EXAMINATION GUIDE
PRE-ALGEBRA: Skills/Problem Solving/Applications

PRE-ALGEBRA: Skills/Problem Solving/Applications is a program of instruction that is designed to accommodate the wide range of abilities and interests of the students in a pre-algebra program. This is evidenced by the following list of features around which *Pre-Algebra* is built. A description of each feature can be found on the indicated pages of the Teacher's Manual portion of this *Teacher's Edition*.

SKILLS LESSONS

Examples (p. M-2)
Self-Check (p. M-2)
Procedures (p. M-2)

Classroom Exercises (p. M-2)
Written Exercises (p. M-2)
Mixed Practice (p. M-2)
Applications (p. M-2)

PROBLEM SOLVING AND APPLICATIONS LESSONS

These lessons are integrated throughout the textbook. That is, each Problem Solving and Applications lesson immediately follows the related skills lesson(s). (p. M-3)
NOTE: Also see the Applications exercises that are included within the Written Exercises of many skills lessons. (p. M-2)

REVIEW — MAINTENANCE — TESTING

Review Capsules (p. M-4)
In-chapter Reviews (p. M-4)
Chapter Reviews (p. M-4)

Chapter Tests (p. M-5)
Additional Practice (p. M-5)
Cumulative Reviews (p. M-5)
Common Errors (p. M-5)

OPTIONAL FEATURES

Consumer Applications (p. M-6)
Career Applications (p. M-6)
Calculator Applications (p. M-6)

Computer Applications (p. M-7)
Enrichment (p. M-7)
More Challenging Exercises (p. M-7)

TEACHER'S EDITION: PART I and PART II

The components of the *Teacher's Edition: Part I* are described on page M-10. The three types of copying masters that are included in the paperback *Teacher's Edition: Part II* are described on pages M-8 and M-9.

SATELLITE PUBLICATIONS

The satellite publications that accompany *PRE-ALGEBRA: Skills/Problem Solving/Applications* are decribed on pages M-8 and M-9.

HBJ Harcourt Brace Jovanovich, Publishers

Teacher's Edition: Part I

Annotated Edition with Teacher's Manual
and Review of Essential Skills

Vincent Brumfiel

Neal Golden

Mary Heins

PRE-ALGEBRA

SKILLS

PROBLEM SOLVING

APPLICATIONS

HBJ Harcourt Brace Jovanovich, Publishers

Orlando   New York   Chicago   San Diego   Atlanta   Dallas

> The Teacher's Edition, Parts I and II, when used with the pupil's textbook, provides a complete pre-algebra program.
>
> A Teacher's Edition, Parts I and II, is not automatically included with each shipment of a classroom sets of textbooks. However, a Teacher's Edition will be forwarded when requested by a teacher, an administrator, or a representative of Harcourt Brace Jovanovich, Inc.
>
> Part II of the Teacher's Edition is available for *Pre-Algebra: Skills/Problem Solving/ Applications.* For information, please contact your sales representative.

Copyright © 1986 by Harcourt Brace Jovanovich, Inc.

All rights reserved. No part of this publication may be reproduced or transmitted in any form or by any means, electronic or mechanical, including photocopy, recording, or any information storage and retrieval system, without permission in writing from the publisher.

Permission is hereby granted to reproduce the Review of Essential Skills pages in this publication in complete pages, with the copyright notice, for instructional use and not for resale by any teacher using classroom quantities of the related student textbook.

Printed in the United States of America

ISBN 0-15-355701-X

CONTENTS

DESCRIPTION OF THE PROGRAM	**M–1**
OVERVIEW: PRE-ALGEBRA: SKILLS/ PROBLEM SOLVING/APPLICATIONS	**M–11**
CONTENT	**M–12**
LESSON PLAN GUIDE	**M–13**
CHAPTER OVERVIEWS AND SECTION-BY-SECTION COMMENTARY	**M–14**
MASTER TIMETABLE	**M–31**
SUGGESTED TIMETABLES OF ASSIGNMENTS	**M–32**
REVIEW OF ESSENTIAL SKILLS	**M–43**
ANSWERS: REVIEW OF ESSENTIAL SKILLS	**M–61**

DESCRIPTION OF THE PROGRAM

Textbook

Pages M–2 through M–7 describe, both verbally and pictorially, the features of the student textbook: skills lessons, problem solving and applications lessons, the approach to review and testing (*Review Capsules*, in-chapter *Reviews, Common Errors, Chapter Reviews, Chapter Tests, Additional Practice, Cumulative Reviews*), and optional features (*Consumer Applications, Career Applications, Computer Applications*, and *Calculator Applications*).

Teacher's Edition: Part I

The four basic components that appear in the margin of the annotated textbook pages for each lesson are described on page M–10 (the *Quick Quiz, Additional Examples, Answers,* and a three-level *Assignment Guide*). Where appropriate, comments to the teacher also appear in the margin. However, section-by-section commentary for each chapter can be found on pages M–14 through M–30 of the *Teacher's Manual*. Page M–43 describes the *Review of Essential Skills*. The 17 pages of copying masters that comprise this section provide survey tests and additional practice on whole numbers, decimals, fractions, percent and metric measures.

Teacher's Edition: Part II

This unique paperback publication is described on pages M–8 and M–9. In addition to receiving a copy of the hardcover *Teacher's Edition: Part I*, upon request, each teacher will also receive a copy of the *Teacher's Edition: Part II*, upon request. This publication consists of copying masters that are perforated and pre-holed. It contains the following components.

1. Quizzes: Two per chapter
2. Chapter Tests: Two forms
3. Cumulative Tests
4. Placement Test
5. Skills Practice: 112 pages of additional examples and practice for each lesson
6. Introduction to Computer Programming
7. Answers for items 1–6

Tests

The four components of the testing program described above are also available as a bound set of spirit duplicating masters. The answers are also included in this publication.

Skills Practice Book

The 112 Skills Practice section of the *Teacher's Edition: Part II* is also available as a separate paperback publication.

The Lesson

Examples and Procedure
The ideas and steps presented in a given lesson are implemented in a clear step-by-step fashion.

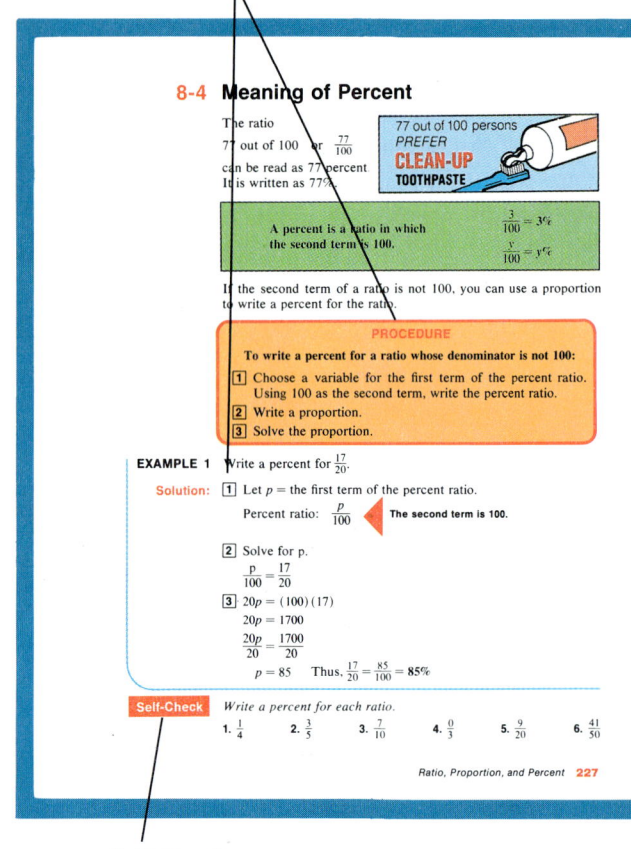

Classroom Exercises
Each set of exercises is referenced to steps in an Example, to an entire Example, to tables, to a definition or to Checks.

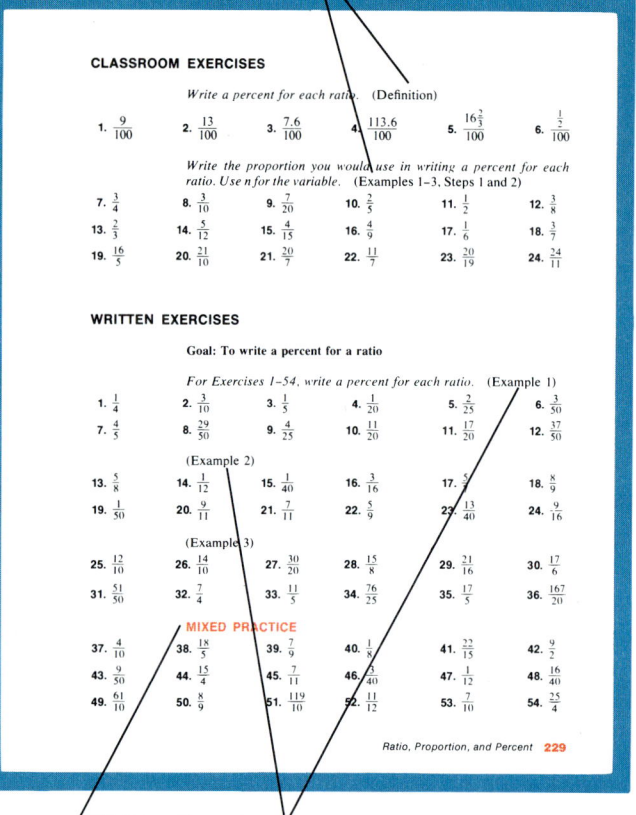

8-4 Meaning of Percent

The ratio 77 out of 100 or $\frac{77}{100}$ can be read as 77 percent. It is written as 77%.

77 out of 100 persons PREFER CLEAN-UP TOOTHPASTE

A percent is a ratio in which the second term is 100.

$\frac{3}{100} = 3\%$

$\frac{y}{100} = y\%$

If the second term of a ratio is not 100, you can use a proportion to write a percent for the ratio.

PROCEDURE

To write a percent for a ratio whose denominator is not 100:
1. Choose a variable for the first term of the percent ratio. Using 100 as the second term, write the percent ratio.
2. Write a proportion.
3. Solve the proportion.

EXAMPLE 1 Write a percent for $\frac{17}{20}$.

Solution:
1. Let p = the first term of the percent ratio.
   Percent ratio: $\frac{p}{100}$ *The second term is 100.*
2. Solve for p.
   $\frac{p}{100} = \frac{17}{20}$
3. $20p = (100)(17)$
   $20p = 1700$
   $\frac{20p}{20} = \frac{1700}{20}$
   $p = 85$ Thus, $\frac{17}{20} = \frac{85}{100} = 85\%$

Self-Check *Write a percent for each ratio.*
1. $\frac{1}{4}$ 2. $\frac{3}{5}$ 3. $\frac{7}{10}$ 4. $\frac{0}{3}$ 5. $\frac{9}{20}$ 6. $\frac{41}{50}$

Ratio, Proportion, and Percent **227**

CLASSROOM EXERCISES

Write a percent for each ratio. (Definition)

1. $\frac{9}{100}$ 2. $\frac{13}{100}$ 3. $\frac{7.6}{100}$ 4. $\frac{113.6}{100}$ 5. $\frac{16\frac{2}{3}}{100}$ 6. $\frac{\frac{1}{2}}{100}$

Write the proportion you would use in writing a percent for each ratio. Use n for the variable. (Examples 1–3, Steps 1 and 2)

7. $\frac{3}{4}$ 8. $\frac{3}{10}$ 9. $\frac{7}{20}$ 10. $\frac{2}{5}$ 11. $\frac{1}{2}$ 12. $\frac{3}{8}$
13. $\frac{2}{3}$ 14. $\frac{5}{12}$ 15. $\frac{4}{15}$ 16. $\frac{4}{9}$ 17. $\frac{1}{6}$ 18. $\frac{3}{7}$
19. $\frac{16}{5}$ 20. $\frac{21}{10}$ 21. $\frac{20}{7}$ 22. $\frac{11}{7}$ 23. $\frac{20}{19}$ 24. $\frac{24}{11}$

WRITTEN EXERCISES

Goal: To write a percent for a ratio

For Exercises 1–54, write a percent for each ratio. (Example 1)

1. $\frac{1}{4}$ 2. $\frac{3}{10}$ 3. $\frac{1}{5}$ 4. $\frac{1}{20}$ 5. $\frac{2}{25}$ 6. $\frac{3}{50}$
7. $\frac{4}{5}$ 8. $\frac{29}{50}$ 9. $\frac{4}{25}$ 10. $\frac{11}{20}$ 11. $\frac{17}{20}$ 12. $\frac{37}{50}$

(Example 2)

13. $\frac{5}{8}$ 14. $\frac{1}{12}$ 15. $\frac{1}{40}$ 16. $\frac{3}{16}$ 17. $\frac{5}{}$ 18. $\frac{8}{9}$
19. $\frac{1}{50}$ 20. $\frac{9}{11}$ 21. $\frac{7}{11}$ 22. $\frac{5}{9}$ 23. $\frac{13}{40}$ 24. $\frac{9}{16}$

(Example 3)

25. $\frac{12}{10}$ 26. $\frac{14}{10}$ 27. $\frac{30}{20}$ 28. $\frac{15}{8}$ 29. $\frac{21}{16}$ 30. $\frac{17}{6}$
31. $\frac{51}{50}$ 32. $\frac{7}{4}$ 33. $\frac{11}{5}$ 34. $\frac{76}{25}$ 35. $\frac{17}{5}$ 36. $\frac{167}{20}$

MIXED PRACTICE

37. $\frac{4}{10}$ 38. $\frac{18}{5}$ 39. $\frac{7}{9}$ 40. $\frac{1}{8}$ 41. $\frac{22}{15}$ 42. $\frac{9}{2}$
43. $\frac{9}{50}$ 44. $\frac{15}{4}$ 45. $\frac{7}{11}$ 46. $\frac{3}{40}$ 47. $\frac{1}{12}$ 48. $\frac{16}{40}$
49. $\frac{61}{10}$ 50. $\frac{8}{9}$ 51. $\frac{119}{10}$ 52. $\frac{11}{12}$ 53. $\frac{7}{10}$ 54. $\frac{25}{4}$

Ratio, Proportion, and Percent **229**

Self-Check
Each Example is immediately followed by a set of *Self-Check* exercises. This gives each student the opportunity to practice the skill just presented.

Written Exercises
Each set of these exercises is referenced to the related Example.

Mixed Practice exercises relate to all the Examples in a lesson.

Applications are often included in the *Written Exercises*, as are *More Challenging Exercise*.

APPLICATIONS: FROM RATIOS TO PERCENTS

Write a percent for each ratio.

55. In 1970, $\frac{2}{5}$ of the population of the United States was under age 21.
56. By 1990, $\frac{8}{}$ of the population of the ... under age 21.

M-2 Teacher's Manual

Problem Solving and Application Lessons

Each *Problem-Solving and Applications* lesson is directly related to the content presented in the given chapter. The applications covered in these sections cover a wide range of practical and mathematical topics.

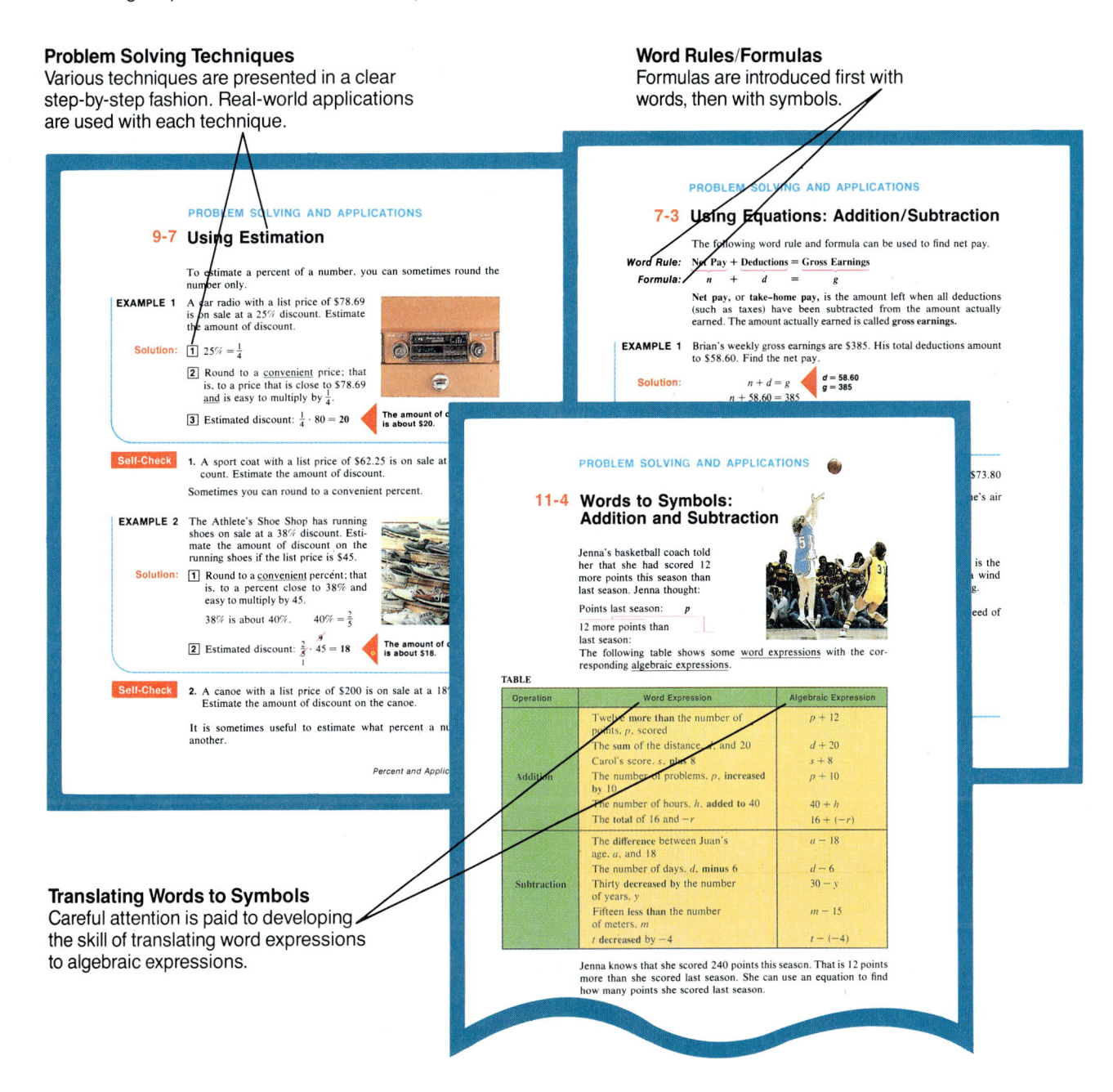

Problem Solving Techniques
Various techniques are presented in a clear step-by-step fashion. Real-world applications are used with each technique.

Word Rules/Formulas
Formulas are introduced first with words, then with symbols.

PROBLEM SOLVING AND APPLICATIONS

9-7 Using Estimation

To estimate a percent of a number, you can sometimes round the number only.

EXAMPLE 1 A car radio with a list price of $78.69 is on sale at a 25% discount. Estimate the amount of discount.

Solution:
1. $25\% = \frac{1}{4}$
2. Round to a <u>convenient</u> price; that is, to a price that is close to $78.69 and is easy to multiply by $\frac{1}{4}$.
3. Estimated discount: $\frac{1}{4} \cdot 80 = 20$

➤ The amount of discount is about $20.

Self-Check 1. A sport coat with a list price of $62.25 is on sale at a ...% discount. Estimate the amount of discount.

Sometimes you can round to a convenient percent.

EXAMPLE 2 The Athlete's Shoe Shop has running shoes on sale at a 38% discount. Estimate the amount of discount on the running shoes if the list price is $45.

Solution:
1. Round to a <u>convenient</u> percent; that is, to a percent close to 38% and easy to multiply by 45.
   38% is about 40%. $40\% = \frac{2}{5}$
2. Estimated discount: $\frac{2}{5} \cdot 45 = 18$

➤ The amount of discount is about $18.

Self-Check 2. A canoe with a list price of $200 is on sale at a 18... Estimate the amount of discount on the canoe.

It is sometimes useful to estimate what percent a n... another.

Percent and Applic...

PROBLEM SOLVING AND APPLICATIONS

7-3 Using Equations: Addition/Subtraction

The following word rule and formula can be used to find net pay.

Word Rule: Net Pay + Deductions = Gross Earnings
Formula: $n\ \ +\ \ d\ \ =\ \ g$

Net pay, or **take-home pay,** is the amount left when all deductions (such as taxes) have been subtracted from the amount actually earned. The amount actually earned is called **gross earnings.**

EXAMPLE 1 Brian's weekly gross earnings are $385. His total deductions amount to $58.60. Find the net pay.

Solution:
$n + d = g$ $d = 58.60$
$n + 58.60 = 385$ $g = 385$

PROBLEM SOLVING AND APPLICATIONS

11-4 Words to Symbols: Addition and Subtraction

Jenna's basketball coach told her that she had scored 12 more points this season than last season. Jenna thought:

Points last season: *p*

12 more points than last season:

The following table shows some <u>word expressions</u> with the corresponding <u>algebraic expressions</u>.

TABLE

Operation	Word Expression	Algebraic Expression
Addition	Twelve more than the number of points, *p*, scored	$p + 12$
	The sum of the distance, *d*, and 20	$d + 20$
	Carol's score, *s*, plus 8	$s + 8$
	The number of problems, *p*, increased by 10	$p + 10$
	The number of hours, *h*, added to 40	$40 + h$
	The total of 16 and $-r$	$16 + (-r)$
Subtraction	The difference between Juan's age, *a*, and 18	$a - 18$
	The number of days, *d*, minus 6	$d - 6$
	Thirty decreased by the number of years, *y*	$30 - y$
	Fifteen less than the number of meters, *m*	$m - 15$
	t decreased by -4	$t - (-4)$

Jenna knows that she scored 240 points this season. That is 12 points more than she scored last season. She can use an equation to find how many points she scored last season.

Translating Words to Symbols
Careful attention is paid to developing the skill of translating word expressions to algebraic expressions.

Teacher's Manual   **M-3**

Review and Testing

Review Capsules
These exercises review prior-taught skills that will be used in the lesson that follows. The exercises are referenced to those pages where the skill was presented.

Review
Each chapter is divided into two parts. Each part is followed by a *Review*. Each set of Exercises in the *Review* is referenced to the related lesson.

Each **Chapter Review** consists of:
1. **Vocabulary** review
2. **Skills** review
3. **Applications** review
Note the help references.

M-4 Teacher's Manual

Review and Testing

The **Chapter Tests** parallel those in the *Tests with Answer Key* (duplicating masters) and in the *Teacher's Edition: Part II* (copying masters).

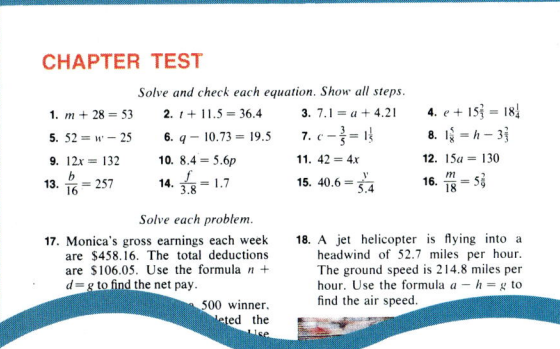

An **Additional Practice** page follows each *Chapter Test*. It provides extra practice for students who did not perform well on the formal *Chapter Test*. Note the help references.

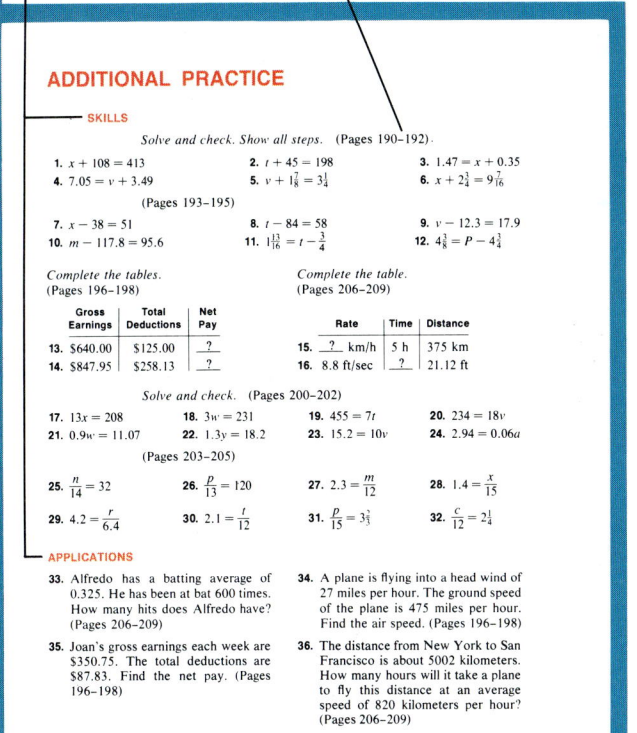

CHAPTER TEST

Solve and check each equation. Show all steps.

1. $m + 28 = 53$
2. $t + 11.5 = 36.4$
3. $7.1 = a + 4.21$
4. $e + 15\frac{2}{3} = 18\frac{1}{4}$
5. $52 = w - 25$
6. $q - 10.73 = 19.5$
7. $c - \frac{3}{5} = 1\frac{1}{5}$
8. $1\frac{5}{8} = h - 3\frac{2}{3}$
9. $12x = 132$
10. $8.4 = 5.6p$
11. $42 = 4x$
12. $15a = 130$
13. $\frac{b}{16} = 257$
14. $\frac{f}{3.8} = 1.7$
15. $40.6 = \frac{v}{5.4}$
16. $\frac{m}{18} = 5\frac{2}{9}$

Solve each problem.

17. Monica's gross earnings each week are $458.16. The total deductions are $106.05. Use the formula $n + d = g$ to find the net pay.
18. A jet helicopter is flying into a headwind of 52.7 miles per hour. The ground speed is 214.8 miles per hour. Use the formula $a - h = g$ to find the air speed.

ADDITIONAL PRACTICE

SKILLS

Solve and check. Show all steps. (Pages 190–192)

1. $x + 108 = 413$
2. $t + 45 = 198$
3. $1.47 = x + 0.35$
4. $7.05 = v + 3.49$
5. $v + 1\frac{7}{8} = 3\frac{1}{4}$
6. $x + 2\frac{3}{4} = 9\frac{7}{16}$

(Pages 193–195)

7. $x - 38 = 51$
8. $t - 84 = 58$
9. $v - 12.3 = 17.9$
10. $m - 117.8 = 95.6$
11. $1\frac{13}{16} = t - \frac{3}{4}$
12. $4\frac{3}{8} = P - 4\frac{3}{4}$

Complete the tables. (Pages 196–198)

	Gross Earnings	Total Deductions	Net Pay
13.	$640.00	$125.00	?
14.	$847.95	$258.13	?

Complete the table. (Pages 206–209)

	Rate	Time	Distance
15.	? km/h	5 h	375 km
16.	8.8 ft/sec	?	21.12 ft

Solve and check. (Pages 200–202)

17. $13x = 208$
18. $3w = 231$
19. $455 = 7t$
20. $234 = 18v$
21. $0.9w = 11.07$
22. $1.3y = 18.2$
23. $15.2 = 10v$
24. $2.94 = 0.06a$

(Pages 203–205)

25. $\frac{n}{14} = 32$
26. $\frac{p}{13} = 120$
27. $2.3 = \frac{m}{12}$
28. $1.4 = \frac{x}{15}$
29. $4.2 = \frac{r}{6.4}$
30. $2.1 = \frac{t}{12}$
31. $\frac{p}{15} = 3\frac{2}{3}$
32. $\frac{c}{12} = 2\frac{1}{4}$

APPLICATIONS

33. Alfredo has a batting average of 0.325. He has been at bat 600 times. How many hits does Alfredo have? (Pages 206–209)
34. A plane is flying into a head wind of 27 miles per hour. The ground speed of the plane is 475 miles per hour. Find the air speed. (Pages 196–198)
35. Joan's gross earnings each week are $350.75. The total deductions are $87.83. Find the net pay. (Pages 196–198)
36. The distance from New York to San Francisco is about 5002 kilometers. How many hours will it take a plane to fly this distance at an average speed of 820 kilometers per hour? (Pages 206–209)

214 Chapter 7

The **Common Errors** feature appears at the end of Chapters 3, 6, 9, 12, 15, and 17. These exercises focus student attention on avoiding some of the most common errors made by students.

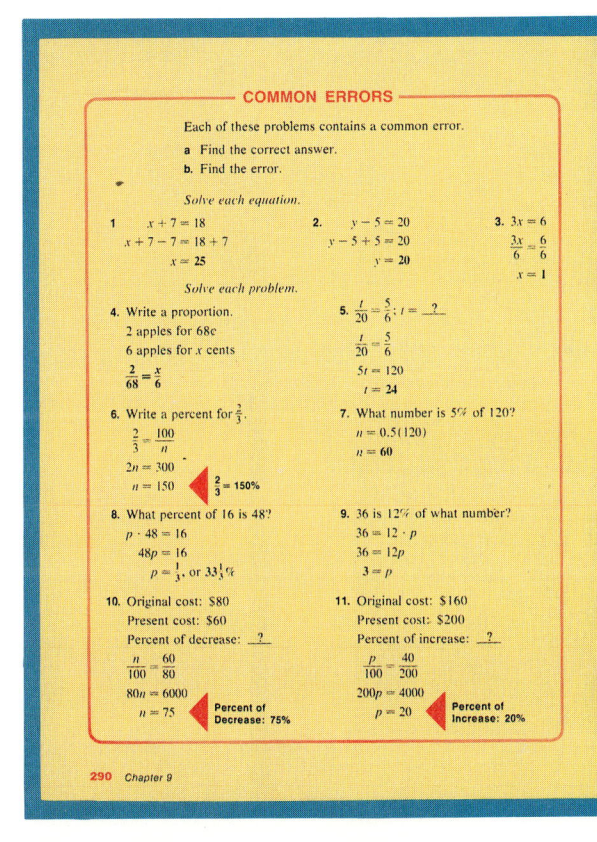

COMMON ERRORS

Each of these problems contains a common error.
 a. Find the correct answer.
 b. Find the error.

Solve each equation.

1. $x + 7 = 18$
   $x + 7 - 7 = 18 + 7$
   $x = 25$

2. $y - 5 = 20$
   $y - 5 + 5 = 20$
   $y = 20$

3. $3x = 6$
   $\frac{3x}{6} = \frac{6}{6}$
   $x = 1$

Solve each problem.

4. Write a proportion.
   2 apples for 68¢
   6 apples for x cents
   $\frac{2}{68} = \frac{x}{6}$

5. $\frac{1}{20} = \frac{5}{6}$; $t = \underline{\ ?\ }$
   $\frac{t}{20} = \frac{5}{6}$
   $5t = 120$
   $t = 24$

6. Write a percent for $\frac{2}{3}$.
   $\frac{2}{3} = \frac{100}{n}$
   $2n = 300$
   $n = 150$ $\frac{2}{3} = 150\%$

7. What number is 5% of 120?
   $n = 0.5(120)$
   $n = 60$

8. What percent of 16 is 48?
   $p \cdot 48 = 16$
   $48p = 16$
   $p = \frac{1}{3}$, or $33\frac{1}{3}\%$

9. 36 is 12% of what number?
   $36 = 12 \cdot p$
   $36 = 12p$
   $3 = p$

10. Original cost: $80
    Present cost: $60
    Percent of decrease: ?
    $\frac{n}{100} = \frac{60}{80}$
    $80n = 6000$
    $n = 75$ Percent of Decrease: 75%

11. Original cost: $160
    Present cost: $200
    Percent of increase: ?
    $\frac{p}{100} = \frac{40}{200}$
    $200p = 4000$
    $p = 20$ Percent of Increase: 20%

290 Chapter 9

A **Cumulative Review** appears after Chapters 3, 6, 9, 12, 15, and 17. Note the **multiple-choice** format.

CUMULATIVE REVIEW: CHAPTERS 7–9

Choose the correct answer. Choose a, b, c, or d.

1. Solve for c: $13.7 = c + 8.9$
   a. 21.6   b. 22.6   c. 5.2   d. 4.8
2. Solve for y: $y - 24 = 33$
   a. 9   b. 57   c. 65   d. 14
3. Last month, Frank's gross earnings were $1158.45. The total deduc... Use the formula $n + d = g$ to find the net pay.
   b. $975.31   c. $875.31

Teacher's Manual   M-5

Optional Features

Consumer Applications
Each consumer topic is an application of the content studied in the related chapter. Each contains a set of exercises.

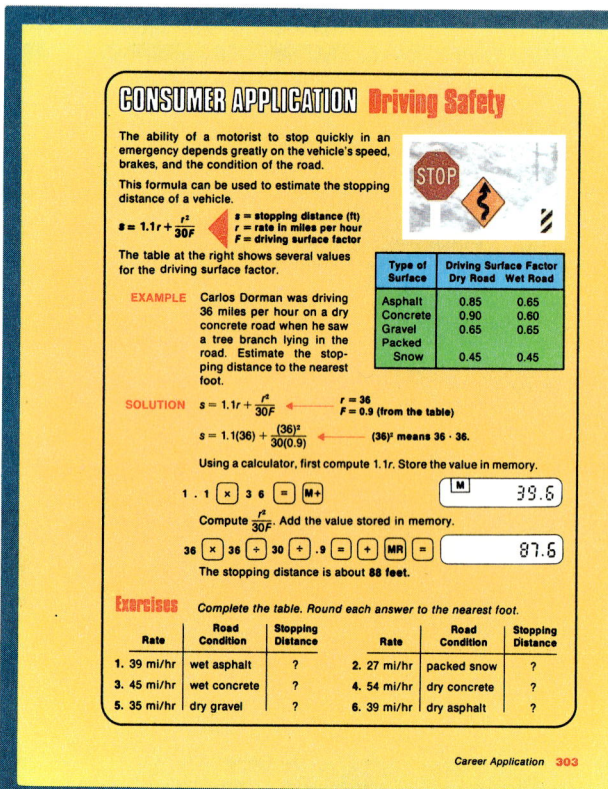

Career Applications
Each career topic is an application of the content studied in the related chapter. Each contains a set of exercises.

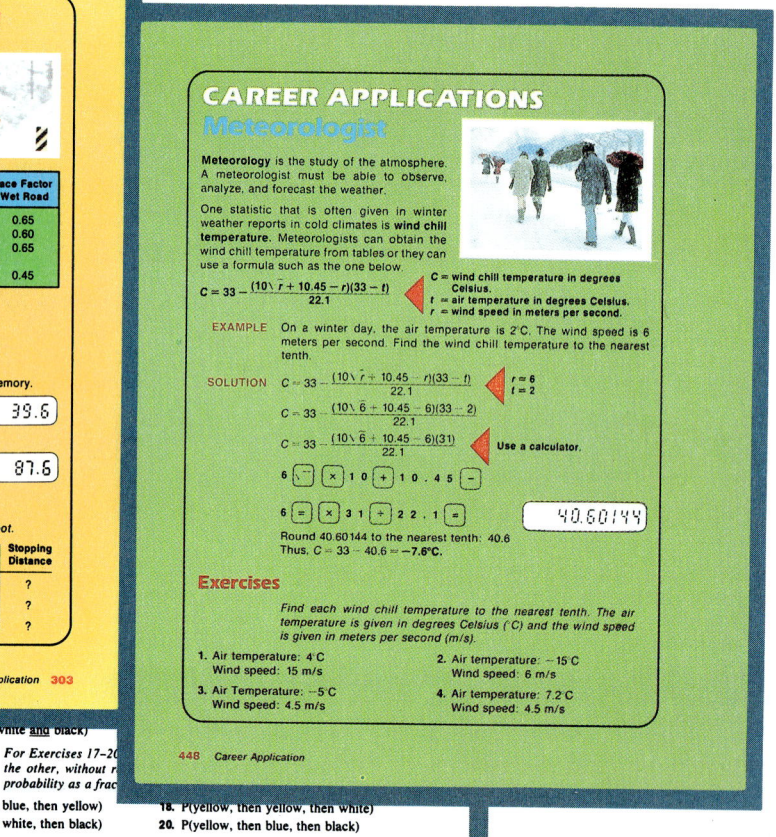

Calculator Applications
Each calculator topic is an application of the content studied in the related chapter. Each contains a set of exercises.

M-6   Teacher's Manual

Optional Features

Computer Applications
Each computer application in the BASIC language is directly related to the content of the related chapter.

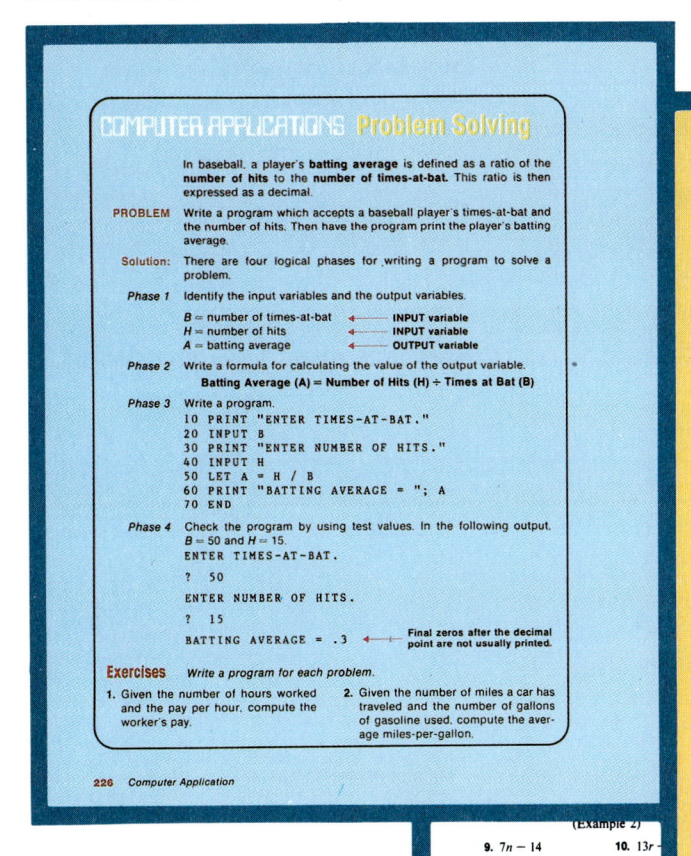

Enrichment
Each Enrichment topic is designed for students who performed well on the formal chapter test.

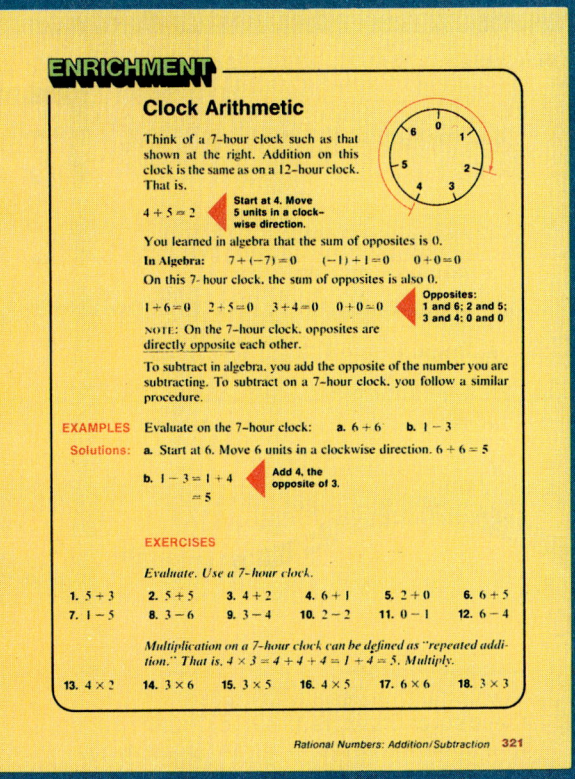

NOTE: The *Teacher's Edition: Part II* contains a five-lesson chapter entitled *Introduction to Computer Programming* on copying masters that is designed to supplement the *Computer Applications*.

More Challenging Exercises
These exercises extend the content of the lesson. They are included in the Written Exercises, where appropriate.

Teacher's Edition: Part II

This paperback consists of perforated and pre-holed copying masters.

It consists of:
1. Testing Program   2. Skills Practice   3. Introduction to Computer Programming   4. Answer Keys

Quizzes
Two quizzes are provided for each chapter. Each quiz covers one half of the content of each chapter.

Chapter Tests: Form A and Form B
Two forms of each Chapter Test are provided. Each parallels the Chapter Test that is included in the student edition.

Name _____ Date _____ Score _____

QUIZ FOR SECTIONS 9-1 THROUGH 9-4

Solve.

1. What percent of 60 is 9?
2. What percent of 72 is 48?
3. What percent of 40 is 15?
4. What percent of 96 is 72?
5. 6 is 8% of what number?
6. 28 is 35% of what number?
7. 54 is 45% of what number?

Name _____ Date _____ Score _____

CHAPTER 11   EQUATIONS: ADDITION/SUBTRACTION

Solve and check.

(11-1)
1. $a + 35 = 16$ 1. _____
2. $t - 12 = -40$ 2. _____
3. $m + 6 = -9$ 3. _____
4. $37.8 = k + 22.9$ 4. _____

(11-2)
5. $8x + 3x = -5 + 27$ 5. _____
6. $-8m + 15m = -2 + 30$ 6. _____
7. $-2c + 4 + 3c = 9 + 6$ 7. _____

Cumulative Tests
Nine Cumulative Tests are included. These cover Chapters 1–3, 4–6, 7–9, 10–13, 14–15, 16–17, 1–13, 1–15, and 1–17.

Placement Test
This four-part Placement Test will help to determine which students should take this course.

NOTE: The testing program is also available separately as a bound set of duplicating masters.

Name _____ Date _____ Score _____

CUMULATIVE TEST: CHAPTERS 1-15

Choose the correct answer. Write the letter of your choice in

(1-4)
1. The low temperatures for one week in a city are listed in the table at the right in degrees Celsius. Find the average low temperature for the week.
   a. 9°   b. 10°   c. 8°   d.

   oblem gives the best estimate of 737
   b. 740 + 4

Name _____ Date _____ Score _____

PLACEMENT TEST   PART 2 Sequences

Each of the following problems has a sequence of numbers that follow a definite pattern. Decide what the next two numbers of each sequence should be. Mark the letter of your choice in the answer column.

Example 1: 5, 9, 13, 17, 21, __?__, __?__
   b. 22, 23   c. 25, 29   d. 25, 27

each number to obtain the next numb

M-8   Teacher's Manual

Teacher's Edition: Part II

Skills Practice
This 112-page *Skills Practice* section provides additional exercises with examples for each section (lesson) in the textbook.

NOTE: Each page is clearly referenced to the related section and pages.

Introduction to Computer Programming
This five-lesson section is designed to supplement the *Computer Applications* in the textbook. It assumes no prior knowledge or experience regarding computers.

11-2 LIKE TERMS/MORE THAN ONE OPERATION   (Pages 326-32)

Solve and check.

Example: $5x - 3x = 7 - 13$ Check: $5x - 3x = 7 - 13$
          $2x = -6$ $5(-3) - 3(-3) \mid 7 - 13$
          $\frac{2x}{2} = \frac{-6}{2}$ $-15 + 9 \mid -6$
          $x = -3$ $-6 \mid -6$

1. $5x - 4x = -7 + 3$ _____
2. $6r - 5r = 3 - 6$ _____
3. $4y + 3y = 14 + 7$ _____
4. $9a + 8a = 5 + 12$ _____
5. $-s - 2s = 12 + 6$ _____
6. $-16e - 15e = -18 - 13$ _____
7. $-3.8p + 4.8p = 5.2 - 7.1$ _____
8. $-7.6m + 10.4m = 21.3 - 4.5$ _____
9. $8\frac{1}{4}t - 7\frac{1}{4}t = 4\frac{1}{2} - 6\frac{1}{2}$ _____
10. $6\frac{1}{8}s - 4\frac{1}{8}s = 6\frac{1}{6} + 3\frac{1}{3}$ _____
11. $3x + 12x = 110 + 70$ _____
12. $9a + 2a = 32 + 89$ _____

Example: $-3x + 4 + 4x = 9 - 15$ Check: $-3x + 4 + 4x = 9 - 15$
          $x + 4 = -6$ $-3(-10) + 4 + 4(-10) \mid 9 - 15$
          $x + 4 - 4 = -6 - 4$ $30 + 4 - 40 \mid -6$
          $x = -10$ $-6 \mid -6$

13. $25y + 4 - 24y = 7 - 14$ _____
14. $35x + 8 - 34x = 3 - 2$ _____
15. $15r + 5 - 14r = -4 + 16$ _____
16. $45x + 18 - 44x = 23 + 15$ _____
17. $4b + 6 - b = 12 - 36$ _____
18. $6h - 7 - h = 13 - 5$ _____
19. $-9n - 5 + 6n = 28 - 12$ _____
20. $-12a - 7 + 4a = 37 - 12$ _____
21. $10m + 3 - 7m = 18 + 9$ _____
22. $8k + 3 - 3k = 18 - 50$ _____
23. $4\frac{1}{2}c - 8 + 2\frac{1}{2}c = -3\frac{1}{4} + 23\frac{1}{4}$ _____
24. $3\frac{1}{3}y - 2 + 2\frac{2}{3}y = -3\frac{1}{6} + 7\frac{1}{6}$ _____

Copyright © 1986 by Harcourt Brace Jovanovich, Inc.
All rights reserved.   W-67   EQUATIONS: ADDITION/SU...

---

5 FOR NEXT Statements

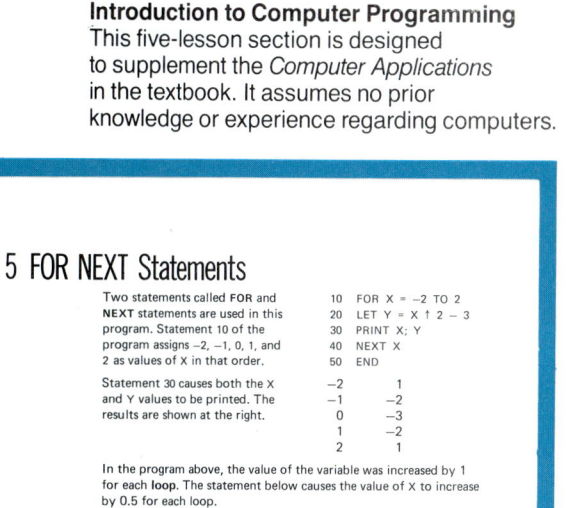

Two statements called **FOR** and **NEXT** statements are used in this program. Statement 10 of the program assigns −2, −1, 0, 1, and 2 as values of X in that order.

Statement 30 causes both the X and Y values to be printed. The results are shown at the right.

```
10 FOR X = -2 TO 2
20 LET Y = X ↑ 2 - 3
30 PRINT X; Y
40 NEXT X
50 END
```

X	Y
−2	1
−1	−2
0	−3
1	−2
2	1

In the program above, the value of the variable was increased by 1 for each *loop*. The statement below causes the value of X to increase by 0.5 for each loop.

   10 FOR X = 1 TO 3 STEP .5

EXAMPLE 1 Write a program with FOR and NEXT statements that will compute and print values of Y for the equation $Y = 10 - X^2$. Use the following input values for x.

   5, 3, 1, −1, −3, −5

PROGRAM:
```
10 FOR X = 5 TO -5 STEP -2
20 LET Y = 10 - X ↑ 2    ← Y = 10 - X²
30 PRINT Y;
40 NEXT X
50 END
```

Output: −15   1   9   9   1   −15

Often an INPUT statement is part of the body of a loop.

EXAMPLE 2 a. The three children of the Palmer family play a video game. Write a program that accepts each player's scores on four games and prints the total score on each player.

Program:
```
10 FOR I = 1 TO 3
20 PRINT "ENTER THE 4 SCORES OF PLAYER "; I
30 INPUT A, B, C, D
40 LET T = A + B + C + D
50 PRINT "TOTAL IS "; T
60 PRINT
70 NEXT I
80 PRINT "END OF PROGRAM"
90 END
```

Copyright © 1986 by Harcourt Brace Jovanovich, Inc.
All rights reserved.   P-13   COMPUTER PROGRAMMING

Skills Practice Book
The *Skills Practice* section is also available as a separate paperback with perforated pages. This version is accompanied by an annotated *Teacher's Edition*.

Answer Keys Each of the 3 components, *Testing Program*, *Skills Practice*, and *Introduction to Computer Programming*, is followed by its *Answer Key*.

Teacher's Manual   **M-9**

Teacher's Edition: Part I

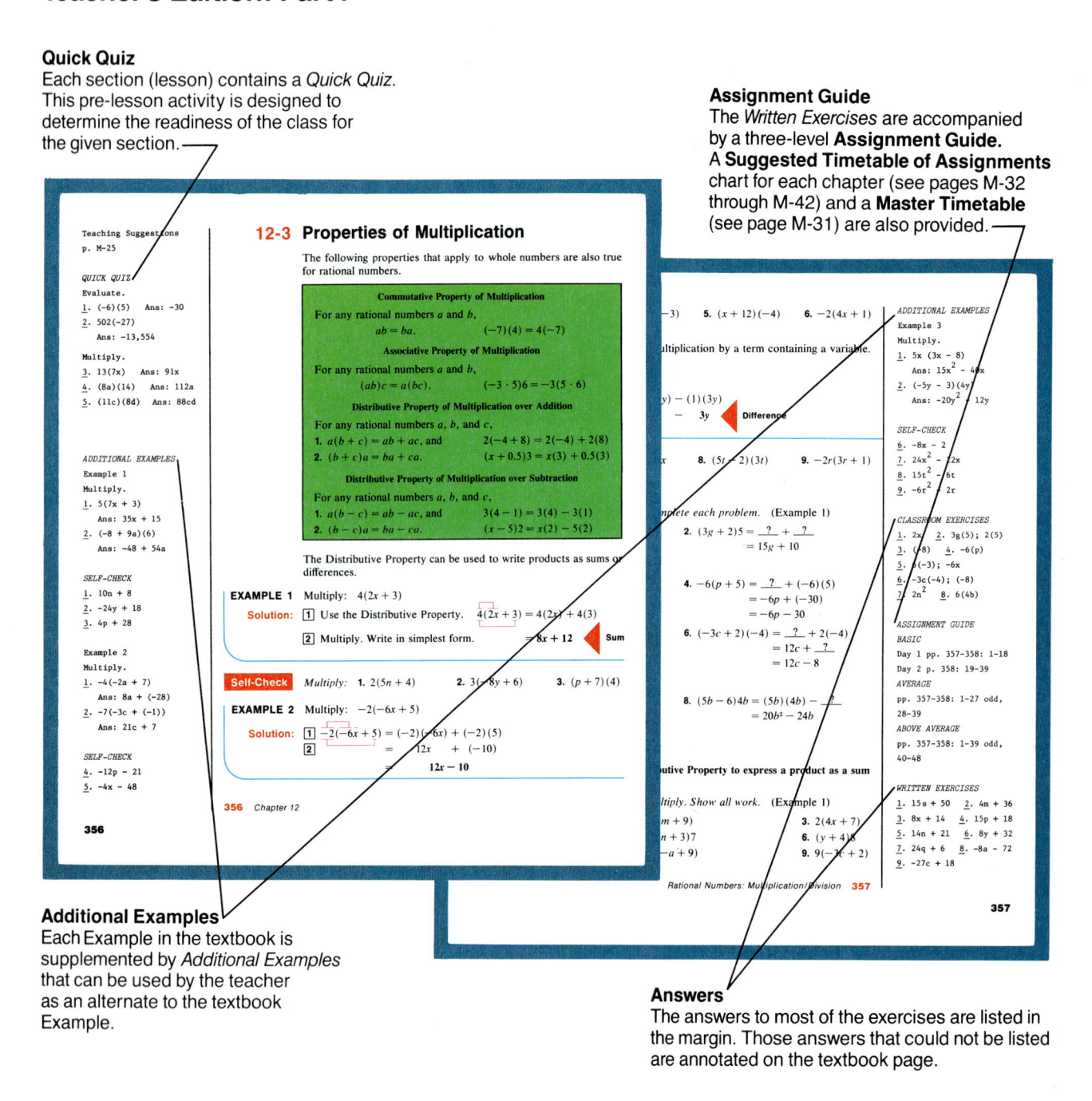

Quick Quiz
Each section (lesson) contains a *Quick Quiz*. This pre-lesson activity is designed to determine the readiness of the class for the given section.

Assignment Guide
The *Written Exercises* are accompanied by a three-level **Assignment Guide**. A **Suggested Timetable of Assignments** chart for each chapter (see pages M-32 through M-42) and a **Master Timetable** (see page M-31) are also provided.

Additional Examples
Each Example in the textbook is supplemented by *Additional Examples* that can be used by the teacher as an alternate to the textbook Example.

Answers
The answers to most of the exercises are listed in the margin. Those answers that could not be listed are annotated on the textbook page.

Review of Essential Skills
See page M-43 for a description of these copying masters.

M-10   Teacher's Manual

OVERVIEW: PRE-ALGEBRA: Skills/Problem Solving/Applications

Philosophy

The major objective of a course in pre-algebra is to prepare students for a course in first-year algebra. In most school systems, pre-algebra is viewed as a ninth-grade course of study that is intended for students who lack the prerequisite skills needed for a course in first-year algebra. Even within the constraints of this definition, there is still a wide range of student abilities and interests in a pre-algebra course. In general, the deficiencies of these students are in two major areas: computation and solving word problems.

In some school systems, a course in pre-algebra is offered to eighth-grade students in place of the traditional eighth-grade course. Students who take this course are able to proceed at a faster pace than the ninth-grade pre-algebra student, and they need material that will challenge their abilities.

Pedagogy

With this understanding of the make-up of a class in pre-algebra, the authors have structured a three-level approach to instruction.

Basic: Chapters 1–13 represent the minimum course. These "basic-level" students will also benefit from the **Review of Essential Skills** (see page M–43), the **Additional Practice** that is included at the end of each chapter in the textbook, and the **Skills Practice Book.**

Average: Chapters 1–15 represent the amount of content that most students should be able to cover. Certain portions of the *Review of Essential Skills* (particularly the content on fractions and percent) and the *Skills Practice Book* will be helpful to the average student. The average student should be encouraged to try some of the **More Challenging Exercises** and **Enrichment** topics.

Above Average: Chapters 1–17 represent the minimum course. Students at this level should be able to handle all of *Enrichment* topics and the *More Challenging Exercises.*

NOTE: The content in the **Consumer, Career, Computer**, and **Calculator** applications is within the ability level of most pre-algebra students. Each of these optional features is directly related to the content of the textbook.

CONTENT

Structure

The seventeen chapters in the textbook are organized into five units as shown in the table of contents. Students who are able to cover Units I–IV will be well prepared for a course in first-year algebra. Recognizing the ability level of most pre-algebra students, the authors believe that short chapters of study are appropriate. As the table of contents shows, the authors have taken an additional step in this direction. Each chapter is divided into two parts and each part is followed by a **Review**. The structure of each section (lesson) also reflects the authors' understanding of these students. Each *Example* is followed by a brief set of **Self-Check** exercises that gives the students the opportunity to determine how well they understand the *Example* before moving on in the lesson.

Computational Skills and Applications

It is the belief of many educators that students will have a more complete understanding of the essential computational skills needed for the later study of first-year algebra if the students have the opportunity to apply these skills at the time they are presented. With this, most skills lessons include a set of **Applications** problems as part of the *Written Exercises*. Each of these *Applications* relates the skill(s) presented in the lesson to real-world situations.

Problem Solving and Applications

Specific techniques for solving word problems are presented in each **Problem Solving and Applications** lesson. Because solving word problems is of particular issue for pre-algebra students, the procedure for solving the various types of word problems is presented in a clear step-by-step fashion. Because each *Problem Solving and Applications* lesson immediately follows the related skills lesson(s), the authors believe that the student's proficiency with computational skills will be improved.

Optional Applications

Each chapter contains other forms of applications that may be considered optional, such as **Consumer, Career, Computer**, and **Calculator** applications. Although these features are optional, they are within the ability of most pre-algebra students. Each of these features is directly related to the content of the textbook.

Enrichment

It is the belief of the authors that pre-algebra students should be given the opportunity to extend the basic content of the course. With this, each chapter includes an **Enrichment** topic. In addition, the *Written Exercises* contain a set of **More Challenging Exercises**, where appropriate.

Review and Maintenance

The authors believe that review should be an ongoing process if it is to be meaningful. Thus, with the exception of the first lesson in each chapter, each lesson is preceeded by a **Review Capsule**. These exercises review those prior-taught skills that will be needed in the lesson that immediately follows. Each chapter is divided into two parts, and each part is followed by a **Review** of that part of the chapter. Prior to the **Chapter Test**, the authors have also included a **Chapter Review**. To help the students help themselves, each set of exercise in the *Review Capsules*, the in-chapter *Reviews*, and the *Chapter Reviews* are referenced to the related pages. The numerous **Cumulative Reviews** present a comprehensive form of review in a multiple-choice format.

LESSON PLAN GUIDE

All of the essential components for structuring effective lesson plans are included in this publication.

Teaching Suggestions Pages M–14 through M–30 contain an **Overview** for each chapter and a brief commentary for each section (lesson) and for each optional feature.

Assignment Guides The **Master Timetable** for three levels of ability on page M–31 is coordinated with the **Suggested Timetables of Assignments** that appear on pages M–32 through M–42 and the **Assignment Guide** that appears in the margin of the annotated pages.

Quick Quiz Because the authors believe that review should be an ongoing process, a **Quick Quiz** is provided in the margin of the annotated pages for each lesson as a pre-lesson activity to determine the readiness of the class for the given lesson.

Additional Examples Since many teachers prefer to use examples other than those in the textbook for teaching a lesson, additional examples are provided in the margin of the annotated pages for each example in the textbook.

Lesson Plan Guide

CHAPTER 1: WHOLE NUMBERS AND APPLICATIONS

OVERVIEW

The purpose of this chapter is to give students practice with all four operations on whole numbers and to introduce some basic algebraic concepts. The basic computational skills are applied to real-world situations through the use of formulas. A section on estimating with whole numbers as a useful skill in everyday life is included. The final two sections of the chapter emphasize the rules for order of operations and their importance in evaluating numerical and algebraic expressions.

SECTION-BY-SECTION COMMENTARY

Section 1-1 (pages 2–4)
The concept of formula is introduced and applied to addition problems involving the buying of a new car. Students read data organized in tables to solve problems that involve more than one step by applying the technique of the "hidden question."

Section 1-2 (pages 5–6)
A formula involving subtraction is applied to finding the number of kilowatt-hours of electricity used by a family or business. An example of an electric bill is provided to clarify the meaning of "present reading" and "previous reading."

Section 1-3 (pages 7–9)
Two distance formulas, one applying to constant speed and the other to acceleration due to gravity, are used to solve problems involving multiplication of whole numbers. Emphasis should be placed on the meaning of an exponent and on the different ways to indicate multiplication.

Section 1-4 (pages 10–12)
Students review addition and division of whole numbers through a variety of applications involving the formula for finding the average (or mean) of a group of numbers or measures.

Section 1-5 (pages 14–16)
Remind students that, when estimating an answer, numbers must be rounded <u>before</u> performing the operation. Point out that estimation is a valuable tool in all computations in determining whether an answer is reasonable.

Section 1-6 (pages 18–21)
Numerous examples of evaluating numerical expressions are provided to demonstrate the rules of order of operations. Show students that the rules for order of operations give unambiguous meaning to numerical expressions.

Section 1-7 (pages 22–24)
This lesson introduces the concept of variable and continues the development of the rules for order of operations. Emphasize that an expression such as $3p$ means to multiply 3 and the given value of p.

COMMENTARY: OPTIONAL FEATURES

Consumer Applications (page 13)
This feature applies the basic computational skills to estimating the amount and cost of floor tile and other materials needed for a room. Thus, it can be used after Section 1-5 (pages 14–16) is taught.

Calculator Applications (page 21)
This feature can be used with Section 1-6 (pages 18–21), as it is an application of the memory and memory recall keys on a calculator to solving problems involving order of operations.

Enrichment (page 29)
This feature challenges students to use the patterns of triangular numbers and square numbers to extend the patterns and to discover relationships between these numbers.

CHAPTER 2: DECIMALS AND APPLICATIONS

OVERVIEW

This chapter introduces the concept of place value and gives students practice with basic operations on decimals. The concept of place value is reinforced through applications involving the comparison of decimals, determining metric units of length, mass, and capacity, and using scientific notation. The skill of evaluating algebraic expressions is developed by including expressions containing variables. Estimation is presented as a useful tool when decimals occur in real-world situations.

SECTION-BY-SECTION COMMENTARY

Section 2-1 (pages 32–34)
The Place Names Chart will help students to give the place name and the value of any digit in a number. Emphasize that the chart can be extended indefinitely in either direction. Remind students of the meaning and use of the symbols ">" and "<."

Section 2-2 (pages 35–37)
In this section, students add and subtract with decimals in both numerical and algebraic expressions. Variables may also be replaced with decimals. Numerical expressions that are written horizontally may also be written vertically. Emphasize the necessity of writing the numbers so that their decimal points are aligned. Annexing zeros is helpful in maintaining place value.

Section 2-3 (pages 38–41)
Multiplication with decimals is practiced and then applied to solving consumer-related word problems involving electricity usage. Emphasize the importance of inserting a decimal point at the right place and maintaining place value in a product.

Section 2-4 (pages 44–46)
This section provides practice in dividing a decimal by a whole number and dividing a decimal by a decimal. These skills are applied to solving word problems. Encourage students to check their answers.

Section 2-5 (pages 47–50)
Estimation is used to solve word problems involving each of the four operations with decimals. Show students that, when multiplying or dividing decimals, estimation helps to know where to insert the decimal point in the product or quotient.

Section 2-6 (pages 51–54)
In this section, students determine equivalent metric units of length, mass, and capacity. Conversion between metric units reinforces the concept of place value.

Section 2-7 (pages 55–57)
Scientific notation is used in many fields of science. The goals of this section are for students to express a decimal number in scientific notation and to express a number in scientific notation as a decimal number.

COMMENTARY: OPTIONAL FEATURES

Consumer/Computer Applications (page 43)
This feature illustrates the use of a computer program for finding the unit price of an item when the cost of the item and the number of units are known. It can be used after Section 2-4 (pages 44–46) is taught, as it applies the skill of dividing a decimal by a whole number.

Calculator Applications (page 50)
This feature can be used with Section 2-5 (pages 47–50), as it is an application of the calculator and of estimation to solving problems.

Enrichment (page 61)
Since this feature uses scientific notation and a calculator to multiply and divide large numbers, it can be used with Section 2-7 (pages 55–57).

CHAPTER 3: NUMBER PROPERTIES/EXPRESSIONS

OVERVIEW

The major goal of this chapter is to simplify algebraic expressions by combining like terms. Students are shown how to use the identity, commutative, and associative properties to simplify algebraic expressions that involve the product of one term or the addition of two or more terms. The two Problem Solving and Applications lessons demonstrate the applicability of this skill. In each of these sections, the students evaluate a formula for finding the area or perimeter of a real-world object.

SECTION-BY-SECTION COMMENTARY

Section 3-1 (pages 64–66)
Mathematical sentences are used to apply the identity properties of zero and one. Students may be interested in knowing why division by zero is undefined.

Section 3-2 (pages 67–69)
In this section, students simplify algebraic expressions that involve multiplication by using the Commutative and Associative Properties of Multiplication. Remind students of the various ways of indicating multiplication.

Section 3-3 (pages 70–73)
The skill of simplifying algebraic expressions that involve multiplication is applied to finding the area of a rectangle and of a square. Students are provided with real-world problems using both metric and customary units.

Section 3-4 (pages 76–78)
Emphasize the fact that terms are separated by an addition or subtraction symbol. Numbers such as 5 and 8 can be considered like terms. Since $x^0 = 1$, they can be expressed as $5x^0$ and $8x^0$.

Section 3-5 (pages 79–80)
The Commutative and Associative Properties of Addition provide the justification for simplifying algebraic expressions by combining like terms. Remind students that terms such as $10xy$ and 10 are not like terms and cannot be combined.

Section 3-6 (pages 81–85)
In this Problem Solving and Applications section, students use the appropriate formula to find the perimeter of a rectangle and of a triangle. This skill is then applied to solving word problems. Point out to students that the formula for finding the perimeter of a square ($P = 4s$) is the result of combining like terms ($P = s + s + s + s$).

COMMENTARY: OPTIONAL FEATURES

Computer Applications (page 75)
This feature illustrates the use of a computer program for evaluating a numerical expression. It applies the skills taught in Section 1-6 (pages 18–21) to writing expressions in the BASIC language.

Calculator Applications (page 86)
This feature can be used with Section 3-3 (pages 70–73) or Section 3-6 (pages 81–85), as it is an application of the calculator to evaluating formulas.

Enrichment (page 90)
Because students use the number properties to simplify numerical computations when evaluating a numerical expression, this *Enrichment* page can be used after Section 3-5 (pages 79–80).

Common Errors (page 92)
The exercises in the *Common Errors* feature focus student attention on avoiding some of the most common errors involved with the skills taught in Chapters 1–3.

CHAPTER 4: NUMBER THEORY

OVERVIEW

The major goal of Chapter 4 is to apply the concept of divisibility to the skills of writing the prime factorization of a composite number and to finding the least common multiple or the greatest common factor of two or more numbers. Students will find these skills useful when working with fractions in Chapters 5 and 6.

SECTION-BY-SECTION COMMENTARY

Section 4-1 (pages 96–98)
Emphasize to students the difference between a factor and a multiple. Remind students to include one and the number itself when listing all factors of a number.

Section 4-2 (pages 99–102)
This section can be an important aid to students. By applying the divisibility tests instead of the division algorithm to determine divisibility, much time can be saved.

Section 4-3 (pages 103–105)
Encourage students to use the appropriate divisibility tests when finding the smallest prime number of a given number. Remind students that the numbers 0 and 1 are neither prime nor composite.

Section 4-4 (pages 108–110)
In this section, students write the prime factorization of a composite number. Point out that, for a given composite number, there is *only one* prime factorization.

Section 4-5 (pages 111–114)
Finding the least common multiple of two or more numbers prepares students to find common denominators when adding or subtracting with fractions.

Section 4-6 (pages 115–117)
Two methods are provided for finding the greatest common factor of two or more numbers. The method to be used is left to the discretion of the teacher.

COMMENTARY: OPTIONAL FEATURES

Calculator Applications (pages 106, 117)
These features can be used with Section 4-3 (pages 103–105) and Section 4-6 (pages 115–117), respectively.

Consumer Applications (page 107)
This feature can be used with Section 3-3 (pages 70–73) and sections 3-6 (pages 81–85), as it applies the skills for finding perimeter and area of a rectangle to solving problems that involve the use of formulas for both.

Enrichment (page 121)
This feature can be used with Section 4-4 (pages 108–110). It applies the use of a factor tree to write the prime factorization for a composite number.

CHAPTER 5: FRACTIONS: MULTIPLICATION/DIVISION

OVERVIEW

This chapter concentrates on the skills of multiplying and dividing with fractions and mixed numerals. The three Problem Solving and Applications sections demonstrate the applicability of these skills to solving word problems involving temperature, area, and volume.

SECTION-BY-SECTION COMMENTARY

Section 5-1 (pages 124–125)
In this section, students are provided with three methods for writing fractions in lowest terms. The choice of method is left to the discretion of the teacher.

Section 5-2 (pages 126–128)
Students are provided with practice in writing mixed numerals through problems that involve addition and subtraction of customary measures. Encourage students to memorize the Table of Customary Measures on page 126.

Section 5-3 (pages 129–131)
Encourage students to divide a numerator and a denominator by a common factor, when possible, before multiplying. Sometimes this process can be carried out more than once.

Section 5-4 (pages 132–134)
Remind students of the difference in the order of operations that parentheses make in the formula for converting temperatures in degrees Fahrenheit to degrees Celsius. Encourage students to memorize these formulas.

Section 5-5 (pages 135–137)
In this section, students express a fraction whose numerator is greater than the denominator as a mixed numeral, write a fraction for a mixed numeral, and multiply with mixed numerals. Point out that writing a fraction as a mixed numeral is not the same as writing a fraction in lowest terms.

Section 5-6 (pages 140–144)
This Problem Solving and Applications lesson provides students with more practice in multiplying with fractions and mixed numerals by using formulas to find the area of a parallelogram and a triangle. Students solve many real-world problems involving parallelograms and triangles.

Section 5-7 (pages 145–147)
The goal of this Problem Solving and Applications section is for students to use a formula to find the volume of a rectangular prism. Practical real-world problems include finding the volume of common objects and solving word problems involving items found in and around the home. The use of customary and metric units provides practice in multiplication with fractions and decimals, respectively.

Section 5-8 (pages 148–151)
Remind students to look for a common factor of the numerator and denominator after writing a division problem as a multiplication problem. The More Challenging Exercises provide students practice with dividing simple algebraic fractions.

COMMENTARY: OPTIONAL FEATURES

Career Applications (page 138)
This feature applies the skills for solving formulas taught in Section 5-7 (pages 145–147) to finding the greatest safe load a beam can bear.

Consumer Application (page 139)
This feature introduces the British thermal unit (BTU) for the measure of cooling capacity of an air conditioner. Since heat and temperature change are directly related, it can be used when Section 5-4 (pages 132–134) is taught.

Enrichment (page 155)
This feature can be used with Section 5-6 (pages 140–144), as it applies the skill of using the formulas for finding the area of a parallelogram and of a triangle to solving problems that involve the use of both formulas to find the area.

CHAPTER 6: FRACTIONS: ADDITION/SUBTRACTION

OVERVIEW

In Chapter 6, students add and subtract with fractions and mixed numerals as well as compare fractions. Section 6-4 uses addition, subtraction, and multiplication with mixed numerals to apply the problem solving technique of the "hidden question" to the solution of two-step problems. Section 6-6 applies multiplication with fractions, mixed numerals, and decimals to finding the circumference and the area of a circle.

SECTION-BY-SECTION COMMENTARY

Section 6-1 (pages 158–160)
In this section, students review the addition and subtraction of like fractions and of mixed numerals containing like fractions.

Section 6-2 (pages 161–164)
The major goal of this section is for students to add and subtract with fractions having unlike denominators. It may be helpful for some students to review Section 4-5 (pages 111–114) on finding the least common multiple of two or more numbers.

Section 6-3 (pages 165–167)
Borrowing from the whole number is an essential skill in subtracting with mixed numerals. The skills of addition and subtraction with mixed numerals are applied to solving word problems that involve track and field sports.

Section 6-4 (pages 169–172)
In this Problem Solving and Applications section, students solve two-step problems involving earnings by applying the technique of the "hidden question." Students read and complete a time card and solve word problems involving total earnings.

Section 6-5 (pages 173–175)
Fractions can be compared by writing a decimal for each fraction and then comparing the decimals. Some students may find it helpful to review the Place Names Chart on page 32.

Section 6-6 (pages 176–179)
This Problem Solving and Applications section applies multiplication with fractions, mixed numerals, and decimals to finding the circumference and area of a circle. Students use customary and metric units to find the circumference and area of real-world objects.

COMMENTARY: OPTIONAL FEATURES

Calculator/Computer Applications (page 168)
This feature illustrates the use of the calculator and a computer program for evaluating powers in which the base and exponent are counting numbers.

Calculator Applications (page 180)
This feature is an application of the calculator to comparing fractions. Therefore, it can be used with Section 6-5 (pages 173–175).

Enrichment (page 184)
This feature applies the skills of adding and subtracting with mixed numerals to the concept of precision used in measurement. Thus, it can be used with Section 6-3 (pages 165–167).

Common Errors (page 186)
The exercises in the *Common Errors* feature focus student attention on avoiding some of the most common errors involved with the skills taught in Chapters 4–6.

Overview and Commentary

CHAPTER 7: SOLVING EQUATIONS

OVERVIEW

In this chapter, students solve equations by subtraction, addition, division, and multiplication. Checking the solutions of all equations serves two purposes: to determine whether any errors were made in solving, and to underscore the meaning of *solution* as a number that makes an equation true. Two Problem Solving and Applications sections provide practice in using equations to solve word problems involving earnings, ground speed and airspeed, and batting average.

SECTION-BY-SECTION COMMENTARY

Section 7-1 (pages 190–192)
Students solve equations containing positive rational numbers by subtraction. Encourage students to check solutions.

Section 7-2 (pages 193–195)
In this section, students solve equations by addition. Emphasize the concept of equivalent equations. The More Challenging Exercises introduce the concept of replacement set.

Section 7-3 (pages 196–198)
This Problem Solving and Applications section provides students with practice in using equations to solve word problems involving a formula for determining net pay and a formula for calculating a plane's airspeed.

Section 7-4 (pages 200–202)
The goal of this section is to solve equations by division. Students practice each subskill involved in solving an equation by division.

Section 7-5 (pages 203–205)
At this point, students should understand the major concepts of inverse operation and equivalent equations presented in this section and in previous sections. Remind students to check each solution.

Section 7-6 (pages 206–209)
In this Problem Solving and Applications section, students use equations to solve word problems involving the formula for the distance traveled by an object at a constant speed and the formula for a baseball player's batting average.

COMMENTARY: OPTIONAL FEATURES

Consumer Applications (page 199)
This feature is an application of the skills of estimating with decimals and solving a formula to finding the number of rolls of wallpaper and the cost of papering rectangular walls. Note the use of the calculator for determining the actual cost of papering the room.

Enrichment (page 213)
This feature applies the skill of solving equations by division to writing a repeating decimal as a fraction. Thus, it can be used with Section 7-4 (pages 200–202).

CHAPTER 8: RATIO, PROPORTION, AND PERCENT

OVERVIEW

The goals of this chapter are to develop the student's understanding and skill in solving proportions and in finding the percent of a number. Applications of percent to word problems involving simple interest and discount are provided in the last two sections of the chapter.

SECTION-BY-SECTION COMMENTARY

Section 8-1 (pages 216–218)
A ratio is a name for a quotient in which a comparison of two numbers is emphasized. The skill of determining whether two ratios are equal is essential for understanding and solving proportions.

Section 8-2 (pages 219–221)
Point out that students can use the method taught in Chapter 7 to solve the resulting multiplication equation. In the applications, students solve word problems involving scales for distances on a map.

Section 8-3 (pages 222–224)
The goal of this section is for students to use proportions to solve word problems. The Classroom Exercises provide students with practice in setting up a proportion. Emphasize the importance of setting up the ratios on each side of the proportion equation correctly.

Section 8-4 (pages 227–230)
In this section, students write a percent for a ratio whose denominator is not 100. In the applications, students apply this skill to solving word problems on population.

Section 8-5 (pages 231–234)
Any rational number can be expressed as either a decimal, a fraction, or a percent. Students practice writing a percent for a decimal and a decimal for a percent.

Section 8-6 (pages 235–239)
In this section, students learn to find a percent of a number. The More Challenging Exercises apply this skill to solving word problems that involve more than one step. Encourage students to memorize the table on page 235.

Section 8-7 (pages 240–243)
Students use the formula for simple interest to solve word problems on borrowing and saving money. Emphasize that in the interest formula, the time (t), is expressed in years.

Section 8-8 (pages 244–247)
The goal of this section is for students to find the amount of discount on an item. The problem solving technique of the "hidden question" is used to solve word problems involving sale price.

COMMENTARY: OPTIONAL FEATURES

Computer Applications (page 226)
This feature illustrates the use of a computer program for solving word problems involving ratios, as introduced in Chapter 7 and related to percent in Section 8-4 (pages 227–230). Two fundamental problem solving steps, identifying the input/output variables and writing the correct formula, are logically extended to writing and checking the BASIC program. The computer solution and printout provides verification and reinforcement of the student's problem solving skills.

Calculator Applications (page 243)
This feature applies the calculator to computing simple interest. Thus, it can be used with Section 8-7 (pages 240–243).

Enrichment (page 251)
In this feature, students write a fraction for a repeating decimal. Thus, it can be used after Section 8-5 (pages 231–234).

Overview and Commentary

CHAPTER 9: PERCENT AND APPLICATIONS

OVERVIEW

Chapter 9 continues the development of solving percent equations begun in the preceding chapter. Section 9-4 presents an alternate method for solving percent problems, the proportion method. In Sections 9-5 and 9-6, students apply percents and the skills of interpreting and constructing bar graphs and circle graphs. Estimation is included as a useful tool in solving problems involving percents.

SECTION-BY-SECTION COMMENTARY

Section 9-1 (pages 254–256)
The goal of this section is for students to determine what percent one number is of another. This skill is applied on two levels: solving numerical problems and solving word problems.

Section 9-2 (pages 257–260)
In this Problem Solving and Applications section, students solve word problems to find the percent of increase in monthly cooling costs and the percent of decrease in monthly heating costs.

Section 9-3 (pages 261–263)
Students solve numerical problems and word problems to find a number when a percent of it is known. In the More Challenging Exercises, students practice this skill with percents greater than 100.

Section 9-4 (pages 264–267)
This section presents the proportions method of solving percent problems as an alternative to the equation method. Teachers who prefer the equation method may wish to use this as a review section.

Section 9-5 (pages 270–273)
In this section, students use data read from bar graphs to solve percent problems and construct bar graphs using data given in tables. Encourage students to use consistent units for the scales and to make their graphs readable.

Section 9-6 (pages 274–278)
Students use data read from circle graphs to solve percent problems and construct circle graphs using data given in tables. A protractor is essential for constructing circle graphs.

Section 9-7 (pages 279–282)
Students will find it helpful to estimate the percent of a number either by rounding the number only or by rounding to a convenient percent. Remind students to follow the rules for rounding.

COMMENTARY: OPTIONAL FEATURES

Consumer Applications (page 269)
This feature applies the skills for solving word problems involving percent to comparing the vitamin or mineral content of selected foods to the adult RDA. Therefore, it can be used after Section 9-4 (pages 264–267) is taught.

Calculator Applications (page 282)
This feature can be used with Section 9-7 (pages 279–282), as it is an application of the calculator to estimating the amount of discount.

Enrichment (page 288)
Because this feature relates sequences to the concept of increase and decrease, it can be used with Section 9-2 (pages 257–260).

Common Errors (page 290)
The exercise in the *Common Errors* feature focus student attention on avoiding some of the most common errors involved with the skills taught in Chapters 7–9.

CHAPTER 10: RATIONAL NUMBERS: ADDITION/SUBTRACTION

OVERVIEW

The purpose of this chapter is to give students practice with addition and subtraction of positive and negative rational numbers. The skill of combining like terms is applied to adding polynomials.

SECTION-BY-SECTION COMMENTARY

Section 10-1 (pages 294–296)
The goals of this section are to introduce the concept of positive and negative numbers, to graph integers on a number line, and to compare integers. The skill of comparing integers is applied to solving word problems involving temperature.

Section 10-2 (pages 297–299)
In this section, the meaning and the symbol for opposite are introduced. The More Challenging Exercises extend the concept of ordering positive and negative integers to include rational numbers.

Section 10-3 (pages 300–302)
Students use a number line to add integers.

Section 10-4 (pages 304–307)
In this section, the rules for adding rational numbers are explained in terms of the concepts of absolute value and opposite. Emphasize that the absolute value of a rational number is always positive.

Section 10-5 (pages 308–310)
Students must be able to express a difference as an equivalent sum as shown in Example 1 and Step 1 of Examples 2 and 3. Practice with this skill is provided in both the Classroom and Written Exercises.

Section 10-6 (pages 312–314)
Students use the properties of addition to add three or more rational numbers. Emphasize the Addition Property of Opposites.

Section 10-7 (pages 315–317)
The goal of this section is for students to add polynomials. Point out that terms should be arranged in order with the greatest exponent first when writing their answers. The More Challenging Exercises cover subtraction of polynomials.

COMMENTARY: OPTIONAL FEATURES

Consumer Applications (page 303)
This feature applies the skills for using formulas to finding the stopping distance of a vehicle when vehicle speed and the road conditions are known. Thus it can be used any place in Chapter 10.

Career Applications (page 311)
This feature applies the concept of positive and negative numbers. Thus, it can be used with Section 10-1 (pages 294–296). It can also be used with Section 10-5 (pages 307–309), as the students find the time difference between two cities in different time zone descriptions.

Calculator Applications (page 314)
As this feature is an application of the calculator to adding rational numbers, it can be used with Section 10-6 (pages 312–314).

Enrichment (page 321)
This feature applies the skills of adding and subtracting with rational numbers to clock arithmetic. Thus, it can be used after Section 10-5 (pages 307–309).

Overview and Commentary **M-23**

CHAPTER 11: EQUATIONS: ADDITION/SUBTRACTION

OVERVIEW

This chapter extends the skills taught in Chapter 7 to include applying the Addition Property of Equations and the Subtraction Property of Equations to solving equations containing like terms. The Problem Solving and Applications section provides students with practice in changing from words to symbols when solving word problems. In the last two sections, students solve and graph inequalities.

SECTION-BY-SECTION COMMENTARY

Section 11-1 (pages 324–325)
The methods used in Sections 1 and 2 of Chapter 7 are formalized to become the Addition Property for Equations and the Subtraction Property for Equations. Students use these properties to solve equations involving positive and negative rational numbers.

Section 11-2 (pages 326–328)
In this section, students solve one- and two-step equations by combining like terms. In each case, the terms containing the variable are on the same side of the equation. Remind students that the goal in solving equations is to get the variable alone on one side of the equation.

Section 11-3 (pages 329–330)
The goal of this section is for students to solve an equation with the variable on both sides. Students must get the terms containing the variable on the left side of the equation before combining like terms.

Section 11-4 (pages 334–337)
Use the table on page 334 to show that key words (underscored) in a word problem can be used as clues to determine which operation is indicated. This section provides practice in writing algebraic equations to solve word problems.

Section 11-5 (pages 338–339)
In this section, students graph inequalities on a number line. Emphasize the meaning of open and closed circles when graphing inequalities.

Section 11-6 (pages 340–342)
Point out, to students, the similarity between the Addition and Subtraction Properties for Inequalities and the Addition and Subtraction Properties for Equations. In the More Challenging Exercises, students solve and graph inequalities with like terms.

COMMENTARY: OPTIONAL FEATURES

Calculator Applications (page 331)
This feature applies the calculator to checking equations with the variable on both sides. Thus, it can be used with Section 11-3 (pages 329–330).

Consumer Applications (page 332)
This feature illustrates the use of a calculator to finding the yearly percentage rate of interest of a bank loan. Because it applies the skills for evaluating an expression by more than one operation, it can be used with Section 11-2 (pages 326–328).

Computer Applications (page 333)
This feature illustrates the use of a computer program for finding the sum of two fractions. Since it applies the skill of adding rational numbers, it can be used after Section 10-4 (pages 304–307).

Calculator Applications (page 342)
Because this feature is an application of the calculator to checking inequalities, it can be used with Section 11-6 (pages 340–342).

Enrichment (page 347)
In this feature, students graph compound inequalities on a number line. Thus, it can be used after Section 11-5 (pages 338–339).

CHAPTER 12: RATIONAL NUMBERS: MULTIPLICATION/DIVISION

OVERVIEW

The purpose of this chapter is to give students practice with multiplication and division of positive and negative numbers. In Section 12-3, the properties of multiplication are introduced. Students use the Distributive Property to write products as sums or differences in preparation for factoring.

SECTION-BY-SECTION COMMENTARY

Section 12-1 (pages 350–352)
In this section, students multiply two rational numbers with unlike signs. This skill is applied to writing algebraic expressions for word expressions and to multiplying monomials.

Section 12-2 (pages 353–355)
Multiplication patterns provide students with a rationale for the rule of multiplying two rational numbers with like signs. In the More Challenging Exercises, students apply this skill to evaluating numerical expressions and multiplying monomials.

Section 12-3 (pages 356–358)
The purpose of this section is for students to use the Distributive Property to express a product as a sum or difference. Emphasize that each term within the parentheses must be multiplied by the term outside the parentheses.

Section 12-4 (pages 360–363)
Students should see that the rules for dividing two rational numbers are the same as the rules for multiplying two rational numbers. The More Challenging Exercises apply the skill of dividing two rational numbers to evaluating algebraic expressions.

Section 12-5 (pages 364–366)
In this section, students factor a sum or difference of two terms by applying the Distributive Property. Point out that the common factor chosen should be the greatest common factor of each term.

COMMENTARY: OPTIONAL FEATURES

Calculator Applications (page 355)
This feature can be used with Section 12-2 (pages 353–355), as it is an application of the calculator to multiplying rational numbers.

Consumer Applications (page 359)
This feature applies the skills for multiplying rational numbers to calculating the rate of heat transfer through a surface. Because this feature requires use of the Distributive Property to express products as a sum or a difference, it can be used after Section 12-3 (pages 356–358) is taught.

Calculator Applications (page 367)
This feature is an application of the calculator to applying the Distributive Property. Thus, it can be used with Section 12-3 (pages 356–358) or Section 12-5 (pages 364–366).

Enrichment (page 371)
Because this feature is an application of the Distributive Property to multiplying two binomials, it can be used after Section 12-3 (pages 356–358).

Overview and Commentary

CHAPTER 13: EQUATIONS: MULTIPLICATION/DIVISION

OVERVIEW

The major goal of this chapter is for students to solve equations and inequalities using the multiplication and division properties for equations and inequalities. Emphasis is placed on solving equations involving two or more steps. These skills are applied throughout the chapter in four Problem Solving and Applications sections. These sections include the following topics: writing equations to solve word problems, angles in a triangle, consecutive numbers, and using inequalities to solve word problems.

SECTION-BY-SECTION COMMENTARY

Section 13-1 (pages 374–376)
The methods used in Sections 7-5 and 7-6 are formalized to become the Multiplication and Division Properties for Equations. Students use these properties to solve equations involving negative as well as positive numbers.

Section 13-2 (pages 377–380)
This section provides practice in translating from word expressions to mathematical expressions. Students apply this skill to solving word problems.

Section 13-3 (pages 381–383)
In this section, students extend the equation-solving skills involved in solving two-step equations introduced in Section 11-3. Emphasis is placed on solving two-step equations in which the coefficient of the variable term can be negative. Encourage students to check their answers.

Section 13-4 (pages 383–385)
In this section, students use the Distributive Property to solve equations containing parentheses. Point out that the parentheses must be removed before using the addition and subtraction properties and before combining like terms.

Section 13-5 (pages 388–389)
This Problem Solving and Applications section applies the skill of solving equations with more than one unknown to finding the measures of the angles of a triangle. Emphasize that the sum of the measures of the angles equals 180°.

Section 13-6 (pages 390–392)
Encourage students to write in words what each variable represents. Emphasize that the problems in this lesson have more than one answer. Thus, students must remember to evaluate each variable in order to include all the answers to a given problem.

Section 13-7 (pages 393–395)
In this section, students solve and graph inequalities using the Multiplication and Division Properties for Inequalities. Emphasize the rule for multiplying or dividing each side of an inequality by the same negative number.

Section 13-8 (pages 396–398)
This Problem Solving and Applications section applies the skills of the previous lesson to solving word problems involving the money, and game and test scores. The Review Capsule on page 395 will help students to review the concept of average.

COMMENTARY: OPTIONAL FEATURES

Computer Applications (page 387)
This feature illustrates the use of a computer program for solving equations. It can be used after Section 13-4 (pages 383–386).

Enrichment (page 402)
This feature introduces students to some symbols and terms used with sets.

Common Errors (page 404)
The exercises in the *Common Errors* feature focus student attention on avoiding some of the most common errors involved with the skills taught in Chapters 10–13.

CHAPTER 14: GRAPHING AND EQUATIONS

OVERVIEW

The three major goals of Chapter 14 are to understand and graph linear equations, to solve systems of linear equations by graphing, and to graph linear inequalities. Students are introduced to the characteristics of the graph of a linear equation (slope, y intercept, x intercept) in Sections 14-4 and 14-5. Each major concept is introduced with a real-world situation. The Problem Solving and Applications section requires students to read and construct a broken line graph.

SECTION-BY-SECTION COMMENTARY

Section 14-1 (pages 408–409)
This section provides students with practice in labeling and graphing a point. Emphasize that the x coordinate is listed first in all ordered pairs.

Section 14-2 (pages 410–412)
Distance/rate/time problems provide a useful application of the algebraic skills of this section. Point out that organizing data in a table helps to identify relationships and patterns. Encourage students to choose a negative value, zero, and a positive value for this variable when constructing data.

Section 14-3 (pages 414–416)
In this section, students read a broken line graph to solve word problems. Again, encourage students to use consistent units for the scales when constructing a broken line graph. Emphasize that points are connected in order by a straight line.

Section 14-4 (pages 417–419)
This section provides students with practice in finding the slope of a line containing two given points. Emphasize the meaning of a line having either a positive, negative, zero, or undefined slope.

Section 14-5 (pages 422–423)
Students are provided with practice in finding the x and y intercepts of a linear equation and with using the x and y intercepts to graph an equation. Since all equations are in the form $y = ax + b$, some students may notice that the numerical term, b, has the same value as the y intercept. You may wish to have the more able students explain why this is so.

Section 14-6 (pages 424–426)
In this section, students are introduced to the concept of direct and indirect variation. Emphasize that the ratio of the variables is constant when the variation is direct, and the product of the variables is constant when the variation is indirect.

Section 14-7 (pages 427–429)
Students are introduced to the concept of solving a system of equations by graphing through an application from business. The major goal is to find the single ordered pair that satisfies both equations. Emphasize that this ordered pair must be checked in <u>both</u> equations of the system.

Section 14-8 (pages 430–431)
This section illustrates how to graph linear inequalities in the coordinate plane. Testing two points, one on each side of the line, will minimize errors and will reinforce the fact that all points on one side of the line satisfy the inequality while no points on the other side satisfy it.

COMMENTARY: OPTIONAL FEATURES

Calculator Applications (page 413)
This feature can be used with Section 14-2 (pages 410–412), as it is an application of the calculator to graphing equations.

Computer Applications (page 421)
This feature illustrates the use of a computer program for finding the slope of a line. It can be used after Section 14-4 (pages 417–419).

Enrichment (page 437)
In this feature, students use Venn diagrams to show how sets are related.

Overview and Commentary

CHAPTER 15: SPECIAL TRIANGLES

OVERVIEW

The work on square roots in this chapter extends student concepts and skills from the rational numbers to the real numbers. Applications with square roots include solving word problems involving formulas with square roots, applying the Pythagorean Theorem, and using the tangent ratio to find the length of a side of a right triangle.

SECTION-BY-SECTION COMMENTARY

Section 15-1 (pages 440–441)
The concept of square root is presented as the inverse operation of squaring a number. Emphasize the notation for the principal square root and the negative square root.

Section 15-2 (pages 442–444)
The goal of this section is for students to identify a number as rational or irrational.

Section 15-3 (pages 445–447)
Students are provided with practice in finding squares and square roots of numbers using the table on page 451. Point out that the table can be used to find the square roots of some numbers greater than 150.

Section 15-4 (pages 451–453)
In this Problem Solving and Applications section, students use formulas to solve word problems related to the sides of a square and the distance to the horizon from an aircraft.

Section 15-5 (pages 454–457)
The Pythagorean Theorem provides an additional application of real numbers. Students apply the skill of finding the measure of a side of a right triangle to solving word problems.

Section 15-6 (pages 458–460)
In this section, students use the tangent ratio and tangent table to find the tangent of an acute angle, and the measure of a leg, of a right triangle.

COMMENTARY: OPTIONAL FEATURES

Career Applications (page 448)
This feature applies the skills for problem solving, using square roots, to calculating the wind chill temperature and illustrates the use of a calculator to evaluate a formula. Thus, it can be used after Section 15-4 (pages 451–453) is taught.

Calculator/Computer Applications (page 449)
This feature illustrates the use of a calculator or a computer program for finding the area of a triangle using Hero's Formula. It can be used after Section 15-3 (pages 445–447) is taught, as it applies the skill of finding the square root of a number.

Calculator Applications (page 461)
As this feature is an application of the calculator to finding tangent values, it can be used with Section 15-6 (pages 458–460).

Enrichment (page 465)
This feature applies the skill of successive approximation to finding the square root of an irrational number to a given number of decimal places. Thus, it can be used after Section 15-3 (pages 445–447).

Common Errors (page 467)
The exercises in the *Common Errors* feature focus student attention on avoiding some of the most common errors involved with the skills taught in Chapters 14–15.

CHAPTER 16: STATISTICS AND PROBABILITY

OVERVIEW

Chapter 16 begins with the statistical concepts of mean, median, and mode. Section 16-2 deals with organizing data in tabular form and displaying it in a histogram. The rest of the chapter develops the concept of mathematical probability. Sections 16-3, 16-4, and 16-5 cover simple probability and the probability of mutually exclusive events. The final section covers the probability of independent and dependent events involving more than one trial.

SECTION-BY-SECTION COMMENTARY

Section 16-1 (pages 470–473)
Point out that "mean" is another word for average. Emphasize that data must be listed in order by size to find the median. Remind students that a group of data may have more than one mode.

Section 16-2 (pages 474–477)
The purpose of this section is for students to list data by using equal intervals in a table and to use a histogram to show data. A histogram should not be confused with a bar graph. A histogram is a graphic representation of a frequency distribution. The width of the vertical rectangles corresponds to a definite range of frequencies, while the height corresponds to the number of frequencies occurring within the range.

Section 16-3 (pages 479–481)
Be sure that students understand the concept of equally likely outcomes and do not confuse it with two or more events having equal probabilities.

Section 16-4 (pages 482–484)
Students may find it helpful to make a table of possible outcomes when finding the probability of two or more mutually exclusive events. Emphasize that probabilities are added when only one trial is involved.

Section 16-5 (pages 485–487)
A tree diagram is another way for students to show all possible outcomes to find the probability of an event. Point out that tree diagrams are most useful when the number of outcomes is relatively small.

Section 16-6 (pages 488–490)
In this section, students find the probability of two or more independent and two or more dependent events. Point out that the probabilities are multiplied because more than one trial is involved.

COMMENTARY: OPTIONAL FEATURES

Career Applications (page 478)
This feature introduces the statistical concept of standard deviation and illustrates a five-step procedure for calculating it. Because standard deviation relates to variations in collected data, it can be used after Section 16-2 (pages 474–477) is taught.

Calculator Applications (page 491)
This feature applies the calculator to computing the probabilities of dependent events. Thus, it can be used with Section 16-6 (pages 488–490).

Enrichment (page 495)
In this feature, students use the concept of a sample, derived from statistical probability, to find the number of defective items in a shipment. Thus, it can be used after Section 16-3 (pages 479–481).

Overview and Commentary

CHAPTER 17: GEOMETRY

OVERVIEW

Chapter 17 can be divided into two parts. The first part introduces students to the description and terminology of some basic geometric figures. In Section 17-4, the relationship of the concept of similar triangles and the skill of solving proportions is used to solve real-world problems.

In the second part of the chapter, students use metric geometry to find the surface area and volume of various solids. The Problem Solving and Applications section provides practice in using a formula to find the surface area of real-world objects having the shape of a rectangular prism.

SECTION-BY-SECTION COMMENTARY

Section 17-1 (pages 498–500)
Labels for lines, segments, and triangles may be given in any order. The labels for rays and angles require greater care because the endpoint of the ray must be named first and the letter at the vertex of the angle must be named in the middle of the three-letter label for an angle.

Section 17-2 (pages 501–503)
In this section, students use a protractor to measure a given angle and to draw a given angle. Students also classify angles according to their measures. Help students to read a protractor accurately.

Section 17-3 (pages 504–507)
Emphasize the meaning of included angle and included side as it relates to the rules of congruence, SAS and ASA.

Section 17-4 (pages 508–511)
Students use proportions to solve problems involving similar triangles. Point out the correspondence of the sides in each ratio.

Section 17-5 (pages 514–516)
Give students extensive practice in recognizing the pairs of congruent angles formed when a transversal intersects two parallel lines. Emphasize that in more complicated figures, the angle relationships hold with respect to a particular transversal.

Section 17-6 (pages 517–519)
In this section, students classify a quadrilateral according to its properties. Some quadrilaterals may have more than one classification.

Section 17-7 (pages 520–523)
This section provides practice in using a formula to find the surface area of real-world objects having the shape of a rectangular solid. Since surface area is a measure of area, it is expressed in square units.

Section 17-8 (pages 524–526)
The volume formula $V = Bh$ for a prism carries over directly to the volume formula for a cylinder. Since the base of a cylinder is a circle, B becomes πr^2 in the volume formula for a cylinder.

Section 17-9 (pages 527–529)
The same relationship described in Section 17-8 for the area of the base, B, applies to the formulas for the volume of a pyramid and of a cone ($V = \frac{1}{3}Bh$). Remind students that volume is measure in cubic units.

COMMENTARY: OPTIONAL FEATURES

Calculator Applications (page 511)
This feature can be used with Section 17-4 (pages 508–511), as it is an application of the calculator to similar triangles.

Computer Applications (page 512)
This feature illustrates the use of a computer program for determining whether two angles are supplementary and for finding the measure of each base angle of an isosceles triangle when the measure of the vertex angle is known. It can be used after Section 17-3 (pages 504–507) is taught.

Enrichment (page 535)
In this feature, students use a formula to find the volume of a sphere. Thus, it can be used after Section 17-9 (pages 527–529).

Common Errors (page 537)
The exercises in the *Common Errors* feature focus student attention on avoiding some of the most common errors involved with the skills taught in Chapters 16–17.

MASTER TIMETABLE

Overview The following *Master Timetable* for three levels of ability is coordinated with the *Suggested Timetables of Assignments* that appear on pages M-32 through M-42 and the *Assignment Guide* that appears in the margin of the annotated pages. Note that the total number of days for each chapter includes additional days for review and testing for each level.

The *Master Timetable*, the *Suggested Timetables of Assignments*, and the *Assignment Guide* should be considered as guidelines only.

MASTER TIMETABLE

	Basic		Average		Above Average	
Chapter	Sections	Days	Sections	Days	Sections	Days
1	All	13	All	11	All	9
2	O (7)	12	All	11	All	10
3	All	13	All	9	All	9
4	All	12	All	11	All	9
5	All	15	All	14	All	11
6	All	14	All	12	All	9
7	All	14	All	12	All	9
8	All	13	All	13	All	11
9	All	12	All	10	All	10
10	O (7)	14	O (7)	11	All	10
11	O (5, 6)	12	O (6)	10	All	9
12	O (5)	11	All	10	All	8
13	O (5-8)	12	O (7, 8)	13	All	11
14	Omit	0	O (8)	11	All	11
15	Omit	0	O (6)	8	All	9
16	Omit	0	Omit	0	All	9
17	Omit	0	Omit	0	All	11
Cumulative Reviews		3		4		5
Total		**170**		**170**		**170**

SUGGESTED TIMETABLES OF ASSIGNMENTS

The *Suggested Timetable of Assignments* for each chapter is structured for three levels of ability: **Basic**, **Average**, and **Above Average** (See page M-11.).

CHAPTER 1: WHOLE NUMBERS AND APPLICATIONS

Section	Pages	Basic	Average	Above Average
1-1	2-4	All 1-12	All 1-12	All 1-12
1-2	5-6	All 1-12	All 1-12	All 1-12
1-3	7-9	All 1-24	Odds 1-23	Odds 1-23
1-4	10-12	Day 1: All 1-11 Day 2: All 12-21	All 1-11 Odds 13-21	Odds 1-21
1-5	14-16	All 1-10	Odds 1-7 All 9-12	Odds 1-7 All 9-12
1-6	18-21	Day 1: Odds 1-39 Day 2: Evens 2-50	Day 1: All 1-32 Day 2: All 33-64	3, 6, 9 . . . , 60 Odds 81-97
1-7	22-24	Odds 1-33	Odds 1-11 All 13-26	Odds 1-33
Review and Testing		3 days	2 days	2 days
Enrichment		1 day	1 day	1 day
Total Days		**13**	**11**	**9**

CHAPTER 2: DECIMALS AND APPLICATIONS

Section	Pages	Basic	Average	Above Average
2-1	32-34	All 1-30	Odds 1-18 All 19-32	Odds 1-18 All 19-32
2-2	35-37	All 1-22	Odds 1-27	Odds 1-25 All 27-29
2-3	38-41	Day 1: All 1-22 Day 2: All 27-33	Day 1: All 1-26 Day 2: All 27-39	Odds 1-33 All 34-43
2-4	44-46	Day 1: Odds 1-25 Day 2: Evens 2-26	All 1-27	Odds 1-23 All 25-30
2-5	47-49	All 1-9	All 1-12	All 1-14
2-6	51-54	All 1-25	All 1-27	Odds 1-23 All 25-30
2-7	55-57	Omit	All 1-24	All 1-24
Review and Testing		3 days	2 days	2 days
Enrichment		1 day	1 day	1 day
Total Days		**12**	**11**	**10**

CHAPTER 3: NUMBER PROPERTIES/EXPRESSIONS

Section	Pages	Basic	Average	Above Average
3-1	64-66	All 1-36	Odds 1-23 All 25-44	Odds 1-43 All 45-54
3-2	67-69	Day 1: All 1-25 Day 2: All 26-45	Odds 1-49	3, 6, 9, . . . , 48 All 50-55
3-3	70-73	Day 1: All 1-18 Day 2: All 19-36	Odds 1-39	Odds 1-33 All 35-44
3-4	76-78	Day 1: All 1-22 Day 2: All 23-40	All 1-10 Odds 11-41	Odds 1-41
3-5	79-80	All 1-22	Odds 1-7 All 8-26	Odds 1-27 All 8-30
3-6	81-85	Odds 1-35	Odds 1-39	Odds 1-35 All 37-48
Review and Testing		3 days	2 days	2 days
Enrichment		1 day	1 day	1 day
Total		**13**	**9**	**9**

Suggested Timetables of Assignments

CHAPTER 4: NUMBER THEORY

Section	Pages	Basic	Average	Above Average
4-1	96-98	Day 1: All 1-21 Day 2: All 22-26	All 1-26	Odds 1-21 All 22-26
4-2	99-102	Odds 1-31	Odds 1-35	3, 6, 9, . . . , 30 All 31-36
4-3	103-105	All 1-24	All 1-24	Odds 1-17 All 19-28
4-4	108-110	Day 1: All 1-20 Day 2: All 31-48	Day 1: All 1-20 Day 2: All 31-54	3, 6, 9, . . . , 60
4-5	111-114	Odds 1-45	Odds 1-45	3, 6, 9, . . . , 47 All 49-52
4-6	115-116	All 1-23	All 1-23	Odds 1-23
Review and Testing		3 days	3 days	2 days
Enrichment		1 day	1 day	1 day
Total		12	11	9

CHAPTER 5: FRACTIONS: MULTIPLICATION/DIVISION

Section	Pages	Basic	Average	Above Average
5-1	124-125	All 1-20	All 1-32	Odds 1-19 All 21-32
5-2	126-128	All 1-28	Odds 1-31	Odds 1-27 All 29-32
5-3	129-131	Day 1: Odds 1-35 Day 2: Evens 2-36	All 1-38	Odds 1-39 All 41-50
5-4	132-133	All 1-12	All 1-15	All 1-15
5-5	135-137	Day 1: Odds 1-33 Day 2: Evens 2-34	Day 1: Odds 1-37 Day 2: Evens 2-38	Odds 1-37 All 39-48
5-6	140-144	Odds 1-27	Odds 1-31	Odds 1-31
5-7	145-147	All 1-16	All 1-18	Odds 1-17 All 19-22
5-8	148-151	Day 1: Odds 1-59 Day 2: Evens 2-58	Day 1: Odds 1-43 Day 2: All 45-67	3, 6, 9, . . . , 63 All 65-77
Review and Testing		3 days	3 days	2 days
Enrichment		1 day	1 day	1 day
Total		15	14	11

Suggested Timetables of Assignments

CHAPTER 6: FRACTIONS: ADDITION/SUBTRACTION

Section	Pages	Basic	Average	Above Average
6-1	158-160	Day 1: Odds 1-33 Day 2: Evens 2-36	Odds 1-21 All 23-36	3, 6, 9, . . . , 36 All 37-51
6-2	161-164	Day 1: Odds 1-47 Day 2: Evens 2-46	Day 1: Odds 1-41 Day 2: All 43-56	Odds 1-71
6-3	165-167	Day 1: Odds 1-39 Day 2: Evens 2-40	Day 1: Odds 1-39 Day 2: Evens 8-42	Odds 1-39 All 41-43
6-4	169-172	All 1-12	All 1-12	Odds 1-7 All 9-16
6-5	173-175	Odds 1-35	Odds 1-39	Odds 1-39
6-6	176-179	Day 1: All 1-16 Day 2: All 17-32	Odds 1-31	Odds 1-31
Review and Testing		3 days	3 days	2 days
Enrichment		1 day	1 day	1 day
Total		14	12	9

CHAPTER 7: SOLVING EQUATIONS

Section	Pages	Basic	Average	Above Average
7-1	190-192	Day 1: All 1-22 Day 2: All 23-45	Day 1: Odds 1-35 Day 2: All 37-51	Odds 1-51
7-2	193-195	Day 1: All 1-24 Day 2: All 25-48	Day 1: Odds 1-35 Day 2: All 37-48	Odds 1-47 All 49-54
7-3	196-198	All 1-16	All 1-18	All 1-18
7-4	200-202	Day 1: All 1-20 Day 2: All 21-44	Odds 1-27 All 29-44	Odds 1-43 All 45-56
7-5	203-205	Day 1: All 1-24 Day 2: All 25-44	Odds 1-31 All 33-44	Odds 1-43 All 45-52
7-6	206-209	All 1-6, 11-18	All 1-20	Odds 1-21 All 22-24
Review and Testing		3 days	3 days	2 days
Enrichment		1 day	1 day	1 day
Total		14	12	9

Suggested Timetables of Assignments

CHAPTER 8: RATIO, PROPORTION, AND PERCENT

Section	Pages	Basic	Average	Above Average
8-1	216-218	All 1-26	Odds 1-43	Odds 1-43
8-2	219-221	Odds 1-29	Odds 1-33	Odds 1-35
8-3	222-225	All 1-10	All 1-14	All 1-14
8-4	227-230	Odds 1-53	Odds 1-57	Odds 1-53 All 55-61
8-5	231-234	All 1-45	Odds 1-57	Odds 1-61
8-6	235-239	Day 1: All 1-14 Day 2: All 15-26	Day 1: Odds 1-25 Day 2: Evens 8-32	Odds 7-25 All 27-38
8-7	240-243	All 1-18	Odds 1-17 All 18-22	Odds 1-17 All 18-22
8-8	244-247	All 1-13	All 1-13	Odds 1-9 All 10-13
Review and Testing		3 days	3 days	2 days
Enrichment		1 day	1 day	1 day
Total		13	13	11

Suggested Timetables of Assignments

CHAPTER 9: PERCENT AND APPLICATIONS

Section	Pages	Basic	Average	Above Average
9-1	254-256	Day 1: All 1-14 Day 2: All 15-26	Odds 1-19 All 21-26	Odds 7-19 All 21-26
9-2	257-260	All 1-16	All 1-19	Odds 1-15 All 17-23
9-3	261-263	All 1-20	All 1-20	Odds 1-21 All 22-26
9-4	264-267	All 1-18	Odds 1-19 All 20-26	Odds 1-19 All 20-34
9-5	270-273	All 1-10	All 1-12	All 1-12
9-6	274-278	All 1-11	All 1-13	All 1-13
9-7	279-282	All 1-19	All 1-25	All 1-25
Review and Testing		3 days	2 days	2 days
Enrichment		1 day	1 day	1 day
Total		**12**	**10**	**10**

CHAPTER 10: RATIONAL NUMBERS: ADDITION/SUBTRACTION

Section	Pages	Basic	Average	Above Average
10-1	294-296	All 1-28	All 1-28	All 1-28
10-2	297-299	All 1-48	All 1-48	Odds 1-49 All 50-56
10-3	300-302	Day 1: All 1-24 Day 2: All 25-48	Odds 1-37 All 38-48	Odds 1-49 All 50-51
10-4	304-307	Day 1: All 1-34 Day 2: All 35-62	Day 1: Odds 1-49 Day 2: All 51-82	3, 6, 9, . . . , 84 All 87-89
10-5	308-310	Day 1: All 1-24 Day 2: All 25-48	Day 1: Odds 1-35 Day 2: All 37-55	3, 6, 9, . . . , 51 All 53-55
10-6	312-314	Day 1: Odds 1-29 Day 2: Evens 2-36	Odds 1-23 All 25-36	Odds 1-35 All 37-44
10-7	315-317	Omit	Omit	Odds 1-23
Review and Testing		3 days	2 days	2 days
Enrichment		1 day	1 day	1 day
Total		**14**	**11**	**10**

Suggested Timetables of Assignments

CHAPTER 11: EQUATIONS: ADDITION/SUBTRACTION

Section	Pages	Basic	Average	Above Average
11-1	324-325	Day 1: Odds 1-33 Day 2: Evens 2-34	Odds 1-39	Evens 2-38
11-2	326-328	Day 1: Odds 1-35 Day 2: Evens 2-36	Odds 1-29 All 30-36	3, 6, 9, . . . , 48
11-3	329-330	Day 1: Odds 1-31 Day 2: Evens 2-32	Day 1: All 1-20 Day 2: All 21-32	Odds 1-31
11-4	334-337	Day 1: All 1-20 Day 2: All 21-36	Day 1: All 1-20 Day 2: All 21-36	Odds 1-35
11-5	338-339	Omit	All 1-25	Odds 1-35
11-6	340-342	Omit	Omit	Odds 1-27 All 37-48
Review and Testing		3 days	2 days	2 days
Enrichment		1 day	1 day	1 day
Total		12	10	9

CHAPTER 12: RATIONAL NUMBERS: MULTIPLICATION/DIVISION

Section	Pages	Basic	Average	Above Average
12-1	350-352	Day 1: All 1-24 Day 2: All 25-44	Day 1: All 1-24 Day 2: All 25-50	3, 6, 9, . . . , 36 All 37-62
12-2	353-354	Day 1: Odds 1-31 Day 2: Evens 2-32	All 1-32	3, 6, 9, . . . , 30 All 33-48
12-3	356-358	Day 1: All 1-18 Day 2: All 19-36	Odds 1-27 All 28-39	Odds 1-39 All 40-48
12-4	360-363	Odds 1-35	All 1-36	Odds 9-31 All 37-56
12-5	364-366	Omit	Odds 1-39	Odds 1-49
Review and Testing		3 days	3 days	2 days
Enrichment		1 day	1 day	1 day
Total		11	10	8

Suggested Timetables of Assignments

CHAPTER 13: EQUATIONS: MULTIPLICATION/DIVISION

Section	Pages	Basic	Average	Above Average
13-1	374-376	Day 1: Odds 1-39 Day 2: Evens 2-40	Odds 1-39	Odds 1-39
13-2	377-380	Day 1: All 1-20 Day 2: All 21-30	Day 1: All 1-20 Day 2: All 21-34	Odds 1-29 All 31-34
13-3	381-382	Day 1: All 1-12 Day 2: All 13-32	Day 1: Odds 1-39 Day 2: Evens 2-40	Odds 1-39
13-4	383-385	Day 1: All 1-16 Day 2: All 17-30	Day 1: Odds 1-33 Day 2: Evens 2-34	3, 6, 9, . . . , 33 All 35-40
13-5	388-389	Omit	All 1-10	All 1-10
13-6	390-392	Omit	All 1-14	All 1-14
13-7	393-395	Omit	Omit	Odds 1-31
13-8	396-398	Omit	Omit	All 1-8
Review and Testing		3 days	3 days	2 days
Enrichment		1 day	1 day	1 day
Total		12	13	11

Suggested Timetables of Assignments

CHAPTER 14: GRAPHING AND EQUATIONS

Section	Pages	Basic	Average	Above Average
14-1	408-409	Omit	All 1-22	All 1-22
14-2	410-412	Omit	Day 1: All 1-10 Day 2: All 11-26	Odds 1-9 All 11-26
14-3	414-416	Omit	All 1-6	All 1-6
14-4	417-419	Omit	All 1-24	Odds 1-23 All 25-36
14-5	422-423	Omit	Odds 1-27	Odds 1-27
14-6	424-426	Omit	All 1-4	All 1-4
14-7	427-429	Omit	All 1-10	All 1-10
14-8	430-431	Omit	Omit	All 1-21
Review and Testing		0 days	2 days	2 days
Enrichment		0 days	1 day	1 day
Total		0	11	11

CHAPTER 15: SPECIAL TRIANGLES

Section	Pages	Basic	Average	Above Average
15-1	440-441	Omit	All 1-48	Odds 1-47
15-2	442-444	Omit	All 1-42	All 1-48
15-3	445-447	Omit	All 1-36	Odds 1-35
15-4	451-453	Omit	All 1-22	All 1-22
15-5	454-457	Omit	All 1-6	All 1-6
15-6	458-460	Omit	Omit	All 1-9
Review and Testing		0 days	2 days	2 days
Enrichment		0 days	1 day	1 day
Total		0	8	9

CHAPTER 16: STATISTICS AND PROBABILITY

Section	Pages	Basic	Average	Above Average
16-1	470-473	Omit	Omit	All 1-13
16-2	474-477	Omit	Omit	All 1-6
16-3	479-481	Omit	Omit	All 1-18
16-4	482-484	Omit	Omit	All 1-16
16-5	485-487	Omit	Omit	All 1-14
16-6	488-490	Omit	Omit	All 1-20
Review and Testing		0 days	0 days	2 days
Enrichment		0 days	0 days	1 day
Total		0	0	9

Suggested Timetables of Assignments

CHAPTER 17: GEOMETRY

Section	Pages	Basic	Average	Above Average
17-1	498-500	Omit	Omit	All 1-14
17-2	501-503	Omit	Omit	All 1-24
17-3	504-507	Omit	Omit	⎡ All 1-10
17-4	508-511	Omit	Omit	⎣ All 1-7
17-5	514-516	Omit	Omit	All 1-12
17-6	517-519	Omit	Omit	All 1-25
17-7	520-523	Omit	Omit	All 1-20
17-8	524-526	Omit	Omit	All 1-18
17-9	527-529	Omit	Omit	All 1-12
Review and Testing		0 days	0 days	2 days
Enrichment		0 days	0 days	1 day
Total		0	0	11

Suggested Timetables of Assignments

REVIEW OF ESSENTIAL SKILLS

Survey Tests

1. Whole Numbers
   Operations — Rounding — Estimation

2. Decimals
   Operations — Rounding — Estimation

3. Fractions
   Operations — Rounding — Estimation

4. Percent
   Operations — Rounding — Estimation

5. Metric Measures
   Units of Length — Units of Mass — Units of Capacity

Practice

1. Whole Numbers
   Operations — Rounding — Estimation

2. Decimals
   Operations — Rounding — Estimation

3. Fractions
   Operations — Rounding — Estimation

4. Percent
   Operations — Rounding — Estimation

5. Metric Measures
   Units of Length — Units of Mass — Units of Capacity

The *Review of Essential Skills* is designed for those students in pre-algebra who have difficulties with the essential skills that are a prerequisite for pre-algebra: whole numbers, decimals, fractions, percent, and metric measures. It consists of two parts: a survey test and corresponding practice exercises. The five survey tests can be used to determine which students require additional practice. Answers for the *Review of Essential Skills* begin on page M-61.

Permission to reproduce pages M-40 through M-60 is granted to users of *Pre-Algebra: Skills/Problem Solving/Applications.*

SURVEY TEST 1: WHOLE NUMBERS

Add.

1. 8
   7
   +9

2. 19
   72
   +57

3. 3074
   969
   + 83

4. 918 + 407 + 21 + 5829

Subtract.

5. 56
   − 9

6. 80
   − 62

7. 907
   − 458

8. 573 − 88

9. 9078 − 4739

Multiply.

10. 57
    × 8

11. 97
    × 30

12. 709
    × 39

13. 6095 × 87

14. 480 × 63

Divide.

15. 7)238

16. 34)238

17. 148 ÷ 7

18. 2244 ÷ 9

19. 19,530 ÷ 8

20. 20)708

21. 24)11,045

22. 252)10,836

Rounding and Estimation

Round each number as indicated.

23. 14; nearest 10

24. 288; nearest hundred

25. 65; nearest 10

26. 7109; nearest thousand

Choose the letter of the response that gives the best estimate. Choose a, b, or c.

27. 85 + 32 + 71   a. 80 + 30 + 70   b. 90 + 40 + 80   c. 90 + 30 + 70

28. 3997 − 1214   a. 4000 − 1000   b. 3000 − 2000   c. 4000 − 2000

29. 195 × 206   a. 100 × 300   b. 200 × 200   c. 200 × 300

30. 89 ÷ 31   a. 80 ÷ 40   b. 90 ÷ 45   c. 90 ÷ 30

SURVEY TEST 2: DECIMALS

Add.

1. 4.7
   + 3.8

2. 0.57
   0.38
   + 0.94

3. 51.169
   3.285
   + 398.426

4. 12.9 + 189.68 + 8.07

Subtract.

5. 8.3
   − 6.7

6. 0.58
   − 0.09

7. 8.032
   − 5.97

8. 60.53 − 15.6

9. 50 − 0.333

Multiply.

10. 9.5
    × 7

11. 39
    × 0.8

12. 0.492
    × 0.12

13. 458.2 × 0.32

14. 0.0513 × 0.007

Divide.

15. 7)18.9

16. 0.06)0.5874

17. 128 ÷ 0.04

18. 820.8 ÷ 0.9

19. 0.08)680.6

20. 10)49

21. 76.7 ÷ 100

22. 317.2 ÷ 1000

Rounding and Estimation

Round each number as indicated.

23. 9.53; nearest whole number
24. 7.05; nearest tenth
25. 0.084; nearest hundredth
26. 1.8036; nearest thousandth

Choose the letter of the response that gives the best estimate. Choose a, b, or c.

27. 8.2 + 19.8 + 34.9   a. 8 + 19 + 34   b. 8 + 20 + 35   c. 10 + 20 + 30
28. 68.07 − 12.96   a. 68 − 13   b. 68 − 12   c. 70 − 10
29. 3.9 × 12.01   a. 3 × 12   b. 4 × 10   c. 4 × 12
30. 89.6 ÷ 3   a. 90 ÷ 3   b. 9 ÷ 3   c. 87 ÷ 3

SURVEY TEST 3: FRACTIONS

Write each fraction in lowest terms.

1. $\frac{4}{10}$
2. $\frac{12}{16}$
3. $\frac{9}{21}$
4. $\frac{1200}{36,000}$

Write a mixed number in lowest terms for each of the following.

5. $\frac{11}{6}$
6. $\frac{19}{12}$
7. $\frac{36}{24}$
8. 15.8
9. 7.25

Write a decimal for each fraction.

10. $\frac{3}{5}$
11. $\frac{9}{10}$
12. $\frac{7}{8}$
13. $\frac{12}{5}$
14. $\frac{11}{50}$

Add or subtract as indicated.

15. $\frac{3}{8} + \frac{10}{16}$
16. $\frac{3}{5} + \frac{4}{7}$
17. $\frac{11}{12} - \frac{3}{8}$
18. $5\frac{2}{5} - 1\frac{4}{5}$
19. $\frac{1}{6} + \frac{3}{4}$
20. $3\frac{3}{10} + 5\frac{5}{8}$
21. $4\frac{1}{3} - 2\frac{5}{9}$
22. $4 - 1\frac{7}{8}$

Multiply or divide as indicated.

23. $\frac{2}{7} \times \frac{3}{5}$
24. $\frac{4}{7} \times \frac{5}{12}$
25. $15 \times \frac{3}{4}$
26. $12 \times 2\frac{1}{5}$
27. $\frac{5}{12} \div \frac{7}{8}$
28. $4\frac{1}{2} \div \frac{3}{8}$
29. $5\frac{1}{4} \div 3$
30. $13 \div \frac{1}{4}$

Rounding and Estimation

Round each fraction to the nearest whole number.

31. $\frac{7}{8}$
32. $\frac{1}{12}$
33. $29\frac{2}{3}$
34. $5\frac{1}{4}$
35. $1\frac{1}{2}$

Choose the letter of the response that gives the best estimate. Choose a, b, or c.

36. $12\frac{3}{4} + 8\frac{1}{6}$ a. 13 + 8   b. 12 + 8   c. 10 + 10
37. $28\frac{11}{12} - 9\frac{4}{5}$ a. 30 − 10   b. 29 − 10   c. 29 − 9
38. $15\frac{1}{8} \times \frac{19}{20}$ a. 16 × 1   b. 15 × 19   c. 15 × 1
39. $2\frac{5}{6} \div \frac{7}{8}$ a. 3 ÷ 1   b. 2 ÷ 1   c. 2 ÷ 2

SURVEY TEST 4: PERCENT

Write a percent for each decimal.

1. 0.64
2. 0.07
3. 0.40
4. 1.02
5. 0.005

Write a decimal for each percent.

6. 42%
7. 3%
8. 10%
9. 125%
10. $3\frac{1}{2}$%

Write a percent for each fraction.

11. $\frac{3}{5}$
12. $\frac{1}{4}$
13. $\frac{8}{10}$
14. $\frac{2}{3}$
15. $\frac{7}{8}$

Find each answer.

16. 24% of 780 = ?
17. 3% of 38 = ?
18. 162% of 70 = ?
19. ? % of 35 = 14
20. ? % of 96 = 12
21. ? % of 72 = 27
22. 6% of ? = 3
23. 160% of ? = 50
24. $15\frac{1}{2}$% of ? = 465

Rounding and Estimation

Round each percent as indicated.

25. 16.5%; nearest whole percent
26. $66\frac{2}{3}$%; nearest whole percent
27. $37\frac{1}{2}$%; nearest ten percent
28. 52.4%; nearest ten percent

Choose the letter of the response that gives the best estimate. Choose a, b, or c.

29. 25% of $399   a. 0.20 × $400   b. 0.30 × $400   c. $\frac{1}{4}$ × $400

30. 12% of $1600   a. $\frac{1}{8}$ × $1600   b. $\frac{1}{10}$ × $1600   c. $\frac{1}{5}$ × $1600

31. $33\frac{1}{3}$% of 295.76   a. $\frac{1}{2}$ × 300   b. $\frac{3}{10}$ × 300   c. $\frac{1}{3}$ × 300

32. $18\frac{3}{4}$% of 10,000   a. $\frac{1}{10}$ × 10,000   b. $\frac{1}{5}$ × 10,000   c. $\frac{1}{8}$ × 10,000

33. 30% of 118.27   a. 0.30 × 120   b. 0.30 × 110   c. $\frac{1}{3}$ × 120

SURVEY TEST 5: METRIC MEASURES

For Exercises 1-4, choose the most suitable unit of measure.
Choose a, b, or c.

1. Length of a pencil   a. millimeter   b. meter   c. centimeter
2. Thickness of a dime   a. millimeter   b. centimeter   c. meter
3. Capacity of a teaspoon   a. milliliter   b. liter   c. kiloliter
4. Weight of an automobile   a. milligram   b. gram   c. kilogram

For Exercises 5-8, choose the most suitable measure.
Choose a, b, c, or d.

5. Thickness of 50 sheets of loose-leaf paper in millimeters
      a. 50 mm   b. 500 mm   c. 0.5 mm   d. 5 mm
6. Height of Niagara Falls in meters
      a. 58 m   b. 5.8 m   c. 580 m   d. 5800 m
7. Weight of a pro football player in kilograms
      a. 12.8 kg   b. 1280 kg   c. 128 kg   d. 1.28 kg
8. Capacity of a can of fruit juice in milliliters
      a. 3.54 mL   b. 354 mL   c. 35.4 mL   d. 3540 mL

For Exercises 9-12, choose the equivalent measure.
Choose a, b, c, or d.

9. 7.8 meters   a. 780 km   b. 0.0078 km   c. 78 mm   d. 78 cm
10. 1800 millimeters   a. 1.8 m   b. 0.018 km   c. 18,000 cm   d. 1.8 cm
11. 84.5 liters   a. 8.45 kL   b. 84,500 mL   c. 8450 mL   d. 84.5 kL
12. 1.6 kilograms   a. 160 g   b. 16 g   c. 1600 g   d. 16,000 g

13. The highway distance from Chicago to Pittsburgh is 728 kilometers. Write this distance in meters.
14. A water pitcher holds 1900 milliliters. Write this capacity in liters.
15. A box of cereal weighs 340 grams. What is this weight in kilograms?

PRACTICE 1: WHOLE NUMBERS

Add.

1.	8	2.	28	3.	927	4.	108	5.	1085	6.	4806
	7		14		348		567		247		7329
	+4		+63		+569		+924		+96		+6194

7. 24 + 5 + 19   8. 38 + 164 + 96   9. 309 + 75 + 54   10. 49 + 3096 + 287

Subtract.

11.	86	12.	90	13.	432	14.	765	15.	408	16.	7283
	− 9		− 24		− 9		− 76		− 129		− 5647

17. 37 − 19   18. 324 − 76   19. 603 − 247   20. 6002 − 2865

Multiply.

21.	47	22.	729	23.	508	24.	4276	25.	593	26.	6008
	× 8		× 6		× 7		× 9		× 72		× 47

27. 52 × 34   28. 46 × 87   29. 308 × 29   30. 690 × 56

31. (75)(84)   32. (476)(100)   33. 907 · 58   34. 6007 × 19

Divide.

35. 6)168   36. 8)1784   37. 9)2736   38. 28)532

Divide. Write each remainder as a fraction in lowest terms.

39. 384 ÷ 9   40. 2516 ÷ 7   41. 4078 ÷ 8   42. 24,046 ÷ 6

43. 18)708   44. 24)1382   45. 72)10,314   46. 208)113,828

Rounding

Rules: To round a whole number, look at the <u>digit</u> to the <u>right</u> of the place to which you are rounding.

1. If the digit is **less than 5**, round **down**.
2. If the digit is **5 or greater than 5**, round up.

Examples: a. **7** rounded to the nearest <u>ten</u> is **10**.

b. **112** rounded to the nearest <u>hundred</u> is **100**.

c. **6806** rounded to the nearest <u>thousand</u> is **7000**.

d. **38,653** rounded to the nearest <u>ten</u> is **38,650**.

e. **38,653** rounded to the nearest <u>hundred</u> is **38,700**.

f. **38,653** rounded to the nearest <u>thousand</u> is **39,000**.

Round to the nearest ten.

47. 72   **48.** 89   **49.** 427   **50.** 893   **51.** 5445   **52.** 6358

Round to the nearest hundred.

53. 562   **54.** 448   **55.** 906   **56.** 8439   **57.** 6752   **58.** 998

Round to the nearest thousand.

59. 6208   **60.** 3647   **61.** 8529   **62.** 7500   **63.** 12,548   **64.** 68,492

Estimation

You can use rounding to estimate answers.

Choose the letter of the response that gives the best estimate.
Choose **a, b,** *or* **c.**

65. 75 + 92 + 48   a. 70 + 90 + 40   b. 80 + 90 + 50   c. 70 + 90 + 50

66. 406 + 712 + 894   a. 400 + 700 + 900   b. 400 + 700 + 800   c. 500 + 800 + 900

67. 803 − 689   a. 900 − 600   b. 800 − 600   c. 800 − 700

68. 6508 − 4078   a. 6500 − 4000   b. 6000 − 4000   c. 6500 − 4100

69. 32 × 28   a. 35 × 30   b. 30 × 30   c. 35 × 25

70. 482 × 811   a. 500 × 800   b. 400 × 800   c. 500 × 900

71. 1887 ÷ 98   a. 2000 ÷ 100   b. 1500 ÷ 100   c. 1900 ÷ 100

72. 3556 ÷ 39   a. 4000 ÷ 40   b. 3600 ÷ 40   c. 3000 ÷ 30

PRACTICE 2: DECIMALS

Add.

1. 3.6
   + 4.9

2. 7.8
   + 4.7

3. 0.25
   0.56
   + 0.84

4. 6.76
   8.68
   + 2.09

5. 65.08
   29.76
   + 35.92

6. 5.08
   267.94
   + 86.47

7. 0.374 + 0.968

8. 14.8 + 9.06 + 186.9

9. 15.76 + 208.49 + 9.06

Subtract.

10. 9.2
    − 4.7

11. 0.86
    − 0.58

12. 12.62
    − 9.75

13. 86.06
    − 57.29

14. 6.182
    − 3.287

15. 44.83
    − 19.7

16. 8.03 − 5.9

17. 24.5 − 13.38

18. 0.3 − 0.186

19. 253 − 19.87

Multiply.

20. 6.2
    × 7

21. 87
    × 0.6

22. 5.9
    × 12

23. 46
    × 0.28

24. 623
    × 4.7

25. 6.37
    × 5.3

26. 0.008 × 0.92

27. 42.8 × 100

28. 7.03 × 0.78

29. 0.089 × 40.6

Divide.

30. 8)194.4

31. 9)63.576

32. 24)114.72

33. 0.07)0.5824

34. 112 ÷ 0.04

35. 319.2 ÷ 0.42

36. 0.4032 ÷ 8.4

37. 76.2 ÷ 0.12

38. 38 ÷ 10

39. 709 ÷ 10

40. 43.6 ÷ 10

41. 6 ÷ 100

42. 73 ÷ 100

43. 187.5 ÷ 100

44. 56 ÷ 1000

45. 28.72 ÷ 1000

Rounding

Rules: To round a decimal, look at the <u>digit</u> to the <u>right</u> of the place to which you are rounding.

1. If the digit is **less than 5**, round **down**.
2. If the digit is **5 or greater than 5**, round **up**.

Examples: a. **8.63** rounded to the nearest tenth is **8.6**.

b. **0.018** rounded to the nearest hundredth is **0.02**.

c. **64.8054** rounded to the nearest whole number is **65**.

d. **64.8054** rounded to the nearest tenth is **64.8**.

e. **64.8054** rounded to the nearest hundredth is **64.81**.

f. **64.8054** rounded to the nearest thousandth is **64.805**.

Round to the nearest whole number.

46. 86.5   **47.** 79.6   **48.** 426.47   **49.** 0.81   **50.** 35.09

Round to the nearest tenth.

51. 8.64   **52.** 4.08   **53.** 12.055   **54.** 0.987   **55.** 65.045

Round to the nearest hundredth.

56. 5.683   **57.** 14.649   **58.** 0.396   **59.** 19.445   **60.** 0.005

Round to the nearest thousandth.

61. 0.1261   **62.** 3.0587   **63.** 0.4064   **64.** 0.1999   **65.** 7.0096

Estimation

You can use rounding to estimate answers.

Choose the letter of the response that gives the best estimate. Choose a, b, or c.

66. 8.75 + 9.10   a. 8 + 9   b. 9 + 9   c. 9 + 10

67. 6.71 + 2.15   a. 7 + 2   b. 6 + 2   c. 5 + 2

68. 0.93 − 0.56   a. 1 − 0.5   b. 1 − 0.6   c. 0.9 − 0.6

69. 68.07 − 13.98   a. 70 − 10   b. 70 − 20   c. 68 − 14

70. 473 × 0.64   a. 473 × 1   b. 500 × 0.6   c. 400 × 0.6

71. 87.3 ÷ 4.8   a. 90 ÷ 5   b. 80 ÷ 4   c. 87 ÷ 5

PRACTICE 3: FRACTIONS

Write in lowest terms.

1. $\dfrac{8}{12}$
2. $\dfrac{15}{18}$
3. $\dfrac{9}{24}$
4. $\dfrac{8}{28}$
5. $\dfrac{14}{21}$
6. $\dfrac{12{,}000}{17{,}000}$
7. $\dfrac{1500}{21{,}000}$
8. $\dfrac{24{,}000}{3600}$
9. $\dfrac{12{,}000{,}000}{33{,}000}$
10. $\dfrac{192{,}000}{39{,}000{,}000}$

Write a mixed number for each of the following.
Write your answer in lowest terms.

11. $\dfrac{12}{7}$
12. $\dfrac{18}{13}$
13. $\dfrac{22}{5}$
14. 15.8
15. 2.125

Write as a decimal.

16. $\dfrac{5}{8}$
17. $\dfrac{7}{16}$
18. $\dfrac{11}{4}$
19. $\dfrac{23}{8}$
20. $\dfrac{85}{32}$

Add or subtract as indicated. Write your answer in lowest terms.

21. $\dfrac{5}{6} + \dfrac{7}{12}$
22. $\dfrac{13}{16} + \dfrac{5}{8}$
23. $\dfrac{2}{5} + \dfrac{3}{4}$
24. $3\dfrac{1}{4} - 1\dfrac{7}{8}$
25. $5\dfrac{1}{6} - 2\dfrac{3}{4}$
26. $\dfrac{1}{8} + \dfrac{5}{12}$
27. $\dfrac{19}{24} - \dfrac{5}{8}$
28. $4\dfrac{5}{12} - 3\dfrac{13}{16}$
29. $6\dfrac{7}{8} + 3\dfrac{5}{6}$

Multiply or divide as indicated.

30. $\dfrac{1}{5} \times \dfrac{3}{4}$
31. $\dfrac{5}{6} \times \dfrac{3}{2}$
32. $\dfrac{5}{16} \div \dfrac{3}{4}$
33. $1\dfrac{5}{6} \div \dfrac{1}{3}$
34. $2\dfrac{1}{2} \times \dfrac{2}{3}$
35. $\dfrac{2}{5} \times 45$
36. $2\dfrac{3}{4} \times 16$
37. $2\dfrac{7}{8} \div \dfrac{1}{4}$
38. $3\dfrac{1}{6} \div 4\dfrac{5}{6}$
39. $6\dfrac{2}{3} \div 2$
40. $14 \div \dfrac{7}{8}$
41. $27 \div \dfrac{9}{16}$
42. $33 \div 1\dfrac{3}{8}$
43. $58 \div 3\dfrac{5}{8}$
44. $16 \div \dfrac{2}{3}$

Rounding

Rules:
1. If a fraction is **less than** $\dfrac{1}{2}$, round **down** to the nearest whole number.
2. If a fraction is **greater than or equal to** $\dfrac{1}{2}$, round **up** to the nearest whole number.

Examples: a. $\frac{4}{5}$ rounded to the nearest whole number is **1**.

b. $5\frac{1}{4}$ rounded to the nearest whole number is **5**.

c. $7\frac{9}{16}$ rounded to the nearest whole number is **8**.

d. $15\frac{1}{2}$ rounded to the nearest whole number is **16**.

e. $2\frac{1}{6}$ rounded to the nearest whole number is **2**.

Round to the nearest whole number.

45. $\frac{7}{8}$ 46. $\frac{5}{6}$ 47. $\frac{1}{9}$ 48. $4\frac{5}{8}$ 49. $7\frac{2}{5}$ 50. $3\frac{3}{4}$

51. $2\frac{5}{12}$ 52. $13\frac{7}{10}$ 53. $6\frac{4}{5}$ 54. $12\frac{15}{16}$ 55. $19\frac{2}{3}$ 56. $3\frac{9}{10}$

Estimation

Choose the letter of the response that gives the best estimate. Choose **a, b,** *or* **c.**

57. $3\frac{1}{4} + 5\frac{3}{8} + 1\frac{15}{16}$ a. 3 + 5 + 1   b. 3 + 6 + 2   c. 3 + 5 + 2

58. $50\frac{9}{16} + 49\frac{7}{8} + 89\frac{5}{6}$ a. 51 + 50 + 90   b. 50 + 49 + 89   c. 40 + 40 + 80

59. $8\frac{7}{8} - 2\frac{1}{4}$ a. 8 − 2   b. 8 − 3   c. 9 − 2

60. $114\frac{2}{3} - 79\frac{5}{6}$ a. 115 − 80   b. 110 − 75   c. 115 − 75

61. $5\frac{1}{4} \times 2\frac{2}{3}$ a. 5 × 3   b. 5 × 4   c. 6 × 2

62. $7\frac{5}{8} \times 5\frac{1}{8}$ a. 8 × 5   b. 7 × 5   c. 7 × 10

63. $19\frac{7}{12} \div 5\frac{1}{8}$ a. 20 ÷ 5   b. 19 ÷ 5   c. 15 ÷ 5

64. $192\frac{1}{8} \div 3\frac{5}{7}$ a. 190 ÷ 4   b. 192 ÷ 4   c. 192 ÷ 3

65. $\frac{11}{12} \div \frac{19}{20}$ a. 2 ÷ 1   b. 2 ÷ $\frac{1}{2}$ c. 1 ÷ 1

66. $89 \div 3\frac{1}{3}$ a. 89 ÷ 3   b. 90 ÷ 3   c. 100 ÷ 3

M-54   Copyright © 1986 by Harcourt Brace Jovanovich, Inc. All rights reserved.

PRACTICE 4: PERCENT

Write a percent for each decimal.

1. 0.16
2. 0.03
3. 0.40
4. 0.785
5. 0.008
6. 1.085

Write a decimal for each percent.

7. 17%
8. 9%
9. 20%
10. $12\frac{1}{2}$%
11. $85\frac{3}{4}$%
12. $9\frac{1}{4}$%
13. 125%
14. 240%
15. $\frac{1}{2}$%
16. $\frac{1}{4}$%
17. $3\frac{1}{8}$%
18. $12\frac{2}{3}$%

Write a percent for each fraction.

19. $\frac{3}{10}$
20. $\frac{4}{5}$
21. $\frac{3}{4}$
22. $\frac{1}{3}$
23. $\frac{1}{8}$
24. $\frac{5}{6}$

Find each answer.

25. 12% of 650 = _?_
26. 5% of 92 = _?_
27. 8% of 98 = _?_
28. $33\frac{1}{3}$% of 96 = _?_
29. 125% of 80 = _?_
30. 0.4% of 16 = _?_
31. 1% of 350 = _?_
32. $87\frac{1}{2}$% of 480 = _?_
33. 4.2% of 800 = _?_
34. _?_ % of 40 = 16
35. _?_ % of 96 = 24
36. _?_ % of 20 = 17
37. ——% of 72 = 18
38. _?_ % of 40 = 12
39. _?_ % of 64 = 48
40. _?_ % of 18 = 27
41. _?_ % of 144 = 60
42. _?_ % of 90 = 72
43. 8% of _?_ = 2
44. 12% of _?_ = 48
45. 20% of _?_ = 7
46. 25% of _?_ = 60
47. 45% of _?_ = 36
48. 0.5% of _?_ = 2
49. 120% of _?_ = 57.6
50. 16.5% of _?_ = 79.2
51. $12\frac{1}{2}$% of _?_ = 9
52. _?_ % of 90 = 120
53. $16\frac{2}{3}$% of 144 = _?_
54. $83\frac{1}{3}$% of _?_ = 55
55. _?_ % of 288 = 16
56. $33\frac{1}{3}$% of _?_ = 15
57. $187\frac{1}{2}$% of 16 = _?_

Rounding

Rule: To round percents, apply the rules for rounding whole numbers, decimals, and fractions.

Examples:
a. 72.5% rounded to the nearest whole percent is 73%.
b. $33\frac{1}{3}$% rounded to the nearest whole number is 33%.
c. 86.3% rounded to the nearest ten percent is 90%.
d. 4.62% rounded to the nearest tenth of a percent is 4.6%.

Round to the nearest whole percent.

58. 15.6%   59. 4.45%   60. $23\frac{1}{2}$%   61. $74\frac{3}{4}$%   62. 0.62%

Round to the nearest ten percent.

63. 48.3%   64. 8.9%   65. $17\frac{5}{8}$%   66. $66\frac{2}{3}$%   67. 326.4%

Round to the nearest tenth of a percent.

68. 8.65%   69. 14.73%   70. 0.16%   71. 0.345%   72. 19.33%.

Estimation

Choose the letter of the response that gives the best estimate. Choose a, b, or c.

73. 20% of $897.60   a. $\frac{1}{4}$ × $900   b. $\frac{1}{4}$ × $800   c. $\frac{1}{5}$ × $900

74. $66\frac{2}{3}$% of $1785   a. $\frac{1}{6}$ × $1800   b. $\frac{2}{3}$ × $1800   c. $\frac{4}{5}$ × $1800

75. 59% of 2095   a. $\frac{1}{2}$ × 3000   b. $\frac{4}{5}$ × 2000   c. 0.60 × 2000

76. $12\frac{1}{2}$% of 512   a. $\frac{1}{12}$ × 512   b. $\frac{1}{10}$ × 512   c. $\frac{1}{8}$ × 512

77. 39% of 289   a. 0.40 × 290   b. 0.40 × 280   c. $\frac{1}{3}$ × 300

PRACTICE 5: METRIC SYSTEM

Units of Length

The **basic unit of length** in the metric system is the **meter**. The following table shows how other metric units of length are related to the meter.

Units of Length	Prefix	Meaning
1 millimeter (mm) = 0.001 meter (m)	milli-	$\frac{1}{1000}$, or 0.001
1 centimeter (cm) = 0.01 meter	centi-	$\frac{1}{100}$, or 0.01
1 decimeter (dm) = 0.1 meter	deci-	$\frac{1}{10}$, or 0.1
1 dekameter (dam) = 10 meters	deka-	10
1 hectometer (hm) = 100 meters	hecto-	100
1 kilometer (km) = 1000 meters	kilo-	1000

Examples:
1. To change to a smaller metric unit, multiply by 10, by 100, by 1000, and so on.
2. To change to a larger metric unit, divide by 10, by 100, by 1000, and so on.

   a. Larger to Smaller: 75 km = 75 × 1000 = 75,000 m

   b. 255 mm = 255 ÷ 1000 = 0.255 m (Smaller to Larger)

   c. 165 cm = 165 ÷ 100 = 1.65 m

   d. 165 cm = 165 × 10 = 1650 mm

The actual sizes of the centimeter and the millimeter are shown on the metric ruler below.

The **millimeter** is suitable for measuring the thickness of a pad of writing paper.

The **centimeter** is suitable for measuring the length and width of a sheet of paper.

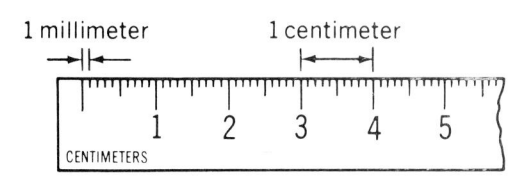

The **meter** is suitable for measuring the height of a building.

The **kilometer** is suitable for measuring the distance between two cities.

EXERCISES

Complete the table. First check whether you multiply or divide. Then check the number by which you multiply or divide. The first one is done for you.

	From	To	Multiply	Divide	By 10	By 100	By 1000
1.	centimeters	meters		✔		✔	
2.	centimeters	millimeters					
3.	meters	kilometers					
4.	kilometers	meters					
5.	decimeters	millimeters					
6.	centimeters	dekameters					

Choose the equivalent measure. Choose a, b, c, or d.

7. The length of a race is 1500 meters. This is equivalent to
   - a. 150 km
   - b. 150,000 cm
   - c. 0.15 km
   - d. 15 cm

8. The bore of an automobile cylinder is 79.5 millimeters. This is equivalent to
   - a. 795 cm
   - b. 7.95 cm
   - c. 0.795 m
   - d. 0.00795 km

9. A calculator is 7.5 millimeters thick. This is equivalent to
   - a. 75 cm
   - b. 0.075 m
   - c. 750 dm
   - d. 0.75 cm

10. The width of some camera film is 3.5 centimeters. This is equivalent to
    - a. 0.35 dm
    - b. 3500 m
    - c. 350 mm
    - d. 0.035 dam

11. The length of scotch tape on a roll is 27.9 meters. This is equivalent to
    - a. 279 dam
    - b. 0.279 km
    - c. 279 dm
    - d. 2790 mm

12. The airline distance from New York to Paris is 5852 kilometers. This is equivalent to
    - a. 585.2 km
    - b. 5,852,000 m
    - c. 5.852 m
    - d. 58,520 dam

13. The thickness of a pad of paper is 1.4 centimeters. This is equivalent to
    - a. 1.4 dm
    - b. 0.0014 dam
    - c. 0.14 dam
    - d. 14 km

Units of Mass

The **basic unit of mass** (weight) in the metric system is the **gram** (g). The other most commonly used units are the **kilogram** (kg) and the **milligram** (mg).

A nickel weighs about 5 grams. Most aspirin tablets weigh 325 milligrams each. A 250-pound football lineman weighs about 113 kilograms.

The following table shows how other metric units of mass are related to the gram.

1 milligram (mg) = 0.001 gram   1 dekagram (dag) = 10 grams
1 centigram (cg) = 0.01 gram   1 hectogram (hg) = 100 grams
1 decigram (dg) = 0.1 gram   1 kilogram (kg) = 1000 grams
                1 metric ton (t) = 1000 kilograms

Choose the most suitable unit of measure.
Choose a, b, or c.

1. Contact lens   a. milligram   b. gram   c. kilogram
2. Airline luggage   a. milligram   b. gram   c. kilogram
3. One serving of cereal   a. milligram   b. gram   c. kilogram
4. Outboard motor for boat a. milligram   b. gram   c. kilogram

Choose the equivalent measure. Choose a, b, c, or d.

5. 325 grams   a. 32.5 kg   b. 3250 kg   c. 0.325 kg   d. 3.25 kg
6. 14.3 kilograms   a. 143 g   b. 14,300 g   c. 1430 g   d. 1.43 g
7. 1850 milligrams   a. 185 g   b. 1.85 g   c. 18.5 g   d. 18,500 g
8. 25,000 kilograms   a. 2.5 t   b. 0.25 t   c. 2500 t   d. 25 t

Choose the most suitable measure. Choose a, b, c, or d.

9. Automobile   a. 100 kg   b. 2 t   c. 5000 g   d. 1,000 mg
10. Baseball   a. 142 g   b. 3500 mg   c. 1.2 kg   d. 500 g
11. Algebra book   a. 85 g   b. 8500 mg   c. 850 g   d. 8.5 kg

M-59

Units of Capacity

The **basic unit of capacity** in the metric system is the **liter** (L). The other most commonly used units are the **milliliter** (mL) and the **kiloliter** (kL).

A quart of milk is about 946 milliliters. A small glass of orange juice is about 125 milliliters. The fuel tank on a compact car holds about 50 liters. A family-size swimming pool holds about 75 kiloliters of water.

The following table shows how other metric units of capacity are related to the liter.

1 milliliter (mL) = 0.001 liter	1 dekaliter (daL) = 10 liters
1 centiliter (cL) = 0.01 liter	1 hectoliter (hL) = 100 liters
1 deciliter (dL) = 0.1 liter	1 kiloliter (kL) = 1000 liters

Choose the most suitable unit of measure. Choose a, b, or c.

1. A bottle of cough syrup   a. milliliter   b. liter   c. kiloliter
2. Capacity of a picnic jug   a. milliliter   b. liter   c. kiloliter
3. Container of fruit juice concentrate   a. milliliter   b. liter   c. kiloliter
4. Amount of water used by a large hotel in a year   a. milliliter   b. liter   c. kiloliter

Choose the equivalent measure. Choose a, b, c, or d.

5. 325 milliliters   a. 3.25 L   b. 32.5 L   c. 3250 L   d. 0.325 L
6. 45.6 liters   a. 4.56 mL   b. 45,600 mL   c. 4560 mL   d. 0.0456 mL
7. 8.25 kiloliters   a. 8250 L   b. 0.00825 L   c. 82.5 L   d. 825 L
8. 90.5 liters   a. 90,500 kL   b. 9.05 kL   c. 0.0905 kL   d. 9050 kL

Choose the most suitable measure. Choose a, b, c, or d.

9. Engine size of a car   a. 500 mL   b. 2.8 L   c. 1.9 kL   d. 14.5 L
10. Half-filled bathtub   a. 25 kL   b. 1000 mL   c. 750 L   d. 160 L
11. Cup of cocoa   a. 250 mL   b. 0.5 L   c. 30 mL   d. 800 mL

ANSWERS: REVIEW OF ESSENTIAL SKILLS

Page M-44 Survey Test 1

1. 24 2. 148 3. 4126 4. 7175 5. 47 6. 18 7. 449 8. 485 9. 4339 10. 456 11. 2910 12. 27,651
13. 530,265 14. 30,240 15. 34 16. 7 17. $21\frac{1}{7}$ 18. $249\frac{1}{3}$ 19. $2441\frac{1}{4}$ 20. $35\frac{2}{5}$ 21. $460\frac{5}{24}$ 22. 43
23. 10 24. 300 25. 70 26. 7000 27. c 28. a 29. b 30. c

Page M-45 Survey Test 2

1. 8.5 2. 1.89 3. 452.880 4. 210.65 5. 1.6 6. 0.49 7. 2.062 8. 44.93 9. 49.667 10. 66.5 11. 31.2
12. 0.05904 13. 146.624 14. 0.0003591 15. 2.7 16. 9.79 17. 3200 18. 912 19. 8507.5 20. 4.9
21. 0.767 22. 0.3172 23. 10 24. 7.1 25. 0.08 26. 1.804 27. b 28. a 29. c 30. a

Page M-46 Survey Test 3

1. $\frac{2}{5}$ 2. $\frac{3}{4}$ 3. $\frac{3}{7}$ 4. $\frac{1}{30}$ 5. $1\frac{5}{6}$ 6. $1\frac{7}{12}$ 7. $1\frac{1}{2}$ 8. $15\frac{4}{5}$ 9. $7\frac{1}{4}$ 10. 0.6 11. 0.9 12. 0.875 13. 2.4
14. 0.22 15. 1 16. $1\frac{6}{35}$ 17. $\frac{13}{24}$ 18. $3\frac{3}{5}$ 19. $\frac{11}{12}$ 20. $8\frac{37}{40}$ 21. $1\frac{7}{9}$ 22. $2\frac{1}{8}$ 23. $\frac{6}{35}$ 24. $\frac{5}{21}$ 25. $11\frac{1}{4}$
26. $26\frac{2}{5}$ 27. $\frac{10}{21}$ 28. 12 29. $1\frac{3}{4}$ 30. 52 31. 1 32. 0 33. 30 34. 5 35. 2 36. a 37. b 38. c
39. a

Page M-47 Survey Test 4

1. 64% 2. 7% 3. 40% 4. 102% 5. 0.5% 6. 0.42 7. 0.03 8. 0.10 9. 1.25 10. 0.035 11. 60%
12. 25% 13. 80% 14. $66\frac{2}{3}$% 15. $87\frac{1}{2}$% 16. 187.2 17. 1.14 18. 113.4 19. 40 20. 12.5 21. 37.5
22. 50 23. 31.25 24. 3000 25. 17% 26. 67% 27. 40% 28. 50% 29. c 30. a 31. c 32. b 33. a

Page M-48 Survey Test 5

1. c 2. a 3. a 4. c 5. d 6. a 7. c 8. b 9. b 10. a 11. b 12. c 13. 728,000 m 14. 1.9 L
15. 0.34 kg

Page M-49 Practice 1

1. 19 2. 105 3. 1844 4. 1599 5. 1428 6. 18,329 7. 48 8. 298 9. 438 10. 3432 11. 77 12. 66
13. 423 14. 689 15. 279 16. 1636 17. 18 18. 248 19. 356 20. 3137 21. 376 22. 4374 23. 3556
24. 38,484 25. 42,696 26. 282,376 27. 1768 28. 4002 29. 8932 30. 38,640 31. 6300 32. 47,600
33. 52,606 34. 114,133 35. 28 36. 223 37. 304 38. 19 39. $42\frac{2}{3}$ 40. $359\frac{3}{7}$ 41. $509\frac{3}{4}$ 42. $4007\frac{2}{3}$
43. $39\frac{1}{3}$ 44. $57\frac{7}{12}$ 45. $143\frac{1}{4}$ 46. $547\frac{1}{4}$ 47. 70 48. 90 49. 430 50. 890 51. 5450 52. 6360 53. 600
54. 400 55. 900 56. 8400 57. 6800 58. 1000 59. 6000 60. 4000 61. 9000 62. 8000 63. 13,000
64. 68,000 65. b 66. a 67. c 68. c 69. b 70. a 71. c 72. b

Answers: Review of Essential Skills   M-61

Page M-51 Practice 2

1. 8.5 2. 12.5 3. 1.65 4. 17.53 5. 130.76 6. 359.49 7. 1.342 8. 210.76 9. 233.31 10. 13.9
11. 0.28 12. 2.87 13. 28.77 14. 2.895 15. 25.13 16. 2.13 17. 11.12 18. 0.114 19. 233.13 20. 43.4
21. 52.2 22. 70.8 23. 12.88 24. 2928.1 25. 33.761 26. 0.00736 27. 4280 28. 5.4834 29. 3.6134
30. 24.3 31. 7.064 32. 4.78 33. 8.32 34. 2800 35. 760 36. 0.048 37. 635 38. 3.8 39. 70.9
40. 4.36 41. 0.06 42. 0.73 43. 1.875 44. 0.056 45. 0.02872 46. 87 47. 80 48. 426 49. 1 50. 35
51. 8.6 52. 4.1 53. 12.1 54. 1.0 55. 65.0 56. 5.68 57. 14.65 58. 0.40 59. 19.45 60. 0.01
61. 0.126 62. 3.059 63. 0.406 64. 0.200 65. 7.010 66. b 67. a 68. c 69. c 70. a 71. a

Page M-53 Practice 3

1. $\frac{2}{3}$ 2. $\frac{5}{6}$ 3. $\frac{3}{8}$ 4. $\frac{2}{7}$ 5. $\frac{2}{3}$ 6. $\frac{12}{17}$ 7. $\frac{1}{14}$ 8. $6\frac{2}{3}$ 9. $363\frac{7}{11}$ 10. $\frac{8}{1625}$ 11. $1\frac{5}{7}$ 12. $1\frac{5}{13}$ 13. $4\frac{2}{5}$
14. $15\frac{4}{5}$ 15. $2\frac{1}{8}$ 16. 0.625 17. 0.4375 18. 2.75 19. 2.875 20. 2.65625 21. $1\frac{5}{12}$ 22. $1\frac{7}{16}$ 23. $1\frac{3}{20}$
24. $1\frac{3}{8}$ 25. $2\frac{5}{12}$ 26. $\frac{13}{24}$ 27. $\frac{1}{6}$ 28. $\frac{29}{48}$ 29. $10\frac{17}{24}$ 30. $\frac{3}{20}$ 31. $1\frac{1}{4}$ 32. $\frac{5}{12}$ 33. $5\frac{1}{2}$ 34. $1\frac{2}{3}$ 35. 18
36. 44 37. $11\frac{1}{2}$ 38. $\frac{19}{29}$ 39. $3\frac{1}{3}$ 40. 16 41. 48 42. 24 43. 16 44. 24 45. 1 46. 1 47. 0 48. 5
49. 7 50. 4 51. 2 52. 14 53. 7 54. 13 55. 20 56. 4 57. c 58. a 59. c 60. a 61. a 62. a
63. a 64. b 65. c 66. b

Page M-55 Practice 4

1. 16% 2. 3% 3. 40% 4. 78.5% 5. 0.8% 6. 108.5% 7. 0.17 8. 0.09 9. 0.2 10. 0.125 11. 0.8575
12. 0.0925 13. 1.25 14. 2.4 15. 0.005 16. 0.0025 17. 0.03125 18. 0.12$\overline{6}$ 19. 30% 20. 80% 21. 75%
22. $33\frac{1}{3}$% 23. $12\frac{1}{2}$% 24. $83\frac{1}{3}$% 25. 78 26. 4.6 27. 7.84 28. 32 29. 100 30. 0.064 31. 3.5 32. 420
33. 33.6 34. 40 35. 25 36. 85 37. 25 38. 30 39. 75 40. 150 41. $41\frac{2}{3}$ 42. 80 43. 25 44. 400
45. 35 46. 240 47. 80 48. 400 49. 48 50. 480 51. 72 52. $133\frac{1}{3}$ 53. 24 54. 66 55. $5\frac{5}{9}$ 56. 45
57. 30 58. 16% 59. 4% 60. 24% 61. 75% 62. 1% 63. 50% 64. 10% 65. 20% 66. 70% 67. 330%
68. 8.7% 69. 14.7% 70. 0.2% 71. 0.3% 72. 19.3% 73. c 74. b 75. c 76. c 77. a

Page M-57 Practice 5

1. Divide by 100. 2. Multiply by 10. 3. Divide by 1000. 4. Multiply by 1000. 5. Multiply by 100.
6. Divide by 1000. 7. b 8. b 9. d 10. a 11. c 12. b 13. b

Page M-59 Units of Mass

1. a 2. c 3. b 4. c 5. c 6. b 7. b 8. d 9. b 10. d 11. c

Page M-60 Units of Capacity

1. a 2. b 3. a 4. c 5. d 6. b 7. a 8. c 9. b 10. d 11. a

TEACHER'S NOTES

TEACHER'S NOTES

TEACHER'S NOTES

TEACHER'S NOTES

TEACHER'S NOTES

TEACHER'S NOTES

Vincent Brumfiel
Neal Golden
Mary Heins

PRE-ALGEBRA

SKILLS

PROBLEM SOLVING

APPLICATIONS

HBJ Harcourt Brace Jovanovich, Publishers
Orlando   New York   Chicago   San Diego   Atlanta   Dallas

ABOUT THE AUTHORS

VINCENT BRUMFIEL
Mathematics Instructor
Albion College
Albion, Michigan

BROTHER NEAL GOLDEN
Chairman, Department of Mathematics and Computer Science
Brother Martin High School
New Orleans, Louisiana

MARY HEINS
Mathematics Teacher
Coronado High School
El Paso, Texas

EDITORIAL ADVISORS

Kozo Nishifue
Mathematics Consultant
Division of Curriculum and Instruction
Oakland Unified School District
Oakland, California

Leo Ramirez
Mathematics Teacher
McAllen High School
McAllen, Texas

Kathy Ross-Warren
Mathematics Consultant K-12
Jefferson Parish Public Schools
Gretna, Louisiana

Pansy Rivenbark Rumley
Mathematics Teacher
D. C. Virgo Junior High School
Wilmington, North Carolina

Laurence R. Wantuck
Mathematics Supervisor K-12
Broward County Schools
Ft. Lauderdale, Florida

Richard J. Wyllie
Mathematics Department Chairman
Downers Grove South High School
Downers Grove, Illinois

Copyright © 1986 by Harcourt Brace Jovanovich, Inc.

All rights reserved. No part of this publication may be reproduced or transmitted in any form or by any means, electronic or mechanical, including photocopy, recording, or any information storage and retrieval system, without permission in writing from the publisher.

Requests for permission to make copies of any part of the work should be mailed to: Permissions, Harcourt Brace Jovanovich, Publishers, Orlando, Florida 32887

Printed in the United States of America.

ISBN 0-15-355700-1

CONTENTS

UNIT 1 ESSENTIAL SKILLS AND APPLICATIONS

Chapter 1: Whole Numbers and Applications — 1

Formulas and Whole Numbers
- 1–1 Using Formulas: Addition — 2
- 1–2 Using Formulas: Subtraction — 5
- 1–3 Using Formulas: Multiplication — 7
- 1–4 Using Formulas: Division — 10
- 1–5 Problem Solving and Applications: *Estimating with Whole Numbers* — 14

Using Whole Numbers and Variables
- 1–6 Numerical Expressions — 18
- 1–7 Using Variables — 22

Features Consumer: 13 ● Calculator 21 ● Enrichment 29

Review/Testing Review: *Sections 1–1—1–5* 17 ● Review: *Sections 1–6—1–7* 24
Chapter Review 25 ● Chapter Test 28 ● Additional Practice 30

Chapter 2: Decimals and Applications — 31

Operations with Decimals and Applications
- 2–1 Decimals and Place Value — 32
- 2–2 Addition and Subtraction — 35
- 2–3 Multiplication — 38
- 2–4 Division — 44

Decimals and Applications
- 2–5 Problem Solving and Applications: *Estimating with Decimals* — 47
- 2–6 Metric Units of Measurement — 51
- 2–7 Problem Solving and Applications: *Scientific Notation* — 55

Features Consumer/Computer Application: 43
Calculator 50 ● Enrichment 61

Review/Testing Review: *Sections 2–1—2–3* 42 ● Review: *Sections 2–4—2–7* 57
Chapter Review 58 ● Chapter Test 60 ● Additional Practice 62

Chapter 3: Number Properties/Expressions — 63

Number Properties and Applications
- 3–1 The Special Numbers 0 and 1 — 64
- 3–2 Simplifying Algebraic Expressions — 67
- 3–3 Problem Solving and Applications: *Formulas: Area* — 70

Algebraic Expressions and Applications
- 3–4 Like Terms — 76
- 3–5 Combining Like Terms — 79
- 3–6 Problem Solving and Applications: *Formulas: Perimeter* — 81

Features	Computer 75 • Calculator 86 • Enrichment 90 • Common Errors 92	
Review/ Testing	Review: *Sections 3–1—3–3* 74 • Review: *Sections 3–4—3–6* 86 Chapter Review 87 • Chapter Test 89 Additional Practice 91 • Cumulative Review: *Chapters 1–3* 93	

Chapter 4: Number Theory — 95

Factors and Divisibility

4–1	Factors, Multiples, and Divisibility	96
4–2	Divisibility Tests	99

Properties of Prime Numbers

4–3	Prime Numbers	103
4–4	Prime Factorization	108
4–5	Least Common Multiple	111
4–6	Greatest Common Factor	115

Features	Calculator 106, 117 • Consumer 107 • Enrichment 121	
Review/ Testing	Review: *Sections 4–1—4–3* 105 • Review: *Sections 4–4—4–6* 117 Chapter Review 118 • Chapter Test 120 • Additional Practice 122	

Chapter 5: Fractions: Multiplication/Division — 123

Fractions and Applications

5–1	Writing Fractions in Lowest Terms	124
5–2	Customary Measures	126
5–3	Multiplying Fractions	129
5–4	Problem Solving and Applications: *Formulas: Temperature*	132

Mixed Numerals and Applications

5–5	Multiplying Mixed Numerals	135
5–6	Problem Solving and Applications: *Area: Parallelograms and Triangles*	140
5–7	Problem Solving and Applications: *Volume: Rectangular Prisms*	145
5–8	Dividing Fractions	148

Features	Career 138 • Consumer 139 • Enrichment 155	
Review/ Testing	Review: *Sections 5–1—5–4* 134 • Review: *Sections 5–5—5–8* 151 Chapter Review 152 • Chapter Test 154 • Additional Practice 156	

Chapter 6: Fractions: Addition/Subtraction — 157

Fractions and Mixed Numerals

6–1	Addition and Subtraction: Like Denominators	158
6–2	Addition and Subtraction: Unlike Denominators	161
6–3	Addition and Subtraction: Mixed Numerals	165

Applications of Addition/Subtraction

6–4	Problem Solving and Applications: *Time Cards*	169
6–5	Fractions and Decimals	173
6–6	Problem Solving and Applications: *Circles: Circumference and Area*	176

Features	Calculator/Computer 168 • Calculator 180 Enrichment 184 • Common Errors 186

Review/ Testing	Review: *Sections 6–1—6–3* 167 • Review: *Sections 6–4—6–6* 180 Chapter Review 181 • Chapter Test 183 Additional Practice 185 • Cumulative Review: *Chapters 4–6* 187	

UNIT 2 PERCENT AND APPLICATIONS

Chapter 7: Solving Equations — 189

Equations: Subtraction and Addition
7–1	Solving Equations by Subtraction	190
7–2	Solving Equations by Addition	193
7–3	Problem Solving and Applications: *Using Equations: Addition/Subtraction*	196

Equations: Division and Multiplication
7–4	Solving Equations by Division	200
7–5	Solving Equations by Multiplication	203
7–6	Problem Solving and Applications: *Using Equations: Multiplication/Division*	206

Features	Consumer 199 • Enrichment 213	
Review/ Testing	Review: *Sections 7–1—7–3* 198 • Review: *Sections 7–4—7–6* 209 Chapter Review 210 • Chapter Test 212 • Additional Practice 214	

Chapter 8: Ratio, Proportion, and Percent — 215

Ratio and Proportion
8–1	Ratio	216
8–2	Proportions	219
8–3	Problem Solving and Applications: *Using Proportions*	222

Introduction to Percent
8–4	Meaning of Percent	227
8–5	Percents and Decimals	231
8–6	Finding a Percent of a Number	235
8–7	Problem Solving and Applications: *Formulas: Simple Interest*	240
8–8	Problem Solving and Applications: *Formulas: Discount*	244

Features	Computer 226 • Calculator 243 • Enrichment 251	
Review/ Testing	Review: *Sections 8–1—8–3* 225 • Review: *Sections 8–4—8–8* 247 Chapter Revew 248 • Chapter Test 250 • Additional Practice 252	

Chapter 9: Percent and Applications — 253

More on Percent
9–1	Finding What Percent One Number Is of Another	254
9–2	Problem Solving and Applications: *Percent of Increase and Decrease*	257
9–3	Finding a Number Given a Percent	261
9–4	Using Proportions to Solve Percent Problems	264

Graphs and Estimation
9–5	Bar Graphs	270
9–6	Circle Graphs	274
9–7	Problem Solving and Applications: *Using Estimation*	279

Features	Consumer 269 ● Calculator 282 Enrichment 288 ● Common Errors 290	
Review/ Testing	Review: *Sections 9–1—9–4* 267 ● Review: *Sections 9–5—9–7* 283 Chapter Review 284 ● Chapter Test 287 Additional Practice 289 ● Cumulative Review: *Chapters 7–9* 291	

UNIT 3 INTEGERS AND RATIONAL NUMBERS

Chapter 10: Rational Numbers: Addition/Subtraction 293

Integers and Rational Numbers

10–1	Integers	294
10–2	Rational Numbers	297
10–3	Addition on the Number Line	300

Addition and Subtraction

10–4	Adding Rational Numbers	304
10–5	Subtracting Rational Numbers	308
10–6	Properties of Addition	312
10–7	Adding Polynomials	315

Features	Consumer 303 ● Career 311 ● Calculator 314 ● Enrichment 321	
Review/ Testing	Review: *Sections 10–1—10–3* 302 ● Review: *Sections 10–4—10–7* 317 Chapter Review 318 ● Chapter Test 320 ● Additional Practice 322	

Chapter 11: Equations: Addition/Subtraction 323

Addition and Subtraction Properties

11–1	Equations: Addition/Subtraction	324
11–2	Like Terms/More Than One Equation	326
11–3	Variable on Both Sides	329

Applications of Equations

11–4	Problem Solving and Applications: *Words to Symbols: Addition/Subtraction*	334
11–5	Inequalities on the Number Line	338
11–6	Using Addition and Subtraction to Solve Inequalities	340

Features	Calculator 331 ● Consumer 332 ● Computer 333 Calculator 342 ● Enrichment 347	
Review/ Testing	Review: *Sections 11–1—11–3* 331 ● Review: *Sections 11–4—11–6* 343 Chapter Review 344 ● Chapter Test 346 ● Additional Practice 348	

Chapter 12: Rational Numbers: Multiplication/Division 349

Multiplication

12–1	Multiplication: Unlike Signs	350
12–2	Multiplication: Like Signs	353
12–3	Properties of Multiplication	356

Division

12–4	Dividing Rational Numbers	360
12–5	Factoring	364

Features	Calculator 355 ● Consumer 359 ● Calculator 367 ● Enrichment 371	
Review/ Testing	Review: *Sections 12–1—12–3* 358 ● Review: *Sections 12–4—12–5* 367 Chapter Review 368 ● Chapter Test 370 ● Additional Practice 372	

Chapter 13: Equations: Multiplication/Division   373

Multiplication and Division Properties

13–1	Equations: Multiplication/Division	374
13–2	Problem Solving and Applications: *Words to Symbols: Multiplication/Division*	377
13–3	Combined Operations	381
13–4	Equations with Parentheses	383

Applications of Equations

13–5	Problem Solving and Applications: *More Than One Unknown: Angles in a Triangle*	388
13–6	Problem Solving and Applications: *More Than One Unknown: Consecutive Numbers*	390
13–7	Inequalities: Multiplication/Division	393
13–8	Problem Solving and Applications: *Using Inequalities*	396

Features	Computer 387 ● Enrichment 402 ● Common Errors 404
Review/ Testing	Review: *Sections 13–1—13–4* 386 ● Review: *Sections 13–5—13–8* 398 Chapter Review 399 ● Chapter Test 401 Additional Practice 403 ● Cumulative Review: *Chapters 10–13* 405

UNIT 4 REAL NUMBERS

Chapter 14: Graphing and Equations   407

The Coordinate Plane

14–1	Graphing Ordered Pairs	408
14–2	Graphing Equations	410
14–3	Problem Solving and Applications: *Broken Line Graphs*	414
14–4	Slope of a Line	417

Using Graphs

14–5	Intercepts	422
14–6	Direct and Indirect Variation	424
14–7	Solving a System of Equations by Graphing	427
14–8	Graphing Linear Inequalities	430

Features	Calculator 413 ● Computer 421 ● Enrichment 437
Review/ Testing	Review: *Sections 14–1—14–4* 420 ● Review: *Sections 14–5—14–8* 432 Chapter Review 433 ● Chapter Test 436 ● Additional Practice 438

Chapter 15: Special Triangles   439

Squares and Square Roots

15–1	Meaning of Square Root	440
15–2	Irrational Numbers	442
15–3	Table of Squares and Square Roots	445
15–4	Problem Solving and Applications: *Formulas: Using Square Roots*	451

Pythagorean Theorem and Special Triangles

15–5	Pythagorean Theorem and Applications	454
15–6	Tangent Ratio and Applications	458

Features Career 448 • Calculator/Computer 449 • Calculator 461 Enrichment 465 • Common Errors 467

Review/Testing Review: *Sections 15–1—15–3* 450 • Review: *Sections 15–4—15–6* 461
Chapter Review 462 • Chapter Test 464
Additional Practice 466 • Cumulative Review: *Chapters 14–15* 468

UNIT 5   OTHER APPLICATIONS

Chapter 16:   Statistics and Probability — 469

Statistics

16–1	Mean, Median, and Mode	470
16–2	Histograms	474

Probability

16–3	Probability	479
16–4	Probability and Tables	482
16–5	Tree Diagrams	485
16–6	Multiplying Probabilities	488

Features Career 478 • Calculator 490 • Enrichment 495

Review/Testing Review: *Sections 16–1—16–2* 477 • Review: *Sections 16–3—16–6* 491
Chapter Review 492 • Chapter Test 494 • Additional Practice 496

Chapter 17:   Geometry — 497

Introduction to Geometry

17–1	Introduction to Geometry	498
17–2	Angles	501
17–3	Properties of Triangles	504
17–4	Problem Solving and Applications: *Using Similar Triangles*	508
17–5	Perpendicular and Parallel Lines	514
17–6	Properties of Quadrilaterals	517

Surface Area and Volume

17–7	Problem Solving and Applications: *Surface Area and Rectangular Prisms*	520
17–8	Volume: Cylinders	524
17–9	Volume: Pyramids and Cones	527

Features Calculator 511 • Computer 512 • Enrichment 535 • Common Errors 537

Review/Testing Review: *Sections 17–1—17–4* 513 • Review: *Sections 17–5—17–9* 530
Chapter Review 531 • Chapter Test 534
Additional Practice 536 • Cumulative Review: *Chapters 16–17* 538

Glossary	540
Index	543
Answers to Selected Exercises	547

CHAPTER 1
Whole Numbers and Applications

SECTIONS
1-1 Using Formulas: Addition
1-2 Using Formulas: Subtraction
1-3 Using Formulas: Multiplication
1-4 Using Formulas: Division
1-5 Problem Solving and Applications: Estimating with Whole Numbers
1-6 Numerical Expressions
1-7 Using Variables

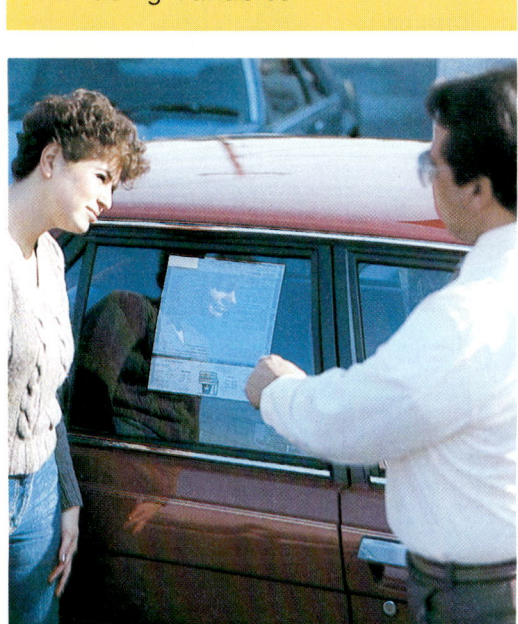

FEATURES
Calculator Application:
Order of Operations
Consumer Application:
Saving $ on Housing Costs
Enrichment:
Geometric Numbers

Teaching Suggestions
p. M-14

QUICK QUIZ

Add.

1. 203 + 387   Ans: 590
2. 39 + 87 + 9250
   Ans: 9376
3. 70 + 85 + 300
   Ans: 455
4. 16,320 + 250 + 1200
   Ans: 17,770
5. 2395 + 482   Ans: 2877
6. 423 + 86 + 1059
   Ans: 1568
7. 9452 + 200 + 1483
   Ans: 11,135

ADDITIONAL EXAMPLES

Example 1

1. The base price of a new car is $12,300. The destination charge is $400. The options total $857. Find the sticker price.
   Ans: $13,557
2. Find the sticker price.
   Base price: $9350
   Options: $1320
   Destination charge: $325
   Ans: $10,995

SELF-CHECK

1. $9875

1-1 Using Formulas: Addition

The following word rule tells you how to find the sticker price of a new car.

Word Rule: The **sticker price** of a new car is the sum of the base price, the destination charge (delivery charges), and the cost of the optional equipment.

A **formula** is a shorthand way of writing a word rule. In a formula, letters and mathematical symbols are used to represent the words.

Formula: $S = B + D + O$

◀ S = sticker price; B = base price;
D = destination charge;
O = cost of optional equipment

PROCEDURE

To use formulas to solve problems:

1. Write the formula.
2. Identify the known values.
3. Replace the known values and solve.

EXAMPLE 1 The base price of a new Larkspur is $8500. The total cost of the options is $927 and the destination charge is $135.
Find the sticker price.

Solution:
1. $S = B + D + O$ ◀ Write the formula.
2. $B = 8500;\ D = 135;\ O = 927$
3. $B = 8500 + 135 + 927$
   $S = 9562$

The sticker price is **$9562.**

Self-Check 1. Find the total cost.
Base price: $7620   Destination charge: $205   Cost of options: $2050

Chapter 1

The table at the right shows the cost of some optional equipment. Thus, to find the total cost of the car, you must first answer the question:

What is the total cost of the options?

This is the **hidden question** in the problem. Then you can use the formula to find the sticker price.

Optional Equipment	
Air conditioning	$648
Automatic transmission	$585
Power brakes	$300
Cruise control	$205
Glass (tinted)	$ 70
Radio, AM/FM	$153
Defogger	$ 85

EXAMPLE 2 Cora Lee bought a new car with a base price of $7980. The destination charge amounted to $197. Cora chose options of air conditioning, automatic transmission and tinted glass.
What was the sticker price?

Solution:
1. $S = B + D + O$
2. $B = 7980$; $D = 197$;
   $O = 648 + 585 + 70 = 1303$
3. $S = 7980 + 197 + 1303$
   $S = 9480$

Hidden Question: Optional Costs = ___?___

Total cost of optional equipment

The sticker price was **$9480**.

Self-Check 2. Find the sticker price. Refer to the Table of Optional Equipment.
Base price: $7125   Destination charge: $155
Options: Automatic transmission, cruise control, AM/FM radio, defogger

CLASSROOM EXERCISES

Add.

1. 5
    6
   +9

2. 23
    51
   +48

3. 109
    561
   + 73

4. 5187
    609
   + 73

5. 2109
    1087
   + 679

6. $723 + 2486 + 1097$

7. $68 + 7418 + 923$

8. $75 + 6541 + 983$

Whole Numbers and Applications **3**

ADDITIONAL EXAMPLE
Example 2
A new car has a base price of $6290. The destination charge is $260. The options (see the Table of Optional Equipment) are power brakes; AM/FM radio; cruise control. Find the sticker price.
Ans: $7208

SELF-CHECK
2. $8308

CLASSROOM EXERCISES
1. 20   2. 122   3. 743
4. 5869   5. 3875
6. 4306   7. 8409
8. 7599

ASSIGNMENT GUIDE
BASIC
p. 4: 1-12
AVERAGE
p. 4: 1-12
ABOVE AVERAGE
p. 4: 1-12

WRITTEN EXERCISES
1. $10,235   2. $8358
3. $9135   4. $10,168
5. $10,640 6. $9350
7. $9880   8. $10,153
9. $8828   10. $8351
11. $9667 12. $10,690

REVIEW CAPSULE
This Review Capsule reviews prior-taught skills used in Section 1-2. The reference is to the pages where the skills were taught.

WRITTEN EXERCISES

Goal: To use a formula to solve problems involving addition

For Exercises 1–7, find the sticker price for each car. (Example 1)

	Base Price	Destination Charge	Total Cost of Options
1.	$7120	$225	$2890
2.	$6800	$108	$1450
3.	$7200	$195	$1740
4.	$9050	$170	$ 948
5.	$8365	$210	$2065

6. The base price of a certain car is $8300. The total cost of optional equipment is $940 and the destination charge is $110.

7. The total cost of the options for a new car is $1875 and the destination charge is $145. The base price is $7860.

Find the sticker price. Refer to the Table of Optional Equipment on page 3. (Example 2)

	Base Price	Destination Charge	Optional Equipment
8.	$8500	$215	Air conditioning; automatic transmission; cruise control
9.	$7600	$188	Automatic transmission; power brakes; tinted glass; defogger
10.	$6950	$158	Cruise control; radio, AM/FM; automatic transmission; power brakes

Solve each problem.

11. Hector plans to buy a car with a base price of $8250 and options of air conditioning, power brakes, cruise control, and tinted glass. The destination charge is $194. What is the sticker price?

12. The base price of the car Luisa wants to buy is $9450. She chooses options of automatic transmission, power brakes, tinted glass, and defogger. The destination charge is $200. Find the sticker price.

REVIEW CAPSULE FOR SECTION 1-2

Subtract.

1. 82 − 38   2. 128 − 9   3. 355 − 8   4. 215 − 34   5. 403 − 78
6. 225 − 156   7. 693 − 427

The answers are on page 27.

4 Chapter 1

1-2 Using Formulas: Subtraction

The amount of an electric bill is based on the number of **kilowatt-hours (kwh)** of electricity used. You use one kilowatt-hour when you use 1 kilowatt (1000 watts) for one hour.

Word rule: The number of kilowatt-hours of electricity used equals the difference between the present reading and the previous reading.

Formula: $n = r - p$

n = number of kwh
r = present reading
p = previous reading

EXAMPLE A portion of the Miles family's electric bill is shown below. Find the number of kilowatt-hours used.

PUBLIC ELECTRIC & POWER COMPANY
025-6100

DATE FROM	READING FROM	DATE TO	READING TO	Kwh USED	AMT.
4 30	6982	5 31	7579		

ACCOUNT NUMBER 012 38 45

Solution:
1. $n = r - p$
2. Present reading, r: 7579   Previous reading, p: 6982
3. $n = 7579 - 6982$
   $n = 597$ Thus, **597 kwh** were used.

Self-Check

1. The present reading on a meter is 9123. The previous reading was 8947. How many kilowatt-hours of electricity were used?

CLASSROOM EXERCISES The answers are on page 6.

Subtract.

1. 3628
   −2416

2. 7142
   −2685

3. 9082
   − 765

4. 7111
   −1236

5. 6255
   −3729

6. 3000 − 1765

7. 2000 − 751

8. 7126 − 4178

Whole Numbers and Applications **5**

Teaching Suggestions
p. M-14

QUICK QUIZ
Find the sticker price.
1. Base price: $8900
   Destination charge: $295
   Optional equipment: $2130
   Ans: $11,325
2. Destination charge: $315
   Optional equipment: $1880
   Base price: $10,830
   Ans: $13,025
3. Base price: $6337
   Destination charge: $219
   Optional equipment:
   AM radio: $103
   4 speed transmission: $425
   bucket seats: $247
   fog lights: $82
   Ans: $7413

ADDITIONAL EXAMPLE
The previous reading on a meter was 937. The present reading is 1720. How many kilowatt-hours were used? Ans: 783

SELF-CHECK
1. 176

CLASSROOM EXERCISES
1. 1212 2. 4457
3. 8317 4. 5875
5. 2526 6. 1235
7. 1249 8. 2948

ASSIGNMENT GUIDE
BASIC
p. 6: 1-12
AVERAGE
p. 6: 1-12
ABOVE AVERAGE
p. 6: 1-12

WRITTEN EXERCISES
1. 817   2. 1767
3. 4204 4. 804
5. 878   6. 878   7. 56
8. 1277 9. 8196
10. 8398 11. 14,543
12. 41,549

WRITTEN EXERCISES

Goal: To use a formula to solve problems involving subtraction

For Exercises 1–8, find how many kilowatt-hours of electricity were used. (Example)

	Previous Reading	Present Reading
1.	1286	2103
2.	2335	4102
3.	5238	9442
4.	3116	3920
5.	8624	9502

6. The reading on a meter on April 5 is 4258. The reading on May 5 is 5136.

7. The reading on a meter on September 15 is 8947. The reading on October 15 is 9003.

8. The present reading on the Gibson's meter is 7705. The previous reading was 6428.

The table at the right below shows the June through October meter readings for Carlin Industries. Refer to the table for Exercises 9–12. (Example)

9. How many kilowatt-hours of electricity were used from July to August?

10. How many kilowatt-hours of electricity were used from August to September?

11. How many kilowatt-hours of electricity did Carlin Industries use from June to August?

12. If the meter reading in November is 51,011, how many kilowatt-hours of electricity were used from June to November?

Month	Meter Reading
June	9462
July	15,809
August	24,005
September	32,403
October	40,016

REVIEW CAPSULE FOR SECTION 1-3

Multiply.

1. 16 × 8
2. 24 × 9
3. 125 × 6
4. 138 × 7
5. 16 × 15
6. 35 × 23
7. 116 × 18
8. 326 × 24
9. 6 × 6 × 6
10. 1 × 1 × 1 × 0

The answers are on page 27.

6 Chapter 1

1-3 Using Formulas: Multiplication

The following word rule relates distance, rate, and time.

Word Rule: Distance equals rate multiplied by time.

Formula: $d = rt$

d = distance
r = rate
t = time

In the formula, rt means "r times t".
Here are several ways to indicate "r times t".

$r \times t \qquad rt \qquad r \cdot t \qquad (r)(t)$ — These have the same meaning.

This table illustrates some units that are used in measuring distance, time, and rate (average speed).

Distance (d)	Time (t)	Rate (r)
meters	seconds	meters per second (m/sec)
kilometers	hours	kilometers per hour (km/hr)
feet	seconds	feet per second (ft/sec)
miles	hours	hour (mi/hr)
miles	minutes	minute (mi/min)

EXAMPLE 1 Light travels at the rate of about 300,000,000 meters per second. It takes 496 seconds for light from the sun to reach the earth. Find the distance from the earth to the sun.

Solution:
[1] $d = rt$
[2] $r = 300,000,000$ m/sec; $t = 496$ sec
[3] $d = 300,000,000 \cdot 496$
$d = 148,800,000,000$

The distance is about **148,800,000,000** meters.

Self-Check
1. Find d when $r = 142$ miles per hour and $t = 4$ hours.
2. Find d when $r = 250$ meters per second and $t = 8$ seconds.

Whole Numbers and Applications

Teaching Suggestions
p. M-14

QUICK QUIZ
Multiply.
1. 23 x 8   Ans: 184
2. 127 x 7   Ans: 889
3. 513 x 26   Ans: 13,338
4. 20,000 x 37
   Ans: 740,000
5. 53 x 12   Ans: 636
6. 217 x 35   Ans: 7595
7. 372 x 100
   Ans: 37,200
8. 5 x 5 x 5   Ans: 125
9. 2 x 2 x 2 x 2
   Ans: 16
10. 16 x 1 x 1   Ans: 16

ADDITIONAL EXAMPLES
Example 1
1. A jet plane has a rate of 680 mph. How far does the jet fly in 8 hours?
   Ans: 5440 miles
2. A car travels 77 feet per second. How many feet will the car travel in 60 seconds?
   Ans: 4620 feet

SELF-CHECK
1. 568 miles
2. 2000 meters

You can use this formula to find the distance in feet that an object will fall in t seconds.

$$d = 16t^2 \quad \begin{array}{l} d = \text{distance} \\ t = \text{time} \end{array}$$

In the formula, t^2 means $t \cdot t$. When a number is multiplied by itself two or more times, the number is <u>raised to a power</u>. The raised "2" is called an <u>exponent</u>. The **exponent** indicates how many times a number is multiplied by itself.

$$3^2 = 3 \cdot 3 = 9$$
$$5^3 = 5 \cdot 5 \cdot 5 = 125$$

3^2 ← **Exponent**
   ← **Base**

3^2 is read "3 squared."
5^3 is read "5 cubed."

EXAMPLE 2 Use $d = 16t^2$ to find how far an object will fall in 4 seconds.

Solution:
[1] $d = 16t^2$
[2] $t = 4; t^2 = 4 \cdot 4$
[3] $d = 16 \cdot 4 \cdot 4$
$d = 16 \cdot 16$
$d = 256$ The object will fall **256 feet**.

Self-Check 3. Use $d = 16t^2$ to find how far an object will fall in 3 seconds.
144 feet

ADDITIONAL EXAMPLES
Example 2
Use the formula $d = 16t^2$ to find the distance an object falls in the given number of seconds.
<u>1</u>. 8   Ans: 1024 feet
<u>2</u>. 11   Ans: 1936 feet

CLASSROOM EXERCISES
<u>1</u>. 15   <u>2</u>. 160   <u>3</u>. 522
<u>4</u>. 11,200   <u>5</u>. 81
<u>6</u>. 343   <u>7</u>. 64   <u>8</u>. 1
<u>9</u>. 1   <u>10</u>. 216

CLASSROOM EXERCISES

Find the distance for the given rate and time. (Example 1)

	Rate	Time	Distance
1.	5 miles per minute	3 minutes	_?_ miles
2.	80 kilometers per hour	2 hours	_?_ kilometers
3.	9 meters per second	58 seconds	_?_ meters
4.	140 feet per second	80 seconds	_?_ feet

Evaluate. (Example 2, step 2)

5. 9^2 **6.** 7^3 **7.** 4^3 **8.** 1^2 **9.** 1^3 **10.** 6^3

Chapter 1

WRITTEN EXERCISES

Goal: To use a formula to solve problems involving multiplication

For Exercises 1–6, find the distance for the given rate and time. (Example 1)

	Rate	Time		Rate	Time
1.	120 feet per second	20 seconds	2.	50 miles per hour	10 hours
3.	75 kilometers per hour	5 hours	4.	25 meters per second	30 seconds

Solve each problem. (Example 1)

5. An elephant runs at a rate of 31 feet per second. How many feet will the elephant run in 15 seconds?

6. Amelia Earhart flew a plane from Newfoundland to Ireland in 15 hours at a rate of 135 miles per hour. What was the distance flown?

7. The speed of sound is about 343 meters per second. How many meters will sound travel in 6 seconds?

8. The Concorde SST flies at a speed of 1550 miles per hour. How many miles can it fly in 6 hours?

Evaluate. (Example 2, step 2)

9. 2^2 10. 2^3 11. 4^2 12. 1^4 13. 1^5 14. 5^2
15. 3^2 16. 3^3 17. 6^2 18. 10^2 19. 10^3 20. 8^2

Solve each problem. Use the formula $d = 16t^2$. (Example 2)

21. How far will an object fall in 6 seconds?

22. How far will an object fall in 12 seconds?

23. How far will an object fall in 4 seconds?

24. How far will an object fall in 8 seconds?

REVIEW CAPSULE FOR SECTION 1–4

Add.

1. $14 + 18 + 22 + 17$
2. $15 + 18 + 23 + 17 + 22$
3. $9 + 8 + 6 + 5 + 4$
4. $12 + 19 + 24 + 32 + 15 + 21$

Divide.

5. $4\overline{)36}$ 6. $6\overline{)42}$ 7. $5\overline{)65}$ 8. $7\overline{)154}$ 9. $5\overline{)235}$ 10. $4\overline{)832}$

The answers are on page 27.

Whole Numbers and Applications **9**

ASSIGNMENT GUIDE
BASIC
p. 9: 1–24
AVERAGE
p. 9: 1–23 odd
ABOVE AVERAGE
p. 9: 1–23 odd

WRITTEN EXERCISES
1. 2400 feet
2. 500 miles
3. 375 kilometers
4. 750 meters
5. 465 feet
6. 2025 miles
7. 2058 meters
8. 9300 miles
9. 4   10. 8   11. 16
12. 1   13. 1   14. 25
15. 9   16. 27   17. 36
18. 100   19. 1000
20. 64   21. 576 feet
22. 2304 feet
23. 256 feet
24. 1024 feet

Teaching Suggestions
p. M-14

QUICK QUIZ
Divide.
1. $3\overline{)375}$ Ans: 125
2. $7\overline{)2135}$ Ans: 305
3. $\frac{432}{2}$ Ans: 216
4. $\frac{112}{14}$ Ans: 8
5. $2926 \div 7$ Ans: 418

ADDITIONAL EXAMPLES
Example 1
Sally's test scores were 80, 95, 87, and 82. What was Sally's average score?
Ans: 86

SELF-CHECK
1. 143   2. 122

Example 2
Barry has a Fahrenheit thermometer outside his window. He takes temperature readings at different times during the day. Yesterday's readings were as follows: 48°; 51°; 60°; 63°; 64°; 62°. Find the mean temperature for the day.
Ans: 58°F

SELF-CHECK
3. 5°

1-4 Using Formulas: Division

Finding an average (or mean) involves finding a quotient.

Word Rule: The **average** of two or more numbers or measures is the sum of the measures divided by the number of measures.

Formula: $a = \frac{S}{n}$

a = average
S = sum of measures
n = number of measures

EXAMPLE 1 The heights of five members of a basketball team are given below in centimeters (cm). What is the average of the players' heights?

Player	A	B	C	D	E
Height	182 cm	190 cm	178 cm	204 cm	191 cm

Solution:
① $a = \frac{S}{n}$
② $S = 182 + 190 + 178 + 204 + 191 = 945$; $n = 5$
③ $a = \frac{945}{5}$
   $a = 189$ The average is **189 centimeters.**

Self-Check *Find the average score for each team.*
1. Team A: 135, 172, 118, 147
2. Team B: 108, 94, 135, 151

EXAMPLE 2 The low temperatures for one week in a city are listed below in degrees Celsius. Find the mean low temperature for the week.

S	M	T	W	T	F	S
5°	3°	4°	3°	1°	2°	3°

Solution:
① $a = \frac{S}{n}$
② $S = 5 + 3 + 4 + 3 + 1 + 2 + 3 = 21$; $n = 7$
③ $a = \frac{21}{7}$
   $a = 3$ The mean low temperature is **3°.**

Mean is another word for average

Self-Check
3. Find the mean low temperature for the week.

S	M	T	W	T	F	S
6°	5°	2°	5°	4°	7°	6°

10   Chapter 1

CLASSROOM EXERCISES

Divide.

1. 600 ÷ 50
2. 432 ÷ 16
3. 882 ÷ 21
4. 6300 ÷ 84
5. 7626 ÷ 62
6. 474 ÷ 79
7. 4185 ÷ 93
8. 3976 ÷ 71

Find the average. (Examples 1 and 2)

9.

Month	Rainy Days
January	2
February	3
March	7
April	20
May	8
June	2

Average number of rainy days: __?__

10.

Age at First Inaugural	
G. Washington	57
J. Adams	61
T. Jefferson	57
J. Madison	57
J. Monroe	58

Average age at first inaugural: __?__

WRITTEN EXERCISES

Goal: To use a formula to solve problems involving division

For Exercises 1–6, find the average score for each team. (Example 1)

1. Team A: 135, 172, 118, 147
2. Team B: 108, 94, 135, 151
3. Team C: 162, 145, 170, 159
4. Team D: 144, 196, 278, 110
5. Team E: 161, 163, 113, 151
6. Team F: 202, 132, 133, 129

Find the average. (Example 1)

7.

Player	Weight in kilograms
Beal	48
Davis	41
Lopez	49
Porter	45
Chang	39
Williams	48

Average weight: __?__

8.

Weekly Pay Before Taxes	
Shoe-factory Worker	$165
Clothing Worker	$165
Laundry Worker	$153
Retail-trade Worker	$149
Leather-goods Worker	$168

Average weekly pay: __?__

Whole Numbers and Applications

CLASSROOM EXERCISES

1. 12   2. 27   3. 42
4. 75   5. 123   6. 6
7. 45   8. 56   9. 7
10. 58

ASSIGNMENT GUIDE
BASIC
Day 1 pp. 11–12: 1–11
Day 2 p. 12: 12–21
AVERAGE
pp. 11–12: 1–11, 13–21 odd
ABOVE AVERAGE
pp. 11–12: 1–21 odd

WRITTEN EXERCISES
1. 143   2. 122   3. 159
4. 182   5. 147   6. 149
7. 45kg   8. $160

WRITTEN EXERCISES

9. 83
10. 73
11. $150,600
12. 359
13. 90
14. 5116
15. 22
16. 7868
17. 76,329
18. 40°
19. 25
20. 73
21. Yes

9. Diane's math test scores are 74, 85, 88, 86, 85, 89, 85 and 72. What is her average score?

10. Brian's pulse rates for a week were 68, 74, 77, 59, 82, 76, and 75. What was his average pulse rate?

11. Annual salaries for starting players on a professional football team are $90,000, $123,000, $225,000, $178,000, and $137,000. What is the average salary for a starter on the team?

Find the mean. (Example 2)

12. 336, 464, 304, 288, 403

13. 98, 85, 90, 92, 96, 85, 84

14. 4608, 5479, 6500, 3877

15. 18, 20, 17, 24, 24, 30, 15, 18, 20, 25, 31

16. 8916, 9214, 5684, 6382, 9144

17. 85, 612, 64, 186, 79, 189

Solve each problem.

18. The daily high temperatures for a week were 43°, 41°, 23°, 54°, 36°, 38°, 45°. What was the mean temperature?

19. A basketball player scored a total of 1875 points in 75 games. Find the mean number of points scored per game.

20. Wilson's department store received a total of 18,250 phone calls during 250 working days. What was the mean number of calls per day?

21. In Mohawk City, new traffic lights are installed when the mean number of vehicles per day using an intersection is 13,000 or more. The traffic safety division counted 102,340 vehicles using an intersection over 7 days. Will the city install a new traffic light?

REVIEW CAPSULE FOR SECTION 1–5

Round each number to the nearest 10.

1. 42 2. 28 3. 55 4. 12 5. 83 6. 96

Round each number to the nearest 100.

7. 225 8. 176 9. 389 10. 622 11. 1254 12. 795

Round each number to the nearest 1000.

13. 12,450 14. 8125 15. 18,505 16. 2050 17. 32,806 18. 13,499

The answers are on page 27.

12 Chapter 1

CONSUMER APPLICATION
Saving $ on Housing Costs

On certain jobs in your home, you can save the cost of labor by doing the work yourself. The drawing below shows a plan for tiling a floor. Each of the larger squares represents a 12" × 12" square.

NOTE: The symbol " represents inches.

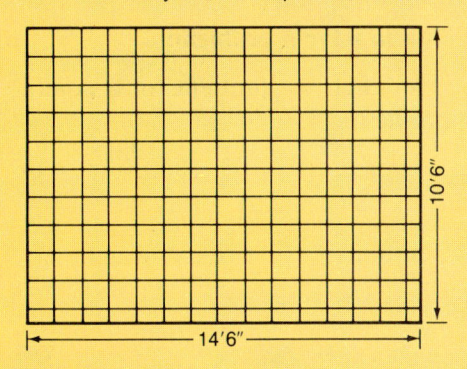

Exercises

1. How many 12" × 12" tiles will be needed?

2. How many 6" × 12" tiles will be needed?

3. How many 6" × 6" tiles will be needed?

4. What is the least number of 12-inch tiles needed? (HINT: you can get two 6" × 12" tiles from one 12" × 12" tile.)

5. There are 45 12-inch tiles in one box. How many boxes will be needed if only full boxes are sold?

6. The tile costs $14 per box. Adhesive costs $8 and a spreader costs $2. What is the total cost of these materials?

7. Molding will be installed around the room at the bottom of the walls. Without allowing for doorways, how many feet of molding are needed?

8. The molding cost 40¢ per foot. What will be the total cost of the molding for the room?

9. What is the total cost of all the materials for the room, excluding sales tax?

10. A carpenter will supply all the materials and do the work for $200. How much can be saved by doing the work yourself?

CONSUMER APPLICATIONS
This feature applies the basic computational skills to estimating the amount and cost of floor tile and other materials needed for a room. Thus, it can be used after Section 1-5 (pages 14-16) is taught.

1. 140
2. 24
3. 1
4. 153
5. 4
6. $66
7. 50
8. $20
9. $86
10. $114

Consumer Application 13

Teaching Suggestions
p. M-14

QUICK QUIZ

Round each number to the nearest ten.

1. 32   Ans: 30
2. 58   Ans: 60

Round each number to the nearest hundred.

3. 851   Ans: 900
4. 4260 Ans: 4300

Round each number to the nearest thousand.

5. 5260 Ans: 4000
6. 6837 Ans: 7000

ADDITIONAL EXAMPLES

Example 1

1. The number of vehicles that crossed an intersection each day of a week was as follows: Mon: 728; Tues: 1077; Wed: 563; Thurs: 440; Fri: 981. Estimate the total to the nearest hundred. Ans: 3800

2. The Orange Bowl seats 75,459 people. The Rose Bowl seats 106,721. Estimate the difference to the nearest hundred. Ans: 31,200

SELF-CHECK

1. 2300 km   2. 300 km

PROBLEM SOLVING AND APPLICATIONS

1-5 Estimating with Whole Numbers

Estimating sums, differences, products, and quotients is a skill that can be very useful in everyday life.

You use the rules for rounding to estimate answers.

> **PROCEDURE**
>
> **To use estimation to solve a problem:**
>
> [1] Write a sum, difference, product, or quotient (or a combination of these) for the problem.
>
> [2] Round each number to the nearest 10, 100, 1000, and so on as indicated.
>
> [3] Perform the indicated operations on the rounded numbers.

EXAMPLE 1 In one month, the number of vehicles that entered a National Park consisted of 556 cars, 429 vans, and 317 trucks.

Estimate the total number of vehicles to the nearest hundred.

Solution: [1] Problem   [2] Rounded to the nearest 100

```
    556  →      600
    429  →      400
   +317  →     +300
              [3] 1300   ◄ Estimate
```

About **1300 vehicles** entered the park in one month.

Self-Check

1. On a car trip, the Rulio family traveled 788 kilometers the first week, 913 kilometers the second week, and 632 kilometers the third week. Estimate the number of kilometers traveled to the nearest hundred.

2. Last year the Rulio family traveled 2576 kilometers. Estimate how many more kilometers they traveled last year than this year.

14   Chapter 1

Example 2 shows how to estimate a product.

EXAMPLE 2 On each of 42 trips to a national park in July, a sightseeing bus was almost filled. The bus carries 28 passengers. Estimate the number of passengers carried by the bus in July.

Solution: ⃞1 Problem   ⃞2 Rounded to the nearest 10

$$\begin{array}{r}42\\ \times 28\\ \hline\end{array} \qquad \begin{array}{r}40\\ \times 30\\ \hline \boxed{3}\ 1200\end{array}$$ ◀ Estimate

The bus carried about **1200 passengers** in July.

Self-Check 3. A campsite with 34 spaces for rental was almost completely filled for 27 days. Estimate the total number of daily rentals.

In problems involving division, round the dividend to the nearest convenient number that is evenly divisible by the rounded divisor.

You can use this formula to find how long it will take to travel a certain distance at a given average speed (rate).

$$t = \frac{d}{r}$$ ◀ t = time
d = distance
r = rate

EXAMPLE 3 Estimate how long it will take to drive 381 miles at an average speed of 48 miles per hour.

Solution: ⃞1 Problem   ⃞2 Rounded Numbers

$$\frac{381}{48} \longrightarrow \frac{400}{50}$$

$$\boxed{3}\ \frac{400}{50} = 8$$ ◀ Estimate

It will take about **8 hours**.

Self-Check 4. William travels 635 kilometers at an average speed of 83 kilometers per hour. Estimate the number of hours he traveled.

5. Amtrak's Silver Meteor travels the 1325 miles from Miami to New York City at an average rate of 53 miles per hour. Estimate the number of hours the trip takes.

Whole Numbers and Applications **15**

ADDITIONAL EXAMPLES
Example 2
For the Labor Day picnic, 53 dozen rolls were ordered. The rolls cost $1.59 per dozen. Estimate the cost of the rolls. Ans: $80

SELF-CHECK
3. 900 rentals

Example 3
A marching band has 237 members. The band travels to football games by bus. Each bus carries 42 people. Estimate the number of busses needed for each game. Ans: 6

SELF-CHECK
4. 8 hr   5. 26 hr

CLASSROOM EXERCISES

1. a
2. c
3. c
4. a

ASSIGNMENT GUIDE
BASIC
p. 16: 1-10
AVERAGE
p. 16: 1-7 odd, 9-12
ABOVE AVERAGE
p. 16: 1-7 odd, 9-12

WRITTEN EXERCISES

1. a
2. b
3. a
4. c
5. a
6. b
7. b
8. c
9. 5000 meters
10. 18 days
11. 2700 seats
12. $5

CLASSROOM EXERCISES

Choose the best estimate. Choose a, b, or c.
(Examples 1, 2, and 3, steps 1 and 2)

1. 112 + 631 a. 100 + 600 b. 200 + 700 c. 100 + 700
2. 2655 − 1878 a. 2700 − 1800 b. 3000 − 2000 c. 2700 − 1900
3. 72 × 48 a. 70 × 40 b. 80 × 40 c. 70 × 50
4. 774 ÷ 68 a. 770 ÷ 70 b. 760 ÷ 60 c. 800 ÷ 60

WRITTEN EXERCISES

Goal: To use estimation to solve word problems

For Exercises 1–8, choose the best estimate. Choose a, b, or c.
(Examples 1, 2, and 3)

1. 738 + 593 a. 1300 b. 1200 c. 1400
2. 4630 + 2397 a. 6000 b. 7000 c. 8000
3. 963 − 857 a. 100 b. 50 c. 120
4. 4103 − 2106 a. 2200 b. 2100 c. 2000
5. 39 × 48 a. 2000 b. 1500 c. 1200
6. 705 × 18 a. 7000 b. 14,000 c. 16,000
7. 328 ÷ 27 a. 15 b. 11 c. 14
8. 2847 ÷ 72 a. 65 b. 25 c. 40

9. Mt. McKinley has an elevation of 8848 meters above sea level. The elevation of Pike's Peak is 4301 meters above sea level. Estimate, to the nearest thousand meters, the difference in their elevations.

10. On a park reserve, 18 kilograms of grain are fed to the animals each day. If there are 357 kilograms of grain on hand, estimate how many days the feed will last (nearest day).

11. In an amusement park amphitheater, there are 87 rows of seats. Each row has 31 seats. Estimate the number of seats in the amphitheater.

12. A hiking club bought T-shirts for 57 members. The total cost was $299.25. Estimate the cost of each shirt to the nearest dollar.

REVIEW: SECTIONS 1-1 — 1-5

Find the sticker price for each car. (Section 1-1)

1. Base Price: $6480
   Destination Charge: $125
   Total Cost of Options: $1048

2. Base Price: $12,350
   Destination Charge: $192
   Total Cost of Options: $1250

Find how many kilowatt-hours of electricity were used. (Section 1-2)

3. Previous Reading: 3819
   Present Reading: 4903

4. Previous Reading: 9426
   Present Reading: 11,054

Find the distance. (Section 1-3)

5. A cheetah can run 31 meters per second. How far can it run in 6 seconds?

6. Using the formula $d = 16t^2$, find how far an object will fall in 7 seconds.

Find the average score. (Section 1-4)

7. Blue Team: 65, 42, 73, 56

8. Red Team: 84, 87, 93, 79, 82

Estimate the answer. (Section 1-5)

9. A speed skater skated the first lap in 44 seconds, the second lap in 46 seconds, and the third lap in 42 seconds. Estimate the total time to the nearest ten seconds.

10. A basketball player scored 1401 points in 68 games. Estimate, to the nearest ten, the average number of points scored per game.

REVIEW CAPSULE FOR SECTION 1-6

Add.

1. 356
   59
   258
   +635

2. 1246
   2389
   5642
   + 384

3. $6 + 8 + 12$
4. $115 + 36 + 84$
5. $325 + 42 + 1456$

Subtract.

6. 56
   −49

7. 2040
   − 528

8. $932 - 62$
9. $5004 - 525$

Multiply.

10. 5×9
11. 25×13
12. $14 \cdot 16$
13. $156(42)$
14. $(45)(1461)$

Divide.

15. $6\overline{)42}$
16. $12\overline{)48}$
17. $62\overline{)434}$
18. $4130 \div 35$
19. $5025 \div 15$

The answers are on page 27.

Whole Numbers and Applications

QUIZ: SECTIONS 1-1—1-5
After completing this Review, you may want to administer a quiz covering the same sections. A Quiz is provided in the *Teacher's Edition: Part II*.

REVIEW: SECTIONS 1-1—1-5
1. $7653   2. $13,792
3. 1084 kwh
4. 1628 kwh
5. 186 m   6. 784 ft
7. 59   8. 85
9. 130 sec
10. 20 points

Teaching Suggestions
p. M-14

QUICK QUIZ

Add.

1. 382 + 79   Ans: 461
2. 406 + 922 + 83
   Ans: 1411
3. 27 + 1026 + 358
   Ans: 1411

Subtract.

4. 270 - 93   Ans: 177
5. 2000 - 1363
   Ans: 637
6. 2271 - 362
   Ans: 1909

Multiply.

7. 213 × 7   Ans: 1491
8. 63 × 22   Ans: 1386
9. (106)(20)   Ans: 2120

Divide.

10. 23)̄391   Ans: 17
11. 1736 ÷ 31   Ans: 56

1-6 Numerical Expressions

Each of the following is a *numerical expression*.

$$19 + 8 - 12$$
$$8 + 3 \times 2$$
$$(44 \times 6) \div 2$$
$$30 \div (3 \times 5)$$

◀ Numerical expressions containing parentheses

As these four examples show, a **numerical expression** contains <u>at least one</u> of the operations of addition, subtraction, multiplication, and division.

To **evaluate** (find the value of) a numerical expression, it is important to know which operation should be done first. For example,

if you add first: $8 + 3 \times 2 = 11 \times 2 = $ **22**
if you multiply first: $8 + 3 \times 2 = 8 + 6 = $ **14**

◀ Both cannot be correct.

To be sure that each numerical expression has only one value, follow the rules for order of operations.

ORDER OF OPERATIONS

1 When the *only operations* are addition and subtraction, perform the operations in order from left to right.

2 When the *only operations* are multiplication and division, perform the operations in order from left to right.

3 When multiplication or division is involved *along with* addition and subtraction, perform the multiplication and division before the addition and subtraction.

4 When numerical expressions contain *parentheses*, perform the operations inside the parentheses first.

The Examples in the table on the following page show how to apply the rules for order of operations.

Chapter 1

Example	Procedure	Computation
1a. $19 + 8 - 12$	① Add. (Rule 1) ② Subtract.	$19 + 8 - 12 = 27 - 12$ $= 15$
1b. $36 - 14 + 7$	① Subtract. (Rule 1) ② Add.	$36 - 14 + 7 = 22 + 7$ $= 29$
2a. $48 \div 12 \times 7$	① Divide. (Rule 2) ② Multiply.	$48 \div 12 \times 7 = 4 \times 7$ $= 28$
2b. $9 \times 4 \div 2$	① Multiply. (Rule 2) ② Divide.	$9 \times 4 \div 2 = 36 \div 2$ $= 18$
3a. $12 \div 2 + 4$	① Divide. (Rule 3) ② Add.	$12 \div 2 + 4 = 6 + 4$ $= 10$
3b. $21 - 6 \times 3$	① Multiply. (Rule 3) ② Subtract.	$21 - 6 \times 3 = 21 - 18$ $= 3$
3c. $24 - 18 \div 3 \times 4$	① Divide. (Rule 3) ② Multiply. (Rule 3) ③ Subtract.	$24 - 18 \div 3 \times 4 = 24 - 6 \times 4$ $24 - 6 \times 4 = 24 - 24$ $= 0$
4a. $10 \div (2 + 3)$	① Add. (Rule 4) ② Divide.	$10 \div (2 + 3) = 10 \div 5$ $= 2$
4b. $5 + (3 - 1) + 4$	① Subtract. (Rule 4) ② Add.	$5 + (3 - 1) + 4 = 5 + 2 + 4$ $= 11$
4c. $(5 + 3) - (1 + 4)$	① Add. (Rule 4) ② Subtract.	$(5 + 3) - (1 + 4) = 8 - 5$ $= 3$

CLASSROOM EXERCISES

Name the rule that was applied in each case. (Examples 1–4, step 1)

1. $15 + 7 - 6 = 22 - 6$
2. $72 \div 36 \times 6 = 2 \times 6$
3. $4 + 9 \div 3 = 4 + 3$
4. $26 - 12 \times 2 = 26 - 24$
5. $16 - (2 + 8) = 16 - 10$
6. $5 \times 8 \div 2 = 40 \div 2$
7. $12 - 3 + 5 = 9 + 5$
8. $(8 + 4) \div 3 = 12 \div 3$
9. $3 \times 12 \div 4 = 36 \div 4$
10. $(8 - 4) \div 2 = 4 \div 2$
11. $14 - 3 \times 4 = 14 - 12$
12. $12 \div 2 \times 8 = 6 \times 8$
13. $6 + 8 \times 9 = 6 + 72$
14. $20 + 1 - 8 = 21 - 8$
15. $6 \times 8 \div 4 = 48 \div 4$

Whole Numbers and Applications

ADDITIONAL EXAMPLES

Example 1
1. $22 + 11 - 10$ Ans: 23
2. $31 - 5 + 6$ Ans: 32

Example 2
1. $36 \div 3 \times 2$ Ans: 24
2. $15 \times 10 \div 5$ Ans: 30

Example 3
1. $15 + 2 \times 3$ Ans: 21
2. $30 \div 10 + 5$ Ans: 8

Example 4
1. $16 \times (8 + 8) \div 2$
 Ans: 128
2. $72 \div 6 \times (12 - 10)$
 Ans: 24

CLASSROOM EXERCISES
1. Rule 1
2. Rule 2
3. Rule 3
4. Rule 3
5. Rule 4
6. Rule 2
7. Rule 1
8. Rule 4
9. Rule 2
10. Rule 4
11. Rule 3
12. Rule 2
13. Rule 3
14. Rule 1
15. Rule 3

ASSIGNMENT GUIDE

BASIC
Day 1 p. 20: 1-39 odd
Day 2 p. 20: 2-50 even

AVERAGE
Day 1 p. 20: 1-32
Day 2 p. 20: 33-64

ABOVE AVERAGE
p. 20: 3, 6, 9, ..., 60
p. 21: 81-97 odd

WRITTEN EXERCISES

1. addition
2. subtraction
3. multiplication
4. multiplication
5. division
6. division
7. multiplication
8. multiplication
9. 8   10. 18   11. 27
12. 28   13. 28   14. 5
15. 8   16. 40   17. 24
18. 2   19. 54   20. 20
21. 100   22. 2   23. 4
24. 56   25. 4   26. 12
27. 47   28. 12   29. 60
30. 23   31. 27   32. 78
33. 36   34. 3   35. 96
36. 390   37. 57   38. 16
39. 50   40. 5   41. 35
42. 21   43. 3   44. 4
45. 19   46. 4   47. 23
48. 61   49. 18   50. 6
51. 56   52. 24   53. 4
54. 6   55. 90   56. 111
57. 21   58. 16   59. 1
60. 51   61. 2   62. 23

WRITTEN EXERCISES

Goal: To use order of operations in evaluating numerical expressions

For Exercises 1–8, name the operation that should be performed first. (Examples 1, 2, and 3, step 1)

1. $7+5-2$
2. $10-3+6$
3. $5\times 4+7$
4. $4\times 6\div 3$
5. $14\div 2+5$
6. $3+24\div 8$
7. $9\times 4\div 3$
8. $14\times 6\div 8-2$

For Exercises 9–80, evaluate each numerical expression.

(Example 1)

9. $5+6-3$
10. $15-3+6$
11. $29+6-8$
12. $23-9+14$
13. $19-7+16$
14. $8+12-15$
15. $16+22-30$
16. $45-41+36$

(Example 2)

17. $12\div 3\cdot 6$
18. $4\cdot 6\div 12$
19. $24\div 4\cdot 9$
20. $4\cdot 15\div 3$
21. $28\div 7\cdot 25$
22. $5\cdot 6\div 15$
23. $7\cdot 8\div 14$
24. $32\div 4\cdot 7$

(Example 3)

25. $12-4\cdot 2$
26. $15\div 3+7$
27. $8\cdot 5+7$
28. $16-20\div 5$
29. $45-2+51\div 3$
30. $29-42\div 2+15$
31. $33-9+21\div 7$
32. $23+18\div 6+52$

(Example 4)

33. $(4+8)\cdot 3$
34. $5+(3\cdot 2)-8$
35. $(27-3)\cdot 4$
36. $(42-3)\cdot (4+6)$
37. $56+4\div (10-6)$
38. $19+(5\div 5)-4$
39. $14+(3\cdot 12)$
40. $(33\div 3)+5-11$

MIXED PRACTICE

41. $15\div 3\cdot 7$
42. $13-4+12$
43. $6\cdot 8\div 16$
44. $3\div 3\cdot (9-5)$
45. $30-9\cdot 3+16$
46. $3+9-8$
47. $(54-8)\cdot 2\div 4$
48. $12\cdot 5+6\div 6$
49. $(16-9)+11$
50. $16\div 4\cdot 3-6$
51. $32\div 4\cdot 7$
52. $16-(8-4)+12$
53. $3\cdot 4\cdot 2\div 6$
54. $8+9-11$
55. $9+7\cdot 8+25$
56. $3+9\cdot 12$
57. $6+7+2\cdot 4$
58. $14\div 2\cdot 3-5$
59. $36\div 3\div 6\div 2$
60. $15\cdot 4-18\div 2$
61. $5\div 5\cdot 7-5$
62. $37-7\cdot 4\div 2$
63. $(4\cdot 8)\div 4\cdot 3$
64. $6\cdot 3+2\cdot 9$
65. $5\cdot 2+4\cdot 7$
66. $(5\cdot 6)\div 3\cdot 2$
67. $21\div 3\cdot 7$
68. $43-6\div 3+4$
69. $(21-15)+4$
70. $20\div (12-2)$
71. $(8+7)-(5+2)$
72. $(47+5)\div 4$
73. $17+(8-5)-6$
74. $25\div (12-7)$
75. $(26-9)+(8-6)$
76. $95-(42+6)-7$
77. $(8+5)-(9+4)$
78. $66\div (2\cdot 3)$
79. $52+(10-7)-30$
80. $(15-9)+(21-8)$

20 Chapter 1

MORE CHALLENGING EXERCISES

Copy each problem. Replace the ● with +, −, ×, or ÷ so that the numeral expression equals the given answer.

81. 6 ● 7 ● 3 = 27
82. 9 ● 9 ● 4 = 4
83. (9 ● 9) ● 4 = 72
84. 50 ● (40 ● 10) = 0
85. 4 ● 4 ● 4 = 64
86. 9 ● 9 ● 9 = 729
87. 8 ● 9 ● 4 ● 9 ● 6 = 73
88. 14 ● 5 ● 7 ● 3 = 13
89. 8 ● 2 ● 9 ● 3 = 43
90. (5 ● 2) ● (20 ● 4) = 15
91. (12 ● 3) ● (21 ● 7) = 45
92. 8 ● (40 ● 10) = 32
93. (6 ● 9) ● 14 = 40
94. (8 ● 6) ● (7 ● 7) = 2
95. 18 ● (35 ● 7) = 13
96. (28 ● 14) ● 16 = 32
97. (9 ● 9) ● (33 ● 11) = 0
98. 29 ● (42 ● 7) = 35

ORDER OF OPERATIONS

You can use the [M+] (memory) and [MR] (memory recall) keys to solve problems involving order of operations.

EXAMPLE: Solve: **a.** $80 - 4 \times 6$ **b.** $98 \times 76 - 189 \div 7$

Solutions:

a. 4 [×] 6 [=] [M+] 8 0 [−] [MR] [=] → M 56.

b. 1 8 9 [÷] 7 [=] [M+] 9 8 [×] 7 6 [=] [−] [MR] [=] → M 7421.

EXERCISES *Use a calculator to evaluate each numerical expression.*

1. $65 - 7 \cdot 8 \div 4$
2. $14 \times 6 \div 12 - 3 \times 2$
3. $80 \div 10 - 5 + 6 \div 3$
4. $11 - 15 \div 3 + 15 \times 4$
5. $20 - 12 \div 4 + 9 \cdot 2$
6. $110 \div 5 - 3 \cdot 7 + 6$
7. $75 - 25 \cdot 2 + 10$
8. $85 \times 41 - 210 \div 3$
9. $18 + 45 \div 9 - 50 \div 5$
10. $22 \times 3 + 18 \div 6 - 30 \div 15$

REVIEW CAPSULE FOR SECTION 1-7

Perform the indicated operations.

1. $6 + 4 + 8$
2. $12 + 9 + 15$
3. $24 - 9 - 6$
4. $35 - 12 - 8$
5. $9 \times 8 \times 7$
6. $13 \times 12 \times 5$
7. $6 \times 5 + 4$
8. $14 + 3 \times 9$
9. $45 \div 5 \div 3$
10. $88 \div 2 \div 2$
11. $32 \div 2 \div 4$
12. $54 \div 3 \div 2$

The answers are on page 27.

Whole Numbers and Applications

WRITTEN EXERCISES

63. 24
64. 36
65. 38
66. 20
67. 49
68. 45
69. 10
70. 2
71. 8
72. 13
73. 8
74. 5
75. 19
76. 40
77. 0
78. 11
79. 25
80. 19
81. 6 + 7 · 3 = 27
82. 9 − 9 + 4 = 4
83. (9 + 9) · 4 = 72
84. 50 − (40 + 10) = 0
85. 4 · 4 · 4 = 64
86. 9 · 9 · 9 = 729
87. 8 · 9 + 4 − 9 + 6 = 73
88. 14 − 5 + 7 − 3 = 13
89. 8 · 2 + 9 · 3 = 43
90. (5 · 2) + (20 ÷ 4) = 15
91. (12 + 3) · (21 ÷ 7) = 45
92. 8 · (40 ÷ 10) = 32
93. (6 · 9) − 14 = 40
94. (8 − 6) − (7 − 7) = 2
95. 18 − (35 ÷ 7) = 13
96. (28 ÷ 14) · 16 = 32
97. (9 − 9) · (33 ÷ 11) = 0
98. 29 + (42 ÷ 7) = 35

CALCULATOR EXERCISES

1. 51
2. 1
3. 5
4. 66
5. 35
6. 7
7. 35
8. 3415
9. 13
10. 67

Teaching Suggestions
p. M–14

QUICK QUIZ
Perform the indicated operations.

<u>1</u>. 13 + 6 + 2   Ans: 21
<u>2</u>. 18 − 7 + 6   Ans: 17
<u>3</u>. 27 − 10 − 7   Ans: 10
<u>4</u>. 13 × 6 × 2   Ans: 156
<u>5</u>. 30 ÷ 3 ÷ 2   Ans: 5
<u>6</u>. 20 + 17 − 12   Ans: 25
<u>7</u>. 2 × 5 + 6   Ans: 16
<u>8</u>. 12 ÷ 2 + 4   Ans: 10
<u>9</u>. 3 + 6 × 5   Ans: 33
<u>10</u>. 30 − 10 ÷ 5   Ans: 28

ADDITIONAL EXAMPLES
Example 1
<u>1</u>. Evaluate 5 + 3x
   when x = 7.   Ans: 26
<u>2</u>. Evaluate 3w − 18
   when w = 6.   Ans: 0

1-7 Using Variables

For the square parking lot at the right, let n represent the length of each side. Since n can represent any length, n is a *variable*. A **variable** is a letter representing one or more numbers.

The sum of the lengths of the sides of a square is the **perimeter**. The perimeter of the square parking lot can be written as follows.

$n + n + n + n$ or $4 \times n$ or $4 \cdot n$ or $4(n)$ or $4n$

Each of the expressions for perimeter is an *algebraic expression*. An **algebraic expression** contains *at least one variable*. Here are some additional examples.

$$19 + c$$
$$z \div 14$$
$$5w \div (r - 3)$$
$$3 \cdot q - (t + 5)$$

▶ Algebraic expressions

To **evaluate** (find the value of) an algebraic expression, follow these steps.

PROCEDURE

Steps for Evaluating an Algebraic Expression

[1] Replace each variable with a given value.

[2] Perform the operations according to the rules for order of operations.

EXAMPLE 1   Evaluate $4n - 3$ when $n = 8$.

Solution: [1] Replace the variable. $4n - 3 = 4(8) - 3$
              [2] Multiply first. (Rule 3) $= 32 - 3$
                  Then subtract. $= \mathbf{29}$

Self-Check   *Evaluate each expression when $t = 9$.*

1. $12t$ **2.** $t - 4$ **3.** $81 \div t - 8$ **4.** $1 + 2t$
    108   5   1   19

22   Chapter 1

An algebraic expression may contain more than one variable.

EXAMPLE 2 Evaluate $(2x + 3) - 5y$ when $x = 12$ and $y = 4$.

Solution:
|1| Replace the variables. $(2x + 3) - 5y = (2 \cdot 12 + 3) - 5(4)$
|2| Evaluate inside parenthesis. $= (24 + 3) - 5(4)$
    (Rule 4) $= 27 - 5(4)$
|3| Multiply first. (Rule 3) $= 27 - 20$
    Then subtract. $= 7$

Self-Check *Evaluate each expression when $w = 12$ and $r = 6$.*

5. $(2w - 3) + 5r$ 6. $(3w \div 9) + r$ 7. $(3w \div 12 + 4r) \div 9$

CLASSROOM EXERCISES

Write the expression after each variable is replaced.
(Examples 1 and 2, Step 1)

1. $15p + 32$ when $p = 5$
2. $29 - 3k$ when $k = 8$
3. $6q + 5r$ when $q = 4$ and $r = 7$
4. $36a - 27b$ when $a = 13$ and $b = 10$
5. $24c + 18d - 30$ when $c = 19$ and $d = 15$
6. $27m + 14 - 11q$ when $m = 15$ and $q = 10$
7. $2xy - y$ when $x = 3$ and $y = 4$
8. $x \cdot x + 5xy$ when $x = 5$ and $y = 6$

Evaluate each expression when the variable equals 2. (Example 1)

9. $8 - 3n$
10. $4 + x$
11. $3y \div 2$
12. $7 - y + 2$
13. $x + x$
14. $3a + a$
15. $11 - 2y + 2$
16. $6n \div 3$
17. $5 - n + 1$
18. $6 - 2n + 2$
19. $3h - 2 + h$
20. $4p + 6 - p$

Evaluate each expression when $x = 4$ and $y = 10$. (Example 2)

21. $x + (2y \div 5)$
22. $(3x - y) + 7$
23. $(7y \div 5) + 2x$
24. $(4x + 8) - 2y$
25. $(12 \div x) + (90 \div y)$
26. $7x + (10y - 15)$
27. $15y - (10 + 5x)$
28. $xy - (x + y)$
29. $xy + (y - x)$
30. $2xy - y$
31. $x \cdot x + 5xy$
32. $x \cdot x - y$

Whole Numbers and Applications   **23**

ASSIGNMENT GUIDE

BASIC
p. 24: 1-33 odd

AVERAGE
p. 24: 1-11 odd, 13-26

ABOVE AVERAGE
p. 24: 1-33 odd

WRITTEN EXERCISES
1. 5 2. 14 3. 8
4. 13 5. 4 6. 1
7. 4 8. 16 9. 7
10. 19 11. 9 12. 9
13. 36 14. 75
15. 51 16. 0 17. 46
18. 2 19. 12 20. 30
21. 29 22. 71
23. 100 24. 69
25. 42 26. 36 27. 4
28. 17 29. 42
30. 650 31. 15
32. 111 33. 48
34. 63

QUIZ: SECTIONS 1-6–1-7
After completing this Review, you may want to administer a quiz covering the same sections. A Quiz is provided in the *Teacher's Edition: Part II*.

REVIEW: SECTIONS 1-6–1-7
1. 6 2. 14 3. 10
See page 25 for the answers to Exercises 4-16.

WRITTEN EXERCISES

Goal: To evaluate (find the value of) an algebraic expression for given replacements of the variable or variables

Evaluate each expression when the variable is equal to 3. (Example 1)

1. $2a - 1$
2. $4x + 2$
3. $2 + 2y$
4. $1 + 4c$
5. $2 + n - 1$
6. $5 - k - 1$
7. $2 + 6 \div x$
8. $12 \div p \cdot 4$
9. $8 - a + 2$
10. $15 - x + 7$
11. $n + 2 \cdot 3$
12. $x + 3 \cdot 2$

For Exercises 9–32, evaluate each expression. (Example 2)

13. $p + 8q$ when $p = 12$ and $q = 3$
14. $r + 12s$ when $r = 3$ and $s = 6$
15. $12c - 3d$ when $c = 5$ and $d = 3$
16. $15x - 8y$ when $x = 8$ and $y = 15$
17. $2v + 3 \cdot x$ when $v = 5$ and $x = 12$
18. $16 \div p - 6 \cdot r$ when $p = 2$ and $r = 1$
19. $8j - (12 + 3k)$ when $j = 6$ and $k = 8$
20. $3h - (2k + 11)$ when $h = 23$ and $k = 14$

MIXED PRACTICE

21. $8p - 19$ when $p = 6$
22. $16x + 3y$ when $x = 2$ and $y = 13$
23. $16 + 12s$ when $s = 7$
24. $12h - 25m$ when $h = 12$ and $m = 3$
25. $x \cdot x + 3x - 12$ when $x = 6$
26. $32r - 92$ when $r = 4$
27. $15 \div a - (b - 1)$ when $a = 3$ and $b = 2$
28. $5 \cdot x - (y + 1)$ when $x = 4$ and $y = 2$
29. $16 + 78 \div q$ when $q = 3$
30. $(14m - m) \cdot n$ when $m = 5$ and $n = 10$
31. $3 \cdot s \div (5 - t)$ when $s = 15$ and $t = 2$
32. $15 + 8z$ when $z = 12$
33. $3f - 8g + g^2$ when $f = 21$ and $g = 5$
34. $q \div (7 + b) \cdot 9$ when $q = 105$ and $b = 8$

REVIEW: SECTIONS 1-6–1-7

Evaluate each expression. (Section 1-6)

1. $8 - 6 + 4$
2. $7 - 5 + 12$
3. $5 \cdot 4 \div 2$
4. $8 \cdot 12 \div 6$
5. $3 + 8 \cdot 5$
6. $15 \div (3 + 2)$
7. $25 \div (9 - 4)$
8. $7 - 5 \cdot 6 \div 10$

Evaluate each expression. (Section 1-7)

9. $12 - 4c$ when $c = 2$
10. $3r + 15$ when $r = 9$
11. $16y - 5z$ when $y = 3$ and $z = 7$
12. $17 - p + 3q$ when $p = 11$ and $q = 16$
13. $k + 3 - 4m$ when $k = 21$ and $m = 6$
14. $a \cdot b \div 3$ when $a = 7$ and $b = 12$
15. $3 \cdot a - (2b - 1)$ when $a = 9$ and $b = 3$
16. $7m - (3n - 2)$ when $m = 5$ and $n = 4$

Chapter 1

CHAPTER REVIEW

PART 1: VOCABULARY

For Exercises 1–8, choose from the box at the right below the word(s) that best corresponds to each description.

1. A letter representing one or more numbers   ?
2. The sum of measures divided by the number of measures   ?
3. A shorthand way of writing a word rule   ?
4. An expression that includes at least one of the operations of addition, subtraction, multiplication, or division   ?
5. Unit for measuring the amount of electricity used   ?
6. The sum of the base price, the destination charge, and the cost of optional equipment   ?
7. To   ?   means to find the value of.
8. An expression that contains at least one variable   ?

> evaluate
> average
> sticker price
> formula
> variable
> algebraic expression
> numerical expression
> kilowatt-hour

PART 2: SKILLS

For Exercises 9–12, find the sticker price for each car. (Section 1-1)

	Base Price	Destination Charge	Total Cost of Options
9.	$4560	$135	$687
11.	$8032	$189	$883

	Base Price	Destination Charge	Total Cost of Options
10.	$4516	$165	$1102
12.	$10,080	$105	$752

For Exercises 13–16, find how many kilowatt-hours of electricity were used. (Section 1-2)

	Previous Reading	Present Reading
13.	1405	1624
14.	6325	8006
15.	3456	4123
16.	12,623	15,010

Whole Numbers and Applications **25**

REVIEW: SECTIONS 1-6–1-7
p. 24
4. 16   5. 43   6. 3
7. 5   8. 4   9. 4
10. 42   11. 13
12. 54   13. 0   14. 28
15. 22   16. 25

CHAPTER REVIEW
1. variable
2. average
3. formula
4. numerical expression
5. kilowatt-hour
6. sticker price
7. evaluate
8. algebraic expression
9. $5382   10. $5783
11. $9104   12. $10,937
13. 219   14. 1681
15. 667   16. 2387

CHAPTER REVIEW

17. 160 hours
18. 243 kilometers
19. 90 meters
20. 525 feet
21. 1   22. 36
23. 1   24. 16
25. 49   26. 144
27. 64   28. 1024
29. 400   30. 16
31. 3600   32. 1936
33. 14°   34. 45°
35. 83°   36. 22°
37. c   38. a
39. b   40. b
41. 12   42. 44
43. 20   44. 18
45. 90   46. 2
47. 30   48. 14
49. 31   50. 7
51. 21   52. 51
53. 8   54. 35
55. 1   56. 53
57. 4   58. 180

For Exercises 17–20, find the distance for the given rate and time. (Section 1-3)

	Rate	Time
17.	40 miles per hour	4 hours
19.	6 meters per second	15 seconds

	Rate	Time
18.	81 kilometers per hour	3 hours
20.	75 feet per second	7 seconds

Evaluate. (Section 1-3)

21. 1^2 22. 6^2 23. 1^3 24. 4^2 25. 7^2 26. 12^2

For Exercises 27–32, use $d = 16t^2$ to find d for the given value of t. (Section 1-3)

27. $t = 2$ 28. $t = 8$ 29. $t = 5$ 30. $t = 1$ 31. $t = 15$ 32. $t = 11$

For Exercises 33–36, find the mean temperature. (Section 1-4)

33. 8°, 12°, 10°, 16°, 24°
34. 41°, 52°, 38°, 50°, 49°, 40°
35. 78°, 81°, 92°, 87°, 76°, 84°
36. 16°, 15°, 24°, 30°, 17°, 24°, 28°

Choose the best estimate. Choose a, b, or c. (Section 1-5)

37. $114 + 381$ a. $100 + 300$ b. $200 + 400$ c. $100 + 400$
38. $262 - 229$ a. $260 - 230$ b. $260 - 220$ c. $270 - 220$
39. 24×38 a. 20×30 b. 20×40 c. 30×40
40. $644 \div 38$ a. $700 \div 40$ b. $640 \div 40$ c. $640 \div 30$

Evaluate each expression. (Section 1-6)

41. $13 + 5 - 6$ 42. $32 - 4 + 16$ 43. $15 \div 3 \cdot 4$ 44. $6 \cdot 9 \div 3$
45. $5 \cdot (12 + 6)$ 46. $20 \div (4 + 6)$ 47. $42 - 18 \cdot 6 \div 9$ 48. $16 \div 2 + 2 \cdot 3$

Evaluate each expression. (Section 1-7)

49. $19 + 3y$ when $y = 4$
50. $4t - 17$ when $t = 6$
51. $12a - 13b$ when $a = 5$ and $b = 3$
52. $3p - q + 2p$ when $p = 15$ and $q = 24$
53. $(h - 3) \cdot (k \div 4)$ when $h = 7$ and $k = 8$
54. $(6r + 2) \cdot s - 5$ when $r = 3$ and $s = 2$
55. $s \cdot s - 3t$ when $s = 2$ and $t = 1$
56. $(8 - k) + r^2$ when $k = 4$ and $r = 7$
57. $(q + 32) \div 5t$ when $q = 8$ and $t = 2$
58. $ac(a - c)$ when $a = 9$ and $c = 4$

PART 3: APPLICATIONS

Solve each problem.

59. The total cost of the options for a new car is $1506 and the destination charge is $185. If the base price is $8,475, find the sticker price. (Section 1-1)

60. The present reading on the Hoelthe's meter is 45,006. The previous reading was 43,100. How many kilowatt-hours of electricity were used? (Section 1-2)

61. In 1927, Charles Lindberg flew the Spirit of St. Louis from New York to Paris in about 33 hours at a rate of 107 miles per hour. Find the distance he traveled. (Section 1-3)

62. Sandra Palmer scored 78, 74, 71, and 73 to win the U.S. Open Golf Tournament. What was her average score? (Section 1-4)

63. A food wholesaler has 35 cases of peas in one warehouse, 71 cases in a second warehouse, and 29 cases in a third warehouse. Estimate the total number of cases (nearest ten). (Section 1-5)

64. A car travels 608 miles on a full tank of fuel. The fuel tank holds 18 gallons when full. Estimate how far the car can travel on one gallon of fuel. (Section 1-5)

CHAPTER REVIEW

59. $10,166
60. 1906 kilowatt-hours
61. 3331 miles
62. 74
63. 140 cases
64. 30 miles

ANSWERS TO REVIEW CAPSULES

Page 4 **1.** 44 **2.** 119 **3.** 347 **4.** 181 **5.** 325 **6.** 69 **7.** 266
Page 6 **1.** 128 **2.** 216 **3.** 750 **4.** 966 **5.** 240 **6.** 805 **7.** 2088 **8.** 7824 **9.** 216 **10.** 0
Page 9 **1.** 71 **2.** 95 **3.** 32 **4.** 123 **5.** 9 **6.** 7 **7.** 13 **8.** 22 **9.** 47 **10.** 208
Page 12 **1.** 40 **2.** 30 **3.** 60 **4.** 10 **5.** 80 **6.** 100 **7.** 200 **8.** 200 **9.** 400 **10.** 600 **11.** 1300 **12.** 800 **13.** 12,000 **14.** 8000 **15.** 19,000 **16.** 2000 **17.** 33,000 **18.** 13,000
Page 17 **1.** 1308 **2.** 9661 **3.** 26 **4.** 235 **5.** 1823 **6.** 7 **7.** 1512 **8.** 870 **9.** 4479 **10.** 45 **11.** 325 **12.** 224 **13.** 6552 **14.** 65,745 **15.** 7 **16.** 4 **17.** 7 **18.** 118 **19.** 335
Page 21 **1.** 18 **2.** 36 **3.** 9 **4.** 15 **5.** 504 **6.** 780 **7.** 34 **8.** 41 **9.** 3 **10.** 22 **11.** 4 **12.** 9

Whole Numbers and Applications   **27**

CHAPTER TEST

Two forms of a chapter test, Form A and Form B, are provided on copying masters in the *Teacher's Edition: Part II*.

1. $9100   2. 779
3. 1440 m   4. 750 km
5. 1936 ft   6. 82
7. b   8. a   9. b
10. c   11. 20   12. 81
13. 7   14. 13   15. 18
16. 16   17. 35
18. 14   19. $5528
20. 22

CHAPTER TEST

Complete.

1. Base price of new car: $7830
   Destination charge: $ 150
   Total cost of options $1120
   Sticker price: ?

2. Previous meter reading: 4825
   Present meter reading: 5604
   Total number of
   kilowatt-hours used: ?

3. Rate: 45 meters per second
   Time: 32 seconds
   Distance: ?

4. Rate: 125 kilometers per hour
   Time: 6 hours
   Distance: ?

Solve each problem.

5. Using the formula $d = 16t^2$, find the distance an object falls in 11 seconds.

6. Marvin's test scores are 75, 82, 95, 85, 63, and 92. What is his average score?

Choose the best estimate. Choose a, b, or c.

7. $824 + 773$ a. $800 + 700$ b. $800 + 800$ c. $820 + 800$
8. $1422 - 629$ a. $1400 - 600$ b. $1400 - 620$ c. $1420 - 600$
9. 38×87 a. 30×80 b. 40×90 c. 40×100
10. $601 \div 27$ a. $600 \div 20$ b. $610 \div 20$ c. $600 \div 30$

Evaluate each expression.

11. $18 + 6 - 4$
12. $36 \div 4 \cdot 9$
13. $(4 + 16) \div 2 - 3$
14. $17 - 4 \cdot 3 + 8$
15. $x + 3y$ when $x = 6$ and $y = 4$
16. $3s - 4t$ when $s = 8$ and $t = 2$
17. $5p \div 3 + 5q$ when $p = 9$ and $q = 4$
18. $7 \cdot (8a - 6) \div b$ when $a = 2$ and $b = 5$

Solve each problem.

19. The total cost of options for Ted's new car is $842 and the destination charge is $106. If the base price is $4580, find the sticker price.

20. Andrea scored 264 points in 12 basketball games. Find the average number of points she scored per game.

28   Chapter 1

ENRICHMENT

Geometric Numbers

Interesting relationships exist between geometric shapes and numbers. The first four **triangular numbers** and the first four **square numbers** are shown below.

Triangular Numbers

1   3   6   10

Square Numbers

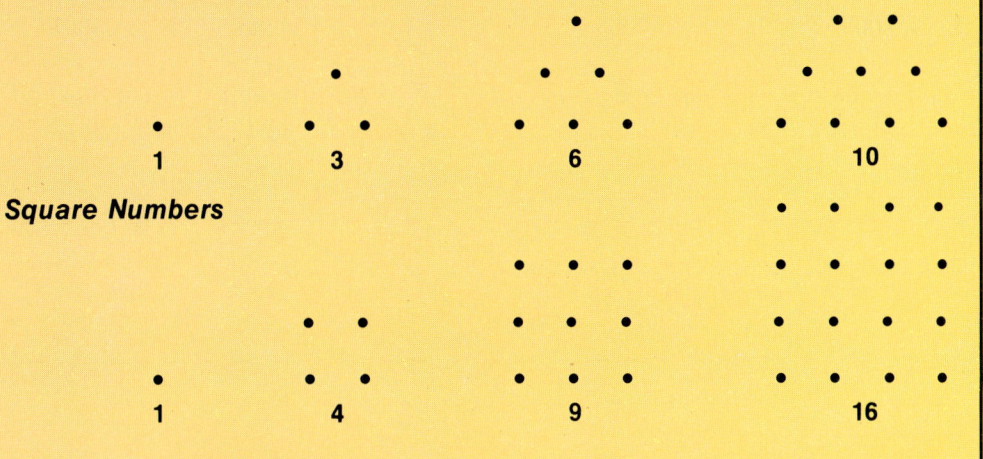

1   4   9   16

EXERCISES

1. Draw the next three triangular numbers after 10.
2. The sums at the right show that the first counting number is the first triangular number, the sum of the first two counting numbers is the second triangular number, and the sum of the first three counting numbers is the third triangular number.

    $1 = 1$
    $1 + 2 = 3$
    $1 + 2 + 3 = 6$

    Write the next six triangular numbers as sums of consecutive counting numbers.
3. Draw the next three square numbers after 16.
4. The diagram at the right shows that the first odd counting number equals the first square number, and the sum of the first two odd counting numbers equals the second square number.

    1    $1 + 3 = 4$

    Write the next four square numbers as sums of consecutive odd counting numbers.

Whole Numbers and Applications **29**

ENRICHMENT
You may wish to use this lesson for students who performed well on the formal Chapter Test.

2. $1 + 2 + 3 + 4 = 10$;
   $1 + 2 + 3 + 4 + 5 = 15$;
   $1 + 2 + 3 + 4 + 5 + 6 = 21$;
   $1 + 2 + 3 + 4 + 5 + 6 + 7 = 28$;
   $1 + 2 + 3 + 4 + 5 + 6 + 7 + 8 = 36$;
   $1 + 2 + 3 + 4 + 5 + 6 + 7 + 8 + 9 = 45$

ENRICHMENT

4. 1 + 3 + 5 + 7 = 16;
1 + 3 + 5 + 7 + 9 = 25;
1 + 3 + 5 + 7 + 9 + 11 = 36;
1 + 3 + 5 + 7 + 9 + 11 + 13 = 45

ADDITIONAL PRACTICE

You may wish to use all or some of these exercises, depending on how well students performed on the formal Chapter Test.

1. $9810   2. $9078
3. 1527 kwh   4. 887 kwh
5. 1425 ft   6. 882 m
7. 576 ft   8. 256 ft
9. 2304 ft   10. 3136 ft
11. 78   12. 12
13. 374   14. 210
15. a   16. b   17. a
18. 34   19. 11
20. 215   21. 40
22. 36   23. 48
24. 79   25. 203 m

ADDITIONAL PRACTICE

SKILLS

Find the sticker price for each car. (Pages 2–4)

	Base Price	Destination Charge	Total Cost of Options
1.	$8340	$330	$1140
2.	$6800	$175	$2103

Find how many kilowatt-hours of electricity were used. (Pages 5–6)

	Previous Reading	Present Reading
3.	1254	2781
4.	3114	4001

Find the distance for the given rate and time. (Pages 7–8)

	Rate	Time
5.	95 feet per second	15 seconds
6.	42 meters per second	21 seconds

Use the formula $d = 16t^2$ to find d for each value of t. (Pages 7–8)

7. $t = 6$ seconds   8. $t = 4$ seconds
9. $t = 12$ seconds   10. $t = 14$ seconds

Find the mean. (Pages 10–12)

11. 64, 86, 70, 92
12. 17, 14, 16, 19, 4, 3, 8, 11, 16
13. 632, 436, 108, 421, 341, 306
14. 97, 421, 68, 309, 155

Choose the best estimate. Choose a, b, or c. (Pages 14–16)

15. 198 + 205 + 486   a. 900   b. 800   c. 1000
16. 797 ÷ 21   a. 30   b. 40   c. 25
17. 71 × 33   a. 2100   b. 2800   c. 21,000

Evaluate each expression. (Pages 18–20)

18. 6 + 4 · 7   19. (42 − 12) ÷ 3 + 1   20. (15 + 12) · 8 − 1   21. 100 ÷ 5 + 20

(Pages 22–24)

22. $8x - 12$ when $x = 6$
23. $9 \cdot (x + y) + y$ when $x = 2$ and $y = 3$

APPLICATIONS

24. Roger had scores of 63, 84, 75, 92, 68, and 92 on six math tests. Find his average score. (Pages 10–12)

25. A cheetah was timed running 29 meters per second. How far would it run in 7 seconds? (Pages 7–9)

30   Chapter 1

CHAPTER 2
Decimals and Applications

SECTIONS
2-1 Decimals and Place Value
2-2 Addition and Subtraction
2-3 Multiplication
2-4 Division
2-5 Problem Solving and Applications: Estimating with Decimals
2-6 Metric Units of Measurement
2-7 Problem Solving and Applications: Scientific Notation

FEATURES
Calculator Application: *Estimation*
Consumer/Computer Application: *Unit Price*
Enrichment: *Using Scientific Notation*

Teaching Suggestions
p. M-15

QUICK QUIZ

Replace the ? with > or <.

1. 40 ? 400   Ans: <
2. 0.4 ? 0.04   Ans: >
3. 73.2 ? 7.32 Ans: >
4. 508 ? 580   Ans: <
5. 0.007 ? 0.07
   Ans: <

ADDITIONAL EXAMPLES
Example 1
Give the value of the "3" in each number.

1. 305.2   Ans: 300
2. 503.2   Ans: 3
3. 2.931   Ans: 0.03
4. 0.0830   Ans: 0.003
5. 1235.06   Ans: 30

SELF-CHECK

1. 50   2. 0.03
3. 0.4   4. 0

2-1 Decimals and Place Value

In our decimal system, the value of each decimal place is *ten* times the value of the place to its right.

PROCEDURE
To give the value of any digit in a number:
1. Identify the place name.
2. Refer to the digit and the place name to give the value.

TABLE

	Hundred Thousands	Ten Thousands	Thousands	Hundreds	Tens	Ones	.	Tenths	Hundredths	Thousandths	Ten-Thousandths	Hundred-Thousandths
a.	2	7	9	4	6	3	.	5	2	8	1	2
b.					9	3	.	7	5			
c.				1	7	2	.	3				
d.						3	.	0	0	7		
e.						0	.	0	0	0	0	7

EXAMPLE 1 Give the value of the "7" for each number in the table.

Solutions:

	1 Place Name	2 Value	
a.	Ten Thousands	a. 70,000	Read: "seventy thousand".
b.	Tenths	b. 0.7	Read: "seven tenths".
c.	Tens	c. 70	Read: "seventy".
d.	Thousandths	d. 0.007	Read: "seven thousandths".
e.	Hundred-Thousandths	e. 0.00007	Read: "seven hundred thousandths".

Self-Check Give the value of the underlined digit in each number.

1. 5̲8.4   2. 6.0̲3   3. 19.4̲   4. 301.82

32 Chapter 2

The symbols > and < are used to compare numbers.

> means is greater than.

< means is less than.

> **PROCEDURE**
>
> **To compare decimals:**
>
> 1. Write the decimals one under the other. Line up the decimal points.
> 2. Starting from left to right, identify and compare the first digits that are <u>not</u> alike. Annex final zeros when necessary.
> 3. Compare the decimals in the <u>same order</u> as the unlike digits in step 2.

NOTE: Annexing final zeros does not change the value of a decimal.

EXAMPLE 2 Replace the __?__ with > or <.

  a. 5.39 __?__ 5.42   b. 0.031 __?__ 0.03

Solutions:

		a.	b.
1	Write the decimals one under the other. Align the decimal points.	5.39 5.42	0.031 0.03
2	Starting from the left, identify the first digits that are <u>not</u> alike. Compare the digits.	5.39 5.42 3 < 4	0.031 0.030 1 > 0	Annex one zero.
3	Compare the decimals.	5.39 < 5.42	0.031 > 0.03

Self-Check *Replace the __?__ with > or <.*

5. 10.83 __?__ 10.8   6. 0.1 __?__ 0.11   7. 61.4 __?__ 60.3

CLASSROOM EXERCISES

Give the <u>place name</u> for the "8" in each number. (Example 1, step 1)

1. 18   2. 843   3. 8762   4. 9.3248   5. 9.058

Replace the __?__ with > or <. (Example 2)

6. 3.2 __?__ 3.5   7. 6.5 __?__ 6.05   8. 0.01 __?__ 0.1   9. 0.4 __?__ 0.38

Decimals and Applications 33

ADDITIONAL EXAMPLES

Example 2

Replace the __?__ with > or <.

1. 3.07 __?__ 3.70
   Ans: <
2. 16.7 __?__ 16.3
   Ans: >
3. 2.07 __?__ 2.17
   Ans: <
4. 41.5 __?__ 41.52
   Ans: <
5. 2.22 __?__ 2.2
   Ans: >

SELF-CHECK

<u>5</u>. >   <u>6</u>. <   <u>7</u>. >

CLASSROOM EXERCISES

<u>1</u>. ones   <u>2</u>. hundreds
<u>3</u>. thousands
<u>4</u>. ten-thousandths
<u>5</u>. thousandths
<u>6</u>. <   <u>7</u>. >
<u>8</u>. <   <u>9</u>. >

ASSIGNMENT GUIDE
BASIC
p. 34: 1-30
AVERAGE
p. 34: 1-18 odd, 19-32
ABOVE AVERAGE
P. 34: 1-18 odd, 19-32

WRITTEN EXERCISES
1. 0.5   2. 0.05   3. 5
4. 0.0005   5. 20
6. 0.002   7. 200
8. 0.0001   9. 2000
10. 70   11. 0.007
12. 0.00001   13. 400
14. 1   15. 0.3   16. 60
17. 0.07   18. 0.005
19. >   20. <   21. <
22. <   23. <   24. >
25. >   26. >   27. >
28. <   29. <   30. <
31. Jane   32. Yes
33. 22.00124

REVIEW CAPSULE
This Review Capsule reviews prior-taught skills used in Section 2-2. The reference is to the pages where the skills were taught.

WRITTEN EXERCISES

Goals: To give the value of any digit in a number

To use place value to compare decimals and to apply this skill to solving word problems

For Exercises 1–12, give the value of the underlined digit. (Example 1)

1. 13.<u>5</u>   2. 13.0<u>5</u>   3. 1<u>5</u>.03   4. 13.176<u>5</u>   5. <u>2</u>3.14   6. 21.63<u>2</u>
7. <u>2</u>56.8   8. 100.025<u>1</u>   9. <u>2</u>001.5   10. 1<u>7</u>5.4   11. 2.51<u>7</u>   12. 46.0387<u>1</u>

For Exercises 13–17, give the value of the indicated digit in the number 2461.375. (Example 1)

13. 4   14. 1   15. 3   16. 6   17. 7   18. 5

For Exercises 19–30, replace the __?__ with > or <. (Example 2)

19. 6.51 __?__ 6.48   20. 0.02 __?__ 0.027   21. 7.01 __?__ 7.89
22. 1.63 __?__ 1.632   23. 0.319 __?__ 0.32   24. 0.72 __?__ 0.7
25. 1.007 __?__ 0.07   26. 3.05 __?__ 3.005 27. 0.628 __?__ 0.625
28. 0.421 __?__ 0.422 29. 0.914 __?__ 0.92 30. 0.00010 __?__ 0.00015

APPLICATIONS: USING PLACE VALUE

31. Luis has a batting average of 0.325. Jane has a batting average of 0.335. Who has the higher batting average?
32. Juanita needs a piece of wire 1.46 meters long for her science project. She buys 1.5 meters of wire. Does she have enough wire for her project?
33. Which is less, 22.00124 or 22.00134?

REVIEW CAPSULE FOR SECTION 2-2

Evaluate each expression. (Pages 22–24)

1. $x + y$ when $x = 15$ and $y = 24$
2. $a + b$ when $a = 280$ and $b = 124$
3. $m - n$ when $m = 2400$ and $n = 1368$
4. $p - q$ when $p = 1564$ and $q = 89$
5. $a + b + c$ when $a = 97$, $b = 259$ and $c = 803$
6. $w + x + y$ when $w = 3162$, $x = 480$ and $y = 1004$

The answers are on page 60.

34   Chapter 2

2-2 Addition and Subtraction

In the 1968 Olympic Games, Jan Henne of the United States won the 100 meter women's freestyle race in 60 seconds. In the 1984 Olympics, Carrie Steinseifer and Nancy Hogshead of the United States tied for the gold medal with winning times of 55.92 seconds. To find the difference in times, you use subtraction with decimals.

PROCEDURE
To add or subtract with decimals:
1. Line up the decimal points one under the other.
2. Annex final zeros when necessary.
3. Add or subtract as with whole numbers.

EXAMPLE 1 Find the difference between the winning times in the 100 meter women's freestyle race in the 1984 and 1968 Olympic Games.

Solution:
1968 winning time in seconds: 60.00   ← **Annex two zeros.**
1984 winning time in seconds: −55.92
Difference: 4.08

Self-Check 1. Find the difference between 87.3 and 55.42.

Variables in an algebraic expression can be replaced with decimals.

EXAMPLE 2 Evaluate $x + y + z$ when $x = 18.2$, $y = 11.94$ and $z = 253.07$.

Solution: $x + y + z = 18.2 + 11.94 + 253.07$
$= 283.21$

$$\begin{array}{r} 18.20 \\ 11.94 \\ +253.07 \\ \hline 283.21 \end{array}$$

◀ **Annex a zero to help line up the decimal points.**

Self-Check 2. Evaluate $t + r + s$ when $t = 9.15$, $r = 0.003$, and $s = 216.5$.
The answer is on page 36.

Decimals and Applications **35**

SELF-CHECK
2. 225.653

CLASSROOM EXERCISES
1. 0.58   2. 9.7
3. 19.94   4. 56.635
5. 214.3   6. 1.18
7. 115.263   8. 21.122
9. 133.706   10. 0.43
11. 2.32   12. 115.4
13. 2.043   14. 110.215
15. 0.40   16. 2.427
17. 5.884   18. 36.159
19. 6.8   20. 147.71
21. 23.7   22. 90.541

ASSIGNMENT GUIDE
BASIC
pp. 36-37: 1-22
AVERAGE
pp. 36-37: 1-27 odd
ABOVE AVERAGE
pp. 36-37: 1-25 odd, 27-29

CLASSROOM EXERCISES

Add or subtract as indicated. (Example 1)

1. 0.01
   0.22
   +0.35

2. 1.6
   2.3
   +5.8

3. 6.26
   5.8
   +7.88

4. 22.44
   8.72
   17.435
   + 8.04

5. 7.8
   182.1
   3.24
   + 21.16

6. $0.08 + 1.07 + 0.03$

7. $0.903 + 1.07 + 100.4 + 12.89$

8. $3.38 + 1.75 + 12.8 + 3.192$

9. $4.105 + 3.2 + 108.241 + 18.16$

10. 0.86
    −0.43

11. 11.67
    − 9.35

12. 265.2
    −149.8

13. 5.24
    −3.197

14. 125.6
    − 15.385

15. $0.79 - 0.39$

16. $11.06 - 8.633$

17. $94.3 - 88.416$

18. $36.5 - 0.341$

Evaluate each expression. (Example 2)

19. $r - s$ when $r = 15.7$ and $s = 8.9$

20. $m - n$ when $m = 180.6$ and $n = 32.89$

21. $x + y + z$ when $x = 4.8$, $y = 6.5$ and $z = 12.4$

22. $a + b + c$ when $a = 45.6$, $b = 36.24$ and $c = 8.701$

WRITTEN EXERCISES

Goals: To add and subtract with decimals and to apply these skills to solving word problems

To apply addition and subtraction of decimals to evaluating algebraic expressions

Add or subtract as indicated. (Example 1)

1. 4.75
   11.04
   + 9.13

2. 6.26
   12.58
   + 3.91

3. 26.1
   3.95
   + 8.492

4. 0.33
   21.56
   38.2
   + 7.351

5. 126.2
   73.08
   1.38
   + 42.261

6. $4.65 + 3.27 + 15.98$

7. $5.93 + 12.41 + 6 + 3.4$

8. $74.656 + 8.9 + 5.013 + 713.85$

9. $57.09 + 878.015 + 6.02 + 5.535$

10. 16.78
    − 9.43

11. 4.968
    −2.375

12. 7.475
    −0.3551

13. 1.7
    −0.87

14. 21.195
    − 7.2325

15. $18.4 - 7.9$

16. $132.4 - 86.51$

17. $9 - 0.048$

18. $11.06 - 8.633$

36 Chapter 2

Evaluate each expression. (Example 2)

19. $x - y$ when $x = 15.05$ and $y = 7.88$
20. $a - b$ when $a = 114.5$ and $b = 38.56$
21. $c - d$ when $c = 164$ and $d = 28.56$
22. $r - s$ when $r = 8.04$ and $s = 0.1564$
23. $x + y + z$ when $x = 5.8$, $y = 12.2$ and $z = 7.3$
24. $r + s + t$ when $r = 0.32$, $s = 1.5$ and $t = 3.16$
25. $a + b + c + d$ when $a = 35.4$, $b = 18.5$, $c = 1.23$, and $d = 16.08$
26. $m + n + p + q$ when $m = 125.8$, $n = 26.5$, $p = 32.06$, and $q = 54$

APPLICATIONS: USING DECIMALS

Solve each problem.

27. In the 800-meter relay, Nathan ran the first leg in 29.2 seconds, Ted ran the second leg in 28.6 seconds, Bill ran the third leg in 30.2 seconds, and Roberto ran the fourth leg in 28.0 seconds. Find their total time.

28. The weather bureau reported that for the five times it rained in July, 1.12 inches, 2 inches, 1.5 inches, 0.46 inches, and 0.3 inches of rain fell. What was the total rainfall for July?

29. Latasha is 1.67 meters tall and Julie is 1.58 meters tall. How much taller is Latasha than Julie?

WRITTEN EXERCISES
1. 24.92 2. 22.75
3. 38.542 4. 67.441
5. 242.921 6. 23.9
7. 27.74 8. 802.419
9. 946.66 10. 7.35
11. 2.593 12. 7.1199
13. 0.83 14. 13.9625
15. 10.5 16. 45.89
17. 8.952 18. 2.427
19. 7.17 20. 75.94
21. 135.44 22. 7.8836
23. 25.3 24. 4.98
25. 71.21 26. 238.36
27. 116 seconds
28. 5.38 inches
29. 0.09 meter

REVIEW CAPSULE FOR SECTION 2-3

Evaluate each expression. (Pages 22-23)

1. $3x$ when $x = 14$
2. $4y$ when $y = 127$
3. $4001p$ when $p = 6$
4. $2xy$ when $x = 506$ and $y = 20$
5. $12ab$ when $a = 16$ and $b = 84$
6. $a \cdot b$ when $a = 7$ and $b = 504$

Round to the nearest cent.

7. $14.759
8. $8.962
9. $25.015
10. $37.104
11. $61.706

The answers are on page 60.

Decimals and Applications

Teaching Suggestions
p. M-15

QUICK QUIZ
Multiply.
1. (327)(18)   Ans: 5886
2. (706)(29)
   Ans: 20,474
3. (420)(100)
   Ans: 42,000

Evaluate.
4. 7a when a = 25
   Ans: 175
5. 300x when x = 52
   Ans: 15,600

ADDITIONAL EXAMPLES
Example 1
1. At a cost of $0.073 per kilowatt-hour, find the average cost of operating a water heater for one month. Round the answer to the nearest cent.
   Ans: $29.20
2. At a cost of $0.068 per kilowatt-hour, find the average cost of operating a radio for one month. Round the answer to the nearest cent.
   Ans: $0.51

SELF-CHECK
1. 73¢

2-3 Multiplication

You use **one kilowatt-hour (kwh)** of electricity when you use 1 kilowatt (1000 watts) for one hour.

The tables below show the average number of kilowatt-hours for operating certain appliances for one month.

Appliance	Kilowatt-Hours per Month
Microwave oven	15.8
Range with oven	97.6
Refrigerator (frostless)	94.7
Refrigerator (defrost)	152.4

Appliance	Kilowatt-Hours per Month
Water heater	400.0
Radio	7.5
Television (black-white)	29.6
Television (color)	55.0

PROCEDURE
To multiply with decimals:
1. Multiply as with whole numbers.
2. Count the number of decimal places in the numbers being multiplied (factors). Starting at the right and moving left, count the same number of decimal places in the product. Insert the decimal point.

EXAMPLE 1 At a cost of $0.097 per kilowatt-hour, find the average cost of operating a refrigerator (defrost) for one month. Round the answer to the nearest cent.

Solution: Multiply the number of kilowatt-hours (refer to the table) and the cost per kilowatt-hour.

Average cost = 152.4 (0.097)   152.4 ← 1 (Decimal Places)
             = 14.7828   ×0.097 ← 3
             = 14.78 Nearest   10668
                      cent   13716
                                    14.7828 ← 1 + 3 = 4

The average monthly cost is **$14.78**.

Self-Check 1. At a cost of $0.097 per kilowatt-hour, find the cost of operating a radio for one month. Round the answer to the nearest cent.

38 Chapter 2

When multiplying, you may have to insert zeros in a product.

EXAMPLE 2 At a cost of $0.029 per hour, how much will it cost to operate a television (black-white) for 0.5 hour? Round the answer to the nearest cent.

Solution: Cost = 0.5 (0.029)
= 0.0145
= 0.01 ◀ **Nearest cent**

It will cost about **$0.01**.

```
            Decimal Places
    0.5   ◀──  1
  ×0.029  ◀──  3
  ─────
     45
     10
  ─────
  0.0145  ◀── Insert 1 zero to make
              4 decimal places.
```

Self-Check 2. At a cost of $0.035 per hour, how much will it cost to operate a black-white television set for 0.5 hour? Round the answer to the nearest cent.

CLASSROOM EXERCISES

Copy each product. Then insert the decimal point. (Example 1)

1. 25 ×0.8 / 200	**2.** 2.5 ×0.8 / 200	**3.** 2.5 × 8 / 200	**4.** 0.25 × 8 / 200	**5.** 0.25 × 0.8 / 200

(Example 2)

6. 0.15 ×0.06 / 90	**7.** 0.015 × 0.06 / 90	**8.** 0.15 ×0.006 / 90	**9.** 0.015 ×0.006 / 90	**10.** 0.15 ×0.060 / 900

Multiply. (Example 1)

11. 45 ×0.6	**12.** 37 ×0.3	**13.** 121 ×0.13	**14.** 1.4 ×0.24	**15.** 32.5 ×0.56

(Example 2)

16. 0.12 × 0.7	**17.** 0.35 ×0.04	**18.** 0.24 ×0.09	**19.** 0.356 ×0.008	**20.** 0.156 ×0.014

ADDITIONAL EXAMPLES
Example 2
1. At a cost of $0.038 per hour, how much will it cost to operate a stereo for 0.5 hr? Round the answer to the nearest cent. Ans: $0.02
2. At a cost of $0.091 per hour, how much will it cost to operate an oven for 0.25 hr? Round the answer to the nearest cent. Ans: $0.02

SELF-CHECK
2. 2¢

CLASSROOM EXERCISES
1. 20.0   2. 2.00
3. 20.0   4. 2.00
5. 0.200   6. 0.0090
7. 0.00090
8. 0.00090
9. 0.000090
10. 0.00900   11. 24.0
12. 11.1   13. 15.73
14. 0.336   15. 18.200
16. 0.084   17. 0.014
18. 0.0216
19. 0.002848
20. 0.002184

Decimals and Applications 39

ASSIGNMENT GUIDE
BASIC
Day 1 p. 40: 1-22
Day 2 p. 40: 27-33
AVERAGE
Day 1 p. 40: 1-26
Day 2 pp. 40-41: 27-39
ABOVE AVERAGE
pp. 40-41: 1-33 odd, 34-43

WRITTEN EXERCISES
1. 9.6   2. 7.50
3. 97.20   4. 148.00
5. 13.6890 6. 0.2944
7. 28.084 8. 13.620
9. 0.58357 10. 0.0490
11. 1.0224 12. 5.7868
13. 0.087528 14. 0.048
15. 0.018 16. 0.01134
17. 0.00800
18. 0.000525
19. 0.0244 20. 0.00874
21. 0.0456 22. 0.00676
23. 0.00175
24. 0.0000384
25. 0.00728
26. 0.001224
27. $3.08, $19.25, $1.54, $3.96
28. $22.40, $140.00, $11.20, $28.80
29. $5.47, $34.16, $2.73, $7.03
30. $.88, $5.53, $.44, $1.14
31. $.42, $2.63, $.21, $.54

WRITTEN EXERCISES

Goals: To multiply with decimals
To apply the skill of multiplying with decimals to solving consumer-related problems

Multiply. (Example 1)

| 1. 12 ×0.8 | 2. 12.5 × 0.6 | 3. 135 ×0.72 | 4. 400 ×0.37 | 5. 210.6 ×0.065 |

6. (12.8)(0.023)   7. (47.6)(0.59)   8. (9.08)(1.5)   9. (8.71)(0.067)
10. (3.5)(0.014) 11. (42.6)(0.024) 12. (125.8)(0.046) 13. (1.042)(0.084)

(Example 2)

| 14. 0.8 ×0.06 | 15. 0.6 ×0.03 | 16. 0.54 ×0.021 | 17. 0.25 ×0.032 | 18. 0.35 ×0.0015 |

19. (0.61)(0.04)   20. (0.23)(0.038)   21. (0.8)(0.057)   22. (0.13)(0.052)
23. (0.35)(0.005) 24. (0.024)(0.0016) 25. (0.56)(0.013) 26. (0.68)(0.0018)

APPLICATIONS: USING MULTIPLICATION WITH DECIMALS

For Exercises 27–33, refer to the table on page 38. Find the monthly cost for each appliance at each rate. Round each answer to the nearest cent. (Example 1)

Appliance	Cost at $0.056 per kwh	Cost at $0.35 per kwh	Cost at $0.028 per kwh	Cost at $0.072 per kwh
27. Color television	?	?	?	?
28. Water heater	?	?	?	?
29. Range with oven	?	?	?	?
30. Microwave oven	?	?	?	?
31. Radio	?	?	?	?
32. Refrigerator (frostless)	?	?	?	?
33. Refrigerator (defrost)	?	?	?	?

40 Chapter 2

For Exercises 34–37, find the cost of operating each appliance at the given rate and time. Round each answer to the nearest cent. (Example 2)

Appliance	Cost per hour	Time Used		Appliance	Cost per hour	Time Used
34. Electric blanket	$0.008	6 hours	35.	Iron	$0.067	0.25 hours
36. Vacuum cleaner	$0.077	0.5 hours	37.	Fan	$0.015	3.5 hours

Solve. Round each answer to the nearest cent.

38. A frostless refrigerator uses an average of 94.7 kilowatt–hours per month. At a cost of $0.065 per kilowatt–hour, find the cost of operating a frostless refrigerator for 2 months.

39. A microwave oven uses an average of 15.8 kilowatt–hours per month. At a cost of $0.034 per kilowatt–hour, find the cost of operating a microwave oven for one year.

40. The cost of operating a washing machine is $0.022 per load. The cost of operating a dryer is $0.035 per load. How much will it cost to wash and dry four loads?

41. In Exercise 40, how much will it cost to wash and dry four loads of clothes for 52 weeks?

MORE CHALLENGING EXERCISES

Solve. Round each answer to the nearest cent.

42. A frostless refrigerator uses an average of 94.7 kilowatt–hours per month. A manual-defrost refrigerator uses an average of 152.4 kilowatt–hours per month. At a cost of $0.057 per kilowatt–hour, find how much more it will cost to operate the manual–defrost refrigerator for 6 months?

43. The average cost of operating a coffee maker is $0.018 per pot. The average cost of operating a toaster is $0.003 per slice. Find the total cost of operating the coffee maker and the toaster to make 1 pot of coffee and 4 slices of toast each day for 365 days.

32. $5.30, $33.15, $2.65, $6.82
33. $8.53, $53.34, $4.27, $10.97
34. $.05   35. $.02
36. $.04   37. $.05
38. $12.31   39. $6.45
40. $.23   41. $11.96
42. $19.73   43. $10.95

Decimals and Applications

QUIZ: SECTIONS 2-1-2-3
After completing this Review, you may want to administer a quiz covering the same sections. A Quiz is provided in the *Teacher's Edition: Part II*.

REVIEW: SECTIONS 2-1-2-3
1. hundredths   2. tens
3. <   4. <   5. <
6. >   7. 19.94
8. 192.7
9. 2.19 inches
10. $3.47   11. $.06
12. 2.05 inches

REVIEW: SECTIONS 2-1—2-3

Complete. (Section 2-1)

1. The place name of the 5 in 22.35 is __?__.
2. The place name of the 6 in 1465.3 is __?__.

For Exercises 3-6, replace the __?__ with > or <. (Section 2-1)

3. 3.28 __?__ 3.31   4. 0.008 __?__ 0.0082   5. 5.003 __?__ 6.003   6. 4.506 __?__ 4.5055

Evaluate each expression. (Section 2-2)

7. $x + y + z$ when $x = 4.2$, $y = 3.04$ and $z = 12.7$
8. $a + b$ when $a = 132.5$ and $b = 60.2$

Solve each problem.

9. The normal annual rainfall in Houston, Texas is 48.19 inches. The normal annual rainfall in Nashville, Tennessee is 46 inches. On the average, how much more rain falls in Houston? (Section 2-2)

10. A color television uses an average of 55 kilowatt-hours per month. At a cost of $0.063 per kilowatt-hour, find the cost of operating the television for 30 days. (Section 2-3)

11. At a cost of $0.077 per hour, how much will it cost to operate a vacuum cleaner for 0.75 hour? (Section 2-3)

12. Last month, 0.02 inch, 1.38 inches, 0.08 inch and 0.57 inch of rain fell in a certain town. What was the total rainfall? (Section 2-2)

REVIEW CAPSULE FOR SECTION 2-4

Evaluate each expression. (Pages 22-24)

1. $x \div 4$ when $x = 24$
2. $90 \div y$ when $y = 6$
3. $248 \div x$ when $x = 4$
4. $2432 \div p$ when $p = 8$
5. $q \div 3$ when $q = 7623$
6. $a \div b$ when $a = 117$ and $b = 3$
7. $m \div n$ when $m = 756$ and $n = 7$
8. $x \div y$ when $x = 72$ and $y = 12$
9. $p \div q$ when $p = 1922$ and $q = 62$
10. $m \div n$ when $m = 3485$ and $n = 17$

The answers are on page 60.

Chapter 2

CONSUMER/COMPUTER APPLICATIONS
Unit Price

You can save money if you use unit price to compare the costs of products. **Unit price** is the price per gram, per ounce, and so on.

The following word rule and the formula tell how to find unit price.

Word Rule: *Unit price* equals the *price of an item* divided by the *number of units*.

Formula: $U = p \div n$, or $U = \dfrac{p}{n}$

U = unit price
P = price of an item
n = number of units

The following is a program which will compute and print the unit price of an item, given the price of the item and the number of units.

Program:

```
100 REM P = TOTAL PRICE
110 REM N = NUMBER OF UNITS
120 REM U = UNIT PRICE OF AN ITEM
130 PRINT "WHAT IS THE TOTAL PRICE";
140 INPUT P
150 PRINT "HOW MANY UNITS";
160 INPUT N
170 LET U = P/N
180 PRINT "THE UNIT PRICE = $"; U
190 PRINT "ANY MORE CALCULATIONS (1 = YES, 0 = NO)";
200 INPUT Z
210 IF Z = 1 THEN 130
220 END
```

REM statements are used to add explanatory comments to a program.

← $U = P \div N$

← $Z = 1$ or $Z = 0$

Exercises

Use the program to compute U to the nearest tenth of a cent.

1. $p = \$1.09$; $n = 24$
2. $p = \$1.99$; $n = 46$
3. $p = \$0.99$; $n = 9$
4. $p = \$3.10$; $n = 15$
5. $p = \$0.85$; $n = 50$
6. $p = \$1.59$; $n = 36$
7. $p = \$1.88$; $n = 25$
8. $p = \$4.51$; $n = 18$
9. $p = \$2.85$; $n = 19$

CONSUMER/COMPUTER APPLICATIONS

This feature illustrates the use of a computer program for finding the unit price of an item when the cost of the item and the number of units are known. It can be used after Section 2-4 (pages 44-46) is taught, as it applies the skill of dividing a decimal by a whole number.

1. $0.045
2. $0.043
3. $0.11
4. $0.207
5. $0.017
6. $0.044
7. $0.075
8. $0.251
9. $0.15

Computer Application

Sidebar

Teaching Suggestions
p. M-15

QUICK QUIZ
Divide.
1. 972 ÷ 27   Ans: 36
2. 8034 ÷ 13   Ans: 618
3. 50,616 ÷ 8
   Ans: 6327
4. 5775 ÷ 11   Ans: 525

ADDITIONAL EXAMPLE
Example 1
A certain car travels 355.5 miles on a full tank of fuel. The tank holds 15 gallons when full. How far does it travel on one gallon?
Ans: 23.7 miles

SELF-CHECK
1. 26.2 miles

2-4 Division

The **fuel economy** of a car refers to how many miles it can travel on one gallon of gasoline. To find fuel economy, you use division skills.

$$\text{Fuel Economy (Miles per gallon)} = \text{Number of Miles} \div \text{Number of Gallons}$$

PROCEDURE

To divide a decimal by a whole number:
1. Place the decimal point in the quotient directly above the decimal point in the dividend.
2. Divide as with whole numbers.

EXAMPLE 1 A certain car travels 248.4 miles on a full tank of fuel. The tank holds 12 gallons when full. How far does it travel on one gallon?

Solution:
1. Place the decimal point in the quotient. → 12)248.4

2.
```
    20.
12)248.4
   24
    8
```
Since 8 is not divisible by 12, write a 0 above the 8. →
```
    20.7
12)248.4
   24
    84
    84
```

The car travels **20.7 miles** on one gallon of fuel.

Self-Check
1. A car travels 366.8 miles on a full tank of fuel. The tank holds 14 gallons when full. How far does it travel on one gallon?

When dividing by a decimal, the first step is to multiply the divisor <u>and</u> the dividend by the <u>same</u> number in order to obtain a whole number divisor.

PROCEDURE

To divide by a decimal:
1. Multiply both the divisor and the dividend by 10, or by 100, or by 1000, and so on, in order to obtain a whole number divisor.
2. Divide.

44 Chapter 2

EXAMPLE 2 Divide: $0.224 \overline{)5.6}$ Check the answer.

Solution: ① Since the divisor is 224 <u>thousandths</u>, multiply the divisor and the dividend by 1000.

$0.224 \overline{)5.600}$ ◄ Annex two zeros.

②
```
        25.
224 )5600.
     448
     1120
     1120
```

Check:
```
    0.224
  ×   25
   1120
    448
   5.600
```
◄ Same as the original dividend.

Self-Check *Divide. Check each answer.*

2. $0.12 \overline{)4.2}$ 3. $0.9 \overline{)657}$ 4. $0.348 \overline{)1009.2}$

CLASSROOM EXERCISES

Divide. (Example 1)

1. $6 \overline{)20.88}$ 2. $3 \overline{)7.23}$ 3. $5 \overline{)3.580}$ 4. $7 \overline{)43.96}$

Multiply the divisor and the dividend by 10, or by 100, or by 1000, to make the divisor a whole number. (Example 2, step 1)

5. $0.32 \overline{)8.96}$ 6. $0.12 \overline{)2.4}$ 7. $0.036 \overline{)7.2}$ 8. $0.64 \overline{)22.4}$

Divide. (Example 2)

9. $0.4 \overline{)3.76}$ 10. $0.68 \overline{)30.6}$ 11. $0.004 \overline{)21.88}$ 12. $0.7 \overline{)238}$

WRITTEN EXERCISES

Goals: To divide a decimal by a whole number
To divide a decimal by a decimal
To apply these skills to solving word problems

Divide. (Example 1)

1. $6 \overline{)17.04}$ 2. $4 \overline{)22.12}$ 3. $6 \overline{)58.86}$ 4. $7 \overline{)29.05}$
5. $9 \overline{)0.585}$ 6. $4 \overline{)0.232}$ 7. $15 \overline{)121.05}$ 8. $17 \overline{)596.7}$
9. $42.63 \div 29$ 10. $75.84 \div 12$ 11. $2.912 \div 52$ 12. $8.112 \div 78$

Decimals and Applications   **45**

ADDITIONAL EXAMPLES
Example 2
Divide.
1. $0.021 \overline{)0.777}$
   Ans: 37
2. $67.945 \div 1.07$
   Ans: 63.5
3. $0.03159 \div 0.81$
   Ans: 0.039

SELF-CHECK
2. 35   3. 730   4. 2900

CLASSROOM EXERCISES
1. 3.48   2. 2.41
3. 0.716   4. 6.28
5. 28   6. 20   7. 200
8. 35   9. 9.4   10. 45
11. 5470   12. 340

ASSIGNMENT GUIDE
BASIC
Day 1 pp. 45-46:
1-25 odd
Day 2 pp. 45-46:
1-26 even
AVERAGE
pp. 45-46: 1-27
ABOVE AVERAGE
pp. 45-46: 1-23 odd
25-30

WRITTEN EXERCISES

1. 2.84   2. 5.53
3. 9.81   4. 4.15
5. 0.065 6. 0.058
7. 8.07   8. 35.1
9. 1.47   10. 6.32
11. 0.056 12. 0.104
13. 6.2   14. 1.59
15. 89   16. 18.3
17. 0.35 18. 20.5
19. 7060 20. 62.6
21. 3670 22. 320.1
23. 8620 24. 9140
25. 16.7 miles
26. 41.2 miles
27. 84.5 miles
28. 18.5 miles
29. first car
30. first car

Divide. Check each answer. (Example 2)

13. $0.8\overline{)4.96}$ 14. $0.9\overline{)1.431}$ 15. $0.06\overline{)5.34}$ 16. $0.05\overline{).915}$
17. $1.2\overline{).42}$ 18. $4.3\overline{)88.15}$ 19. $0.38\overline{)2682.8}$ 20. $0.81\overline{)50.706}$
21. 14.68 ÷ 0.004   22. 2.5608 ÷ 0.008   23. 94.82 ÷ 0.011   24. 511.84 ÷ 0.056

APPLICATIONS: USING DIVISION WITH DECIMALS

25. The fuel tank of a certain luxury car holds 24 gallons when full. The car travels 400.8 miles on a full tank. How many miles does it travel on one gallon?

26. A sub-compact car travels 535.6 miles on a full tank of fuel. A full tank holds 13 gallons. How far does it travel on one gallon?

27. A certain motorcycle travels 135.2 miles on a full tank of fuel. The tank holds 1.6 gallons when full. How far does it travel on one gallon?

28. The fuel tank in a certain sports car holds 14.2 gallons when full. The car travels 262.7 miles on a full tank. How far will it travel on one gallon?

MORE CHALLENGING EXERCISES

Determine which car has the greater fuel economy.

29. One compact car travels 500 miles on 12.5 gallons of fuel. Another compact car travels 462 miles on 13.2 gallons of fuel.

30. One sports car travels 789 kilometers on 52.6 liters of fuel. Another sports car travels 843.9 kilometers on 58.2 liters of fuel.

REVIEW CAPSULE FOR SECTION 2-5

Round to the nearest whole number.

1. 4.3   2. 8.7   3. 0.9   4. 16.82   5. 11.38

Round to the nearest tenth.

6. 3.27   7. 1.32   8. 17.84   9. 11.409   10. 0.987

Round to the nearest hundredth.

11. 5.863   12. 0.396   13. 14.649   14. 0.005   15. 19.442

The answers are on page 60.

Chapter 2

PROBLEM SOLVING AND APPLICATIONS

2-5 Estimating with Decimals

At the local supermarket, Carlos bought items for $0.43, $1.69, $0.79, $2.29, $0.67, and $3.82. To estimate the total cost, Carlos rounded the cost of each item to the nearest dollar.

You can round decimals to the nearest whole number, to the nearest tenth, to the nearest hundredth, and so on. You can use the rules for rounding to estimate answers.

PROCEDURE

To use estimation to solve a problem:

1. Write a sum, difference, product, or quotient (or a combination of these) for the problem.
2. Round each number to the nearest whole number, or to the nearest tenth, or to the nearest hundredth, and so on.
3. Perform the indicated operations on the rounded numbers.

EXAMPLE 1 Estimate Carlos' total bill at the supermarket. Round to the nearest dollar.

Solution:

1 Problem	2 Round to the nearest dollar.
0.43	→ 0.00
1.69	→ 2.00
0.79	→ 1.00
2.29	→ 2.00
0.67	→ 1.00
3.82	→ +4.00
	3 10.00

Estimate: $10.00

SOUP ** 0.43
CATSUP ** 1.69
CORN ** 0.79
MILK ** 2.29
FRUIT ** 0.67
CHICKEN ** 3.82
TOTAL
THANK YOU

Self-Check

1. Emily bought school supplies for $3.69, $0.72, $1.37, $4.19, $0.35 and $2.55. Estimate the total cost to the nearest dollar.

Decimals and Applications 47

Teaching Suggestions
p. M–15

QUICK QUIZ

Round to the nearest whole number.

1. 15.26   Ans: 15
2. 27.603   Ans: 28
3. 19.362   Ans: 19

Round to the nearest tenth.

4. 27.083   Ans: 27.1
5. 27.038   Ans: 27.0

ADDITIONAL EXAMPLES
Example 1

1. Marge bought hardware supplies for $6.75, $14.32, $2.68, $0.49, $9.51, and $4.17. Estimate the total cost to the nearest dollar.   Ans: $38

2. Burt bought books for $3.95, $2.95, $4.15, $16.25, $11.50, and $8.95. Estimate the total cost to the nearest dollar.   Ans: $48

SELF-CHECK

1. $13.00

47

ADDITIONAL EXAMPLE

Example 2

Joe is paid $0.032 for each bolt he makes. He received $57.60 for six hours work. Estimate how many bolts he made in six hours.   Ans: 300

SELF-CHECK

2. 50 pounds

CLASSROOM EXERCISES

1. c   2. a   3. a
4. c   5. b   6. a
7. b   8. a   9. b
10. a 11. c 12. a

When estimating the answer to a problem that involves division of decimals or whole numbers, round the dividend and the divisor to the nearest whole numbers that divide evenly.

EXAMPLE 2 Mae is paid $0.045 for each envelope she addresses. She received $14.85 for five hours work. Estimate how many envelopes Mae addressed in the five hours.

Solution:

[1] Problem

$0.045 \overline{)14.85}$ ⟶

[2] Round to convenient numbers.

$0.05 \overline{)15.00}$

[3] $1500 \div 5 = 300$

Mae addressed about **300** envelopes.

Self-Check 2. Karl is paid $0.042 for each pound of beans he picks. He received $2.10 for one bag. Estimate how many pounds of beans were in the bag.

CLASSROOM EXERCISES

Choose the best estimate. Choose a, b, or c.
(Examples 1 and 2, steps 1 and 2)

1. $3.70 + $5.20   a. $3.00 + $5.00   b. $4.00 + $6.00   c. $4.00 + $5.00
2. $14.80 + $6.75   a. $15.00 + $7.00   b. $14.00 + $7.00   c. $14.00 + $6.00
3. $10.20 − $7.80   a. $10 − $8   b. $10 − $9   c. $11 − $7
4. $24.60 − $19.90   a. $24 − $18   b. $25 − $19   c. $25 − $20
5. 3.8 (10.1)   a. 3 · 10   b. 4 · 10   c. 3 · 11
6. 10.3 (38.9)   a. 10 · 39   b. 11 · 39   c. 10 · 38
7. 60.3 ÷ 29.8   a. 65 ÷ 20   b. 60 ÷ 30   c. 60 ÷ 20
8. 19.9 ÷ 1.8   a. 20 ÷ 2   b. 20 ÷ 1   c. 18 ÷ 2
9. 2.1(29.6)   a. 3 · 30   b. 2 · 30   c. 2 · 20
10. 81.8 ÷ 19.6   a. 80 ÷ 20   b. 90 ÷ 20   c. 80 ÷ 15
11. 148.5 ÷ 52.2   a. 100 ÷ 50   b. 200 ÷ 50   c. 150 ÷ 50
12. 38.6(3.9)   a. 40 · 4   b. 30 · 4   c. 30 · 3

48 Chapter 2

WRITTEN EXERCISES

Goal: To use estimation to solve word problems involving decimals

Choose the best estimate. Choose a, b, or c. (Examples 1 and 2)

1. $16.70 + $3.90 + $21.20   **a.** $40   **b.** $42   **c.** $45
2. $9.10 + $8.60 + $11.90   **a.** $30   **b.** $35   **c.** $28
3. $11.22 − $4.85   **a.** $4   **b.** $6   **c.** $7
4. $22.70 − $19.80   **a.** $5   **b.** $4   **c.** $3
5. 2.7 (3.9)   **a.** 6   **b.** 8   **c.** 12
6. 11.08 (1.02)   **a.** 11   **b.** 15   **c.** 14
7. 17.5 ÷ 6.3   **a.** 6   **b.** 3   **c.** 5
8. 14.88 ÷ 3.01   **a.** 5   **b.** 3   **c.** 2

APPLICATIONS: USING ESTIMATION WITH DECIMALS

9. At Games Galore, Tim paid $2.79 for ping pong balls, $6.39 for dominoes, and $2.40 for a book of rules for games. Estimate the total cost to the nearest dollar.

10. The band that played for the Homecoming Dance was paid $84 per hour. The band played for 3.75 hours. Estimate to the nearest dollar how much the band was paid for the evening.

11. The regular price for a set of golf clubs was $178.95. The sale price was $113.25. Estimate the amount saved to the nearest dollar.

12. Claude was paid $6.25 for each page of manuscript he typed. His total pay amounted to $300. About how many pages did he type?

13. For a pep rally in the school gym, pennants were used to decorate the walls. Each pennant cost $2.19 and the total bill was $100.74. About how many pennants were used?

14. The French Club sold corsages to raise money for a trip. They charged $5.20 for each corsage and sold 289 corsages. Estimate, to the nearest dollar, the total receipts for the sale of corsages.

Decimals and Applications   **49**

ASSIGNMENT GUIDE
BASIC
p. 49: 1–9
AVERAGE
p. 49: 1–12
ABOVE AVERAGE
p. 49: 1–14

WRITTEN EXERCISES
1. b   2. a   3. b
4. c   5. c   6. a
7. b   8. a   9. $11
10. $336   11. $66
12. 50 pages
13. 50 pennants
14. $1450

CALCULATOR EXERCISES

1. 280, 281
2. 3600, 3619
3. 12,000, 12,139
4. 800, 807.94
5. 70, 68.6
6. 10, 9.9

ESTIMATION

You can use the rules for rounding and estimation and the calculator to solve problems.

EXAMPLE 54.6 ÷ 10.4

Solution Use paper and pencil to estimate the answer: 50 ÷ 10 = 5

Use the calculator to find the exact answer.

5 4 . 6 ÷ 1 0 . 4 =   5.25

Since the estimate is 5, 5.25 is a reasonable answer.

EXERCISES First estimate each answer. Then use a calculator to find the exact answer.

Problem	Estimated Answer	Exact Answer
1. 29 + 141 + 38 + 73	?	?
2. 5418 − 1799	?	?
3. 199 × 61	?	?
4. 39.8 × 20.3	?	?
5. 3498.6 ÷ 51	?	?
6. 90.09 ÷ 9.1	?	?

REVIEW CAPSULE FOR SECTION 2-6

Evaluate each expression for m = 10. (Pages 22–24)

1. $3.4m$ 2. $15m$ 3. $0.5m$ 4. $4.06m$ 5. $65.2m$ 6. $80m$
7. $130 \div m$ 8. $12.4 \div m$ 9. $6 \div m$ 10. $0.8 \div m$ 11. $3000 \div m$ 12. $9.6 \div m$

Evaluate each expression for p = 100. (Pages 22–24)

13. $2.8p$ 14. $74p$ 15. $0.6p$ 16. $0.013p$ 17. $20.4p$ 18. $7.56p$
19. $2000 \div p$ 20. $200 \div p$ 21. $62 \div p$ 22. $428.6 \div p$ 23. $45000 \div p$ 24. $3.5 \div p$

Evaluate each expression for t = 1000. (Pages 22–24)

25. $6.2t$ 26. $8.5t$ 27. $40.7t$ 28. $0.3t$ 29. $0.07t$ 30. $38t$
31. $4000 \div t$ 32. $405 \div t$ 33. $32 \div t$ 34. $78,000 \div t$ 35. $3840 \div t$ 36. $6 \div t$

The answers are on page 60.

Chapter 2

2-6 Metric Units of Measurement

The base unit of length in the metric system is the **meter**. The other most commonly used units of length are the **centimeter**, the **millimeter**, and the **kilometer**.

The following table shows how other metric units of length are related to the meter. It also shows how each prefix is related to place value.

TABLE 1

Units of Length	Prefix	Meaning of Prefix
1 millimeter (mm) = **0.001** meter (m)	milli-	thousandths
1 centimeter (cm) = **0.01** meter	centi-	hundredths
1 decimeter (dm) = **0.1** meter	deci-	tenths
1 dekameter (dam) = **10** meters	deka-	tens
1 hectometer (hm) = **100** meters	hecto-	hundreds
1 kilometer (km) = **1000** meters	kilo-	thousands

NOTE: The value of each metric unit is 10 times the value of the next smaller unit.

PROCEDURE

1 To change to a smaller metric unit, multiply by 10, or by 100, or by 1000, and so on.

2 To change to a larger metric unit, divide by 10 or by 100, or by 1000, and so on.

EXAMPLE 1 Complete.

a. 75 km = __?__ m   b. 255 mm = __?__ m   c. 165 cm = __?__ mm

Larger to smaller: Multiply.

Solutions: a. 75 km = 75 · 1000 = 75,000 m

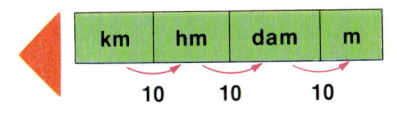

Smaller to larger: Divide.

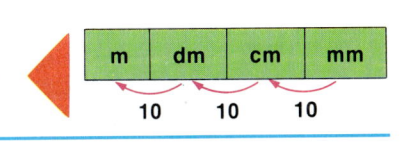

b. 255 mm = 255 ÷ 1000 = 0.255 m

km	hm	dam	m
	10	10	10

m	dm	cm	mm
	10	10	10

Decimals and Applications   **51**

Teaching Suggestions
p. M-15

QUICK QUIZ

Evaluate each expression.

1. 3.97 x when x = 100
   Ans: 397
2. 21.6y when y = 100
   Ans: 2160
3. 16.97a when a = 10
   Ans: 169.7
4. 19.03d when d = 100
   Ans: 1903
5. 3.9h when h = 1000
   Ans: 3900
6. 18.4 k when k = 1000
   Ans: 18,400
7. 6.3 ÷ a when a = 100
   Ans: 0.063
8. 1492.4 ÷ x when x = 100   Ans: 14.924
9. 27.63 ÷ b when b = 10
   Ans: 2.763
10. 0.38 ÷ y when y = 1000   Ans: 0.00038

ADDITIONAL EXAMPLES
Example 1
Complete.
1. 372 m = __?__ km
   Ans: 0.372 km
2. 84 m = __?__ cm
   Ans: 8400 cm
3. 5000 cm = __?__ km
   Ans: 0.05 km

SELF-CHECK

1. 25,000 m
2. 1.5 cm
3. 790 cm

ADDITIONAL EXAMPLES

Example 2

Complete.

1. 273 cg = ___?___ g
   Ans: 2.73 g
2. 70 kg = ___?___ g
   Ans: 70,000 g
3. 46.2 g = ___?___ mg
   Ans: 46,200 mg

SELF-CHECK

4. 3600 cg
5. 0.350 g
6. 53.1 dg

Self-Check *Complete.*

1. 25 km = ___?___ m   2. 15 mm = ___?___ cm   3. 7.9 m = ___?___ cm

The prefixes used for other metric units are the same as those used for units of length.

The most commonly used units of mass in the metric system are the **milligram**, the **gram**, and the **kilogram**.

TABLE 2   **Units of Mass**

kilogram (kg)	hectogram (hg)	dekagram (dag)	gram (g)	decigram (dg)	centigram (cg)	milligram (mg)

NOTE: A metric ton (t) equals 1000 kilograms.

EXAMPLE 2 Karen's doctor wants her to take one iron capsule per day. Each capsule contains 63 milligrams of iron.

a. How many milligrams is this per week?

b. How many grams is this per week?

Solution: a. 63 · 7 = 441 mg per week.

Smaller to larger: Divide.

b. 441 mg = 441 ÷ 1000 = 0.441 g

Karen takes **0.441 g** of iron per week.

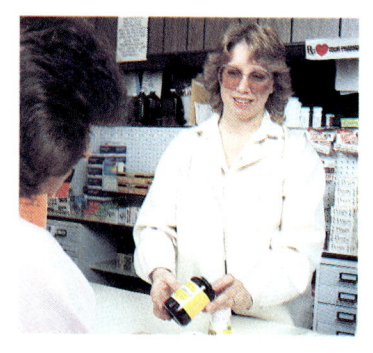

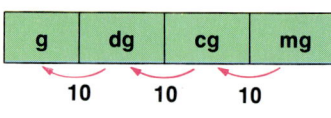

g	dg	cg	mg
	10	10	10

Self-Check *Complete.*

4. 36 g = ___?___ cg   5. 350 mg = ___?___ g   6. 5.31 g = ___?___ dg

The most commonly used units of capacity are the **kiloliter**, the **liter**, and the **milliliter**.

TABLE 3   **Units of Capacity**

kiloliter (kL)	hectoliter (hL)	dekaliter (daL)	liter (L)	deciliter (dL)	centiliter (cL)	milliliter (mL)

52 Chapter 2

EXAMPLE 3 The gasoline tank of a car holds 45.6 liters. How many kiloliters is this?

Smaller to larger: Divide.

Solution: 45.6 L = 45.6 ÷ 1000 = 0.0456 kL

The tank holds **0.0456 kL**.

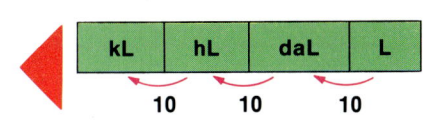

kL	hL	daL	L

Self-Check

Complete.

7. 22 mL = _?_ L   8. 37 L = _?_ mL   9. 792 kL = _?_ L

CLASSROOM EXERCISES

Complete. Use the words "multiply" or "divide."
(Tables 1, 2, and 3)

1. To change from millimeters to centimeters, _?_ by 10.
2. To change from meters to millimeters, _?_ by 1000.
3. To change from grams to centigrams, _?_ by 100.
4. To change from grams to kilograms, _?_ by 1000.
5. To change from milliliters to liters, _?_ by 1000.
6. To change from kiloliters to liters, _?_ by 1000.
7. To change from meters to decimeters, multiply by _?_.
8. To change from decimeters to centimeters, multiply by _?_.
9. To change from grams to kilograms, divide by _?_.
10. To change from centigrams to grams, divide by _?_.

WRITTEN EXERCISES

Goal: To determine equivalent metric units of length, mass, and capacity

Complete. (Example 1)

1. 3 cm = _?_ mm
2. 7.2 m = _?_ cm
3. 54 km = _?_ m
4. 65,000 cm = _?_ km

	km	m	cm
5.	3.5	?	?
7.	?	256	?

	m	cm	mm
6.	?	8.4	?
8.	?	?	32.5

Decimals and Applications **53**

ADDITIONAL EXAMPLES
Example 3
Complete.
1. 31.7 L = _?_ kL
   Ans: 0.0317 kL
2. 7.62 L = _?_ mL
   Ans: 7620 mL
3. 0.38 dL = _?_ L
   Ans: 0.038 L

SELF-CHECK
7. 0.022 L   8. 37,000 mL
9. 792,000 L

CLASSROOM EXERCISES
1. divide
2. multiply
3. multiply
4. divide   5. divide
6. multiply   7. 10
8. 10   9. 1000   10. 100

ASSIGNMENT GUIDE
BASIC
pp. 53-54: 1-25
AVERAGE
pp. 53-54: 1-27
ABOVE AVERAGE
pp. 53-54: 1-23 odd, 25-30

WRITTEN EXERCISES

1. 30 mm
2. 720 cm
3. 54,000 m
4. 0.65 km
5. 3500 m; 350,000 cm
6. 0.084m; 84 mm
7. 0.256 km; 25,600 cm
8. 0.0325 m; 3.25 cm
9. 5.100 kg
10. 21,800 mg
11. 900 kg
12. 320 g
13. 3500 g; 3,500,000 mg
14. 420 kg; 420,000 g
15. 0.72 kg; 720,000 mg
16. 0.0346 t; 34.6 kg
17. 1000 L
18. 3.250 L
19. 450 mL
20. 1785 L
21. 23,000 mL
22. 0.4 L
23. 620 L
24. 0.043 kL
25. 4200 m
26. 41,000,000 kg
27. 0.946 mi
28. 4000 k
29. 1000 mg
30. 255,000 g

Complete. (Example 2)

9. 5100 g = ? kg 10. 21.8 g = ? mg 11. 0.9 t = ? kg 12. 0.32 kg = ? g

	kg	g	mg
13.	3.5	?	?
15.	?	720	?

	t	kg	g
14.	0.42	?	?
16.	?	?	34,600

Complete. (Example 2)

17. 1 kL = ? L   18. 3250 mL = ? L   19. 0.45 L = ? mL   20. 1.785 kL = ? L
21. 23 L = ? mL 22. 400 mL = ? L   23. 0.62 kL = ? L   24. 43,000 mL = ? kL

APPLICATIONS: USING METRIC MEASURES

25. The jogging track in Central Park is 4.2 kilometers long. How many meters is this?

26. A recent estimate states that 41,000 metric tons of gold still exist in the earth. How many kilograms is this?

27. A quart of milk is about 946 milliliters. How many liters is this?

28. The polar continent Antarctica is covered by an ice dome that is almost 4 kilometers high. Express this height in meters.

29. One milliliter of water weighs (has a mass of) one gram. Express the weight of 1 mL of water in terms of milligrams.

30. A weight-lifting record of 255 kilograms was set in the 1976 Olympics. Express 255 kilograms in terms of grams.

REVIEW CAPSULE FOR SECTION 2-7

Replace the ? with the correct exponent. (Pages 7-9)

1. $10 \cdot 10 = 10^?$
2. $10 \cdot 10 \cdot 10 \cdot 10 \cdot 10 = 10^?$
3. $10 \cdot 10 \cdot 10 = 10^?$
4. $10 \cdot 10 \cdot 10 \cdot 10 \cdot 10 \cdot 10 = 10^?$
5. $10 \cdot 10 \cdot 10 \cdot 10 \cdot 10 \cdot 10 \cdot 10 = 10^?$
6. $10 \cdot 10 \cdot 10 \cdot 10 = 10^?$

Evaluate each expression. (Pages 7-9)

7. 10^3 8. 10^5 9. 10^6 10. 10^1 11. 10^4 12. 10^7

The answers are on page 60.

Chapter 2

PROBLEM SOLVING AND APPLICATIONS

2-7 Scientific Notation

The distance from the earth to the moon is approximately 360,000,000 meters. Very large numbers such as this are often expressed in **scientific notation**.

$$360{,}000{,}000 = 3.6 \times 10^8$$

Number between 1 and 10 — Power of 10 — Scientific Notation

PROCEDURE

To write a number in scientific notation:

1. Count the number of places the decimal point must be moved in order to arrive at a number between 1 and 10.
2. Express it as a product in this form:
(Number between 1 and 10) × (Power of 10)

EXAMPLE 1 The distance from the sun to the earth is about 150,000,000 kilometers. Express this number in scientific notation.

Solution:
1. Count the number of places the decimal point must be moved to arrive at a number between 1 and 10.   150,000,000   8 places
2. Write the number as a product. 1.5×10^8

Number between 1 and 10 — Power of 10

Self-Check 1. The Puerto Rico Trench is the deepest known point in the Atlantic Ocean. It is about 2600 meters below sea level. Express this number in scientific notation.

A number written in scientific notation can be expressed as a decimal numeral.

Decimals and Applications **55**

Teaching Suggestions
p. M-15

QUICK QUIZ
Replace the ? with the correct exponent.
1. $10 \cdot 10 \cdot 10 = 10^?$
   Ans: 10^3
2. $10 \cdot 10 = 10^?$
   Ans: 10^2

Evaluate each expression.
3. 10^1 Ans: 10
4. 10^7 Ans: 10,000,000
5. 10^4 Ans: 10,000

ADDITIONAL EXAMPLE
Example 1
The number of square feet in the Pentagon at Washington, D.C., is 3,708,000. Express this number in scientific notation.
Ans: 3.708×10^6

SELF-CHECK
1. 2.6×10^3

ADDITIONAL EXAMPLE

Example 2

The distance of the Great Galaxy from the earth is about 1.5×10^6 light years. Express the distance as a decimal numeral.

Ans: 1,500,000

SELF-CHECK

2. 864,000

CLASSROOM EXERCISES

1. 1   2. 2   3. 3
4. 2   5. 4   6. 7
7. 4 x 10   8. 2 x 10^5
9. 4 x 10^2
10. 2 x 10^6
11. 4 x 10^3
12. 2 x 10^8 13. 160
14. 38,625
15. 38,000

ASSIGNMENT GUIDE
BASIC
pp. 56-57: Omit
AVERAGE
pp. 56-57: 1-24
ABOVE AVERAGE
pp. 56-57: 1-24

PROCEDURE

To express a number written in scientific notation as a decimal numeral:

1. Evaluate the power of 10.
2. Multiply.

EXAMPLE 2 The diameter of the planet Mars is approximately 4.214×10^3 miles. Express the diameter as a decimal numeral.

Solution:
1. $4.214 \times 10^3 = 4.214 \times 1000$ $10^3 = 10 \cdot 10 \cdot 10 = 1000$
2. $ = 4214$

Self-Check   2. The diameter of the sun is 8.64×10^5 miles. Express the diameter as a decimal numeral.

NOTE: $10^1 = 10$ and $10^0 = 1$.

CLASSROOM EXERCISES

Complete. (Example 1, step 2)

1. $39 = 3.9 \times 10^?$
2. $395 = 3.95 \times 10^?$
3. $3950 = 3.950 \times 10^?$
4. $573 = 5.73 \times 10^?$
5. $57,300 = 5.73 \times 10^?$
6. $57,300,000 = 5.73 \times 10^?$

Write each number in scientific notation. (Example 1)

7. 40   8. 200,000   9. 400   10. 2,000,000   11. 4000   12. 200,000,000

Express as a decimal numeral. (Example 2)

13. 1.6×10^2 14. 3.8625×10^4 15. 3.8×10^4

WRITTEN EXERCISES

Goal: To express a number in scientific notation

To express a number written in scientific notation as a number in decimal notation

Express each number in scientific notation. (Example 1)

1. 3400   2. 8000   3. 90,000   4. 64,000   5. 1,050,000   6. 4,000,000
7. 875   8. 24,100   9. 80,000,000   10. 302   11. 76,000   12. 40,300,000

56   Chapter 2

Express as a decimal numeral. (Example 2)

13. 3.2×10^3
14. 5.8×10^6
15. 2.51×10^5
16. 6.05×10^9
17. 9.123×10^6
18. 1.104×10^3
19. 5.2×10^7
20. 4.0×10^5
21. 1.42×10^1

22. The Space Shuttle's orbit is about 22,240 miles above the equator. Express this number in scientific notation.

23. The assembled Space Shuttle weighs about 2,000,000 kilograms at liftoff. Express this number in scientific notation.

24. The three main engines of the Space Shuttle orbiter burn about 242,000 liters of fuel per minute. Express this number in scientific notation.

REVIEW: SECTIONS 2-4 — 2-7

Solve each problem. (Section 2-4)

1. Arlita has a piece of fabric 2.8 meters long. She wants to cut it into 4 equal pieces. How long should each piece be?

2. Harold can travel 27.6 miles in 1.5 hours on his bicycle. Find how far he can travel in one hour.

Choose the best estimate. Choose a, b, or c. (Section 2-5)

3. $3.70 + $8.20 + $0.90   a. $11   b. $13   c. $14
4. $22.85 − $13.20   a. $10   b. $11   c. $8
5. 6.2 (4.8)   a. 31   b. 27   c. 30
6. 21.04 ÷ 6.83   a. 3   b. 5   c. 2

Complete. (Section 2-6)

7. 4.3 m = __?__ km
8. 0.4 kg = __?__ g
9. 320.4 L = __?__ kL

Complete. (Section 2-7)

10. $3200 = 3.2 \times 10^?$
11. $940,000 = 9.4 \times 10^?$
12. $8,250,000 =$ __?__ $\times 10^6$
13. $45,000,000 =$ __?__ $\times 10^7$

WRITTEN EXERCISES
1. 3.4×10^3
2. 8×10^3
3. 9×10^4
4. 6.4×10^4
5. 1.05×10^6
6. 4×10^6
7. 8.75×10^2
8. 2.41×10^4
9. 8×10^7
10. 3.02×10^2
11. 7.6×10^4
12. 4.03×10^7 13. 3200
14. 5,800,000
15. 251,000
16. 6,050,000,000
17. 9,123,000   18. 1104
19. 52,000,000
20. 400,000   21. 14.2
22. 2.224×10^4
23. 2×10^6
24. 2.42×10^5

QUIZ: SECTIONS 2-4-2-7
After completing this Review, you may want to administer a quiz covering the same sections. A Quiz is provided in the *Teacher's Edition: Part II.*

REVIEW: SECTIONS 2-4-2-7
1. 0.7 m   2. 18.4 mi
3. b   4. a   5. c
6. a   7. 0.0043

Decimals and Applications

REVIEW: SECTIONS 2-4—2-7
<u>8.</u> 400   <u>9.</u> 0.3204
<u>10.</u> 3   <u>11.</u> 5
<u>12.</u> 8.25   <u>13.</u> 4.5

CHAPTER REVIEW
<u>1.</u> liter   <u>2.</u> milli-
<u>3.</u> deka-   <u>4.</u> kilowatt
<u>5.</u> meter   <u>6.</u> kilo-
<u>7.</u> metric ton   <u>8.</u> centi-
<u>9.</u> scientific notation
<u>10.</u> deci-   <u>11.</u> gram
<u>12.</u> hecto-   <u>13.</u> 0.4
<u>14.</u> 40   <u>15.</u> 0.04
<u>16.</u> 5000   <u>17.</u> 0.005
<u>18.</u> 0.00003   <u>19.</u> >
<u>20.</u> <   <u>21.</u> <   <u>22.</u> >
<u>23.</u> >   <u>24.</u> <
<u>25.</u> 10.84
<u>26.</u> 230.505   <u>27.</u> 1.669
<u>28.</u> 271.987
<u>29.</u> 23.595   <u>30.</u> 0.06
<u>31.</u> 105.792
<u>32.</u> 44.739   <u>33.</u> 40.81
<u>34.</u> 166.78

CHAPTER REVIEW

PART 1: VOCABULARY

In Exercises 1–12, choose from the box at the right below the word(s) that best corresponds to each description.

1. A unit of capacity in the metric system ?
2. A prefix meaning thousandths ?
3. A prefix meaning tens ?
4. One thousand watts ?
5. The base unit of length in the metric system ?
6. A prefix meaning thousands ?
7. One thousand kilograms ?
8. A prefix meaning hundredths ?
9. A way of expressing very large numbers ?
10. A prefix meaning tenths ?
11. The base unit of mass in the metric system ?
12. A prefix meaning hundreds ?

milli-
centi-
deci-
deka-
hecto-
kilo-
kilowatt
meter
gram
liter
metric ton
scientific notation

PART 2: SKILLS

Give the value of each underlined digit.   (Section 2-1)

13. 25.<u>4</u>   14. <u>4</u>8.12   15. 15.1<u>4</u>6   16. <u>5</u>140.27   17. 9.02<u>5</u>   18. 2.5471<u>3</u>

For Exercises 19–24, replace the ? with > or <.   (Section 2-1)

19. 2.46 ? 2.39
20. 5.03 ? 5.11
21. 1.76 ? 1.761
22. 0.628 ? 0.627
23. 0.03 ? 0.009
24. 2.007 ? 2.018

Add or subtract as indicated.   (Section 2-2)

25. 2.02
     5.68
    +3.14

26.   1.9
      57.21
       3.025
    +168.37

27.   6.46
     −4.791

28.   371.082
     − 99.095

29. $0.025 + 1.76 + 21.81$
30. $0.64 - 0.58$
31. $91.23 + 4.8 + 0.062 + 9.7$
32. $84.3 - 39.561$

Evaluate each expression.   (Section 2-2)

33. $x + y + z$ when $x = 8.9$, $y = 31.4$ and $z = 0.51$
34. $a - b$ when $a = 215.3$ and $b = 48.52$

58   Chapter 2

For Exercises 35–36, find the monthly cost for each appliance at the given rate. Round each answer to the nearest cent. (Section 2-3)

Appliance	Kilowatt-hours per month	Cost at $0.098 per kwh	Cost at $0.075 per kwh	Cost at $0.047 per kwh
35. Refrigerator	152.4	?	?	?
36. Color television	55.0	?	?	?

Divide. (Section 2-4)

37. $4\overline{)23.56}$
38. $25\overline{)308.75}$
39. $341.6 \div 2.8$
40. $298.08 \div 0.048$

Choose the best estimate. Choose a, b, or c. (Section 2-5)

41. $18.90 + $4.30 + $35.80   a. 57   b. 62   c. 59
42. $45.70 − $13.35   a. 33   b. 31   c. 30
43. (14.8)(5.3)   a. 200   b. 75   c. 50
44. $19.75 \div 4.05$ a. 6   b. 4   c. 5

Complete. (Section 2-6)

45. 2 cm = __?__ mm
46. 1500 m = __?__ km
47. 4.51 g = __?__ mg
48. 8350 mL = __?__ L
49. 0.55 kL = __?__ L
50. 64,000 mL = __?__ kL

Express each number in scientific notation. (Section 2-7)

51. 250
52. 34,500
53. 5,000,000
54. 6,030,000,000

PART 3: APPLICATIONS

55. During a Women's Nordic Skiing Event, Andrea recorded runs of 5.52 minutes, 7.87 minutes, and 6.14 minutes. Estimate to the nearest minute the total time for the runs. Choose a, b, or c. (Section 2-5)
    a. 18   b. 22   c. 20

56. In 1933, the World's Land Speed Record for 1 mile was 272.109 miles per hour. In 1963, the speed record set by Craig Breedlove was 407.45 miles per hour. By how many miles per hour did the speed record increase? (Section 2-2)

57. A washing machine uses about 208.08 kilowatt-hours per month. At a cost of $0.12 per kilowatt-hour, find the cost of operating a washing machine for 12 months. (Section 2-3)

58. A car travels 343.35 miles on a full tank of fuel. The tank holds 10.5 gallons when full. How far does the car travel on one gallon? (Section 2-4)

CHAPTER REVIEW

35. $14.94, $11.43, $7.16
36. $5.39, $4.13, $2.59
37. 5.89   38. 12.35
39. 122   40. 6210
41. c   42. a   43. b
44. c   45. 20   46. 1.5
47. 4510   48. 8.35
49. 550   50. 0.064
51. 2.5×10^2
52. 3.45×10^4
53. 5×10^6
54. 6.03×10^9
55. c   56. 135.341 mph
57. $299.64
58. 32.7 miles

Decimals and Applications

CHAPTER TEST

Two forms of a chapter test, Form A and Form B, are provided on copying masters in the Teacher's Edition: Part II.

1. hundredths   2. tens
3. >   4. <   5. <
6. 106.921   7. 191.83
8. 37.02   9. 37.35
10. 54.75   11. 476.37
12. 56.7   13. 3.26
14. 9.7   15. 0.002832
16. c   17. b
18. 0.515   19. 2600
20. 9.5   21. 3   22. 5
23. 6   24. 2.05 inches
25. 28.2 miles

CHAPTER TEST

Complete.

1. The place name of the 2 in 49.02 is ___?___.
2. The place name of the 7 in 573.1 is ___?___.

For Exercises 3–5, replace the ___?___ with > or <.

3. 4.15 ___?___ 4.09
4. 1.001 ___?___ 1.015
5. 25.03 ___?___ 25.033

6. $51.9 + 0.52 + 4.83 + 49.671 = $ ___?___
7. $253.41 - 61.58 = $ ___?___
8. $2.01 + 34.18 + 0.83 = $ ___?___
9. $41.2 - 3.85 = $ ___?___

10. Evaluate $x + y + z$ when $x = 1.07$, $y = 0.28$, and $z = 53.4$
11. Evaluate $a - b$ when $a = 512.01$ and $b = 35.64$

12. 63×0.9
13. $21 \overline{)68.46}$
14. $3.88 \div 0.4$
15. $(0.048)(0.059)$

Choose the best estimate. Choose a, b, or c.

16. $4.15 + $11.75 + $8.90   **a.** 26   **b.** 23   **c.** 25
17. $(15.015)(9.7)$ **a.** 135   **b.** 150   **c.** 160

Complete.

18. $51.5 \text{ cm} = $ ___?___ m
19. $2.6 \text{ kg} = $ ___?___ g
20. $9500 \text{ mL} = $ ___?___ L
21. $4100 = 4.1 \times 10^{?}$
22. $570{,}000 = 5.7 \times 10^{?}$
23. $8{,}200{,}000 = 8.2 \times 10^{?}$

24. Last month, 0.02 inch, 1.38 inches, 0.08 inch, and 0.57 inch of rain fell. What was the total rainfall?
25. A car travels 352.5 miles on a full tank (12.5 gallons) of fuel. How far does it travel on one gallon?

ANSWERS TO REVIEW CAPSULES

Page 34   **1.** 39   **2.** 404   **3.** 1032   **4.** 1475   **5.** 1159   **6.** 4646

Page 37   **1.** 42   **2.** 508   **3.** 24,006   **4.** 20,240   **5.** 16,128   **6.** 3528   **7.** $14.76   **8.** $8.96   **9.** $25.02   **10.** $37.10   **11.** $61.71

Page 42   **1.** 6   **2.** 15   **3.** 62   **4.** 304   **5.** 2541   **6.** 39   **7.** 108   **8.** 6   **9.** 31   **10.** 205

Page 46   **1.** 4   **2.** 9   **3.** 1   **4.** 17   **5.** 11   **6.** 3.3   **7.** 1.3   **8.** 17.8   **9.** 11.4   **10.** 1.0   **11.** 5.86   **12.** 0.40   **13.** 14.65   **14.** 0.01   **15.** 19.44

Page 50   **1.** 34   **2.** 150   **3.** 5   **4.** 40.6   **5.** 652   **6.** 800   **7.** 13   **8.** 1.24   **9.** 0.6   **10.** 0.08   **11.** 300   **12.** 0.96   **13.** 280   **14.** 7400   **15.** 60   **16.** 1.3   **17.** 2040   **18.** 756   **19.** 20   **20.** 2   **21.** 0.62   **22.** 4.286   **23.** 450   **24.** 0.035   **25.** 6200   **26.** 8500   **27.** 40,700   **28.** 300   **29.** 70   **30.** 38,000   **31.** 4   **32.** 0.405   **33.** 0.032   **34.** 78   **35.** 3.840   **36.** 0.006

Page 54   **1.** 2   **2.** 5   **3.** 3   **4.** 7   **5.** 8   **6.** 4   **7.** 1000   **8.** 100,000   **9.** 1,000,000   **10.** 10   **11.** 10,000   **12.** 10,000,000

60   Chapter 2

ENRICHMENT

Using Scientific Notation

Space ships of the future called galactic cruisers may be able to travel at the speed of light. Light travels at the rate of 1.86×10^5 miles per second. This can also be expressed as 2.976×10^5 kilometers per second.

You can use scientific notation to multiply and divide with large numbers such as 1.86×10^5.

EXAMPLE How long would it take a galactic cruiser to travel 3.91×10^8 miles from Earth to Jupiter? Round your answer to the nearest second.

Solution: Use the distance formula.

$\boxed{1}$ $t = \dfrac{d}{r}$ $\quad d = 3.91 \times 10^8$ $\quad \boxed{2}$ $t = \dfrac{3.91 \times 10^8}{1.86 \times 10^5} = \dfrac{3.91}{1.86} \times \dfrac{10^8}{10^5}$
$\quad\quad\quad\quad\quad r = 1.86 \times 10^6$

$\boxed{3}$ Use a calculator to find $3.91 \div 1.86$.

3 . 9 1 [÷] 1 . 8 6 [=]   **2.1021505**

$\boxed{4}$ Find $\dfrac{10^8}{10^5}$. $\quad \dfrac{10^8}{10^5} = \dfrac{10^5 \times 10^3}{10^5} = \dfrac{\overset{1}{\cancel{10^5}} \times 10^3}{\underset{1}{\cancel{10^5}}} = 10^3$ $\quad\quad \dfrac{10^5}{10^5} = 1$

$\boxed{5}$ Thus, $\dfrac{3.91 \times 10^8}{1.86 \times 10^5} = 2.1021505 \times 10^3$, or **2102.1505**

It would take **2102 seconds** (Nearest second)

EXERCISES

How long would it take a galactic cruiser to travel from the Earth to each given location? Round your answer to the nearest second.

1. Sun: 1.488×10^8 kilometers
2. Pluto: 3.68×10^9 miles
3. Today, space ships travel at a speed of about 2.4×10^4 miles per hour.
   a. At this rate, how many hours would it take to travel 3.91×10^8 miles to Jupiter? Round your answer to the nearest <u>hour</u>.
   b. How many days would it take?
4. The closest star (other than the Sun) to the Earth is 4×10^{13} kilometers away. To the nearest <u>year</u>, how long would it take a galactic cruiser to travel to this star? (HINT: First find the number of seconds it would take.)

Decimals and Applications

ENRICHMENT
You may wish to use this lesson for students who performed well on the formal Chapter Test.

<u>1</u>. 500 sec
<u>2</u>. 19,785 sec
<u>3</u>. <u>a</u>. 16,292 hr
   <u>b</u>. 679 days
<u>4</u>. 4 yr

ADDITIONAL PRACTICE

You may wish to use all or some of these exercises, depending on how well students performed on the formal Chapter Test.

1. 0.8   2. 3000   3. 60
4. 0.07   5. 0.0003
6. 0.00008   7. <   8. >
9. >   10. >
11. 82.993   12. 28.3
13. 1.26   14. 0.2754
15. 0.000912
16. 819.84   17. 13.2
18. 3.5   19. 660
20. 1684   21. b   22. a
23. b   24. a
25. 4,300,000
26. 1.384   27. 14,000
28. 4.3×10^3
29. 7.7×10^4
30. 1.3×10^6
31. 1.3×10^2
32. 9×10^7 33. 310
34. 420,000
35. 35,000,000
36. 40,040   37. $.01
38. 35 miles

ADDITIONAL PRACTICE

SKILLS

Give the value of the underlined digit. (Pages 32–34)

1. 423.<u>8</u>   2. <u>3</u>071   3. 5<u>6</u>2.04   4. 4.0<u>7</u>4   5. 3.164<u>3</u>   6. 0.9371<u>8</u>

Replace each __?__ with > or <. (Pages 32–34)

7. 0.347 __?__ 0.359   8. 14.2 __?__ 14.199   9. 0.84 __?__ 0.8   10. 1.883 __?__ 0.9

11. Evaluate $a + b + c$ when $a = 37.4$, $b = 38.79$ and $c = 6.803$. (Pages 35–37)
12. Evaluate $x - y$ when $x = 37.4$ and $y = 9.1$. (Pages 35–37)

Multiply. (Pages 38–41)

13. 14(0.09)   14. (0.34)(0.81)   15. (0.012)(0.076)   16. (97.6)(8.4)

Divide. (Pages 44–46)

17. $79.2 \div 6$ 18. $2.8 \div 0.8$ 19. $429 \div 0.65$ 20. $10.104 \div 0.006$

Choose the best estimate. Choose a, b, or c. (Pages 47–49)

21. 10.3(59.8)   a. 5980   b. 600   c. 458
22. 13.21 + 6.9   a. 20   b. 19   c. 21
23. 74.9 ÷ 24.8   a. 3.5   b. 3   c. 4
24. 120.21 − 19.8   a. 100   b. 102   c. 99

Complete. (Pages 51–54)

25. 43 km = __?__ cm   26. 1384 g = __?__ kg   27. 14 L = __?__ mL

Express each number in scientific notation. (Pages 55–57)

28. 4300   29. 77,000   30. 1,300,000   31. 130   32. 90,000,000

Express as a decimal numeral. (Pages 55–57)

33. 3.1×10^2 34. 4.2×10^5 35. 3.5×10^7 36. 4.004×10^4

APPLICATIONS

37. At a cost of $0.007 per hour, how much will it cost to operate a 60-watt light bulb for 1.5 hours? (Pages 38–41)

38. A certain compact travels 497 miles on a full tank of fuel. A full tank holds 14.2 gallons. How far does it travel on one gallon? (Pages 44–46)

Chapter 2

CHAPTER 3
Number Properties/Expressions

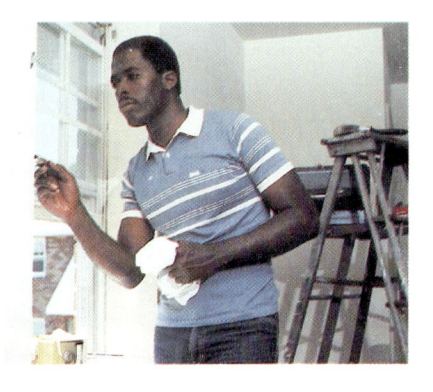

SECTIONS
3-1 The Special Numbers 0 and 1
3-2 Simplifying Algebraic Expressions
3-3 Problem Solving and Applications: Formulas: Area
3-4 Like Terms
3-5 Combining Like Terms
3-6 Problem Solving and Applications: Formulas: Perimeter

FEATURES
Calculator Application:
Using Formulas

Computer Application:
Evaluating Expressions

Enrichment:
Number Properties

Common Errors

3-1 The Special Numbers 0 and 1

Teaching Suggestions p. M-16

QUICK QUIZ

Evaluate.
1. x + 7 when x = 0
   Ans: 7
2. 7y when y = 1
   Ans: 7
3. $\frac{6a}{6}$ when a = 7
   Ans: 7
4. 8b when b = 0
   Ans: 0
5. 5h + 7 when h = 0
   Ans: 7

ADDITIONAL EXAMPLES

Example 1

Write the value of x that makes each sentence true. Name the property that gives the reason for your choice.
1. 6.3x = 6.3
   Ans: x = 1,
   Mult. Prop. of One
2. 6.3x = 0
   Ans: x = 0;
   Mult. Prop. of Zero

SELF-CHECK
1. t = 0;
   Add. Prop. of Zero
2. s = 0;
   Mult. Prop. of Zero
3. r = 58;
   Mult. Prop. of One

The numbers 0 and 1 have special properties.
The sum of any number and 0 is the number.
The product of any number and 0 is 0.
The product of any number (except 0) and 1 is the number.

Properties of 0 and 1

	Addition Property of Zero	
Arithmetic		**Algebra**
$8 + 0 = 0 + 8 = 8$		$a + 0 = 0 + a = a$
	Multiplication Property of Zero	
Arithmetic		**Algebra**
$(79.3)(0) = (0)(79.3) = 0$		$a \cdot 0 = 0 \cdot a = 0$
	Multiplication Property of One	
Arithmetic		**Algebra**
$257 \cdot 1 = 1 \cdot 257 = 257$		$a \cdot 1 = 1 \cdot a = a$

Expressions such as $8 + 0 = 0$ and $a \cdot 1 = a$ are **mathematical sentences.**

EXAMPLE 1 Write the value of the variable that makes each sentence true. Name the property that gives the reason for your choice.

	Sentence	Solution	Property
a.	$(4.81)(1) = r$	$r = 4.81$	Multiplication Property of One
b.	$p + 18.9 = 18.9$	$p = 0$	Addition Property of Zero
c.	$210 \cdot q = 0$	$q = 0$	Multiplication Property of Zero

Self-Check Write the value of the variable that makes each sentence true.

1. $t + 16 = 16$
2. $(115.6)(s) = 0$
3. $1 \cdot 58 = r$

Since $a \cdot 1 = a$, $a \div a = 1$, $a \neq 0$. ◀ ≠ means "is not equal to."

That is, **any number divided by itself equals 1.**
This is true for all numbers except 0.

Arithmetic		Algebra
$10 \div 10 = 1 \quad \frac{18}{18} = 1$		$\frac{a}{a} = 1, a \neq 0$

64 Chapter 3

EXAMPLE 2 Find the value of t that makes each sentence true.

a. $2\left(\dfrac{9}{9}\right) = t$ b. $\dfrac{23}{23}(t) = 45$ c. $6.5\left(\dfrac{t}{19}\right) = 6.5$

Solutions: a. $t = 2$ b. $t = 45$ c. $t = 19$

Self-Check Find the value of r that makes each sentence true.

4. $r = \left(\dfrac{5}{5}\right)(21)$ 5. $7\left(\dfrac{r}{15}\right) = 7$ 6. $\dfrac{10}{10}(r) = 21$

CLASSROOM EXERCISES

Name the property that makes each sentence true. (Example 1)

1. $4 + 0 = 4$
2. $1.3(1) = 1.3$
3. $m(0) = 0$
4. $3y = 0 + 3y$

For Exercises 5–20, find the value of the variable that makes each sentence true. (Example 1)

5. $12 + x = 12$
6. $(0.8)y = 0$
7. $(1)(c) = 7$
8. $m + 0 = 2.7$
9. $3.6(p) = 3.6$
10. $84(n) = 0$
11. $0.7(0) = t$
12. $124 + s = 124$

(Example 2)

13. $\dfrac{5}{5}(m) = 24$
14. $\dfrac{18}{18}(11) = x$
15. $\dfrac{3}{3}(0.3) = k$
16. $q = 1.5\left(\dfrac{8}{8}\right)$
17. $\left(\dfrac{x}{7}\right)(5) = 5$
18. $16\left(\dfrac{91}{w}\right) = 16$
19. $(2.5)\left(\dfrac{31}{31}\right) = a$
20. $\left(\dfrac{5}{5}\right)(0.2) = b$

WRITTEN EXERCISES

Goal: To apply the properties of 0 and 1

For Exercises 1–44, find the value of the variable that makes each sentence true. (Example 1)

1. $6 + n = 6$
2. $t + 11 = 11$
3. $0.45 = 0 + x$
4. $0 + y = 0.8$
5. $12y = 0$
6. $0 = 3c$
7. $(m)(3.9) = 0$
8. $0 = (n)(14.1)$
9. $225n = 225$
10. $46 = 46t$
11. $(1)(4.5) = k$
12. $2.75 = 2.75m$

(Example 2)

13. $x\left(\dfrac{3}{3}\right) = 13$
14. $15 = y\left(\dfrac{10}{10}\right)$
15. $z\left(\dfrac{7}{7}\right) = 19$
16. $28 = k\left(\dfrac{15}{15}\right)$
17. $21.3\left(\dfrac{4}{4}\right) = n$
18. $y = 0.8\left(\dfrac{7}{7}\right)$
19. $\dfrac{8}{8}(p) = 23$
20. $\left(\dfrac{2}{2}\right)39 = t$

Number Properties/Expressions **65**

WRITTEN EXERCISES

1. 0 2. 0 3. 0.45
4. 0.8 5. 0 6. 0
7. 0 8. 0 9. 1
10. 1 11. 4.5 12. 1
13. 13 14. 15
15. 19 16. 28
17. 21.3 18. 0.8
19. 23 20. 39
21. 3 22. 10 23. 5
24. 11 25. 1 26. 1
27. 0 28. 0 29. 0
30. 18 31. 1 32. 0
33. 16 34. 46 35. 0
36. 75 37. 3 38. 2
39. 7 40. 8 41. 3
42. 12 43. 13 44. 8
45. 5 46. 4 47. 0
48. 1 49. 6 50. 12
51. 36 52. 5 53. 5
54. 2

21. $16 = 16\left(\dfrac{3}{c}\right)$
22. $\left(\dfrac{z}{10}\right)21 = 21$
23. $9 = 9\left(\dfrac{5}{m}\right)$
24. $\left(\dfrac{t}{11}\right)34 = 34$

MIXED PRACTICE

25. $8.3n = 8.3$
26. $4.9 = (t)(4.9)$
27. $0.6 + r = 0.6$
28. $7.3 = x + 7.3$
29. $256t = 0$
30. $18 + 0 = v$
31. $0.9c = 0.9$
32. $93s = 0$
33. $16\left(\dfrac{8}{8}\right) = n$
34. $n = \left(\dfrac{9}{9}\right)(46)$
35. $\left(\dfrac{17}{17}\right)0 = x$
36. $\left(\dfrac{19}{19}\right)r = 75$
37. $11 = 11\left(\dfrac{n}{3}\right)$
38. $8 = 8\left(\dfrac{2}{x}\right)$
39. $\dfrac{n}{7}(21) = 21$
40. $\dfrac{x}{8}(35) = 35$
41. $n\left(\dfrac{7}{7}\right) = 3$
42. $12 = \dfrac{5}{5}(p)$
43. $\left(\dfrac{17}{17}\right)x = 13$
44. $n = \left(\dfrac{14}{14}\right)(8)$

MORE CHALLENGING EXERCISES

Write the value of the variable that makes each sentence true.

45. $25 \div 5b + b(0) = 1$
46. $(11 - 2 \cdot 5)(3x + x \cdot 0) = 12$
47. $(5 \cdot z + 5 \cdot 0)(4 + z \cdot 0) = 0$
48. $(18 \cdot k - 18 \cdot 0)(5 + k \cdot 0) = 90$
49. $12 + n\left(\dfrac{3}{3}\right) - 42(0) = 18$
50. $(14 - 3 \cdot 4)\left(\dfrac{5}{5} \cdot x\right) = 24$
51. $\left(\dfrac{r}{36} \cdot 14\right) + 39(0) = 14$
52. $45 \div 3a + \left(\dfrac{a}{3}\right)(0) = 3$
53. $\dfrac{5n}{5} + (5n \cdot 0) = 5$
54. $(5 \cdot x + 2)\left(3 + \dfrac{8}{8}\right) = 48$

REVIEW CAPSULE

This Review Capsule reviews prior-taught skills used in Section 3-2. The reference is to the pages where the skills were taught.

REVIEW CAPSULE FOR SECTION 3-2

Multiply. (Pages 18–21)

1. a. 4×5
   b. 5×4
2. a. 3×8
   b. 8×3
3. a. 6×12
   b. 12×6
4. a. $2 \times 3 \times 5$
   b. $2 \times 5 \times 3$
5. a. $4 \times (3 \times 2)$
   b. $(4 \times 3) \times 2$
6. a. $6 \times (9 \times 7)$
   b. $(9 \times 7) \times 6$
7. a. $(7 \times 10) \times 12$
   b. $(10 \times 12) \times 7$
8. a. $5 \times 7 \times 5$
   b. $5 \times 5 \times 7$
9. a. $10 \times (11 \times 5)$
   b. $11 \times (10 \times 5)$
10. a. $(3 \times 6)(4 \times 7)$
    b. $(6 \times 4)(3 \times 7)$
11. a. $(5 \times 9)(2 \times 2)$
    b. $(2 \times 2)(5 \times 9)$
12. a. $(3 \times 4)(5 \times 6)$
    b. $(3 \times 6)(5 \times 4)$

The answers are on page 89.

3-2 Simplifying Algebraic Expressions

Multiplying two numbers in either order gives the same product. This is the **Commutative Property of Multiplication**.

$312 \cdot 9 = 2808$
$9 \cdot 312 = 2808$

When you multiply three or more numbers, you may group them in any way. This is the **Associative Property of Multiplication**.

$27(4 \cdot 2) = 27 \cdot 8 = 216$
$(27 \cdot 4) \cdot 2 = (108) \cdot 2 = 216$

Properties of Multiplication

> **Commutative Property of Multiplication**
>
> *Any two numbers can be multiplied in either order.*
>
> **Arithmetic**   **Algebra**
> $4 \cdot 10 = 10 \cdot 4$ $a \cdot b = b \cdot a$
>
> **Associative Property of Multiplication**
>
> *The way three or more numbers are grouped for multiplication does not affect the product.*
>
> **Arithmetic**   **Algebra**
> $(16 \cdot 3) \cdot 4 = 16 \cdot (3 \cdot 4)$ $(a \cdot b) \cdot c = a \cdot (b \cdot c)$

EXAMPLE 1 Write the value of the variable that makes each sentence true. Name the property that gives the reason for your choice.

Sentence	Solution	Property
a. $3 \cdot 8 = 8(c)$	$c = 3$	Commutative Property of Multiplication
b. $(14 \cdot 2) \cdot 5 = 14 \cdot (2 \cdot n)$	$n = 5$	Associative Property of Multiplication
c. $(d \cdot 7) \cdot 9 = 8 \cdot (7 \cdot 9)$	$d = 8$	Associative Property of Multiplication

Self-Check *Write the value of the variable that makes each sentence true.*

1. $(2 \cdot 77) \cdot 50 = k \cdot (77 \cdot 50)$
2. $(6.9)(d) = 7(6.9)$

The Commutative and Associative Properties of Multiplication can be used to simplify algebraic expressions. Recall that there are several ways to indicate multiplication.

$5m \cdot p \qquad (5m)(p) \qquad 5m(p) \qquad (5m)p \qquad 5mp$

Number Properties/Expressions **67**

ADDITIONAL EXAMPLES

Example 2
Multiply.
1. 17(20a)   Ans: 340a
2. 17a(20b) Ans: 340ab

SELF-CHECK
3. 30x   4. 62xy
5. 33cz 6. 11xy

Example 3
Multiply
1. 5a(7ab)   Ans: $35a^2b$
2. $(12a^2)(5ab)$
   Ans: $60a^3b$

SELF-CHECK
7. $50c^2n$ 8. $30a^3b$
9. $48xy^3$

CLASSROOM EXERCISES
1. 5   2. 10   3. 30
4. 13   5. 2.5   6. 4
7. 7   8. 4   9. 3
10. 50y   11. 48t
12. 2t   13. 5.5k
14. 18xy 15. 8ab
16. 30rs 17. 8rs
18. $56a^2$ 19. $6x^2y$
20. $48m^3$ 21. $21p^2q^2r$
22. $24b^3c^2$
23. $90f^2g^2h$
24. $6r^2s^3t^2$
25. $49h^2n^3$

EXAMPLE 2   Multiply: $(5e)(8g)$

Solution: $(5e)(8g)$ means $5 \cdot e \cdot 8 \cdot g$.

[1] Change the order. Write the numbers first. $5 \cdot e \cdot 8 \cdot g = 5 \cdot 8 \cdot e \cdot g$
[2] Group the numbers and the letters. $= (5 \cdot 8) \cdot (e \cdot g)$
Multiply. $= \mathbf{40eg}$

◂ Comm. and Assoc. Properties

Self-Check   *Multiply.*

3. $3(10x)$ 4. $(15.5x)(4y)$ 5. $(3c)(11z)$ 6. $5x(2.2y)$

You can also simplify algebraic expressions involving exponents.

EXAMPLE 3   Multiply: $(3a)(2b)(7b^2)$

Solution: $(3a)(2b)(7b^2) = 3 \cdot a \cdot 2 \cdot b \cdot 7 \cdot b \cdot b$
  $= 3 \cdot 2 \cdot 7 \cdot a \cdot b \cdot b \cdot b$
  $= 42 \cdot a \cdot b^3$
  $= \mathbf{42ab^3}$

◂ $b^2 = b \cdot b$

Self-Check   *Multiply.*

7. $(10c)(5cn)$ 8. $(2a)(3b)(5a^2)$ 9. $(8x)(2y)(3y^2)$

CLASSROOM EXERCISES

Find the value of the variable that makes each sentence true. (Example 1)

1. $8 \cdot x = 5 \cdot 8$
2. $7 \cdot 10 = b \cdot 7$
3. $20(m) = 30 \cdot 20$
4. $7 \cdot 13 = (t)7$
5. $2.5(3.8) = 3.8(p)$
6. $3.8(q) = 4(3.8)$
7. $0.5(1.2y) = (0.5 \cdot 1.2)7$
8. $3(4 \cdot 5) = (3 \cdot m)5$
9. $(20 \cdot t) \cdot 5 = 20(3 \cdot 5)$

Multiply.   (Example 2)

10. $10(5y)$ 11. $4(12t)$ 12. $20(0.1t)$ 13. $(1.1)(5k)$
14. $(6x)(3y)$ 15. $(2a)(4b)$ 16. $(100r)(0.3s)$ 17. $(2r)(4s)$

(Example 3)

18. $(8a)(7a)$ 19. $(2x)(3xy)$ 20. $(3m^2)(8)(2m)$ 21. $(3pq)(qr)(7p)$
22. $(2b^2)(3bc)(4c)$ 23. $(6fg)(5gh)(3f)$ 24. $(2rt)(3s^2)(rst)$ 25. $(7n)(kn)(7kn)$

Chapter 3

WRITTEN EXERCISES

Goal: To simplify algebraic expressions that involve multiplication

Find the value of the variable that makes each sentence true. (Example 1)

1. $7 \cdot 6 = 6(a)$
2. $x(9) = 9 \cdot 4$
3. $(11 \cdot 2)7 = 11(n \cdot 7)$
4. $13(r \cdot 6) = (13 \cdot 4)6$
5. $3.4n = (6.2)(3.4)$
6. $(8.9)(6.2) = 6.2a$
7. $(4 \cdot 3.8)k = 4(3.8 \cdot 5)$
8. $3(t \cdot 7) = 7(15 \cdot 3)$
9. $(8.4)(1.5)x = 3(1.5)(8.4)$

Multiply. (Example 2)

10. $5(7x)$
11. $6(4y)$
12. $7(6a)$
13. $9(10b)$
14. $(8t)7$
15. $(1.2n)7$
16. $(2.3k)4$
17. $(12r)8$
18. $(9a)(7b)$
19. $(8s)(12t)$
20. $(1.3x)(3y)$
21. $(6p)(1.4q)$
22. $(5r)(20s)$
23. $(3.2a)(6b)$
24. $(56p)(2r)$
25. $(1.9r)(3.2s)(t)$

(Example 3)

26. $(13x)(3x)$
27. $(6y)(14y)$
28. $(12mn)(3n)$
29. $(3k)(12kt)$
30. $(9x)(4xy)$
31. $(7m)(6m^2n)$
32. $(10p)(0.2np)$
33. $(15st)(0.4t^2)$
34. $(3x^2)(8xy)$
35. $(20mn^2)(4mn)$
36. $(5k)(2kt)(3t^2)$
37. $(9dp)(3a^2p)(3d)$

MIXED PRACTICE

38. $(2x)(9y)$
39. $(6m)(12n)$
40. $7(12t)$
41. $11(27r)$
42. $(0.5x^2)(18xy)$
43. $(12m)(0.5mn^2)$
44. $(15w)0.5$
45. $(28a)(0.2)$
46. $(2r)(3s)(st)$
47. $(4a)(3b)(2c)$
48. $(rs)(2st)(3r^2)$
49. $(2mn)(m^2p)(3p)$

MORE CHALLENGING EXERCISES

Multiply. Then evaluate the expression when $x = 2$ and $y = 1$.

50. $(x^2)(xy)(3y)$
51. $(2xy)(4x)(3y)$
52. $x \cdot x \cdot y \cdot y^2 \cdot x \cdot y$
53. $y \cdot y \cdot x \cdot x \cdot y^3$
54. $(2x)(3xy)(4y^2)$
55. $(5y)(2x^2y)(9xy^3)$

REVIEW CAPSULE FOR SECTION 3-3

Evaluate each expression. (Pages 7–9)

1. xy when $x = 7$ and $y = 4$
2. ab when $a = 19$ and $b = 21$
3. mx when $m = 6.2$ and $x = 3.8$
4. rs when $r = 3.6$ and $s = 1.2$
5. $n \cdot n$ when $n = 6$
6. $a \cdot a$ when $a = 13.4$
7. x^2 when $x = 15$
8. s^2 when $s = 8.04$

The answers are on page 89.

Number Properties/Expressions **69**

ASSIGNMENT GUIDE
BASIC
Day 1 p. 69: 1-25
Day 2 p. 69: 26-45
AVERAGE
p. 69: 1-49 odd
ABOVE AVERAGE
p. 69: 3, 6, 9, ..., 48, 50-55

WRITTEN EXERCISES
1. a = 7   2. x = 4
3. n = 2   4. r = 4
5. n = 6.2   6. a = 8.9
7. k = 5   8. t = 15
9. x = 3   10. 35x
11. 24y   12. 42a
13. 90b   14. 56t
15. 8.4n   16. 9.2k
17. 96r   18. 63ab
19. 96st   20. 3.9xy
21. 8.4pq   22. 100rs
23. 19.2ab   24. 112pr
25. 6.08rst   26. $39x^2$
27. $84y^2$ 28. $36mn^2$
29. $36k^2t$ 30. $36x^2y$
31. $42m^3n$ 32. $2np^2$
33. $6st^3$ 34. $24x^3y$
35. $80m^2n^3$ 36. $30k^2t^3$
37. $81a^2d^2p^2$ 38. 18xy
39. 72mn   40. 84t
41. 297r   42. $9x^3y$
43. $6m^2n^2$ 44. 7.5w
45. 5.6a   46. $6rs^2t$
47. 24abc   48. $6r^3s^2t$
See page 70 for the answers to Exercises 49-55.

Teaching Suggestions
p. M-16

QUICK QUIZ
Multiply.
1. (23)(27) Ans: 621
2. 18^2 Ans: 324
3. (7.8)(12.3)
   Ans: 95.94
4. $(6.1)^2$ Ans: 37.21
Evaluate.
5. hk when h = 9
   and k = 32 Ans: 288

ADDITIONAL EXAMPLES
Example 1
Find the area of
each rectangle.
1. l = 5 ft; w = 3 ft
   Ans: 15 ft^2
2. l = 17 m; w = 8 m
   Ans: 136 m^2

SELF-CHECK
1. 1.25 m^2

WRITTEN EXERCISES p. 69
49. $6m^3np^2$
50. $3x^3y^2$; 24
51. $24x^2y^2$; 96
52. x^3y^4; 8
53. x^2y^5; 4
54. $24x^2y^3$; 96
55. $90x^3y^5$; 720

70

PROBLEM SOLVING AND APPLICATIONS

3-3 Formulas: Area

Each side of this figure is 1 centimeter long. The area of the figure is 1 · 1 or **1 square centimeter (1 cm²)**. **Area is the number of square units needed to cover a surface.**

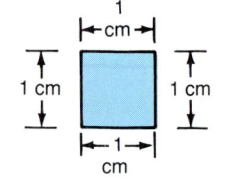

A rectangle has four right angles.
Opposite sides of a rectangle are equal.

This word rule and formula tell how to find the area of a rectangle.

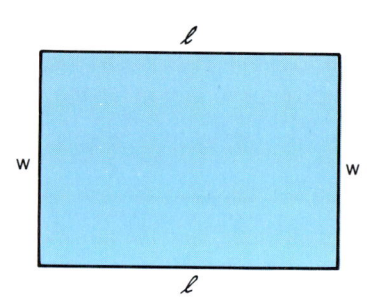

Word Rule: To find the area of a rectangle, multiply the length and the width.

Formula: $A = lw$

◁ A = area
l = length; w = width

Small areas are measured in square centimeters (cm²) or square inches (in²). Larger areas are measured in square meters (m²), square miles (mi²), square yards (yd²), or square feet (ft²).

EXAMPLE 1 A soccer field is 119 meters long and 91 meters wide. Find the area.

Solution:
1. Write the formula. $A = lw$
2. Identify known values. $l = 119$ m; $w = 91$ m
3. Replace the variables. Solve. $A = 119(91)$
   $A = 10,829$ The area is **10,829 m²**.

Self-Check 1. A desk is 1.25 meters long and 1 meter wide. Area = ___?___

A **square** is a rectangle with four equal sides. You can use this word rule and formula to find the area of a square.

Word Rule: The area of a square equals the product of the lengths of any two sides.

Formula: $A = s^2$

◁ A = area
s = length of a side

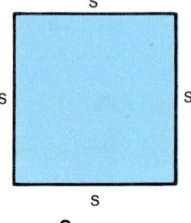

Square

70 Chapter 3

EXAMPLE 2 Find the area of this square storm-warning flag.

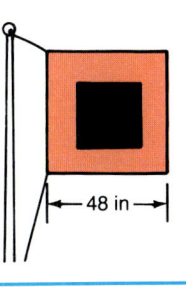

Solution:
1 Formula: $A = s^2$
2 Known value: $s = 48$
3 Solve. $A = (48)(48)$
$A = 2304$

The area of the flag is **2304 in²**.

Self-Check 2. Each side of a square rug is 3 yards long. Find the area of the rug.

CLASSROOM EXERCISES

Find the area of each rectangle. (Example 1)

1. $l = 10$ cm; $w = 5$ cm
2. $l = 5$ m; $w = 3$ m
3. $l = 10$ km; $w = 0.5$ km
4. $l = 13$ ft; $w = 10$ ft
5. $l = 7$ in; $w = 3$ in
6. $l = 5$ yd; $w = 2$ yd

Find the area of each square. (Example 2)

7. $s = 8$ cm
8. $s = 12$ km
9. $s = 0.5$ m
10. $s = 1.2$ m
11. $s = 15$ in
12. $s = 12$ yd
13. $s = 25$ ft
14. $s = 8$ mi

WRITTEN EXERCISES

Goals: To find the area of a rectangle and of a square
To apply the skill of finding area to solving word problems

Find the area of each rectangle. (Example 1)

METRIC MEASURES

1. **Gym Floor** — 42 m, 27 m
2. **Dollar Bill** — 6.5 cm, 15.5 cm
3. **Swimming Pool** — 3 m, 4 m

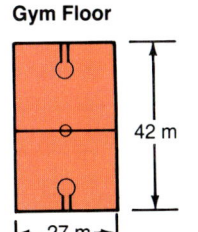

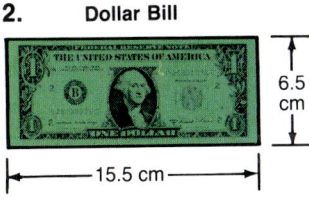

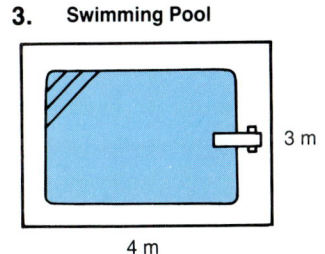

ADDITIONAL EXAMPLES

Example 2
Find the area of each square.

1. $s = 90$ ft
   Ans: 8100 ft²

2. $s = 0.51$ m
   Ans: 0.2601 m²

SELF-CHECK

2. 9 yd²

CLASSROOM EXERCISES

1. 50 cm²   2. 15 m²
3. 5 km²   4. 130 ft²
5. 21 in²   6. 10 yd²
7. 64 cm²   8. 144 km²
9. 0.25 m² 10. 1.44 m²
11. 225 in² 12. 144 yd²
13. 625 ft² 14. 64 mi²

ASSIGNMENT GUIDE

BASIC
Day 1 pp. 71-72: 1-18
Day 2 pp. 72-73: 19-36
AVERAGE
pp. 71-73: 1-39 odd
ABOVE AVERAGE
pp. 71-73: 1-33 odd, 35-44

WRITTEN EXERCISES

1. 1134 m²
2. 100.75 cm²
3. 12 m²

Number Properties/Expressions **71**

WRITTEN EXERCISES

4. 12 cm^2 5. 3 m^2
6. 23 m^2 7. 36.4 mm^2
8. 48 km^2 9. 21 m^2
10. 35 in^2 11. 308 yd^2
12. 15 ft^2 13. 104 in^2
14. 24 ft^2 15. 48 yd^2
16. 896 in^2 17. 3 mi^2
18. 216 ft^2 19. 961 m^2
20. 408.04 mm^2
21. 1.21 m^2 22. 16 cm^2
23. 100 m^2
24. 38.44 m^2
25. 190.44 mm^2

Find the area of each rectangle. (Example 1)

	Length	Width	Area
4.	4 cm	3 cm	?
6.	5 m	4.6 m	?
8.	8 km	6 km	?

	Length	Width	Area
5.	2 m	1.5 m	?
7.	13 mm	2.8 mm	?
9.	7.5 m	2.8 m	?

CUSTOMARY MEASURES

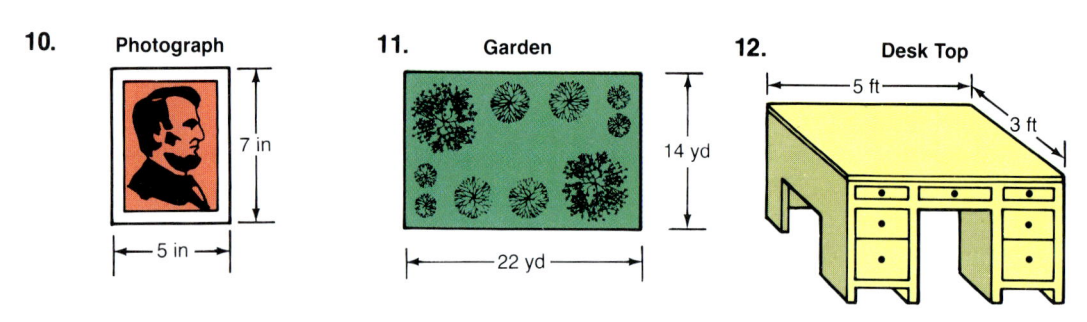

10. Photograph — 7 in by 5 in
11. Garden — 14 yd by 22 yd
12. Desk Top — 5 ft by 3 ft

	Length	Width	Area
13.	13 in	8 in	?
15.	8 yd	6 yd	?
17.	3 mi	1 mi	?

	Length	Width	Area
14.	6 ft	4 ft	?
16.	32 in	28 in	?
18.	12 ft	18 ft	?

Find the area of each square. (Example 2)

METRIC MEASURES

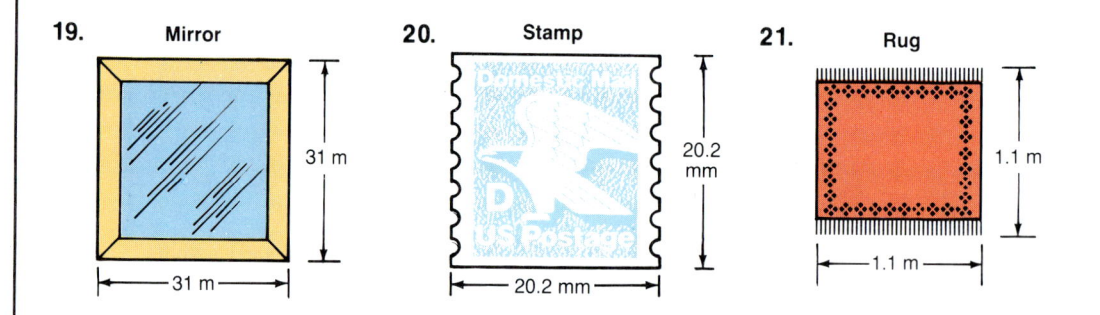

19. Mirror — 31 m by 31 m
20. Stamp — 20.2 mm by 20.2 mm
21. Rug — 1.1 m by 1.1 m

	Length of Side	Area
22.	4 cm	?
24.	6.2 m	?

	Length of Side	Area
23.	10 m	?
25.	13.8 mm	?

72 Chapter 3

CUSTOMARY MEASURES

26. Road Sign — 15 ft, 15 ft

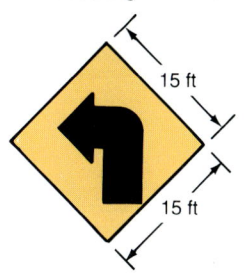

27. City Square — 2 mi × 2 mi

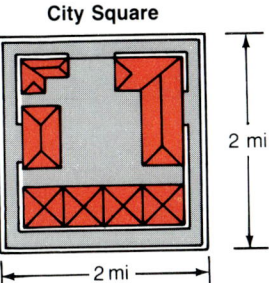

28. Baseball Diamond — 30 yd, 30 yd

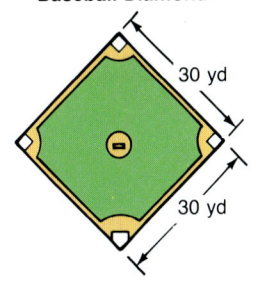

	Length of Side	Area			Length of Side	Area
29.	3 ft	?		**30.**	16 in	?
31.	16 yd	?		**32.**	23 ft	?
33.	4 mi	?		**34.**	5 yd	?

MIXED PRACTICE

35. A movie screen is 125 centimeters long and 100 centimeters wide. Find the area of the screen.

36. Each base of the twin towers in New York City's World Trade Center is a square. Each base measures 62.7 meters on a side. Find the area of one base.

37. A rectangular swimming pool is 164 feet long and 80 feet wide. Find the area of the pool.

38. The top of a rectangular table is 1 meter long and 0.8 meter wide. Find the area of the table top.

39. The length of each side of a square box is 15 inches. Find the area of the top of the box.

40. The length of a tennis court is 21.95 meters and the width is 11.4 meters. Find the area.

41. A farmer's field is 83.5 meters long and 250.4 meters wide. What is the area of the field?

42. A piece of carpet is 12 yards wide and 12 yards long. Find the area of the carpet.

43. A room has the shape of a rectangle. It is 3.1 meters long and 2.9 meters wide. Find the area.

44. A standard sheet of plywood is 48 inches wide and 96 inches long. What is the area of the sheet?

---

26. 225 ft^2 **27.** 4 mi^2
28. 900 yd^2 **29.** 9 ft^2
30. 256 in^2 **31.** 256 yd^2
32. 529 ft^2 **33.** 16 mi^2
34. 25 yd^2
35. 12,500 cm^2
36. 3931.29 m^2
37. 13,120 ft^2
38. 0.8 m^2 **39.** 225 in^2
40. 250.23 m^2
41. 20,908.4 m^2
42. 144 yd^2
43. 8.99 m^2
44. 4608 in^2

Number Properties/Expressions **73**

QUIZ: SECTIONS 3-1-3-3
After completing this Review, you may wish to administer a quiz covering the same sections. A Quiz is provided in the Teacher's Edition: Part II.

REVIEW: SECTIONS 3-1-3-3
1. Mult. Prop. of Zero
2. Add. Prop. of Zero
3. Mult. Prop. of One
4. $t = 3.8$ 5. $r = 32$
6. $x = 0$
7. Assoc. Prop. of Mult.
8. Comm. Prop. of Mult.
9. Assoc. Prop. of Mult.
10. $x = 8$ 11. $g = 6$
12. $y = 9$ 13. $24y$
14. $24t$ 15. $12a^2b$
16. 5670.09 m^2
17. 1008 in^2
18. 2914 cm^2
19. 324 in^2

REVIEW: SECTIONS 3-1-3-3

Name the property illustrated in each sentence. (Section 3-1)

1. $2.7(0) = 0$
2. $x + 0 = x$
3. $(4.52)(1) = 4.52$

Find the value of the variable that makes each sentence true. (Section 3-1)

4. $3.8 + 0 = t$
5. $\left(\frac{12}{12}\right)r = 32$
6. $5x = 0$

Name the property illustrated in each sentence. (Section 3-2)

7. $2(10 \cdot 8) = (2 \cdot 10)8$
8. $3 \cdot x = x \cdot 3$
9. $(26 \cdot 4)5 = 26(4 \cdot 5)$

Find the value of the variable that makes each sentence true. (Section 3-2)

10. $4.2(3.5 \cdot x) = (4.2 \cdot 3.5)8$
11. $(8.7)(g) = 6(8.7)$
12. $(15 \cdot 21)y = 15(21 \cdot 9)$

Multiply. (Section 3-3)

13. $6(4y)$
14. $(8t)(3)$
15. $(3a)(4ab)$

16. Each side of a square parking lot is 75.3 meters long. Find the area of the parking lot. (Section 3-3)

17. A rectangular drawing board is 36 inches long and 28 inches wide. Find the area of the drawing board. (Section 3-3)

18. A television screen is 62 centimeters long and 47 centimeters wide. What is the area of the screen? (Section 3-3)

19. Eric and Beth put together a square jig-saw puzzle. The puzzle is 18 inches on each side. Find the area of the puzzle. (Section 3-3)

REVIEW CAPSULE FOR SECTION 3-4

Add or subtract as indicated. (Pages 35-37)

1. $3.2 + 1.8$
2. $4.2 + 0.5$
3. $9.7 + 3$
4. $5 + 8.9$
5. $10.6 - 4.9$
6. $13.4 - 8.5$
7. $7 - 0.6$
8. $5 - 3.1$
9. $31.1 - 2.2$
10. $11 - 3.2$
11. $16 + 2.6$
12. $4.3 + 5.7$

The answers are on page 89.

Chapter 3

COMPUTER APPLICATIONS
Evaluating Expressions

These are the symbols for arithmetic operations in BASIC.

Exponentiate means "raise to a power."

Operation	Symbol	Arithmetic	BASIC
Add.	+	9 + 7	9 + 7
Subtract.	−	8 − 3	8 − 3
Multiply.	*	6 × 5	6 * 5
Divide.	/	7 ÷ 2	7 / 2
Exponentiate.	↑, or ^	3^2	3 ↑ 2

When two or more operations appear in a BASIC expression or formula, the computer follows this order of operations.

1. Exponentiation (raising to a power) is done first.
2. Next multiplication and division are done from left to right.
3. Then addition and subtraction are done left to right.
4. When there are parentheses, do the operations inside the parentheses first, beginning with the innermost parentheses. Thus, the order listed in 1–3 is changed.

The computer will evaluate an expression if you type it as a direct command, such as **PRINT**.

EXAMPLE Type each direct command into a computer.
    a. PRINT 20 − 3 * 6   **b.** PRINT 5 ↑ 2 * (4 + 2)

Output: **a.** 2   **b.** 150

Exercises

Use a computer to evaluate each expression.

1. PRINT 4 + 5
2. PRINT 2 * 8 + 3
3. PRINT 3*(9 − 4)
4. PRINT 30 − 5 * 6
5. PRINT 5*11/2
6. PRINT 40/2*3
7. PRINT (14 + 7)/3
8. PRINT 14/(9 − 5)
9. PRINT 50/5*2
10. PRINT 50/(5*2)
11. PRINT 6 ↑ 2
12. PRINT (8 − 1) ↑ 2
13. PRINT 100 − 4 ↑ 2
14. PRINT 2 ↑ 3
15. PRINT (7 + 5)*9

COMPUTER APPLICATIONS
This feature illustrates the use of a computer program for evaluating a numerical expression. It applies the skills taught in Section 1-6 (pages 18-21) to writing expressions in the BASIC language.

1. 9
2. 19
3. 15
4. 0
5. 27.5
6. 60
7. 7
8. 3.5
9. 20
10. 5
11. 36
12. 49
13. 84
14. 8
15. 108

Computer Application

Teaching Suggestions
p. M-16

QUICK QUIZ
Add or subtract.
1. 23 + 18   Ans: 41
2. 102 − 46   Ans: 56
3. 7.9 − 3.2   Ans: 4.7
4. 16.2 + 5   Ans: 21.2
5. 30.71 − 5.9
   Ans: 24.81

ADDITIONAL EXAMPLES
Example 1
Does each pair show like terms? Answer *Yes* or *No*.
1. 16x and $5x^2$ Ans: No
2. $9a^2b$ and $7a^2b$
   Ans: Yes

SELF-CHECK
1. Yes   2. Yes
3. No   4. No

3-4 Like Terms

A number such as 257 can be represented by this algebraic expression.

$$2h + 5t + 7o$$

This expression contains three terms: $2h$, $5t$, and $7o$. Since the terms have different variables, h, t, and o, the terms are unlike.

Hundreds	+	Tens	+	Ones
2h	+	5t	+	7o

Like terms have the *same variable* and the *same exponent* for each variable.

Two or more **like terms** have	**Like:** $5a$ and $7a$
1. the *same* variables	**Like:** $2x^2$ and $9x^2$
2. the *same powers* of these variables.	**Unlike:** $5a$ and $3a^2$

EXAMPLE 1 For each term in List 1 select one or more like terms from List 2.

List 1
1. $0.5d$
2. $6rs$
3. $2x^2$
4. $3xy$

List 2
a. $4rs$ b. $30x^2$
c. $1.5r$ d. $0.9xy$
e. $2s$ f. $18d$
g. $11d$ h. $1.8xy$

Solutions: 1. f, g   2. a   3. b   4. d, h

Self-Check Which pairs show like terms? Answer *Yes* or *No*.
1. $3x$ and $1.5x$
2. $6rt$ and $21rt$
3. $9r^2$ and $5s^2$
4. $0.5d$ and $1.5cd$

You can add or subtract like terms. This is called **combining like terms**. In the expression $5x^2$, 5 is the **numerical coefficient** of x^2.

PROCEDURE
To add or subtract like terms:
[1] Add or subtract the numerical coefficients.
[2] Write the same variable(s) and exponent(s).

Chapter 3

EXAMPLE 2 Combine like terms: $5x + 3.1x$

Solution:
1. Add the numerical coefficients. $\quad 5 + 3.1 = 8.1$
2. Use the same variable(s) and exponent(s). $\quad 5x + 3.1x = \mathbf{8.1x}$

Self-Check *Combine like terms.*

5. $3y + 7y$ **6.** $4rs + 9rs$ **7.** $6.1z^2 + 3.4z^2$

NOTE: $13n^2 - n^2$ is the same as $13n^2 - 1n^2$.

EXAMPLE 3 Combine like terms: $13n^2 - n^2$

Solution:
1. Subtract the numerical coefficients. $\quad 13 - 1 = 12$
2. Use the same variable(s) and exponent(s). $\quad 13n^2 - n^2 = \mathbf{12n^2}$

Self-Check *Combine like terms.*

8. $15p - 9p$ **9.** $13r - r$ **10.** $5.9t^2 - 1.8t^2$

CLASSROOM EXERCISES

Identify each pair of terms as like terms or unlike terms. Give a reason for each answer. (Example 1)

1. $3x$ and $4x$ **2.** $6m$ and $3r$ **3.** $2.4y$ and $3.5y^2$ **4.** $8b$ and $5b$
5. $7xy$ and xy **6.** $4c^2$ and $2.2c^2d$ **7.** $4.2b$ and $2ab$ **8.** $5a^2$ and $6a^2$

Add or subtract the numerical coefficients in each expression. (Step 1, Examples 2 and 3)

9. $7m + 15m$ **10.** $13r - 5r$ **11.** $0.9t - 0.4t$ **12.** $4.8x + 1.3x$
13. $5k - 4k$ **14.** $26m^2 + 17m^2$ **15.** $32z^2 - 19z^2$ **16.** $80ab + 23ab$

For Exercises 17–32, combine like terms. (Example 2)

17. $8x + 3x$ **18.** $6a + 2a$ **19.** $0.3d + 0.6d$ **20.** $1.2t + 2.5t$
21. $13r^2 + 8r^2$ **22.** $9m^2 + 22m^2$ **23.** $15ad + 19ad$ **24.** $bc + 99bc$

(Example 3)

25. $9y - 4y$ **26.** $12x - x$ **27.** $40b - 15b$ **28.** $19g - 6g$
29. $4.5m^2 - 1.2m^2$ **30.** $5.9t^2 - 1.3t^2$ **31.** $35st - 15st$ **32.** $80jk - 79jk$

Number Properties/Expressions **77**

ADDITIONAL EXAMPLES

Example 2
Combine like terms.
1. $16ab + 7ab$ Ans: $23ab$
2. $12.6y^2 + 3.7y^2$
   Ans: $16.3y^2$

SELF-CHECK
5. $10y$ 6. $13rs$
7. $9.5z^2$

Example 3
Combine like terms.
1. $20x - x$ Ans: $19x$
2. $500a - 27.3a$
   Ans: $472.7a$

SELF-CHECK
8. $6p$ 9. $12r$
10. $4.1t^2$

CLASSROOM EXERCISES
1. like; same variable
2. unlike;
   different variable
3. unlike;
   different exponents
4. like; same variables
5. like; same variables
6. unlike;
   different variables
7. unlike;
   different variables
8. like; same variables

See page 78 for the answers to Exercises 19-30.

CLASSROOM EXERCISES p. 77

<u>9</u>. 22 <u>10</u>. 8 <u>11</u>. 0.5
<u>12</u>. 6.1 <u>13</u>. 1 <u>14</u>. 43
<u>15</u>. 13 <u>16</u>. 103
<u>17</u>. 11x <u>18</u>. 8a
<u>19</u>. 0.9d <u>20</u>. 3.7t
<u>21</u>. 21r^2 <u>22</u>. 31m^2
<u>23</u>. 34ad <u>24</u>. 100bc
<u>25</u>. 5y <u>26</u>. 11x
<u>27</u>. 25b <u>28</u>. 13g
<u>29</u>. 3.3m^2 <u>30</u>. 4.6t^2
<u>31</u>. 20st <u>32</u>. jk

ASSIGNMENT GUIDE
BASIC
Day 1 p. 78: 1-22
Day 2 p. 78: 23-40
AVERAGE
p. 78: 1-10, 11-41 odd
ABOVE AVERAGE
p. 78: 1-41 odd

WRITTEN EXERCISES

<u>1</u>. H <u>2</u>. B <u>3</u>. D, R
<u>4</u>. E, S <u>5</u>. O <u>6</u>. A
<u>7</u>. C <u>8</u>. M <u>9</u>. Q
<u>10</u>. N, T <u>11</u>. 16a
<u>12</u>. 14m <u>13</u>. 13b
<u>14</u>. 10r <u>15</u>. 39y
<u>16</u>. 50n <u>17</u>. 3.9x
<u>18</u>. 5.7y <u>19</u>. 12m^2
<u>20</u>. 16b^2 <u>21</u>. 14xy
<u>22</u>. 10ab <u>23</u>. 7s
<u>24</u>. 6t <u>25</u>. 5a <u>26</u>. 8x
<u>27</u>. 5.9t <u>28</u>. 4.5b

See page 83 for the answers to Exercises 29-42.

WRITTEN EXERCISES

Goals: To identify like terms
To combine like terms

For each expression in Exercises 1–10, select one or more like terms from A–T.

1. $5x$	2. $3a$	A. $4y^2$	B. $12a$	C. $3rs$	D. $5y$
3. $2y$	4. x^2	E. $2x^2$	F. r^2t	G. $2mn$	H. $3x$
5. $0.7m$	6. y^2	I. $9ab$	J. $4r$	K. $2x^2y$	L. $0.7b$
7. rs	8. $7t^2$	M. $0.4t^2$	N. $6xy$	O. $3.2m$	P. $3xy^2$
9. $1.8a^2$	10. xy	Q. $3a^2$	R. $8y$	S. $8x^2$	T. $3.5xy$

For Exercises 11–34, combine like terms. (Example 2)

11. $7a + 9a$ 12. $6m + 8m$ 13. $12b + b$ 14. $r + 9r$
15. $35y + 4y$ 16. $41n + 9n$ 17. $2.4x + 1.5x$ 18. $3.2y + 2.5y$
19. $6m^2 + 6m^2$ 20. $7b^2 + 9b^2$ 21. $8xy + 6xy$ 22. $8.4ab + 1.6ab$

(Example 3)

23. $12s - 5s$ 24. $13t - 7t$ 25. $6a - a$ 26. $9x - x$
27. $7.8t - 1.9t$ 28. $4.9b - 0.4b$ 29. $7a - 3.2a$ 30. $9s - 2.5s$
31. $4x^2 - 2x^2$ 32. $5.7m^2 - 2.9m^2$ 33. $6ab - 3ab$ 34. $1.4xy - 0.9xy$

MIXED PRACTICE

Combine like terms.

35. $16n - 14n$ 36. $24a^2 - 9a^2$ 37. $22st + 23st$ 38. $19r^2 - 8r^2$
39. $5.3f^2 + 2.9f^2$ 40. $6cp - 5cp$ 41. $3.83m + 12.19m$ 42. $122.8p^2 - 23.6p^2$

REVIEW CAPSULE FOR SECTION 3-5

Add. (Pages 35–37)

1. a. $12 + 9$
   b. $9 + 12$
2. a. $4.6 + 9.8$
   b. $9.8 + 4.6$
3. a. $12 + 7.8$
   b. $7.8 + 12$
4. a. $12.6 + 5.9 + 7.4$
   b. $7.4 + 5.9 + 12.6$
5. a. $26 + 18 + 35$
   b. $18 + 26 + 35$
6. a. $9 + 13 + 17$
   b. $9 + 17 + 13$

The answers are on page 89.

78 Chapter 3

3-5 Combining Like Terms

Properties of addition are similar to those of multiplication.

Properties of Addition

Commutative Property of Addition

Any two numbers can be added in either order.

Arithmetic	Algebra
$3.5 + 1.9 = 1.9 + 3.5$	$a + b = b + a$

Associative Property of Addition

The way numbers are grouped for addition does not affect the sum.

Arithmetic	Algebra
$(7 + 8) + 15 = 7 + (8 + 15)$	$(a + b) + c = a + (b + c)$

EXAMPLE 1 Write the value of the variable that makes each sentence true. Name the property or properties that give the reason for your choice.

	Sentence	Solution	Property
a.	$(6 + 11) + w = 6 + (11 + 8)$	$w = 8$	Associative Property of Addition
b.	$5.1 + 7.6 = 7.6 + p$	$p = 5.1$	Commutative Property of Addition
c.	$16 + (9 + 4) = (16 + n) + 9$	$n = 4$	Commutative and Associative Properties of Addition

Self-Check Write the value of the variable that makes each sentence true.

1. $16.9 + q = 11.5 + 16.9$
2. $3 + (1 + 5) = (t + 1) + 5$

The Commutative and Associative Properties can be used to simplify algebraic expressions.

EXAMPLE 2 Combine like terms: $15rx + 92 + 36rx$

Solution:

1. Change the order. Write the like terms first. $15rx + 36rx + 92$ ◄ **Commutative Property**

2. Group like terms and add. $(15rx + 36rx) + 92$ ◄ **Associative Property**
   $51rx + 92$

Number Properties/Expressions 79

ADDITIONAL EXAMPLE
Example 2
2. 9.2 + 7.3y + 0.2y + 17
   Ans: 26.2 + 7.5y

SELF-CHECK
3. 17a + 7   4. 11t
5. 3.9x + 13

CLASSROOM EXERCISES
1. 11; Comm. and Assoc. Prop. of Add.
2. 50; Assoc. Prop. of Add.
3. 5; Comm. Prop. of Add.
4. 4.8; Comm. and Assoc. Prop. of Add..
5. 52r + 10   6. 19b + 5
7. 42st + 25   8. 9x
9. 1.5t + 1.8
10. 2.3s + 25
11. 87n² + 9   12. 12a²

ASSIGNMENT GUIDE
BASIC
p. 80: 1–22
AVERAGE
p. 80: 1–7 odd, 8–26
ABOVE AVERAGE
p. 80: 1–7 odd, 8–30

WRITTEN EXERCISES
1. r = 2.7   2. g = 3.5
3. x = 4   4. b = 8
5. x = 6   6. m = 15
See page 84 for the answers to Exercises 7–30.

Self-Check *Combine like terms.*
3. $4a + 7 + 13a$ 4. $6t + 4t + t$ 5. $1.8x + 13 + 2.1x$

CLASSROOM EXERCISES

Write the value of the variable that makes each sentence true. Name the property or properties that give the reason for your choice. (Example 1)

1. $7 + (9 + 11) = (9 + 7) + r$ 2. $(c + 100) + 25 = 50 + (100 + 25)$
3. $(3 + 4) + x = 5 + (3 + 4)$ 4. $(3.2 + 1.5) + 4.8 = (b + 1.5) + 3.2$

Combine like terms. (Example 2)

5. $36r + 10 + 16r$ 6. $12b + 5 + 7b$ 7. $14st + 25 + 28st$ 8. $3x + 2x + 4x$
9. $0.6t + 1.8 + 0.9t$ 10. $1.4s + 25 + 0.9s$ 11. $71n^2 + 9 + 16n^2$ 12. $4a^2 + 3a^2 + 5a^2$

WRITTEN EXERCISES

Goal: To simplify algebraic expressions that involve addition of more than two terms

Write the value of the variable that makes each sentence true. (Example 1)

1. $2.7 + 11.9 = 11.9 + r$ 2. $4.5 + q = 3.5 + 4.5$
3. $(7 + 9) + x = 7 + (4 + 9)$ 4. $(12 + 15) + 8 = 12 + (b + 15)$
5. $(41 + x) + 90 = (41 + 90) + 6$ 6. $(m + 14) + 25 = 15 + (14 + 25)$

Combine like terms. (Example 2)

7. $3a + 2a + 4$ 8. $8x + 3x + 9$ 9. $12a + 15 + 17a$ 10. $23t + 9 + 18t$
11. $0.9m + 4.2 + 3m$ 12. $8.7r + 5 + 2.8r$ 13. $21k + 7 + 34k$ 14. $33n + 16 + 39n$
15. $3x + 6x + 9x$ 16. $4a + 2a + 8a$ 17. $1.6t + 3.8t + 4.1t$ 18. $8.5p + 0.4p + 9.1p$
19. $7a^2 + 4 + 6a^2$ 20. $5x^2 + 6 + 12x^2$ 21. $9xy + 4 + xy$ 22. $1.2mn + 3.8mn + 2.9$
23. $3p + 1.9 + 0.9p$ 24. $3x^2 + 4x^2 + 9x^2$ 25. $7p + 8p + 16$ 26. $3.2 + 8.5t + 4.9t$

MORE CHALLENGING EXERCISES

Combine like terms.

27. $24r + 19r + r + 29r + 18r$ 28. $43a^2 + 6a^2 + 9 + 8a^2 + 15$
29. $77 + 114p + 98p + 113 + 42p$ 30. $112.1z^2 + 31.6z^2 + 0.9 + 101.8z^2$

80   Chapter 3

REVIEW CAPSULE FOR SECTION 3-6

Evaluate each expression. (Pages 22-24)

1. $x + y + z$ when $x = 4$, $y = 3$, and $z = 5$
2. $2x + 2y$ when $x = 3$ and $y = 8$
3. $m + r + s$ when $m = 8.4$, $r = 12$, and $s = 9.5$
4. $4a$ when $a = 9.7$
5. $2a + 2b$ when $a = 5.4$ and $b = 4.2$
6. $4x$ when $x = 112$

The answers are on page 89.

PROBLEM SOLVING AND APPLICATIONS

3-6 Formulas: Perimeter

The **perimeter** of a figure is the distance around it. To find perimeter, add the lengths of the sides.

The following word rule and formula tell how to find the perimeter of any rectangle.

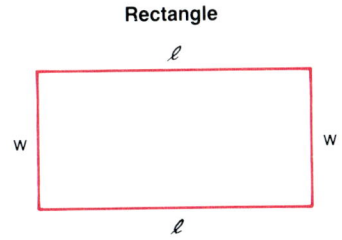

Rectangle

$P = l + l + w + w$
$P = 2l + 2w$

Word Rule: The perimeter of a rectangle equals twice the length plus twice the width.

Formula: $P = 2l + 2w$

P = perimeter
l = length
w = width

Perimeter is expressed in linear units such as kilometers (km), meters (m), centimeters (cm), miles (mi), yards (yd), feet (ft), and so on.

EXAMPLE 1 The sign at the right is 25 centimeters wide and 43.8 centimeters long. Find the perimeter.

Solution:

1. Write the formula. $P = 2l + 2w$
2. Identify known values. $l = 43.8$ cm; $w = 25$ cm
3. Replace the known values and solve.
$P = 2(43.8) + 2(25)$
$P = 87.6 + 50$
$P = 137.6$

The perimeter is **137.6 centimeters**.

Self-Check   *Find the perimeter of each rectangle.*

1. $l = 24$ m; $w = 17$ m
2. $l = 22$ ft; $w = 15$ ft

Number Properties/Expressions

This word rule tells how to find the perimeter of any triangle.

Word Rule: The perimeter of a triangle equals the sum of the lengths of the sides.

Formula: $P = a + b + c$

P = perimeter
a, b, c = lengths of the sides

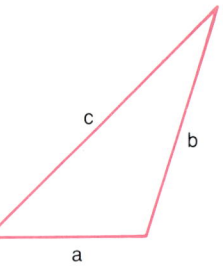

EXAMPLE 2 The course for a boat race is triangular in shape. One leg of the race is 50 miles long. The second leg is 46 miles long and the third leg is 62 miles long. Find the total length of the race.

Solution: Since the course has the shape of a triangle, use the formula for the perimeter of a triangle.

1. Formula: $P = a + b + c$
2. Known values: $a = 50; b = 46; c = 62$
3. $P = 50 + 46 + 62$
   $P = 158$ The total length of the race is **158 miles.**

Self-Check *Find the perimeter of each triangle.*

3. $a = 4$ m; $b = 7$ m; $c = 9$ m
4. $a = 2$ yd; $b = 5$ yd; $c = 6$ yd

Since a square is a rectangle with four equal sides, you can write a formula for the perimeter of a square by adding like terms.

$P = s + s + s + s$

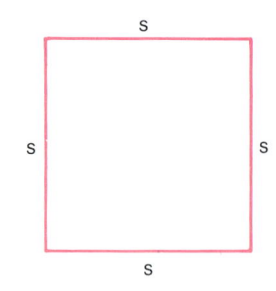

Formula: $P = 4s$

P = perimeter
s = length of a side

CLASSROOM EXERCISES

Find the perimeter of each rectangle. (Example 1)

1. $l = 7$ cm; $w = 4$ cm
2. $l = 12.4$ m; $w = 8.9$ m
3. $l = 15$ m; $w = 12$ m
4. $l = 3$ ft; $w = 1$ ft
5. $l = 8$ in; $w = 4$ in
6. $l = 6$ yd; $w = 4$ yd

Chapter 3

ADDITIONAL EXAMPLES

Example 2

Find the perimeter of each triangle.

1. $a = 19.6$ in;
   $b = 8.7$ in; $c = 13$ in
   Ans: 41.3 in
2. $a = 2.16$ cm;
   $b = 0.8$ cm;
   $c = 2.507$ cm
   Ans: 5.467 cm

SELF-CHECK

3. 20 m   4. 13 yd

CLASSROOM EXERCISES

1. 22 cm   2. 42.6 cm
3. 54 m   4. 8 ft
5. 24 in   6. 20 yd

Find the perimeter of each triangle. (Example 2)

7. $a = 6$ cm; $b = 9$ cm, $c = 4$ cm
8. $a = 2.5$ m; $b = 5.1$ m; $c = 3$ m
9. $a = 12$ cm; $b = 13.8$ cm; $c = 8$ cm
10. $a = 4$ ft; $b = 5$ ft; $c = 2$ ft
11. $a = 4$ in; $b = 10$ in; $c = 8$ in
12. $a = 22$ yd; $b = 18$ yd; $c = 13$ yd

WRITTEN EXERCISES

Goals: To find the perimeter of a rectangle and of a triangle
To apply the skill of finding perimeter to solving word problems

Find the perimeter of each rectangle. (Example 1)

METRIC MEASURES

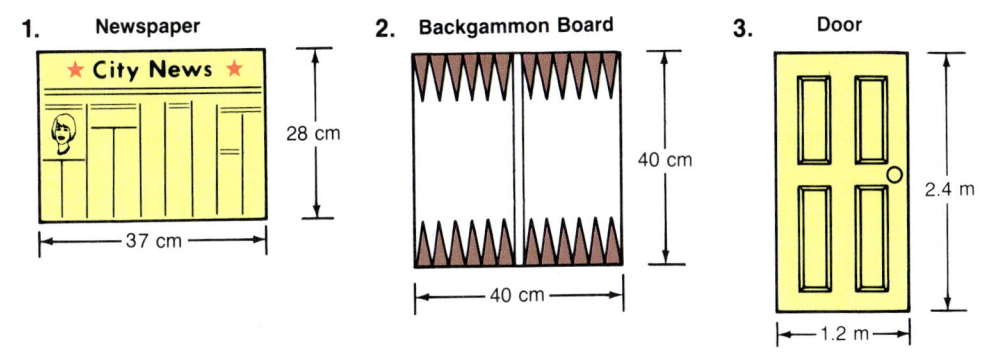

1. Newspaper — 37 cm × 28 cm
2. Backgammon Board — 40 cm × 40 cm
3. Door — 1.2 m × 2.4 m

	Length	Width	Perimeter
4.	8 cm	5 cm	?
6.	3.4 cm	2.1 cm	?
8.	4.2 m	4.2 m	?

	Length	Width	Perimeter
5.	25.1 m	9.8 m	?
7.	3.2 cm	1.9 cm	?
9.	16 mm	16 mm	?

CUSTOMARY MEASURES

10. Book Cover — 7 in × 9 in
11. Table Top — 6 ft × 2 ft
12. Checker Board — 15 in × 15 in

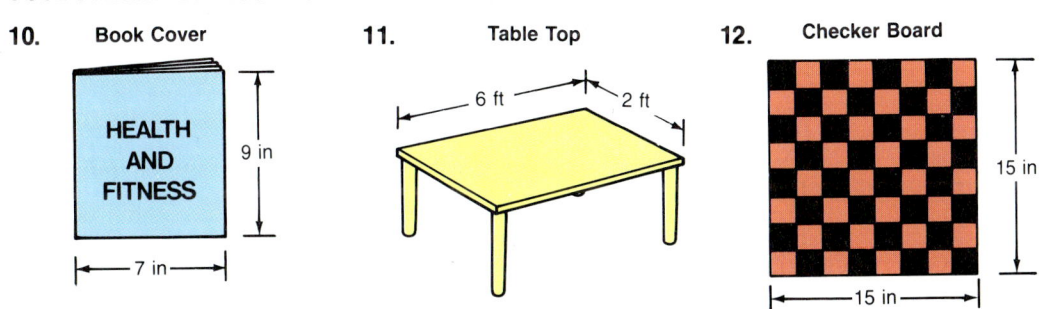

CLASSROOM EXERCISES
7. 19 cm   8. 10.6 m
9. 33.8 cm 10. 11 ft
11. 22 in   12. 53 yd

ASSIGNMENT GUIDE
BASIC
pp. 83-84: 1-35 odd
AVERAGE
pp. 83-85: 1-39 odd
ABOVE AVERAGE
pp. 83-85: 1-35 odd, 37-48

WRITTEN EXERCISES
1. 130 cm   2. 160 cm
3. 7.2 m   4. 26 cm
5. 69.8 m   6. 11 cm
7. 10.2 cm 8. 16.8 m
9. 64 mm   10. 32 in
11. 16 ft   12. 60 in

WRITTEN EXERCISES p. 78
29. 3.8a   30. 6.5s
31. $2x^2$ 32. $2.8m^2$
33. 3ab   34. 0.5xy
35. 2n   36. $15a^2$
37. 45st   38. $11r^2$
39. $8.2f^2$ 40. cp
41. 16.02m 42. $99.2p^2$

Number Properties/Expressions **83**

WRITTEN EXERCISES

13. 42 in   14. 10 ft
15. 24 yd   16. 14 in
17. 20 ft   18. 40 yd
19. 2.9 m   20. 96 cm
21. 25.1 m 22. 26 cm
23. 9.8 m   24. 12 cm
25. 32 cm   26. 14.1 m
27. 9 m   28. 13 yd
29. 36 ft   30. 44 ft
31. 12 ft   32. 27 in
33. 15 yd   34. 14 ft
35. 49 in   36. 9 in

	Length	Width	Perimeter
13.	14 in	7 in	?
15.	6 yd	6 yd	?
17.	5 ft	5 ft	?

	Length	Width	Perimeter
14.	2 ft	3 ft	?
16.	3 in	4 in	?
18.	12 yd	8 yd	?

Find the perimeter of each triangle. (Example 2)

METRIC MEASURES

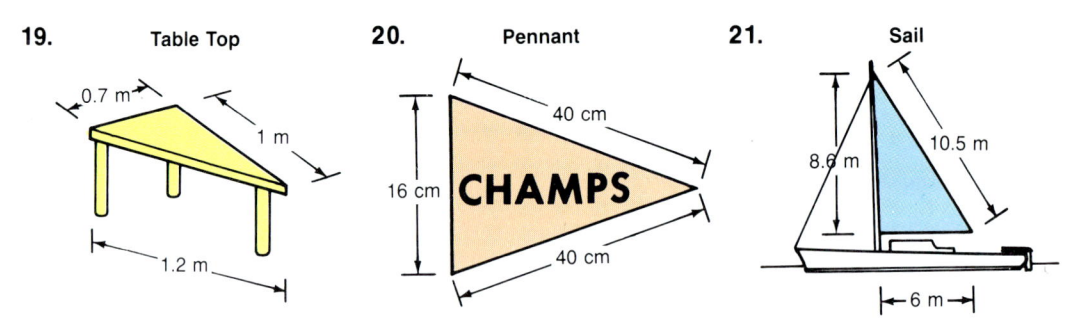

19. Table Top — 0.7 m, 1 m, 1.2 m
20. Pennant — 40 cm, 16 cm, 40 cm (CHAMPS)
21. Sail — 8.6 m, 10.5 m, 6 m

	a	b	c	Perimeter
22.	8 cm	12 cm	6 cm	?
24.	4 cm	4 cm	4 cm	?
26.	4.2 m	3.5 m	6.4 m	?

	a	b	c	Perimeter
23.	3.2 m	4.1 m	2.5 m	?
25.	12 cm	12 cm	8 cm	?
27.	3 m	3 m	3 m	?

CUSTOMARY MEASURES

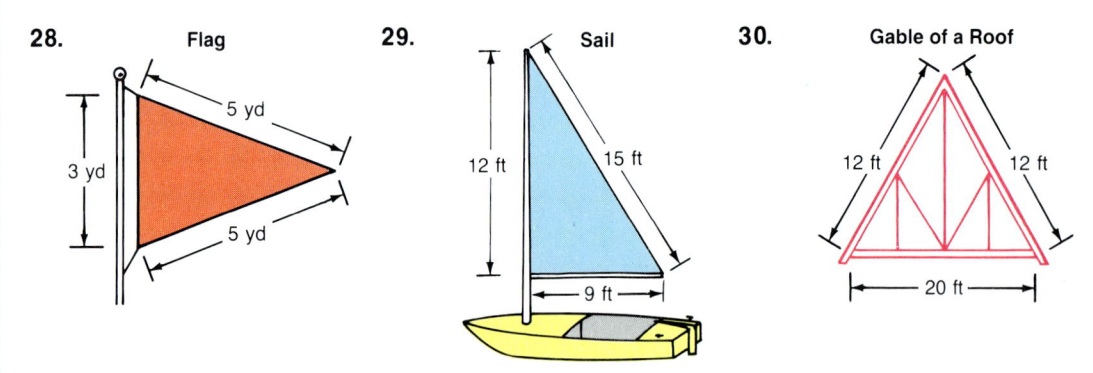

28. Flag — 5 yd, 3 yd, 5 yd
29. Sail — 12 ft, 15 ft, 9 ft
30. Gable of a Roof — 12 ft, 12 ft, 20 ft

	a	b	c	Perimeter
31.	3 ft	4 ft	5 ft	?
33.	5 yd	5 yd	5 yd	?
35.	22 in	11 in	16 in	?

	a	b	c	Perimeter
32.	9 in	11 in	7 in	?
34.	4 ft	4 ft	6 ft	?
36.	2 in	3 in	4 in	?

WRITTEN EXERCISES p. 80

7. 5a + 4   8. 11x + 9
9. 39a + 15
10. 41t + 9
11. 3.9m + 4.2
12. 11.5r + 5
13. 55k + 7
14. 72n + 16   15. 18x
16. 14a   17. 9.5t
18. 18p   19. $13a^2 + 4$
20. $17x^2 + 6$
21. 10xy + 4
22. 5.0mn + 2.9
23. 3.9p + 1.9
24. $16x^2$ 25. 15p + 16
26. 3.2 + 13.4t
27. 91r
28. $57a^2 + 24$
29. 190 + 254p
30. $245.5z^2 + 0.9$

Chapter 3

MIXED PRACTICE

37. The rectangular building in which spacecraft are assembled at Cape Canaveral in Florida is 654 feet long and 474 feet wide. Find its perimeter.

38. The base of the Great Pyramid of Egypt is a square. Each side of the square is 236.4 meters long. Find the perimeter of the base.

39. A triangular lot has sides having lengths of 32 yards 45 yards and 27 yards. Find the perimeter of the lot.

40. A football field has a rectangular shape. It is 110 meters long and 49 meters wide. Find the perimeter of the field.

MORE CHALLENGING EXERCISES

Find the perimeter of each figure.

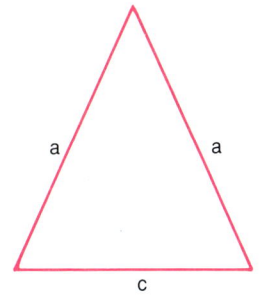

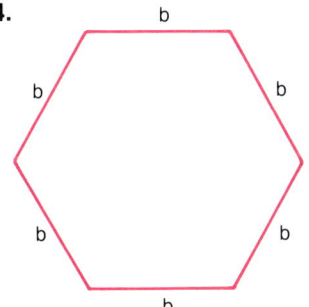

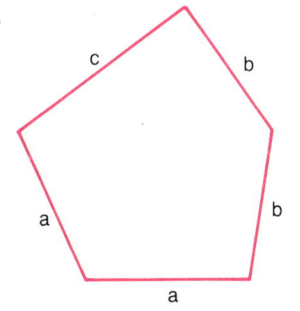

41. (square with sides 2b)
42. (triangle with sides a, a, c)
43. (triangle with sides a, a, a)
44. (hexagon with sides b)
45. (pentagon with sides a, b, c, b, a)
46. (octagon with sides a)

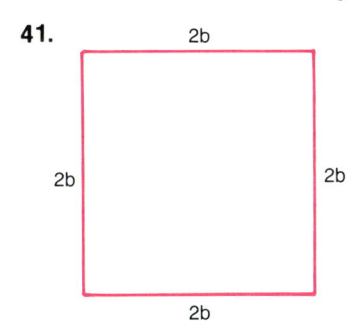

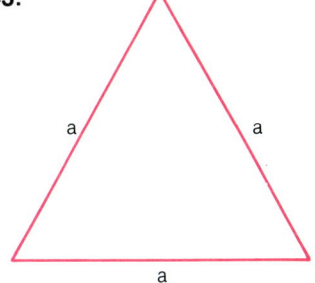

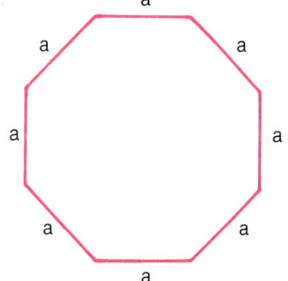

47. The length of a rectangle is 4x and the width is 2x. Find the perimeter of the rectangle.

48. The lengths of the sides of a triangle are 4s, 6s, and 7s. Find the perimeter of the triangle.

WRITTEN EXERCISES
37. 2256 ft
38. 945.6 m
39. 104 yd   40. 318 m
41. 8b   42. 2a + c
43. 3a   44. 6b
45. 2a + 2b + c
46. 8a   47. 12x
48. 17s

Number Properties/Expressions

QUIZ: SECTIONS 3-4–3-6
After completing this Review, you may wish to administer a quiz covering the same sections. A Quiz is provided in the Teacher's Edition: Part II.

REVIEW: SECTIONS 3-4–3-6
1. Yes   2. No   3. Yes
4. 5x   5. 7.2b
6. 1.9ef 7. $8n^2$
8. a = 3.4   9. c = 4
10. m = 3   11. 12b
12. 14g + 17
13. 11.3t + 10
14. $8.2x^2 + 2.5$
15. 180 mi
16. 18.4 km

CALCULATOR EXERCISES
1. 153   2. 276
3. 2346   4. 121,771
5. 183,315
6. 856,086

REVIEW: SECTIONS 3-4 — 3-6

Which pairs show like terms? Answer Yes or No. (Section 3-4)

1. $4x$ and $5x$
2. $2.5c^2d$ and $7.1cd^2$
3. $2xy$ and $1.5xy$

Combine like terms. (Section 3-4)

4. $2x + 3x$
5. $6.2b + b$
6. $4.7ef - 2.8ef$
7. $9n^2 - n^2$

Write the value of the variable that makes each sentence true. (Section 3-5)

8. $3.4 + 5 = 5 + a$
9. $(7 + c) + 9 = 7 + (4 + 9)$
10. $9 + (m + 6) = (9 + 6) + 3$

Combine like terms. (Section 3-5)

11. $2b + 6b + 4b$
12. $9g + 5g + 17$
13. $7.2t + 10 + 4.1t$
14. $5x^2 + 2.5 + 3.2x^2$

Solve each problem. (Section 3-6)

15. A boat race is run over a triangular course. The first leg is 72 miles long; the second leg is 30 miles long; and the third is 78 miles long. Find the perimeter. (Section 3-6)

16. A rectangular pasture needs to be fenced. The pasture is 2.5 kilometers wide and 6.7 kilometers long. Find the perimeter of the pasture. (Section 3-6)

USING FORMULAS

You can use a calculator and the following formula to find the sum of a given number of counting numbers. (The counting numbers are 1, 2, 3, 4, 5, ···)

$$S = \frac{n(n + 1)}{2}$$

S represents the sum.
n represents how many numbers are to be added.

EXAMPLE Find the sum of the first 230 counting numbers.

Solution $S = \dfrac{n(n + 1)}{2}$ Replace *n* with 230. Then $n + 1 = 231$.

$$S = \frac{230(230 + 1)}{2} = \frac{230(231)}{2}$$

2 3 0 [•] 2 3 1 [÷] 2 [=] $\boxed{26565.}$

EXERCISES Find the sum of the first *n* counting numbers for the given value of *n*.

1. 17
2. 23
3. 68
4. 493
5. 605
6. 1308

CHAPTER REVIEW

PART 1: VOCABULARY

For Exercises 1–11, choose from the box at the right the word(s) or numerical expression(s) that best corresponds to each description.

1. The number of square units needed to cover a surface __?__
2. Addition Property of Zero __?__
3. Commutative Property of Addition __?__
4. Multiplication Property of One __?__
5. A rectangle with four equal sides __?__
6. Associative Property of Addition __?__
7. Multiplication Property of Zero __?__
8. Terms that have the same variable and the same exponents for each variable __?__
9. Commutative Property of Multiplication __?__
10. Associative Property of Multiplication __?__
11. The distance around a figure __?__

$7 + 0 = 7$
$25 \cdot 0 = 0$
$(8)(1) = 8$
$5 \cdot 12 = 12 \cdot 5$
$4 + 3 = 3 + 4$
$3 \cdot (6 \cdot 8) = (3 \cdot 6) \cdot 8$
$(2 + 6) + 9 = 2 + (6 + 9)$
perimeter
like terms
square
area

PART 2: SKILLS

Name the property that makes each sentence true.
(Sections 3-1 – 3-2)

12. $2.4(1) = 2.4$
13. $(46)(0) = 0$
14. $7x = 0 + 7x$
15. $(8)(y) = (y)(8)$
16. $(23 \cdot 9)4 = 23(9 \cdot 4)$
17. $5(16 \cdot 32) = (5 \cdot 16)32$

Find the value of the variable that makes each sentence true.
(Sections 3-1 – 3-2)

18. $0 = 5e$
19. $1.9 = 0 + y$
20. $3.2 \cdot t = 3.2$
21. $6 \cdot x = 3 \cdot 6$
22. $4(q \cdot 7) = 7(9 \cdot 4)$
23. $(2 \cdot 8.5)k = 4(8.5 \cdot 2)$

Multiply. (Section 3-2)

24. $5(4x)$
25. $(4.2m)6$
26. $(2x)(12y)$
27. $(10r)(1.7t)$
28. $(12x)(3x)$
29. $(5a)(9a)$
30. $(2x)(5xy)$
31. $(17pq)(4p)$
32. $(15cd)(2.4c^2)$
33. $(20ab^2)(5ab)$
34. $(14e^2f)(3ef)$
35. $(9x^2y^2)(6y)$

Number Properties/Expressions **87**

CHAPTER REVIEW
1. area   2. $7 + 0 = 7$
3. $4 + 3 = 3 + 4$
4. $(8)(1) = 8$
5. square
6. $(2 + 6) + 9 = 2 + (6 + 9)$
7. $25 \cdot 0 = 0$
8. like terms
9. $5 \cdot 12 = 12 \cdot 5$
10. $3 \cdot (6 \cdot 8) = (3 \cdot 6) \cdot 8$
11. perimeter
12. Mult. Prop. of One
13. Mult. Prop. of Zero
14. Add. Prop. of Zero
15. Comm. Prop. of Mult.
16. Assoc. Prop. of Mult.
18. $e = 0$ 19. $y = 1.9$
20. $t = 1$ 21. $x = 3$
22. $q = 9$ 23. $k = 4$
24. $20x$ 25. $25.2m$
26. $24xy$ 27. $17rt$
28. $36x^2$ 29. $45a^2$
30. $10x^2y$ 31. $68p^2q$
32. $36c^3d$ 33. $100a^2b^3$
34. $42e^3f^2$
35. $54x^2y^3$

87

CHAPTER REVIEW

36. 432 m^2
37. 70.56 cm^2
38. 44.16 mm^2
39. 441 m^2
40. unlike terms
41. like   42. like
43. unlike   44. like
45. unlike   46. 30x
47. 13n   48. 9.5y
49. 11a   50. 5s
51. 12.1t   52. 19x
53. 2.8n^2 54. m = 3.1
55. y = 3   56. 13x
57. 6.7a + 3
58. 9y + 2
59. 13ab + 5
60. 14c^2 + 12
61. 9t + 8   62. 1 + 6u
63. 9x^2 64. 22 cm
65. 84 in   66. 42.4 m
67. 29.5 cm   68. 117 in
69. 108 ft^2
70. 5.76 m^2
71. 136.6 m
72. 30 ft

Find the area of each rectangle or square. (Section 3-3)

	Length	Width	Area
36.	27 m	16 m	?
37.	8.4 cm	8.4 cm	?
38.	9.2 mm	4.8 mm	?
39.	21 m	21 m	?

Combine like terms. (Section 3-4)

46. $18x + 12x$
47. $7n + 6n$
48. $2.7y + 6.8y$
49. $a + 10a$
50. $9s - 4s$
51. $18t - 5.9t$
52. $20x - x$
53. $8.4n^2 - 5.6n^2$

Find the value of the variable that makes each sentence true. (Section 3-5)

54. $3.1 + 12.4 = 12.4 + m$
55. $(5 + 8) + y = 5 + (8 + 3)$

Combine like terms. (Section 3-5)

56. $4x + 7x + 2x$
57. $5a + 1.7a + 3$
58. $6y + 2 + 3y$
59. $4ab + 5 + 9ab$
60. $13c^2 + 12 + c^2$
61. $7t + 2t + 8$
62. $1 + u + 5u$
63. $2x^2 + x^2 + 6x^2$

Find the perimeter of each rectangle. (Section 3-6)

64. $l = 6$ cm; $w = 5$ cm
65. $l = 10$ in; $w = 32$ in
66. $l = 8.7$ m; $w = 12.5$ m

Find the perimeter of each triangle. (Section 3-6)

67. $a = 14.2$ cm; $b = 9.7$ cm; $c = 5.6$ cm
68. $a = 24$ in; $b = 55$ in; $c = 38$ in

Identify each pair of terms as like terms or unlike terms. (Section 3-4)

40. 1.4^2 and $2x$
41. a and $7a$
42. $25s$ and $1.3s$
43. $2c$ and cx
44. $16xy$ and $8xy$
45. z and z^2

Combine like terms. (Section 3-4)

46. $18x + 12x$
47. $7n + 6n$
48. $2.7y + 6.8y$
49. $a + 10a$
50. $9s - 4s$
51. $18t - 5.9t$
52. $20x - x$
53. $8.4n^2 - 5.6n^2$

PART 3: **APPLICATIONS**

69. A rectangular wing on a single-engine airplane is 3 feet wide and 36 feet long. Find the area of the wing. (Section 3-3)

70. The sides of a square Oriental rug are 2.4 meters long. Find the area of the rug. (Section 3-3)

71. A triangular courtyard has sides measuring 50.4 meters, 47.2 meters and 39 meters. Find the perimeter of the courtyard. (Section 3-6)

72. A rectangular picture window is 6 feet long and 9 feet wide. Find the perimeter of the picture window. (Section 3-6)

CHAPTER TEST

Find the value of the variable that makes each sentence true.

1. $x + 0 = 5.2$
2. $3(6 \cdot y) = (3 \cdot 6)7$
3. $(48)(a) = 48$
4. $5 \cdot b = 9 \cdot 5$

Multiply.

5. $4.1\,(3.6x)$
6. $(13s)(5t)$
7. $(4.5x)(8x)$
8. $(7cd)(8cd^2)$

Find the area of each rectangle or square.

9. length: 3.4 m; width: 3.4 m
10. length: 9 ft; width: 10 ft
11. length: 41 in; width: 8 in
12. length: 2.3 cm; width: 2.3 cm

Combine like terms.

13. $3x + 12x$
14. $5.2y - y$
15. $4a + 3 + 10a$
16. $7m^2 - 3.9m^2$

Find the perimeter of each rectangle.

17. $l = 14$ m; $w = 6.2$ m
18. $l = 25$ in; $w = 19$ in
19. $l = 8.9$ cm; $w = 4.6$ cm

Find the perimeter of each triangle.

20. $a = 2$ m; $b = 6.5$ m; $c = 7.3$ m
21. $a = 48$ ft; $b = 27$ ft; $c = 36$ ft
22. $a = 45$ in; $b = 60$ in; $c = 75$ in
23. $a = 9.8$ cm; $b = 5.4$ cm; $c = 6.7$ cm

24. Maria Gala plans to wallpaper her bedroom. In finding the area of the walls she must allow for the area of the doorway. A rectangular doorway is 3 feet wide and 7 feet long. Find the area of the doorway.

25. A rectangular sign is 0.75 meters wide and 3.4 meters long. Find the perimeter of the sign.

ANSWERS TO REVIEW CAPSULES

Page 66 **1.** 20; 20 **2.** 24; 24 **3.** 72; 72 **4.** 30; 30 **5.** 24; 24 **6.** 378; 378 **7.** 840; 840 **8.** 175; 175 **9.** 550; 550 **10.** 504; 504 **11.** 180; 180 **12.** 360; 360

Page 69 **1.** 28 **2.** 399 **3.** 23.56 **4.** 4.32 **5.** 36 **6.** 179.56 **7.** 225 **8.** 64.6416

Page 74 **1.** 5.0 **2.** 4.7 **3.** 12.7 **4.** 13.9 **5.** 5.7 **6.** 4.9 **7.** 6.4 **8.** 1.9 **9.** 28.9 **10.** 7.8 **11.** 18.6 **12.** 10.0

Page 78 **1.** 21; 21 **2.** 14.4; 14.4 **3.** 19.8; 19.8 **4.** 25.9; 25.9 **5.** 79; 79 **6.** 39; 39

Page 81 **1.** 12 **2.** 22 **3.** 29.9 **4.** 38.8 **5.** 19.2 **6.** 448

Number Properties/Expressions **89**

CHAPTER TEST

Two forms of a chapter test, Form A and Form B, are provided on copying masters in the *Teacher's Edition: Part II.*

1. x = 5.2 2. y = 7
3. a = 1 4. b = 9
5. 14.76x 6. 65st
7. $36x^2$ 8. $56c^2d^3$
9. 11.56 m^2 10. 90 ft^2
11. 328 in^2
12. 5.29 cm^2
13. 15x 14. 4.2y
15. 14a + 3 16. 3.1 m^2
17. 40.4 m 18. 88 in
19. 27 cm 20. 15.8 m
21. 111 ft 22. 180 in
23. 21.9 cm 24. 21 ft^2
25. 8.3 m

ENRICHMENT

You may wish to use this lesson for students who performed well on the formal Chapter Test.

1. 190   2. 520
3. 21,000   4. 742,000
5. 37,100   6. 12,000
7. 162   8. 9,000
9. 500   10. 1,286
11. 16,100
12. 19,000

ENRICHMENT

Number Properties

Number properties can be used to simplify numerical computations.

EXAMPLES Use number properties to evaluate each expression.

Solutions:

a. $(93 + 8) + (7 + 72) = (93 + 7) + (8 + 72)$
$= 100 + 80 = \mathbf{180}$

◀ By the Commutative and Associative Properties for Addition

b. $(6 \cdot 25) \cdot (7 \cdot 4) = (6 \cdot 7) \cdot (25 \cdot 4)$
$= 42 \cdot 100 = \mathbf{4200}$

◀ By the Commutative and Associative Properties for Multiplication

c. $997 \cdot 42 + 3 \cdot 42 = 997x + 3x$

◀ Think of $997 \cdot 42$ as $997x$.
Think of $3 \cdot 42$ as $3x$.

$= 1000x$

◀ Now replace x with 42.

$= 1000 \cdot 42 = \mathbf{42{,}000}$

d. $17 \cdot 54 - 7 \cdot 54 = 17y - 7y$

◀ Think of $17 \cdot 54$ as $17y$.
Think of $7 \cdot 54$ as $7y$.

$= 10y$

◀ Replace y with 54.

$= 10 \cdot 54 = \mathbf{540}$

EXERCISES

Use number properties to evaluate each expression.

1. $(64 + 78) + (36 + 12)$
2. $(450 + 19) + (50 + 1)$
3. $(50 \cdot 42) \cdot (2 \cdot 5)$
4. $(250 \cdot 371) \cdot (4 \cdot 2)$
5. $51 \cdot 371 + 49 \cdot 371$
6. $822 \cdot 12 + 178 \cdot 12$
7. $109 \cdot 9 - 9 \cdot 91$
8. $1364 \cdot 9 - 364 \cdot 9$
9. $(10 \cdot 4 + 5) + (90 \cdot 4 + 95)$
10. $(37 \cdot 12 + 32) + (63 \cdot 12 + 54)$
11. $88 \cdot 161 + 11 \cdot 161 + 161$
12. $960 \cdot 19 + 39 \cdot 19 + 19$

ADDITIONAL PRACTICE

SKILLS *Find the value of the variable that makes each sentence true.* (Pages 64–66)

1. $72 + x = 72$
2. $0 = 4v$
3. $7.1x = 7.1$
4. $43 = y\left(\dfrac{12}{12}\right)$
5. $\left(\dfrac{6}{x}\right)134 = 134$

(Pages 67–69)

6. $17x = 33(17)$
7. $(5 \cdot 4)y = 5 \cdot (4 \cdot 13)$
8. $3(t \cdot 6) = (3 \cdot 6) \cdot 12$

Multiply. (Pages 67–69)

9. $4(9x)$
10. $(3xy)(2xy)$
11. $(10p)(2p^2)$
12. $(4.3a)(2.1c)$
13. $(4x^2y^2)(5xy)$

Find the area of each rectangle or square. (Pages 70–73)

14. Rectangle: $l = 13$ cm; $w = 8$ cm
15. Square: $s = 9.2$ cm
16. Square: $s = 27$ m
17. Rectangle: $l = 12.4$ m; $w = 11.1$ m

Combine like terms. (Pages 76–78)

18. $6x + 8x$
19. $4cd - cd$
20. $13.2x^2y - 4.7x^2y$
21. $16.8x^2y^2 + 19.9x^2y^2$

(Pages 79–80)

22. $10c + 12 + c$
23. $4xy + 3 + xy$
24. $12ab^2 + 4ab^2 + 6$
25. $9x + 4 + 3.8x$
26. $4.1xy + 11.9xy + 2$
27. $9.6m^2n + 13 + 8.7m^2n$

Find the perimeter of each figure. (Pages 81–85)

28. Rectangle: $l = 15$ in; $w = 6$ in
29. Square: $s = 52$ yd
30. Triangle: $a = 13.5$ m; $b = 13.5$ m; $c = 13.5$ m
31. Triangle: $a = 17$ cm; $b = 25$ cm; $c = 30$ cm

APPLICATIONS

32. A desk top has a length of 1.5 meters and a width of 0.75 meter. Find the perimeter and the area of the desk top. (Pages 70–73; 81–85)

33. Each side of a triangular lot is 17 yards long. Each side of a square lot is 13 yards long. Which lot has the greater perimeter? How much greater? (Pages 81–85)

34. A boat race is run over a triangular course. The first leg is 82 kilometers long, the second leg is 27 kilometers long, and the third leg is 91 kilometers long. Find the perimeter of the course. (Pages 81–85)

35. A rectangular room is 5.3 yards long and 4.4 yards wide. Find the number of square yards of carpet needed to carpet the room. (Pages 70–73)

ADDITIONAL PRACTICE

You may wish to use all or some of these exercises, depending on how well students performed on the formal Chapter Test.

1. $x = 0$
2. $v = 0$
3. $x = 1$
4. $y = 43$
5. $x = 6$
6. $x = 33$
7. $y = 13$
8. $t = 12$
9. $36x$
10. $6x^2y^2$
11. $20p^3$
12. $9.03ac$
13. $20x^3y^3$
14. 104 cm^2
15. 84.64 cm^2
16. 729 m^2
17. 137.64 m^2
18. $14x$
19. $3cd$
20. $8.5x^2y$
21. $36.7x^2y^2$
22. $11c + 12$
23. $5xy + 3$
24. $16ab^2 + 6$
25. $12.8x + 4$
26. $16xy + 2$
27. $18.3\ m^2n + 13$
28. 42 in
29. 208 yd
30. 40.5 m
31. 72 cm
32. perimeter: 4.5 m; area: 1.125 m^2
33. square; 1 yd greater
34. 200 km
35. 23.32 yd^2

Number Properties/Expressions

COMMON ERRORS

In preparation for the Cumulative Review, these exercises focus the student's attention on the most common errors to be avoided.

1. 9   2. 14   3. 11
4. 29   5. 21.39
6. 1.125   7. 4000 mm
8. $28a^2$ 9. 9b
10. $15y + 6$
11. 32 cm
12. 64 ft^2

COMMON ERRORS

Each of these problems contains a common error.

a. Find the correct answer.
b. Find the error.

Evaluate each expression.

1. $3^2 = 3 \cdot 2$
   $= 6$

2. $6 + 4 \cdot (5 - 3) = 6 + 4 \cdot 2$
   $= 10 \cdot 2 = 20$

3. $3 + 6 \cdot 2 - 4 = 9 \cdot 2 - 4$
   $= 18 - 4$
   $= 14$

4. $xy + 5$ when $x = 6$ and $y = 4$
   $xy + 5 = 6(4) + 5$
   $= 64 + 5 = 69$

Complete.

5. $3.24 + 1.5 + 16.65 = \underline{\ ?\ }$

   3.24
   1.5
   $\underline{+16.65}$
   20.04

6. $2.5(0.45) = \underline{\ ?\ }$

   2.5
   $\underline{\times 0.45}$
   125
   100
   $\overline{11.25}$

7. $4 \text{ m } \underline{\ ?\ } \text{ mm}$
   $4 \text{ m} = (4 \div 1000) \text{ mm}$
   $= 0.004 \text{ mm}$

8. $(4a)(7a) = \underline{\ ?\ }$
   $(4a)(7a) = 4 \cdot 7 \cdot a$
   $= 28a$

Combine like terms.

9. $2b + 7b = \underline{\ ?\ }$
   $2 + 7 = 9$
   Thus, $2b + 7b = 9b^2$.

10. $12y + 6 + 3y = \underline{\ ?\ }$
    $12 + 6 + 3 = 21$
    Thus, $12y + 6 + 3y = 21y$.

11. Find the perimeter of this rectangle.
    l: 10 cm; w: 6 cm
    $P = l + w$
    $P = 10 + 6$
    $P = 16$ cm

12. Find the area of a square with a side 8 feet in length.
    $A = s^2$
    $A = 8^2$
    $A = 16$ feet

Chapter 3

CUMULATIVE REVIEW: CHAPTERS 1–3

Choose the correct answer.
Choose a, b, c, or d.

NEW CAR	
BASE PRICE	$9475
OPTIONAL EQUIPMENT	$ 655
DESTINATION CHARGE	$ 160

1. Find the sticker price of the car.
   a. $10,300 **b.** $10,280 **c.** $10,290 **d.** $10,190

2. The reading on a meter on June 10 is 8146. The reading on May 10 was 7359. How many kilowatt-hours of electricity were used?
   a. 887 **b.** 797 **c.** 793 **d.** 787

3. An airplane is flying at a rate of 358 kilometers per hour. How many kilometers will the plane fly in 12 hours?
   a. 4286 **b.** 4196 **c.** 4296 **d.** 29.8

4. The formula $d = 16t^2$ gives the distance in feet that an object will fall in t seconds. How many feet will the object fall in 7 seconds?
   a. 784 **b.** 224 **c.** 12,544 **d.** 512

5. Erica's science test scores are 68, 74, 82, 96, 62, and 74. Find the average score.
   a. 76 **b.** 78 **c.** 75 **d.** 74

6. Choose the best estimate: 179×299
   a. 51,000 **b.** 54,000 **c.** 49,300 **d.** 53,280

7. Evaluate: $6 + 4 \times 17$
   a. 170 **b.** 74 **c.** 160 **d.** 64

8. Evaluate: $(20 + 60) \div 5 + 5$
   a. 37 **b.** 26 **c.** 8 **d.** 21

9. Evaluate $13x - 11y$ when $x = 14$ and $y = 12$.
   a. 40 **b.** 50 **c.** 30 **d.** 60

10. Evaluate $(18 + 72) \div p$ when $p = 3$.
    a. 32 **b.** 42 **c.** 30 **d.** 40

11. Which decimal has the greatest value?
    a. 0.511 **b.** 0.509 **c.** 0.4978 **d.** 0.5059

12. Subtract: $96.05 - 3.48$
    a. 92.57 **b.** 92.67 **c.** 93.57 **d.** 93.53

CUMULATIVE REVIEW

A cumulative test is provided on copying masters in the *Teacher's Edition: Part II.*

1. c 2. d 3. c
4. a 5. a 6. b
7. b 8. d 9. b
10. c 11. a 12. a

Cumulative Review: Chapters 1–3

CUMULATIVE REVIEW

13. b 14. b
15. c 16. d
17. d 18. c
19. d 20. d
21. c 22. d
23. d 24. c
25. d 26. b

13. Evaluate $x + y + z$ when $x = 7.8$, $y = 14.7$, and $z = 6.9$.
    a. 28.4
    b. 29.4
    c. 2.4
    d. 19.4

14. Multiply: 2.84×0.008
    a. 2.272
    b. 0.02272
    c. 22.72
    d. 0.2272

15. Divide: $.81 \overline{)51.03}$
    a. 630
    b. .63
    c. 63
    d. 6.3

16. Harold's subcompact car travels 518.4 miles on 12 gallons of gas. How many miles does it travel on one gallon?
    a. 432
    b. 44.2
    c. 6220.8
    d. 43.2

17. Choose the best estimate: $99.8 \div 24.9$
    a. 4.3
    b. 5
    c. 3.9
    d. 4

18. A bottle holds 1250 milliliters of water. How many liters is this?
    a. 12.50
    b. 125
    c. 1.25
    d. 0.1250

19. Saturn is 887,000,000 miles from the sun. Express this number in scientific notation.
    a. 88.7×10^7
    b. 887×10^6
    c. 0.887×10^9
    d. 8.87×10^8

20. Multiply: $(4x)(3x)$
    a. $7x^2$
    b. $7x$
    c. $12x$
    d. $12x^2$

21. Multiply: $(3x)(4xy)(3x^2)$
    a. $108x^4y$
    b. $36x^3y$
    c. $36x^4y$
    d. $108x^3y$

22. The length of each side of a square parking lot is 72 meters. Find the area of the parking lot.
    a. 288 m²
    b. 5084 m²
    c. 358 m²
    d. 5184 m²

23. Combine like terms: $4.1xy - 2.5xy$
    a. $2.6xy$
    b. $1.6x^2y^2$
    c. $2.6x^2y^2$
    d. $1.6xy$

24. Combine like terms: $1.4pt + 8.1 + 6.7pt + 9.2$
    a. $8.1p^2t^2 + 17.3$
    b. $7.1p^2t^2 + 17.3$
    c. $8.1pt + 17.3$
    d. $7.1pt + 17.3$

25. Write the value of the variable that makes this sentence true.
    $(4.6 + 9.4) + 13 = 13 + (9.4 + x)$
    a. 0
    b. 13
    c. 9.4
    d. 4.6

26. The cover of a book is 10 inches long and 8 inches wide. Find the perimeter of the cover.
    a. 80 in
    b. 36 in
    c. 800 in
    d. 26 in

Cumulative Review: Chapters 1-3

CHAPTER 4
Number Theory

SECTIONS
- **4-1** Factors, Multiples, and Divisibility
- **4-2** Divisibility Tests
- **4-3** Prime Numbers
- **4-4** Prime Factorization
- **4-5** Least Common Multiple
- **4-6** Greatest Common Factor

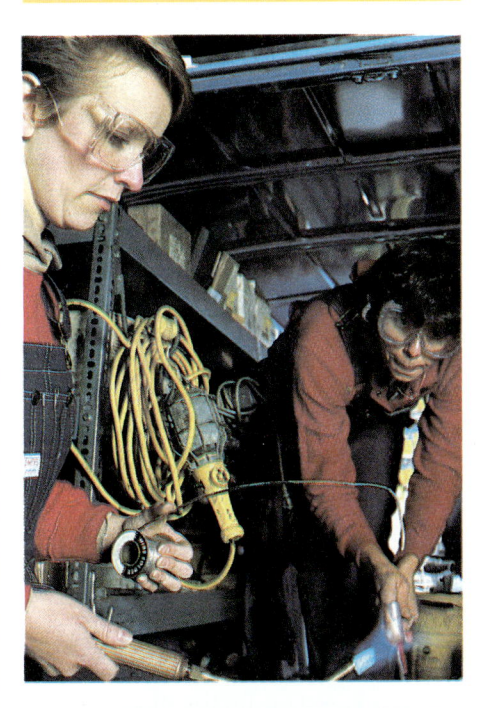

FEATURES
Calculator Application: *Prime Numbers and Exponents*
Calculator Application: *Prime Numbers*
Consumer Application: *Painting and Estimation*
Enrichment: *Factor Trees*

Teaching Suggestions
p M-17

QUICK QUIZ
Divide. Write the remainder.
1. 480 ÷ 5   Ans: 96 R 0
2. 342 ÷ 6   Ans: 57 R 0
3. 5119 ÷ 3
   Ans: 1706 R 1

Multiply.
4. 4 x 150   Ans: 600
5. 10 x 60   Ans: 600

ADDITIONAL EXAMPLES
Example 1
Write True (T) or False (F) for each statement. Give a reason.
1. 6 is a factor of 3114.
   Ans: T; 6 · 519 = 3114.
2. 7 is a multiple of 378.
   Ans: F; 378 ÷ 7 = 54.

SELF-CHECK
1. False; since 5 · 7 = 35.
2. True; since 28 ÷ 4 = 7.

4-1 Factors, Multiples, and Divisibility

> A number is **divisible** by another number if the remainder is 0.

$$4\overline{)24} \quad \underline{24} \quad 0$$
24 is divisible by 4.

$$6\overline{)32} \quad \underline{30} \quad 2$$
32 is **not** divisible by 6.

Since 24 is divisible by 4, 4 is a **factor** of 24. Here are all the factors of 24.

$$1, 2, 3, 4, 6, 8, 12, 24$$
Factors of 24

NOTE: The only factors that we consider are counting numbers.

Counting numbers: 1, 2, 3, 4, 5, . . .   These are also called **natural** numbers.

The number 24 is a **multiple** of each of its factors.

$$1 \cdot 24 = 24 \quad 2 \cdot 12 = 24 \quad 3 \cdot 8 = 24 \quad 4 \cdot 6 = 24$$

The table below shows several related statements using the words is divisible by, is a factor of, and is a multiple of.

TABLE: Related Statements

Statement 1	Statement 2	Statement 3
24 is divisible by 8.	8 is a factor of 24.	24 is a multiple of 8.
9 is divisible by 3.	3 is a factor of 9.	9 is a multiple of 3.
7 is divisible by 1.	1 is a factor of 7.	7 is a multiple of 1.

EXAMPLE 1   Write True or False for each statement. Give a reason for each answer.

a. 50 is a multiple of 5.
b. 28 is divisible by 9.
c. 9 is a factor of 27.
d. 17 is a multiple of 17.

Solutions:
a. True, since $5 \cdot 10 = 50$.
b. False, since $28 \div 9 = 3$ R 1.
c. True, since $27 \div 9 = 3$.
d. True, since $1 \cdot 17 = 17$.

Self-Check   Write True or False for each statement. Give the reason.

1. 36 is a multiple of 5.
2. 4 is a factor of 28.

96   Chapter 4

> **PROCEDURE**
>
> **To determine the factors of a counting number, n:**
>
> [1] Starting with $1 \cdot n$, list all the ways of writing n as the product of two numbers. Stop when the factors begin to repeat.
>
> [2] Use the list to write the factors of n. Start with the least.

EXAMPLE 2 List all the factors of 30.

Solution: [1] List all the pairs of numbers whose product is 30.

$1 \cdot 30 \qquad 2 \cdot 15 \qquad 3 \cdot 10 \qquad 5 \cdot 6$ ◀ **Stop!** $6 \cdot 5$ has the same factors as $5 \cdot 6$.

[2] Write the factors in order.
Start with the least. Factors of 30: **1, 2, 3, 5, 6, 10, 15, 30**

Self-Check List all the factors of each number.

3. 12   **4.** 48   **5.** 3   **6.** 11   **7.** 76   **8.** 105

CLASSROOM EXERCISES

Write a true statement about each pair of numbers. Use the words is divisible by. (Definition)

1. 40 and 20   **2.** 4 and 4   **3.** 2 and 6   **4.** 6 and 12
5. 5 and 40   **6.** 35 and 5   **7.** 7 and 7   **8.** 7 and 56

Write a true statement about each pair of numbers. Use the words is a factor of. (Table)

9. 55 and 11   **10.** 1 and 8   **11.** 8 and 40   **12.** 12 and 24
13. 7 and 1   **14.** 39 and 13 **15.** 33 and 33 **16.** 47 and 47

Write a true statement about each pair of numbers. Use the words is a multiple of. (Table)

17. 24 and 8   **18.** 13 and 1   **19.** 6 and 6   **20.** 16 and 4
21. 1 and 9   **22.** 5 and 30   **23.** 3 and 15 **24.** 11 and 11

Show all the ways of writing each number as a product of two numbers. (Example 2, Step 1)

25. 13   **26.** 9   **27.** 18   **28.** 28   **29.** 30   **30.** 80

Number Theory

Example 2
List all the factors.
1. 45   Ans: 1, 3, 5, 9, 15, 45
2. 108 Ans: 1, 2, 3, 4, 6, 9, 12, 18, 27, 36, 54, 108

SELF-CHECK
3. 1, 2, 3, 4, 6, 12
4. 1, 2, 3, 4, 6, 8, 12, 16, 24, 48
5. 1, 3   6. 1, 11
7. 1, 2, 4, 19, 38, 76
8. 1, 3, 5, 7, 15, 21, 35, 105

CLASSROOM EXERCISES
1. 40 is divisible by 20.
2. 4 is divisible by 4.
3. 6 is divisible by 2.
4. 12 is divisible by 6.
5. 40 is divisible by 5.
6. 35 is divisible by 5.
7. 7 is divisible by 7.
8. 56 is divisible by 7.
9. 11 is a factor of 55.
10. 1 is a factor of 8.
11. 8 is a factor of 40.
12. 12 is a factor of 24.
13. 1 is a factor of 7.
See page 98 for the answers to Exercises 14–30.

ASSIGNMENT GUIDE
BASIC
Day 1 p. 98: 1-21
Day 2 p. 98: 22-26
AVERAGE
p. 98: 1-26
ABOVE AVERAGE
p. 98: 1-21 odd, 22-26

WRITTEN EXERCISES
1. T; since 18 ÷ 6 = 3
2. T; since 36 ÷ 6 = 6
3. F; since 21 ÷ 8 = 2 R5
4. F; since 58 ÷ 17 = 3 R7
5. T; since 29 · 1 = 29
6. T; since 7 · 4 = 28
7. T; since 5 · 15 = 75
8. T; since 132 ÷ 12 = 11
9. T; since 143 ÷ 13 = 11
10. 1, 2, 4, 8
11. 1, 13
12. 1, 3, 5, 15
See page 99 for the answers to Exercises 13-21.

CLASSROOM EXERCISES p.97
14. 13 is a factor of 39.
15. 33 is a factor of 33.
16. 47 is a factor of 47.
17. 24 is a multiple of 8.
18. 13 is a multiple of 1.
19. 6 is a multiple of 6.
20. 16 is a multiple of 4.
21. 9 is a multiple of 1.
22. 30 is a multiple of 5.
23. 15 is a multiple of 3.
24. 11 is a multiple of 11.
25. 1 · 13
26. 1 · 9; 3 · 3
27. 1 · 18; 2 · 9; 3 · 6
28. 1 · 28; 2 · 14; 4 · 7
29. 1 · 30; 2 · 15; 3 · 10; 5 · 6
30. 1 · 80; 2 · 40; 4 · 20; 5 · 16; 8 · 10

WRITTEN EXERCISES

Goal: To identify the factors of a number

Write True or False for each sentence. Write a reason for each answer. (Example 1)

1. 6 is a factor of 18.
2. 6 is a factor of 36.
3. 21 is divisible by 8.
4. 58 is divisible by 17.
5. 29 is a multiple of 29.
6. 28 is a multiple of 7.
7. 75 is a multiple of 5.
8. 132 is divisible by 12.
9. 143 is divisible by 13.

List all the factors of each number. (Example 2)

10. 8
11. 13
12. 15
13. 23
14. 27
15. 31
16. 40
17. 60
18. 24
19. 46
20. 72
21. 49

MIXED PRACTICE

For Exercises 22-26, write Yes or No to answer the questions in a, b, and c. For d in each row, list the factors of n.

n	Is n Divisible by This Number?	Is This Number a Factor of n?	Is n a Multiple of This Number?	List of Factors of n
22. 72	a. 24 _Yes_	b. 18 _Yes_	c. 48 _No_	d. _1,2,3,4,6,8,9,12, 18,24,36,72_
23. 15	a. 30 _No_	b. 30 _No_	c. 30 _Yes_	d. _1,3,5,15_
24. 77	a. 27 _No_	b. 11 _Yes_	c. 7 _Yes_	d. _1,7,11,77_
25. 56	a. 8 _Yes_	b. 7 _Yes_	c. 4 _Yes_	d. _1,2,4,7,8,14,28,56_
26. 60	a. 12 _Yes_	b. 9 _No_	c. 15 _Yes_	d. _1,2,3,4,5,6,10,12,15, 20,30,60_

REVIEW CAPSULE FOR SECTION 4-2

This Review Capsule reviews prior-taught skills used in Section 4-2. The reference is to the pages where the skills were taught.

For each number:
a. *Write the remainder when the number is divided by 3.*
b. *Write the remainder when the number is divided by 9.*

1. 17
2. 12
3. 18
4. 11
5. 31
6. 44

For each number:
a. *Write the remainder when the number is divided by 4.*
b. *Write the remainder when the number is divided by 5.*

7. 21
8. 18
9. 39
10. 108
11. 226
12. 198

The answers are on page 120.

98 Chapter 4

4-2 Divisibility Tests

For a family reunion dinner, the Gills would like the same number of persons seated at each table and all the tables completely filled. For the 157 expected guests, the caterer used the test for divisibility by 9 to determine whether each table could seat exactly nine persons.

The tests for divisibility involve the **digits** 0, 1, 2, 3, 4, 5, 6, 7, 8, and 9 and the set of whole numbers.

Whole numbers: 0, 1, 2, 3, 4, 5, . . .

Whole Numbers: 0 + Counting Numbers

TESTS FOR DIVISIBILITY

A whole number is divisible by:
2 if it ends in 0, 2, 4, 6, or 8.
3 if the sum of its digits is divisible by 3.
4 if its last two digits are divisible by 4.
5 if it ends in 0 or 5.
9 if the sum of its digits is divisible by 9.
10 if it ends in 0.

Any whole number that does not meet one of these tests is not divisible by the number shown. Recall that the following statements are related (see page 96).

38 is divisible by 2.   38 is a multiple of 2.   2 is a factor of 38.

EXAMPLE 1 Write True or False for each statement. Give the reason.

Solutions:

	Statement	True or False? Why?
a.	456 is a multiple of 2.	True. 456 ends in 6.
b.	522 is divisible by 4.	False. 22 is not divisible by 4.
c.	5 is a factor of 723.	False. 723 does not end in 0 or 5.
d.	10 is a factor of 520.	True. 520 ends in 0.

Number Theory **99**

Teaching Suggestions p. M-17

QUICK QUIZ

Divide. Write the remainder.

1. 13,224 ÷ 4   Ans: 0
2. 7560 ÷ 4   Ans: 0
3. 4106 ÷ 4   Ans: 2
4. 7560 ÷ 9   Ans: 0

ADDITIONAL EXAMPLES

Example 1

Write True (T) or False (F) for each statement. Give a reason.

1. 5 is a factor of 73,124.   Ans: F; 73,124 ends in 4.
2. 75,120 is divisible by 10.   Ans: T; 75,120 ends in 0.

WRITTEN EXERCISES p. 98

13. 1, 23
14. 1, 3, 9, 27
15. 1, 31
16. 1, 2, 4, 5, 8, 10, 20, 40
17. 1, 2, 3, 4, 5, 6, 10, 12, 15, 20, 30, 60
18. 1, 2, 3, 4, 6, 8, 12, 24
19. 1, 2, 23, 46
20. 1, 2, 3, 4, 6, 8, 9, 12, 18, 24, 36, 72
21. 1, 7, 49

99

SELF-CHECK
1. 2, 4   2. 2, 4
3. 2, 4, 5, 10
4. 5   5. 2, 4, 5, 10

ADDITIONAL EXAMPLES
Example 2
Write True (T) or False (F) for each statement. Give the digit sum.
1. 3 is a factor of 17,622   Ans: T; digit sum: 18
2. 2176 is divisible by 3   Ans: F; digit sum: 16

SELF-CHECK
6. 3, 9   7. 3   8. 3
9. neither

Example 3
Jim has 247 trees. He plans to plant the trees so that there are 10 trees in each row. Will the 247 trees completely fill the rows? If not, how many more trees will be needed to do this?
Ans: No; 3

SELF-CHECK
10. No   11. 2

Self-Check *Which of these numbers are divisible by 2? by 4? by 5? by 10?*

1. 48   **2.** 412   **3.** 260   **4.** 1005   **5.** 1340

> A number that is divisible by 2 is an **even number**.
> A number that is <u>not</u> divisible by 2 is an **odd number**.

EXAMPLE 2 Write <u>True</u> or <u>False</u> for each statement. Give the reason.

Solutions:

	Statements	Digit Sum	True or False? Why?
a.	113 is divisible by 3.	1 + 1 + 3 = 5	False. 5 is not divisible by 3.
b.	3 is a factor of 546.	5 + 4 + 6 = 15	True. 15 is divisible by 3.
c.	468 is divisible by 9.	4 + 6 + 8 = 18	True. 18 is divisible by 9.

Self-Check *Which of these numbers are divisible by 3? by 9?*

6. 58,014   **7.** 768   **8.** 2346   **9.** 40,693

The tests for divisibility can be used to solve problems.

EXAMPLE 3 At a reunion of the Gill Family, 157 persons were present.

  a. Can the guests be seated so that there are <u>exactly</u> 9 persons at <u>each</u> table?
  b. If not, how many more persons are needed to exactly fill the tables?

Solutions:

  a. Determine whether 157 is divisible by 9.   1 + 5 + 7 = 13   ◀ 13 is not divisible by 9.
     Since 13 is not divisible by 9, 157 is not divisible by 9.
     Therefore, 157 persons **cannot be seated** with exactly 9 persons at each table.

  b. Find the nearest whole number greater than 13 that is divisible by 9.   13 + 5 = 18   ◀ 18 is divisible by 9.

     Thus, **5 more persons** would be needed to fill the tables with exactly 9 persons per table.

Self-Check *Audrey has 148 books. She plans to build shelves holding 10 books each.*

  10. Will the 148 books completely fill the shelves?
  11. If not, how many more books will be needed to do this?

CLASSROOM EXERCISES

For Exercises 1-18, write True or False for each statement. Write a reason for each answer. (Example 1)

1. 71 is divisible by 2.
2. 375 is divisible by 5.
3. 116 is a multiple of 4.
4. 622 is a multiple of 4.
5. 385 is divisible by 10.
6. 321 is divisible by 2.
7. 5 is a factor of 4130.
8. 10 is a factor of 4130.

(Example 2)

9. 126 is divisible by 3.
10. 3 is a factor of 372.
11. 72,631 is a multiple of 9.
12. 576 is divisible by 9.
13. 648 is a multiple of 9.
14. 9 is a factor of 71,173.
15. 9 is a factor of 3429.
16. 1403 is divisible by 3.
17. 2463 is a multiple of 3.
18. 3 is a factor of 81,273.
19. 9 is a factor of 9.
20. 3 is a factor of 81.

WRITTEN EXERCISES

**Goals: To test a whole number for divisibility by 2, 3, 4, 5, 9, or 10
To apply these tests to solving word problems**

For Exercises 1-32, write True or False for each statement. Write a reason for each answer. (Example 1)

1. 428 is divisible by 2.
2. 57 is divisible by 5.
3. 95 is divisible by 5.
4. 936 is a multiple of 4.
5. 717 is a multiple of 4.
6. 9020 is divisible by 10.
7. 428 is divisible by 4.
8. 2 is a factor of 313.
9. 10 is a factor of 926.
10. 7032 is divisible by 2.

(Example 2)

11. 825 is divisible by 3.
12. 657 is divisible by 3.
13. 3 is a factor of 4321.
14. 3 is a factor of 15,423.
15. 90,426 is a multiple of 3.
16. 72,646 is a multiple of 3.
17. 74,613 is divisible by 9.
18. 35,226 is divisible by 9.
19. 671,346 is a multiple of 9.
20. 1,301,206 is a multiple of 9.

Number Theory **101**

CLASSROOM EXERCISES

1. F; does not end in 0, 2, 4, 6, 8
2. T; ends in 5
3. T; is divisible by 4
4. F; not divisible by 4
5. F; does not end in 0
6. F; does not end in 0, 2, 4, 6, 8
7. T; ends in 0
8. T; ends in 0
9. T; is divisible by 3
10. T; is divisible by 3
11. F; not divisible by 9
12. T; is divisible by 9
13. T; is divisible by 9
14. F; not divisible by 9
15. T; is divisible by 9
16. F; not divisible by 3
17. T; is divisible by 3
18. T; is divisible by 3
19. T; is divisible by 9
20. T; is divisible by 3

ASSIGNMENT GUIDE
BASIC
pp. 101-102: 1-31 odd
AVERAGE
pp. 101-102: 1-35 odd
ABOVE AVERAGE
pp. 101-102: 3, 6, 9, ..., 30, 31-36

WRITTEN EXERCISES

1. T; ends in 8
2. F; does not end in 5
3. T; ends in 5

101

WRITTEN EXERCISES

4. T; is divisible by 4
5. F; not divisible by 4
6. T; ends in 0
7. T; is divisible by 4
8. F; does not end in 0, 2, 4, 6, or 8
9. F; does not end in 0
10. T; ends in 2
11. T; is divisible by 3
12. T; is divisible by 3
13. F; not divisible by 3
14. T; is divisible by 3
15. T; is divisible by 3
16. F; not divisible by 3
17. F; not divisible by 9
18. T; is divisible by 9
19. T; is divisible by 9
20. F; not divisible by 9
21. F; does not end in 0
22. T; is divisible by 9
23. T; is divisible by 3
24. F; does not end in 5
25. T; is divisible by 4
26. T; ends in 4
27. F; does not end in 0, 2, 4, 6, or 8
28. F; not divisible by 3
29. T; ends in 5
30. T; ends in 0
31. F; not divisible by 9
32. F; not divisible by 4
33. No; 8   34. No; 2
35. No; 2   36. No; 7

MIXED PRACTICE

21. 52,374 is divisible by 10.
22. 648 is a multiple of 9.
23. 62,235 is a multiple of 3.
24. 5 is a factor of 323.
25. 4 is a factor of 7044.
26. 324 is divisible by 2.
27. 67,237 is divisible by 2.
28. 7,324,675 is a multiple of 3.
29. 5 is a factor of 675.
30. 7150 is divisible by 10.
31. 7630 is a multiple of 9.
32. 4 is a factor of 655.

APPLICATIONS: USING THE RULES FOR DIVISIBILITY

Solve. (Example 3)

33. Sheila built shelves for her collection of 982 books. Each shelf will hold 10 books.
    a. Will the 982 books completely fill the shelves?
    b. If not, how many more books are needed to do this?

34. Mr. Garcia wishes to give $70,622 in equal amounts to four different charities.
    a. Can $70,622 be divided into four equal whole-dollar amounts?
    b. If not, how many dollars must be added so that this can be done?

35. A farmer has 3788 kilograms of potatoes. He plans to sell the potatoes in bags that hold 5 kilograms each.
    a. Will all the bags be completely filled?
    b. If not, how many additional kilograms are needed to completely fill the bags?

36. To make money, a bicycle club with 9 members plans to sell 326 boxes of cookies.
    a. Can each member sell the same number of boxes?
    b. If not, how many more boxes are needed for this to be possible?

REVIEW CAPSULE FOR SECTION 4-3

Complete. (Pages 7-9)

1. In the numerical expression 3^2, 3 is called the base and the raised 2 is called the __?__.

2. In Exercise 1, the raised 2 in 3^2 indicates how many times to multiply 3 by __?__.

Evaluate. (Pages 7-9)

3. 3^2 4. 2^3 5. 4^2 6. 2^4 7. 1^2 8. 3^3

The answers are on page 120.

102   Chapter 4

4-3 Prime Numbers

Each number listed below is a *prime number*.

2, 3, 5, 7, 11, 13, 17, 19, 23

> A **prime number** is a counting number greater than 1 that has exactly two counting-number factors, the number itself and 1.

$3 = 3 \cdot 1$
$59 = 59 \cdot 1$

Numbers that are not prime are *composite numbers*.

> A **composite number** is a counting number greater than 1 that is not a prime number.

$32 = 2 \cdot 16$
$75 = 3 \cdot 5 \cdot 5$

The numbers 0 and 1 are neither prime nor composite.

PROCEDURE

To determine the smallest prime-number factor of a given number:

1. Starting with 2, write the first several prime numbers in order.
2. Test whether each prime is a factor of the given number. If none of these primes is a factor, the smallest prime factor is the number itself.

EXAMPLE 1 What is the smallest prime number that is a factor of 119?

Solution:

1. Write several prime numbers in order.   2, 3, 5, 7, 11, 13, 17, 19, 23, . . .
2. **Think**

Is 119 divisible by 2?	No. 119 is an odd number.
Is 119 divisible by 3?	No. $1 + 1 + 9 = 11$. Since 11 is not divisible by 3, 119 is not divisible by 3.
Is 119 divisible by 5?	No. 119 does not end in 0 or 5.
Is 119 divisible by 7?	$\begin{array}{r} 17 \\ 7\overline{)119} \\ \underline{7} \\ 49 \\ \underline{49} \end{array}$ Since the remainder is 0, 119 is divisible by 7.

Thus, the smallest prime factor of 119 is 7.

Number Theory **103**

SELF-CHECK
1. 3   2. 2   3. 2
4. 5   5. 31

ADDITIONAL EXAMPLES
Example 2
Use exponents to rewrite each expression.
1. 5 · 2 · 2 · 3 · 5 · 7 · 2
   Ans: $2^3 \cdot 3 \cdot 5^2 \cdot 7$
2. 7 · 7 · 7 · 2 · 3 · 2 · 3
   Ans: $2^2 \cdot 3^2 \cdot 7^3$

SELF-CHECK
6. $3^2 \cdot 5^2 \cdot 7^2 \cdot 11$
7. $2 \cdot 3^2 \cdot 5^3 \cdot 19$

CLASSROOM EXERCISES
1. prime   2. composite
3. composite   4. prime
5. prime   6. prime
7. 2   8. 11   9. 3
10. 2   11. 7   12. 2
13. $2 \cdot 3 \cdot 5^2$
14. $2^2 \cdot 3^2 \cdot 7^3$
15. $2^3 \cdot 3^2 \cdot 7 \cdot 11^2$
16. $3 \cdot 5^3 \cdot 7^2$
17. $2^4 \cdot 11^2$
18. $2 \cdot 3 \cdot 17^2 \cdot 23^3$
19. $3^4 \cdot 4^2 \cdot 7^3$
20. $3 \cdot 5^3 \cdot 11^4$

Self-Check *What is the smallest prime factor of each number?*

1. 51   2. 550   3. 62   4. 365   5. 31

Exponents are often useful in expressing a product of prime numbers.

PROCEDURE
To use exponents to express a product of prime numbers:

1. Write each prime number once as a factor.
   Write the factors in order from least to greatest.
2. For each prime number, write the exponent that indicates how many times it is a factor.

EXAMPLE 2 Use exponents to rewrite this expression: 3 · 7 · 2 · 5 · 2 · 2 · 13 · 5 · 3

Solution:
1. Starting with the smallest, list each prime factor once.   2 · 3 · 5 · 7 · 13
2. Use exponents to indicate how many times each prime number is a factor. $2^3 \cdot 3^2 \cdot 5^2 \cdot 7 \cdot 13$

Self-Check *Use exponents to rewrite each expression.*

6. 5 · 5 · 3 · 7 · 3 · 7 · 11   7. 3 · 5 · 3 · 5 · 2 · 19 · 5

NOTE: $2^3 \cdot 3^2 \cdot 5^2 \cdot 7 \cdot 13$ is read: "two cubed times three squared times five squared times seven times 13."

CLASSROOM EXERCISES

Identify each number as prime or composite. (Definitions)

1. 17   2. 55   3. 12   4. 41   5. 59   6. 29

Write the smallest prime factor of each number. (Example 1)

7. 10   8. 121   9. 9   10. 80   11. 77   12. 162

Use exponents to write each product. (Example 2)

13. 2 · 5 · 3 · 5
14. 3 · 7 · 2 · 2 · 3 · 7 · 7
15. 2 · 11 · 3 · 2 · 2 · 3 · 7 · 11
16. 7 · 5 · 5 · 5 · 3 · 7
17. 2 · 11 · 11 · 2 · 2 · 2
18. 17 · 17 · 23 · 2 · 3 · 23 · 23
19. 3 · 3 · 3 · 3 · 7 · 4 · 7 · 7 · 4
20. 11 · 11 · 11 · 3 · 5 · 5 · 5 · 11

104   Chapter 4

WRITTEN EXERCISES

Goal: To rewrite a product of prime factors with exponents

Find the smallest prime factor of each number. (Example 1)

1. 6	2. 8	3. 15	4. 21	5. 25	6. 91
7. 175	8. 696	9. 187	10. 343	11. 169	12. 84
13. 29	14. 23	15. 219	16. 1573	17. 1859	18. 2187

Use exponents to rewrite each product. (Example 2)

19. $2 \cdot 2 \cdot 3 \cdot 3 \cdot 2$
20. $3 \cdot 3 \cdot 5 \cdot 5 \cdot 3 \cdot 7 \cdot 7$
21. $2 \cdot 2 \cdot 3 \cdot 3 \cdot 2 \cdot 5 \cdot 11 \cdot 11$
22. $7 \cdot 7 \cdot 11 \cdot 2 \cdot 2 \cdot 7 \cdot 7$
23. $11 \cdot 11 \cdot 13 \cdot 13 \cdot 2 \cdot 2 \cdot 11$
24. $17 \cdot 11 \cdot 17 \cdot 13 \cdot 17 \cdot 2 \cdot 2 \cdot 2$
25. $2 \cdot 3 \cdot 3 \cdot 2 \cdot 2 \cdot 3 \cdot 5 \cdot 7 \cdot 5$
26. $3 \cdot 3 \cdot 5 \cdot 3 \cdot 5 \cdot 7 \cdot 7 \cdot 3 \cdot 5$
27. $2 \cdot 2 \cdot 3 \cdot 2 \cdot 2 \cdot 2 \cdot 3 \cdot 2 \cdot 3$
28. $2 \cdot 3 \cdot 2 \cdot 2 \cdot 3 \cdot 2 \cdot 3 \cdot 7$

REVIEW: SECTIONS 4-1 — 4-3

List all the factors of each number. (Section 4-1)

| 1. 33 | 2. 10 | 3. 40 | 4. 26 | 5. 15 | 6. 24 |
| 7. 20 | 8. 12 | 9. 7 | 10. 13 | 11. 30 | 12. 32 |

Write True or False for each statement. Give a reason for each answer. (Section 4-2)

13. 3 is a factor of 31,206.
14. 72,429 is divisible by 3.
15. 31,867 is divisible by 9.
16. 4 is a multiple of 8024.
17. 644 is a factor of 4.
18. 2 is a factor of 1379.
19. 5 is a factor of 2160.
20. 13,370 is a multiple of 10.

Find the smallest prime factor of each number. (Section 4-3)

21. 6012
22. 37,055
23. 1263
24. 539

Use exponents to rewrite each product. (Section 4-3)

25. $3 \cdot 3 \cdot 3 \cdot 5 \cdot 5$
26. $2 \cdot 2 \cdot 2 \cdot 3 \cdot 7 \cdot 7$
27. $2 \cdot 2 \cdot 3 \cdot 3 \cdot 11 \cdot 11 \cdot 11$
28. $2 \cdot 3 \cdot 3 \cdot 3 \cdot 5 \cdot 5 \cdot 5 \cdot 7 \cdot 7$
29. $3 \cdot 3 \cdot 3 \cdot 5 \cdot 5 \cdot 5$
30. $2 \cdot 2 \cdot 3 \cdot 3 \cdot 4 \cdot 4 \cdot 5 \cdot 5$

Number Theory 105

ASSIGNMENT GUIDE
BASIC
p. 105: 1-24
AVERAGE
p. 105: 1-24
ABOVE AVERAGE
p. 105: 1-17 odd, 19-28

WRITTEN EXERCISES
1. 2   2. 2   3. 3
4. 3   5. 5   6. 7
7. 5   8. 2   9. 11
10. 7 11. 13 12. 2
13. 29 14. 23 15. 3
16. 11 17. 11 18. 3
19. $2^3 \cdot 3^2$
20. $3^3 \cdot 5^2 \cdot 7^2$
21. $2^3 \cdot 3^2 \cdot 5 \cdot 11^2$
22. $2^2 \cdot 7^4 \cdot 11$
23. $2^2 \cdot 11^3 \cdot 13^2$
24. $2^3 \cdot 11 \cdot 13 \cdot 17^3$
25. $2^3 \cdot 3^3 \cdot 5^2 \cdot 7$
26. $3^4 \cdot 5^3 \cdot 7^2$
27. $2^6 \cdot 3^3$
28. $2^5 \cdot 3^3 \cdot 7$

QUIZ: SECTIONS 4-1-4-3
After completing this Review, you may want to administer a quiz covering the same sections. A Quiz is provided in the *Teacher's Edition: Part II.*

105

REVIEW: SECTIONS 4-1–4-3

1. 1, 3, 11, 33
2. 1, 2, 5, 10
3. 1, 2, 4, 5, 8, 10, 20, 40
4. 1, 2, 13, 26
5. 1, 3, 5, 15
6. 1, 2, 3, 4, 6, 8, 12, 24
7. 1, 2, 4, 5, 10, 20
8. 1, 2, 3, 4, 6, 12
9. 1, 7   10. 1, 13
11. 1, 2, 3, 5, 6, 10, 15, 30
12. 1, 2, 4, 8, 16, 32
13. T; is divisible by 3
14. T; is divisible by 3
15. F; not divisible by 9
16. T; is divisible by 4
17. T; is divisible by 4
18. F; does not end in 0, 2, 4, 6, or 8
19. T; ends in 0
20. T; ends in 0
21. 2   22. 5   23. 3
24. 7   25. $3^3 \cdot 5^2$
26. $2^3 \cdot 3 \cdot 7^2$
27. $2^2 \cdot 3^2 \cdot 11^3$
28. $2 \cdot 3^3 \cdot 5^3 \cdot 7^2$
29. $3^3 \cdot 5^3$
30. $2^2 \cdot 3^2 \cdot 4^2 \cdot 5^2$
31. No; 3   32. No; 1

CALCULATOR EXERCISES
1. 7   2. 127   3. 1023
4. 63   5. 32,767
6. 131,071

Solve. (Section 4-3)

31. A grocer has 1347 apples to pack in boxes. Each box holds 10 apples.
  a. Will all the boxes be completely filled?
  b. If not, how many more apples are needed to fill the boxes?

32. Maria wishes to spend $71 to buy gifts for three friends.
  a. Can the $71 be divided into three equal whole-dollar amounts?
  b. If not, how many more dollars are needed in order to do this?

PRIME NUMBERS AND EXPONENTS

Some prime numbers can be found by using this formula.

$$P = 2^n - 1$$

EXAMPLE: Use the formula above to find P when $n = 5$.

Solution: Use a calculator. On some calculators, you can keep pressing the $=$ key to evaluate powers.

2 [×] [=] [=] [=] [=] [−] 1 [=]   31.

EXERCISES Use the formula $P = 2^n - 1$ to find P for each value of n.

1. 3   2. 7   3. 10   4. 6   5. 15   6. 17

REVIEW CAPSULE FOR SECTION 4-4

Which of these numbers are divisible by 2? (Pages 99–102)

1. 105   2. 328   3. 400   4. 63   5. 225   6. 630

Which of these numbers are divisible by 3? (Pages 99–102)

7. 603   8. 128   9. 516   10. 71   11. 543   12. 89

Which of these numbers are divisible by 5? (Pages 99–102)

13. 125   14. 374   15. 600   16. 95   17. 101   18. 558

Find each quotient.

19. $7\overline{)161}$ 20. $3\overline{)102}$ 21. $5\overline{)1015}$ 22. $17\overline{)323}$ 23. $13\overline{)221}$

The answers are on page 120.

106   Chapter 4

CONSUMER APPLICATION
Painting and Estimation

To estimate the amount of paint needed to paint a room, you must first answer these questions.

What is the combined area of the walls and ceiling?
How much area must be subtracted for the doors and windows?

These are the **hidden questions** in the problem.

EXAMPLE A rectangular-shaped room is 4.6 meters long and 3.4 meters wide. It is 2.6 meters high. The combined area of windows and doors is 7.1 square meters. One liter of paint will cover 9 square meters. How many liters of paint are needed to paint the room?

SOLUTION
1. Find the area of the walls. Multiply the perimeter of the room by the height of the room.
   Perimeter: $2(4.6) + 2(3.4) = 16$ m
   Area of walls: $16 \times 2.6 = 41.6$ m²
2. Area of ceiling: $4.6 \times 3.4 = 15.64$ m²
3. Combined area of walls and ceiling: $41.6 + 15.64 = 57.24$ m²
4. Subtract the area of doors and windows: $57.24 - 7.1 = 50.14$ m²
5. Divide the area to be painted by the area covered by one liter of paint.
   $50.14 \div 9 = 5.57\overline{1}$ or **6 liters.** ◄ Round up to the next whole liter.

Exercises

Find the number of liters of paint needed to paint each room. Assume that one liter of paint covers 9 square meters.

Room	Length	Width	Height	Combined Area Windows–Doors
1. Kitchen	4.1 m	3.7 m	2.5 m	7.1 m²
2. Bedroom	5.4 m	4.8 m	2.7 m	10.8 m²
3. Den	3.2 m	3.6 m	2.7 m	4.5 m²
4. Family Room	6.7 m	4.2 m	2.8 m	5.3 m²

Consumer Application

CONSUMER APPLICATIONS
This feature can be used with Section 3-3 (pages 70-73) and Section 3-6 (pages 81-85), as it applies the skills for finding perimeter and area of a rectangle to solving problems that involve the use of formulas for both.

1. 6 liters
2. 9 liters
3. 5 liters
4. 10 liters

Teaching Suggestions
p. M-17

QUICK QUIZ
Evaluate.
<u>1</u>. 8 · 15   Ans: 120
<u>2</u>. 6 · 20   Ans: 120
<u>3</u>. 8 · 3 · 5   Ans: 120

Write <u>True</u> or <u>False</u> for each statement.
<u>4</u>. 2015 is divisible by 2.
  Ans: False
<u>5</u>. 3 is a factor of 1911.
  Ans: True

ADDITIONAL EXAMPLES
Example 1
Write the prime factorization of each number.
<u>1</u>. 110   Ans: 2 · 5 · 11
<u>2</u>. 195   Ans: 3 · 5 · 13
<u>3</u>. 483   Ans: 3 · 7 · 23

SELF-CHECK
<u>1</u>. 7 · 11   <u>2</u>. 3 · 11
<u>3</u>. 2 · 3 · 5
<u>4</u>. 2 · 3 · 5 · 7

4-4 Prime Factorization

Every composite number can be expressed as a product of prime factors.

$75 = 3 \cdot 25$ ← **25 is a composite number.**

$ = 3 \cdot 5 \cdot 5$ ← **Each factor is a prime number.**

The product, 3 · 5 · 5, is called the **prime factorization** of 75. In a prime factorization, the factors are usually written in order from least to greatest.

The prime numbers less than 24 are listed below. Refer to them in the Examples and Exercises.

2, 3, 5, 7, 11, 13, 17, 19, 23

EXAMPLE 1   Write the prime factorization of 105.

Solution: [1] Is 105 divisible by 2? No; 105 is an odd number.

[2] Is 105 divisible by 3? Yes, by the rule for divisibility by 3. So divide.

$3\overline{)105} \atop 35$ **Not a prime number**   The quotient is **35**.

[3] Is 35 divisible by 3? No

[4] Is 35 divisible by 5? Yes, by the rule for divisibility by 5. So divide.

$3\overline{)105} \atop 5\overline{)35} \atop 7$ **Prime number**   The quotient is **7**.

Since 7 is a prime number, all the prime factors of 105 have been determined.

[5] Write the prime factorization of 105.
$105 = 3 \cdot 5 \cdot 7$

Self-Check   Write the prime factorization of each number.

1. 77   **2.** 33   **3.** 30   **4.** 210

When the same prime number appears more than once in a prime factorization, exponents are often used to indicate the number of times each prime factor occurs.

108   Chapter 4

EXAMPLE 2 Write the prime factorization of 126.

Solution:

[1] Is 126 divisible by 2? Yes. Divide. The quotient is 63.

$\begin{array}{r}2\underline{|126}\\63\end{array}$ ◄ Not a prime number

[2] Is 63 divisible by 2? No.

[3] Is 63 divisible by 3? Yes. Divide. The quotient is 21.

$\begin{array}{r}2\underline{|126}\\3\underline{|63}\\21\end{array}$ ◄ Not a prime number

[4] Is 21 divisible by 3? Yes. Divide. The quotient is 7. Since 7 is a prime number, all the prime factors of 126 have been found.

$\begin{array}{r}2\underline{|126}\\3\underline{|63}\\3\underline{|21}\\7\end{array}$ ◄ Prime number

[5] Write the prime factorization: $126 = 2 \cdot 3 \cdot 3 \cdot 7$, or $2 \cdot 3^2 \cdot 7$.

Self-Check *Write the prime factorization of each number.*

5. 20 **6.** 18 **7.** 100 **8.** 242

CLASSROOM EXERCISES

Write the prime factorization of each number.
(Step 5, Examples 1 and 2)

1. 455
$\begin{array}{r}5\underline{|455}\\7\underline{|91}\\13\end{array}$

2. 261
$\begin{array}{r}3\underline{|261}\\3\underline{|87}\\29\end{array}$

3. 150
$\begin{array}{r}2\underline{|150}\\3\underline{|75}\\5\underline{|25}\\5\end{array}$

4. 204
$\begin{array}{r}2\underline{|204}\\2\underline{|102}\\3\underline{|51}\\17\end{array}$

5. 492
$\begin{array}{r}2\underline{|492}\\2\underline{|246}\\3\underline{|123}\\41\end{array}$

6. 315
$\begin{array}{r}3\underline{|315}\\3\underline{|105}\\5\underline{|35}\\7\end{array}$

7. 198
$\begin{array}{r}?\underline{|198}\\?\underline{|99}\\?\underline{|33}\\11\end{array}$

8. 220
$\begin{array}{r}?\underline{|220}\\?\underline{|110}\\?\underline{|55}\\11\end{array}$

9. 261
$\begin{array}{r}?\underline{|261}\\?\underline{|87}\\29\end{array}$

10. 396
$\begin{array}{r}?\underline{|396}\\?\underline{|198}\\?\underline{|99}\\?\underline{|33}\\11\end{array}$

11. 450
$\begin{array}{r}?\underline{|450}\\?\underline{|225}\\?\underline{|75}\\?\underline{|25}\\5\end{array}$

12. 312
$\begin{array}{r}?\underline{|312}\\?\underline{|156}\\?\underline{|78}\\?\underline{|39}\\13\end{array}$

(Example 1)

13. 35 **14.** 70 **15.** 15 **16.** 165 **17.** 130 **18.** 182

(Example 2)

19. 36 **20.** 135 **21.** 140 **22.** 525 **23.** 594 **24.** 375

Number Theory

ADDITIONAL EXAMPLES
Example 2
Write the prime factorization of each number.
1. 60   Ans: $2^2 \cdot 3 \cdot 5$
2. 1089 Ans: $3^2 \cdot 11^2$

SELF-CHECK
5. $2^2 \cdot 5$ 6. $2 \cdot 3^2$
7. $2^2 \cdot 5^2$ 8. $2 \cdot 11^2$

CLASSROOM EXERCISES
1. $5 \cdot 7 \cdot 13$
2. $3^2 \cdot 29$
3. $2 \cdot 3 \cdot 5^2$
4. $2^2 \cdot 3 \cdot 17$
5. $2^2 \cdot 3 \cdot 41$
6. $3^2 \cdot 5 \cdot 7$
7. $2 \cdot 3^2 \cdot 11$
8. $2^2 \cdot 5 \cdot 11$
9. $3^2 \cdot 29$
10. $2^2 \cdot 3^2 \cdot 11$
11. $2 \cdot 3^2 \cdot 5^2$
12. $2^3 \cdot 3 \cdot 13$
13. $5 \cdot 7$
14. $2 \cdot 5 \cdot 7$
15. $3 \cdot 5$
16. $3 \cdot 5 \cdot 11$
17. $2 \cdot 5 \cdot 13$
18. $2 \cdot 7 \cdot 13$
19. $2^2 \cdot 3^2$
20. $3^3 \cdot 5$
21. $2^2 \cdot 5 \cdot 7$
22. $3 \cdot 5^2 \cdot 7$
23. $2 \cdot 3^3 \cdot 11$
24. $3 \cdot 5^3$

ASSIGNMENT GUIDE

BASIC
Day 1 p. 110: 1–20
Day 2 p. 110: 31–48

AVERAGE
Day 1 p. 110: 1–20
Day 2 p. 110: 31–54

ABOVE AVERAGE
p. 110: 3, 6, 9, ..., 60

WRITTEN EXERCISES

1. $2 \cdot 3$ 2. $5 \cdot 11$
3. $2 \cdot 17$ 4. $3 \cdot 13$
5. $5 \cdot 11$ 6. $5 \cdot 17$
7. $2 \cdot 5 \cdot 11$
8. $11 \cdot 13$ 9. $5 \cdot 17$
10. $2 \cdot 19$ 11. $7 \cdot 13$
12. $3 \cdot 7 \cdot 11$
13. $2^3 \cdot 7$ 14. $2 \cdot 3^3$
15. $2 \cdot 5^2$ 16. $2^4 \cdot 3$
17. $2 \cdot 3 \cdot 13$
18. $3 \cdot 5 \cdot 7$
19. $2^2 \cdot 3 \cdot 7$
20. $2^2 \cdot 3 \cdot 5$
21. 2^7 22. $2^3 \cdot 3 \cdot 7$
23. 2^6 24. 5^3
25. $2 \cdot 3^3 \cdot 7$
26. $2^2 \cdot 7 \cdot 13$
27. $2^2 \cdot 3^4$
28. $2^5 \cdot 3^2$
29. $2^2 \cdot 3^3 \cdot 5$
30. $2^2 \cdot 3 \cdot 5 \cdot 7$
31. $2^2 \cdot 3^2 \cdot 5$
32. $3 \cdot 7 \cdot 19$
33. $7 \cdot 17$ 34. $5 \cdot 23$

See page 111 for the answers to Exercises 35–60.

WRITTEN EXERCISES

Goal: To write the prime factorization of a composite number

Write the prime factorization of each number. (Example 1)

| 1. 6 | 2. 55 | 3. 34 | 4. 39 | 5. 55 | 6. 85 |
| 7. 110 | 8. 143 | 9. 85 | 10. 38 | 11. 91 | 12. 231 |

(Example 2)

13. 56	14. 54	15. 50	16. 48	17. 78	18. 105
19. 84	20. 60	21. 128	22. 168	23. 64	24. 125
25. 378	26. 364	27. 324	28. 288	29. 540	30. 420

MIXED PRACTICE

31. 180	32. 399	33. 119	34. 115	35. 468	36. 133
37. 187	38. 504	39. 512	40. 729	41. 484	42. 418
43. 1768	44. 1071	45. 2431	46. 221	47. 323	48. 1547
49. 3185	50. 4998	51. 437	52. 391	53. 1150	54. 1425
55. 4655	56. 546	57. 42,588	58. 250,965	59. 12,675	60. 17,850

REVIEW CAPSULE FOR SECTION 4-5

For Exercises 1–5, list the first six multiples of each number. (Pages 96–98)

| 1. 2 | 2. 3 | 3. 4 | 4. 5 | 5. 6 |

6. List any numbers that are in the answers to <u>both</u> Exercises 1 and 2.
7. List any numbers that are in the answers to <u>both</u> Exercises 3 and 5.
8. List any numbers that are in the answers to <u>both</u> Exercises 3 and 6.

Use exponents to rewrite each expression. (Pages 103–105)

9. $2 \cdot 2 \cdot 5 \cdot 3 \cdot 3 \cdot 3$
10. $2 \cdot 2 \cdot 5 \cdot 5 \cdot 2 \cdot 5$
11. $2 \cdot 2 \cdot 2 \cdot 3 \cdot 3 \cdot 5 \cdot 5 \cdot 2 \cdot 3$
12. $5 \cdot 11 \cdot 11 \cdot 5 \cdot 5 \cdot 11 \cdot 17 \cdot 5$
13. $2 \cdot 3 \cdot 4 \cdot 3 \cdot 2 \cdot 2 \cdot 3 \cdot 4$
14. $6 \cdot 9 \cdot 9 \cdot 6 \cdot 6 \cdot 9 \cdot 12 \cdot 12$

The answers are on page 120.

110 Chapter 4

4-5 Least Common Multiple

Brian and his sister Sara have part-time jobs after school. Brian works every third day and Sara works every fourth day. If both start to work on the same day, what is the first working day after this when they *both* will be working at their part-time jobs?

Sara solved this problem by using *common multiples* of 3 and 4.

12 is a **common multiple** of 3 and 4.

24 is a **common multiple** of 3 and 4.

Multiples of 3: 3, 6, 9, **12**, 15, 18, 21, **24**, 27, ...
Multiples of 4: 4, 8, **12**, 16, 20, **24**, 28, 32, 36, ...

Answer: Since 12 is the smallest common multiple of 3 and 4, they would both be working again at their part-time jobs on the **twelfth day**.

The number 12 is the *least common multiple* of 3 and 4.

| The **least common multiple (LCM)** of two or more counting numbers is the smallest counting number that is divisible by the given numbers. | LCM of 2 and 3: **6**. |

You can use prime factors to find the LCM of two, three, or more counting numbers.

PROCEDURE
To find the LCM of two or more counting numbers:

1. Write the prime factorization of each number.
2. Write a product using each prime factor only once.
3. For each prime factor, write the highest exponent used for that factor in any of the prime factorizations of the numbers.
4. Multiply these factors.

Number Theory

Teaching Suggestions p. M-17

QUICK QUIZ
List the first six multiples.
1. 10   Ans: 10, 20, 30, 40, 50, 60
2. 15   Ans: 15, 30, 45, 60, 75, 90
3. 6   Ans: 6, 12, 18, 24, 30, 36
4. 4   Ans: 4, 8, 12, 16, 20, 24

WRITTEN EXERCISES p. 110
35. $2^2 \cdot 3^2 \cdot 13$
36. $7 \cdot 19$ 37. $11 \cdot 17$
38. $2^3 \cdot 3^2 \cdot 7$ 39. 2^9
40. 3^6 41. $2^2 \cdot 11^2$
42. $2 \cdot 11 \cdot 19$
43. $2^3 \cdot 13 \cdot 17$
44. $3^2 \cdot 7 \cdot 17$
45. $11 \cdot 13 \cdot 17$
46. $13 \cdot 17$
47. $17 \cdot 19$
48. $7 \cdot 13 \cdot 17$
49. $5 \cdot 7^2 \cdot 13$
50. $2 \cdot 3 \cdot 7^2 \cdot 17$
51. $19 \cdot 23$ 52. $17 \cdot 23$
53. $2 \cdot 5^2 \cdot 23$
54. $3 \cdot 5^2 \cdot 19$
55. $5 \cdot 7^2 \cdot 19$
56. $2 \cdot 3 \cdot 7 \cdot 13$
57. $2^2 \cdot 3^2 \cdot 7 \cdot 13^2$
58. $3^3 \cdot 5 \cdot 11 \cdot 13^2$
59. $3 \cdot 5^2 \cdot 13^2$
60. $2 \cdot 3 \cdot 5^2 \cdot 7 \cdot 17$

ADDITIONAL EXAMPLES

Example 1

Find the LCM.

1. 8 and 10   Ans: 40
2. 25 and 15 Ans: 75
3. 20 and 70 Ans: 140

SELF-CHECK

1. 8   2. 84   3. 63

Example 2

Write the LCM of each group of numbers.

1. 6, 10, and 15
   Ans: 30
2. 4, 12 and 25
   Ans: 300
3. 36, 27, and 225
   Ans: 2700

SELF-CHECK

4. 12   5. 54   6. 300

EXAMPLE 1 Find the least common multiple (LCM).

　　　　　　　a 4 and 6　　　　　b. 12 and 18

Solutions:　　　　　　　a.　　　　　　　　　　b

[1] Prime factorization:　　$4 = 2 \cdot 2 = 2^2$　　　　[1] $12 = 2 \cdot 2 \cdot 3 = 2^2 \cdot 3$
　　Prime factorization:　　$6 = 2 \cdot 3$　　　　　　　$18 = 2 \cdot 3 \cdot 3 = 2 \cdot 3^2$

[2] Write a product. Use　$2 \cdot 3$　　　　　　　　[2] $2 \cdot 3$
　　each prime factor once.

[3] For each prime fac-　　$2^2 \cdot 3$　　　　　　　　[3] $2^2 \cdot 3^2$
　　tor, write the highest
　　exponent used in any
　　of the prime factoriza-
　　tions.

[4] Multiply.　　　　　　　$4 \cdot 3 = 12$ ◀ LCM of 4 and 6　　[4] $4 \cdot 9 = 36$ ◀ LCM of 12 and 18

Self-Check Write the LCM of each pair of numbers.

　　1. 4 and 8　　　2. 12 and 14　　　3. 9 and 21

NOTE: The LCM of two numbers is always equal to, or larger than, either of the numbers.

Follow the same procedure to find the LCM of three or more counting numbers.

EXAMPLE 2 Find the least common multiple of 8, 36, and 45.

Solution: [1] Prime factorization:　　　　　　$8 = 2 \cdot 2 \cdot 2 = 2^3$
　　　　　　　Prime factorization:　　　　　　$36 = 2 \cdot 2 \cdot 3 \cdot 3 = 2^2 \cdot 3^2$
　　　　　　　Prime factorization:　　　　　　$45 = 3 \cdot 3 \cdot 5 = 3^2 \cdot 5$

[2] Product that includes each prime　　$2 \cdot 3 \cdot 5$
　　factor once.

[3] Product that shows the highest　　$2^3 \cdot 3^2 \cdot 5$
　　exponent for each factor.

[4] Multiply.　　　　　　　　　　　　$2^3 \cdot 3^2 \cdot 5 = 8 \cdot 9 \cdot 5 = 360$

　　　　　　　　LCM of 8, 36, and 45: **360**

Self-Check Write the LCM of each group of numbers.

　　4. 4, 6, and 12　　　5. 6, 18, and 27　　　6. 15, 20, and 25

112 Chapter 4

CLASSROOM EXERCISES

The prime factorizations of two or three numbers are given. Write the prime factorization of their least common multiple.
(Step 3, Examples 1 and 2)

1. $2 \cdot 3^2$
   $2^2 \cdot 3$
2. $2^2 \cdot 3^2 \cdot 5$
   $2^3 \cdot 3$
3. $2^2 \cdot 3^3 \cdot 5^2$
   $2^3 \cdot 3^2 \cdot 7$
4. $11 \cdot 17$
   11^2
5. $2^3 \cdot 3^2 \cdot 5^3$
   $2^2 \cdot 3^3 \cdot 5$
6. $2 \cdot 3^2 \cdot 7$
   2^4
7. $2 \cdot 3 \cdot 5$
   $3 \cdot 5$
   $2^3 \cdot 3^2$
8. $2 \cdot 3 \cdot 5^2$
   $2^2 \cdot 3^3$
   5

Find the least common multiple. (Example 1)

9. 4 and 10
10. 3 and 27
11. 3 and 5
12. 2 and 3
13. 7 and 27
14. 6 and 24
15. 27 and 18
16. 4 and 18
17. 3 and 7
18. 2 and 11
19. 15 and 25
20. 6 and 14
21. 120 and 16
22. 135 and 54
23. 240 and 16
24. 216 and 45

(Example 2)

25. 3, 5, and 15
26. 9, 5, and 14
27. 3, 15, and 27
28. 6, 27, and 10
29. 3, 9, and 27
30. 5, 10, and 15
31. 3, 15, and 25
32. 2, 27, and 18
33. 7, 14, and 21
34. 15, 27, and 25
35. 9, 6, and 4
36. 2, 16, and 24

WRITTEN EXERCISES

Goals: To find the least common multiple of two or more numbers

To apply the skill of finding the least common multiple to solving word problems

For Exercises 1-48, write the least common multiple. (Example 1)

1. 15 and 75
2. 8 and 40
3. 6 and 20
4. 45 and 28
5. 3 and 11
6. 5 and 13
7. 28 and 40
8. 36 and 405
9. 12 and 56
10. 45 and 27
11. 5 and 27
12. 45 and 7
13. 49 and 42
14. 18 and 168
15. 42 and 96
16. 120 and 150

(Example 2)

17. 8, 28, and 21
18. 20, 30, and 50
19. 36, 32, and 12
20. 50, 35, and 28
21. 10, 15, and 24
22. 54, 45, and 36
23. 15, 10, and 16
24. 15, 35, and 49
25. 135, 225, and 405
26. 27, 45, and 100
27. 24, 30, and 42
28. 16, 20, and 48

Number Theory   113

CLASSROOM EXERCISES

1. $2^2 \cdot 3^2$
2. $2^3 \cdot 3^2 \cdot 5$
3. $2^3 \cdot 3^3 \cdot 5^2 \cdot 7$
4. $11^2 \cdot 17$
5. $2^3 \cdot 3^3 \cdot 5^3$
6. $2^4 \cdot 3^2 \cdot 7$
7. $2^3 \cdot 3^2 \cdot 5$
8. $2^2 \cdot 3^3 \cdot 5^2$
9. 20   10. 27   11. 15
12. 6   13. 189   14. 24
15. 54   16. 36   17. 21
18. 22   19. 75   20. 42
21. 240   22. 270
23. 240   24. 1080
25. 15   26. 630
27. 135   28. 270
29. 27   30. 30   31. 75
32. 54   33. 42
34. 675   35. 36
36. 48

ASSIGNMENT GUIDE
BASIC
pp. 113-14: 1-45 odd
AVERAGE
pp. 113-114: 1-45 odd
ABOVE AVERAGE
pp. 113-114: 3, 6, 9, ..., 47, 49-52

WRITTEN EXERCISES

1. 75   2. 40   3. 60
4. 1260   5. 33   6. 65
7. 280   8. 1620   9. 168
10. 135   11. 135
12. 315   13. 294
14. 504   15. 672

WRITTEN EXERCISES

16. 600   17. 168
18. 300   19. 288
20. 700   21. 120
22. 540   23. 240
24. 735   25. 2025
26. 2700 27. 840
28. 240   29. 200
30. 80   31. 180
32. 240   33. 2100
34. 60   35. 216
36. 336   37. 187
38. 81   39. 319
40. 128   41. 120
42. 720   43. 630
44. 540   45. 144
46. 360   47. 210
48. 2520
49. 12th day
50. 60 days
51. 360 in
52. $144

MIXED PRACTICE

29. 5, 25, and 40   **30.** 16 and 40   **31.** 6, 36, and 30   **32.** 16, 6, and 30
33. 42, 100, and 14   **34.** 6 and 20   **35.** 54 and 8   **36.** 24, 28, and 16
37. 11 and 17   **38.** 9, 27, and 81   **39.** 11 and 29   **40.** 8, 16, and 128
41. 30, 24, and 20   **42.** 9, 20, and 48   **43.** 30 and 126   **44.** 108 and 90
45. 48, 12, and 36   **46.** 72 and 45   **47.** 21 and 70   **48.** 18, 72, and 105

APPLICATIONS: USING LEAST COMMON MULTIPLES

49. Jill, Jean, and Alex have part-time jobs after school. Jill works every third day, Jean every fourth day, and Alex every other day. If all three start to work on the same day, on what working day after that will <u>all three</u> again be working at their part-time jobs?

50. In a grocery store, produce is delivered every other business day, milk every third business day, and paper goods every twentieth business day. All three of these items are delivered on Tuesday. How many business days will pass before <u>all three</u> are again delivered on the same day?

51. In a warehouse, cartons 36, 40, and 24 inches tall are being stacked next to each other in three separate piles. What is the shortest height possible for all three piles to have the same height?

52. May earns $16 per hour and Joan earns $18 per hour. On Thursday, both girls worked a whole number of hours and they both earned the same amount. What is the least amount they could have earned?

REVIEW CAPSULE FOR SECTION 4-6

List all the factors of each number. (Pages 96–98)

1. 8   **2.** 10   **3.** 15   **4.** 24   **5.** 36   **6.** 42

Write the prime factorization of each number. (Pages 108–110)

7. 6   **8.** 16   **9.** 20   **10.** 32   **11.** 45   **12.** 50

The answers are on page 120.

114 Chapter 4

4-6 Greatest Common Factor

Martin made a list of the factors of 12 and 18. He noticed that 12 and 18 have some of the same factors. These are called **common factors**.

Factors of 12: ①, ②, ③, 4, ⑥, 12
Factors of 18: ①, ②, ③, ⑥, 9, 18

Common factors: **1, 2, 3, 6** ◀ The greatest common factor (GCF) is 6.

> The **greatest common factor (GCF)** of two or more counting numbers is the greatest number that is a factor of each number.

The table below shows how to find the GCF by listing the factors of the numbers.

TABLE

Numbers	Factors	Common Factors	GCF
16	①, ②, ④, ⑧, 16	1, 2, 4, 8	8
24	①, ②, 3, ④, 6, ⑧, 12, 24		
9	①, ③, 9	1, 3	3
15	①, ③, 5, 15		
10	①, ②, ⑤, ⑩	1, 2, 5, 10	10
20	①, ②, 4, ⑤, ⑩, 20		

Another way to find the GCF is to write the prime factorization of each number.

> ### PROCEDURE
> To determine the greatest common factor (GCF) of two or more numbers:
>
> **1** Write the prime factorization of each number. Identify the common factors.
>
> **2** Write the common factors as a product.

Number Theory **115**

Teaching Suggestions
p. M-17

QUICK QUIZ

List all the factors of each number.

1. 20 Ans: 1, 2, 4, 5, 10, 20
2. 35 Ans: 1, 5, 7, 35
3. 56 Ans: 1, 2, 4, 7, 8, 14, 28, 56

Write the prime factorization of each number.

4. 44 Ans: $2^2 \cdot 11$
5. 36 Ans: $2^2 \cdot 3^2$

Margin Notes

ADDITIONAL EXAMPLE
Find the GCF of 18 and 54.   Ans: 18

SELF-CHECK
1. 2   2. 4   3. 3

CLASSROOM EXERCISES
1. 2   2. 4   3. 2
4. 9   5. 2 · 3 · 3; 2

ASSIGNMENT GUIDE
BASIC
p. 116: 1–23
AVERAGE
p. 116: 1–23
ABOVE AVERAGE
p. 116: 1–23 odd

WRITTEN EXERCISES
1. 3; 5; 2   2. 5; 5; 15
3. 3; 3; 6   4. 3   5. 2
6. 10   7. 3   8. 15
9. 8   10. 12   11. 4
12. 20   13. 15   14. 1
15. 15   16. 2   17. 6
18. 3   19. 5   20. 3
21. 4   22. 11   23. 9

EXAMPLE   Find the GCF of 24 and 36.

Solution:
1. 24 = ②·②·2·③
    36 = ②·②·③·3   Common factors: 2, 2, and 3
2. GCF: 2 · 2 · 3 = **12**

Self-Check   *Find the GCF of each pair of numbers.*

1. 4 and 10   2. 16 and 28   3. 15 and 48

CLASSROOM EXERCISES

Complete. (Table)

1. Factors of 6: 1, 2, 3, 6
   Factors of 8: 1, 2, 4, 8
   Common factors: 1, 2
   GCF: __?__

2. Factors of 12: 1, 2, 3, 4, 6, 12
   Factors of 16: 1, 2, 4, 8, 16
   Common factors: 1, 2, 4
   GCF: __?__

(Example)

3. 16 = 2 · 2 · 2 · 2
   30 = 2 · 3 · 5
   GCF: __?__

4. 27 = 3 · 3 · 3
   36 = 2 · 2 · 3 · 3
   GCF: __?__

5. 8 = 2 · 2 · 2
   18 = __?__
   GCF: __?__

WRITTEN EXERCISES

Goal: To find the greatest common factor (GCF) of two or more numbers

Complete. Find the GCF of each pair of numbers.
(Table and Example)

1. 6 = 2 × __?__
   10 = 2 × __?__
   GCF: __?__

2. 30 = 2 × 3 × __?__
   45 = 3 × 3 × __?__
   GCF: 3 × __?__ = __?__

3. 18 = 2 × 3 × __?__
   24 = 2 × 2 × 2 × __?__
   GCF: 2 × __?__ = __?__

Find the GCF. (Table and Example)

4. 15 and 18
5. 8 and 14
6. 20 and 30
7. 9 and 15
8. 30 and 45
9. 40 and 24
10. 36 and 24
11. 16 and 60
12. 60 and 100
13. 45 and 75
14. 9 and 25
15. 15 and 30
16. 4, 8, 10
17. 12, 18, 24
18. 6, 9, 15
19. 10, 15, 25
20. 12, 15, 21
21. 16, 20, 24
22. 22, 66, 88
23. 18, 27, 36

116   Chapter 4

REVIEW: SECTIONS 4-4 — 4-6

Write the prime factorization of each number. (Section 4-4)

1. 48	2. 40	3. 180	4. 200	5. 125	6. 66
7. 75	8. 340	9. 95	10. 363	11. 288	12. 50

Write the least common multiple. (Section 4-5)

13. 8 and 40	14. 18, 20, and 24	15. 12, 14, and 16	16. 50 and 60
17. 18, 36, and 210	18. 42 and 70	19. 13 and 19	20. 14, 20, and 35

Solve. (Section 4-5)

21. Tim and Michelle have part-time jobs after school. Tim works every third day and Michelle works every fourth day. If they both work on Wednesday, how many days will pass before they both will again be working on the same day? (Section 4-5)

22. In a meat market, pork is delivered every third business day, lamb every eighth business day, and beef every other business day. All three of these items are delivered on Monday. How many business days will pass before all three items are again delivered on the same day? (Section 4-5)

Find the GCF of each pair of numbers. (Section 4-6)

23. 18 and 72	24. 40 and 24	25. 12 and 108	26. 22 and 77
27. 63 and 15	28. 20 and 25	29. 28 and 36	30. 24 and 42

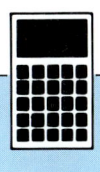

PRIME NUMBERS

Some prime numbers can be determined by evaluating the formula $n^2 - n + 41$ for values of n from 1 through 40.

EXAMPLE What prime number is produced by evaluating $n^2 - n + 41$ for $n = 29$?

Solution $n^2 - n + 41 = 29^2 - 29 + 41$

Use a calculator with a $\boxed{x^2}$ key.

29 $\boxed{x^2}$ $\boxed{-}$ 29 $\boxed{+}$ $\boxed{=}$ 853.

EXERCISES Determine the prime number produced when $n^2 - n + 41$ is evaluated for each value of n.

1. 10   2. 17   3. 26   4. 35   5. 38   6. 40

QUIZ: SECTIONS 4-4-4-6
After completing this Review, you may want to administer a quiz covering the same sections. A Quiz is provided in the *Teacher's Edition: Part II.*

REVIEW: SECTIONS 4-4-4-6
1. $2^4 \cdot 3$ 2. $2^3 \cdot 5$
3. $2^2 \cdot 3^2 \cdot 5$
4. $2^3 \cdot 5^2$
5. 5^3 6. $2 \cdot 3 \cdot 11$
7. $3 \cdot 5^2$
8. $2^2 \cdot 5 \cdot 17$
9. $5 \cdot 19$ 10. $3 \cdot 11^2$
11. $2^5 \cdot 3^2$
12. $2 \cdot 5^2$ 13. 40
14. 360 15. 336
16. 300 17. 1260
18. 210 19. 247
20. 140 21. 12 days
22. 24 days 23. 18
24. 8 25. 12 26. 11
27. 3 28. 5 29. 4
30. 6

CALCULATOR EXERCISES
1. 131 2. 313
3. 691 4. 1231
5. 1447 6. 1601

CHAPTER REVIEW

1. digits
2. prime number
3. least common multiple
4. divisible by
5. composite numbers
6. prime factorization
7. even number
8. factor
9. counting numbers
10. whole numbers
11. multiple
12. odd number
13. 1, 2, 7, 14
14. 1, 2, 3, 4, 6, 8, 12, 24
15. 1, 2, 3, 4, 6, 9, 12, 18, 36
16. 1, 2, 4, 5, 8, 10, 20, 40
17. 1, 2, 3, 6, 9, 18
18. 1, 2, 4, 8, 16, 32
19. 1, 2, 5, 10
20. 1, 2, 3, 6
21. 1, 17   22. 1, 13
23. 1, 2, 3, 4, 6, 8, 9, 12, 18, 24, 36, 70
24. 1, 2, 4, 5, 8, 10, 16, 20, 40, 80
25. T; ends in 5
26. T; is divisible by 3
27. F; not divisible by 9
28. T; ends in 0
29. F; does not end in 0, 2, 4, 6, or 8
30. T; is divisible by 4

118

CHAPTER REVIEW

PART 1: VOCABULARY

For Exercises 1–12, choose from the box at the right the word(s) that best correspond to each description.

1. The name of the symbols 0, 1, 2, 3, 4, 5, 6, 7, 8, and 9 __?__
2. A counting number greater than 1 that has exactly two factors __?__
3. The smallest counting number divisible by two or more counting numbers __?__
4. If the remainder is zero when a first number is divided by a second, the first number is said to be __?__ the second.
5. Counting numbers greater than 1 that are not prime __?__
6. Expressing a counting number as a product of primes __?__
7. A number that is divisible by 2 __?__
8. Since 24 is divisible by 4, 4 is a __?__ of 24.
9. The numbers 1, 2, 3, 4, . . . __?__
10. The counting numbers and zero __?__
11. A counting number is a __?__ of each of its factors.
12. A number not divisible by 2 __?__

> divisible by
> factor
> counting numbers
> multiple
> whole numbers
> digits
> even number
> odd number
> prime number
> composite numbers
> prime factorization
> least common multiple

PART 2: SKILLS

Write the factors of each number. (Section 4–1)

| 13. 14 | 14. 24 | 15. 36 | 16. 40 | 17. 18 | 18. 32 |
| 19. 10 | 20. 6 | 21. 17 | 22. 13 | 23. 72 | 24. 80 |

Write True or False for each statement. Give a reason based on the rules on page 99. (Section 4-2)

25. 435 is divisible by 5.
26. 9723 is divisible by 3.
27. 9 is a factor of 103,421.
28. 6420 is a multiple of 10.
29. 531 is divisible by 2.
30. 1024 is divisible by 4.

118   Chapter 4

Use exponents to rewrite each product. (Section 4-3)

31. 2 · 3 · 5 · 3 · 3 · 7 · 5 · 11 · 3
32. 3 · 5 · 2 · 2 · 3 · 2 · 5
33. 5 · 27 · 3 · 5 · 27 · 5 · 5
34. 2 · 5 · 11 · 11 · 11 · 2 · 2 · 2 · 5 · 7 · 7

Write the prime factorization of each number. (Section 4-4)

35. 105   **36.** 44   **37.** 300   **38.** 1210   **39.** 88   **40.** 240

The prime factorizations of two or three numbers are given below. Write the prime factorization of their least common multiple. (Section 4-5)

41. 2 · 3² · 5³ · 13; 2 · 3 · 5 · 13²
42. 2⁵ · 3 · 7³ · 11; 2³ · 5² · 11⁴
43. 2 · 3 · 5³; 3⁴ · 5 · 7; 2⁴ · 5³ · 11
44. 2⁴ · 5 · 17; 2³ · 5⁴ · 11; 2² · 5² · 11³

Find the least common multiple of the given numbers. (Section 4-5)

45. 8 and 30   **46.** 15 and 14   **47.** 36, 42, and 18   **48.** 15, 18, and 50

Find the GCF. (Section 4-6)

49. 35 and 49   **50.** 16 and 34   **51.** 9 and 24   **52.** 20 and 8
53. 5, 15, 25   **54.** 8, 10, 16   **55.** 9, 12, 18   **56.** 21, 35, 14

PART 3: APPLICATIONS

57. A manager of a grocery store has 121 pounds of onions that are to be put into 3 pound bags.
  a. Will all the bags be completely filled?
  b. If not, how many additional pounds of onions are needed to completely fill the bags? (Section 4-1)

58. Nine high school students earned $4967 one summer.
  a. Can the $4967 be divided into nine equal whole-dollar amounts?
  b. If not, how many more dollars are needed in order to do this? (Section 4-1)

59. In a grocery store, boxes 20 inches tall, 24 inches tall, and 45 inches tall are being stacked next to each other in three piles. What is the shortest height possible so that all three piles will have the same height? (Section 4-5)

CHAPTER REVIEW

31. 2 · 3⁴ · 5² · 7 · 11
32. 2³ · 3² · 5²
33. 3 · 5⁴ · 27²
34. 2⁴ · 5² · 7² · 11³
35. 3 · 5 · 7
36. 2² · 11
37. 2² · 3 · 5²
38. 2 · 5 · 11²
39. 2³ · 11
40. 2⁴ · 3 · 5
41. 2 · 3² · 5³ · 13²
42. 2⁵ · 3 · 5² · 7³ · 11⁴
43. 2⁴ · 3⁴ · 5³ · 7 · 11
44. 2⁴ · 5⁴ · 11³ · 17
45. 120   46. 210
47. 252   48. 450
49. 7   50. 2   51. 3
52. 4   53. 5   54. 2
55. 3   56. 7
57. a. No   b. 2
58. a. No   b. 1
59. 360

Number Theory

CHAPTER TEST

Two forms of a chapter test, Form A and Form B, are provided on copying masters in the *Teacher's Edition: Part II*.

1. 1, 2, 3, 5, 6, 10, 15, 30
2. 1, 11
3. 1, 2, 3, 4, 6, 8, 12, 24
4. 1, 2, 4, 5, 8, 10, 20, 40
5. T; is divisible by 3
6. T; is divisible by 4
7. F; not divisible by 15
8. T; is divisible by 9
9. 7
10. $2^4 \cdot 3^3 \cdot 5^2 \cdot 11$
11. $2^4 \cdot 3^3 \cdot 5^2$
12. $2^4 \cdot 3^4 \cdot 5^2$
13. $5^5 \cdot 7^2 \cdot 11 \cdot 13$
14. $2 \cdot 3 \cdot 7$
15. $2 \cdot 3^2 \cdot 5$
16. $3^2 \cdot 5^2$
17. $2^3 \cdot 3^2 \cdot 5$
18. 1400 19. 1
20. 60 days
21. No; 1

CHAPTER TEST

List all of the factors of each number.

1. 30
2. 11
3. 24
4. 40

For Exercises 5–8, write True or False for each sentence. Give a reason for each answer.

5. 47,209,131 is divisible by 3.
6. 130,244 is a multiple of 4.
7. 15 is a factor of 13,324.
8. 6,734,385 is divisible by 9.
9. What is the smallest prime factor of 847?

Use exponents to rewrite each product.

10. $2 \cdot 2 \cdot 3 \cdot 3 \cdot 2 \cdot 5 \cdot 2 \cdot 3 \cdot 5 \cdot 11$
11. $2 \cdot 2 \cdot 3 \cdot 2 \cdot 3 \cdot 2 \cdot 3 \cdot 5 \cdot 5$
12. $2 \cdot 2 \cdot 2 \cdot 3 \cdot 3 \cdot 3 \cdot 3 \cdot 5 \cdot 5 \cdot 2$
13. $5 \cdot 7 \cdot 7 \cdot 5 \cdot 5 \cdot 11 \cdot 5 \cdot 13 \cdot 5$

For Exercises 14–17, write the prime factorization of each number.

14. 42
15. 90
16. 225
17. 360

18. Find the least common multiple of 100 and 56.
19. Find the greatest common factor of 12, 63, and 98.

20. In a neighborhood grocery, milk is delivered every third business day, bread every fifth business day, and paper goods every twelfth business day. All three of these items are delivered on Monday. How many more business days will pass before all three items are delivered on the same day?

21. A rock band with three members was offered $245 to play at a prom.
    a. Could the $245 be divided evenly so that each member received a whole-dollar amount?
    b. If not, how many more dollars would the band need to earn for this to be possible?

ANSWERS TO REVIEW CAPSULES

Page 98 1. 2; 8 2. 0; 3 3. 0; 0 4. 2; 2 5. 1; 4 6. 2; 8 7. 1; 1 8. 2; 3 9. 3; 4 10. 0; 3 11. 2; 1 12. 2; 3

Page 102 1. exponent 2. 3, or itself 3. 9 4. 8 5. 16 6. 16 7. 1 8. 27

Page 106 1. no 2. yes 3. yes 4. no 5. no 6. yes 7. yes 8. no 9. yes 10. no 11. yes 12. no 13. yes 14. no 15. yes 16. yes 17. no 18. no 19. 23 20. 34 21. 203 22. 19 23. 17

Page 110 1. 2, 4, 6, 8, 10, 12 2. 3, 6, 9, 12, 15, 18 3. 4, 8, 12, 16, 20, 24 4. 5, 10, 15, 20, 25, 30 5. 6, 12, 18, 24, 30, 36 6. 6, 12 7. 12, 24 8. 12 9. $2^2 \cdot 3^3 \cdot 5$ 10. $2^3 \cdot 5^3$ 11. $2^4 \cdot 3^3 \cdot 5^2$ 12. $5^4 \cdot 11^3 \cdot 17$ 13. $2^3 \cdot 3^3 \cdot 4^2$ 14. $6^3 \cdot 9^3 \cdot 12^2$

Page 114 1. 1, 2, 4, 8 2. 1, 2, 5, 10 3. 1, 3, 5, 15 4. 1, 2, 3, 4, 6, 8, 12, 24 5. 1, 2, 3, 4, 6, 9, 12, 18, 36 6. 1, 2, 3, 6, 7, 14, 21, 42 7. $2 \cdot 3$ 8. 2^4 9. $2^2 \cdot 5$ 10. 2^5 11. $3^2 \cdot 5$ 12. $2 \cdot 5^2$

Chapter 4

ENRICHMENT

Factor Trees

Another way to write the prime factorization of a composite number is to use a **factor tree**.

EXAMPLE Use a factor tree to factor 180. Then use exponents to write the prime factors.

Solution: A factor tree can begin with any pair of factors of the number.

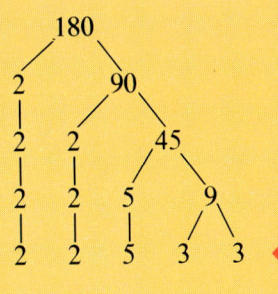

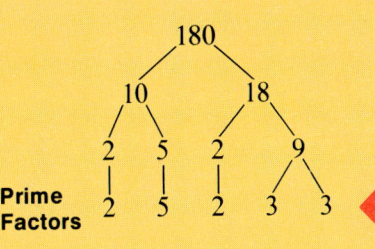

$180 = 2 \times 2 \times 5 \times 3 \times 3$
$= 2^2 \times 3^2 \times 5$

$180 = 2 \times 5 \times 2 \times 3 \times 3$
$= 2^2 \times 3^2 \times 5$

EXERCISES

Copy and complete the factor trees below. Then use exponents to write the prime factorization of the given number.

1. 294
2. 1200

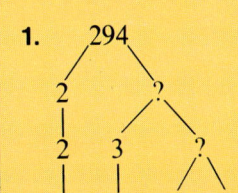

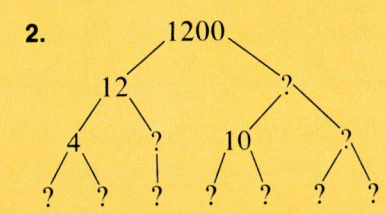

Use a factor tree to find the prime factorization of each number. Then use exponents to write each prime factorization.

3. 36
4. 40
5. 100
6. 80
7. 150
8. 200
9. 1000
10. 1260
11. 2025
12. 2079

Number Theory **121**

ENRICHMENT

You may wish to use this lesson for students who performed well on the formal Chapter Test.

1. $2 \cdot 3 \cdot 7^2$
2. $2^4 \cdot 3 \cdot 5^2$
3. $2^2 \cdot 3^2$
4. $2^3 \cdot 5$
5. $2^2 \cdot 5^2$
6. $2^4 \cdot 5$
7. $2 \cdot 3 \cdot 5^2$
8. $2^3 \cdot 5^2$
9. $2^3 \cdot 5^3$
10. $2^2 \cdot 3^2 \cdot 5 \cdot 7$
11. $3^4 \cdot 5^2$
12. $3^3 \cdot 7 \cdot 11$

ADDITIONAL PRACTICE

You may wish to use all or some of these exercises, depending on how well students performed on the formal Chapter Test.

1. 1, 3, 9 2. 1, 19
3. 1, 2, 3, 4, 6, 8, 12, 16, 24, 48 4. 1, 2, 3, 4, 5, 6, 10, 12, 15, 20, 30, 60 5. 1, 2, 4, 8, 16, 32 6. 1, 2, 4, 5, 8, 10, 20, 25, 40, 50, 100, 200
7. T; is divisible by 9
8. F; 97 ÷ 7 = 13 R 6
9. T; 6 · 11 = 66
10. T; 92 ÷ 4 = 23
11. T; ends in 5
12. F; not divisible by 4
13. F; does not end in 0
14. T; is divisible by 9
15. 37 16. 11 17. 3
18. 7 19. 3 20. 11
21. $2^3 \cdot 3^2 \cdot 5$
22. $2^4 \cdot 3^2 \cdot 5^3$
23. $2^3 \cdot 3^2 \cdot 11^2 \cdot 17$
24. $2^3 \cdot 3^3 \cdot 5^2 \cdot 17^2$
25. 5 · 13
26. 2 · 3 · 7
27. 2 · 7 · 13
28. $2^3 \cdot 3^2 \cdot 5$
29. $2^3 \cdot 3^3$
30. $2^3 \cdot 5^2 \cdot 7$
31. 450; 2 32. 90; 9
33. 630; 1 34. 540; 2
35. No; 2 36. 240 min

122

ADDITIONAL PRACTICE

SKILLS

List all the factors of each number. (Pages 96–98)

1. 9 **2.** 19 **3.** 48 **4.** 60 **5.** 32 **6.** 200

Write True *or* False *for each sentence. Write a reason for each answer.* (Pages 99–102)

7. 9 is a factor of 54.
8. 97 is divisible by 7.
9. 66 is a multiple of 11.
10. 4 is a factor of 92.
11. 135 is divisible by 5.
12. 1642 is divisible by 4.
13. 1342 is a multiple of 10.
14. 9 is a factor of 47,358.

Find the smallest prime factor of each number. (Pages 103–105)

15. 37 **16.** 121 **17.** 4035 **18.** 343 **19.** 7419 **20.** 1331

Use exponents to rewrite each product. (Pages 103–105)

21. 2 · 2 · 3 · 2 · 3 · 5
22. 2 · 2 · 3 · 2 · 2 · 5 · 5 · 3 · 5
23. 2 · 11 · 2 · 3 · 11 · 2 · 3 · 17
24. 17 · 2 · 3 · 3 · 2 · 5 · 3 · 2 · 5 · 17

Write the prime factorization of each number. (Pages 108–110)

25. 65 **26.** 42 **27.** 182 **28.** 360 **29.** 216 **30.** 1400

For Exercises 31–34, write the LCM and the GCF. (Pages 111–116)

31. 18 and 50 **32.** 18 and 45 **33.** 6, 45, and 35 **34.** 20, 54, and 180

APPLICATIONS

35. The members of a student council want to divide $874 evenly between 3 clubs.
  a. Can each club receive a whole-number dollar amount?
  b. If not, how many additional dollars are needed for this to be possible? (Pages 99–102)

36. Three satellites are in orbit around the earth. The satellites make one complete orbit every 60, 80, and 120 minutes, respectively. If all three satellites are directly over Los Angeles, California, how many minutes will pass before they are all over Los Angeles again? (Pages 111–114)

122 Chapter 4

CHAPTER 5
Fractions: Multiplication/Division

FEATURES
Career Application: *Engineering Technician*
Consumer Application: *Saving $ on Cooling Costs*
Enrichment: *Special Areas*

SECTIONS
5-1 Writing Fractions in Lowest Terms
5-2 Customary Measures
5-3 Multiplying Fractions
5-4 Problem Solving and Applications: Formulas: Temperature
5-5 Multiplying Mixed Numerals
5-6 Problem Solving and Applications: Area: Parallelograms and Triangles
5-7 Problem Solving and Applications: Volume: Rectangular Prisms
5-8 Dividing Fractions

Teaching Suggestions
p. M-18

QUICK QUIZ

Write the prime factorization of each number.

1. 70   Ans: $2 \cdot 5 \cdot 7$
2. 231 Ans: $3 \cdot 7 \cdot 11$
3. 20   Ans: $2^2 \cdot 5$
4. 48   Ans: $2^4 \cdot 3$
5. 35   Ans: $5 \cdot 7$
6. 90   Ans: $2 \cdot 3^2 \cdot 5$
7. 168 Ans: $2^3 \cdot 3 \cdot 7$
8. 54   Ans: $2 \cdot 3^3$
9. 13   Ans: 13
10. 200 Ans: $2^3 \cdot 5^2$

ADDITIONAL EXAMPLES

Example 1
Find the GCF.
1. 30 and 36   Ans: 6
2. 500 and 300   Ans: 100
3. 27 and 63   Ans: 9

Example 2
Write the lowest terms.
1. $\frac{22}{66}$ Ans: $\frac{1}{3}$
2. $\frac{56}{70}$ Ans: $\frac{4}{5}$
3. $\frac{25}{48}$ Ans: $\frac{25}{48}$

5-1 Writing Fractions in Lowest Terms

Six out of every 300 gallons of water used in a home are used for cleaning purposes. Six out of 300 can be written as a fraction.

$\frac{6}{300}$ ← **Numerator** / **Denominator**

The fraction $\frac{6}{300}$ can be simplified; that is, it can be written in *lowest terms*.

> A fraction is in **lowest terms** when its numerator and denominator have no common prime factor.
>
> $\frac{12}{35} = \frac{2 \cdot 2 \cdot 3}{5 \cdot 7}$ ← Lowest Terms

Here are three methods for writing fractions in lowest terms.

TABLE

Method 1 (GCF)	$\frac{6}{300} = \frac{6 \div 6}{300 \div 6}$ $= \frac{1}{50}$	Divide the numerator and denominator by the GCF. Lowest terms
Method 2 (Prime factorization)	$\frac{12}{20} = \frac{2 \cdot 2 \cdot 3}{2 \cdot 2 \cdot 5}$ $= \frac{\cancel{2} \cdot \cancel{2} \cdot 3}{\cancel{2} \cdot \cancel{2} \cdot 5} = \frac{3}{5}$	Write the prime factorizations. Lowest terms
Method 3 (Common factors)	$\frac{12}{18} = \frac{\cancel{12}^{6}}{\cancel{18}_{9}}$ $= \frac{6}{9}$ $= \frac{\cancel{6}^{2}}{\cancel{9}_{3}} = \frac{2}{3}$	2 is a common factor. Divide the numerator and denominator by 2. Not in lowest terms. 3 is also a common factor. Lowest terms

Chapter 5

CLASSROOM EXERCISES

Determine whether each fraction is in lowest terms. Answer Yes or No. (Table)

1. $\frac{8}{24}$ 2. $\frac{12}{48}$ 3. $\frac{9}{25}$ 4. $\frac{24}{51}$ 5. $\frac{15}{40}$ 6. $\frac{21}{52}$

WRITTEN EXERCISES

Goal: To write a fraction in lowest terms

Write each fraction in lowest terms. (Table)

1. $\frac{6}{9}$ 2. $\frac{6}{12}$ 3. $\frac{16}{28}$ 4. $\frac{12}{20}$ 5. $\frac{2}{16}$ 6. $\frac{15}{48}$
7. $\frac{21}{24}$ 8. $\frac{4}{10}$ 9. $\frac{6}{45}$ 10. $\frac{24}{64}$ 11. $\frac{100}{300}$ 12. $\frac{25}{55}$
13. $\frac{9}{16}$ 14. $\frac{22}{88}$ 15. $\frac{16}{36}$ 16. $\frac{16}{49}$ 17. $\frac{57}{90}$ 18. $\frac{12}{30}$

APPLICATIONS: USING FRACTIONS

19. Sheri cut a cantaloupe into 10 slices. She then ate 4 of the slices. What fractional part (lowest terms) of the cantaloupe did she eat?

20. The string section of a certain orchestra has 42 instruments. Twenty of these instruments are violins. What fractional part (lowest terms) of the string section is not violins?

MORE CHALLENGING EXERCISES

Simplify. No variable equals zero.

EXAMPLE: $\frac{3a}{6} = \frac{\overset{1}{\cancel{3}} \cdot a}{\underset{2}{\cancel{6}}} = \frac{1 \cdot a}{2} = \frac{a}{2}$

21. $\frac{5t}{10}$ 22. $\frac{6p}{8}$ 23. $\frac{4}{6y}$ 24. $\frac{9}{6x}$ 25. $\frac{6p}{9pn}$ 26. $\frac{8st}{10t}$
27. $\frac{6y}{14y^2}$ 28. $\frac{4a^2}{7a}$ 29. $\frac{3pn}{5p^2}$ 30. $\frac{8x^2y}{12xy^2}$ 31. $\frac{27cde}{60def}$ 32. $\frac{12rst}{8st^2}$

Fractions: Multiplication/Division **125**

CLASSROOM EXERCISES

<u>1</u>. No <u>2</u>. No <u>3</u>. Yes
<u>4</u>. No <u>5</u>. No <u>6</u>. Yes

ASSIGNMENT GUIDE
BASIC
p. 125: 1-20
AVERAGE
p. 125: 1-32
ABOVE AVERAGE
p. 125: 1-19 odd, 21-32

WRITTEN EXERCISES

<u>1</u>. $\frac{2}{3}$ <u>2</u>. $\frac{1}{2}$ <u>3</u>. $\frac{4}{7}$
<u>4</u>. $\frac{3}{5}$ <u>5</u>. $\frac{1}{8}$ <u>6</u>. $\frac{5}{16}$
<u>7</u>. $\frac{7}{8}$ <u>8</u>. $\frac{2}{5}$ <u>9</u>. $\frac{2}{15}$
<u>10</u>. $\frac{3}{8}$ <u>11</u>. $\frac{1}{3}$ <u>12</u>. $\frac{5}{11}$
<u>13</u>. $\frac{9}{16}$ <u>14</u>. $\frac{1}{4}$ <u>15</u>. $\frac{4}{9}$
<u>16</u>. $\frac{16}{49}$ <u>17</u>. $\frac{19}{30}$
<u>18</u>. $\frac{2}{5}$ <u>19</u>. $\frac{2}{5}$ <u>20</u>. $\frac{11}{21}$
<u>21</u>. $\frac{t}{2}$ <u>22</u>. $\frac{3p}{4}$
<u>23</u>. $\frac{2}{3y}$ <u>24</u>. $\frac{3}{2x}$
<u>25</u>. $\frac{2}{3n}$ <u>26</u>. $\frac{4s}{5}$
<u>27</u>. $\frac{3}{7y}$ <u>28</u>. $\frac{4a}{7}$
<u>29</u>. $\frac{3n}{5p}$ <u>30</u>. $\frac{2x}{3y}$
<u>31</u>. $\frac{9c}{20f}$ <u>32</u>. $\frac{3r}{2t}$

125

Teaching Suggestions
p. M-18

QUICK QUIZ

Add or subtract.

1. 1007 + 286   Ans: 1293
2. 271 − 182   Ans: 89
3. 1007 − 286   Ans: 721
4. 8146 + 8089
   Ans: 16,235
5. 3000 − 2708 Ans: 292

ADDITIONAL EXAMPLES

Example 1
Write a mixed numeral for each measure.

1. 8 ft 10 in
   Ans: $8\frac{5}{6}$ ft
2. 17 lb 8 oz
   Ans: $17\frac{1}{2}$ lb
3. 23 yd 1 ft
   Ans: $23\frac{1}{3}$ yd
4. 12 T 500 lb
   Ans: $12\frac{1}{4}$ T

SELF-CHECK

1. $3\frac{3}{4}$ lb   2. $4\frac{2}{3}$ ft
3. $2\frac{2}{3}$ yd

Example 2

1. 5 qt 1 pt
   +3 qt 1 pt
   Ans: 9 qt
2. 9 ft 7 in
   −4 ft 9 in
   Ans: 4 ft 10 in

5-2 Customary Measures

This table shows some commonly used customary units.

TABLE

Length	Capacity
12 inches (in) = 1 foot (ft)	2 cups (c) = 1 pint (pt)
3 feet = 1 yard (yd)	2 pints = 1 quart (qt)
36 inches = 1 yd	4 quarts = 1 gallon (gal)
5280 feet = 1 mile (mi)	**Weight**
	16 ounces (oz) = 1 pound (lb)
	2000 pounds = 1 ton (T)

A measure such as 3 feet 9 inches can be written as a mixed numeral. **A mixed numeral** (shown at the right), represents the sum of a whole number and a fraction.

Mixed Numeral

$3\frac{3}{4}$

Whole Number — Fraction

EXAMPLE 1 Write a mixed numeral for each measure.

a. 3 ft 9 in   b. 5 lb 2 oz

Solutions: a. Since 12 in = 1 ft, 9 in = $\frac{9}{12}$, or $\frac{3}{4}$ ft. ◀ Lowest terms

Thus, 3 ft 9 in = $3\frac{3}{4}$ **ft.**

b. Since 16 oz = 1 lb, 2 oz = $\frac{2}{16}$, or $\frac{1}{8}$ lb.

Thus, 5 lb 2 oz = $5\frac{1}{8}$ **lb.**

Self-Check Write a mixed numeral for each measure.

1. 3 lb 12 oz   2. 4 ft 8 in   3. 2 yd 2 ft

Example 2 shows how to add and subtract with customary measures.

EXAMPLE 2 Add or subtract as indicated. Express the answer as a mixed numeral.

a. 3 ft 9 in   b. 4 gal 2 qt   ◀ 4 gal 2 qt =
   +2 ft 8 in   −1 gal 3 qt   3 gal + 4 qt + 2 qt =
                                              3 gal 6 qt

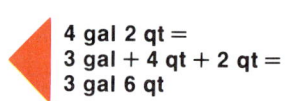

126 Chapter 5

Solutions: **a.** 3 ft 9 in
+2 ft 8 in
5 ft 17 in = 5 ft + 1 ft 5 in
= 6 ft 5 in
= $6\frac{5}{12}$ ft

b. Since you cannot subtract 3 qt from 2 qt, write 4 gal 2 qt as 3 gal 6 qt.

4 gal 2 qt → 3 gal 6 qt
−1 gal 3 qt   −1 gal 3 qt
2 gal 3 qt,
or $2\frac{3}{4}$ gal

Self-Check

Add or subtract as indicated.

4. 6 yd 2 ft
+1 yd 2 ft

5. 8 lb 6 oz
−3 lb 8 oz

6. 9 gal 1 qt
−4 gal 3 qt

CLASSROOM EXERCISES

Write a mixed numeral for each measure. (Example 1)

1. 4 gal 2 qt   **2.** 3 yd 1 ft   **3.** 5 qt 1 pt   **4.** 6 ft 9 in
5. 6 lb 8 oz   **6.** 1 T 200 lb   **7.** 3 qt 1 pt   **8.** 3 lb 10 oz

Add or subtract as indicated. Express the answer as a mixed numeral. (Example 2)

9. 2 ft 3 in
+4 ft 5 in

10. 5 gal 2 qt
+2 gal 1 qt

11. 12 lb 8 oz
+ 8 lb 12 oz

12. 4 yd 2 ft
+8 yd 2 ft

13. 9 ft 8 in
−4 ft 10 in

14. 15 lb 12 oz
−10 lb 4 oz

15. 7 yd 1 ft
−3 yd 2 ft

16. 13 gal 1 qt
− 2 gal 3 qt

WRITTEN EXERCISES

Goals To add and subtract customary measures
To apply these skills to solving word problems

Write a mixed numeral for each measure. (Example 1)

1. 6 gal 3 qt   **2.** 3 ft 5 in   **3.** 6 yd 2 ft   **4.** 7 yd 12 in
5. 4 gal 3 qt   **6.** 24 lb 6 oz   **7.** 2 pt 1 c   **8.** 3 T 500 lb
9. 1 mi 528 ft   **10.** 1 qt 5 pt   **11.** 1 yd 5 ft   **12.** 98 lb 12 oz

Fractions: Multiplication/Division   **127**

---

3.   9 yd
     −2 yd 2 ft
Ans: 6 yd 1 ft

SELF-CHECK

4. $8\frac{1}{3}$ yd   **5.** $4\frac{7}{8}$ lb

6. $4\frac{1}{2}$ gal

CLASSROOM EXERCISES

1. $4\frac{1}{2}$ gal   **2.** $3\frac{1}{3}$ yd
3. $5\frac{1}{2}$ qt   **4.** $6\frac{3}{4}$ ft
5. $6\frac{1}{2}$ lb   **6.** $1\frac{1}{10}$ T
7. $3\frac{1}{2}$ qt   **8.** $3\frac{5}{8}$ lb
9. $6\frac{2}{3}$ ft   **10.** $7\frac{3}{4}$ gal
11. $21\frac{1}{4}$ lb **12.** $13\frac{1}{3}$ yd
13. $4\frac{5}{6}$ ft   **14.** $5\frac{1}{2}$ lb
15. $3\frac{2}{3}$ yd   **16.** $10\frac{1}{2}$ gal

ASSIGNMENT GUIDE
BASIC
pp. 127–128: 1–28
AVERAGE
pp. 127–128: 1–31 odd
ABOVE AVERAGE
pp. 127–128: 1–27 odd, 29–32

WRITTEN EXERCISES

1. $6\frac{3}{4}$ gal   **2.** $3\frac{5}{12}$ ft
3. $6\frac{2}{3}$ yd   **4.** $7\frac{1}{3}$ yd
5. $4\frac{3}{4}$ gal **6.** $24\frac{3}{8}$ lb
7. $2\frac{1}{2}$ pt   **8.** $3\frac{1}{4}$ T

WRITTEN EXERCISES

9. $1\frac{1}{10}$ mi   10. $3\frac{1}{2}$ qt
11. $2\frac{2}{3}$ yd   12. $98\frac{3}{4}$ lb
13. $3\frac{11}{12}$ ft   14. 8 ft
15. $10\frac{1}{3}$ yd   16. $11\frac{2}{3}$ yd
17. $11\frac{3}{4}$ gal   18. $20\frac{1}{4}$ gal
19. $13\frac{1}{4}$ lb   20. $29\frac{13}{16}$ lb
21. $2\frac{1}{4}$ ft   22. $4\frac{5}{6}$ ft
23. $6\frac{2}{3}$ yd   24. $3\frac{1}{3}$ yd
25. $7\frac{1}{4}$ gal   26. $3\frac{1}{2}$ gal
27. $4\frac{1}{2}$ lb   28. $81\frac{1}{2}$ lb
29. $3\frac{1}{4}$ gal   30. $9\frac{2}{3}$ ft
31. $1\frac{7}{12}$ ft   32. $12\frac{1}{4}$ lb

Add or subtract as indicated. Write each answer as a mixed numeral. (Example 2)

13. 1 ft 7 in
    +2 ft 4 in

14. 5 ft 9 in
    +2 ft 3 in

15. 1 yd 2 ft
    +8 yd 2 ft

16. 6 yd 1 ft
    +5 yd 1 ft

17. 8 gal 2 qt
    +3 gal 1 qt

18. 19 gal 3 qt
    +   2 qt

19. 3 lb 12 oz
    +9 lb 8 oz

20. 16 lb 14 oz
    +12 lb 15 oz

21. 3 ft 8 in
    −1 ft 5 in

22. 9 ft 6 in
    −4 ft 8 in

23. 9 yd 1 ft
    −2 yd 2 ft

24. 4 yd 2 ft
    −1 yd 1 ft

25. 10 gal 3 qt
    − 3 gal 2 qt

26. 6 gal 1 qt
    −2 gal 3 qt

27. 18 lb 4 oz
    −13 lb 12 oz

28. 98 lb
    −16 lb 8 oz

APPLICATIONS: **USING CUSTOMARY MEASURES**

Solve. Write each answer as a mixed numeral.

29. May measured the amount of water collected from a dripping faucet.

    Monday: 1 gallon 2 quarts
    Tuesday: 1 gallon 3 quarts

    Find the total amount collected in the two days.

30. Larry wishes to replace the weather-stripping at the top and on one side of a door. The height of the door is 6 feet 11 inches and the width 2 feet 9 inches. How many feet of weather-stripping will be needed?

31. Juan is 6 feet 3 inches tall. His younger sister is 4 feet 8 inches tall. How much taller is Juan than his sister?

32. An airline passenger is allowed to bring luggage weighing 60 pounds. Coleen's luggage weighs 72 pounds 4 ounces. How many pounds over the limit is the weight of her luggage?

This Review Capsule reviews prior-taught skills used in Section 5-3. The reference is to the pages where the skills were taught.

REVIEW CAPSULE FOR SECTION 5-3

Multiply. (Pages 7–9)

1. 8 · 22   2. 42 · 9   3. 10 · 15   4. 12 · 13   5. 16 · 21   6. 18 · 32

Write each fraction in lowest terms. (Pages 124–125)

7. $\frac{4}{6}$ 8. $\frac{6}{14}$ 9. $\frac{12}{18}$ 10. $\frac{28}{30}$ 11. $\frac{50}{75}$ 12. $\frac{18}{81}$

The answers are on page 154.

Chapter 5

5-3 Multiplying Fractions

Two-thirds of Grant City Park is covered with grass. The Director of City Parks plans to replace $\frac{1}{5}$ of this grass with shrubs.

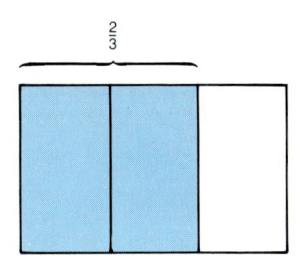

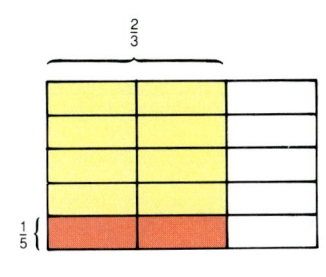

15 equal parts
2 of the 15 will have shrubs.

Grass covers $\frac{2}{3}$ of the park. $\frac{1}{5} \cdot \frac{2}{3} = \frac{1 \cdot 2}{5 \cdot 3} = \frac{2}{15}$

One way of multiplying with fractions is to multiply the numerators <u>and</u> multiply the denominators. Then write the answer in lowest terms.

EXAMPLE 1   Multiply: **a.** $\frac{2}{3} \cdot \frac{7}{10}$ **b.** $\frac{3}{16} \cdot 5$

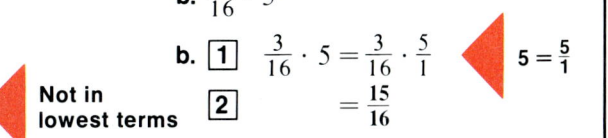

Solution: **a.** ① $\frac{2}{3} \cdot \frac{7}{10} = \frac{2 \cdot 7}{3 \cdot 10}$
$= \frac{14}{30}$ ◀ Not in lowest terms
② $= \frac{14 \div 2}{30 \div 2}$
$= \frac{7}{15}$ ◀ Lowest terms

b. ① $\frac{3}{16} \cdot 5 = \frac{3}{16} \cdot \frac{5}{1}$ $5 = \frac{5}{1}$
② $= \frac{15}{16}$

Self-Check   Multiply:

1. $\frac{3}{4} \cdot \frac{3}{10}$ 2. $\frac{4}{15} \cdot 2$ 3. $\frac{1}{18} \cdot 11$ 4. $\frac{4}{5} \cdot \frac{3}{12}$

It is sometimes easier to divide a numerator <u>and</u> a denominator by a common factor <u>before</u> multiplying.

EXAMPLE 2   Multiply: **a.** $\frac{5}{7} \cdot \frac{4}{15}$ **b.** $\frac{2}{9} \cdot \frac{3}{4}$

Solutions: **a.** $\frac{5}{7} \cdot \frac{4}{15} = \frac{\cancel{5}^1}{7} \cdot \frac{4}{\cancel{15}_3}$ $5 \div 5 = 1$
$15 \div 5 = 3$
$= \frac{1 \cdot 4}{7 \cdot 3} = \frac{4}{21}$

b. $\frac{2}{9} \cdot \frac{3}{4} = \frac{\cancel{2}^1}{\cancel{9}_3} \cdot \frac{\cancel{3}^1}{\cancel{4}_2}$
$= \frac{1 \cdot 1}{3 \cdot 2} = \frac{1}{6}$

Fractions: Multiplication/Division   **129**

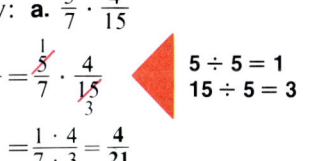

Teaching Suggestions
p. M-18

QUICK QUIZ
Multiply.
1. 16 • 20   Ans: 320
2. 75 • 15   Ans: 1125

Write in lowest terms.
3. $\frac{45}{72}$ Ans: $\frac{5}{8}$
4. $\frac{96}{120}$ Ans: $\frac{4}{5}$
5. $\frac{102}{126}$ Ans: $\frac{17}{21}$

ADDITIONAL EXAMPLES
Example 1
Multiply.
1. $2 \cdot \frac{3}{8}$ Ans: $\frac{3}{4}$
2. $\frac{7}{8} \cdot \frac{5}{6}$ Ans: $\frac{35}{48}$
3. $\frac{4}{9} \cdot \frac{12}{25}$ Ans: $\frac{16}{75}$

SELF-CHECK
1. $\frac{9}{40}$ 2. $\frac{8}{15}$ 3. $\frac{11}{18}$
4. $\frac{1}{5}$

Example 2
1. $\frac{2}{25} \cdot 5$ Ans: $\frac{2}{5}$
2. $\frac{10}{21} \cdot \frac{14}{45}$ Ans: $\frac{4}{27}$
3. $\frac{12}{20} \cdot \frac{25}{49}$ Ans: $\frac{15}{49}$

SELF-CHECK

5. $\frac{4}{5}$ 6. $\frac{2}{3}$ 7. $\frac{3}{5}$
8. $\frac{1}{4}$

CLASSROOM EXERCISES

1. $\frac{1}{14}$ 2. $\frac{1}{20}$ 3. $\frac{3}{16}$
4. $\frac{4}{35}$ 5. $\frac{16}{45}$ 6. $\frac{9}{20}$
7. $\frac{5}{21}$ 8. $\frac{1}{15}$ 9. $\frac{15}{7}$
10. $\frac{2}{5}$ 11. $\frac{6}{7}$ 12. $\frac{8}{9}$
13. $\frac{4}{7}$ 14. $\frac{3}{5}$ 15. $\frac{2}{5}$
16. $\frac{3}{7}$ 17. $\frac{1}{5}$ 18. $\frac{7}{9}$
19. $\frac{2}{5}$ 20. $\frac{2}{9}$ 21. $\frac{7}{18}$
22. $\frac{1}{3}$ 23. $\frac{3}{10}$ 24. $\frac{1}{6}$

ASSIGNMENT GUIDE
BASIC
Day 1 p. 130: 1-35 odd
Day 2 p. 130: 2-36 even
AVERAGE
pp. 130-131: 1-38
ABOVE AVERAGE
pp. 130-131: 1-39 odd, 41-50

WRITTEN EXERCISES

1. $\frac{1}{8}$ 2. $\frac{1}{15}$ 3. $\frac{3}{10}$

Self-Check *Multiply.*

5. $\frac{6}{7} \cdot \frac{14}{15}$ 6. $\frac{7}{9} \cdot \frac{18}{21}$ 7. $\frac{12}{15} \cdot \frac{3}{4}$ 8. $\frac{2}{3} \cdot \frac{3}{8}$

> For the fractions $\frac{a}{b}$ and $\frac{c}{d}$, $b \neq 0$, $d \neq 0$,
> $\frac{a}{b} \cdot \frac{c}{d} = \frac{a \cdot c}{b \cdot d} = \frac{ac}{bd}$. $\frac{2}{5} \cdot \frac{3}{7} = \frac{2 \cdot 3}{5 \cdot 7} = \frac{6}{35}$

CLASSROOM EXERCISES

Multiply. Write each answer in lowest terms. (Examples 1 and 2)

1. $\frac{1}{2} \cdot \frac{1}{7}$ 2. $\frac{1}{4} \cdot \frac{1}{5}$ 3. $\frac{3}{4} \cdot \frac{1}{4}$ 4. $\frac{4}{5} \cdot \frac{1}{7}$ 5. $\frac{2}{5} \cdot \frac{8}{9}$ 6. $\frac{3}{4} \cdot \frac{3}{5}$
7. $\frac{5}{6} \cdot \frac{2}{7}$ 8. $\frac{2}{5} \cdot \frac{1}{6}$ 9. $5 \cdot \frac{3}{7}$ 10. $2 \cdot \frac{1}{5}$ 11. $\frac{2}{7} \cdot 3$ 12. $\frac{2}{9} \cdot 4$
13. $\frac{4}{5} \cdot \frac{5}{7}$ 14. $\frac{3}{4} \cdot \frac{4}{5}$ 15. $\frac{3}{5} \cdot \frac{2}{3}$ 16. $\frac{4}{7} \cdot \frac{3}{4}$ 17. $\frac{1}{3} \cdot \frac{3}{5}$ 18. $\frac{7}{8} \cdot \frac{8}{9}$
19. $\frac{1}{2} \cdot \frac{4}{5}$ 20. $\frac{1}{4} \cdot \frac{8}{9}$ 21. $\frac{5}{9} \cdot \frac{7}{10}$ 22. $\frac{4}{5} \cdot \frac{5}{12}$ 23. $\frac{9}{10} \cdot \frac{1}{3}$ 24. $\frac{3}{4} \cdot \frac{2}{9}$

WRITTEN EXERCISES

Goals: To multiply with fractions
To apply the skill of multiplying with fractions to solving word problems

Multiply. Write each answer in lowest terms. (Examples 1 and 2)

1. $\frac{1}{4} \cdot \frac{1}{2}$ 2. $\frac{1}{3} \cdot \frac{1}{5}$ 3. $\frac{3}{5} \cdot \frac{1}{2}$ 4. $\frac{3}{5} \cdot \frac{2}{7}$ 5. $\frac{2}{9} \cdot \frac{4}{5}$ 6. $\frac{3}{11} \cdot \frac{5}{7}$
7. $4 \cdot \frac{1}{5}$ 8. $5 \cdot \frac{1}{8}$ 9. $3 \cdot \frac{1}{6}$ 10. $\frac{3}{10} \cdot 3$ 11. $\frac{2}{15} \cdot 4$ 12. $6 \cdot \frac{3}{20}$
13. $\frac{4}{5} \cdot \frac{7}{12}$ 14. $\frac{10}{11} \cdot \frac{3}{5}$ 15. $\frac{5}{9} \cdot \frac{3}{7}$ 16. $\frac{3}{4} \cdot \frac{2}{5}$ 17. $\frac{4}{5} \cdot \frac{3}{8}$ 18. $\frac{5}{6} \cdot \frac{1}{5}$
19. $\frac{2}{3} \cdot \frac{3}{8}$ 20. $\frac{4}{5} \cdot \frac{5}{8}$ 21. $\frac{9}{10} \cdot \frac{5}{6}$ 22. $\frac{2}{3} \cdot \frac{9}{10}$ 23. $\frac{5}{8} \cdot \frac{4}{15}$ 24. $\frac{3}{4} \cdot \frac{8}{15}$
25. $\frac{5}{8} \cdot \frac{11}{25}$ 26. $\frac{5}{6} \cdot \frac{4}{5}$ 27. $\frac{10}{21} \cdot \frac{14}{15}$ 28. $\frac{9}{10} \cdot \frac{8}{15}$ 29. $\frac{4}{3} \cdot \frac{3}{4}$ 30. $\frac{1}{9} \cdot 6$
31. $2 \cdot \frac{3}{8}$ 32. $\frac{3}{8} \cdot \frac{5}{7}$ 33. $\frac{8}{9} \cdot \frac{9}{8}$ 34. $\frac{3}{4} \cdot \frac{4}{9}$ 35. $\frac{2}{3} \cdot \frac{4}{5}$ 36. $\frac{3}{8} \cdot \frac{8}{9}$

130 Chapter 5

APPLICATIONS: USING MULTIPLICATION WITH FRACTIONS

Solve. Write each answer in lowest terms.

37. Angelo spent $\frac{5}{6}$ of an hour practicing basketball. He spent $\frac{2}{3}$ of that time practicing lay-up shots. What fractional part of an hour did he spend practicing lay-up shots?

38. Susan can mow her front lawn in $\frac{3}{4}$ hour. Her brother can mow the lawn in $\frac{4}{5}$ that time. How long does it take Susan's brother to mow the lawn?

39. Mr. Mitchell's lawn mower will hold $\frac{9}{10}$ gallon of gasoline. The tank is $\frac{1}{3}$ full. How much gasoline is in the tank?

40. Band practice lasted 2 hours. One third of that time was spent marching. What fractional part of an hour was spent marching?

MORE CHALLENGING EXERCISES

Multiply. No variable equals zero.

EXAMPLES: a. $\dfrac{3}{t} \cdot \dfrac{t}{9} = \dfrac{\overset{1}{\cancel{3}} \cdot \overset{1}{\cancel{t}}}{\underset{1}{\cancel{t}} \cdot \underset{3}{\cancel{9}}} = \dfrac{1 \cdot 1}{1 \cdot 3} = \dfrac{1}{3}$

b. $10 \cdot \dfrac{1}{b} = \dfrac{10}{1} \cdot \dfrac{1}{b} = \dfrac{10 \cdot 1}{1 \cdot b} = \dfrac{10}{b}$

41. $\dfrac{2}{a} \cdot \dfrac{a}{6}$ **42.** $\dfrac{5}{x} \cdot \dfrac{x}{10}$ **43.** $\dfrac{a^2}{5} \cdot \dfrac{3}{a}$ **44.** $\dfrac{2m}{3} \cdot \dfrac{5}{6m}$ **45.** $\dfrac{3st^2}{5} \cdot \dfrac{5}{3t}$

46. $6 \cdot \dfrac{1}{x}$ **47.** $\dfrac{2}{b} \cdot 3$ **48.** $9 \cdot \dfrac{1}{3t}$ **49.** $12 \cdot \dfrac{1}{4m}$ **50.** $8 \cdot \dfrac{3}{6p}$

REVIEW CAPSULE FOR SECTION 5-4

Evaluate each expression. (Pages 18–21)

1. $2 \cdot 10 + 1$ **2.** $8 \cdot 15 + 16$ **3.** $9 \cdot 32 - 25$ **4.** $13 \cdot 25 - 47$

Evaluate each expression. (Pages 22–24)

5. $\frac{2}{3}x + 15$ when $x = 6$ **6.** $\frac{4}{9}y + 24$ when $y = 36$ **7.** $\frac{4}{5}(a + 10)$ when $a = 75$

8. $\frac{5}{6}s - 45$ when $s = 90$ **9.** $\frac{3}{4}(d - 8)$ when $d = 120$ **10.** $\frac{3}{8}m - 51$ when $m = 424$

The answers are on page 154.

Fractions: Multiplication/Division

WRITTEN EXERCISES

4. $\frac{6}{35}$ 5. $\frac{8}{45}$ 6. $\frac{15}{77}$
7. $\frac{4}{5}$ 8. $\frac{5}{8}$ 9. $\frac{1}{2}$
10. $\frac{9}{10}$ 11. $\frac{8}{15}$ 12. $\frac{9}{10}$
13. $\frac{7}{15}$ 14. $\frac{6}{11}$ 15. $\frac{5}{21}$
16. $\frac{3}{10}$ 17. $\frac{3}{10}$ 18. $\frac{1}{6}$
19. $\frac{1}{4}$ 20. $\frac{1}{2}$ 21. $\frac{3}{4}$
22. $\frac{3}{5}$ 23. $\frac{1}{6}$ 24. $\frac{2}{5}$
25. $\frac{11}{40}$ 26. $\frac{2}{3}$ 27. $\frac{4}{9}$
28. $\frac{12}{25}$ 29. 1   30. $\frac{2}{3}$
31. $\frac{3}{4}$ 32. $\frac{15}{56}$ 33. 1
34. $\frac{1}{3}$ 35. $\frac{8}{15}$ 36. $\frac{1}{3}$
37. $\frac{5}{9}$ hr   38. $\frac{3}{5}$ hr
39. $\frac{3}{10}$ gal   40. $\frac{2}{3}$ hr
41. $\frac{1}{3}$ 42. $\frac{1}{2}$ 43. $\frac{3a}{5}$
44. $\frac{5}{9}$ 45. st   46. $\frac{6}{x}$
47. $\frac{6}{b}$ 48. $\frac{3}{t}$ 49. $\frac{3}{m}$
50. $\frac{4}{p}$

Teaching Suggestions
p. M-18

QUICK QUIZ

Evaluate.

1. 6 · 12 + 5   Ans: 77
2. 6(12 + 5)   Ans: 102
3. 8(25 − 10)   Ans: 120
4. 7 · 12 − 9   Ans: 75
5. $\frac{2}{3}$ · 15 + 6   Ans: 16
6. 8x + 2 if x = $\frac{3}{4}$
   Ans: 8
7. $\frac{2}{3}$(x − 5) if x = 17
   Ans: 8
8. $\frac{4}{9}$y + 26 if y = 198
   Ans: 114
9. $\frac{1}{5}$a − 92 if a = 735
   Ans: 55
10. $\frac{5}{7}$(b + 18) if b = 59
    Ans: 55

ADDITIONAL EXAMPLES

Example 1

Express each of the following as a Fahrenheit reading.

1. 275° C   Ans: 527° F
2. 40° C   Ans: 104° F
3. 100° C   Ans: 212° F

SELF-CHECK

1. 41° F   2. 221° F
3. 1472° F   4. 2192° F

PROBLEM SOLVING AND APPLICATIONS

5-4 Formulas: Temperature

The **Celsius (C)** scale is used to measure temperature. In the United States, temperatures are also given in degrees **Fahrenheit (°F)**.

It may be helpful to memorize the information in this table.

TABLE

Temperature in °C	Meaning	Temperature in °F
100°C	Boiling point of water	212°F
37°C	Normal body temperature	98.6°F
35°C	Hot summer day	95°F
20°C	Comfortable room temperature	68°F
0°C	Freezing point of water	32°F

You can use this formula to write temperatures in °C (read: degrees Celsius) as temperatures in °F.

$$F = \frac{9}{5}C + 32$$

Given: °C
Find: Corresponding °F

EXAMPLE 1   The temperature at which copper boils is 2300°C. Express this as a Fahrenheit reading.

Solution: $F = \frac{9}{5}C + 32$ Replace C with 2300.

$F = \frac{9}{5}(2300) + 32$

$F = 9(460) + 32$

$F = 4140 + 32 = 4172$ 2300 °C corresponds to 4172 °F.

Self-Check   *Express each of the following as a Fahrenheit reading.*

1. 5°C   2. 105°C   3. 800°C   4. 1200°C

You can use the following formula to write temperatures in °F as temperature in °C.

$$C = \frac{5}{9}(F - 32)$$

Given: °F
Find: Corresponding °C

132   Chapter 5

When using the formula at the bottom of page 132, be sure to do the subtraction <u>inside the parentheses first</u>. Then multiply by $\frac{5}{9}$.

EXAMPLE 2 The temperature of a cup of hot tea is 176°F. What is the corresponding Celsius reading?

Solution:
$C = \frac{5}{9}(F - 32)$ ◀ Replace F with 176.
$C = \frac{5}{9}(176 - 32)$
$C = \frac{5}{9} \cdot 144$ $C = 80$ ◀ 176°F corresponds to 80°C

Self-Check *Express each of the following as a Celsius reading.*

5. 41°F   **6.** 221°F   **7.** 986°F   **8.** 185°F

CLASSROOM EXERCISES

Express each of the following as a Fahrenheit reading. (Example 1)

1. 50°C   **2.** 25°C   **3.** 15°C   **4.** 90°C   **5.** 200°C   **6.** 1000°C

Express each of the following as a Celsius reading. (Example 2)

7. 68°F   **8.** 86°F   **9.** 140°F   **10.** 77°F   **11.** 131°F   **12.** 338°F

WRITTEN EXERCISES

Goal: To use the formulas for Celsius–Fahrenheit temperatures

Express each of the following as a Fahrenheit reading. (Example 1)

1. 10°C   **2.** 45°C   **3.** 100°C   **4.** 400°C   **5.** 225°C   **6.** 85°C

Express each of the following as a Celsius reading. (Example 2)

7. 50°F   **8.** 257°F   **9.** 113°F   **10.** 59°F   **11.** 464°F   **12.** 932°F

13. The average high temperature on the moon's surface is about 248°F. Express this as a Celsius reading.

14. The average surface temperature of Venus is about 860°F. Express this as a Celsius reading.

15. The average surface temperature of the sun is about 5525°C. Express this as a Fahrenheit reading.

Fractions: Multiplication/Division

ADDITIONAL EXAMPLES
Example 2
Express each of the following as a Celsius reading.
1. 167° F   Ans: 75° C
2. 32° F   Ans: 0° C
3. 50° F   Ans: 10° C

SELF-CHECK
5. 5° C   6. 105° C
7. 530° C 8. 85° C

CLASSROOM EXERCISES
1. 122° F   2. 77° F
3. 59° F   4. 194° F
5. 392° F   6. 1832° F
7. 20° C   8. 30° C
9. 60° C   10. 25° C
11. 55° C   12. 170° C

ASSIGNMENT GUIDE
BASIC
p. 133: 1-12
AVERAGE
p. 133: 1-15
ABOVE AVERAGE
p. 133: 1-15

WRITTEN EXERCISES
1. 50° F   2. 113° F
3. 212° F 4. 752° F
5. 437° F 6. 185° F
7. 10° C   8. 125° C
9. 45° C   10. 15° C
11. 240° C 12. 500° C

WRITTEN EXERCISES
13. 120° C   14. 460° C
15. 9977° F

QUIZ: SECTIONS 5-1–5-4
After completing this Review, you may wish to administer a quiz covering the same sections. A Quiz is provided in the *Teacher's Edition: Part II.*

REVIEW SECTIONS 5-1–5-4
1. $\frac{3}{4}$ 2. $\frac{5}{8}$ 3. $\frac{6}{7}$
4. $\frac{2}{5}$ 5. $\frac{4}{5}$ 6. $\frac{2}{3}$
7. $8\frac{11}{12}$ ft   8. $19\frac{5}{16}$ lb
9. $3\frac{2}{3}$ yd   10. $2\frac{3}{4}$ gal
11. $\frac{5}{12}$ 12. $\frac{1}{3}$ 13. $\frac{6}{7}$
14. $\frac{2}{15}$ 15. 20   16. 122
17. 180   18. 932
19. $1\frac{7}{12}$ ft   20. $\frac{3}{4}$

REVIEW: SECTIONS 5-1–5-4

Express each fraction in lowest terms. (Section 5-1)

1. $\frac{9}{12}$ 2. $\frac{20}{32}$ 3. $\frac{18}{21}$ 4. $\frac{200}{500}$ 5. $\frac{12}{15}$ 6. $\frac{36}{54}$

Add or subtract as indicated. Write each answer as a mixed numeral. (Section 5-2)

7. 5 ft 4 in
  +3 ft 7 in

8. 10 lb 9 oz
  + 8 lb 12 oz

9. 7 yd 1 ft
  −3 yd 2 ft

10. 8 gal 2 qt
  −5 gal 3 qt

Multiply. Write each answer in lowest terms. (Section 5-3)

11. $\frac{1}{2} \cdot \frac{5}{6}$ 12. $\frac{3}{8} \cdot \frac{8}{9}$ 13. $\frac{2}{7} \cdot 3$ 14. $\frac{4}{9} \cdot \frac{3}{10}$

Complete. (Section 5-4)

15. 68°F = ___?___ °C
16. 50°C = ___?___ °F
17. 356°F = ___?___ °C
18. 500°C = ___?___ °F

19. David's sailboat is 13 feet 10 inches long. Ellie's sailboat is 15 feet 5 inches long. How much longer is Ellie's boat than David's? (Section 5-2)

20. Tennis practice lasted 3 hours. One-fourth of that time was spent practicing the serve. What fractional part of an hour was spent practicing the serve? (Section 5-3)

REVIEW CAPSULE FOR SECTION 5-5

Write the remainder.

1. $2\overline{)3}$ 2. $3\overline{)5}$ 3. $5\overline{)8}$ 4. $7\overline{)20}$ 5. $8\overline{)15}$ 6. $3\overline{)20}$

Multiply. Write each answer in lowest terms. (Pages 124–125)

7. $\frac{1}{3} \cdot \frac{3}{4}$ 8. $\frac{4}{7} \cdot \frac{3}{8}$ 9. $\frac{5}{9} \cdot \frac{7}{10}$ 10. $\frac{4}{5} \cdot \frac{7}{12}$ 11. $\frac{3}{10} \cdot 2$ 12. $\frac{2}{15} \cdot 6$

The answers are on page 154.

134 Chapter 5

5-5 Multiplying Mixed Numerals

The length of the bolt at the right can be given as $\frac{7}{4}$ inches or as $1\frac{3}{4}$ inches.

$$\frac{7}{4} = 1\frac{3}{4}$$

- Fraction
- Whole Number

Recall from page 126 that $1\frac{3}{4}$ represents the sum of a whole number and a fraction. It is called a **mixed numeral**. Any fraction whose numerator is greater than its denominator can be expressed as a mixed numeral.

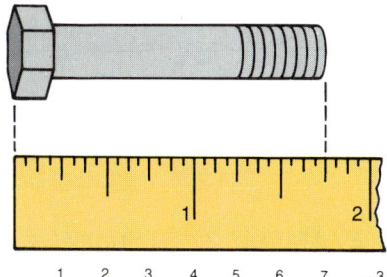

$\frac{1}{4}$ $\frac{2}{4}$ $\frac{3}{4}$ $\frac{4}{4}$ $\frac{5}{4}$ $\frac{6}{4}$ $\frac{7}{4}$, or $1\frac{3}{4}$

PROCEDURE

To write a mixed numeral for a fraction whose numerator is greater than the denominator:

1 Divide the numerator by the denominator.
2 Write a fraction for the remainder.

EXAMPLE 1 Write a mixed numeral for $\frac{37}{5}$.

Solution: 1 $\frac{37}{5} = 37 \div 5$

$5\overline{)37}$ → 7 R 2, or $7\frac{2}{5}$
35
2

2 $= 7\frac{2}{5}$ ◀ Mixed numeral

Self-Check Write a mixed numeral for each fraction.

1. $\frac{23}{7}$ 2. $\frac{41}{5}$ 3. $\frac{16}{3}$ 4. $\frac{81}{6}$

This table shows how to write a fraction for a mixed numeral.

Mixed Numeral	Procedure		
	1 Multiply the denominator and the whole number.	2 Add this product to the numerator.	3 Write this sum over the denominator.
$5\frac{3}{4}$	$4 \times 5 = 20$	$20 + 3 = 23$	$\frac{23}{4}$ $5\frac{3}{4} = \frac{23}{4}$

Fractions: Multiplication/Division

Teaching Suggestions
p. M-18

QUICK QUIZ
Divide.
1. $6 \div 5$ Ans: 1 R 1
2. $18 \div 7$ Ans: 2 R 4
3. $29 \div 9$ Ans: 3 R 2
4. $73 \div 12$ Ans: 6 R 1

Multiply.
5. $\frac{2}{3} \cdot \frac{6}{7}$ Ans: $\frac{4}{7}$
6. $\frac{10}{49} \cdot \frac{21}{25}$ Ans: $\frac{6}{35}$
7. $\frac{4}{9} \cdot 27$ Ans: 12
8. $\frac{8}{18} \cdot \frac{9}{4}$ Ans: 1

ADDITIONAL EXAMPLES
Example 1
Write as a mixed numeral.
1. $\frac{17}{6}$ Ans: $2\frac{5}{6}$
2. $\frac{100}{9}$ Ans: $11\frac{1}{9}$
3. $\frac{95}{4}$ Ans: $23\frac{3}{4}$

SELF-CHECK
1. $3\frac{2}{7}$ 2. $8\frac{1}{5}$ 3. $5\frac{1}{3}$
4. $13\frac{1}{2}$

ADDITIONAL EXAMPLES
Example 2
Multiply.

1. $5\frac{2}{3} \cdot \frac{6}{7}$ Ans: $4\frac{6}{7}$
2. $18\frac{1}{2} \cdot 10$ Ans: 185
3. $7\frac{3}{5} \cdot 3\frac{3}{4}$ Ans: $28\frac{1}{2}$

SELF-CHECK

5. 10   6. 60   7. $5\frac{1}{2}$
8. $5\frac{11}{16}$

CLASSROOM EXERCISES

1. $4\frac{1}{2}$ 2. $2\frac{1}{5}$ 3. $5\frac{2}{5}$
4. $9\frac{5}{6}$ 5. $6\frac{4}{7}$ 6. $9\frac{1}{4}$
7. $1\frac{3}{5}$ 8. $\frac{7}{2}$ 9. $\frac{9}{4}$

See page 137 for the answers to Exercises 10–19.

ASSIGNMENT GUIDE
BASIC
Day 1 pp. 136–137:
1–33 odd
Day 2 pp. 136–137:
2–34 even
AVERAGE
Day 1 pp. 136–137:
1–37 odd
Day 2 pp. 136–137:
2–38 even
ABOVE AVERAGE
pp. 136–137: 1–37 odd, 39–48

PROCEDURE
To multiply with mixed numerals:
1 Write a fraction for each mixed numeral.
2 Multiply the fractions.
3 Write the product in lowest terms.

EXAMPLE 2 Multiply. **a.** $4 \cdot 3\frac{1}{16}$ **b.** $6\frac{2}{3} \cdot 2\frac{3}{4}$

Solutions:

a. 1 $4 \cdot 3\frac{1}{16} = \frac{4}{1} \cdot \frac{49}{16}$

2 $= \frac{\cancel{4}}{1} \cdot \frac{49}{\cancel{16}}$ ← $\frac{1 \cdot 49}{1 \cdot 4}$

$= \frac{49}{4}$

3 $= 12\frac{1}{4}$ ← Lowest terms →

b. 1 $6\frac{2}{3} \cdot 2\frac{3}{4} = \frac{20}{3} \cdot \frac{11}{4}$

2 $= \frac{\cancel{20}^{5}}{3} \cdot \frac{11}{\cancel{4}_{1}}$

$= \frac{55}{3}$

3 $= 18\frac{1}{3}$

Self-Check *Multiply.*

5. $2\frac{1}{2} \cdot 4$ 6. $9 \cdot 6\frac{2}{3}$ 7. $1\frac{1}{3} \cdot 4\frac{1}{8}$ 8. $1\frac{5}{8} \cdot 3\frac{1}{2}$

CLASSROOM EXERCISES

Write a mixed numeral for each fraction. (Example 1)

1. $\frac{9}{2}$ 2. $\frac{11}{5}$ 3. $\frac{27}{5}$ 4. $\frac{59}{6}$ 5. $\frac{46}{7}$ 6. $\frac{37}{4}$ 7. $\frac{8}{5}$

Write a fraction for each mixed numeral. (Table)

8. $3\frac{1}{2}$ 9. $2\frac{1}{4}$ 10. $5\frac{3}{4}$ 11. $9\frac{3}{8}$ 12. $2\frac{2}{3}$ 13. $3\frac{7}{16}$ 14. $5\frac{3}{7}$

Multiply. Write each answer in lowest terms. (Example 2)

15. $7 \cdot 2\frac{1}{2}$ 16. $8 \cdot 6\frac{1}{3}$ 17. $2\frac{2}{3} \cdot 1\frac{1}{2}$ 18. $1\frac{5}{8} \cdot 1\frac{1}{4}$ 19. $2\frac{2}{5} \cdot 3\frac{1}{3}$

WRITTEN EXERCISES

Goals: To multiply with mixed numerals
To apply multiplication with mixed numerals to solving word problems

Write a mixed numeral for each fraction. (Example 1)

1. $\frac{10}{7}$ 2. $\frac{11}{4}$ 3. $\frac{12}{5}$ 4. $\frac{22}{3}$ 5. $\frac{15}{8}$ 6. $\frac{25}{12}$ 7. $\frac{27}{10}$

Chapter 5

Write a fraction for each mixed numeral. (Table)

8. $4\frac{7}{8}$ 9. $3\frac{2}{3}$ 10. $4\frac{3}{4}$ 11. $7\frac{2}{5}$ 12. $6\frac{5}{7}$ 13. $3\frac{4}{9}$ 14. $3\frac{7}{10}$

Multiply. Write each answer in lowest terms. (Example 2)

15. $3\frac{1}{2} \cdot 2$ 16. $5\frac{1}{3} \cdot 4$ 17. $4 \cdot 1\frac{1}{2}$ 18. $6 \cdot 2\frac{1}{3}$ 19. $2\frac{1}{4} \cdot 4$
20. $2\frac{1}{2} \cdot 9$ 21. $1\frac{3}{4} \cdot 4\frac{1}{2}$ 22. $2\frac{1}{4} \cdot 3\frac{1}{3}$ 23. $4\frac{1}{2} \cdot 2\frac{3}{5}$ 24. $5\frac{2}{3} \cdot 2\frac{1}{5}$
25. $5\frac{1}{2} \cdot 6\frac{1}{3}$ 26. $8\frac{2}{3} \cdot 5\frac{1}{6}$ 27. $4\frac{1}{8} \cdot 4\frac{3}{5}$ 28. $3\frac{1}{2} \cdot 4\frac{2}{3}$ 29. $1\frac{1}{2} \cdot 1\frac{5}{9}$
30. $1\frac{7}{8} \cdot 3\frac{2}{3}$ 31. $3\frac{2}{3} \cdot 3\frac{2}{3}$ 32. $4\frac{1}{2} \cdot 2\frac{1}{3}$ 33. $10 \cdot 1\frac{5}{6}$ 34. $6\frac{1}{2} \cdot 12$

APPLICATIONS: USING MIXED NUMERALS

Solve. Write each answer in lowest terms.

35. A sports car averages $12\frac{1}{2}$ miles to one gallon of fuel. The tank holds $9\frac{1}{3}$ gallons. How far will the car travel on a full tank of fuel?

36. Vivian works $5\frac{1}{6}$ hours every Saturday. How many hours does she work in a month if there are 4 Saturdays in the month?

37. A casserole recipe calls for $1\frac{1}{4}$ cups of chicken. How much chicken is needed to make 3 times that recipe?

38. Jules is putting carpet in his hall. The hall is $8\frac{2}{3}$ feet long and $4\frac{1}{2}$ feet wide. How many square feet of carpet will he need to cover the hall floor?

MORE CHALLENGING EXERCISES

Multiply. No variable equals zero.

EXAMPLE: $2\frac{1}{4} \cdot \frac{x}{3} = \frac{9}{4} \cdot \frac{x}{3}$

$$= \frac{\overset{3}{\cancel{9}} \cdot x}{4 \cdot \underset{1}{\cancel{3}}} = \frac{3 \cdot x}{4 \cdot 1} = \frac{3x}{4}$$

39. $1\frac{1}{2} \cdot \frac{p}{3}$ 40. $6\frac{2}{3} \cdot \frac{y}{5}$ 41. $2\frac{1}{4} \cdot \frac{2}{a}$ 42. $5\frac{1}{3} \cdot \frac{9}{b}$ 43. $2\frac{1}{3} \cdot \frac{m}{3}$
44. $\frac{3d}{2} \cdot 1\frac{2}{3}$ 45. $\frac{5t}{6} \cdot 4\frac{1}{5}$ 46. $3\frac{1}{3} \cdot \frac{6t}{10}$ 47. $5\frac{1}{4} \cdot \frac{2}{7p}$ 48. $1\frac{2}{5} \cdot \frac{3}{5x}$

CLASSROOM EXERCISES
p. 136

10. $\frac{23}{4}$ 11. $\frac{75}{8}$ 12. $\frac{8}{3}$
13. $\frac{55}{16}$ 14. $\frac{38}{7}$ 15. $17\frac{1}{2}$
16. $50\frac{2}{3}$ 17. 4 18. $2\frac{1}{32}$
19. 8

WRITTEN EXERCISES

1. $1\frac{3}{7}$ 2. $2\frac{3}{4}$ 3. $2\frac{2}{5}$
4. $7\frac{1}{3}$ 5. $1\frac{7}{8}$ 6. $2\frac{1}{12}$
7. $2\frac{7}{10}$ 8. $\frac{39}{8}$ 9. $\frac{11}{3}$
10. $\frac{19}{4}$ 11. $\frac{37}{5}$ 12. $\frac{47}{7}$
13. $\frac{31}{9}$ 14. $\frac{37}{10}$ 15. 7
16. $21\frac{1}{3}$ 17. 6 18. 14
19. 9 20. $22\frac{1}{2}$ 21. $7\frac{7}{8}$
22. $7\frac{1}{2}$ 23. $11\frac{7}{10}$
24. $12\frac{7}{15}$ 25. $34\frac{5}{6}$
26. $44\frac{7}{9}$ 27. $18\frac{39}{40}$
28. $16\frac{1}{3}$ 29. $2\frac{1}{3}$
30. $6\frac{7}{8}$ 31. $13\frac{4}{9}$
32. $10\frac{1}{2}$ 33. $18\frac{1}{3}$
34. 78 35. $116\frac{2}{3}$ mi

See page 140 for the answers to Exercises 36–48.

Fractions: Multiplication/Division

CAREER APPLICATIONS

This feature applies the skills for solving formulas taught in Section 5-7 (pages 145-147) to finding the greatest safe load a beam can bear.

1. 8975 pounds
2. 5744 pounds
3. 34,464 pounds
4. 15,078 pounds
5. 13,350 pounds

CAREER APPLICATIONS
Engineering Technician

Engineering technicians often work with engineers and perform some of the same tasks as engineers.

In construction work, it is important to know the greatest safe load that a beam can bear. The formula below applies to a steel I-beam with the load in the middle of the beam. The load is in pounds.

$$S = \frac{1795Ad}{l}$$

S = greatest safe load (lbs)
A = cross-sectional area (in^2)
d = depth of beam (in)
l = length of beam (ft)

EXAMPLE Calculate the greatest safe load in pounds for a steel I-beam 24 feet long, 6 inches deep, and with a cross-sectional area of 28 square inches.

SOLUTION
$$S = \frac{1795Ad}{l}$$
$$S = \frac{1795(28)(6)}{24}$$

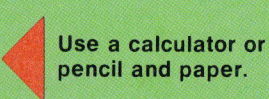

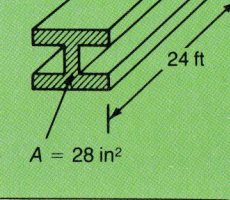

Use a calculator or pencil and paper.

1795 × 28 × 6 ÷ 24 =

`12565.`

The greatest safe load is **12,565 pounds**.

Exercises

Calculate the greatest safe load in pounds of a steel I-beam that has the given dimensions.

1. $A = 15$ in^2
   $d = 4$ in
   $l = 12$ ft

2. $A = 8$ in^2
   $d = 4$ in
   $l = 10$ ft

3. $A = 48$ in^2
   $d = 8$ in
   $l = 20$ ft

4. $A = 21$ in^2
   $d = 6$ in
   $l = 15$ ft

5. The formula $S = \frac{1780Ad}{l}$ is used to calculate the greatest safe load in pounds of a solid steel beam with the load distributed along the beam. Find the greatest safe load of such a beam with a length of 24 feet, a depth of 6 inches, and a cross-sectional area of 30 square inches.

CONSUMER APPLICATION
Saving $ on Cooling Costs

When buying an air conditioner, it is important to select a model that is the right size. The size or **cooling capacity** of an air conditioner is measured in **British Thermal units (BTU's)**. One **BTU** is the energy needed to raise the temperature of one pound of water one degree Fahrenheit.

The graph shown at the right can be used to estimate how much cooling capacity is needed for a given room of a house or apartment. (The graph can be used for any room that is about eight feet high provided that it is not a kitchen and not directly below an attic floor.) The exposure of the room's exterior wall determines which line you should use.

EXAMPLE The room of a house has a northern exposure. Its floor area is 300 square feet. What must be the size (cooling capacity) of the air conditioner for this room?

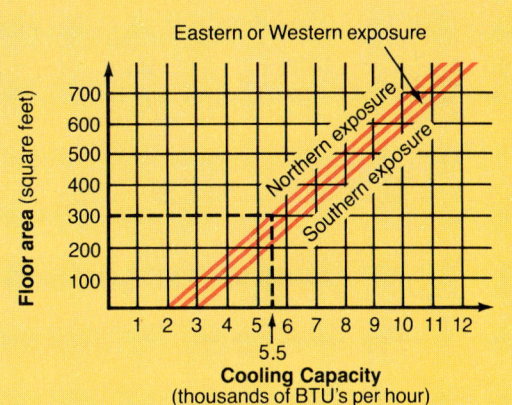

SOLUTION
1. On the Floor Area scale, find the floor area, 300 square feet.
2. Since the room has a northern exposure, find the point on the northern exposure line directly to the right of the 300 reading.
3. From this point on the line read directly <u>down</u> on the Cooling Capacity scale to find the correct cooling capacity, **5500 BTU's per hour.**

Exercises
Estimate the size of an air conditioner needed to cool each room described. Round each answer to the nearest 500 BTU's.

1. Southern exposure; 300 square feet
2. Eastern exposure; 300 square feet
3. Southern exposure; 500 square feet
4. Northern exposure; 500 square feet
5. Western exposure; 100 square feet
6. Northern exposure; 100 square feet

CONSUMER APPLICATION
This feature introduces the British thermal unit (BTU) for the measure of cooling capacity of an air conditioner. Since heat and temperature change are directly related, it can be used when Section 5-4 (pages 132-134) is taught.
1. 6500 BTU's
2. 6000 BTU's
3. 9000 BTU's
4. 8000 BTU's
5. 3500 BTU's
6. 3000 BTU's

Teaching Suggestions
p. M-18

QUICK QUIZ
Multiply.
1. $18 \cdot \frac{1}{6}$ Ans: 3
2. $\frac{1}{3} \cdot 5 \cdot 6$ Ans: 10
3. $\frac{3}{5}(7)(1.6)$ Ans: 6.72

Evaluate.
4. 7h when h = 22
   Ans: 154
5. xy when x = 8
   and y = 7.6
   Ans: 60.8
6. 16ab when a = 5
   and b = $\frac{3}{4}$
   Ans: 60
7. $\frac{1}{2}$ck when c = $\frac{3}{5}$
   and k = 10
   Ans: 3

WRITTEN EXERCISES p. 137
36. $20\frac{2}{3}$ hr   37. $3\frac{3}{4}$ cups
38. 39 ft^2 39. $\frac{p}{2}$
40. $\frac{4y}{3}$ 41. $\frac{9}{2a}$ 42. $\frac{48}{b}$
43. $\frac{7m}{9}$ 44. $\frac{5d}{2}$ 45. $\frac{7t}{2}$
46. 2t   47. $\frac{3}{2p}$ 48. $\frac{21}{25x}$

REVIEW CAPSULE FOR SECTION 5-6

Evaluate each expression   (Pages 22-24)

1. $(4.5)(a)$ when $a = 2.7$
2. $(3.1)(y)$ when $y = 6$
3. $2n$ when $n = 3.5$
4. $\frac{1}{2}bc$ when $b = 8$ and $c = 5$
5. $\frac{1}{2}pq$ when $p = \frac{2}{3}$ and $q = 4$
6. mt when $m = \frac{8}{9}$ and $t = 5$

Multiply. (Pages 38-41)

7. $3.4(8.1)$ 8. $1.4(3.5)$ 9. $4(2.3)$ 10. $(1.6)(2.1)$ 11. $12.4(3.5)$ 12. $0.5(8.1)$

Multiply. Write each answer in lowest terms. (Pages 124-125)

13. $\frac{1}{2} \cdot \frac{4}{5}$ 14. $\frac{1}{4} \cdot \frac{8}{9}$ 15. $\frac{2}{3} \cdot 6$ 16. $\frac{3}{5} \cdot 15$ 17. $\frac{4}{9} \cdot \frac{1}{4}$ 18. $16 \cdot \frac{5}{6}$

The answers are on page 154.

PROBLEM SOLVING AND APPLICATIONS

5-6 Area: Parallelograms and Triangles

A rectangle can be changed into a **parallelogram** of equal area.

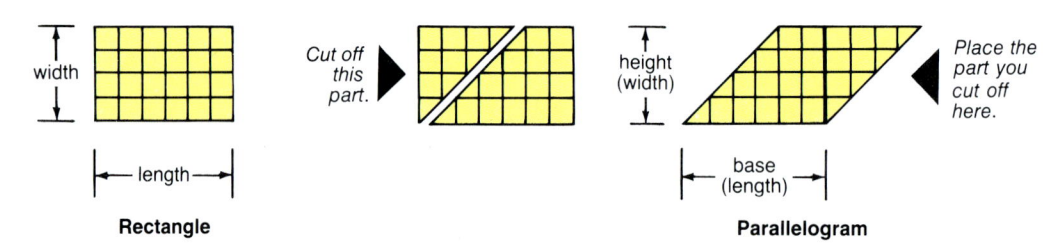

Thus, the formula for finding the area of a parallelogram is similar to the formula for finding the area of a rectangle.

Word Rule: The area of a parallelogram is the product of the base and the height.

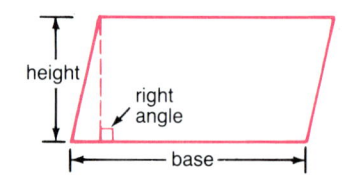

Formula: $A = bh$

A = area
b = length of a base
h = corresponding height

The base of a parallelogram can be any side.
The height of a parallelogram is measured along a line at right angles to the base.

EXAMPLE 1 An address plate on an office building has the shape of a parallelogram. The plate is 32 centimeters long and 16.5 centimeters high. Find the area.

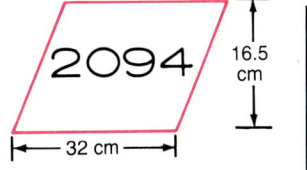

2094   16.5 cm
32 cm

Solution:
1. Formula: $A = bh$
2. Known values: $b = 32$ cm; $h = 16.5$ cm
3. Replace b with 32 and h with 16.5. $A = 32 \times 16.5$
4. Multiply. $A = 528.0$ The area is **528 cm²**.

Self-Check Find the area of each parallelogram.
1. $b = 7\frac{1}{2}$ in; $h = 3$ in
2. $b = 12.8$ cm; $h = 5.6$ cm

A parallelogram can be changed into two triangles with equal area. The area of each triangle is one-half the area of the parallelogram.

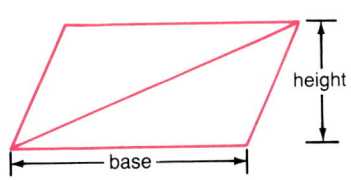

Word Rule: The area of a triangle is one-half the product of the base and the height.

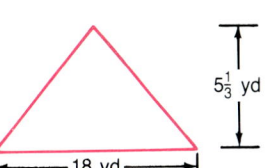

Formula: $A = \frac{1}{2}bh$

A = area
b = length of a base
h = corresponding height

EXAMPLE 2 The gable of a roof has the shape of a triangle. The gable is 18 yards long and $5\frac{1}{3}$ yards high. Find the area.

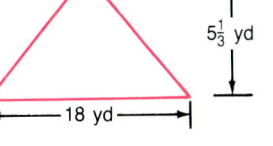

Solution:
1. Formula: $A = \frac{1}{2}bh$
2. Known values: $b = 18$ yd; $h = 5\frac{1}{3}$ yd
3. Replace b with 18 and h with $5\frac{1}{3}$. Write a fraction for $5\frac{1}{3}$. $A = \frac{1}{2} \times 18 \times \frac{16}{3}$
4. Multiply. $A = \frac{1}{2} \times \frac{\overset{9}{\cancel{18}}}{1} \times \frac{16}{\cancel{3}}$

$A = \frac{48}{1} = 48$

The area is **48 yd²**.

Self-Check Find the area of each triangle.
3. $b = 4\frac{1}{2}$ ft; $h = 6$ ft
4. $b = 8.6$ m; $h = 10$ m

ADDITIONAL EXAMPLES

Example 1
Find the area of each parallelogram.
1. b = 12 yd; h = 12 yd
   Ans: 144 yd²
2. b = $3\frac{1}{2}$ ft; h = $7\frac{1}{3}$ ft
   Ans: $25\frac{2}{3}$ ft²
3. b = 8.1 m; h = 3.6 m
   Ans: 29.16 m²

SELF-CHECK
1. $22\frac{1}{2}$ in² 2. 71.68 cm²

Example 2
Find the area of each triangle.
1. b = 12 yd; h = 12 yd
   Ans: 72 yd²
2. b = $3\frac{1}{2}$ ft; h = $7\frac{1}{3}$ ft
   Ans: $12\frac{5}{6}$ ft²
3. b = 8.1 m; h = 3.6 m
   Ans: 14.58 m²

SELF-CHECK
3. $13\frac{1}{2}$ ft² 4. 43 m²

Fractions: Multiplication/Division

CLASSROOM EXERCISES

<u>1</u>. 28 in^2 <u>2</u>. $4\frac{2}{3}$ ft^2
<u>3</u>. 91 yd^2 <u>4</u>. 24.8 cm^2
<u>5</u>. 98.8 m^2
<u>6</u>. 225.4 cm^2 <u>7</u>. 6 ft^2
<u>8</u>. $12\frac{1}{2}$ in^2 <u>9</u>. 21 yd^2
<u>10</u>. 10 m^2 <u>11</u>. 36.6 cm^2
<u>12</u>. 42.64 mm^2

ASSIGNMENT GUIDE
BASIC
pp. 142-144: 1-27 odd
AVERAGE
pp. 142-144: 1-31 odd
ABOVE AVERAGE
pp. 142-144: 1-31 odd

WRITTEN EXERCISES
<u>1</u>. 225 in^2 <u>2</u>. $4\frac{1}{2}$ in^2
<u>3</u>. $1\frac{3}{4}$ ft^2 <u>4</u>. 20 in^2
<u>5</u>. 96 ft^2 <u>6</u>. $9\frac{4}{5}$ yd^2
<u>7</u>. 200 in^2

142

CLASSROOM EXERCISES

Find the area of each parallelogram. (Example 1)

CUSTOMARY MEASURES

1. $b = 8$ in; $h = 3\frac{1}{2}$ in   **2.** $b = 2\frac{2}{3}$ ft; $h = 1\frac{3}{4}$ ft   **3.** $b = 13$ yd; $h = 7$ yd

METRIC MEASURES

4. $b = 6.2$ cm; $h = 4$ cm   **5.** $b = 13$ m; $h = 7.6$ m   **6.** $b = 24.5$ cm; $h = 9.2$ cm

Find the area of each triangle. (Example 2)

CUSTOMARY MEASURES

7. $b = 4$ ft; $h = 3$ ft   **8.** $b = 6\frac{1}{4}$ in; $h = 4$ in   **9.** $b = 9$ yd; $h = 4\frac{2}{3}$ yd

METRIC MEASURES

10. $b = 5$ m; $h = 4$ m   **11.** $b = 12.2$ cm; $h = 6$ cm   **12.** $b = 10.4$ mm; $h = 8.2$ mm

WRITTEN EXERCISES

Goals: To use a formula to find the area of a parallelogram and of a triangle
To apply these formulas to solving word problems

Find the area of each parallelogram. (Example 1)

CUSTOMARY MEASURES

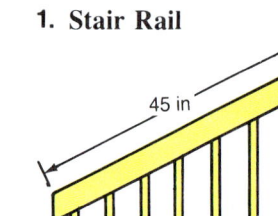

1. Stair Rail — 45 in, 5 in

2. Product Label — 3 in, $1\frac{1}{2}$ in

3. Address Plate — $2\frac{1}{3}$ ft, $\frac{3}{4}$ ft

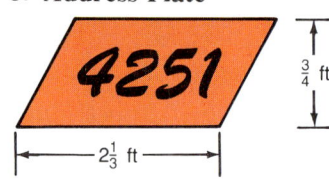

	Base	Height	Area
4.	6 in	$3\frac{1}{3}$ in	?
6.	$4\frac{1}{5}$ yd	$2\frac{1}{3}$ yd	?

	Base	Height	Area
5.	12 ft	8 ft	?
7.	16 in	$12\frac{1}{2}$ in	?

142   Chapter 5

METRIC MEASURES

8. Parking Lot

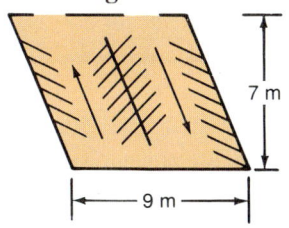

7 m, 9 m

9. Decorative Tile

13 cm, 17.5 cm

10. Product Label

1.5 cm, 4 cm

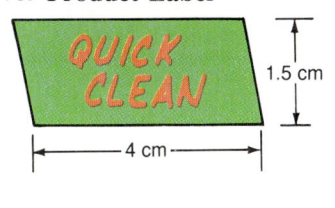

	Base	Height	Area
11.	3.2 cm	4 cm	?
13.	7 m	12 m	?

	Base	Height	Area
12.	15.5 mm	18 mm	?
14.	35 cm	14 cm	?

Find the area of each triangle. (Example 2)

CUSTOMARY MEASURES

15. Table Top

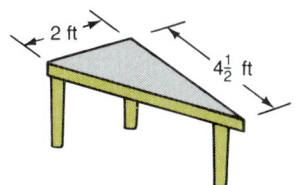

2 ft, $4\frac{1}{2}$ ft

16. Sail

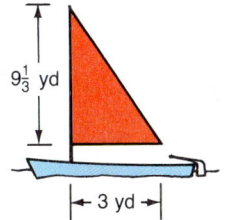

$9\frac{1}{3}$ yd, 3 yd

17. Tie Back

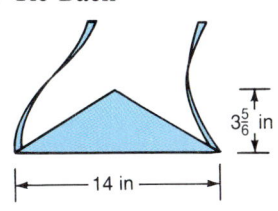

$3\frac{5}{6}$ in, 14 in

	Base	Height	Area
18.	4 in	6 in	?
20.	$8\frac{1}{6}$ ft	4 ft	?

	Base	Height	Area
19.	10 ft	$18\frac{1}{4}$ ft	?
21.	5 yd	$3\frac{1}{9}$ yd	?

METRIC MEASURES

22. Gable of a Roof

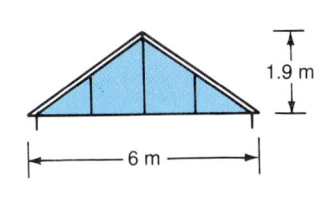

1.9 m, 6 m

23. Flag

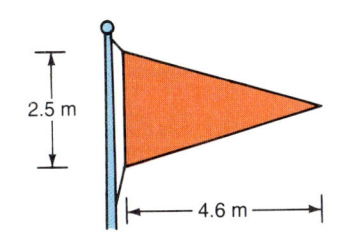

2.5 m, 4.6 m

24. Sail

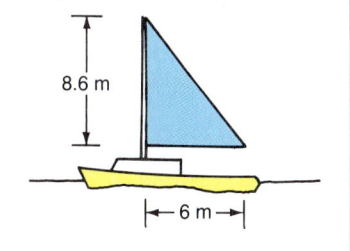

8.6 m, 6 m

WRITTEN EXERCISES

8. 63 m^2 9. 227.5 cm^2
10. 6 cm^2 11. 12.8 cm^2
12. 279 mm^2 13. 84 m^2
14. 490 cm^2
15. $4\frac{1}{2}$ ft^2 16. 14 yd^2
17. $26\frac{5}{6}$ in^2
18. 12 in^2
19. $91\frac{1}{4}$ ft^2
20. $16\frac{1}{3}$ ft^2
21. $7\frac{7}{9}$ yd^2 22. 5.7 m^2
23. 5.75 m^2
24. 25.8 m^2

Fractions: Multiplication/Division

WRITTEN EXERCISES

25. 12.6 cm^2
26. 3.75 m^2
27. 4.55 cm^2
28. 112 mm^2
29. 135 ft^2
30. 33.6 m^2
31. $15\frac{3}{4}$ in^2
32. 3.6 m^2

Find the area of each triangle. (Example 2)

	Base	Height	Area
25.	6 cm	4.2 cm	?
27.	3.5 cm	2.6 cm	?

	Base	Height	Area
26.	3 m	2.5 m	?
28.	16 mm	14 mm	?

MIXED PRACTICE

Solve each problem.

29. The mainsail on Jim's catamaran is in the shape of a triangle. The base is $6\frac{2}{3}$ feet and the height is $20\frac{1}{4}$ feet. Find the area of the sail.

30. Ellen's garden has the shape of a parallelogram. The base is 6 meters and the height is 5.6 meters. Find the area of the garden.

31. A wing on a model rocket has the shape of a parallelogram. The base is $2\frac{1}{3}$ inches and the height is $6\frac{3}{4}$ inches. Find the area of the wing.

32. Each side of a tent is triangular in shape. The base of the triangle is 2.4 meters and the height is 3 meters. Find the area of one side of the tent.

REVIEW CAPSULE FOR SECTION 5-7

Evaluate each expression. (Pages 22–24)

1. xyz when $x = 2$, $y = 4$, and $z = 12$
2. ab when $a = 3$ and $b = 9$
3. kmn when $k = 6$, $m = 1.7$, and $n = 3.2$
4. st when $s = 0.6$ and $t = 4.9$
5. cd when $c = 8$ and $d = 3\frac{1}{3}$
6. rst when $r = 2$, $s = 3\frac{1}{2}$, and $t = 1\frac{1}{3}$

Multiply. (Pages 38–41)

7. 4(3.2)
8. 7.1(5)
9. 2.8(1.6)
10. 1.9(3.5)
11. (5.2)(1.8)
12. (3.6)(0.7)
13. (12)(3.8)
14. (6)(2.9)

Multiply. Write each answer in lowest terms. (Pages 129–131)

15. $1\frac{1}{2} \cdot 2$
16. $3 \cdot 2\frac{1}{4}$
17. $1\frac{1}{3} \cdot 4\frac{1}{2}$
18. $2\frac{2}{5} \cdot 1\frac{1}{8}$
19. $1\frac{1}{6} \cdot 2\frac{1}{7}$
20. $8\frac{1}{3} \cdot 2\frac{2}{5}$
21. $4 \cdot 2\frac{3}{4}$
22. $6\frac{1}{4} \cdot 2\frac{3}{5}$

The answers are on page 154.

Chapter 5

PROBLEM SOLVING AND APPLICATIONS

5-7 Volume: Rectangular Prisms

A **rectangular prism** is a solid such as the shoe box at the right. The **volume** of a rectangular prism is the amount of space it contains. Volume is measured in **cubic units** such as cubic centimeters (cm³), cubic meters (m³), cubic inches (in³), cubic feet (ft³), and so on.

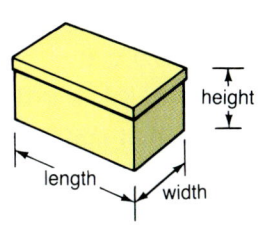

Word Rule: The volume of a rectangular prism is the product of the length, the width, and the height.

Formula: $V = lwh$ V = volume; l = length; w = width; h = height

EXAMPLE A shoe box has a length of 35 centimeters, a width of 15 centimeters, and a height of 10 centimeters. Find the volume of the shoe box.

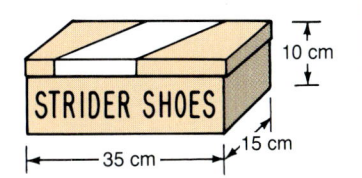

Solution:
1. Formula: $V = lwh$
2. Known values: $l = 35$ cm; $w = 15$ cm; $h = 10$ cm
3. Replace l with 35, w with 15, and h with 10.
   $V = 35 \times 15 \times 10$
   $V = 5250$

The volume is **5250 cm³**.

Self-Check *Find the volume of each rectangular prism.*

1. $l = 6\frac{1}{2}$ ft; $w = 3\frac{1}{3}$ ft; $h = 2$ ft
2. $l = 12$ m; $w = 4.8$ m; $h = 5.2$ m

NOTE: The formula for the volume of a rectangular prism can also be written as follows.

$V = l \cdot w \cdot h$

$V = B \cdot h$, or

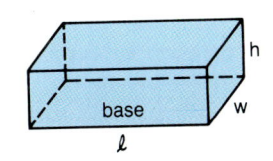

$V = Bh$ B = area of base
           h = height

Fractions: Multiplication/Division **145**

Teaching Suggestions
p. M-18

QUICK QUIZ
Multiply.
1. 21 · 13 · 6
   Ans: 1638
2. $2\frac{1}{2}$ · 8 · 3   Ans: 60
3. 2 · $5\frac{1}{4}$ · 12   Ans: 126

Evaluate.
4. 5cd when c = 2
   and d = 18   Ans: 180
5. 6ax when a = $2\frac{3}{4}$
   and x = $\frac{1}{5}$ Ans: $3\frac{3}{10}$
6. hjk when h = 7.1;
   j = 1.5; k = 0.2
   Ans: 2.13

ADDITIONAL EXAMPLES
Example
Find the volume of each rectangular prism.
1. l = 18 ft; w = 12 ft;
   h = 7 ft
   Ans: 1512 ft³
2. l = $2\frac{1}{3}$ in; w = $\frac{5}{6}$ in;
   h = 3 in
   Ans: $5\frac{5}{6}$ in³
3. l = 6 km; w = 2.8 km;
   h = 5.02 km
   Ans: 84.336 km³

SELF-CHECK
1. $43\frac{1}{3}$ ft³
2. 299.52 m³

145

CLASSROOM EXERCISES

Find the volume of each rectangular prism. (Example)

1. $l = 9$ ft; $w = 6$ ft; $h = 2\frac{1}{2}$ ft
2. $l = 10$ in; $w = 5\frac{1}{4}$ in; $h = 3$ in
3. $l = 0.5$ cm; $w = 1$ cm; $h = 1.7$ cm
4. $l = 2.7$ m; $w = 3.1$ m; $h = 5$ m
5. $l = 8\frac{1}{2}$ yd; $w = 3$ yd; $h = 2\frac{1}{3}$ yd
6. $l = 4\frac{1}{4}$ ft; $w = 3$ ft; $h = 5\frac{1}{7}$ ft
7. $l = 16.4$ mm; $w = 9.2$ mm; $h = 5$ mm
8. $l = 2.5$ cm; $w = 1.4$ cm; $h = 4$ cm

WRITTEN EXERCISES

Goals: To use a formula to find the volume of a rectangular prism
To apply this formula to solving word problems

Find the volume of each rectangular prism. (Example)

CUSTOMARY MEASURES

1. Shed

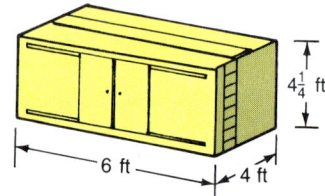

6 ft, 4 ft, $4\frac{1}{4}$ ft

2. Truck

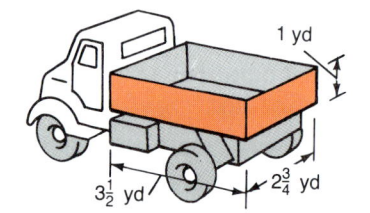

$3\frac{1}{2}$ yd, $2\frac{3}{4}$ yd, 1 yd

3. Refrigerator

$3\frac{3}{4}$ ft, 3 ft, 6 ft

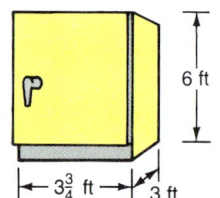

	Length	Width	Height
4.	10 ft	5 ft	4 ft
6.	5 yd	$6\frac{1}{5}$ yd	$3\frac{1}{2}$ yd

	Length	Width	Height
5.	$16\frac{1}{2}$ in	8 in	$2\frac{1}{4}$ in
7.	12 ft	$3\frac{1}{8}$ ft	$3\frac{1}{3}$ ft

METRIC MEASURES

8. Planter

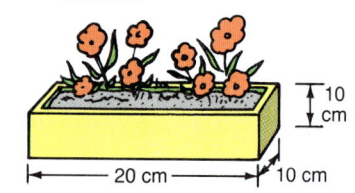

20 cm, 10 cm, 10 cm

9. Paper Weight

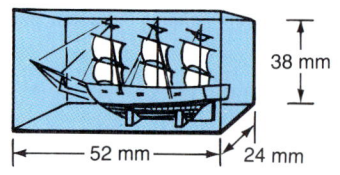

52 mm, 24 mm, 38 mm

10. Stereo Speaker

26 cm, 40 cm

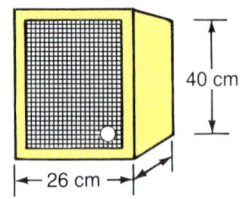

	Length	Width	Height
11.	5 cm	2 cm	4 cm
13.	9 m	6 m	5.1 m

	Length	Width	Height
12.	8 m	6 m	3.9 m
14.	2.1 mm	1.9 mm	5.4 mm

146 Chapter 5

MIXED PRACTICE

15. Stephanie McCloud is building a kitchen cabinet that is 3 feet long, $1\frac{3}{4}$ feet wide, and $2\frac{1}{2}$ feet high. Find the volume of the cabinet.

16. A wall oven is 100 centimeters wide, 85 centimeters deep, and 64 centimeters high. Find the volume of the oven.

17. A fish tank is 45 centimeters long, 30 centimeters wide, and 25.4 centimeters high. Find the volume.

18. A wading pool is 12 feet long, $8\frac{1}{2}$ feet wide, and $4\frac{1}{6}$ feet high. Find the volume of the wading pool.

MORE CHALLENGING EXERCISES

Find the volume of each figure.

19.

20.

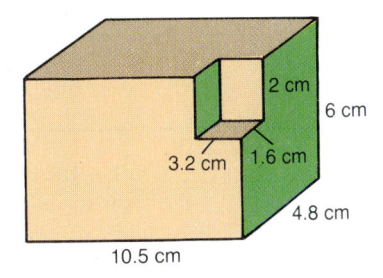

21. The term "1 inch of rainfall" means a rainfall 1 inch high covering 1 acre of ground. An acre equals 43,560 square feet. Find the volume of water in cubic feet for 1 inch of rainfall. (HINT: 1 inch = $\frac{1}{12}$ foot)

22. Twenty-one people crowded into a telephone booth together. The telephone booth was $3\frac{1}{2}$ feet long, $3\frac{1}{2}$ feet wide, and $7\frac{1}{5}$ feet high. How many cubic feet of space did each person have in the booth?

WRITTEN EXERCISES

15. $13\frac{1}{8}$ ft^3
16. 544,000 cm^3
17. 34,290 cm^3
18. 425 ft^3
19. 161.4 m^3
20. 292.16 cm^3
21. 3630 ft^3
22. $4\frac{1}{5}$ ft^3

REVIEW CAPSULE FOR SECTION 5-8

Write a fraction for each mixed numeral. (Pages 135–137)

1. $1\frac{1}{2}$ 2. $2\frac{3}{4}$ 3. $5\frac{1}{3}$ 4. $4\frac{2}{5}$ 5. $3\frac{1}{6}$ 6. $1\frac{2}{7}$

Multiply. Write each answer in lowest terms. (Pages 129–131)

7. $\frac{1}{3} \cdot \frac{9}{1}$ 8. $6 \cdot \frac{1}{2}$ 9. $\frac{6}{5} \cdot \frac{10}{9}$ 10. $\frac{7}{4} \cdot \frac{8}{3}$ 11. $8 \cdot \frac{1}{4}$ 12. $\frac{3}{2} \cdot \frac{4}{9}$

The answers are on page 154.

Fractions: Multiplication/Division

Teaching Suggestions
p. M-18

QUICK QUIZ
Write a fraction for each.
1. $5\frac{2}{3}$ Ans: $\frac{17}{3}$
2. $10\frac{3}{8}$ Ans: $\frac{83}{8}$

Multiply.
3. $\frac{2}{3} \cdot \frac{5}{7}$ Ans: $\frac{10}{21}$
4. $\frac{7}{8} \cdot \frac{10}{3}$ Ans: $2\frac{11}{12}$
5. $12 \cdot 7\frac{2}{3}$ Ans: 92
6. $1\frac{2}{5} \cdot 3\frac{1}{3}$ Ans: $4\frac{2}{3}$

ADDITIONAL EXAMPLES
Example 1
Divide.
1. $\frac{5}{8} \div 2$ Ans: $\frac{5}{16}$
2. $10 \div \frac{1}{5}$ Ans: 50
3. $\frac{3}{7} \div \frac{1}{6}$ Ans: $2\frac{4}{7}$

SELF-CHECK
1. $2\frac{1}{4}$ 2. $\frac{1}{24}$ 3. $\frac{2}{25}$
4. $\frac{5}{12}$

148

5-8 Dividing Fractions

When the product of two numbers is 1, the numbers are **reciprocals** of each other. Reciprocals are sometimes called **multiplicative inverses**.

TABLE

Number	Reciprocal	Reason
$\frac{5}{16}$	$\frac{16}{5}$	$\frac{5}{16} \cdot \frac{16}{5} = 1$
$\frac{1}{12}$	12	$\frac{1}{12} \cdot 12 = 1$
$\frac{10}{3}$	$\frac{3}{10}$	$\frac{10}{3} \cdot \frac{3}{10} = 1$
$3\frac{1}{2}$, or $\frac{7}{2}$	$\frac{2}{7}$	$\frac{7}{2} \cdot \frac{2}{7} = 1$

Dividing by a number gives the same result as multiplying by its reciprocal.

$$18 \div 2 = 9 \qquad 18 \cdot \frac{1}{2} = 9$$

PROCEDURE

To divide with fractions:

[1] Using the reciprocal of the divisor, write the division problem as a multiplication problem.
[2] Multiply.
[3] Write the answer in lowest terms.

EXAMPLE 1   Divide.

a. $\frac{5}{16} \div \frac{3}{4}$ b. $\frac{1}{12} \div 9$

Solutions:

a. [1] $\frac{5}{16} \div \frac{3}{4} = \frac{5}{16} \cdot \frac{4}{3}$ ◄ The reciprocal of $\frac{3}{4}$ is $\frac{4}{3}$.

[2] $= \frac{5}{\cancel{16}4} \cdot \frac{\cancel{4}^1}{3}$

[3] $= \frac{5}{12}$ ◄ Lowest terms

b. [1] $\frac{1}{12} \div 9 = \frac{1}{12} \cdot \frac{1}{9}$ ◄ The reciprocal of 9 is $\frac{1}{9}$.

[2] $= \frac{1 \cdot 1}{12 \cdot 9}$

[3] $= \frac{1}{108}$

Self-Check   Divide.

1. $\frac{3}{5} \div \frac{4}{15}$ 2. $\frac{7}{8} \div 21$ 3. $\frac{4}{5} \div 10$ 4. $\frac{2}{9} \div \frac{24}{45}$

Chapter 5

> **PROCEDURE**
> **To divide with mixed numerals:**
> 1. Write a fraction for each mixed numeral.
> 2. Divide the fractions.
> 3. Write the answer in lowest terms.

EXAMPLE 2 Divide.

a. $2\frac{1}{4} \div 3\frac{3}{8}$

b. $24 \div 5\frac{1}{3}$

Solution:

a. $\boxed{1}$ $2\frac{1}{4} \div 3\frac{3}{8} = \frac{9}{4} \div \frac{27}{8}$

$\boxed{2}$ $= \frac{9}{4} \cdot \frac{8}{27}$ ◀ The reciprocal of $\frac{27}{8}$ is $\frac{8}{27}$.

$= \frac{\cancel{9}^1}{\cancel{4}_1} \cdot \frac{\cancel{8}^2}{\cancel{27}_3}$

$= \frac{1 \cdot 2}{1 \cdot 3}$

$\boxed{3}$ $= \frac{2}{3}$

b. $\boxed{1}$ $24 \div 5\frac{1}{3} = 24 \div \frac{16}{3}$

$\boxed{2}$ $= \frac{24}{1} \cdot \frac{3}{16}$

$= \frac{\cancel{24}^3}{1} \cdot \frac{3}{\cancel{16}_2}$

$= \frac{3 \cdot 3}{1 \cdot 2}$

$\boxed{3}$ $= \frac{9}{2} = 4\frac{1}{2}$

Self-Check *Divide.*

5. $3\frac{1}{4} \div 6\frac{1}{2}$
6. $12 \div 2\frac{4}{7}$
7. $5\frac{1}{3} \div 8$
8. $4\frac{1}{2} \div 2\frac{3}{4}$

> For the fractions $\frac{a}{b}$ and $\frac{c}{d}$, $b \ne 0$, $c \ne 0$, $d \ne 0$,
>
> $\frac{a}{b} \div \frac{c}{d} = \frac{a}{b} \cdot \frac{d}{c} = \frac{ad}{bc}$ $\qquad$ $\frac{1}{2} \div \frac{3}{5} = \frac{1}{2} \cdot \frac{5}{3} = \frac{5}{6}$

CLASSROOM EXERCISES

Write the reciprocal of each number. (Table)

1. $\frac{2}{3}$
2. $\frac{3}{5}$
3. $\frac{9}{7}$
4. 2
5. 4
6. $\frac{10}{3}$
7. $\frac{3}{2}$

Divide. Write each answer in lowest terms. (Example 1)

8. $\frac{1}{4} \div \frac{1}{3}$
9. $\frac{3}{8} \div 3$
10. $\frac{2}{7} \div \frac{3}{8}$
11. $\frac{4}{7} \div \frac{2}{3}$
12. $\frac{3}{5} \div \frac{15}{20}$
13. $\frac{7}{10} \div 14$
14. $\frac{4}{9} \div 16$
15. $\frac{9}{10} \div 15$
16. $\frac{1}{7} \div 21$
17. $\frac{3}{5} \div 12$

Fractions: Multiplication/Division **149**

ADDITIONAL EXAMPLES
Example 2
Divide.
1. $15 \div 2\frac{1}{2}$ Ans: 6
2. $1\frac{3}{4} \div \frac{5}{6}$ Ans: $2\frac{1}{10}$
3. $3\frac{1}{6} \div 2\frac{2}{3}$ Ans: $1\frac{3}{16}$

SELF-CHECK
5. $\frac{1}{2}$ 6. $4\frac{2}{3}$ 7. $\frac{2}{3}$
8. $1\frac{7}{11}$

CLASSROOM EXERCISES
1. $\frac{3}{2}$ 2. $\frac{5}{3}$ 3. $\frac{7}{9}$
4. $\frac{1}{2}$ 5. $\frac{1}{4}$ 6. $\frac{3}{10}$
7. $\frac{2}{3}$ 8. $\frac{3}{4}$ 9. $\frac{1}{8}$
10. $\frac{16}{21}$ 11. $\frac{6}{7}$ 12. $\frac{4}{5}$
13. $\frac{1}{20}$ 14. $\frac{1}{36}$
15. $\frac{3}{50}$ 16. $\frac{1}{147}$
17. $\frac{1}{20}$

CLASSROOM EXERCISES

18. $\frac{3}{4}$ 19. $2\frac{8}{25}$ 20. $\frac{2}{3}$
21. 1   22. $1\frac{6}{13}$ 23. $\frac{1}{2}$
24. $\frac{3}{4}$ 25. 3   26. $\frac{2}{5}$
27. $6\frac{1}{8}$

ASSIGNMENT GUIDE
BASIC
Day 1 p. 150: 1-59 odd
Day 2 p. 150: 2-58 even
AVERAGE
Day 1 p. 150: 1-43 odd
Day 2 pp. 150-151: 45-67
ABOVE AVERAGE
pp. 150-151: 3, 6, 9, ..., 63, 65-77

WRITTEN EXERCISES

1. 3   2. $\frac{5}{3}$ 3. $\frac{7}{4}$
4. $\frac{18}{5}$ 5. $\frac{1}{8}$ 6. $\frac{1}{4}$
7. $\frac{1}{12}$ 8. $\frac{5}{11}$ 9. $\frac{11}{18}$
10. $\frac{4}{21}$ 11. $\frac{4}{13}$ 12. $\frac{1}{21}$
13. $\frac{9}{7}$ 14. $\frac{1}{9}$ 15. $\frac{2}{3}$
16. $\frac{3}{4}$ 17. $\frac{3}{4}$ 18. $\frac{3}{4}$
19. $\frac{5}{6}$ 20. $\frac{48}{49}$ 21. $\frac{3}{5}$
22. $\frac{1}{2}$ 23. $\frac{20}{33}$ 24. $\frac{1}{18}$
25. $\frac{1}{30}$ 26. $\frac{1}{30}$ 27. 6
28. 32   29. $\frac{2}{15}$ 30. $4\frac{3}{4}$
31. $3\frac{7}{25}$ 32. $\frac{1}{2}$ 33. $\frac{1}{2}$
34. $1\frac{1}{2}$ 35. $3\frac{1}{5}$ 36. $1\frac{3}{5}$
37. 4   38. $2\frac{5}{6}$

150

(Example 2)

18. $2\frac{1}{2} \div 3\frac{1}{3}$ 19. $9\frac{2}{3} \div 4\frac{1}{6}$ 20. $2\frac{1}{4} \div 3\frac{3}{8}$ 21. $2\frac{1}{2} \div 2\frac{1}{2}$ 22. $4\frac{3}{4} \div 3\frac{1}{4}$

23. $7\frac{1}{2} \div 15$ 24. $9\frac{3}{4} \div 13$ 25. $4 \div 1\frac{1}{3}$ 26. $3 \div 7\frac{1}{2}$ 27. $21 \div 3\frac{3}{7}$

WRITTEN EXERCISES

Goals: To divide with fractions and mixed numerals
To apply this skill to solving word problems

Write the reciprocal of each number. (Table)

1. $\frac{1}{3}$ 2. $\frac{3}{5}$ 3. $\frac{4}{7}$ 4. $\frac{5}{18}$ 5. 8   6. 4   7. 12

8. $\frac{11}{5}$ 9. $\frac{18}{11}$ 10. $\frac{21}{4}$ 11. $\frac{13}{4}$ 12. 21   13. $\frac{7}{9}$ 14. 9

Divide. Write each answer in lowest terms. (Example 1)

15. $\frac{1}{2} \div \frac{3}{4}$ 16. $\frac{1}{4} \div \frac{1}{3}$ 17. $\frac{3}{5} \div \frac{4}{5}$ 18. $\frac{5}{8} \div \frac{5}{6}$ 19. $\frac{2}{3} \div \frac{4}{5}$

20. $\frac{4}{7} \div \frac{7}{12}$ 21. $\frac{1}{3} \div \frac{5}{9}$ 22. $\frac{1}{8} \div \frac{1}{4}$ 23. $\frac{5}{9} \div \frac{11}{12}$ 24. $\frac{7}{9} \div 14$

25. $\frac{1}{6} \div 5$ 26. $\frac{1}{5} \div 6$ 27. $4 \div \frac{2}{3}$ 28. $8 \div \frac{1}{4}$ 29. $\frac{2}{3} \div 5$

(Example 2)

30. $6\frac{1}{3} \div 1\frac{1}{3}$ 31. $8\frac{1}{5} \div 2\frac{1}{2}$ 32. $3\frac{1}{4} \div 6\frac{1}{2}$ 33. $1\frac{3}{5} \div 3\frac{1}{5}$ 34. $3\frac{3}{8} \div 2\frac{1}{4}$

35. $6\frac{6}{7} \div 2\frac{1}{7}$ 36. $2 \div 1\frac{1}{4}$ 37. $6 \div 1\frac{1}{2}$ 38. $5\frac{2}{3} \div 2$ 39. $4\frac{4}{5} \div 6$

40. $4 \div 6\frac{1}{4}$ 41. $6\frac{2}{3} \div 1\frac{3}{4}$ 42. $9\frac{1}{7} \div 8\frac{1}{2}$ 43. $3\frac{7}{8} \div 4$ 44. $5\frac{1}{4} \div 6\frac{7}{8}$

MIXED PRACTICE

Divide. Write each answer in lowest terms.

45. $6 \div \frac{1}{2}$ 46. $\frac{8}{9} \div \frac{2}{3}$ 47. $3\frac{1}{2} \div 7$ 48. $4\frac{2}{3} \div 1\frac{7}{9}$ 49. $3 \div \frac{2}{5}$

50. $1\frac{1}{2} \div 3\frac{3}{8}$ 51. $\frac{2}{5} \div \frac{4}{15}$ 52. $\frac{3}{4} \div \frac{1}{8}$ 53. $1\frac{7}{10} \div 3\frac{2}{5}$ 54. $2 \div 4\frac{1}{2}$

55. $1\frac{1}{5} \div 2\frac{1}{5}$ 56. $\frac{4}{9} \div 4$ 57. $5\frac{1}{4} \div 4\frac{3}{8}$ 58. $7 \div 3\frac{2}{3}$ 59. $\frac{11}{12} \div \frac{5}{6}$

60. $\frac{7}{8} \div \frac{1}{6}$ 61. $3\frac{3}{8} \div 2\frac{1}{2}$ 62. $7\frac{1}{3} \div 4\frac{2}{3}$ 63. $\frac{3}{7} \div \frac{9}{14}$ 64. $8 \div 4\frac{1}{2}$

150 Chapter 5

APPLICATIONS: USING FRACTIONS AND MIXED NUMERALS

65. A stack of magazines weighs 10 pounds. Each magazine weighs $\frac{5}{8}$ pound. How many magazines are in the stack?

66. Jason rides his bicycle 25 kilometers in $1\frac{1}{4}$ hours. How far does he ride in 1 hour?

67. How many boards $2\frac{1}{4}$ feet long can be cut from a board $20\frac{1}{4}$ feet long?

MORE CHALLENGING EXERCISES

Divide. No variable equals zero.

EXAMPLE: $\frac{x}{5} \div x = \frac{x}{5} \cdot \frac{1}{x}$ ◀ The reciprocal of x is $\frac{1}{x}$.

$$= \frac{\cancel{x} \cdot 1}{5 \cdot \cancel{x}} = \frac{1 \cdot 1}{5 \cdot 1} = \frac{1}{5}$$

68. $\frac{y}{4} \div y$ **69.** $\frac{p}{6} \div p$ **70.** $\frac{2}{5} \div \frac{n}{5}$ **71.** $\frac{1}{4} \div \frac{t}{4}$ **72.** $\frac{r}{4} \div \frac{r}{5}$

73. $\frac{x}{3} \div \frac{x}{6}$ **74.** $1\frac{1}{3} \div a$ **75.** $4\frac{1}{2} \div 3p$ **76.** $3\frac{1}{3} \div \frac{y}{3}$ **77.** $6\frac{1}{4} \div \frac{5}{t}$

REVIEW: SECTIONS 5-5—5-8

Multiply. Write each answer in lowest terms. (Section 5-5)

1. $1\frac{2}{5} \cdot 2\frac{1}{7}$ **2.** $3\frac{1}{3} \cdot 1\frac{1}{2}$ **3.** $4\frac{1}{6} \cdot 5\frac{3}{5}$ **4.** $2\frac{2}{9} \cdot 6\frac{1}{10}$

Find the area of each triangle. (Section 5-6)

5. $b = 5$ ft; $h = 6$ ft   **6.** $b = 8.6$ m; $h = 3.5$ m   **7.** $b = 4\frac{1}{8}$ in; $h = 6$ in

Find the volume of each rectangular prism. (Section 5-7)

8. $l = 6$ yd; $w = 8$ yd; $h = 2\frac{1}{8}$ yd   **9.** $l = 3.7$ cm; $w = 5.2$ cm; $h = 2.5$ cm

Divide. Write each answer in lowest terms. (Section 5-8)

10. $\frac{1}{2} \div \frac{1}{3}$ **11.** $\frac{3}{4} \div 6$ **12.** $3 \div 2\frac{1}{7}$ **13.** $1\frac{1}{2} \div 1\frac{3}{5}$

14. The base of a parallelogram is $3\frac{1}{2}$ inches. The height is $2\frac{1}{4}$ inches. Find the area. (Section 5-6)

15. A car travels 120 miles on $6\frac{2}{5}$ gallons of fuel. How far does it travel on one gallon of fuel? (Section 5-8)

Fractions: Multiplication/Division

WRITTEN EXERCISES

39. $\frac{4}{5}$ **40.** $\frac{16}{25}$ **41.** $3\frac{17}{21}$

42. $1\frac{9}{119}$ **43.** $\frac{31}{32}$

44. $\frac{42}{55}$ **45.** 12   **46.** $1\frac{1}{3}$

47. $\frac{1}{2}$ **48.** $2\frac{5}{8}$ **49.** $7\frac{1}{2}$

50. $\frac{4}{9}$ **51.** $1\frac{1}{2}$ **52.** 6

53. $\frac{1}{2}$ **54.** $\frac{4}{9}$ **55.** $\frac{6}{11}$

56. $\frac{1}{9}$ **57.** $1\frac{1}{5}$ **58.** $1\frac{10}{11}$

See page 155 for the answers to Exercises 59–77.

QUIZ: SECTIONS 5-5–5-8
After completing this Review, you may wish to administer a quiz covering the same sections. A Quiz is provided in the *Teacher's Edition: Part II.*

REVIEW: SECTIONS 5-5–5-8

1. 3   **2.** 5   **3.** $23\frac{1}{3}$
4. $13\frac{5}{9}$ **5.** 15 ft^2
6. 15.05 m^2
7. $12\frac{3}{8}$ in^2
8. 102 yd^3
9. 48.1 cm^3
10. $1\frac{1}{2}$ **11.** $\frac{1}{8}$
12. $1\frac{2}{5}$ **13.** $\frac{15}{16}$
14. $7\frac{7}{8}$ in^2 **15.** $18\frac{3}{4}$ mi

CHAPTER REVIEW

1. Fahrenheit
2. reciprocals
3. mixed numeral
4. Celsius
5. lowest terms
6. greatest common factor

7. $\frac{1}{2}$ 8. $\frac{2}{9}$ 9. $\frac{1}{4}$
10. $\frac{14}{15}$ 11. $\frac{7}{16}$ 12. $\frac{6}{17}$
13. $3\frac{11}{12}$ ft   14. 15 ft
15. $10\frac{1}{2}$ gal   16. $11\frac{7}{16}$
17. $4\frac{2}{3}$ yd   18. $\frac{3}{4}$ ft
19. $6\frac{1}{4}$ gal   20. $28\frac{9}{16}$ lb
21. $\frac{3}{16}$ 22. $\frac{3}{8}$ 23. $\frac{2}{5}$
24. $\frac{3}{10}$ 25. $\frac{2}{3}$ 26. $\frac{3}{5}$
27. $\frac{1}{4}$ 28. $\frac{1}{14}$ 29. $\frac{1}{2}$
30. $\frac{1}{12}$ 31. 104   32. 40
33. 50   34. $1\frac{2}{3}$ 35. $3\frac{1}{2}$
36. $2\frac{1}{12}$ 37. $3\frac{3}{4}$
38. $1\frac{6}{7}$ 39. $2\frac{5}{8}$

CHAPTER REVIEW

PART 1: VOCABULARY

For Exercises 1–6, choose from the box at the right the word(s) that best corresponds to each description.

1. The boiling point of water is 212° on this temperature scale. __?__
2. Two numbers whose product is 1 __?__
3. Represents the sum of a whole number and a fraction __?__
4. The freezing point of water is 0° on this temperature scale. __?__
5. A numerator and denominator having no common prime factors __?__

> lowest terms
> greatest common factor
> mixed numeral
> Celsius
> Fahrenheit
> reciprocals

6. The greatest number that is a factor of each of two or more numbers __?__

PART 2: SKILLS

Write each fraction in lowest terms. (Section 5-1)

7. $\frac{3}{6}$ 8. $\frac{4}{18}$ 9. $\frac{6}{24}$ 10. $\frac{28}{30}$ 11. $\frac{14}{32}$ 12. $\frac{18}{51}$

Add or subtract as indicated. Write each answer as a mixed numeral. (Section 5-2)

13. 1 ft 8 in
    +2 ft 3 in

14. 5 ft 6 in
    +9 ft 6 in

15. 4 gal 3 qt
    +5 gal 3 qt

16. 7 lb 15 oz
    +3 lb 8 oz

17. 8 yd 1 ft
    −3 yd 2 ft

18. 7 ft 5 in
    −5 ft 8 in

19. 10 gal 2 qt
    − 4 gal 1 qt

20. 57 lb
    −28 lb 7 oz

Multiply. Write each answer in lowest terms. (Section 5-3)

21. $\frac{1}{2} \cdot \frac{3}{8}$ 22. $\frac{3}{4} \cdot \frac{1}{2}$ 23. $\frac{4}{5} \cdot \frac{1}{2}$ 24. $\frac{2}{5} \cdot \frac{3}{4}$ 25. $4 \cdot \frac{1}{6}$

26. $\frac{2}{3} \cdot \frac{9}{10}$ 27. $\frac{5}{6} \cdot \frac{3}{10}$ 28. $\frac{4}{7} \cdot \frac{1}{8}$ 29. $\frac{5}{8} \cdot \frac{4}{5}$ 30. $\frac{8}{9} \cdot \frac{3}{32}$

Complete. (Section 5-4)

31. 40°C = __?__ °F
32. 104°F = __?__ °C
33. 10°C = __?__ °F

Write a mixed numeral for each fraction. (Section 5-5)

34. $\frac{5}{3}$ 35. $\frac{7}{2}$ 36. $\frac{25}{12}$ 37. $\frac{15}{4}$ 38. $\frac{13}{7}$ 39. $\frac{21}{8}$

Write a fraction for each mixed numeral. (Section 5-5)

40. $3\frac{1}{5}$ 41. $1\frac{1}{4}$ 42. $6\frac{1}{3}$ 43. $4\frac{3}{8}$ 44. $7\frac{3}{4}$ 45. $2\frac{4}{7}$

Multiply. Write each answer in lowest terms. (Section 5-5)

46. $3\frac{3}{5} \cdot 5$ 47. $4\frac{1}{3} \cdot 9$ 48. $6\frac{1}{2} \cdot 3$ 49. $\frac{5}{6} \cdot 3\frac{3}{10}$ 50. $\frac{5}{8} \cdot 2\frac{1}{5}$

51. $1\frac{1}{6} \cdot 1\frac{4}{5}$ 52. $\frac{1}{8} \cdot 3\frac{3}{4}$ 53. $3\frac{1}{2} \cdot 4\frac{2}{3}$ 54. $1\frac{1}{4} \cdot 1\frac{1}{4}$ 55. $1\frac{7}{8} \cdot 6\frac{2}{5}$

Find the area of each parallelogram. (Section 5-6)

	Base	Height	Area
56.	52.7 m	15 m	?
57.	4 in	$2\frac{1}{2}$ in	?

Find the area of each triangle. (Section 5-6)

	Base	Height	Area
58.	20 ft	$16\frac{1}{10}$ ft	?
59.	3.5 cm	2.6 cm	?

Find the volume of each rectangular prism. (Section 5-7)

	Length	Width	Height
60.	5 ft	2 ft	$5\frac{1}{4}$ ft
62.	4.6 m	1.2 m	0.6 m

	Length	Width	Height
61.	6.2 m	4.1 m	5 m
63.	$12\frac{1}{2}$ in	$8\frac{2}{5}$ in	$4\frac{1}{3}$ in

Divide. Write each answer in lowest terms. (Section 5-8)

64. $\frac{5}{6} \div \frac{2}{3}$ 65. $3\frac{1}{3} \div 1\frac{1}{5}$ 66. $6 \div \frac{3}{8}$ 67. $1\frac{1}{8} \div 3$ 68. $2\frac{1}{4} \div 3\frac{3}{8}$

69. $\frac{3}{8} \div \frac{3}{4}$ 70. $2\frac{1}{5} \div 4$ 71. $2 \div \frac{4}{5}$ 72. $3\frac{1}{2} \div \frac{3}{4}$ 73. $4\frac{1}{4} \div \frac{1}{4}$

PART 3: APPLICATIONS

74. A Lunar Rover carrying an astronaut wearing a space suit and backpack weighs 880 pounds 5 ounces. The astronaut wearing a spacesuit and backpack weighs 379 pounds 13 ounces. Find the weight of the Lunar Rover. (Section 5-2)

75. At sunset, the temperature on the moon is about 59°F. Express this as a Celsius reading. (Section 5-4)

76. The wing on a hang glider is in the shape of a triangle. The base is 5.4 meters and the height is 1.5 meters. Find the area. (Section 5-6)

77. A rectangular decompression chamber is $10\frac{1}{2}$ feet long, 5 feet wide, and $9\frac{1}{3}$ feet high. Find the volume. (Section 5-7)

CHAPTER REVIEW

40. $\frac{16}{5}$ 41. $\frac{5}{4}$ 42. $\frac{19}{3}$
43. $\frac{35}{8}$ 44. $\frac{31}{4}$ 45. $\frac{18}{7}$
46. 18 47. 39
48. $19\frac{1}{2}$ 49. $2\frac{3}{4}$
50. $1\frac{3}{8}$ 51. $2\frac{1}{10}$
52. $\frac{15}{32}$ 53. $16\frac{1}{3}$
54. $1\frac{9}{16}$ 55. 12
56. 790.5 m^2
57. 10 in^2
58. 161 ft^2
59. 4.55 cm^2
60. 52.5 ft^3
61. 127.1 m^3
62. 3.312 m^3
63. 455 in^3 64. $1\frac{1}{4}$
65. $2\frac{7}{9}$ 66. 16 67. $\frac{3}{8}$
68. $\frac{2}{3}$ 69. $\frac{1}{2}$ 70. $\frac{11}{20}$
71. $2\frac{1}{2}$ 72. $4\frac{2}{3}$
73. 17 74. $500\frac{1}{2}$ lb
75. 15° C 76. 4.05 m^2
77. 490 ft^3

Fractions: Multiplication/Division

CHAPTER TEST

Two forms of a chapter test, Form A and Form B, are provided on copying masters in the *Teacher's Edition: Part II.*

1. $\frac{1}{2}$ 2. $\frac{1}{6}$ 3. $\frac{3}{8}$
4. $\frac{3}{7}$ 5. $8\frac{3}{4}$ ft
6. $1\frac{1}{2}$ gal 7. $\frac{5}{12}$
8. $\frac{1}{3}$ 9. 14 10. $8\frac{5}{8}$
11. 130.5 m^2
12. 46 ft^2
13. $26\frac{2}{3}$ in^2
14. 53.24 cm^2 15. 2
16. $1\frac{1}{9}$ 17. $6\frac{3}{4}$
18. $\frac{16}{25}$ 19. 55° C
20. 1.6 m^3

CHAPTER TEST

Write each fraction in lowest terms.

1. $\frac{7}{14}$
2. $\frac{6}{36}$
3. $\frac{15}{40}$
4. $\frac{12}{28}$

Add or subtract as indicated. Write each answer as a mixed numeral.

5. 3 ft 7 in
   +5 ft 2 in

6. 5 gal 1 qt
   −3 gal 3 qt

Multiply. Write each answer in lowest terms.

7. $\frac{1}{2} \cdot \frac{5}{6}$
8. $\frac{4}{5} \cdot \frac{5}{12}$
9. $2\frac{1}{3} \cdot 6$
10. $2\frac{1}{4} \cdot 3\frac{5}{6}$

Find the area of each parallelogram.

11. $b = 15$ m; $h = 8.7$ m
12. $b = 14\frac{3}{8}$ ft; $h = 3\frac{1}{5}$ ft

Find the area of each triangle.

13. $b = 10$ in; $h = 5\frac{1}{3}$ in
14. $b = 12.1$ cm; $h = 8.8$ cm

Divide. Write each answer in lowest terms.

15. $\frac{1}{2} \div \frac{1}{4}$
16. $\frac{5}{6} \div \frac{3}{4}$
17. $3 \div \frac{4}{9}$
18. $2\frac{2}{5} \div 3\frac{3}{4}$

19. On a certain day in Death Valley, California, the temperature reached 131°F. Write this as a Celsius reading.

20. A storage closet for linen is 2 meters long, 1 meter wide, and 0.8 meters high. Find the volume of the closet.

ANSWERS TO REVIEW CAPSULES

Page 128 1. 176 2. 378 3. 150 4. 156 5. 336 6. 576 7. $\frac{2}{3}$ 8. $\frac{3}{7}$ 9. $\frac{2}{3}$ 10. $\frac{14}{15}$ 11. $\frac{2}{3}$ 12. $\frac{2}{9}$
Page 131 1. 21 2. 136 3. 263 4. 278 5. 19 6. 40 7. 68 8. 30 9. 84 10. 108
Page 134 1. 1 2. 2 3. 3 4. 6 5. 7 6. 2 7. $\frac{1}{8}$ 8. $\frac{3}{14}$ 9. $\frac{7}{18}$ 10. $\frac{7}{15}$ 11. $\frac{3}{5}$ 12. $\frac{4}{5}$
Page 140 1. 12.15 2. 18.6 3. 7 4. 20 5. $1\frac{1}{3}$ 6. $4\frac{4}{9}$ 7. 27.54 8. 4.9 9. 9.2 10. 3.36 11. 43.4 12. 4.05 13. $\frac{2}{5}$ 14. $\frac{2}{9}$ 15. 4 16. 9 17. $\frac{1}{9}$ 18. $13\frac{1}{3}$
Page 144 1. 96 2. 27 3. 32.64 4. 2.94 5. $26\frac{2}{3}$ 6. $9\frac{1}{3}$ 7. 12.8 8. 35.5 9. 4.48 10. 6.65 11. 9.36 12. 2.52 13. 45.6 14. 17.4 15. 3 16. $6\frac{3}{4}$ 17. 6 18. $2\frac{7}{10}$ 19. $2\frac{1}{2}$ 20. 20 21. 11 22. $16\frac{1}{4}$
Page 147 1. $\frac{3}{2}$ 2. $\frac{11}{4}$ 3. $\frac{16}{3}$ 4. $\frac{22}{5}$ 5. $\frac{19}{6}$ 6. $\frac{9}{7}$ 7. 3 8. 3 9. $1\frac{1}{3}$ 10. $4\frac{2}{3}$ 11. 2 12. $\frac{2}{3}$

Chapter 5

ENRICHMENT

Special Areas

You sometimes have to use more than one formula to find area.

Remember: Opposite sides of rectangles, parallelograms, and squares have the same length.

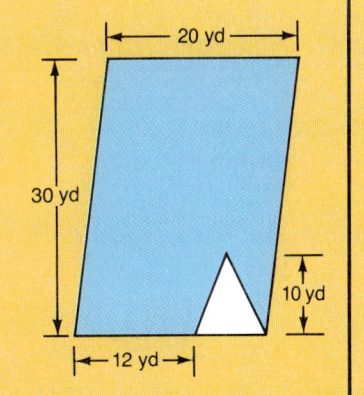

EXAMPLE Find the area of the shaded region.

Solution:

Area of Shaded Region	=	Area of Parallelogram	–	Area of Triangle
A	=	bh	–	$\frac{1}{2}bh$
A	=	$30 \cdot 20$	–	$\frac{1}{2} \cdot 8 \cdot 10$
A	=	$600 - 40$, or **560 yd²**		

EXERCISES

Find the area of the shaded region.

1. 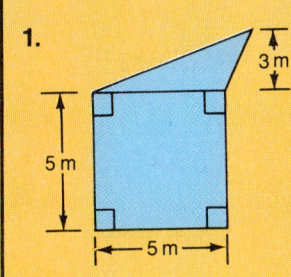
2. 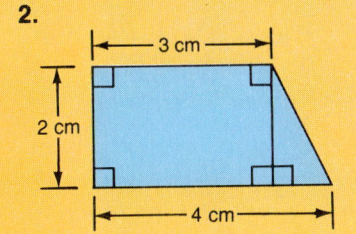
3. 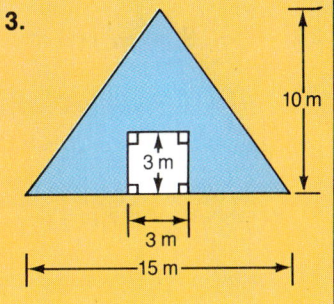
4. 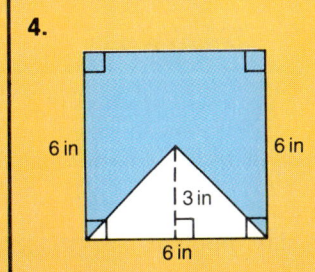
5. 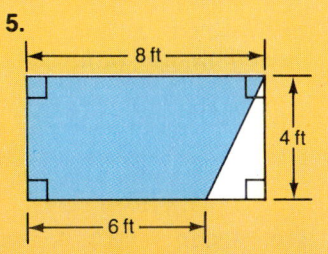
6.

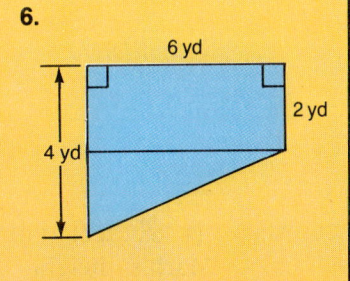

Fractions: Multiplication/Division   **155**

ENRICHMENT

You may wish to use this lesson for students who performed well on the formal Chapter Test.

1. $32\frac{1}{2}$ m²   2. 7 cm²
3. 66 m²   4. 27 in²
5. 28 ft²   6. 18 yd²

WRITTEN EXERCISES
pp. 150–151

59. $1\frac{1}{10}$ 60. $5\frac{1}{4}$
61. $1\frac{7}{20}$ 62. $1\frac{4}{7}$
63. $\frac{2}{3}$ 64. $1\frac{7}{9}$
65. 16   66. 20 km
67. 9   68. $\frac{1}{4}$
69. $\frac{1}{6}$ 70. $\frac{2}{n}$
71. $\frac{1}{t}$ 72. $1\frac{1}{4}$
73. 2   74. $\frac{4}{3a}$
75. $\frac{3}{2p}$ 76. $\frac{10}{y}$
77. $\frac{5t}{4}$

ADDITIONAL PRACTICE

You may wish to use all or some of these exercises, depending on how well students performed on the formal Chapter Test.

1. $\frac{1}{2}$ 2. $\frac{1}{3}$ 3. $\frac{2}{7}$
4. $\frac{2}{3}$ 5. $\frac{3}{7}$ 6. $\frac{5}{6}$
7. $28\frac{1}{4}$ lb 8. $4\frac{3}{4}$ gal
9. $1\frac{3}{4}$ ft 10. $13\frac{1}{3}$ yd
11. $\frac{1}{4}$ 12. $\frac{6}{7}$ 13. $\frac{6}{35}$
14. $\frac{1}{12}$ 15. $\frac{6}{11}$
16. $\frac{2}{15}$ 17. 95° F
18. 40° C 19. 0° C
20. 10 21. $22\frac{1}{2}$
22. $3\frac{1}{3}$ 23. 6 24. $1\frac{1}{4}$
25. 28 cm^2
26. $10\frac{5}{8}$ ft^2
27. 12 m^2
28. 10.5 ft^2
29. 60 m^3
30. 100.86 ft^3
31. $1\frac{1}{4}$ 32. $\frac{4}{5}$ 33. $\frac{9}{20}$
34. $1\frac{1}{8}$ 35. $5\frac{1}{3}$
36. $2\frac{3}{5}$ ft 37. 7500 m^3

ADDITIONAL PRACTICE

SKILLS

Write each fraction in lowest terms. (Pages 124–125)

1. $\frac{15}{30}$ 2. $\frac{7}{21}$ 3. $\frac{12}{42}$ 4. $\frac{30}{45}$ 5. $\frac{42}{98}$ 6. $\frac{15}{18}$

Add or subtract as indicated. Write each answer as a mixed numeral. (Pages 126–128)

7. 13 lb 8 oz
   +14 lb 12 oz

8. 16 gal 2 qt
   −11 gal 3 qt

9. 4 ft 6 in
   −2 ft 9 in

10. 5 yd 2 ft
    +7 yd 2 ft

Multiply. Write each answer in lowest terms. (Pages 129–131)

11. $\frac{2}{5} \cdot \frac{5}{8}$ 12. $\frac{3}{7} \cdot 2$ 13. $\frac{2}{7} \cdot \frac{3}{5}$ 14. $\frac{3}{8} \cdot \frac{2}{9}$ 15. $\frac{2}{11} \cdot 3$ 16. $\frac{4}{9} \cdot \frac{3}{10}$

Complete. (Pages 132–134)

17. 35°C = __?__ F 18. 104°F = __?__ C 19. 32°F = __?__ C

Multiply. Write each answer in lowest terms. (Pages 135–137)

20. $3\frac{1}{3} \cdot 3$ 21. $2\frac{1}{4} \cdot 10$ 22. $1\frac{1}{3} \cdot 2\frac{1}{2}$ 23. $4\frac{1}{5} \cdot 1\frac{3}{7}$ 24. $1\frac{5}{6} \cdot \frac{15}{22}$

Find the area of each parallelogram. (Pages 140–144)

25. Base: 8 cm; Height: $3\frac{1}{2}$ cm
26. Base: $2\frac{1}{2}$ ft; Height: $4\frac{1}{4}$ ft

Find the area of each triangle. (Pages 141–144)

27. Base: 4 m; Height: 6 m
28. Base: 8.4 ft; Height: 2.5 ft

Find the volume of each rectangular prism. (Pages 145–147)

29. $l = 8$ m; $w = 4\frac{1}{2}$ m; $h = 1\frac{2}{3}$ m
30. $l = 8.2$ ft; $w = 4.1$ ft; $h = 3$ ft

Divide. Write each answer in lowest terms. (Pages 148–151)

31. $\frac{2}{3} \div \frac{8}{15}$ 32. $\frac{2}{3} \div \frac{5}{6}$ 33. $1\frac{1}{5} \div 2\frac{2}{3}$ 34. $4\frac{1}{2} \div 4$ 35. $1\frac{1}{3} \div \frac{1}{4}$

APPLICATIONS

36. Mark is $6\frac{1}{2}$ feet tall. His sister is $\frac{3}{5}$ Mark's height. How much taller is Mark than his sister? (Pages 129–131)

37. A rectangular pool is 50 meters long, 25 meters wide, and 6 meters deep. Find the volume of the pool. (Pages 145–147)

156 Chapter 5

CHAPTER 6
Fractions: Addition/Subtraction

SECTIONS
6-1 Addition and Subtraction: Like Denominators
6-2 Addition and Subtraction: Unlike Denominators
6-3 Addition and Subtraction: Mixed Numerals
6-4 Problem Solving and Applications: Time Cards
6-5 Fractions and Decimals
6-6 Problem Solving and Applications Circles: Circumference and Area

FEATURES
Calculator Application: *Comparing Fractions*
Calculator/Computer Application: *Evaluating Powers*
Enrichment: *Precision of Measurement*
Common Errors

Teaching Suggestions
p. M-19

QUICK QUIZ

Write in lowest terms.

1. $\frac{18}{30}$ Ans: $\frac{3}{5}$
2. $\frac{72}{84}$ Ans: $\frac{6}{7}$
3. $\frac{27}{153}$ Ans: $\frac{3}{17}$

Write each fraction as a mixed numeral.

4. $\frac{9}{8}$ Ans: $1\frac{1}{8}$
5. $\frac{37}{5}$ Ans: $7\frac{2}{5}$

ADDITIONAL EXAMPLES
Example 1
Add or subtract as indicated.

1. $\frac{3}{5} + \frac{4}{5}$ Ans: $1\frac{2}{5}$
2. $\frac{17}{20} - \frac{5}{20}$ Ans: $\frac{3}{5}$

SELF-CHECK

1. $1\frac{1}{3}$ 2. $\frac{2}{5}$ 3. $\frac{2}{5}$
4. $1\frac{3}{5}$

6-1 Addition and Subtraction: Like Denominators

Susan is making a pocketbook. She needs $\frac{3}{8}$ yard of felt for the pocketbook and $\frac{1}{8}$ yard for the strap. To find the total number of yards, Susan adds $\frac{3}{8}$ and $\frac{1}{8}$.

$$\frac{3}{8} + \frac{1}{8} = \frac{3+1}{8} = \frac{4}{8}, \text{ or } \frac{1}{2}$$ ◀ Lowest terms

Thus, she needs $\frac{1}{2}$ yard of felt in all.

Like fractions, such as $\frac{3}{8}$ and $\frac{1}{8}$, have **like denominators**.

PROCEDURE
To add or subtract like fractions:
1. Add or subtract the numerators. Write the sum or difference over the <u>like</u> denominator.
2. Write the answer in lowest terms.

EXAMPLE 1 Add or subtract as indicated: a. $\frac{7}{12} + \frac{11}{12}$ b. $\frac{13}{15} - \frac{7}{15}$

Solutions: a. [1] $\frac{7}{12} + \frac{11}{12} = \frac{7+11}{12}$ b. [1] $\frac{13}{15} - \frac{7}{15} = \frac{13-7}{15}$
                        $= \frac{18}{12}$ ◀ Write a mixed numeral for $\frac{18}{12}$. $= \frac{6}{15}$
             [2] $= 1\frac{6}{12}$, or $1\frac{1}{2}$ [2] $= \frac{2}{5}$

Self-Check Add or subtract as indicated.

1. $\frac{5}{12} + \frac{11}{12}$ 2. $\frac{14}{15} - \frac{8}{15}$ 3. $\frac{19}{20} - \frac{11}{20}$ 4. $\frac{9}{10} + \frac{7}{10}$

Mixed numerals such as $1\frac{1}{6}$ and $3\frac{5}{6}$ also have like denominators.

PROCEDURE
To add or subtract mixed numerals:
1. Add or subtract the whole numbers.
2. Add or subtract the fractional parts.
3. Write the answer in lowest terms.

EXAMPLE 2 Add or subtract as indicated: **a.** $4\frac{1}{5}+3\frac{2}{5}$ **b.** $10\frac{7}{8}-5\frac{3}{8}$

Solutions:
a. $4\frac{1}{5}+3\frac{2}{5}=7\frac{1+2}{5}$ (4+3)
$=7\frac{3}{5}$

b. $10\frac{7}{8}-5\frac{3}{8}=5\frac{7-3}{8}$ (10−5)
$=5\frac{4}{8}$, or $5\frac{1}{2}$

Self-Check Add or subtract as indicated.

5. $2\frac{1}{4}+5\frac{3}{4}$ **6.** $5\frac{3}{4}-2\frac{1}{4}$ **7.** $8\frac{9}{10}-3\frac{7}{10}$ **8.** $5\frac{1}{5}+2\frac{3}{5}$

For fractions $\frac{a}{b}$ and $\frac{c}{b}$, $b \neq 0$,
$\frac{a}{b}+\frac{c}{b}=\frac{a+c}{b}$ and $\frac{a}{b}-\frac{c}{b}=\frac{a-c}{b}$. $\frac{8}{9}-\frac{4}{9}=\frac{8-4}{9}$, or $\frac{4}{9}$

CLASSROOM EXERCISES

Add or subtract. Write each answer in lowest terms. (Example 1)

1. $\frac{1}{4}+\frac{1}{4}$ **2.** $\frac{7}{16}-\frac{3}{16}$ **3.** $\frac{9}{7}-\frac{2}{7}$ **4.** $\frac{3}{10}+\frac{3}{10}$ **5.** $\frac{11}{16}+\frac{13}{16}$

(Example 2)

6. $3\frac{1}{3}+2\frac{1}{3}$ **7.** $4\frac{7}{8}-2\frac{3}{8}$ **8.** $5\frac{3}{5}+2\frac{2}{5}$ **9.** $4\frac{3}{4}-3\frac{1}{4}$ **10.** $2\frac{7}{8}-1\frac{1}{8}$

WRITTEN EXERCISES

Goals: To add and subtract fractions and mixed numerals (like denominators)

To apply the skill of adding and subtracting fractions and mixed numerals (like denominators) to solving word problems.

Add or subtract. Write each answer in lowest terms. (Example 1)

1. $\frac{1}{3}+\frac{1}{3}$ **2.** $\frac{1}{5}+\frac{2}{5}$ **3.** $\frac{1}{6}+\frac{1}{6}$ **4.** $\frac{1}{8}+\frac{1}{8}$ **5.** $\frac{3}{4}-\frac{1}{4}$

6. $\frac{7}{15}-\frac{4}{15}$ **7.** $\frac{11}{13}+\frac{2}{13}$ **8.** $\frac{11}{25}+\frac{12}{25}$ **9.** $\frac{7}{10}+\frac{7}{10}$ **10.** $\frac{19}{20}+\frac{11}{20}$

(Example 2)

11. $3\frac{1}{7}+2\frac{1}{7}$ **12.** $4\frac{1}{5}+9\frac{2}{5}$ **13.** $11\frac{7}{9}-2\frac{2}{9}$ **14.** $13\frac{11}{12}-9\frac{4}{12}$

15. $5\frac{3}{6}+1\frac{1}{6}$ **16.** $4\frac{7}{8}-2\frac{1}{8}$ **17.** $9\frac{11}{12}-4\frac{3}{12}$ **18.** $17\frac{11}{20}+13\frac{5}{20}$

19. $5\frac{3}{7}+4\frac{2}{7}$ **20.** $6\frac{5}{8}+4\frac{2}{8}$ **21.** $6\frac{9}{11}+3\frac{3}{11}$ **22.** $4\frac{11}{13}+9\frac{2}{13}$

Fractions: Addition/Subtraction **159**

WRITTEN EXERCISES

9. $1\frac{2}{5}$ 10. $1\frac{1}{2}$ 11. $5\frac{2}{7}$
12. $13\frac{3}{5}$ 13. $9\frac{5}{9}$
14. $4\frac{7}{12}$ 15. $6\frac{2}{3}$
16. $2\frac{3}{4}$ 17. $5\frac{2}{3}$
18. $30\frac{4}{5}$ 19. $9\frac{5}{7}$
20. $10\frac{7}{8}$ 21. $10\frac{1}{11}$
22. 14   23. $\frac{3}{5}$ 24. $1\frac{1}{4}$
25. $\frac{3}{25}$ 26. $\frac{1}{7}$ 27. $8\frac{2}{5}$
28. 8   29. $2\frac{1}{4}$ 30. $2\frac{1}{2}$
31. $14\frac{1}{4}$ 32. $7\frac{3}{5}$
33. $13\frac{1}{4}$ 34. 5
35. $6\frac{1}{2}$ hr   36. $3\frac{1}{5}$
37. $\frac{8}{a}$ 38. $\frac{9}{n}$ 39. $\frac{9}{x}$
40. $\frac{3}{n}$ 41. $\frac{2x}{3}$ 42. $\frac{3a}{4}$
43. $\frac{2s}{9}$ 44. $\frac{y}{2}$ 45. $\frac{m}{5}$
46. $\frac{2t}{7}$ 47. $\frac{3x}{y}$ 48. $\frac{2a}{c}$
49. $\frac{5a}{b}$ 50. $\frac{2n}{s}$ 51. $\frac{5a}{3f}$

REVIEW CAPSULE

This Review Capsule reviews prior-taught skills used in Section 6–2. The reference is to the pages where the skills were taught.

MIXED PRACTICE

23. $\frac{1}{10} + \frac{5}{10}$ 24. $\frac{7}{8} + \frac{3}{8}$ 25. $\frac{6}{25} - \frac{3}{25}$ 26. $\frac{3}{14} - \frac{1}{14}$

27. $6\frac{1}{5} + 2\frac{1}{5}$ 28. $4\frac{1}{7} + 3\frac{6}{7}$ 29. $3\frac{3}{8} - 1\frac{1}{8}$ 30. $5\frac{7}{8} - 3\frac{3}{8}$

31. $11\frac{5}{32} + 3\frac{3}{32}$ 32. $2\frac{3}{20} + 5\frac{9}{20}$ 33. $27\frac{7}{8} - 14\frac{5}{8}$ 34. $9\frac{3}{4} - 4\frac{3}{4}$

APPLICATIONS: USING ADDITION AND SUBTRACTION

35. Stan is flying from Washington, D.C. to Seattle with a stop in San Francisco. It takes $4\frac{5}{8}$ hours for the flight to San Francisco and $1\frac{7}{8}$ hours for the flight to Seattle. What is the total time? Write the answer in lowest terms.

36. In 1978, petroleum production in the United States was $18\frac{4}{5}$ million barrels per day. By 1982, production dropped to $15\frac{3}{5}$ million barrels per day. How many millions of barrels per day did oil production drop?

MORE CHALLENGING EXERCISES

Add or subtract. Write each answer in lowest terms.

EXAMPLES: a. $\frac{3}{x} + \frac{5}{x} = \frac{3+5}{x}$ b. $\frac{11a}{9} - \frac{5a}{9} = \frac{11a - 5a}{9}$ $11a - 5a = 6a$

$= \frac{8}{x}$ $= \frac{\overset{2}{\cancel{6a}}}{\underset{3}{\cancel{9}}} = \frac{2a}{3}$

37. $\frac{3}{a} + \frac{5}{a}$ 38. $\frac{2}{n} + \frac{7}{n}$ 39. $\frac{12}{x} - \frac{3}{x}$ 40. $\frac{5}{n} - \frac{2}{n}$ 41. $\frac{x}{3} + \frac{x}{3}$

42. $\frac{3a}{16} + \frac{9a}{16}$ 43. $\frac{4s}{9} - \frac{2s}{9}$ 44. $\frac{5y}{8} - \frac{y}{8}$ 45. $\frac{3m}{10} - \frac{m}{10}$ 46. $\frac{5t}{7} - \frac{3t}{7}$

47. $\frac{7x}{3y} + \frac{2x}{3y}$ 48. $\frac{2a}{3c} + \frac{4a}{3c}$ 49. $\frac{7a}{b} - \frac{2a}{b}$ 50. $\frac{5n}{2s} - \frac{n}{2s}$ 51. $\frac{7a}{3f} - \frac{2a}{3f}$

REVIEW CAPSULE FOR SECTION 6–2

Write the least common multiple of each pair of numbers. (Pages 111–114)

1. 4 and 6   2. 3 and 8   3. 2 and 12   4. 2 and 3   5. 15 and 20

Write each fraction in lowest terms. (Pages 124–125)

6. $\frac{5}{10}$ 7. $\frac{8}{20}$ 8. $\frac{14}{18}$ 9. $\frac{12}{30}$ 10. $\frac{24}{56}$

The answers are on page 183.

160   Chapter 6

6-2 Addition and Subtraction: Unlike Denominators

Unlike fractions have unlike denominators. To add or subtract fractions having unlike denominators, you first write *equivalent fractions*.

Equivalent fractions name the same number.

$$\frac{4}{6} = \frac{\cancel{2} \cdot 2}{\cancel{2} \cdot 3} \qquad \frac{6}{9} = \frac{2 \cdot \cancel{3}}{3 \cdot \cancel{3}} \qquad \frac{10}{15} = \frac{2 \cdot \cancel{5}}{3 \cdot \cancel{5}}$$

$$\frac{4}{6} = \frac{2}{3} \qquad \frac{6}{9} = \frac{2}{3} \qquad \frac{10}{15} = \frac{2}{3}$$

Thus, $\frac{4}{6}$, $\frac{6}{9}$, and $\frac{10}{15}$ are **equivalent fractions**.

PROCEDURE

To write equivalent fractions, multiply or divide <u>both</u> the numerator and denominator of a fraction by the <u>same</u> number.

EXAMPLE 1 Write a fraction that is equivalent to $\frac{5}{6}$ and has a denominator of 30.

Solution: Think: $\frac{5}{6} = \frac{?}{30}$

$\frac{5}{6} = \frac{5}{6} \cdot \frac{5}{5}$ ◀ Multiply both the numerator and denominator by the same number.

$= \frac{25}{30}$

Self-Check *Complete by writing an equivalent fraction as indicated.*

1. $\frac{1}{2} = \frac{?}{10}$ 2. $\frac{2}{5} = \frac{?}{20}$ 3. $\frac{7}{12} = \frac{?}{60}$ 4. $\frac{5}{16} = \frac{?}{64}$

To add or subtract fractions having unlike denominators, you write equivalent fractions having like denominators.

PROCEDURE

To add or subtract with fractions having unlike denominators:

1. Find the LCM of the denominators.
2. Write each fraction as an equivalent fraction having the least common multiple as its new denominator.
3. Add or subtract as with like fractions.
4. Write the answer in lowest terms.

Fractions: Addition/Subtraction **161**

Teaching Suggestions
p. M-19

QUICK QUIZ
Write the value of
x that makes each
sentence true.

1. $3(\frac{x}{7}) = 3$ Ans: 7
2. $(\frac{8}{x}) \cdot 9 = 9$ Ans: 8
3. $19(\frac{6}{6}) = x$ Ans: 19

Write in lowest terms.

4. $\frac{45}{60}$ Ans: $\frac{3}{4}$
5. $\frac{9}{12}$ Ans: $\frac{3}{4}$

ADDITIONAL EXAMPLES
Example 1
Complete. Write an
equivalent fraction
as indicated.

1. $\frac{2}{3} = \frac{?}{12}$ Ans: 8
2. $\frac{4}{5} = \frac{?}{25}$ Ans: 20

SELF-CHECK

1. $\frac{5}{10}$ 2. $\frac{8}{20}$ 3. $\frac{35}{60}$
4. $\frac{20}{64}$

ADDITIONAL EXAMPLES
Example 2
Add.

1. $\frac{3}{5} + \frac{1}{4}$ Ans: $\frac{17}{20}$
2. $\frac{5}{12} + \frac{3}{4}$ Ans: $1\frac{1}{6}$

SELF-CHECK

5. $1\frac{1}{2}$ 6. $1\frac{1}{5}$ 7. $\frac{19}{20}$
8. $\frac{19}{24}$

Example 3
Subtract.

1. $\frac{7}{10} - \frac{1}{5}$ Ans: $\frac{1}{2}$
2. $\frac{2}{3} - \frac{1}{2}$ Ans: $\frac{1}{6}$

SELF-CHECK

9. $\frac{1}{4}$ 10. $\frac{1}{2}$ 11. $\frac{1}{40}$
12. $\frac{1}{16}$

CLASSROOM EXERCISES

1. $\frac{3}{9}$ 2. $\frac{6}{15}$ 3. $\frac{12}{16}$
4. $\frac{3}{24}$ 5. $\frac{15}{35}$ 6. $\frac{8}{18}$
7. $\frac{7}{8}$ 8. $1\frac{1}{6}$

EXAMPLE 2 Add: $\frac{1}{4} + \frac{7}{8}$

Solution:
[1] Find the LCM of 4 and 8. $\left. \begin{array}{l} 4 = 2 \cdot 2 = 2^2 \\ 8 = 2 \cdot 2 \cdot 2 = 2^3 \end{array} \right\}$ LCM: $2^3 = 8$

[2] Write an equivalent fraction for $\frac{1}{4}$ having 8 as a denominator.
$$\frac{1}{4} + \frac{7}{8} = \frac{1}{4} \cdot \frac{2}{2} + \frac{7}{8}$$

[3] $\qquad = \frac{2}{8} + \frac{7}{8}$ ◀ $\frac{2+7}{8}$

$\qquad = \frac{9}{8}$, or $1\frac{1}{8}$

Self-Check Add.

5. $\frac{5}{6} + \frac{2}{3}$ 6. $\frac{1}{2} + \frac{7}{10}$ 7. $\frac{3}{4} + \frac{1}{5}$ 8. $\frac{3}{8} + \frac{5}{12}$

Example 3 shows how to subtract fractions having unlike denominators.

EXAMPLE 3 Subtract: $\frac{5}{8} - \frac{7}{12}$

Solution:
[1] Find the LCM of 8 and 12.
$8 = 2 \cdot 2 \cdot 2 = 2^3$
$12 = 2 \cdot 2 \cdot 3 = 2^2 \cdot 3$
LCM: $2^3 \cdot 3 = 8 \cdot 3 = 24$

[2] Write equivalent fractions with a denominator of 24 for $\frac{5}{8}$ and $\frac{7}{12}$.
$$\frac{5}{8} - \frac{7}{12} = \frac{5}{8} \cdot \frac{3}{3} - \frac{7}{12} \cdot \frac{2}{2}$$

[3] $\qquad = \frac{15}{24} - \frac{14}{24}$ ◀ $\frac{15-14}{24}$

[4] $\qquad = \frac{1}{24}$

Self-Check Subtract.

9. $\frac{7}{12} - \frac{1}{3}$ 10. $\frac{5}{6} - \frac{1}{3}$ 11. $\frac{5}{8} - \frac{3}{5}$ 12. $\frac{7}{16} - \frac{3}{8}$

CLASSROOM EXERCISES

Complete. Write an equivalent fraction as indicated. (Example 1)

1. $\frac{1}{3} = \frac{?}{9}$ 2. $\frac{2}{5} = \frac{?}{15}$ 3. $\frac{3}{4} = \frac{?}{16}$ 4. $\frac{1}{8} = \frac{?}{24}$ 5. $\frac{3}{7} = \frac{?}{35}$ 6. $\frac{4}{9} = \frac{?}{18}$

Add. Write each answer in lowest terms. (Example 2)

7. $\frac{5}{8} + \frac{1}{4}$ 8. $\frac{5}{6} + \frac{1}{3}$ 9. $\frac{1}{6} + \frac{2}{9}$ 10. $\frac{1}{4} + \frac{5}{6}$ 11. $\frac{1}{3} + \frac{3}{8}$

12. $\frac{2}{5} + \frac{1}{2}$ 13. $\frac{3}{7} + \frac{11}{14}$ 14. $\frac{1}{2} + \frac{5}{6}$ 15. $\frac{4}{5} + \frac{3}{8}$ 16. $\frac{9}{10} + \frac{7}{20}$

162   Chapter 6

Subtract. Write each answer in lowest terms. (Example 3)

17. $\frac{2}{3} - \frac{1}{6}$ 18. $\frac{3}{4} - \frac{1}{8}$ 19. $\frac{5}{6} - \frac{3}{8}$ 20. $\frac{7}{10} - \frac{4}{15}$ 21. $\frac{3}{4} - \frac{1}{5}$

22. $\frac{4}{5} - \frac{1}{3}$ 23. $\frac{5}{12} - \frac{1}{4}$ 24. $\frac{7}{8} - \frac{1}{4}$ 25. $\frac{1}{2} - \frac{3}{8}$ 26. $\frac{6}{10} - \frac{3}{20}$

WRITTEN EXERCISES

Goals: To add and subtract fractions with unlike denominators

To apply the skill of adding and subtracting fractions with unlike denominators to solving word problems

Complete. Write an equivalent fraction as indicated. (Example 1)

1. $\frac{1}{3} = \frac{?}{6}$ 2. $\frac{2}{5} = \frac{?}{15}$ 3. $\frac{5}{8} = \frac{?}{40}$ 4. $\frac{3}{4} = \frac{?}{20}$ 5. $\frac{5}{6} = \frac{?}{36}$ 6. $\frac{1}{9} = \frac{?}{18}$

7. $\frac{3}{7} = \frac{?}{42}$ 8. $\frac{9}{10} = \frac{?}{100}$ 9. $\frac{2}{3} = \frac{?}{15}$ 10. $\frac{1}{2} = \frac{?}{8}$ 11. $\frac{7}{12} = \frac{?}{36}$ 12. $\frac{8}{15} = \frac{?}{45}$

Add. Write each answer in lowest terms. (Example 2)

13. $\frac{5}{8} + \frac{1}{4}$ 14. $\frac{1}{4} + \frac{1}{3}$ 15. $\frac{7}{15} + \frac{1}{3}$ 16. $\frac{3}{20} + \frac{2}{5}$ 17. $\frac{2}{9} + \frac{1}{6}$

18. $\frac{5}{12} + \frac{3}{8}$ 19. $\frac{2}{9} + \frac{4}{15}$ 20. $\frac{1}{8} + \frac{1}{6}$ 21. $\frac{3}{5} + \frac{1}{3}$ 22. $\frac{2}{7} + \frac{1}{5}$

23. $\frac{5}{8} + \frac{1}{3}$ 24. $\frac{2}{3} + \frac{1}{10}$ 25. $\frac{5}{6} + \frac{2}{3}$ 26. $\frac{3}{4} + \frac{7}{12}$ 27. $\frac{11}{14} + \frac{5}{21}$

Subtract. Write each answer in lowest terms. (Example 3)

28. $\frac{7}{12} - \frac{1}{2}$ 29. $\frac{4}{5} - \frac{3}{10}$ 30. $\frac{9}{14} - \frac{3}{7}$ 31. $\frac{7}{15} - \frac{1}{3}$ 32. $\frac{9}{10} - \frac{1}{4}$

33. $\frac{5}{12} - \frac{2}{9}$ 34. $\frac{11}{15} - \frac{3}{10}$ 35. $\frac{5}{6} - \frac{2}{9}$ 36. $\frac{3}{5} - \frac{1}{3}$ 37. $\frac{5}{9} - \frac{2}{5}$

38. $\frac{6}{7} - \frac{1}{4}$ 39. $\frac{11}{13} - \frac{2}{3}$ 40. $\frac{4}{5} - \frac{3}{4}$ 41. $\frac{1}{2} - \frac{3}{8}$ 42. $\frac{5}{7} - \frac{1}{3}$

MIXED PRACTICE

Add or subtract. Write each answer in lowest terms. (Examples 1–3)

43. $\frac{5}{6} + \frac{1}{2}$ 44. $\frac{3}{5} - \frac{1}{6}$ 45. $\frac{5}{8} - \frac{1}{10}$ 46. $\frac{3}{7} + \frac{1}{4}$ 47. $\frac{9}{11} - \frac{1}{4}$

48. $\frac{1}{9} + \frac{2}{3}$ 49. $\frac{4}{5} + \frac{2}{9}$ 50. $\frac{3}{5} - \frac{1}{10}$ 51. $\frac{9}{14} + \frac{2}{7}$ 52. $\frac{7}{8} - \frac{1}{6}$

Fractions: Addition/Subtraction   **163**

CLASSROOM EXERCISES

9. $\frac{7}{18}$ 10. $1\frac{1}{12}$

11. $\frac{17}{24}$ 12. $\frac{9}{10}$

13. $1\frac{3}{14}$ 14. $1\frac{1}{3}$

15. $1\frac{7}{40}$ 16. $1\frac{1}{4}$ 17. $\frac{1}{2}$

18. $\frac{5}{8}$ 19. $\frac{11}{24}$ 20. $\frac{13}{30}$

21. $\frac{11}{20}$ 22. $\frac{7}{15}$ 23. $\frac{1}{6}$

24. $\frac{5}{8}$ 25. $\frac{1}{8}$ 26. $\frac{9}{20}$

ASSIGNMENT GUIDE
BASIC
Day 1 p. 163: 1–47 odd
Day 2 p. 163: 2–46 even
AVERAGE
Day 1 p. 163: 1–41 odd
Day 2 pp. 163–164: 43–56
ABOVE AVERAGE
pp. 163–164: 1–51 odd,
53–71 odd

WRITTEN EXERCISES

1. $\frac{2}{6}$ 2. $\frac{6}{15}$ 3. $\frac{25}{40}$

4. $\frac{15}{20}$ 5. $\frac{30}{36}$ 6. $\frac{2}{18}$

7. $\frac{18}{42}$ 8. $\frac{90}{100}$ 9. $\frac{10}{15}$

10. $\frac{4}{8}$ 11. $\frac{21}{36}$ 12. $\frac{24}{45}$

13. $\frac{7}{8}$ 14. $\frac{7}{12}$ 15. $\frac{4}{5}$

16. $\frac{11}{20}$

WRITTEN EXERCISES

17. $\frac{7}{18}$ 18. $\frac{19}{24}$ 19. $\frac{22}{45}$
20. $\frac{7}{24}$ 21. $\frac{14}{15}$ 22. $\frac{17}{35}$
23. $\frac{23}{24}$ 24. $\frac{23}{30}$ 25. $1\frac{1}{2}$
26. $1\frac{1}{3}$ 27. $1\frac{1}{42}$
28. $\frac{1}{12}$ 29. $\frac{1}{2}$ 30. $\frac{3}{14}$
31. $\frac{2}{15}$ 32. $\frac{13}{20}$ 33. $\frac{7}{36}$
34. $\frac{13}{30}$ 35. $\frac{11}{18}$ 36. $\frac{4}{15}$
37. $\frac{7}{45}$ 38. $\frac{17}{28}$ 39. $\frac{7}{39}$
40. $\frac{1}{20}$ 41. $\frac{1}{8}$ 42. $\frac{8}{21}$
43. $1\frac{1}{3}$ 44. $\frac{13}{30}$ 45. $\frac{21}{40}$
46. $\frac{19}{28}$ 47. $\frac{25}{44}$ 48. $\frac{7}{9}$
49. $1\frac{1}{45}$ 50. $\frac{1}{2}$ 51. $\frac{13}{14}$
52. $\frac{17}{24}$ 53. $1\frac{3}{8}$ in
54. $\frac{9}{20}$ lb 55. $\frac{7}{30}$ mi
56. $1\frac{13}{30}$ hr 57. $\frac{6x}{25}$
58. $\frac{n}{2}$ 59. $\frac{a}{18}$ 60. $\frac{3y}{10}$
61. $\frac{5x}{8}$ 62. $\frac{9y}{8}$
63. $\frac{13}{2y}$ 64. $\frac{11}{2m}$
65. $\frac{7}{3x}$ 66. $\frac{4n}{3t}$
67. $\frac{7a}{2n}$ 68. $\frac{13x}{2y}$
69. $\frac{9y}{16z}$ 70. $\frac{29n}{6p}$
71. $\frac{61s}{15t}$

APPLICATIONS: **UNLIKE DENOMINATORS**

53. On Tuesday, the rainfall was $\frac{5}{8}$ inch. On Wednesday, the rainfall was $\frac{3}{4}$ inch. What was the total rainfall for the two days?

54. Carlton is making beef stew. The recipe calls for $\frac{3}{4}$ of a pound of beef. He has $\frac{3}{10}$ of a pound. How much more beef does he need?

55. Yolanda lives $\frac{9}{10}$ of a mile from school. Eric lives $\frac{2}{3}$ of a mile from school. How much farther from the school does Yolanda live?

56. Elena spent $\frac{5}{6}$ of an hour doing her math homework and $\frac{3}{5}$ of an hour doing her science homework. How much time did she spend altogether?

MORE CHALLENGING EXERCISES

Add or subtract. Write each answer in lowest terms. No variable equals zero.

EXAMPLE: a. $\frac{t}{6} + \frac{t}{2} = \frac{t+3t}{6}$
$= \frac{4t}{6} = \frac{2t}{3}$

b. $\frac{7}{f} - \frac{5}{2f} = \frac{14}{2f} - \frac{5}{2f}$
$= \frac{14-5}{2f} = \frac{9}{2f}$ LCM: $2f$

57. $\frac{x}{5} + \frac{x}{25}$ 58. $\frac{n}{6} + \frac{n}{3}$ 59. $\frac{a}{6} - \frac{a}{9}$ 60. $\frac{y}{2} - \frac{y}{5}$ 61. $\frac{3x}{8} + \frac{x}{4}$

62. $\frac{5y}{8} + \frac{y}{2}$ 63. $\frac{9}{2y} + \frac{2}{y}$ 64. $\frac{1}{2m} + \frac{5}{m}$ 65. $\frac{3}{x} - \frac{2}{3x}$ 66. $\frac{7n}{3t} - \frac{n}{t}$

67. $\frac{3a}{n} + \frac{a}{2n}$ 68. $\frac{5x}{y} + \frac{3x}{2y}$ 69. $\frac{21y}{16z} - \frac{3y}{4z}$ 70. $\frac{11n}{2p} - \frac{2n}{3p}$ 71. $\frac{17s}{5t} + \frac{2s}{3t}$

REVIEW CAPSULE FOR SECTION 6-3

Write the least common multiple of each pair of numbers. (Pages 111–114)

1. 3 and 11   2. 6 and 20   3. 8 and 32   4. 10 and 15   5. 4 and 9

Write each fraction in lowest terms.   (Pages 124–125)

6. $\frac{10}{20}$ 7. $\frac{5}{30}$ 8. $\frac{12}{26}$ 9. $\frac{8}{24}$ 10. $\frac{15}{40}$

The answers are on page 183.

6-3 Addition and Subtraction: Mixed Numerals

The record for the high jump at Doug's school is $6\frac{2}{3}$ feet. Doug's best jump at the last intramural meet was $6\frac{1}{4}$ feet. To find how much short of the record he is, Doug found this difference.

$6\frac{2}{3} - 6\frac{1}{4}$ ◀ **Mixed numerals with unlike denominators**

Adding and subtracting mixed numerals with unlike denominators is similar to adding and subtracting fractions having unlike denominators.

EXAMPLE 1 Add or subtract as indicated: **a.** $2\frac{5}{8} + 5\frac{3}{4}$ **b.** $6\frac{2}{3} - 6\frac{1}{4}$

Solutions:

a. LCM: 8

$2\frac{5}{8} + 5\frac{3}{4} = 2\frac{5}{8} + 5\frac{6}{8}$ ◀ $\begin{array}{l}2 + 5 = 7\\ \frac{5}{8} + \frac{6}{8} = \frac{11}{8}\end{array}$

$= 7\frac{11}{8}$

$= 7 + 1\frac{3}{8}$

$= 8\frac{3}{8}$

b. LCM: 12

$6\frac{2}{3} - 6\frac{1}{4} = 6\frac{8}{12} - 6\frac{3}{12}$ ◀ $\begin{array}{l}6 - 6 = 0\\ \frac{8}{12} - \frac{3}{12} = \frac{5}{12}\end{array}$

$= \frac{5}{12}$ ◀ Doug's jump was $\frac{5}{12}$ ft short of the record

Self-Check *Add or subtract as indicated.*

1. $8\frac{4}{5} + 5\frac{1}{3}$ **2.** $5\frac{3}{4} - 2\frac{1}{2}$ **3.** $7\frac{3}{4} - 5\frac{1}{3}$ **4.** $6\frac{7}{8} + 1\frac{1}{4}$

When subtracting with mixed numerals, it is sometimes necessary to "borrow" from the whole number.

EXAMPLE 2 Subtract: **a.** $12\frac{1}{4} - 2\frac{2}{5}$ **b.** $14 - 8\frac{5}{6}$

Solutions: **a.** LCM: 20

$12\frac{1}{4} - 2\frac{2}{5} = 12\frac{5}{20} - 2\frac{8}{20}$ ◀ Since $\frac{5}{20}$ is less than $\frac{8}{20}$, borrow 1, or $\frac{20}{20}$, from 12.

$= 11\frac{25}{20} - 2\frac{8}{20}$

$= 9\frac{17}{20}$

b. LCM: 6 $14 - 8\frac{5}{6} = 13\frac{6}{6} - 8\frac{5}{6}$ ◀ $14 = 13\frac{6}{6}$

$= 5\frac{1}{6}$

Fractions: Addition/Subtraction **165**

Teaching Suggestions p. M-19

QUICK QUIZ

Write the LCM.

<u>1</u>. 6 and 7   Ans: 42

<u>2</u>. 10 and 5   Ans: 10

<u>3</u>. 12 and 18   Ans: 36

Add or subtract as indicated.

<u>4</u>. $\frac{13}{16} + \frac{5}{16}$ Ans: $1\frac{1}{8}$

<u>5</u>. $\frac{13}{16} - \frac{5}{16}$ Ans: $\frac{1}{2}$

ADDITIONAL EXAMPLES

Example 1

Add or subtract as indicated.

<u>1</u>. $3\frac{5}{6} + 4\frac{2}{3}$ Ans: $8\frac{1}{2}$

<u>2</u>. $9\frac{11}{12} - \frac{3}{8}$ Ans: $9\frac{13}{24}$

SELF-CHECK

<u>1</u>. $14\frac{2}{15}$ <u>2</u>. $3\frac{1}{4}$

<u>3</u>. $2\frac{5}{12}$ <u>4</u>. $8\frac{1}{8}$

Example 2

Subtract.

<u>1</u>. $2\frac{1}{3} - 1\frac{4}{9}$ Ans: $\frac{8}{9}$

SELF-CHECK

<u>5</u>. $1\frac{1}{2}$ <u>6</u>. $2\frac{1}{2}$ <u>7</u>. $5\frac{2}{5}$

<u>8</u>. $2\frac{2}{3}$

Self-Check Subtract. The answers are on page 165.

5. $5\frac{1}{4} - 3\frac{3}{4}$ **6.** $7\frac{1}{6} - 4\frac{2}{3}$ **7.** $9 - 3\frac{3}{5}$ **8.** $5 - 2\frac{1}{3}$

CLASSROOM EXERCISES

Complete. (Example 1)

1. $3\frac{1}{4} + 2\frac{1}{2} = 3\frac{1}{4} + 2\frac{?}{4}$ **2.** $6\frac{3}{10} + 2\frac{2}{5} = 6\frac{3}{10} + 2\frac{?}{10}$
3. $6\frac{4}{5} - 2\frac{1}{4} = 6\frac{?}{20} - 2\frac{5}{20}$ **4.** $5\frac{2}{3} - 1\frac{1}{4} = 5\frac{?}{12} - 1\frac{3}{12}$

Complete. (Example 2)

5. $7\frac{1}{6} - 2\frac{2}{3} = 6\frac{?}{6} - 2\frac{4}{6}$ **6.** $11\frac{1}{2} - 2\frac{5}{8} = 10\frac{?}{8} - 2\frac{5}{8}$
7. $6 - 3\frac{1}{4} = 5\frac{?}{4} - 3\frac{1}{4}$ **8.** $3 - 1\frac{5}{6} = 2\frac{?}{6} - 1\frac{5}{6}$

WRITTEN EXERCISES

Goals: To add and subtract with mixed numerals
To apply the skill of adding and subtracting with mixed numerals to solving word problems

Add or subtract. Write each answer in lowest terms. (Example 1)

1. $5\frac{1}{2} + 2\frac{1}{4}$ **2.** $2\frac{3}{8} + 1\frac{1}{2}$ **3.** $1\frac{1}{6} + \frac{3}{4}$ **4.** $8\frac{4}{5} + 5\frac{1}{3}$ **5.** $6\frac{3}{4} + 1\frac{2}{8}$
6. $12\frac{1}{2} + 11\frac{1}{8}$ **7.** $6\frac{2}{5} + \frac{3}{4}$ **8.** $12\frac{1}{12} + 13\frac{1}{4}$ **9.** $5\frac{3}{4} - 2\frac{1}{2}$ **10.** $5\frac{11}{12} - 4\frac{3}{8}$
11. $2\frac{5}{6} - \frac{3}{4}$ **12.** $7\frac{7}{10} - \frac{2}{5}$ **13.** $8\frac{7}{8} - 6\frac{3}{4}$ **14.** $12\frac{1}{2} - 3\frac{1}{3}$ **15.** $6\frac{8}{15} - \frac{2}{5}$

Subtract. Write each answer in lowest terms. (Example 2)

16. $5\frac{1}{4} - 1\frac{3}{4}$ **17.** $7\frac{1}{6} - 2\frac{1}{2}$ **18.** $9\frac{2}{3} - 4\frac{11}{12}$ **19.** $14 - 12\frac{1}{2}$ **20.** $6 - 4\frac{2}{3}$
21. $11\frac{1}{2} - \frac{5}{6}$ **22.** $2\frac{1}{6} - \frac{3}{8}$ **23.** $7\frac{1}{5} - 2\frac{5}{12}$ **24.** $3\frac{1}{4} - 1\frac{2}{3}$ **25.** $8 - 2\frac{9}{10}$
26. $9\frac{1}{12} - 3\frac{2}{3}$ **27.** $7\frac{1}{8} - 3\frac{1}{4}$ **28.** $5 - 3\frac{1}{3}$ **29.** $9 - 2\frac{5}{16}$ **30.** $15\frac{1}{4} - 8\frac{3}{8}$

MIXED PRACTICE

Add or subtract as indicated. Write each answer in lowest terms.

31. $7\frac{3}{5} + 2\frac{1}{3}$ **32.** $2\frac{1}{2} - 1\frac{1}{8}$ **33.** $8\frac{1}{3} + 7\frac{3}{4}$ **34.** $4\frac{11}{15} + 6\frac{3}{5}$ **35.** $8\frac{1}{8} - 2\frac{1}{4}$
36. $9\frac{1}{2} - 3\frac{1}{5}$ **37.** $3 - 1\frac{7}{12}$ **38.** $4\frac{6}{7} + 2\frac{5}{6}$ **39.** $20\frac{2}{3} - 8\frac{1}{4}$ **40.** $11\frac{3}{11} + 2\frac{1}{3}$

166 Chapter 6

APPLICATIONS: USING MIXED NUMERALS

Solve. Write each answer in lowest terms.

41. The school record for running the mile is $4\frac{3}{4}$ minutes. Rick's record is $5\frac{3}{10}$ minutes. Find the difference between the two records.

42. Rose can throw the shot put $34\frac{3}{8}$ feet. Julie can throw the shot put $39\frac{1}{4}$ feet. How much farther can Julie throw the shot put than Rose?

43. Arthur is $5\frac{2}{3}$ feet tall. His record in the long jump is $10\frac{1}{2}$ feet more than his height. What is his long jump record?

REVIEW: SECTIONS 6-1—6-3

Add or subtract. Write each answer in lowest terms. (Section 6-1)

1. $\frac{1}{8} + \frac{5}{8}$
2. $3\frac{1}{3} + 4\frac{1}{3}$
3. $\frac{8}{13} - \frac{5}{13}$
4. $15\frac{9}{20} - 4\frac{3}{20}$
5. $6\frac{7}{8} + 3\frac{5}{8}$

Complete. Write an equivalent fraction as indicated. (Section 6-2)

6. $\frac{1}{2} = \frac{?}{20}$
7. $\frac{3}{7} = \frac{?}{28}$
8. $\frac{3}{4} = \frac{?}{12}$
9. $\frac{5}{6} = \frac{?}{30}$
10. $\frac{7}{8} = \frac{?}{48}$

Add or subtract. Write each answer in lowest terms. (Section 6-2)

11. $\frac{1}{2} + \frac{2}{3}$
12. $\frac{1}{5} + \frac{3}{10}$
13. $\frac{5}{6} - \frac{1}{8}$
14. $\frac{4}{5} - \frac{3}{10}$
15. $\frac{1}{2} - \frac{1}{6}$

(Section 6-3)

16. $1\frac{5}{6} + 2\frac{1}{4}$
17. $4\frac{2}{3} + 5\frac{1}{5}$
18. $5\frac{2}{3} - 1\frac{1}{6}$
19. $8\frac{3}{5} - 5\frac{3}{4}$
20. $15 - 4\frac{1}{8}$

Solve each problem. Write answers in lowest terms.

21. Tamara swam $2\frac{1}{5}$ miles on Wednesday and $1\frac{3}{5}$ miles on Friday. How many miles did she swim in all? (Section 6-1)

22. Willie spent $\frac{3}{4}$ of an hour mowing his front yard. He spent $\frac{2}{3}$ of an hour edging the driveway and front curb. How much time did he spend altogether on mowing and edging? (Section 6-2)

Fractions: Addition/Subtraction **167**

WRITTEN EXERCISES

25. $5\frac{1}{10}$ 26. $5\frac{5}{12}$
27. $3\frac{7}{8}$ 28. $1\frac{2}{3}$
29. $6\frac{11}{16}$ 30. $6\frac{7}{8}$
31. $9\frac{14}{15}$ 32. $1\frac{3}{8}$
33. $16\frac{1}{12}$ 34. $11\frac{1}{3}$
35. $5\frac{7}{8}$ 36. $6\frac{3}{10}$
37. $1\frac{5}{12}$ 38. $7\frac{29}{42}$
39. $12\frac{5}{12}$ 40. $13\frac{20}{33}$
41. $\frac{11}{20}$ min 42. $4\frac{7}{8}$ ft
43. $16\frac{1}{6}$ ft

QUIZ: SECTIONS 6-1-6-3
After completing this Review, you may want to administer a quiz covering the same sections. A Quiz is provided in the *Teacher's Edition: Part II*.

REVIEW: SECTIONS 6-1-6-3

1. $\frac{3}{4}$ 2. $7\frac{2}{3}$ 3. $\frac{3}{13}$
4. $11\frac{3}{10}$ 5. $10\frac{1}{2}$
6. $\frac{10}{20}$ 7. $\frac{12}{28}$ 8. $\frac{9}{12}$
9. $\frac{25}{30}$ 10. $\frac{42}{48}$ 11. $1\frac{1}{6}$
12. $\frac{1}{2}$

CALCULATOR/COMPUTER APPLICATIONS

This feature illustrates the use of the calculator and a computer program for evaluating powers in which the base and exponent are counting numbers.

1. 49
2. 729
3. 4096
4. 531,441
5. 1024
6. 1331
7. 19,487,171
8. 43,046,721
9. 62,748,517
10. 33,554,432
11. 4,100,625
12. 57,289,761

CALCULATOR/COMPUTER APPLICATIONS
Evaluating Powers

To evaluate a formula such as

$$x = b^n - 2$$

you first have to evaluate b^n. For $b = 3$ and $n = 4$,

$$x = 3^4 - 2$$
$$= 3 \cdot 3 \cdot 3 \cdot 3 - 2$$
$$= 81 - 2 = \mathbf{79}$$

You can use a calculator to evaluate a power.

Calculator: Evaluate 15^3.

Solution: Press " = " *once* for the b^2, *twice* for the b^3, *three* times for b^4, and so on. Since the exponent in 15^3 is three, press the = key twice.

15 [×] [=] [=] | 3375.

Computer: You can use the following program to evaluate b^n, where b and n are counting numbers. Note that this program does not use the ↑ operator of BASIC.

Program:
```
10 INPUT B,N
20 LET X = 1
30 FOR I = 1 TO N
40 LET X = X * B
50 NEXT I
60 PRINT "ANSWER IS"; X
70 END
```

Statements 30–50 represent a **FOR-NEXT** loop. This loop multiplies B times itself N times. However, before the loop begins, X is set to 1 in statement 20.

Exercises

Use a calculator or a computer to evaluate each of the following.

1. 7^2
2. 9^3
3. 8^4
4. 9^6
5. 4^5
6. 11^3
7. 11^7
8. 9^8
9. 13^7
10. 32^5
11. 45^4
12. 87^4

168   Computer Application

REVIEW CAPSULE FOR SECTION 6-4

Multiply. Write each answer in lowest terms. (Pages 135–137)

1. $\frac{3}{4} \cdot \frac{9}{10}$
2. $\frac{5}{6} \cdot \frac{8}{15}$
3. $2\frac{1}{4} \cdot 4\frac{2}{5}$
4. $1\frac{1}{3} \cdot 3\frac{1}{6}$
5. $5 \cdot 1\frac{1}{2}$
6. $9 \cdot 2\frac{2}{3}$
7. $24 \cdot 3\frac{1}{8}$
8. $150 \cdot 2\frac{3}{10}$

Complete. (Pages 161–164)

9. $15 \text{ min} = \frac{15}{60} \text{ hr} = \underline{\ ?\ } \text{ hr}$
10. $\frac{3}{4} \text{ hr} = \frac{?}{60} \text{ hr} = \underline{\ ?\ } \text{ min}$
11. $10 \text{ min} = \frac{?}{60} \text{ hr} = \frac{1}{6} \text{ hr}$
12. $\frac{1}{3} \text{ hr} = \frac{?}{60} \text{ hr} = \underline{\ ?\ } \text{ min}$

The answers are on page 183.

PROBLEM SOLVING AND APPLICATIONS

6-4 Time Cards

Rosemarie and Karl have part time jobs at a supermarket near their home. At the supermarket, time cards are used to record how many hours each employee works per week.

EXAMPLE 1 A copy of Rosemarie's time card for one week is shown at the right. Find the total number of hours she worked that week.

DAYS	IN	OUT	IN	OUT
1	—	—	5:30	8:00
2	—	—	3:00	6:15
3	—	—	4:00	7:20
4	7:00	11:30	1:00	4:00
5	9:00	12:00	1:30	5:15

Solution:

1 Find the hours worked each day.

Day 1	Day 2	Day 3	Day 4	Day 5
$2\frac{1}{2}$ hr	$3\frac{1}{4}$ hr	$3\frac{1}{3}$ hr	$7\frac{1}{2}$ hr	$6\frac{3}{4}$ hr

2 Find the total hours for the week. LCM: 12

$2\frac{1}{2} + 3\frac{1}{4} + 3\frac{1}{3} + 7\frac{1}{2} + 6\frac{3}{4} = 2\frac{6}{12} + 3\frac{3}{12} + 3\frac{4}{12} + 7\frac{6}{12} + 6\frac{9}{12}$

$= 21\frac{28}{12}$ ◀ $\frac{28}{12} = 2\frac{4}{12} = 2\frac{1}{3}$

$= 23\frac{1}{3}$ hours

Self-Check

1. Karl worked $3\frac{1}{2}$ hours on Tuesday, $6\frac{1}{4}$ hours on Friday, and $5\frac{2}{3}$ hours on Saturday. Find the total number of hours worked.

Fractions: Addition/Subtraction   **169**

REVIEW: SECTIONS 6-1–6-3

13. $\frac{17}{24}$
14. $\frac{1}{2}$
15. $\frac{1}{3}$
16. $4\frac{1}{12}$
17. $9\frac{13}{15}$
18. $4\frac{1}{2}$
19. $2\frac{17}{20}$
20. $10\frac{7}{8}$
21. $3\frac{4}{5}$ mi
22. $1\frac{5}{12}$ hr

Teaching Suggestions
p. M-19

QUICK QUIZ

Perform the indicated operation.
Write answers in lowest terms.

1. $\frac{7}{16} + \frac{5}{16}$ Ans: $\frac{3}{4}$
2. $\frac{5}{8} + \frac{5}{8}$ Ans: $1\frac{1}{4}$
3. $\frac{2}{3} \cdot \frac{5}{9}$ Ans: $\frac{10}{27}$
4. $\frac{4}{5} \cdot \frac{3}{8}$ Ans: $\frac{3}{10}$

ADDITIONAL EXAMPLE

Example 1
Find the hours worked each day.

Mon.	Tues.	Wed.
$7\frac{2}{3}$	8	$8\frac{1}{4}$

Ans: $23\frac{11}{12}$ hr

SELF-CHECK

1. $15\frac{5}{12}$ hr

Rosemarie is paid $4.35 per hour and Karl is paid $4.15 per hour. To determine how much Rosemarie and Karl earn each week, you must first answer the question:

How many hours do they work each week?

This is the **hidden question** in the problem. Then you can find the weekly earnings.

EXAMPLE 2 Rosemarie is paid $4.35 per hour. How much did she earn for the week discussed in Example 1?

Solutions:

① Find the total number of hours worked.

Total hours: $23\frac{1}{3}$ ◀ **From Example 1**

② Weekly Earnings = (Pay per Hour) × (Total Hours)

$$= (4.35)\left(23\frac{1}{3}\right) \quad 23\frac{1}{3} = \frac{70}{3}$$

$$= (\cancel{4.35})\left(\frac{\cancel{70}}{\cancel{3}}\right)^{1.45}_{1}$$

$$= \$101.50$$

Rosemarie's weekly earnings amounted to **$101.50**.

Self-Check 2. Karl is paid $4.15 per hour. How much did he earn for the week discussed in the Self-Check that follows Example 1?

CLASSROOM EXERCISES

For Exercises 1–2:

a. *Find the number of hours worked each day.*

b. *Find the total number of hours for the week.* (Example 1)

1.

DAYS	IN	OUT	IN	OUT	DAILY TOTALS
1	—	—	1:00	5:30	?
2	9:00	12:15	1:15	5:00	?
3	8:30	12:00	1:00	4:30	?
4	8:45	12:00	1:00	4:15	?
5	9:15	12:00	—	—	?
WEEKLY TOTAL					?

2.

DAYS	IN	OUT	IN	OUT	DAILY TOTALS
1	9:00	12:30	1:30	5:00	?
2	9:00	12:20	1:00	5:15	?
3	8:45	11:30	—	—	?
4	9:00	12:45	1:45	5:00	?
5	8:40	12:00	1:00	5:30	?
WEEKLY TOTAL					?

ADDITIONAL EXAMPLE

Example 2

Max is paid $5.70 per hour. How much did he earn for $18\frac{1}{3}$ hours work?

Ans: $104.50

SELF-CHECK

2. $63.98

CLASSROOM EXERCISES

1. $4\frac{1}{2}$; 7; 7; $6\frac{1}{2}$; $2\frac{3}{4}$; $27\frac{3}{4}$ hr

2. 7; $7\frac{7}{12}$; $2\frac{3}{4}$; 7; $7\frac{5}{6}$; $32\frac{1}{6}$ hr

WRITTEN EXERCISES

Goal: To solve two-step problems involving earnings by applying the technique of the "hidden question"

For Exercises 1–4:
a. *Find the total number of hours worked each day.*
b. *Find the total number of hours worked for the week.* (Example 1)

1.

Name: Sonya Wilson				
DAYS	IN	OUT	IN	OUT	DAILY TOTALS
1	8:00	12:00	1:00	4:15	?
2	8:00	12:00	1:00	4:30	?
3	8:00	12:30	1:00	5:15	?
4	8:00	12:00	1:00	4:30	?
5	8:00	12:15	1:15	4:30	?
WEEKLY TOTAL				?

2.

Name: Paul Chan				
DAYS	IN	OUT	IN	OUT	DAILY TOTALS
1	9:00	11:30	12:30	5:30	?
2	8:30	11:30	12:00	5:20	?
3	8:15	11:30	12:30	5:15	?
4	9:00	12:30	1:30	5:30	?
5	8:30	11:30	12:30	5:40	?
WEEKLY TOTAL				?

3.

Name: Ted Schultz				
DAYS	IN	OUT	IN	OUT	DAILY TOTALS
1	9:00	1:00	2:00	5:30	?
2	9:00	1:00	2:00	5:45	?
3	8:15	12:00	—	—	?
4	8:30	12:00	1:00	5:15	?
5	9:00	12:30	1:30	5:30	?
WEEKLY TOTAL				?

4.

Name: Lisa Kotlowski				
DAYS	IN	OUT	IN	OUT	DAILY TOTALS
1	9:00	12:20	1:20	5:00	?
2	—	—	1:15	5:00	?
3	8:30	1:00	2:00	4:20	?
4	8:40	1:00	2:00	5:00	?
5	9:00	12:30	1:30	5:15	?
WEEKLY TOTAL				?

For Exercises 5–8, use the information from Exercises 1–4 to solve each problem.

5. Sonya Wilson is paid $5.20 per hour (see Exercise 1). Find her total weekly earnings.

6. Paul Chan is paid $4.50 per hour (see Exercise 2). Find his total weekly earnings.

7. Ted Schultz is paid $6.20 per hour (see Exercise 3). Find his total weekly earnings.

8. Lisa Kotlowski is paid $3.78 per hour (see Exercise 4). Find her total weekly earnings.

Fractions: Addition/Subtraction **171**

ASSIGNMENT GUIDE
BASIC
pp. 171–172: 1–12
AVERAGE
pp. 171–172: 1–12
ABOVE AVERAGE
pp. 171–172: 1–7 odd, 9–16

WRITTEN EXERCISES

<u>1.</u> $7\frac{1}{4}$; $7\frac{1}{2}$; $8\frac{3}{4}$; $7\frac{1}{2}$; $7\frac{1}{2}$; $38\frac{1}{2}$ hr

<u>2.</u> $7\frac{1}{2}$; $8\frac{1}{3}$; 8; $7\frac{1}{2}$; $8\frac{1}{6}$; $39\frac{1}{2}$ hr

<u>3.</u> $7\frac{1}{2}$; $7\frac{3}{4}$; $3\frac{3}{4}$; $7\frac{3}{4}$; $7\frac{1}{2}$; $34\frac{1}{4}$ hr

<u>4.</u> 7; $3\frac{3}{4}$; $6\frac{5}{6}$; $7\frac{1}{3}$; $7\frac{1}{4}$; $32\frac{1}{6}$ hr

<u>5.</u> $200.20   <u>6.</u> $177.75
<u>7.</u> $212.35   <u>8.</u> $121.59

WRITTEN EXERCISES

9. $93.80 10. $74.40
11. $134.20
12. $150.50 13. $22.95
14. $4.30 per hour
15. $12.90 per hour
16. $392.36

Solve each problem.

9. Tom Hastings worked $6\frac{1}{2}$ hours on Monday, $5\frac{3}{4}$ hours on Wednesday, and $4\frac{1}{2}$ hours on Friday. He is paid $5.60 per hour. Find his weekly earnings.

10. Julie Pendleton worked $3\frac{1}{2}$ hours on Tuesday, $3\frac{3}{4}$ hours on Friday, and $8\frac{1}{4}$ hours on Saturday. She is paid $4.80 per hour. Find her weekly earnings.

11. Robert Felch worked from 9:00 A.M. to 4:15 P.M. on Monday, Wednesday, and Friday, and from 8:30 A.M. to 5:15 P.M. on Saturday. He is paid $4.40 per hour. Find his weekly earnings.

12. Doris Youngblood worked from 3:00 P.M. to 9:30 P.M. on Monday and Tuesday, from 3:00 P.M. to 8:20 P.M. on Thursday, and from 3:15 P.M. to 10:00 P.M. on Friday. She is paid $6.00 per hour. Find her weekly earnings.

MORE CHALLENGING EXERCISES

13. Tim Murray works $38\frac{1}{4}$ hours each week. He is paid $4.60 per hour. After his raise he will be paid $5.20 per hour. How much more money will Tim earn each week?

14. Denise Turner is paid $165.55 per week for $38\frac{1}{2}$ hours of work. How much does Denise earn in an hour?

15. Mike Fu is paid $344.00 for 40 hours of work. He is paid $1\frac{1}{2}$ times the hourly rate for time over 40 hours. How much is Mike paid per hour for time over 40 hours?

16. Marjorie Russell was paid $344.00 for 40 hours of work. She was paid $9.30 per hour for time over 40 hours. She worked $45\frac{1}{5}$ hours last week. How much did Marjorie earn?

REVIEW CAPSULE FOR SECTION 6-5

Divide to the hundredths place. (Pages 44–46)

1. $2\overline{)1.7}$ 2. $8\overline{)4.4}$ 3. $4\overline{)2.6}$ 4. $5\overline{)1.3}$ 5. $6\overline{)4.5}$

Replace the __?__ with > or <. (Pages 32–34)

6. 0.421 __?__ 0.422   7. 0.914 __?__ 0.92   8. 0.021 __?__ 0.039
9. 0.77 __?__ 0.68   10. 0.628 __?__ 0.625   11. 0.01 __?__ 0.15
12. 0.163 __?__ 0.1603   13. 0.199 __?__ 0.1991   14. 0.004 __?__ 0.014

The answers are on page 183.

Chapter 6

6-5 Fractions and Decimals

The San Andreas Fault extends $\frac{3}{4}$ of the length of California's coastal region.

To write a decimal for $\frac{3}{4}$, divide 3 by 4.

PROCEDURE

To write a decimal for a fraction, divide the numerator of the fraction by the denominator.

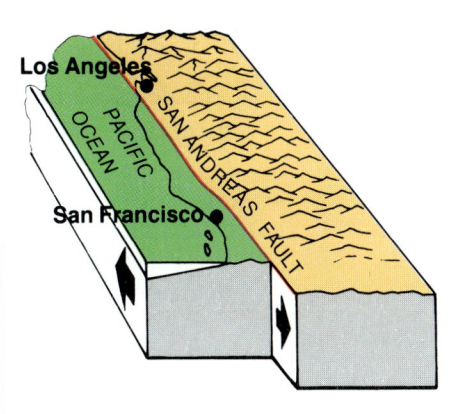

EXAMPLE 1 Write a decimal for $\frac{3}{4}$.

Solution: $\frac{3}{4}$ means $3 \div 4$.

$$\begin{array}{r} 0.75 \\ 4\overline{)3.00} \\ \underline{2\ 8} \\ 20 \\ \underline{20} \\ 0 \end{array}$$

The remainder is 0. The decimal **terminates**.

Self-Check Write a decimal for each fraction.

1. $\frac{1}{2}$ 2. $\frac{2}{5}$ 3. $\frac{5}{8}$ 4. $\frac{5}{2}$ 5. $\frac{9}{5}$

The fraction $\frac{3}{4} = 0.75$. The decimal 0.75 is called a **terminating decimal**. Some decimals do not terminate.

EXAMPLE 2 About $\frac{2}{3}$ of the earth's surface is covered with water. Write a decimal for $\frac{2}{3}$. Carry the division to three decimal places.

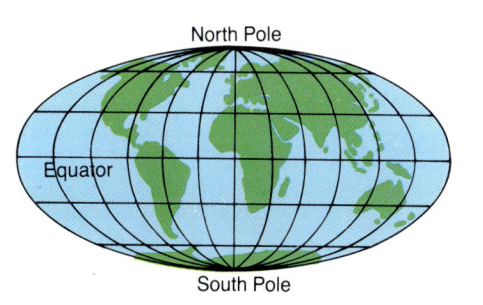

Solution:
$$\begin{array}{r} 0.666 \\ 3\overline{)2.000} \\ \underline{1\ 8} \\ 20 \\ \underline{18} \\ 20 \\ \underline{18} \\ 2 \end{array}$$

The digit, 6, repeats indefinitely.

$\frac{2}{3} = 0.666\cdots$ The three dots mean that "6" repeats indefinitely.

Fractions: Addition/Subtraction

Teaching Suggestions p. M-19

QUICK QUIZ

Divide to the hundredths place.

1. $3\overline{)7.86}$ Ans: 2.62
2. $6\overline{)3.9}$ Ans: 0.65
3. $21.6 \div 9$ Ans: 2.40

Replace the ? with > or <.

4. 29 ? 290   Ans: <
5. 2.9 ? 0.29   Ans: >

ADDITIONAL EXAMPLES

Example 1

Write as a decimal.

1. $\frac{3}{8}$ Ans: 0.375
2. $\frac{4}{5}$ Ans: 0.8

SELF-CHECK

1. 0.5   2. 0.4
3. 0.625   4. 2.5
5. 1.8

Example 2

Write as a decimal.

1. $\frac{5}{12}$ Ans: 0.41666$\cdots$
2. $\frac{7}{11}$ Ans: 0.636363$\cdots$
3. $\frac{23}{111}$

Ans: 0.207207$\cdots$

SELF-CHECK

6. $0.333\cdots$

7. $0.833\cdots$

8. $0.133\cdots$

9. $0.555\cdots$

10. $0.277\cdots$

ADDITIONAL EXAMPLES

Example 3

Complete.

Write >, <, or =.

1. $\frac{7}{16}$? $\frac{2}{5}$ Ans: >

2. $\frac{5}{6}$? $\frac{7}{8}$ Ans: <

3. $\frac{3}{4}$? $\frac{5}{7}$ Ans: >

SELF-CHECK

11. <

12. <

13. >

14. =

CLASSROOM EXERCISES

1. =   2. ≠   3. ≠

4. =   5. ≠   6. =

7. =   8. ≠   9. T

10. T   11. R   12. R

13. T   14. R   15. <

16. <   17. <   18. <

Self-Check *Write a decimal for each fraction. Use dots to show repeating digits.*

6. $\frac{1}{3}$ 7. $\frac{5}{6}$ 8. $\frac{2}{15}$ 9. $\frac{5}{9}$ 10. $\frac{5}{18}$

The decimal $0.666\cdots$ is called a **repeating decimal** because the digit 6 repeats indefinitely. The decimal can be written in another way.

$$0.666\cdots = 0.\overline{6}$$

◀ **The bar indicates the repeating digit(s).**

NOTE: Repeating decimals may have one, two, or several repeating digits. Also, the digits may start to repeat only after a certain point.

$\frac{5}{11} = 0.\overline{45}$ ◀ $0.454545\cdots$ $\frac{1}{12} = 0.08\overline{3}$ ◀ $0.083333\ldots$

One way of comparing two fractions is to write a decimal for each fraction. Then compare the decimals (see Example 2 on page 33).

EXAMPLE 3 Suellen wanted to buy $\frac{3}{4}$ of a yard of material. The clerk told her there was $\frac{5}{8}$ of a yard in stock. Was there enough material in stock?

Solution: ① Write a decimal for each fraction.

$\frac{3}{4} = 0.75$ $\frac{5}{8} = 0.625$ ◀ **See Example 1 on page 173.**

② Compare the decimals. $0.75 > 0.625$

Therefore, $\frac{3}{4} > \frac{5}{8}$

Since $\frac{3}{4} > \frac{5}{8}$, there is **not enough material.**

Self-Check *Complete. Write >, <, or =.*

11. $\frac{2}{5}$? $\frac{1}{2}$ 12. $\frac{5}{8}$? $\frac{3}{4}$ 13. $\frac{7}{8}$? $\frac{3}{5}$ 14. $\frac{2}{3}$? $\frac{6}{9}$

CLASSROOM EXERCISES

Replace the __?__ with = or ≠ (is not equal to).   (Example 1)

1. $\frac{1}{25}$ __?__ 0.04 2. $\frac{4}{5}$ __?__ 0.6 3. $\frac{3}{20}$ __?__ 0.17 4. $\frac{7}{50}$ __?__ 0.14

5. $\frac{7}{5}$ __?__ 1.25 6. $\frac{9}{4}$ __?__ 2.25 7. $\frac{19}{10}$ __?__ 1.9 8. $\frac{27}{8}$ __?__ 3.75

State whether each of the following is a repeating or a terminating decimal. Write R for repeating and T for terminating.   (Definitions)

9. 0.32 10. 0.10 11. $0.\overline{2}$ 12. $0.5\overline{2}$ 13. 1.5 14. $3.8\overline{61}$

174 Chapter 6

Compare each pair of fractions. Write >, <, or =. (Example 3)

15. $\frac{1}{2}$ _?_ $\frac{3}{5}$ 16. $\frac{1}{4}$ _?_ $\frac{3}{10}$ 17. $\frac{1}{5}$ _?_ $\frac{1}{3}$ 18. $\frac{3}{8}$ _?_ $\frac{3}{4}$

WRITTEN EXERCISES

Goals: To write a decimal for a fraction

To apply the skill of writing a decimal for a fraction to comparing fractions

Write a decimal for each fraction. (Example 1)

1. $\frac{11}{20}$ 2. $\frac{7}{4}$ 3. $\frac{19}{5}$ 4. $\frac{9}{25}$ 5. $\frac{3}{20}$
6. $\frac{3}{50}$ 7. $\frac{5}{8}$ 8. $\frac{7}{8}$ 9. $\frac{13}{40}$ 10. $\frac{7}{16}$

Write a decimal for each fraction. Use a bar to show any digit or digits that repeat. (Example 2)

11. $\frac{1}{3}$ 12. $\frac{5}{9}$ 13. $\frac{7}{9}$ 14. $\frac{7}{15}$ 15. $\frac{11}{15}$
16. $\frac{17}{3}$ 17. $\frac{10}{9}$ 18. $\frac{9}{11}$ 19. $\frac{7}{12}$ 20. $\frac{10}{33}$
21. $\frac{5}{6}$ 22. $\frac{4}{15}$ 23. $\frac{7}{3}$ 24. $\frac{15}{11}$ 25. $\frac{1}{6}$

Compare each pair of fractions. Write >, <, or =. (Example 3)

26. $\frac{3}{5}$ _?_ $\frac{5}{8}$ 27. $\frac{7}{8}$ _?_ $\frac{3}{4}$ 28. $\frac{9}{20}$ _?_ $\frac{7}{10}$ 29. $\frac{3}{10}$ _?_ $\frac{2}{5}$ 30. $\frac{1}{5}$ _?_ $\frac{3}{10}$
31. $\frac{2}{3}$ _?_ $\frac{5}{8}$ 32. $\frac{5}{6}$ _?_ $\frac{5}{8}$ 33. $\frac{7}{8}$ _?_ $\frac{9}{10}$ 34. $\frac{18}{25}$ _?_ $\frac{7}{8}$ 35. $\frac{2}{3}$ _?_ $\frac{17}{20}$

APPLICATIONS: COMPARING DECIMALS

36. Pearl and Jane are reading the same book. Pearl has read $\frac{7}{8}$ of the book and Jane has read $\frac{7}{12}$ of the book. Who has read fewer pages, Pearl or Jane?

37. Earl and Miyako are painting a fence. Earl has completed $\frac{3}{8}$ of the job and Miyako has completed $\frac{1}{3}$ of the job. Who has completed the greater part?

38. Mario needs $1\frac{7}{10}$ yards of chain to hang a light. He ordered $1\frac{7}{8}$ yards. Did he order enough?

39. Phyllis and Flora are running in the marathon. Phyllis has completed $\frac{11}{12}$ of the distance and Flora has completed $\frac{13}{15}$ of the distance. Who is ahead?

Fractions: Addition/Subtraction

ASSIGNMENT GUIDE

BASIC
p. 175: 1-25, 27-35 odd

AVERAGE
p. 175: 1-39 odd

ABOVE AVERAGE
p. 175: 1-39 odd

WRITTEN EXERCISES
1. 0.55   2. 1.75
3. 3.8   4. 0.36
5. 0.15   6. 0.06
7. 0.625   8. 0.875
9. 0.325   10. 0.4375
11. $0.\overline{3}$ 12. $0.\overline{5}$
13. $0.\overline{7}$ 14. $0.4\overline{6}$
15. $0.7\overline{3}$ 16. $5.\overline{6}$
17. $1.\overline{1}$ 18. $0.\overline{81}$
19. $0.58\overline{3}$ 20. $0.\overline{30}$
21. $0.8\overline{3}$ 22. $0.2\overline{6}$
23. $2.\overline{3}$ 24. $1.\overline{36}$
25. $0.1\overline{6}$ 26. <
27. >   28. <   29. <
30. <   31. >   32. >
33. <   34. <   35. <
36. Jane   37. Earl
38. Yes   39. Phyllis

Teaching Suggestions
p. M-19

QUICK QUIZ
Multiply.
1. (4)(3.14)(10)
   Ans: 125.6
2. (2)(3.14)(5.5)
   Ans: 34.54
3. (3)(3.14)(6.5)
   Ans: 61.23
4. (2)(14)($\frac{22}{7}$)
   Ans: 88
5. (2)(175)($\frac{22}{7}$)
   Ans: 1100
6. ($\frac{22}{7}$)($3\frac{1}{2}$)($3\frac{1}{2}$)
   Ans: $38\frac{1}{2}$
7. ($\frac{22}{7}$)($4\frac{1}{5}$)($4\frac{1}{5}$)
   Ans: $55\frac{11}{25}$

Round to the nearest whole number.
8. 35.2   Ans: 35
9. 17.61 Ans: 18
10. $6\frac{2}{3}$ Ans: 7

REVIEW CAPSULE FOR SECTION 6-6

Multiply. (Pages 38–42)

1. $2 \times 3.14 \times 9.5$
2. $2 \times 3.14 \times 27.8$
3. $3.14 \times 4 \times 4$
4. $3.14 \times 10 \times 10$

Write a fraction for each mixed numeral. (Pages 135–137)

5. $3\frac{1}{7}$ 6. $1\frac{3}{4}$ 7. $5\frac{1}{4}$ 8. $10\frac{7}{8}$ 9. $3\frac{1}{10}$ 10. $2\frac{1}{8}$ 11. $11\frac{1}{2}$ 12. $9\frac{3}{5}$

Multiply. (Pages 135–137)

13. $2 \times 49 \times \frac{22}{7}$
14. $2 \times 280 \times \frac{22}{7}$
15. $\frac{22}{7} \times 8\frac{2}{3} \times 8\frac{2}{3}$
16. $\frac{22}{7} \times 5\frac{1}{4} \times 5\frac{1}{4}$

Round to the nearest whole number. (Pages 14–16)

17. 21.3   18. 73.5   19. 34.72   20. $\frac{3}{4}$ 21. $5\frac{1}{8}$ 22. $63\frac{2}{5}$

The answers are on page 183.

PROBLEM SOLVING AND APPLICATIONS

6-6 Circles: Circumference and Area

The diameter and radius of a circle such as a bicycle wheel are shown at the right. In any circle, the diameter is twice the radius. That is,

$d = 2 \cdot r$, or $d = 2r$

d = diameter
r = radius

The circumference, C, of a circle, is the distance around the circle.

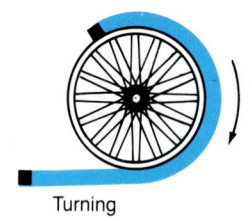

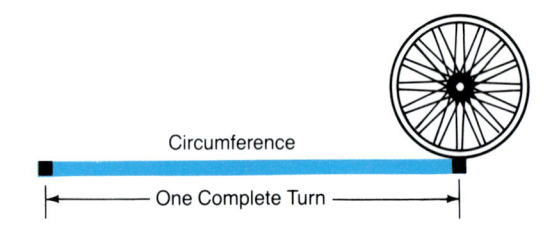

Start   Turning   Circumference — One Complete Turn

For all circles, the quotient $\frac{\text{circumference}}{\text{diameter}}$ is the same. It is the number π (Greek letter "pi"). The value of π is approximately 3.14 or $\frac{22}{7}$.

176 Chapter 6

PROCEDURE

To find the circumference, C, of a circle, find the product of the radius and 2π **or** the product of the diameter and π.

$$C = 2 \cdot \pi \cdot r \text{ or } C = \pi \cdot d$$

EXAMPLE 1 The radius of this bicycle wheel is 13 inches. Find the circumference. Round your answer to the nearest whole number.

13 in

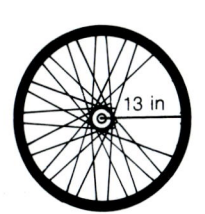

Solutions:

Method 1

$C = 2 \cdot \pi \cdot r$ ◄ $r = 13$ in

$C = 2 \cdot \frac{22}{7} \cdot 13$

$C = 81\frac{5}{7}$

Method 2

$C = \pi \cdot d$

$C = \frac{22}{7} \cdot 26$ ◄ $d = 2r$; $d = 2 \cdot 13$

$C = 81\frac{5}{7}$

The circumference is about **82 inches**.

Self-Check Find the circumference of each circle. Round each answer to the nearest whole number.

1. $r = 3$ in; $\pi = \frac{22}{7}$
2. $r = 10$ ft; $\pi = \frac{22}{7}$
3. $d = 5$ cm; $\pi = 3.14$
4. $d = 25$ m; $\pi = 3.14$

The formula for the area of a circle can be developed from the formula for the area of a parallelogram.

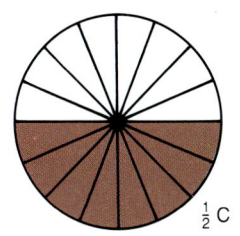

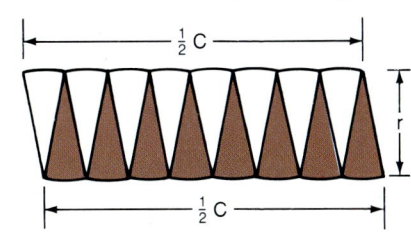

$\frac{1}{2}C$

$A = b \cdot h$ ◄ **Area of parallelogram**

$A = \frac{1}{2} \cdot C \cdot r$

$A = \frac{1}{2} \cdot 2 \cdot \pi \cdot r \cdot r$

$A = \pi \cdot r \cdot r$

PROCEDURE

To find the area of a circle, multiply π times the radius times the radius.

$$A = \pi \cdot r \cdot r \text{ or } A = \pi r^2$$ ◄ r^2 **means** $r \cdot r$.

ADDITIONAL EXAMPLE
Example 1
The diameter of a bicycle wheel is 27 inches. Find the circumference. Round your answer to the nearest whole number.
Ans: 85 inches

SELF-CHECK
1. 19 in   2. 63 ft
3. 16 cm   4. 79 m

Fractions: Addition/Subtraction **177**

ADDITIONAL EXAMPLE
Example 2
A round sign has a radius of 1.4 meters. Find the area. Use 3.14 for π. Round your answer to the nearest whole number.
Ans: 6 meters

SELF-CHECK
5. 314 cm²   6. 39 ft²

CLASSROOM EXERCISES
1. 88 in   2. 66 cm
3. 94 m   4. 707 cm²
5. 28 yd²   6. 98 m²

ASSIGNMENT GUIDE
BASIC
Day 1 pp. 178-179: 1-16
Day 2 pp. 178-179: 17-32
AVERAGE
pp. 178-179: 1-31 odd
ABOVE AVERAGE
pp. 178-179: 1-31 odd

WRITTEN EXERCISES
1. 188 cm   2. 207 cm
3. 20 cm   4. 80 cm
5. 4 m   6. 40,085 km
7. 76 cm
8. 4,370,880 km

EXAMPLE 2 The top of a round table has a radius of 0.5 meter. Find the area. Use 3.14 for π. Round your answer to the nearest whole number.

Solution: $A = \pi \cdot r \cdot r$
$A = 3.14 \cdot 0.5 \cdot 0.5$
$A = 0.785$ The area is about **1 m²**.

Self-Check *Find the area of each circle. Round answers to the nearest whole number.*

5. $r = 10$ cm; $\pi = 3.14$ **6.** $r = 3\frac{1}{2}$ ft; $\pi = \frac{22}{7}$

CLASSROOM EXERCISES

For Exercises 1–3, find the circumference of each circle. Round answers to the nearest whole number. (Example 1)

1. $r = 14$ in; $\pi = \frac{22}{7}$ **2.** $r = 10.5$ cm; $\pi = 3.14$ **3.** $d = 30$ m; $\pi = 3.14$

For Exercises 4–6, find the area of each circle. Round answers to the nearest whole number. (Example 2)

4. $r = 15$ cm; $\pi = 3.14$ **5.** $r = 3$ yd; $\pi = \frac{22}{7}$ **6.** $r = 5.6$ m; $\pi = 3.14$

WRITTEN EXERCISES

Goal: To find the circumference and area of a circle

METRIC MEASURES   *Find the circumference. Round answers to the nearest whole number. For Exercises 1–8, use 3.14 for π.*

1. Auto Tire   **2. Bicycle Wheel**   **3. Tennis Ball**   **4. Record**

$r = 30$ cm $r = 33$ cm $r = 3.2$ cm $r = 12.7$ cm

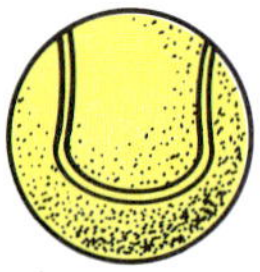

5. Tractor Tire   **6. Earth**   **7. Basketball**   **8. Sun**
$d = 1.3$ m $d = 12,766$ km $r = 12.1$ cm $r = 696,000$ km

178   Chapter 6

CUSTOMARY MEASURES Find the circumference. Round answers to nearest whole number. For Exercises 9–16 use $\frac{22}{7}$ for π.

9. Fountain Base 
$r = 7\frac{1}{2}$ ft

10. Bicycle Gear 
$r = 4\frac{2}{5}$ in

11. Clock 
$d = 13\frac{1}{2}$ in

12. Moon 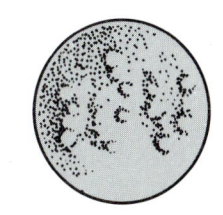
$d = 2250$ mi

13. Douglas Fir
$r = 2$ yd

14. Mars
$d = 11,152$ mi

15. Softball
$d = 12$ in

16. Reel of Tape
$r = 3\frac{1}{2}$ in

METRIC MEASURES Find the area. Round answers to the nearest whole number. For Exercises 17–24, use 3.14 for π. (Example 2)

17. Tray 
$r = 20$ cm

18. Mirror 
$r = 23$ cm

19. Rug 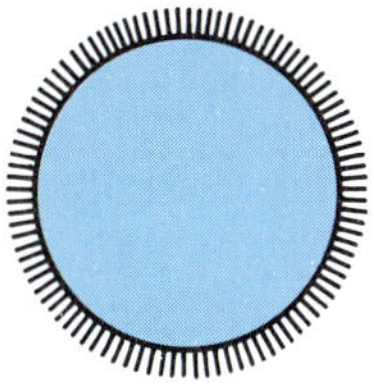
$r = 0.8$ m

20. Camera Lens 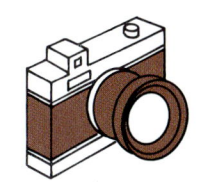
$r = 25$ mm

21. Plate
$r = 11.4$ cm

22. Pool Cover Top
$r = 3$ m

23. Garden
$r = 2.7$ m

24. Top of a Jar
$r = 3.5$ cm

CUSTOMARY MEASURES Use $\frac{22}{7}$ for π.

25. Top of a Drum 
$r = 8\frac{1}{2}$ in

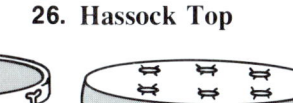

26. Hassock Top 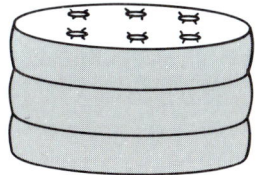
$r = 9$ in

27. Emblem 
$r = 2\frac{1}{4}$ in

28. Window 
$r = 8$ ft

29. Paper Doily
$r = 4$ in

30. Target
$r = 1\frac{1}{2}$ ft

31. Flower Center
$r = 2$ in

32. Clock Face
$r = 2\frac{3}{4}$ in

WRITTEN EXERCISES

9. 47 ft   10. 28 in
11. 42 in   12. 7071 mi
13. 13 yd
14. 35,049 mi
15. 38 in   16. 22 in
17. 1256 cm^2
18. 1661 cm^2 19. 2 m^2
20. 1963 mm^2
21. 408 cm^2 22. 28 m^2
23. 23 m^2 24. 38 cm^2
25. 227 in^2
26. 255 in^2
27. 16 in^2 28. 201 ft^2
29. 50 in^2 30. 7 ft^2
31. 13 in^2 32. 24 in^2

Fractions: Addition/Subtraction

QUIZ: SECTIONS 6-4–6-6

After completing this Review, you may want to administer a quiz covering the same sections. A Quiz is provided in the *Teacher's Edition: Part II.*

REVIEW: SECTIONS 6-4–6-6

1. a. $7\frac{1}{4}$; 7; $7\frac{1}{3}$; $7\frac{1}{4}$; $4\frac{2}{3}$;
   b. $33\frac{1}{2}$
2. $180.90   3. <   4. >
5. =   6. <
7. $37\frac{5}{7}$ in; $113\frac{1}{7}$ in^2
8. 26.38 m; 55.39 m^2
9. Tara's
10. $234\frac{1}{7}$ in^2

CALCULATOR EXERCISES
31. >   32. >   33. <
34. <   35. <

REVIEW: SECTIONS 6-4 — 6-6

For Exercises 1–2, use the information in the time card at the right to solve each problem. (Section 6-4)

1. a. Find the number of hours Mark Thomas worked each day.
   b. Find the total number of hours he worked for the week.
2. Mark Thomas is paid $5.40 per hour. Find his total weekly earnings.

Name:	Mark Thomas			
DAYS	IN	OUT	IN	OUT
1	7:00	11:00	12:00	3:15
2	7:00	11:15	12:15	3:00
3	7:00	11:20	12:20	3:20
4	6:45	11:00	12:00	3:00
5	7:00	11:00	12:00	2:40

Compare each pair of fractions. Write >, <, or =. (Section 6-5)

3. $\frac{1}{4}$? $\frac{2}{5}$
4. $\frac{1}{2}$? $\frac{3}{8}$
5. $\frac{4}{10}$? $\frac{2}{5}$
6. $\frac{8}{5}$? $\frac{5}{3}$

Find the circumference and area of each circle. Round answers to the nearest whole number. (Section 6-6)

7. $r = 6$ in; $\pi = \frac{22}{7}$
8. $r = 4.2$ m; $\pi = 3.14$

9. Joan caught a fish that weighed $1\frac{1}{3}$ pounds. Tara caught a fish that weighed $1\frac{2}{5}$ pounds. Whose fish weighed more, Joan's or Tara's? (Section 6-5)

10. Cathy cut four circle posters for the bulletin board at her school. She cut two circles 7 inches in diameter and two circles 10 inches in diameter. How many square inches of paper did she use?

COMPARING FRACTIONS

You can use a calculator to compare fractions.

EXAMPLE   Which is greater, $\frac{5}{6}$ or $\frac{7}{8}$?

Solution   Write a decimal for each fraction.

5 ÷ 6 = | 0.8333333 | 7 ÷ 8 = | 0.875 |

Since $0.875 > 0.8\overline{3}$, $\frac{7}{8} > \frac{5}{6}$.

EXERCISES   *Use a calculator to compare each pair of fractions in Exercises 31–35 on page 175.*

180   Chapter 6

CHAPTER REVIEW

PART 1: VOCABULARY

For Exercises 1–5, choose from the box at the right the word(s) that best correspond to each description.

1. The number 0.25 is a __?__ decimal.
2. The number $1.1\overline{6}$ is a __?__ decimal.
3. Fractions such as $\frac{3}{5}$ and $\frac{1}{5}$ have __?__ denominators.
4. Fractions that name the same number are __?__.
5. Fractions such as $\frac{2}{3}$ and $\frac{5}{8}$ have __?__ denominators.

> like
> equivalent fractions
> repeating
> terminating
> unlike

PART 2: SKILLS

Add or subtract as indicated. Write each answer in lowest terms. (Section 6-1)

6. $\frac{3}{5} + \frac{1}{5}$
7. $\frac{2}{7} + \frac{4}{7}$
8. $\frac{5}{8} + \frac{2}{8}$
9. $\frac{9}{16} - \frac{1}{16}$
10. $\frac{11}{12} - \frac{5}{12}$
11. $2\frac{1}{3} + 1\frac{1}{3}$
12. $6\frac{1}{4} + 2\frac{3}{4}$
13. $6\frac{7}{8} - 2\frac{3}{8}$
14. $12\frac{9}{10} - 3\frac{7}{10}$
15. $3\frac{3}{5} - 3\frac{2}{5}$

Complete. Write an equivalent fraction as indicated. (Section 6-2)

16. $\frac{3}{5} = \frac{?}{15}$
17. $\frac{4}{9} = \frac{?}{18}$
18. $\frac{1}{7} = \frac{?}{35}$
19. $\frac{7}{12} = \frac{?}{36}$
20. $\frac{2}{3} = \frac{?}{30}$

Add or subtract as indicated. Write each answer in lowest terms. (Section 6-2)

21. $\frac{3}{5} + \frac{1}{3}$
22. $\frac{5}{8} + \frac{1}{2}$
23. $\frac{7}{16} + \frac{11}{24}$
24. $\frac{4}{7} + \frac{3}{14}$
25. $\frac{2}{9} + \frac{1}{3}$
26. $\frac{7}{12} - \frac{1}{6}$
27. $\frac{11}{15} - \frac{3}{10}$
28. $\frac{7}{9} - \frac{1}{4}$
29. $\frac{6}{7} - \frac{2}{21}$
30. $\frac{17}{24} - \frac{3}{8}$

(Section 6-3)

31. $2\frac{1}{2} + 1\frac{1}{3}$
32. $3\frac{1}{6} + 5\frac{2}{5}$
33. $9\frac{2}{7} + 6\frac{2}{3}$
34. $10\frac{3}{14} + 4\frac{1}{2}$
35. $4\frac{5}{6} + 3\frac{1}{8}$
36. $7\frac{2}{3} - 3\frac{1}{8}$
37. $10 - 4\frac{5}{9}$
38. $8\frac{2}{5} - 5\frac{3}{4}$
39. $17\frac{1}{8} - 10\frac{5}{6}$
40. $4\frac{1}{2} - 2\frac{1}{6}$

For Exercises 41–42, complete the table. (Section 6-4)

	Hours Per Day				Total Hours	Rate Per Hour	Weekly Earnings
	1	2	3	4	5		
41.	$7\frac{1}{2}$	8	$7\frac{1}{3}$	$8\frac{1}{4}$	$7\frac{3}{4}$	?	$4.80	?
42.	$6\frac{1}{2}$	5	$8\frac{1}{3}$	$7\frac{1}{2}$	$6\frac{3}{4}$	?	$7.20	?

Fractions: Addition/Subtraction   **181**

CHAPTER REVIEW

1. terminating
2. repeating
3. like
4. equivalent fractions
5. unlike
6. $\frac{4}{5}$ 7. $\frac{6}{7}$ 8. $\frac{7}{8}$
9. $\frac{1}{2}$ 10. $\frac{1}{2}$ 11. $3\frac{2}{3}$
12. 9   13. $4\frac{1}{2}$ 14. $9\frac{1}{5}$
15. $\frac{1}{5}$ 16. $\frac{9}{15}$ 17. $\frac{8}{18}$
18. $\frac{5}{35}$ 19. $\frac{21}{36}$ 20. $\frac{20}{30}$
21. $\frac{14}{15}$ 22. $1\frac{1}{8}$ 23. $\frac{43}{48}$
24. $\frac{11}{14}$ 25. $\frac{5}{9}$ 26. $\frac{5}{12}$
27. $\frac{13}{30}$ 28. $\frac{19}{36}$ 29. $\frac{16}{21}$
30. $\frac{1}{3}$ 31. $3\frac{5}{6}$
32. $8\frac{17}{30}$ 33. $15\frac{20}{21}$
34. $14\frac{5}{7}$ 35. $7\frac{23}{24}$
36. $4\frac{13}{24}$ 37. $5\frac{4}{9}$
38. $2\frac{13}{20}$ 39. $6\frac{7}{24}$
40. $2\frac{1}{3}$
41. $38\frac{5}{6}$; $186.40
42. $34\frac{1}{12}$; $245.40
43. 0.125   44. $0.\overline{4}$
45. $0.\overline{54}$ 46. 1.4
47. 0.15   48. >

181

CHAPTER REVIEW

49. <   50. >   51. >
52. >   53. 24 m; 45 m²
54. 33 km; 88 km²
55. 76 ft; 453 ft²
56. 50 in; 201 in²
57. $\frac{1}{5}$ hr   58. $\frac{7}{8}$ lb
59. $1\frac{17}{20}$ mi   60. Hiram
61. 100 m   62. 1386 in²

Write a decimal for each fraction. (Section 6–5)

43. $\frac{1}{8}$ 44. $\frac{4}{9}$ 45. $\frac{6}{11}$ 46. $\frac{7}{5}$ 47. $\frac{3}{20}$

Compare each pair of fractions. Write >, <, or =. (Section 6–5)

48. $\frac{2}{5}$ __?__ $\frac{1}{4}$ 49. $\frac{5}{6}$ __?__ $\frac{7}{8}$ 50. $\frac{7}{12}$ __?__ $\frac{1}{2}$ 51. $\frac{9}{4}$ __?__ $\frac{11}{5}$ 52. $\frac{1}{3}$ __?__ $\frac{2}{9}$

Find the circumference and area of each circle. Round answers to the nearest whole number. (Section 6–6)

53. $r = 3.8$ m; $\pi = 3.14$ 54. $r = 5.3$ km; $\pi = 3.14$

55. $d = 24$ ft; $\pi = \frac{22}{7}$ 56. $d = 16$ in; $\pi = \frac{22}{7}$

PART 3: APPLICATIONS

57. It takes Diane $\frac{3}{5}$ of an hour to walk to school. It takes Nara $\frac{4}{5}$ of an hour to walk to school. How much longer does it take Nara to walk to school than Diane? (Section 6–1)

58. Martin bought $\frac{1}{2}$ pound of pecans and $\frac{3}{8}$ pound of walnuts. How many pounds of nuts did he buy in all? (Section 6–2)

59. Grant bicycled $8\frac{2}{5}$ miles on Saturday and $10\frac{1}{4}$ miles on Sunday. How much farther did he bicycle on Sunday than on Saturday? (Section 6–3)

60. Rafael worked $\frac{4}{5}$ of his math problems correctly. Hiram worked $\frac{7}{8}$ of his math problems correctly. Who worked more problems correctly, Rafael or Hiram? (Section 6–5)

61. The diameter of a round exercise track on a space station is 32 meters. To the nearest meter, how far must an astronaut walk to go around the track? (Section 6–6)

62. Jane painted the top of a round picnic table. The radius of the table is 21 inches. To the nearest square inch, what is the area she painted? (Section 6–6)

CHAPTER TEST

Complete. Write an equivalent fraction as indicated.

1. $\frac{4}{5} = \frac{?}{10}$
2. $\frac{2}{9} = \frac{?}{45}$
3. $\frac{5}{12} = \frac{?}{48}$
4. $\frac{3}{4} = \frac{?}{24}$

Add or subtract as indicated. Write each answer in lowest terms.

5. $\frac{5}{12} + \frac{1}{12}$
6. $\frac{3}{5} - \frac{1}{4}$
7. $\frac{8}{9} + \frac{1}{6}$
8. $2\frac{1}{2} - 1\frac{1}{3}$
9. $6 - 2\frac{3}{7}$
10. $3\frac{2}{5} + 2\frac{3}{8}$
11. $8\frac{3}{8} + 4\frac{5}{6}$
12. $12\frac{1}{8} - 4\frac{2}{3}$

Write a decimal for each fraction.

13. $\frac{3}{5}$
14. $\frac{2}{3}$
15. $\frac{5}{6}$
16. $\frac{7}{20}$

17. Find the circumference and area of a circle with a radius of 2.3 meters. Use $\frac{22}{7}$ for π. Round your answer to the nearest whole number.

18. California produces $\frac{1}{4}$ of the cotton crop for the United States. Texas produces $\frac{9}{40}$ of the cotton crop for the United States. Which state produces more cotton, California of Texas?

19. Tanya is paid $8.40 per hour. On Monday she worked from 9:00 A.M. to 11:30 A.M.; on Thursday she worked from 10:00 A.M. to 3:15 P.M.; on Friday she worked from 9:00 A.M. to 12:20 P.M. Find her total earnings.

20. The Johnson's bought 30 feet of fence to put around a round flower bed. The diameter of the flower bed is 8 feet. To the nearest foot, how many feet of fence did they have left?

ANSWERS TO REVIEW CAPSULES

Page 160 1. 12 2. 24 3. 12 4. 6 5. 60 6. $\frac{1}{2}$ 7. $\frac{2}{5}$ 8. $\frac{7}{9}$ 9. $\frac{2}{5}$ 10. $\frac{3}{7}$

Page 164 1. 33 2. 60 3. 32 4. 30 5. 36 6. $\frac{1}{2}$ 7. $\frac{1}{6}$ 8. $\frac{6}{13}$ 9. $\frac{1}{3}$ 10. $\frac{3}{8}$

Page 169 1. $\frac{27}{40}$ 2. $\frac{4}{9}$ 3. $9\frac{9}{10}$ 4. $4\frac{2}{9}$ 5. $7\frac{1}{2}$ 6. 24 7. 75 8. 345 9. $\frac{1}{4}$ 10. $\frac{45}{60}$; 45 11. $\frac{10}{60}$ 12. $\frac{20}{60}$; 20

Page 172 1. 0.85 2. 0.55 3. 0.65 4. 0.26 5. 0.75 6. < 7. < 8. < 9. > 10. > 11. < 12. > 13. < 14. <

Page 176 1. 59.66 2. 174.584 3. 50.24 4. 314 5. $\frac{22}{7}$ 6. $\frac{7}{4}$ 7. $\frac{21}{4}$ 8. $\frac{87}{8}$ 9. $\frac{31}{10}$ 10. $\frac{17}{8}$ 11. $\frac{23}{2}$ 12. $\frac{48}{5}$ 13. 308 14. 1760 15. $236\frac{4}{63}$ 16. $86\frac{5}{8}$ 17. 21 18. 74 19. 35 20. 1 21. 5 22. 63

Fractions: Addition/Subtraction

ENRICHMENT

You may wish to use this lesson for students who performed well on the formal Chapter Test.

1. 0.1 cm; ± 0.05 cm; 4.35 cm; 4.25 cm
2. 0.01 cm; ± 0.005 cm; 61.715 cm; 61.705 cm
3. $\frac{1}{2}$ in; ± $\frac{1}{4}$ in; $2\frac{3}{4}$ in; $2\frac{1}{4}$ in
4. $\frac{1}{8}$ in; ± $\frac{1}{16}$ in; $5\frac{7}{16}$ in; $5\frac{5}{16}$ in

ENRICHMENT

Precision of Measurement

In measurement, the word **precision** refers to the *smallest measurement* that can be made on the instrument used to measure. The last digit written in a measurement may be in error by as much as one-half the unit of precision. This is the **maximum possible error**.

EXAMPLE Find the maximum possible error for a measurement of 1.6 cm.

Solution: Unit of precision: 0.1 cm ◀ The last digit of 1.6 cm is in the tenths place.

Maximum possible error: $\frac{1}{2} \times 0.1 = 0.05$

Since all measurements are approximations, a maximum possible error of 0.05 means that a measurement can vary by ±0.05. That is, a measurement of 4.8 centimeters can be as long as (4.8 + 0.05), or 4.85 centimeters and as short as (4.8 − 0.05), or 4.75 centimeters.

This table shows other examples.

Measurement	Precision	Maximum Possible Error	Greatest Length	Shortest Length
4.7 m	0.1 m	±0.05 m	4.75 m	4.65 m
3.41 cm	0.01 cm	±0.005 cm	3.415 cm	3.405 cm
$4\frac{3}{8}$ in	$\frac{1}{8}$ in	±$\frac{1}{16}$ in	$4\frac{7}{16}$ in	$4\frac{5}{16}$ in

EXERCISES

Complete.

	Measurement	Precision	Maximum Possible Error	Greatest Length	Shortest Length
1.	4.3 cm	?	?	?	?
2.	61.71 cm	?	?	?	?
3.	$2\frac{1}{2}$ in	?	?	?	?
4.	$5\frac{3}{8}$ in	?	?	?	?

184 Chapter 6

ADDITIONAL PRACTICE

SKILLS

Add or subtract. Write each answer in lowest terms.
(Pages 158–160)

1. $\frac{3}{10} + \frac{3}{10}$
2. $4\frac{5}{7} + 3\frac{2}{7}$
3. $\frac{11}{12} - \frac{5}{12}$
4. $27\frac{3}{4} - 18\frac{1}{4}$

(Pages 161–164)

5. $\frac{1}{2} + \frac{1}{4}$
6. $\frac{1}{3} + \frac{3}{5}$
7. $\frac{2}{7} + \frac{1}{5}$
8. $\frac{5}{8} + \frac{9}{10}$
9. $\frac{2}{3} - \frac{1}{6}$
10. $\frac{4}{5} - \frac{2}{9}$
11. $\frac{11}{14} - \frac{1}{7}$
12. $\frac{3}{4} - \frac{7}{10}$

(Pages 165–167)

13. $3\frac{1}{6} + 5\frac{1}{3}$
14. $4\frac{1}{3} + 2\frac{1}{2}$
15. $5\frac{7}{8} + 3\frac{3}{4}$
16. $17\frac{7}{8} + 4\frac{9}{10}$
17. $5\frac{4}{9} - 2\frac{1}{6}$
18. $17\frac{1}{8} - 9\frac{3}{4}$
19. $14\frac{1}{5} - 8\frac{1}{3}$
20. $16\frac{7}{10} - 8\frac{3}{4}$

Write a decimal for each fraction. Use a bar to indicate any digit or digits that repeat. (Pages 173–175)

21. $\frac{7}{10}$
22. $\frac{5}{9}$
23. $\frac{4}{15}$
24. $\frac{1}{16}$
25. $\frac{5}{12}$
26. $\frac{5}{3}$

Compare each pair of fractions. Write >, <, or =. Give a reason.
(Pages 173–175)

27. $\frac{4}{5}$? $\frac{3}{4}$
28. $\frac{7}{10}$? $\frac{4}{5}$
29. $\frac{2}{5}$? $\frac{3}{8}$
30. $\frac{2}{3}$? $\frac{4}{6}$
31. $\frac{3}{8}$? $\frac{1}{3}$

Find the circumference and area of each circle. Use 3.14 for π. Round answers to the nearest whole number. (Pages 176–179)

32. $r = 40$ cm
33. $d = 25$ m
34. $r = 11$ km
35. $d = 30.4$ cm

Use $\frac{22}{7}$ for π. (Pages 176–179)

36. $r = 9\frac{1}{2}$ ft
37. $d = 4$ mi
38. $r = 3\frac{3}{4}$ in
39. $d = 18$ yds

APPLICATIONS

40. David Yaw works from 1:00 A.M. to 3:15 P.M. on Monday, Wednesday, and Friday and from 8:30 A.M. to 1:00 P.M. on Tuesday and Thursday. He is paid $4.80 per hour. Find his weekly earnings. (Pages 169–172)

41. About $\frac{1}{10}$ of the households in America have a computer. About $\frac{9}{100}$ of American households own a cordless telephone. Are there more computers or cordless telephones in American households?

Fractions: Addition/Subtraction

ADDITIONAL PRACTICE

You may wish to use all or some of these exercises, depending on how well students performed on the formal Chapter Test.

1. $\frac{3}{5}$
2. 8
3. $\frac{1}{2}$
4. $9\frac{1}{2}$
5. $\frac{3}{4}$
6. $\frac{14}{15}$
7. $\frac{17}{35}$
8. $1\frac{21}{40}$
9. $\frac{1}{2}$
10. $\frac{26}{45}$
11. $\frac{9}{14}$
12. $\frac{1}{20}$
13. $8\frac{1}{2}$
14. $6\frac{5}{6}$
15. $9\frac{5}{8}$
16. $22\frac{31}{40}$
17. $3\frac{5}{18}$
18. $7\frac{3}{8}$
19. $5\frac{13}{15}$
20. $7\frac{19}{20}$
21. 0.7
22. $0.\overline{5}$
23. $0.2\overline{6}$
24. 0.0625
25. $0.41\overline{6}$
26. $1.\overline{6}$
27. >; 0.8 > 0.75
28. <; 0.7 < 0.8
29. >; 0.4 > 0.375
30. =; $0.\overline{6} = 0.\overline{6}$
31. >; $0.375 > 0.\overline{3}$
32. 251 cm; 5024 cm²
33. 79 m; 491 m²
34. 69 km; 380 km²
35. 95 cm; 725 cm²
36. 60 ft; 284 ft²
37. 13 mi; 13 mi²
38. 24 in; 44 in²
39. 57 yd; 255 yd²
40. $75.60
41. More computers

COMMON ERRORS

In preparation for the Cumulative Review, these exercises focus the student's attention on the most common errors to be avoided.

1. 1, 2, 3, 6, 9, 18
2. $2^3 \cdot 3$ 3. 24
4. 2   5. $\frac{12}{25}$ 6. $1\frac{1}{6}$
7. $2\frac{4}{25}$ 8. $9\frac{4}{5}$ 9. $1\frac{5}{12}$
10. $\frac{1}{40}$ 11. $9\frac{1}{3}$
12. $6\frac{13}{20}$ 13. $2\frac{1}{4}$ yd^2
14. 200.96 cm^2

COMMON ERRORS

Each of these problems contains a common error.
  a. Find the correct answer.
  b. Find the error.

1. List all the factors of 18.
   Factors of 18: **2, 3, 6, 9**

2. Write the prime factorization of 24.
   $24 = 1 \cdot 2 \cdot 2 \cdot 2 \cdot 3 = \mathbf{1 \cdot 2^3 \cdot 3}$

3. Find the LCM of 6 and 8.
   $6 = 2 \cdot 3$
   $8 = 2 \cdot 2 \cdot 2$
   LCM of 6 and 8: $2 \cdot 3 = 6$

4. Find the GCF of 4 and 10.
   $4 = ② \cdot 2$
   $10 = ② \cdot 5$
   GCF of 4 and 10: $2 \cdot 2 = 4$

5. $\frac{3}{5} \cdot \frac{4}{5} = $?
   $\frac{3}{5} \cdot \frac{4}{5} = \frac{12}{5}$, or $\mathbf{2\frac{2}{5}}$

6. $\frac{2}{3} \div \frac{4}{7} = $?
   $\frac{2}{3} \div \frac{4}{7} = \frac{2}{3} \cdot \frac{4}{7} = \frac{8}{21}$

7. $3\frac{3}{5} \div 1\frac{2}{3} = $?
   $3\frac{3}{5} \div 1\frac{2}{3} = 3\frac{3}{5} \cdot 1\frac{3}{2} = \mathbf{3\frac{9}{10}}$

8. $2\frac{1}{3} \cdot 4\frac{1}{5} = $?
   $2\frac{1}{3} \cdot 4\frac{1}{5} = \mathbf{8\frac{2}{15}}$

9. $\frac{2}{3} + \frac{3}{4} = $?
   $\frac{2}{3} + \frac{3}{4} = \frac{2}{12} + \frac{3}{12} = \frac{5}{12}$

10. $\frac{5}{8} - \frac{3}{5} = $?
    $\frac{5}{8} - \frac{3}{5} = \frac{2}{3}$

11. $4\frac{1}{6} + 5\frac{1}{6} = $?
    $4\frac{1}{6} + 5\frac{1}{6} = 9\frac{2}{12}$
    $= \mathbf{9\frac{1}{6}}$

12. $10\frac{1}{4} - 3\frac{3}{5} = $?
    $10\frac{1}{4} - 3\frac{3}{5} = 10\frac{5}{20} - 3\frac{12}{20}$
    $= 9\frac{15}{20} - 3\frac{12}{20} = \mathbf{6\frac{3}{20}}$

13. Find the area of this triangle.
    $b = 3$ yd; $h = 1\frac{1}{2}$ yd
    $A = bh$
    $A = 3\left(\frac{3}{2}\right)$
    $A = \frac{9}{2}$, or $\mathbf{4\frac{1}{2}}$ yd^2

14. Find the area of this circle.
    $r = 8$ cm; $\pi = 3.14$
    $A = \pi r^2$
    $A = 3.14(8)(2)$
    $A = \mathbf{50.24}$ cm^2

186 Chapter 6

CUMULATIVE REVIEW: CHAPTERS 4–6

Choose the correct answer. Choose a, b, c, or d.

1. Which number is <u>not</u> a factor of 27?
   a. 1 b. 6 c. 3 d. 9

2. Which number is <u>not</u> divisible by 9?
   a. 315 b. 756 c. 664 d. 558

3. Which number is <u>not</u> a multiple of 10?
   a. 770 b. 660 c. 1300 d. 995

4. Find the smallest prime factor of 1925.
   a. 3 b. 2 c. 5 d. 7

5. Rewrite this product using exponents: $2 \cdot 2 \cdot 3 \cdot 2 \cdot 3 \cdot 5 \cdot 5 \cdot 2 \cdot 3$
   a. $2^3 \cdot 3^3 \cdot 5^2$ b. $2^4 \cdot 3^3 \cdot 5^2$ c. $2^4 \cdot 3^2 \cdot 5^3$ d. $2^3 \cdot 3^2 \cdot 5^2$

6. Find the prime factorization of 540.
   a. $2^2 \cdot 3^2 \cdot 5$ b. $2^2 \cdot 3^3 \cdot 5$ c. $2^3 \cdot 3^3 \cdot 5$ d. $2^2 \cdot 3^2 \cdot 5^2$

7. Find the least common multiple of 15, 30, and 50.
   a. 150 b. 60 c. 300 d. 90

8. Find the greatest common factor of 12 and 20.
   a. 4 b. 2 c. 24 d. 60

9. Write this fraction in lowest terms: $\frac{18}{54}$
   a. $\frac{9}{27}$ b. $\frac{6}{18}$ c. $\frac{1}{3}$ d. $\frac{1}{6}$

10. A basketball player made 25 out of 55 attempted shots. What fractional part of the attempted shots did the player make?
    a. $\frac{2}{5}$ b. $\frac{5}{11}$ c. $\frac{3}{5}$ d. $\frac{3}{10}$

11. Subtract. Express the answer as a mixed numeral. 17 lb 4 oz
    $-$12 lb 12 oz

    a. $5\frac{1}{2}$ lb b. $5\frac{1}{8}$ lb c. $4\frac{1}{2}$ lb d. $4\frac{1}{8}$ lb

12. Multiply: $\frac{4}{15} \cdot \frac{3}{4}$
    a. $\frac{2}{5}$ b. $\frac{4}{5}$ c. $\frac{3}{5}$ d. $\frac{1}{5}$

13. On the planet Mercury, the temperature can climb to 100°C. Use the formula $F = \frac{9}{5}C + 32$ to write this temperature in degrees Fahrenheit.
    a. 124° b. 224° c. 112° d. 212°

CUMULATIVE REVIEW

A cumulative test is provided on copying masters in the *Teacher's Edition: Part II.*

1. b
2. c
3. d
4. c
5. b
6. b
7. a
8. a
9. c
10. b
11. c
12. d
13. d

Cumulative Review: Chapters 4–6 **187**

CUMULATIVE REVIEW

14. a
15. a
16. b
17. a
18. a
19. b
20. b
21. c
22. a
23. d
24. b

14. Multiply: $1\frac{1}{3} \cdot 1\frac{7}{8}$
    a. $2\frac{1}{2}$
    b. $1\frac{7}{24}$
    c. $1\frac{1}{4}$
    d. $2\frac{1}{4}$

15. Sam earns $3.80 per hour. How much will Sam earn in 37 hours?
    a. $140.60
    b. $142.50
    c. $145.00
    d. $172.50

16. Find the area of the gable of a roof that has a base of 6.8 meters and a height of 4.4 meters.
    a. 74.4 m²
    b. 14.96 m²
    c. 7.44 m²
    d. 149.6 m²

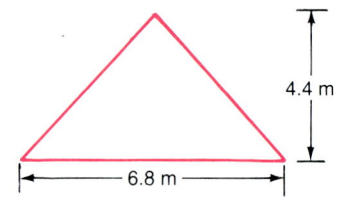

17. Find the volume of a box of detergent 22 centimeters long, 10 centimeters wide, and 30 centimeters high.
    a. 6600 cm³
    b. 62 cm³
    c. 6660 cm³
    d. 124 cm³

18. Divide: $1\frac{1}{3} \div 2\frac{2}{3}$
    a. $\frac{1}{2}$
    b. $\frac{24}{9}$
    c. 2
    d. $\frac{1}{3}$

19. Maria and Andrea ride their bicycles 20 kilometers in $1\frac{1}{4}$ hours. How many kilometers could they ride in one hour?
    a. 25
    b. 16
    c. 18
    d. 20

20. Subtract: $\frac{7}{15} - \frac{3}{10}$
    a. $\frac{4}{5}$
    b. $\frac{1}{6}$
    c. $1\frac{1}{5}$
    d. $\frac{11}{30}$

21. Subtract: $4\frac{1}{6} - 2\frac{1}{2}$
    a. $2\frac{1}{4}$
    b. $2\frac{1}{3}$
    c. $1\frac{2}{3}$
    d. $1\frac{1}{2}$

22. Yoshiro worked $5\frac{1}{2}$ hours on Monday, $2\frac{3}{4}$ hours on Tuesday, and $7\frac{1}{4}$ hours on Wednesday. At $3.40 per hour, how much did Yoshiro earn for the three days?
    a. $52.70
    b. $51.40
    c. $53.80
    d. $55.20

23. Write a decimal for $\frac{7}{25}$.
    a. 0.24
    b. 0.26
    c. 0.25
    d. 0.28

24. Find the circumference of a wheel that has a diameter of 40 centimeters. Use 3.14 for π.
    a. 47.10 cm
    b. 125.60 cm
    c. 1412.50 cm
    d. 5024 cm

Cumulative Review: Chapters 4–6

CHAPTER

7 Solving Equations

SECTIONS
7-1 Solving Equations by Subtraction
7-2 Solving Equations by Addition
7-3 Problem Solving and Applications: Using Equations: Addition/Subtraction
7-4 Solving Equations by Division
7-5 Solving Equations by Multiplication
7-6 Problem Solving and Applications: Using Equations: Multiplication/Division

FEATURES
Consumer Application: *Wallpaper and Estimation*
Enrichment: *Repeating Decimals*

Teaching Suggestions
p. M-20

QUICK QUIZ

Evaluate.

1. 17 + 9 - 9   Ans: 17
2. 3.17 - 1.6   Ans: 1.57
3. $13\frac{5}{6} - 2\frac{1}{3}$ Ans. $11\frac{1}{2}$
4. $8 - 5\frac{3}{4}$ Ans: $2\frac{1}{4}$
5. x + 7 - 7   Ans: x

ADDITIONAL EXAMPLES

Example 1

Solve and check.

1. x + 27 = 83
    Ans: x = 56
2. 17 + y = 21
    Ans: y = 4

SELF-CHECK

1. 36   2. 3   3. 0

7-1 Solving Equations by Subtraction

This table shows that subtraction "undoes" addition.

① Choose a number.	② Add any number to it.	③ Subtract the same number.	④ Result: The original number
12	12 + 8	12 + 8 − 8	12
3.2	3.2 + 8.1	3.2 + 8.1 − 8.1	3.2
$5\frac{1}{3}$	$5\frac{1}{3} + \frac{1}{2}$	$5\frac{1}{3} + \frac{1}{2} - \frac{1}{2}$	$5\frac{1}{3}$
n	$n + 4$	$n + 4 - 4$	n

Thus, subtraction is the **opposite**, or **inverse operation**, of addition. This idea can be applied to solving equations. An **equation** is a mathematical sentence that contains the equality symbol "=".

To **solve an equation** means to find the number(s) that can replace the variable to make the equation true. To solve an addition equation such as

$$x + 9 = 21,$$

subtract 9 from each side of the equation.

The goal in solving an equation is <u>to get the variable alone on one side of the equation.</u>

PROCEDURE

To solve the equation $x + b = c$ for x, subtract b from each side of the equation.

EXAMPLE 1 Solve and check: $x + 9 = 21$

Solution: To get x alone, subtract 9 from each side.

$x + 9 = 21$ **Check:** $x + 9 = 21$
$x + 9 - 9 = 21 - 9$ $12 + 9$
$x = 12$ 21

Replace x with 12.

Self-Check *Solve and check.*

1. $x + 7 = 43$ 2. $t + 9 = 12$ 3. $a + 13 = 13$

190 Chapter 7

As Example 1 shows, you check an equation by replacing the variable with the value you obtained. If this value makes the equation true, it is a **solution** of the equation.

EXAMPLE 2 Solve and check.

　　a. $12.6 = t + 1.3$　　　　b. $g + 1\frac{1}{4} = 2\frac{5}{8}$

Solutions:

a. $12.6 = t + 1.3$　　◀ Subtract 1.3 from both sides.

$12.6 - 1.3 = t + 1.3 - 1.3$

$11.3 = t$

b. $g + 1\frac{1}{4} = 2\frac{5}{8}$　　◀ Subtract $1\frac{1}{4}$ from both sides.

$g + 1\frac{1}{4} - 1\frac{1}{4} = 2\frac{5}{8} - 1\frac{1}{4}$

$g = 2\frac{5}{8} - 1\frac{2}{8}$

$g = \mathbf{1\frac{3}{8}}$

The checks are left for you.

Self-Check *Solve and check.*

4. $19.6 = n + 17.8$　　5. $y + \frac{1}{4} = \frac{5}{8}$　　6. $p + 1\frac{1}{2} = 2\frac{3}{4}$

You can also use the following method to solve equations.

$$\begin{array}{r} x + 9 = 21 \\ -9 = -9 \\ \hline x = 12 \end{array} \qquad \begin{array}{r} 12.6 = t + 1.3 \\ -1.3 = -1.3 \\ \hline 11.3 = t \end{array}$$

To be sure that your answer is a solution of the equation, always use the original equation to check.

CLASSROOM EXERCISES

Evaluate each expression. (Table)

1. $8 + 9 - 9$　　2. $12 + 3 - 3$　　3. $n + 7 - 7$　　4. $m + 5 - 5$
5. $x + \frac{1}{2} - \frac{1}{2}$　　6. $a + \frac{3}{4} - \frac{3}{4}$　　7. $n + 112 - 112$　　8. $c + \frac{2}{3} - \frac{2}{3}$

Write the equation formed as the first step in solving each equation. (Step 1, Examples 1–2)

9. $x + 5 = 19$　　10. $a + 12 = 14$　　11. $n + 5.7 = 10.3$　　12. $14.6 = p + 9.7$
13. $b + \frac{1}{4} = \frac{1}{2}$　　14. $y + 1\frac{3}{4} = 3\frac{7}{8}$　　15. $6\frac{7}{8} = y + 2\frac{3}{4}$　　16. $x + 12.5 = 21.3$
17. $3.2 + r = 9$　　18. $27 = 5 + a$　　19. $6 + c = 9$　　20. $5.6 + z = 8.2$

Solving Equations　**191**

ADDITIONAL EXAMPLES
Example 2
Solve and check.
1. $3.7 + x = 9.2$
　Ans: 5.5
2. $2\frac{1}{3} = x + \frac{1}{6}$
　Ans: $2\frac{1}{6}$
3. $\frac{1}{10} + y = 3\frac{2}{5}$
　Ans: $3\frac{3}{10}$

SELF-CHECK
4. 1.8　　5. $\frac{3}{8}$　　6. $1\frac{1}{4}$

CLASSROOM EXERCISES
1. 8　　2. 12　　3. n
4. m　　5. x　　6. a
7. n　　8. c
9. $x + 5 - 5 = 19 - 5$
10. $a + 12 - 12 =$
　　$14 - 12$
11. $n + 5.7 - 5.7 =$
　　$10.3 - 5.7$
12. $14.6 - 9.7 =$
　　$p + 9.7 - 9.7$
13. $b + \frac{1}{4} - \frac{1}{4} = \frac{1}{2} - \frac{1}{4}$
14. $y + 1\frac{3}{4} - 1\frac{3}{4} =$
　　$3\frac{7}{8} - 1\frac{3}{4}$
15. $6\frac{7}{8} - 2\frac{3}{4} =$
　　$y + 2\frac{3}{4} - 2\frac{3}{4}$
16. $x + 12.5 - 12.5 =$
　　$21.3 - 12.5$
See page 192 for the answers to Exercises 17–20.

CLASSROOM EXERCISES
p. 191.
17. $3.2 - 3.2 + r = 9 - 3.2$
18. $27 - 5 = 5 - 5 + a$
19. $6 - 6 + c = 9 - 6$
20. $5.6 - 5.6 + z = 8.2 - 5.6$

ASSIGNMENT GUIDE
BASIC
Day 1 p. 192: 1-22
Day 2 p. 192: 23-45
AVERAGE
Day 1 p. 192: 1-35 odd
Day 2 p. 192: 37-51
ABOVE AVERAGE
p. 192: 1-51 odd

WRITTEN EXERCISES
1. 7   2. 17   3. 17
4. 19   5. 24   6. 17
7. 17   8. 26   9. 27
10. 30 11. 49   12. 54
13. 1.6 14. 4.2
15. 3.5 16. 2.4
17. 2.9 18. 2.3
19. 11.5 20. 0.47
21. 0.38 22. 0.5
23. 21.74 24. 13.88

REVIEW CAPSULE
This Review Capsule reviews prior-taught skills used in Section 7-2. The reference is to the pages where the skills were taught.

WRITTEN EXERCISES

Goal: To use subtraction to solve equations

For Exercises 1–51, solve and check each equation. Show all steps. (Example 1)

1. $n + 14 = 21$
2. $x + 18 = 35$
3. $46 = y + 29$
4. $42 = r + 23$
5. $t + 19 = 43$
6. $s + 34 = 51$
7. $a + 39 = 56$
8. $b + 45 = 71$
9. $64 = q + 37$
10. $120 = t + 90$
11. $h + 48 = 97$
12. $n + 17 = 71$

(Example 2)

13. $k + 5.8 = 7.4$
14. $x + 3.4 = 7.6$
15. $5.8 = y + 2.3$
16. $6.9 = n + 4.5$
17. $t + 6.8 = 9.7$
18. $b + 12.8 = 15.1$
19. $r + 23.7 = 35.2$
20. $s + 0.43 = 0.9$
21. $0.85 = y + 0.47$
22. $0.74 = x + 0.24$
23. $m + 1.05 = 22.79$
24. $p + 4.31 = 18.19$
25. $r + \frac{3}{8} = 1\frac{5}{8}$ $1\frac{1}{4}$
26. $t + \frac{7}{12} = 3\frac{11}{12}$ $3\frac{1}{3}$
27. $5\frac{3}{4} = x + 3\frac{1}{2}$ $2\frac{1}{4}$
28. $9\frac{1}{2} = n + 5\frac{1}{4}$ $4\frac{1}{4}$
29. $c + 3\frac{3}{4} = 7\frac{1}{4}$ $3\frac{1}{2}$
30. $a + 10\frac{1}{2} = 15\frac{1}{4}$ $4\frac{3}{4}$
31. $c + \frac{1}{2} = \frac{5}{6}$ $\frac{1}{3}$
32. $\frac{3}{5} + p = \frac{9}{10}$ $\frac{3}{10}$
33. $8\frac{1}{3} = 4\frac{7}{9} + m$ $3\frac{5}{9}$
34. $9\frac{3}{4} = y + 5\frac{1}{2}$ $4\frac{1}{4}$
35. $15\frac{1}{5} + q = 17\frac{1}{3}$ $2\frac{2}{15}$
36. $9\frac{1}{2} + n = 17\frac{3}{8}$ $7\frac{7}{8}$

MIXED PRACTICE

37. $5.19 = a + 2.65$ 2.54
38. $t + 3\frac{1}{8} = 5\frac{3}{16}$ $2\frac{1}{16}$
39. $x + 4\frac{3}{4} = 8\frac{7}{12}$ $3\frac{5}{6}$
40. $7.03 = g + 4.28$ 2.75
41. $y + 190 = 403$ 213
42. $s + 8417 = 9378$ 961
43. $x + 0.39 = 1.46$ 1.07
44. $4204 = k + 3787$ 417
45. $205 = m + 198$ 7
46. $v + 5\frac{2}{3} = 12\frac{1}{5}$ $6\frac{8}{15}$
47. $4\frac{3}{16} = r + 2\frac{5}{16}$ $1\frac{7}{8}$
48. $3\frac{1}{8} = s + 1\frac{7}{8}$ $1\frac{1}{4}$
49. $y + 0.57 = 2.15$ 1.58
50. $145 = b + 145$ 0
51. $w + 5.23 = 5.23$ 0

REVIEW CAPSULE FOR SECTION 7-2

Complete. (Pages 5-6; 35-37; 165-167)

1. $7 = 6 - 6 + \underline{}$
2. $1.5 - 1.5 + \underline{} = 6$
3. $9\frac{1}{2} = 1\frac{1}{4} - 1\frac{1}{4} + \underline{}$
4. $58.6 - 58.6 + \underline{} = 0.9$
5. $q = 18 - 18 + \underline{}$
6. $\frac{3}{8} - \frac{3}{8} + c = \underline{}$
7. $1.9 - 1.9 + \underline{} = f$
8. $p + 1800 - p = \underline{}$

The answers are on page 212.

192 Chapter 7

7-2 Solving Equations by Addition

This table shows that addition "undoes" subtraction.

1 Choose a number.	2 Subtract any number.	3 Add the same number.	4 Result: the original number
18	$18 - 7$	$18 - 7 + 7$	18
23.5	$23.5 - 11.8$	$23.5 - 11.8 + 11.8$	23.5
$12\frac{1}{2}$	$12\frac{1}{2} - 3\frac{3}{4}$	$12\frac{1}{2} - 3\frac{3}{4} + 3\frac{3}{4}$	$12\frac{1}{2}$
n	$n - 12$	$n - 12 + 12$	n

Thus, subtraction is the inverse operation of addition. To solve a subtraction equation such as

$$r - 9 = 15$$

add 9 **to each side** of the equation.

PROCEDURE

To solve the equation $x - b = c$ for x, add b **to each side** of the equation.

EXAMPLE 1 Solve and check: $r - 9 = 15$

Solution: To get r alone, add 9 to each side.

$r - 9 = 15$
$r - 9 + 9 = 15 + 9$
$r = 24$

Check: $r - 9 = 15$
$24 - 9$
15

Self-Check *Solve and check.*

1. $x - 5 = 18$ 2. $y - 7 = 11$ 3. $t - 4 = 1$

NOTE: In solving an equation, you write equivalent equations. **Equivalent equations** have the same solution(s). Thus, in the solution of Example 1, these equations are equivalent.

$r - 9 = 15$
$r - 9 + 9 = 15 + 9$ ◀ **Equivalent equations.**
$r = 24$

The three equations have the same solution, **24**.

Solving Equations   **193**

Teaching Suggestions
p. M-20

QUICK QUIZ
Evaluate.
1. 127 - 85 + 85
   Ans: 127
2. 5.7 + 0.63
   Ans: 6.33
3. 18.1 - 9.6 + 9.6
   Ans: 18.1
4. $10\frac{3}{4} + 8\frac{1}{2}$ Ans: $19\frac{1}{4}$

ADDITIONAL EXAMPLES
Example 1
Solve and check.
1. x - 38 = 109
   Ans: 147
2. 53 = a - 81
   Ans: 134

SELF-CHECK
1. 23 2. 18 3. 5

193

ADDITIONAL EXAMPLES

Example 2

Solve and check.

1. $15 = x - 1.8$
   Ans: 16.8
2. $y - \frac{1}{2} = 7\frac{3}{10}$
   Ans: $7\frac{4}{5}$

SELF-CHECK

4. 5.9 5. 1 6. $4\frac{1}{2}$

CLASSROOM EXERCISES

1. a   2. t   3. r
4. s   5. x   6. n
7. a   8. p   9. $3\frac{1}{4}$
10. $x - 7 + 7 = 19 + 7$
11. $a - 12 + 12 =$
    $5 + 12$
12. $23 + 27 = s -$
    $27 + 27$
13. $47 + 83 = t -$
    $83 + 83$
14. $q - 2.7 + 2.7 =$
    $3.9 + 2.7$
15. $12.9 + 5.6 =$
    $b - 5.6 + 5.6$
16. $w - \frac{3}{4} + \frac{3}{4} =$
    $2\frac{1}{2} + \frac{3}{4}$
17. $4\frac{1}{4} + 1\frac{7}{8} =$
    $n - 1\frac{7}{8} + 1\frac{7}{8}$
18. $q - \frac{3}{4} + \frac{3}{4} =$
    $6\frac{1}{2} + \frac{3}{4}$

EXAMPLE 2 Solve and check.

a. $19.7 = t - 8.2$ b. $\frac{1}{2} = r - \frac{3}{4}$

Solutions:

a. $19.7 = t - 8.2$ *Add 8.2 to each side*

$19.7 + 8.2 = t - 8.2 + 8.2$

$27.9 = t$

b. $\frac{1}{2} = r - \frac{3}{4}$ *Add $\frac{3}{4}$ to each side.*

$\frac{1}{2} + \frac{3}{4} = r - \frac{3}{4} + \frac{3}{4}$

$\frac{2}{4} + \frac{3}{4} = r$

$\frac{5}{4} = r$

The checks are left for you.

Self-Check Solve and check.

4. $x - 1.1 = 4.8$ 5. $\frac{5}{6} = r - \frac{1}{6}$ 6. $y - 1\frac{1}{8} = 3\frac{3}{8}$

CLASSROOM EXERCISES

Evaluate each expression. (Table)

1. $a - 5 + 5$
2. $t - 10 + 10$
3. $r - 12 + 12$
4. $s - 37 + 37$
5. $x - 2.8 + 2.8$
6. $n - 0.8 + 0.8$
7. $a - \frac{1}{4} + \frac{1}{4}$
8. $p - 6\frac{1}{2} + 6\frac{1}{2}$
9. $3\frac{1}{4} - x + x$

Write the equivalent equation formed as the first step in solving each equation. (Step 1, Examples 1–3)

10. $x - 7 = 19$
11. $a - 12 = 5$
12. $23 = s - 27$
13. $47 = t - 83$
14. $q - 2.7 = 3.9$
15. $12.9 = b - 5.6$
16. $w - \frac{3}{4} = 2\frac{1}{2}$
17. $4\frac{1}{4} = n - 1\frac{7}{8}$
18. $q - \frac{3}{4} = 6\frac{1}{2}$

WRITTEN EXERCISES

Goal: To use addition to solve equations

For Exercises 1–51, solve and check each equation. Show all steps. (Example 1)

1. $x - 32 = 26$
2. $r - 23 = 36$
3. $t - 48 = 53$
4. $p - 59 = 47$
5. $75 = n - 128$
6. $87 = b - 136$
7. $185 = a - 207$
8. $95 = p - 235$
9. $m - 169 = 169$
10. $378 = k - 378$
11. $y - 4021 = 86$
12. $x - 17 = 8013$

Chapter 7

(Example 2)

13. $s - 4.2 = 5.9$
14. $q - 7.9 = 6.7$
15. $0.86 = w - 0.27$
16. $0.54 = n - 0.89$
17. $19.7 = g - 28.3$
18. $28.5 = h - 42.7$
19. $r - 9.08 = 15.29$
20. $x - 12.37 = 9.09$
21. $c - 10.8 = 9.7$
22. $d - 90.6 = 7.9$
23. $6.08 = p - 104.1$
24. $113.2 = t - 8.4$
25. $b - 2\frac{1}{2} = 5\frac{1}{2}$
26. $a - 3\frac{1}{4} = 2\frac{1}{2}$
27. $3\frac{3}{4} = n - 10\frac{3}{4}$
28. $5\frac{1}{8} = q - 12\frac{3}{8}$
29. $r - \frac{7}{16} = 1\frac{3}{16}$
30. $t - \frac{11}{12} = \frac{7}{12}$
31. $2\frac{1}{3} = m - 3\frac{1}{6}$
32. $c - 7\frac{1}{4} = 3\frac{1}{8}$
33. $11\frac{1}{7} = d - 2\frac{6}{7}$
34. $s - \frac{4}{9} = \frac{1}{3}$
35. $\frac{5}{8} = q - 3\frac{1}{4}$
36. $g - \frac{8}{11} = 3\frac{7}{22}$

MIXED PRACTICE

37. $49 = r - 78$
38. $g - 24.9 = 13.9$
39. $k - 263 = 105$
40. $57 = t - 96$
41. $4\frac{5}{6} = n - 1\frac{5}{12}$
42. $5\frac{5}{8} = p - 2\frac{3}{4}$
43. $t - \frac{3}{4} = 2\frac{15}{16}$
44. $m - 135.8 = 94.7$
45. $2.97 = y - 0.08$
46. $s - \frac{4}{5} = \frac{13}{15}$
47. $b - 276.5 = 57.9$
48. $1356 = f - 867$

MORE CHALLENGING EXERCISES

The set or collection of numbers that a variable represents is called the **replacement set**.

EXAMPLE: Solve the equation $n + 9 = n$ when the replacement set is {whole numbers}.

Solution: Since 0 is the only whole number for which $n + 9 = 9$, the solution is **0**.

Find the solution(s). The replacement set is {2, 4, 6, 8, 10}.

49. $s + 2 = 10$
50. $t + 2 = 2 + t$
51. $r + r = 2r$
52. $p = 2$
53. $q - 5 = 3$
54. $21 - a = 15$

REVIEW CAPSULE FOR SECTION 7-3

Solve each equation. (Pages 190-192)

1. $x + 4 = 18$
2. $p + 3.5 = 16.2$
3. $36 = x + 21$
4. $n + 24 = 86$
5. $m + 0.4 = 3.8$
6. $t + \frac{4}{5} = 2\frac{1}{2}$
7. $s + 2\frac{1}{2} = 4\frac{2}{3}$
8. $25.4 = p + 15.5$

The answers are on page 212.

Solving Equations 195

ASSIGNMENT GUIDE
BASIC
Day 1 pp. 194-195: 1-24
Day 2 p. 195: 25-48
AVERAGE
Day 1 pp. 194-195:
1-35 odd
Day 2 p. 195: 37-48
ABOVE AVERAGE
pp. 194-195: 1-47 odd,
49-54

WRITTEN EXERCISES
1. 58   2. 59   3. 101
4. 106   5. 203
6. 223   7. 392
8. 330   9. 338
10. 756 11. 4107
12. 8030 13. 10.1
14. 14.6 15. 1.13
16. 1.43 17. 48
18. 71.2 19. 24.37
20. 21.46 21. 20.5
22. 98.5 23. 110.18
24. 121.6 25. 8
26. $5\frac{3}{4}$ 27. $14\frac{1}{2}$
28. $17\frac{1}{2}$ 29. $1\frac{5}{8}$
30. $1\frac{1}{2}$ 31. $5\frac{1}{2}$
32. $10\frac{3}{8}$ 33. 14
34. $\frac{7}{9}$ 35. $3\frac{7}{8}$
36. $4\frac{1}{22}$ 37. 127
38. 38.8   39. 368
40. 153   41. $6\frac{1}{4}$

See page 200 for the answers to Exercises 42-54

Teaching Suggestions
p. M-20

QUICK QUIZ
Solve and check.

1. y + 15 = 20.31
   Ans: 5.31
2. $2\frac{1}{2} = x + 1\frac{1}{3}$
   Ans: $1\frac{1}{6}$
3. $5\frac{3}{4} + y = 12$
   Ans: $6\frac{1}{4}$

Evaluate.

4. a + b when a = 29
   and b = 106
   Ans: 135
5. x − h when x = 5.8
   and h = 1.61
   Ans: 4.19
6. k − p when $k = 5\frac{1}{7}$
   and $p = 2\frac{1}{2}$
   Ans: $2\frac{9}{14}$

ADDITIONAL EXAMPLES
Example 1
Find the net pay.
1. Gross earnings:
   $8231;
   Deductions: $529
   Ans: $7702.
2. Gross earnings:
   $56.13;
   Deductions: $5.83
   Ans: $50.30

SELF-CHECK
1. $336.20

PROBLEM SOLVING AND APPLICATIONS

7-3 Using Equations: Addition/Subtraction

The following word rule and formula can be used to find net pay.

Word Rule: Net Pay + Deductions = Gross Earnings
Formula: $n + d = g$

Net pay, or **take-home pay,** is the amount left when all deductions (such as taxes) have been subtracted from the amount actually earned. The amount actually earned is called **gross earnings.**

EXAMPLE 1 Brian's weekly gross earnings are $385. His total deductions amount to $58.60. Find the net pay.

Solution:
$$n + d = g$$
$$n + 58.60 = 385$$
$$n + 58.60 - 58.60 = 385 - 58.60$$
$$n = 326.40$$

d = 58.60
g = 385

Brian's net pay is **$326.40.**

Check: Does $326.40 + 58.60 = $385? Yes ✔

Self-Check 1. Find the net pay: Gross earnings: $410; Deductions: $73.80

The word rule and formula below can be used to find a plane's air speed in miles per hour (mi/h).

Word Rule: Air Speed − Head Wind Speed = Ground Speed
Formula: $a - h = g$

Air speed is the speed of the plane in still air. **Ground speed** is the speed of the plane relative to the ground. A **head wind** is a wind blowing in the opposite direction in which the plane is heading.

EXAMPLE 2 A plane is flying into a head wind of 30 mi/h. The ground speed of the plane is 615 mi/h. Find the air speed.

Solution:
$$a - h = g$$
$$a - 30 = 615$$
$$a - 30 + 30 = 615 + 30$$
$$a = 645$$

h = 30 mi/h;
g = 615 mi/h

The air speed is **645 mi/h.**

Check: Does 645 − 30 = 615? Yes ✔

196 Chapter 7

Self-Check   2. Find the air speed: Head Wind: 25 mi/h; Ground speed: 630 mi/h

CLASSROOM EXERCISES

Use the formula $n + d = g$ to find the net pay. (Example 1)

1. Gross earnings: $285
   Deductions: $52
2. Gross earnings: $152
   Deductions: $27
3. Gross earnings: $600
   Deductions: $137.89

Use the formula $a - h = g$ to find the air speed. (Example 2)

4. g: 430 mi/h
   h: 30 mi/h
5. g: 350 mi/h
   h: 25 mi/h
6. g: 512 mi/h
   h: 15.4 mi/h

WRITTEN EXERCISES

Goal: To use equations to solve word problems

Use the formula $n + d = g$ to find the net pay. (Example 1)

	Gross Earnings	Total Deductions	Net Pay
1.	$560.00	$121.00	?
3.	$600.00	$108.32	?
5.	$392.50	$55.27	?

	Gross Earnings	Total Deductions	Net Pay
2.	$80.00	$9.18	?
4.	$370.00	$49.00	?
6.	$295.82	$54.48	?

7. Last week, Scott's gross earnings were $386.15. The total deductions were $94.38. Find the net pay.

8. Lisa's gross earnings each week are $460.85. The total deductions are $125.09. Find the net pay.

Use the formula $a - h = g$ to find the air speed. (Example 2)

	Ground Speed	Head Wind	Air Speed
9.	320 mi/h	17 mi/h	?
11.	865 mi/h	10.4 mi/h	?
13.	356.8 mi/h	20 mi/h	?

	Ground Speed	Head Wind	Air Speed
10.	650 mi/h	24 mi/h	?
12.	212 mi/h	8.7 mi/h	?
14.	1042 mi/h	15 mi/h	?

15. A plane is flying into a head wind of 23 miles per hour. The ground speed is 452 mi/h. Find the air speed.

16. A plane is flying into a head wind of 42.6 mi/h. The ground speed is 1030.4 mi/h. Find the air speed.

Solving Equations   **197**

ADDITIONAL EXAMPLES
Example 2
Find the airspeed.
1. Head wind: 82 mph;
   Ground speed: 520 mph
   Ans: 602 mph
2. Head wind: 46.5 mph;
   Ground speed: 351.7 mph   Ans: 398.2 mph

SELF-CHECK
2. 655 mph

CLASSROOM EXERCISES
1. $233   2. $125
3. $462.11   4. 460 mph
5. 375 mph
6. 527.4 mph

ASSIGNMENT GUIDE
BASIC
p. 197: 1-16
AVERAGE
pp. 197-198: 1-18
ABOVE AVERAGE
pp. 197-198: 1-18

WRITTEN EXERCISES
1. $439.00   2. $70.82
3. $491.68   4. $321.00
5. $337.23   6. $241.34
7. $291.77   8. $335.76
See page 198 for the answers to Exercises 9-16.

WRITTEN EXERCISES p. 197

9. 337 mph
10. 674 mph
11. 875.4 mph
12. 220.7 mph
13. 376.8 mph
14. 1057 mph
15. 475 mph
16. 1073 mph
17. a + t = g
18. a + 30 = 425; 395

QUIZ: SECTIONS 7-1–7-3
After completing this Review, you may wish to administer a quiz covering the same sections. A Quiz is provided in the Teacher's Edition: Part II.

REVIEW: SECTIONS 7-1–7-3

1. 15   2. 22   3. 8.6
4. 0.33   5. $\frac{1}{3}$ 6. $2\frac{5}{8}$
7. 4.3   8. $\frac{1}{6}$ 9. $4\frac{3}{10}$
10. 38   11. 130
12. 13.6   13. 108
14. 8   15. $2\frac{1}{8}$
16. 10   17. $12\frac{1}{12}$
18. $5\frac{17}{20}$ 19. $257.69
20. $148.80
21. 569.9 mph
22. 597.5 mph

MORE CHALLENGING EXERCISES

Solve each problem.

17. A **tail wind** is a wind blowing in the same direction in which a plane is heading. Write a formula to find the ground speed, using the air speed (a) and the tail wind speed (t).

18. A plane is flying with a tail wind of 30 miles per hour. The ground speed is 425 miles per hour. Write an addition equation to find the air speed (see Exercise 17). Then solve the equation.

REVIEW: SECTIONS 7-1–7-3

Solve and check each equation. Show all steps. (Section 7-1)

1. $x + 12 = 27$
2. $43 = n + 21$
3. $b + 10.2 = 18.8$
4. $0.61 + y = 0.94$
5. $r + \frac{5}{6} = 1\frac{1}{6}$
6. $1\frac{1}{2} + c = 4\frac{1}{8}$
7. $n + 6.7 = 11$
8. $p + \frac{2}{3} = \frac{5}{6}$
9. $7\frac{1}{2} = t + 3\frac{1}{5}$

Solve and check each equation. Show all steps. (Section 7-2)

10. $y - 28 = 10$
11. $h - 65 = 65$
12. $7.2 = s - 6.4$
13. $x - 21.9 = 86.1$
14. $4\frac{1}{2} = t - 3\frac{1}{2}$
15. $m - 1\frac{3}{4} = \frac{3}{8}$
16. $v - 8.25 = 1.75$
17. $6\frac{1}{3} = w - 5\frac{3}{4}$
18. $d - \frac{3}{5} = 5\frac{1}{4}$

For Exercises 19–20, use the formula $n + d = g$. (Section 7-3)

19. Issac's gross earnings each week are $325.86. The total deductions are $68.17. Find his net pay.

20. Sonya's gross earnings are $160 each week. The total deductions are $11.20. What amount does Sonya take home each week?

For Exercises 21–22, use the formula $a - h = g$. (Section 7-3)

21. Airline Flight 101 is flying into a head wind of 94.7 miles per hour. The ground speed is 475.2 miles per hour. Find the air speed of Flight 101.

22. Airline Flight 102 has a ground speed of 525.5 miles per hour. The plane is flying into a head wind of 72 miles per hour. What is the air speed of Flight 102?

Chapter 7

CONSUMER APPLICATION
Wallpaper and Estimation

To find the number of rolls of wallpaper needed to paper a room, use these rules.

RULES
1. One roll of wallpaper covers 3 square meters of wall area.
2. Subtract one roll of paper for every door and one roll of paper for every two windows.

This formula can be used to estimate the amount of wallpaper needed.

$R = \dfrac{Ph}{3} - W$

- R = number of rolls of paper
- P = perimeter of room
- h = height of room
- W = door and window allowance

EXAMPLE A rectangular room is 3.4 meters long, 2.6 meters wide, and 2.9 meters high. The room has 1 door and 2 windows. At $6.45 per roll, how much will it cost to paper the four walls?

SOLUTION

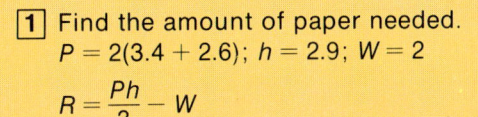

1 Find the amount of paper needed.
$P = 2(3.4 + 2.6)$; $h = 2.9$; $W = 2$

$R = \dfrac{Ph}{3} - W$

$R = \dfrac{(3.4 + 2.6)(2)(2.9)}{3} - 2$ ← Use a calculator.

3 . 4 [+] 2 . 6 [=] [×] 2 [×]

2 . 9 [÷] 3 [−] 2 [=] | 9.6 |

$R = $ **10 rolls** ← Round up to the next whole roll.

2 Find the total cost: $6.45(10) = **$64.50**

Exercises
Complete the table to find the cost of papering each room.

	Length	Width	Height	Doors	Windows	Number of Rolls	Cost Per Roll	Total Cost
1.	4.1 m	3.2 m	2.7 m	2	2	?	$5.79	?
2.	3.6 m	2.4 m	2.7 m	1	2	?	$6.50	?
3.	4.4 m	3.4 m	2.8 m	2	4	?	$7.99	?
4.	5.3 m	3.7 m	2.6 m	1	1	?	$9.43	?

CONSUMER APPLICATIONS
This feature is an application of the skills of estimating with decimals and solving a formula to finding the number of rolls of wallpaper and the cost of papering rectangular walls. Note the use of the calculator for determining the actual cost of papering the room.

1. 11; $63.69
2. 9; $58.50
3. 11; $87.89
4. 15; $141.45

Consumer Application

Teaching Suggestions
p. M-20

QUICK QUIZ
Evaluate.
1. $87 \div 3$ Ans: 29
2. $1290 \div 15$ Ans: 86
3. $7.2 \div 0.9$ Ans: 8
4. $6.63 \div 1.3$
    Ans: 5.1
5. $12 \div \frac{1}{3}$ Ans: 36

ADDITIONAL EXAMPLES
Example 1
Solve and check.
1. $18x = 378$
    Ans: $x = 21$
2. $14{,}035 = 7y$
    Ans: $y = 2005$

SELF-CHECK
1. 14   2. 86
3. 16   4. 32

WRITTEN EXERCISES p. 195
42. $8\frac{3}{8}$ 43. $3\frac{11}{16}$
44. 230.5   45. 3.05
46. $1\frac{2}{3}$ 47. 334.4
48. 2223   49. 8
50. 2, 4, 6, 8, 10
51. 2, 4, 6, 8, 10
52. 2   53. 8   54. 6

REVIEW CAPSULE FOR SECTION 7-4

Evaluate each expression.   (Pages 64–66)

1. $8 \cdot 5 \div 5$
2. $6 \cdot 3 \div 3$
3. $9 \cdot 6 \div 6$
4. $10 \cdot 2 \div 2$
5. $0.4 \times 4 \div 4$
6. $3.7 \times 8 \div 8$
7. $1.6 \times 2 \div 2$
8. $19.5 \times 7 \div 7$
9. $6\frac{1}{2} \cdot 3\frac{1}{3} \div 3\frac{1}{3}$
10. $8 \cdot 2\frac{1}{2} \div 2\frac{1}{2}$
11. $\frac{1}{3} \cdot \frac{2}{5} \div \frac{2}{5}$
12. $4\frac{3}{8} \cdot 2\frac{2}{3} \div 2\frac{2}{3}$

The answers are on page 212.

7-4 Solving Equations by Division

The following table shows that division "undoes" multiplication.

① Choose a number.	② Multiply by any number.	③ Divide by the same number.	④ Result: The original number
7	$7 \cdot 3$	$7 \cdot 3 \div 3$	7
25	$25 \cdot 4$	$25 \cdot 4 \div 4$	25
n	$8n$	$\frac{8n}{8}$, or $\frac{8}{8} \cdot n$	n

Thus, division is the **opposite**, or **inverse operation**, of multiplication. To solve an equation such as

$$12x = 180,$$

divide **each side** of the equation by 12.

> **PROCEDURE**
> To solve the equation $ax = c$, $a \neq 0$, divide **each side** of the equation by a.

EXAMPLE 1   Solve and check: $12x = 180$

Solution: To get x alone, divide each side by 12.

$12x = 180$
$\frac{12x}{12} = \frac{180}{12}$ $\frac{12}{12} = 1$
$x = 15$

Check: $12x = 180$
$12 \cdot 15$
180

Self-Check   *Solve and check.*

1. $8x = 112$
2. $3t = 258$
3. $3n = 48$
4. $4y = 128$

200   Chapter 7

NOTE: In Example 1, these equations are equivalent because they have the <u>same</u> solution, **15**.

$$12x = 180, \quad \frac{12x}{12} = \frac{180}{12}, \quad \text{and} \quad x = 15$$

EXAMPLE 2 Solve and check.

a. $21 = 0.7w$
b. $16w = 200$

Solutions:

a. $21 = 0.7w$ ◀ Divide each side by 0.7.

$\frac{21}{0.7} = \frac{0.7w}{0.7}$

$30 = w$

b. $16w = 200$ ◀ Divide each side by 16.

$\frac{16w}{16} = \frac{200}{16}$

$w = \frac{25}{2}$

$w = 12\frac{1}{2}$, or **12.5**

The checks are left for you.

Self-Check *Solve and check.*

5. $147 = 2.1p$
6. $5w = 79$
7. $1.5q = 3$
8. $100 = 8p$

CLASSROOM EXERCISES

Evaluate each expression. (Step 1, Examples 1–2)

1. $\frac{10x}{10}$
2. $\frac{27y}{27}$
3. $\frac{0.9n}{0.9}$
4. $\frac{128k}{128}$

Write the equivalent equation formed as the first step in solving each equation. (Step 1, Examples 1–2)

5. $4n = 32$
6. $9x = 72$
7. $44 = 11y$
8. $0.7w = 21$
9. $2.6a = 33.8$
10. $63.6 = 53t$
11. $96 = 19.2x$
12. $108b = 1296$

Determine whether the value of x shown is the solution for the given equation. Answer <u>Yes</u> or <u>No</u>. (Check, Example 1)

13. $4x = 8$
    $x = \frac{1}{2}$
14. $12 = 36x$
    $x = \frac{1}{3}$
15. $8x = 30$
    $x = 3\frac{3}{4}$
16. $12.4 = 0.2x$
    $x = 6.2$
17. $14x = 2$
    $x = \frac{1}{7}$
18. $9x = 55$
    $x = 6\frac{1}{9}$
19. $13x = 10$
    $x = 1\frac{3}{10}$
20. $15x = 45$
    $x = \frac{1}{3}$

Solving Equations **201**

ADDITIONAL EXAMPLES

Example 2

Solve and check.

1. $1.3x = 8.45$
   Ans: 6.5
2. $0.8y = 11.24$
   Ans: 14.05
3. $16a = 12$
   Ans: $\frac{3}{4}$ or 0.75

SELF-CHECK

5. 70
6. $15\frac{4}{5}$
7. 2
8. $12\frac{1}{2}$

CLASSROOM EXERCISES

1. x   2. y   3. n
4. k
5. $\frac{4n}{4} = \frac{32}{4}$
6. $\frac{9x}{9} = \frac{72}{9}$
7. $\frac{44}{11} = \frac{11y}{11}$
8. $\frac{0.7w}{0.7} = \frac{21}{0.7}$
9. $\frac{2.6a}{2.6} = \frac{33.8}{2.6}$
10. $\frac{63.6}{53} = \frac{53t}{53}$
11. $\frac{96}{19.2} = \frac{19.2x}{19.2}$
12. $\frac{108b}{108} = \frac{1296}{108}$
13. No   14. No
15. Yes 16. No
17. Yes 18. Yes
19. No   20. No

201

ASSIGNMENT GUIDE

BASIC
Day 1 p. 202: 1–20
Day 2 p. 202: 21–44

AVERAGE
p. 202: 1–27 odd, 29–44

ABOVE AVERAGE
p. 202: 1–43 odd, 45–56

WRITTEN EXERCISES
1. 9   2. 7   3. 5
4. 60   5. 32   6. 23
7. 5   8. 4   9. 14
10. 25   11. 41
12. 23   13. 20
14. 60   15. 120
16. 240   17. 40
18. 120   19. $9\frac{1}{5}$
20. 18   21. $16\frac{4}{5}$
22. $8\frac{2}{5}$ 23. $10\frac{5}{8}$
24. $27\frac{3}{4}$ 25. $\frac{3}{4}$
26. $\frac{1}{3}$ 27. 32
28. 36   29. 12
30. 25   31. 53.5
32. $5\frac{1}{3}$ 33. 125
34. 135   35. $12\frac{1}{2}$
36. $\frac{4}{9}$ 37. 24.5
38. $2\frac{1}{8}$ 39. $\frac{2}{5}$
40. 29   41. 22
42. 8   43. $15\frac{1}{3}$
44. $\frac{5}{8}$ 45. $\frac{5}{8}$ 46. $\frac{2}{3}$
47. 12   48. 6

See page 203 for the answers to Exercises 49–56.

202

WRITTEN EXERCISES

Goal: To use division to solve equations

For Exercises 1–44, solve and check. Show all steps. (Example 1)

1. $6x = 54$
2. $9x = 63$
3. $60 = 12n$
4. $2p = 120$
5. $256 = 8y$
6. $7a = 161$
7. $80 = 16q$
8. $18t = 72$

(Example 2)

9. $8.4 = 0.6r$
10. $22.5 = 0.9n$
11. $5.2k = 213.2$
12. $2.8m = 64.4$
13. $1.9x = 38$
14. $2.7y = 162$
15. $9.6 = 0.08t$
16. $7.2 = 0.03p$
17. $88 = 2.2w$
18. $0.04z = 4.8$
19. $92 = 10c$
20. $108 = 6e$
21. $5x = 84$
22. $15y = 126$
23. $85 = 8x$
24. $111 = 4r$
25. $32x = 24$
26. $45y = 15$
27. $416 = 13s$
28. $504 = 14t$

MIXED PRACTICE

29. $168 = 14x$
30. $8.6t = 215$
31. $2.4p = 128.4$
32. $15s = 80$
33. $375 = 3n$
34. $2.9k = 391.5$
35. $12t = 150$
36. $18x = 8$
37. $1.47 = 0.06a$
38. $16x = 34$
39. $25y = 10$
40. $17y = 493$
41. $7b = 154$
42. $4.3d = 34.4$
43. $138 = 9u$
44. $48v = 30$

MORE CHALLENGING EXERCISES

Solve and check each equation. Show all steps.

45. $4x = 2\frac{1}{2}$
46. $8p = 5\frac{1}{3}$
47. $\frac{3}{4}m = 9$
48. $1\frac{2}{3}n = 10$
49. $\frac{2}{3}y = \frac{3}{5}$
50. $\frac{7}{8}t = 2\frac{1}{2}$
51. $3\frac{2}{3}s = 8\frac{1}{4}$
52. $6\frac{2}{3}w = 6\frac{1}{4}$
53. $1\frac{3}{4}r = 49$
54. $2x = 4\frac{4}{5}$
55. $2\frac{1}{5}q = 3\frac{3}{10}$
56. $\frac{4}{5}v = 5\frac{3}{5}$

REVIEW CAPSULE FOR SECTION 7–5

Evaluate each expression.
(Pages 18–21, 38–41, 44–46, 129–131, 148–151)

1. $16 \div 4 \cdot 4$
2. $84 \div 12 \cdot 12$
3. $\frac{10}{8}(8)$
4. $\frac{240}{3}(3)$
5. $21 \div 4.2 \times 4.2$
6. $\frac{31.6}{1.6}(1.6)$
7. $\frac{136}{6.8}(6.8)$
8. $22.2 \div 3.7 \times 3.7$

The answers are on page 212.

Chapter 7

7-5 Solving Equations by Multiplication

The following table shows that multiplication "undoes" division.

① Choose a number.	② Divide by any number (not zero).	③ Multiply by the same number.	④ Result: The original number
24	$24 \div 6$, or $\frac{24}{6}$	$\frac{24}{6} \cdot 6$	24
63	$\frac{63}{5}$	$\frac{63}{5} \cdot 5$	63
n	$\frac{n}{4}$	$\frac{n}{4} \cdot 4$	n

Thus, multiplication is the **inverse operation** of division. To solve an equation such as

$$\frac{x}{5} = 24,$$

multiply each side by 5.

PROCEDURE

To solve the equation $\frac{x}{b} = c$, $b \neq 0$, multiply each side by b.

EXAMPLE 1 Solve and check: $\frac{x}{5} = 24$

Solution: To get x alone, multiply each side by 5.

$$\frac{x}{5} = 24$$
$$\frac{x}{5} \cdot 5 = 24 \cdot 5$$
$$x = 120$$

Check: $\frac{x}{5} = 24$
$$\frac{120}{5}$$
$$24$$

Self-Check Solve and check.

1. $\frac{v}{3} = 9$ 2. $\frac{z}{8} = 1$ 3. $\frac{q}{16} = 21$ 4. $\frac{w}{9} = 12$

NOTE: In Example 1, these equations are equivalent because they have the <u>same</u> solution, **120**.

$$\frac{x}{5} = 24, \quad \frac{x}{5} \cdot 5 = 24 \cdot 5, \quad \text{and} \quad x = 120$$

Solving Equations **203**

Teaching Suggestions
p. M-20

QUICK QUIZ
Evaluate.
1. $17 \div 9 \cdot 9$ Ans: 17
2. $\frac{15}{7} \cdot 7$ Ans: 15
3. $5\frac{1}{8} \div 3.2 \times 3.2$
   Ans: $5\frac{1}{8}$
4. $\frac{5.21}{6.85}(6.85)$
   Ans: 5.21
5. $2\frac{1}{3} \div 5\frac{3}{7} \times 5\frac{3}{7}$
   Ans: $2\frac{1}{3}$

ADDITIONAL EXAMPLES
Example 1
Solve and check.
1. $\frac{x}{13} = 20$ Ans: 260
2. $15 = \frac{y}{15}$ Ans: 225

SELF-CHECK
1. 27   2. 8
3. 336   4. 108

WRITTEN EXERCISES p. 202
49. $\frac{9}{10}$ 50. $2\frac{6}{7}$
51. $2\frac{1}{4}$ 52. $\frac{15}{16}$
53. 28   54. $2\frac{2}{5}$
55. $1\frac{1}{2}$ 56. 7

203

ADDITIONAL EXAMPLES

Example 2

Solve and check.

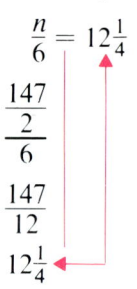

1. $8.6 = \dfrac{x}{1.5}$ Ans: 12.9
2. $\dfrac{a}{0.75} = 12$ Ans: 9
3. $\dfrac{x}{10} = 3\dfrac{1}{5}$ Ans: 32

SELF-CHECK

5. 126   6. 18
7. 10   8. 12.3

CLASSROOM EXERCISES

1. q   2. z   3. a
4. s   5. r   6. n
7. p   8. y
9. $\dfrac{n}{5} \cdot 5 = 6 \cdot 5$
10. $\dfrac{x}{20} \cdot 20 = 9 \cdot 20$
11. $\dfrac{x}{10} \cdot 10 = 3 \cdot 10$
12. $\dfrac{m}{8} \cdot 8 = 15 \cdot 8$
13. $\dfrac{s}{5} \cdot 5 = 1.1(5)$
14. $4.7(7) = \dfrac{t}{7} \cdot 7$
15. $2\dfrac{1}{2} \cdot 4 = \dfrac{y}{4} \cdot 4$
16. $\dfrac{p}{9} \cdot 9 = 3\dfrac{1}{3} \cdot 9$
17. $\dfrac{b}{17} \cdot 17 = 5 \cdot 17$
18. $6.2(3) = \dfrac{c}{3} \cdot 3$
19. $3\dfrac{3}{4} \cdot 8 = \dfrac{d}{8} \cdot 8$
20. $\dfrac{e}{10} \cdot 10 = 5\dfrac{2}{5} \cdot 10$
21. $\dfrac{f}{3} \cdot 3 = 4\dfrac{5}{6} \cdot 3$
22. $7.8(11) = \dfrac{h}{11} \cdot 11$
23. $\dfrac{r}{7} \cdot 7 = 12 \cdot 7$
24. $\dfrac{x}{8} \cdot 8 = 6\dfrac{1}{2} \cdot 8$

EXAMPLE 2   Solve and check: **a.** $10 = \dfrac{d}{12.5}$ **b.** $\dfrac{n}{6} = 12\dfrac{1}{4}$

Solutions: **a.** $10 = \dfrac{d}{12.5}$ ◀ Multiply each side by 12.5.

$10(12.5) = \dfrac{d}{12.5}(12.5)$

$125 = d$

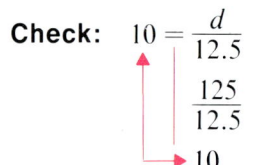

Check: $10 = \dfrac{d}{12.5}$

$\dfrac{125}{12.5}$

10

b. $\dfrac{n}{6} = 12\dfrac{1}{4}$ ◀ Multiply each side by 6.

$\dfrac{n}{6} \cdot 6 = 12\dfrac{1}{4} \cdot 6$

$\dfrac{n}{\cancel{6}} \cdot \cancel{6} = \dfrac{49}{\cancel{4}} \cdot \cancel{6}^{3}$

$n = \dfrac{147}{2}$, or $73\dfrac{1}{2}$

Check: $\dfrac{n}{6} = 12\dfrac{1}{4}$

$\dfrac{\frac{147}{2}}{6}$

$\dfrac{147}{12}$

$12\dfrac{1}{4}$

Self-Check   *Solve and check.*

5. $\dfrac{t}{6.3} = 20$ 6. $2\dfrac{1}{4} = \dfrac{c}{8}$ 7. $\dfrac{m}{3} = 3\dfrac{1}{3}$ 8. $\dfrac{r}{1.5} = 8.2$

CLASSROOM EXERCISES

Evaluate each expression.   (Table)

1. $q \div 3 \cdot 3$ 2. $\dfrac{z}{2} \cdot 2$ 3. $a \div 19 \cdot 19$ 4. $\dfrac{s}{89} \cdot 89$
5. $\dfrac{r}{9.3}(9.3)$ 6. $n \div 8 \cdot 8$ 7. $\dfrac{p}{11} \cdot 11$ 8. $\left(\dfrac{y}{4.6}\right)4.6$

Write the equivalent equation formed as the first step in solving each equation.   (Step 1, Examples 1–2)

9. $\dfrac{n}{5} = 6$ 10. $\dfrac{x}{20} = 9$ 11. $\dfrac{y}{10} = 3$ 12. $\dfrac{m}{8} = 15$
13. $\dfrac{s}{5} = 1.1$ 14. $4.7 = \dfrac{t}{7}$ 15. $2\dfrac{1}{2} = \dfrac{y}{4}$ 16. $\dfrac{p}{9} = 3\dfrac{1}{3}$
17. $\dfrac{b}{17} = 5$ 18. $6.2 = \dfrac{c}{3}$ 19. $3\dfrac{3}{4} = \dfrac{d}{8}$ 20. $\dfrac{e}{10} = 5\dfrac{2}{5}$
21. $\dfrac{f}{3} = 4\dfrac{5}{6}$ 22. $7.8 = \dfrac{h}{11}$ 23. $\dfrac{r}{7} = 12$ 24. $\dfrac{x}{8} = 6\dfrac{1}{2}$

WRITTEN EXERCISES

Goal: To use multiplication to solve equations

Solve and check each equation. Show all steps. (Example 1)

1. $\dfrac{t}{8} = 17$
2. $\dfrac{r}{12} = 9$
3. $12 = \dfrac{x}{23}$
4. $15 = \dfrac{x}{34}$
5. $\dfrac{w}{7} = 123$
6. $\dfrac{k}{9} = 208$
7. $346 = \dfrac{p}{13}$
8. $529 = \dfrac{a}{16}$
9. $\dfrac{x}{2} = 5$
10. $4 = \dfrac{y}{3}$
11. $\dfrac{n}{8} = 2$
12. $5 = \dfrac{r}{6}$

(Example 2)

13. $\dfrac{a}{1.8} = 70$
14. $\dfrac{n}{2.3} = 9$
15. $83 = \dfrac{s}{0.09}$
16. $56 = \dfrac{t}{0.07}$
17. $\dfrac{w}{3.7} = 0.8$
18. $\dfrac{p}{5.2} = 0.4$
19. $23.5 = \dfrac{x}{2.4}$
20. $40.4 = \dfrac{y}{2.5}$
21. $\dfrac{a}{2.5} = 4$
22. $\dfrac{k}{1.3} = 2.3$
23. $9 = \dfrac{m}{1.1}$
24. $2 = \dfrac{w}{3.1}$
25. $\dfrac{x}{8} = 3\dfrac{1}{2}$
26. $\dfrac{k}{18} = 6\dfrac{1}{4}$
27. $4\dfrac{3}{8} = \dfrac{n}{16}$
28. $3\dfrac{5}{6} = \dfrac{m}{18}$
29. $\dfrac{p}{10} = \dfrac{2}{5}$
30. $\dfrac{r}{30} = \dfrac{5}{6}$
31. $12\dfrac{1}{2} = \dfrac{y}{13}$
32. $1\dfrac{1}{4} = \dfrac{a}{5}$

MIXED PRACTICE

33. $2.8 = \dfrac{r}{12}$
34. $\dfrac{p}{16} = 3\dfrac{1}{4}$
35. $\dfrac{n}{26} = 35$
36. $3.6 = \dfrac{t}{15}$
37. $\dfrac{q}{12} = 5\dfrac{1}{6}$
38. $\dfrac{w}{42} = 93$
39. $102 = \dfrac{a}{47}$
40. $1\dfrac{3}{5} = \dfrac{b}{5}$
41. $\dfrac{m}{9.6} = 0.5$
42. $\dfrac{r}{7.8} = 9.2$
43. $2\dfrac{2}{3} = \dfrac{d}{9}$
44. $\dfrac{p}{11} = 122$

MORE CHALLENGING EXERCISES

Solve and check each equation. Show all steps.

45. $\dfrac{x}{\frac{3}{4}} = 12$
46. $\dfrac{x}{\frac{5}{6}} = 3$
47. $\dfrac{x}{\frac{2}{5}} = 8$
48. $\dfrac{x}{1\frac{1}{4}} = \dfrac{3}{10}$
49. $\dfrac{x}{2\frac{1}{3}} = \dfrac{2}{7}$
50. $\dfrac{x}{\frac{1}{6}} = 1\dfrac{1}{2}$
51. $\dfrac{x}{1\frac{1}{3}} = 2\dfrac{3}{4}$
52. $\dfrac{x}{3\frac{1}{9}} = 4\dfrac{1}{5}$

ASSIGNMENT GUIDE
BASIC
Day 1 p. 205: 1–24
Day 2 p. 205: 25–44
AVERAGE
p. 205: 1–31 odd, 33–44
ABOVE AVERAGE
p. 205: 1–43 odd, 45–52

WRITTEN EXERCISES
1. 136 2. 108 3. 276
4. 510 5. 861
6. 1872 7. 4498
8. 8464 9. 10
10. 12 11. 16
12. 30 13. 126
14. 20.7 15. 7.47
16. 3.92 17. 2.96
18. 2.08 19. 56.4
20. 101 21. 10.0
22. 2.99 23. 9.9
24. 6.2 25. 28
26. $112\dfrac{1}{2}$ 27. 70
28. 69 29. 4 30. 25
31. $162\dfrac{1}{2}$ 32. $6\dfrac{1}{4}$
33. 33.6 34. 52
35. 910 36. 54
37. 62 38. 3906
39. 4794 40. 8
41. 4.8 42. 71.76
43. 24 44. 1342
45. 9 46. $2\dfrac{1}{2}$
47. $3\dfrac{1}{5}$ 48. $\dfrac{3}{8}$ 49. $\dfrac{2}{3}$
50. $\dfrac{1}{4}$ 51. $3\dfrac{2}{3}$
52. $13\dfrac{1}{15}$

Solving Equations

Teaching Suggestions
p. M-20

QUICK QUIZ
Solve and check.

<u>1</u>. 3y = 74.1   Ans: 24.7

<u>2</u>. 0.6a = 96   Ans: 160

Evaluate.

<u>3</u>. ab when a = 18
and b = $\frac{2}{3}$ Ans: 12

<u>4</u>. cd when c = 5.1
and d = 0.6
Ans: 3.06

<u>5</u>. $\frac{h}{k}$ when h = $7\frac{1}{2}$
and k = 4   Ans: $1\frac{7}{8}$

ADDITIONAL EXAMPLES
Example 1
Use the formula d = rt.

<u>1</u>. d = 195 ft; r = 32.5 ft/sec. Find t.
Ans: 6 sec

<u>2</u>. d = $42\frac{1}{2}$ mi; t = 6 hr.
Find r.
Ans: $7\frac{1}{12}$ mph

SELF-CHECK
<u>1</u>. 255 sec

REVIEW CAPSULE FOR SECTION 7-6

Solve and check each equation. (Pages 200–202)

1. $2x = 56$
2. $7y = 4.2$
3. $3p = 5$
4. $12n = 1440$
5. $4.5q = 15$
6. $17r = 52$
7. $36 = 1.8t$
8. $12.4 = 31s$

Evaluate each expression. (Pages 22–24)

9. $x \div y$ when $x = 12$ and $y = 0.5$
10. pq when $p = 4.5$ and $q = 0.6$
11. mr when $m = 7\frac{1}{3}$ and $r = \frac{1}{2}$
12. $\frac{s}{t}$ when $s = 3\frac{1}{3}$ and $t = 4$

The answers are on page 212.

PROBLEM SOLVING AND APPLICATIONS

7-6 Using Equations: Multiplication/Division

In Chapter 1, you used this formula to find distance.

$$d = rt$$

d = distance
r = rate
t = time

You can use the same formula to find rate or time.

EXAMPLE 1 A diving bell moves up and down in the water at the rate of 1.4 meters per second. How long would it take for the diving bell to descend 655.2 meters?

Solution: ① Formula: $d = rt$

② Known value: $d = 655.2$ m; $r = 1.4$ m/sec

③ $655.2 = 1.4t$ ◀ **Divide each side by 1.4.**

$\frac{655.2}{1.4} = \frac{1.4t}{1.4}$

$468 = t$ It will take **468 seconds**.

Check: Does $655.2 = 1.4\,(468)$?   Yes ✓

Self-Check 1. How long will it take the diving bell in Example 1 to rise 357 meters?

206   Chapter 7

The following word rule and formula can be used to find a baseball player's batting average.

Word Rule: Batting Average = Number of Hits ÷ Number of Times at Bat

Formula: $A = \dfrac{h}{t}$

A = Batting Average
h = number of hits
t = times at bat

When you know a player's batting average and the number of times at bat, you can use the formula to find the number of hits.

EXAMPLE 2 A player's batting average is 0.400 for 30 times at bat. Find the number of hits.

Solution:
1. Formula: $A = \dfrac{h}{t}$
2. Known values: $A = .400$; $t = 30$
3. $0.400 = \dfrac{h}{30}$ Multiply each side by 30.

$0.400(30) = \dfrac{h}{30}(30)$

$12 = h$ The player had **12** hits.

Check: Does $0.400 = \dfrac{12}{30}$?   Yes ✔

Self-Check 2. Diana Lopez has a batting average of 0.250 for 104 times at bat. Find the number of hits.

CLASSROOM EXERCISES

Use the formula $d = rt$ to find the time t. (Example 1)

1. Distance: 400 miles
   Rate: 20 miles per hour

2. Distance: 750 meters
   Rate: 25 meters per minute

Use the formula $d = rt$ to find the rate, r. (Example 1)

3. Distance: 48 kilometers
   Time: 2 hours

4. Distance: 500 miles
   Time: 10 hours

Use the formula $A = \dfrac{h}{t}$ to find the number of hits, h. (Example 2)

5. Times at Bat: 100 times
   Batting Average: 0.280

6. Times at Bat: 50 times
   Batting Average: 0.180

Solving Equations **207**

Example 2
Use the formula $A = \dfrac{h}{t}$.
1. A = 0.525; t = 120.
   Find h.   Ans: h = 63
2. A = 0.150; t = 240.
   Find h.   Ans: h = 36

SELF-CHECK
2. 26 hits

CLASSROOM EXERCISES
1. 20 hr   2. 30 min
3. 24 km/h 4. 50 mph
5. 28 hits 6. 9 hits

ASSIGNMENT GUIDE
BASIC
p. 208: 1–6, 11–18
AVERAGE
p. 208: 1–20
ABOVE AVERAGE
pp. 208–209: 1–21 odd, 22–24

WRITTEN EXERCISES
1. 6 hr   2. 11.2 km/h
3. 5.24 m/sec
4. $5\frac{1}{3}$ sec   5. $8\frac{1}{5}$ sec
6. 80 km/h
7. 1083.75 mph
8. 312.5 ft/sec
9. 15 hr   10. 35 sec
11. 50 hits
12. 95 hits
13. 30 hits
14. 119 hits
15. 39 hits
16. 57 hits
17. 176 hits
18. 144 hits
19. 26 hits
20. 31 hits

WRITTEN EXERCISES

Goal: To use equations to solve word problems

For Exercises 1–10, use the formula $d = rt$ to find the rate or the time as indicated. (Example 1)

	Rate	Time	Distance
1.	50 mph	? h	300 mi
3.	? m/sec	65 sec	340.6 m
5.	2 ft/sec	? sec	$16\frac{2}{5}$ ft

	Rate	Time	Distance
2.	? km/h	5 h	56.0 km
4.	8 ft/sec	? sec	$42\frac{2}{3}$ ft
6.	? km/h	16 h	1280 km

7. The Concorde SST flies 3468 miles from London, England to New York. The flight takes 3.2 hours. What is the average speed of the Concorde?

8. A peregrine falcon in a vertical fall takes about 16 seconds to dive 5000 feet. What is the falcon's average speed?

9. The distance from St. Louis to Dallas is about 1050 kilometers. How many hours will it take to drive this distance at an average speed of 70 kilometers per hour?

10. A roller skater set a world's speed record by skating 1320 feet at a rate of $37\frac{5}{7}$ feet per second. How long did it take the skater to complete the distance?

For Exercises 11–20, use the formula $A = \frac{h}{t}$ to find the number of hits. (Example 2)

	Times at Bat	Batting Average	Number of Hits
11.	250	0.200	?
13.	100	0.300	?
15.	150	0.260	?
17.	550	0.320	?

	Times at Bat	Batting Average	Number of Hits
12.	250	0.380	?
14.	340	0.350	?
16.	200	0.285	?
18.	605	0.238	?

19. Gilda Sanchez has a batting average of .250. She has been at bat 104 times. Find her number of hits.

20. Nick Demarco has been at bat 124 times. He has a batting average of .250. Find his number of hits.

208   Chapter 7

MORE CHALLENGING EXERCISES

21. In a triathalon, Sue Lin must swim $1\frac{1}{2}$ miles, bicycle 25 miles, and run 6 miles. She hopes to average 2 miles per hour swimming, 20 miles per hour bicycling, and 10 miles per hour running. How long should it take her to complete the triathalon?

22. Southern Skies Flight 224 flies north from Miami to Boston and back. The distance each way is about 2400 miles. The plane maintains a constant airspeed of 570 miles per hour in a 70 mile-per-hour wind blowing from north to south. How much longer does it take to fly from Miami to Boston than from Boston to Miami?

23. During the softball season, Norma had 57 hits and a batting average of 0.285. How many times was she at bat during the season?

24. The earth travels 934,240,000 kilometers around the sun in 8767 hours. How fast (nearest whole number) does the earth travel?

REVIEW: SECTIONS 7-4—7-6

Solve and check each equation. Show all steps. (Section 7-4)

1. $8x = 56$
2. $14b = 84$
3. $1.2m = 7.2$
4. $0.6p = 9$
5. $7q = 30$
6. $8s = 3$
7. $156 = 6t$
8. $4.4 = 0.4y$

Solve and check each equation. Show all steps. (Section 7-5)

9. $\frac{x}{9} = 13$
10. $\frac{y}{4} = 25$
11. $\frac{p}{15} = 1.6$
12. $\frac{t}{3.5} = 4.82$
13. $\frac{s}{8} = 1\frac{1}{3}$
14. $\frac{m}{7} = 5\frac{2}{3}$
15. $104 = \frac{n}{16}$
16. $2.6 = \frac{r}{0.04}$

Solve each problem. (Section 7-6)

17. The deepest point in the Pacific Ocean is 11,023 meters. Sound waves take 7.55 seconds to reach the bottom at this point. Use the formula $d = rt$ to find the speed of sound in water.

18. Michael has a batting average of 0.425 for 40 times at bat. Use
$$A = \frac{h}{t}$$
to find the number of hits.

Solving Equations

CHAPTER REVIEW

PART 1: VOCABULARY

For Exercises 1–10, choose from the box at the right the word(s) that best correspond to each description.

1. Amount of pay left after deductions __?__
2. The speed of a plane in still air __?__
3. An equation such as $n + 4 = 8$ is an __?__ equation.
4. The number of hits divided by the number of times at bat __?__
5. To solve a multiplication equation, you __?__ both sides by the same number.
6. Amount of pay actually earned __?__
7. A mathematical sentence containing the symbol "=" __?__
8. A wind blowing in the opposite direction in which a plane is heading __?__
9. Equations having the same solution or solutions __?__
10. To solve a division equation, you __?__ both sides by the same number.

> multiply
> equation
> gross earnings
> addition
> head wind
> equivalent equations
> batting average
> net pay
> air speed
> divide

PART 2: SKILLS

Solve and check each equation. Show all steps. (Section 7-1)

11. $x + 8 = 42$
12. $r + 28 = 73$
13. $32.1 = y + 19.3$
14. $t + 4.39 = 6.2$
15. $p + 3\frac{1}{2} = 14\frac{1}{4}$
16. $15\frac{1}{4} = n + 7\frac{3}{5}$

Solve and check each equation. Show all steps. (Section 7-2)

17. $a - 9 = 13$
18. $w - 109 = 58$
19. $k - 7.8 = 28.6$
20. $47.8 = b - 23.4$
21. $6\frac{1}{8} = m - 5\frac{1}{4}$
22. $t - 6\frac{3}{4} = 13\frac{1}{3}$

For Exercises 23–26, use the formula $n + d = g$ to find the net pay, n. (Section 7-3)

	Gross Earnings	Total Deductions	Net Pay		Gross Earnings	Total Deductions	Net Pay
23.	$280.00	$56.00	?	24.	$185.00	$37.15	?
25.	$312.30	$72.85	?	26.	$524.00	$108.09	?

210   Chapter 7

For Exercises 27–30, use the formula $a - h = g$ to find the airspeed, a. (Section 7-3)

	Ground-speed	Head Wind	Airspeed		Ground-speed	Head Wind	Airspeed
27.	425 mi/h	15 mi/h	?	28.	708 mi/h	8.4 mi/h	?
29.	256.4 mi/h	21 mi/h	?	30.	384.2 mi/h	31.6 mi/h	?

Solve and check each equation. Show all steps. (Section 7-4)

31. $126 = 14p$
32. $0.8a = 344$
33. $8t = 70$
34. $105 = 9c$

Solve and check each equation. Show all steps. (Section 7-5)

35. $\dfrac{d}{14} = 13$
36. $\dfrac{n}{9.7} = 4.8$
37. $\dfrac{b}{20} = 6\dfrac{1}{3}$
38. $27 = \dfrac{r}{1.3}$

For Exercises 39–44, complete as indicated. (Section 7-6)

	Rate	Time	Distance
39.	6 ft/sec	?	360 ft
40.	?	18 sec	82.8 m
41.	?	6.4 hr	332.8 km

	Times at Bat	Batting Average	Number of Hits
42.	72	0.250	?
43.	240	0.425	?
44.	586	0.314	?

PART 3: APPLICATIONS

45. Albert Chin's gross earnings last week were $382. The total deductions were $82.55. Use the formula $n + d = g$ to find the net pay. (Section 7-4)

46. A plane is flying into a head wind of 27.2 miles per hour. The ground speed is 456 miles per hour. Use the formula $a - h = g$ to find the air speed. (Section 7-4)

47. The longest recorded distance for a woman in a 24 hour walking race is $122\dfrac{1}{2}$ miles. Use the formula $d = rt$ to find the rate. Round the answer to the nearest tenth of a mile per hour. (Section 7-6)

48. The highest lifetime batting average in professional baseball is 0.367. The record is held by Ty Cobb who was at bat 11,420 times. Use the formula $A = \dfrac{h}{t}$ to find the number of hits. Round the answer to the nearest whole number. (Section 7-6)

---

27. 440 mph
28. 716.4 mph
29. 277.4 mph
30. 415.8 mph
31. 9
32. 430
33. $8\dfrac{3}{4}$
34. $11\dfrac{2}{3}$
35. 182
36. 46.56
37. $126\dfrac{2}{3}$
38. 35.1
39. 60 sec
40. 4.6 m/sec
41. 52 km/h
42. 18 hits
43. 102 hits
44. 184 hits
45. $299.45
46. 483.2 mph
47. 5.1 mph
48. 4191 hits

Solving Equations

CHAPTER TEST

Two forms of a chapter test, Form A and Form B, are provided on copying masters in the *Teacher's Edition: Part II.*

1. 25   2. 24.9
3. 2.89   4. $2\frac{7}{12}$
5. 77   6. 30.23
7. $1\frac{4}{5}$ 8. $5\frac{7}{24}$
9. 11   10. 1.5
11. $10\frac{1}{2}$ 12. $8\frac{2}{3}$
13. 4112   14. 6.46
15. 219.24   16. 94
17. $352.11
18. 267.5 mph
19. 156.25 mph
20. 27

CHAPTER TEST

Solve and check each equation. Show all steps.

1. $m + 28 = 53$
2. $t + 11.5 = 36.4$
3. $7.1 = a + 4.21$
4. $e + 15\frac{2}{3} = 18\frac{1}{4}$
5. $52 = w - 25$
6. $q - 10.73 = 19.5$
7. $c - \frac{3}{5} = 1\frac{1}{5}$
8. $1\frac{5}{8} = h - 3\frac{2}{3}$
9. $12x = 132$
10. $8.4 = 5.6p$
11. $42 = 4x$
12. $15a = 130$
13. $\frac{b}{16} = 257$
14. $\frac{f}{3.8} = 1.7$
15. $40.6 = \frac{y}{5.4}$
16. $\frac{m}{18} = 5\frac{2}{9}$

Solve each problem.

17. Monica's gross earnings each week are $458.16. The total deductions are $106.05. Use the formula $n + d = g$ to find the net pay.

18. A jet helicopter is flying into a headwind of 52.7 miles per hour. The ground speed is 214.8 miles per hour. Use the formula $a - h = g$ to find the air speed.

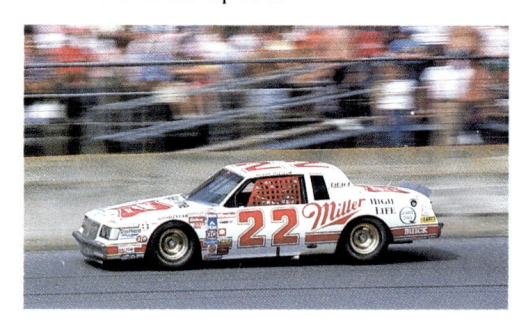

19. In 1984, the Daytona 500 winner, Cale Yarborough, completed the 500 miles in about 3.2 hours. Use the formula $d = rt$ to find his average speed.

20. Maria Hernandez has a batting average of 0.300. She has been at bat 90 times. Use the formula $A = \frac{h}{t}$ to find the number of hits.

ANSWERS TO REVIEW CAPSULES

Page 192 1. 7 2. 6 3. $9\frac{1}{2}$ 4. 0.9 5. q 6. c 7. f 8. 1800
Page 195 1. 14 2. 12.7 3. 15 4. 62 5. 3.4 6. $1\frac{7}{10}$ 7. $2\frac{1}{6}$ 8. 9.9
Page 200 1. 8 2. 6 3. 9 4. 10 5. 0.4 6. 3.7 7. 1.6 8. 19.5 9. $6\frac{1}{2}$ 10. 8 11. $\frac{1}{3}$ 12. $4\frac{3}{8}$
Page 202 1. 16 2. 84 3. 10 4. 240 5. 21 6. 31.6 7. 136 8. 22.2
Page 206 1. 28 2. 0.6 3. $1\frac{2}{3}$ 4. 120 5. $3\frac{1}{3}$ 6. $3\frac{1}{17}$ 7. 20 8. 0.4 9. 24 10. 2.7 11. $3\frac{2}{3}$ 12. $\frac{5}{6}$

Chapter 7

ENRICHMENT

Repeating Decimals

You learned in Chapter 6 that every fraction can be represented by an infinite repeating decimal or by a terminating decimal.

$\frac{2}{3} = 0.666\ldots$, or $0.\overline{6}$ ◄— The bar indicates that "6" repeats

$\frac{3}{4} = 0.75$ ◄— Terminating decimal

This Example shows how to write an infinite repeating decimal in the form $\frac{a}{b}$.

EXAMPLE Write each infinite repeating decimal in the form $\frac{a}{b}$.

a. $0.\overline{1}$ b. $9.\overline{27}$

Solutions: a. Let $n = 0.\overline{1}$ ◄— Multiply each side by 10^1, or 10.
$10n = 1.\overline{1}$ ◄— Subtract $n = 0.\overline{1}$ from this equation.
$\underline{n = 0.\overline{1}}$
$9n = 1.0$
$\frac{9n}{9} = \frac{1}{9}$
$n = \frac{1}{9}$ Thus, $0.\overline{1} = \frac{1}{9}$.

b. Let $n = 9.\overline{27}$ ◄— Multiply each side by 10^2, or 100.
$100n = 927.\overline{27}$ ◄— Subtract $n = 9.\overline{27}$ from this equation.
$\underline{n = 9.\overline{27}}$
$99n = 918$
$\frac{99n}{99} = \frac{918}{99}$
$n = \frac{918}{99}$, or $\frac{102}{11}$ Thus, $9.\overline{27} = \frac{102}{11}$.

EXERCISES

Write each infinite repeating decimal in the form $\frac{a}{b}$.

1. $0.\overline{3}$
2. $0.\overline{5}$
3. $0.\overline{18}$
4. $0.\overline{81}$
5. $0.\overline{72}$
6. $0.\overline{4}$
7. $1.\overline{6}$
8. $2.\overline{36}$
9. $3.\overline{45}$
10. $4.\overline{7}$
11. $0.\overline{918}$
12. $0.\overline{369}$
13. $6.\overline{25}$
14. $3.\overline{16}$
15. $7.\overline{32}$

Solving Equations

ENRICHMENT

You may wish to use this lesson for students who performed well on the formal Chapter Test.

1. $\frac{1}{3}$ 2. $\frac{5}{9}$ 3. $\frac{2}{11}$
4. $\frac{9}{11}$ 5. $\frac{8}{11}$ 6. $\frac{4}{9}$
7. $\frac{5}{3}$ 8. $\frac{26}{11}$ 9. $\frac{38}{11}$
10. $\frac{43}{9}$ 11. $\frac{34}{37}$
12. $\frac{41}{111}$ 13. $\frac{619}{99}$
14. $\frac{313}{99}$ 15. $\frac{725}{99}$

ADDITIONAL PRACTICE

You may wish to use all or some of these exercises, depending on how well students performed on the formal Chapter Test.

1. 305   2. 153
3. 1.12   4. 3.56
5. $1\frac{3}{8}$ 6. $6\frac{11}{16}$
7. 89   8. 142
9. 30.2   10. 213.4
11. $2\frac{9}{16}$ 12. $9\frac{1}{8}$
13. $515.00
14. $589.82
15. 75 km/h
16. 2.4 sec
17. 16   18. 77
19. 65   20. 13
21. 12.3 22. 14
23. 1.52 24. 49
25. 448   26. 1560
27. 27.6 28. 21
29. 26.88 30. 25.2
31. 55   32. 27
33. 195 hits
34. 502 mph
35. $262.92
36. 6.1 hr

ADDITIONAL PRACTICE

SKILLS

Solve and check. Show all steps. (Pages 190–192).

1. $x + 108 = 413$
2. $t + 45 = 198$
3. $1.47 = x + 0.35$
4. $7.05 = v + 3.49$
5. $v + 1\frac{7}{8} = 3\frac{1}{4}$
6. $x + 2\frac{3}{4} = 9\frac{7}{16}$

(Pages 193–195)

7. $x - 38 = 51$
8. $t - 84 = 58$
9. $v - 12.3 = 17.9$
10. $m - 117.8 = 95.6$
11. $1\frac{13}{16} = t - \frac{3}{4}$
12. $4\frac{3}{8} = P - 4\frac{3}{4}$

Complete the tables.
(Pages 196–198)

	Gross Earnings	Total Deductions	Net Pay
13.	$640.00	$125.00	?
14.	$847.95	$258.13	?

Complete the table.
(Pages 206–209)

	Rate	Time	Distance
15.	? km/h	5 h	375 km
16.	8.8 ft/sec	?	21.12 ft

Solve and check. (Pages 200–202)

17. $13x = 208$
18. $3w = 231$
19. $455 = 7t$
20. $234 = 18v$
21. $0.9w = 11.07$
22. $1.3y = 18.2$
23. $15.2 = 10v$
24. $2.94 = 0.06a$

(Pages 203–205)

25. $\frac{n}{14} = 32$
26. $\frac{p}{13} = 120$
27. $2.3 = \frac{m}{12}$
28. $1.4 = \frac{x}{15}$
29. $4.2 = \frac{r}{6.4}$
30. $2.1 = \frac{t}{12}$
31. $\frac{p}{15} = 3\frac{2}{3}$
32. $\frac{c}{12} = 2\frac{1}{4}$

APPLICATIONS

33. Alfredo has a batting average of 0.325. He has been at bat 600 times. How many hits does Alfredo have? (Pages 206–209)

34. A plane is flying into a head wind of 27 miles per hour. The ground speed of the plane is 475 miles per hour. Find the air speed. (Pages 196–198)

35. Joan's gross earnings each week are $350.75. The total deductions are $87.83. Find the net pay. (Pages 196–198)

36. The distance from New York to San Francisco is about 5002 kilometers. How many hours will it take a plane to fly this distance at an average speed of 820 kilometers per hour? (Pages 206–209)

214   Chapter 7

CHAPTER

8 Ratio, Proportion, and Percent

FEATURES
Calculator Application: *Simple Interest*
Computer Application: *Problem Solving*
Enrichment: *Compound Interest*

SECTIONS
8-1 Ratio
8-2 Proportions
8-3 Problem Solving and Applications: Using Proportions
8-4 Meaning of Percent
8-5 Percents and Decimals
8-6 Finding a Percent of a Number
8-7 Problem Solving and Applications: Formulas: Simple Interest
8-8 Problem Solving and Applications: Formulas: Discount

Teaching Suggestions
p. M-21

QUICK QUIZ

Write in lowest terms.

1. $\frac{18}{32}$ Ans: $\frac{9}{16}$

2. $\frac{100}{180}$ Ans: $\frac{5}{9}$

Complete.

3. $\frac{3}{4} = \frac{?}{12}$ Ans: 9

4. $\frac{2}{5} = \frac{?}{20}$ Ans: 8

5. $\frac{1}{2} = \frac{?}{120}$ Ans: 60

ADDITIONAL EXAMPLES

Example 1

Write a fraction in lowest terms for each ratio.

1. 18 to 42 Ans: $\frac{3}{7}$

2. 150 to 180 Ans: $\frac{5}{6}$

SELF-CHECK

1. $\frac{1}{3}$ 2. $\frac{4}{3}$ 3. $\frac{1}{7}$

4. $\frac{5}{1}$

8-1 Ratio

In 1956, the cost for the first ounce of a first-class letter was 3¢. In 1985, the cost was 22¢. You can use this ratio to compare these costs.

$\frac{3}{22}$ ← Cost in 1956
← Cost in 1985

A **ratio** uses division to compare two numbers.

3 to 22, 3 : 22, or $\frac{3}{22}$

22 to 3, 22 : 3, or $\frac{22}{3}$

3 : 22 or $\frac{3}{22}$ is read "3 is to 22."

NOTE: The ratios 3 : 22 and 22 : 3 do not have the same meaning. They are not equal.

Just as with fractions, ratios can be written in lowest terms.

EXAMPLE 1 In 1968, the cost of the first ounce of a first-class letter was 6¢. In 1974, the cost was 10¢. Write a ratio in lowest terms to compare the costs.

Solution: $\frac{\text{Cost in 1968}}{\text{Cost in 1974}} = \frac{6}{10} = \frac{3}{5}$ ◄ Lowest terms

Self-Check Write a fraction in lowest terms for each ratio.

1. 5 to 15   2. 12 to 9   3. 3 : 21   4. 25 : 5

The ratios $\frac{6}{10}$ and $\frac{3}{5}$ are equivalent.

Equivalent ratios are equal.

PROCEDURE

To determine whether two ratios are equal:

1. Find the cross-products.
2. Compare the cross-products. **Equivalent ratios have equal cross-products.**

216 Chapter 8

EXAMPLE 2 Determine whether each pair of ratios is equivalent.

a. $\frac{3}{20}$ and $\frac{9}{40}$ b. $\frac{7}{12}$ and $\frac{21}{36}$

The loops show the cross-products.

Solutions:

a. $\frac{3}{20} \stackrel{?}{=} \frac{9}{40}$

$3 \cdot 40 \stackrel{?}{=} 20 \cdot 9$

$120 \stackrel{?}{=} 180$ **No**

The ratios are **not equivalent**.

b. $\frac{7}{12} \stackrel{?}{=} \frac{21}{36}$

$7 \cdot 36 \stackrel{?}{=} 12 \cdot 21$

$252 \stackrel{?}{=} 252$ **Yes**

The ratios are **equivalent**.

Self-Check Determine whether each pair of ratios is equivalent.

5. $\frac{2}{3}$ and $\frac{3}{4}$ 6. $\frac{1}{2}$ and $\frac{3}{6}$ 7. $\frac{2}{25}$ and $\frac{9}{58}$ 8. $\frac{4}{5}$ and $\frac{7}{9}$

Variables can be used in one, or both, terms of a ratio. However, the variables cannot have a value that will make the second term equal zero.

> In the ratio $\frac{a}{b}$, $b \neq 0$, a is the **first term** and b is the **second term**.

EXAMPLE 3 Write each ratio in lowest terms.

a. $\frac{9q}{12q}$, $q \neq 0$ b. $\frac{3}{12d}$, $d \neq 0$

Solutions:

a. $\frac{9q}{12q} = \frac{\overset{3}{\cancel{9}}}{\underset{4}{\cancel{12}}} \cdot \frac{\overset{1}{\cancel{q}}}{\underset{1}{\cancel{q}}}$

$= \frac{3 \cdot 1}{4 \cdot 1}$

$= \frac{3}{4}$

b. $\frac{3}{12d} = \frac{\overset{1}{\cancel{3}}}{\underset{4}{\cancel{12}}} \cdot \frac{1}{d}$

$= \frac{1 \cdot 1}{4 \cdot d}$

$= \frac{1}{4d}$

Self-Check Write each ratio in lowest terms. No variable equals 0.

9. $\frac{6y}{15y}$ 10. $\frac{25t}{35t}$ 11. $\frac{4}{16n}$ 12. $\frac{9}{21p}$

Ratio, Proportion, and Percent

ADDITIONAL EXAMPLES

Example 2
Determine whether each pair of ratios is equivalent.

1. $\frac{12}{20}$ and $\frac{18}{30}$ Ans: Yes
2. $\frac{6}{21}$ and $\frac{9}{28}$ Ans: No

SELF-CHECK

5. No 6. Yes
7. No 8. No

Example 3
Write in lowest terms.

1. $\frac{80x}{16xy}$ Ans: $\frac{5}{y}$
2. $\frac{132a}{154a}$ Ans: $\frac{6}{7}$

SELF-CHECK

9. $\frac{2}{5}$ 10. $\frac{5}{7}$ 11. $\frac{1}{4n}$
12. $\frac{3}{7p}$

CLASSROOM EXERCISES

1. $\frac{3}{4}$ 2. $\frac{10}{7}$ 3. $\frac{5}{1}$
4. $\frac{12}{5}$ 5. $\frac{4}{7}$ 6. $\frac{3}{1}$
7. Yes 8. No 9. No
10. Yes 11. $\frac{3}{5}$ 12. $\frac{2}{1}$
13. $\frac{7p}{2}$ 14. $\frac{4}{7c}$ 15. $\frac{3}{2}$
16. $\frac{4}{1}$ 17. $\frac{1}{2q}$ 18. $\frac{1}{9t}$
19. $\frac{1}{3}$ 20. $\frac{7}{1}$ 21. $\frac{1}{1}$
22. $\frac{1}{1}$

ASSIGNMENT GUIDE
BASIC
p. 218: 1-26
AVERAGE
p. 218: 1-43 odd
ABOVE AVERAGE
p. 218: 1-43 odd

WRITTEN EXERCISES

1. $\frac{1}{3}$ 2. $\frac{16}{25}$ 3. $\frac{3}{8}$
4. $\frac{1}{5}$ 5. $\frac{1}{10}$ 6. $\frac{9}{5}$
7. $\frac{1}{3}$ 8. $\frac{1}{5}$ 9. $\frac{1}{4}$
10. $\frac{1}{4}$ 11. $\frac{1}{3}$ 12. $\frac{1}{9}$
13. Yes 14. No
15. Yes 16. Yes
17. No 18. No
See page 221 for the answers to Exercises 19-44.

CLASSROOM EXERCISES

Write each ratio in lowest terms. (Example 1)

1. $\frac{9}{12}$ 2. $\frac{50}{35}$ 3. 20 to 4 4. 36 to 15 5. 8 : 14 6. 300 : 100

Determine whether each pair of ratios is equivalent. Answer Yes or No. (Example 2)

7. $\frac{1}{2}$ and $\frac{6}{12}$ 8. $\frac{3}{4}$ and $\frac{15}{22}$ 9. $\frac{2}{7}$ and $\frac{21}{63}$ 10. $\frac{2}{3}$ and $\frac{14}{21}$

Write each ratio in lowest terms. No variable equals 0. (Example 3)

11. $\frac{3x}{5x}$ 12. $\frac{16t}{8t}$ 13. $\frac{42p}{12}$ 14. $\frac{8}{14c}$ 15. $\frac{24d}{16d}$ 16. $\frac{8m}{2m}$
17. $\frac{7}{14q}$ 18. $\frac{4}{36t}$ 19. $\frac{16a}{48a}$ 20. $\frac{77d}{11d}$ 21. $\frac{s}{s}$ 22. $\frac{r}{r}$

WRITTEN EXERCISES

Goals: To write ratios in lowest terms
To determine whether pairs of ratios are equivalent

Write each ratio in lowest terms. (Example 1)

1. 14 to 42 2. 16 to 25 3. 30 : 80 4. 5 : 25 5. 10 to 100 6. 36 to 20
7. $\frac{7}{21}$ 8. $\frac{4}{20}$ 9. $\frac{15}{60}$ 10. $\frac{16}{64}$ 11. $\frac{17}{51}$ 12. $\frac{6}{54}$

Write the cross-product to determine whether each pair of ratios is equivalent. Answer Yes or No. (Example 2)

13. $\frac{4}{1}$ and $\frac{28}{7}$ 14. $\frac{12}{1}$ and $\frac{132}{13}$ 15. $\frac{3}{16}$ and $\frac{9}{48}$ 16. $\frac{3}{2}$ and $\frac{27}{18}$
17. $\frac{7}{10}$ and $\frac{54}{80}$ 18. $\frac{7}{12}$ and $\frac{36}{60}$ 19. $\frac{6}{5}$ and $\frac{17}{12}$ 20. $\frac{1}{3}$ and $\frac{16}{51}$

Write each ratio in lowest terms. No variable equals 0. (Example 3)

21. $\frac{10a}{12a}$ 22. $\frac{21y}{14y}$ 23. $\frac{15r}{6}$ 24. $\frac{8}{20k}$ 25. $\frac{40d}{28d}$ 26. $\frac{12}{30p}$
27. $\frac{15t}{35t}$ 28. $\frac{48c}{18}$ 29. $\frac{5a}{12a}$ 30. $\frac{20c}{5}$ 31. $\frac{15a}{8}$ 32. $\frac{18x}{60x}$
33. $\frac{9r}{9r}$ 34. $\frac{3d}{3d}$ 35. $\frac{8t}{6t}$ 36. $\frac{14a}{49a}$ 37. $\frac{5r}{10}$ 38. $\frac{60y}{15}$
39. $\frac{21r}{21}$ 40. $\frac{47r}{47}$ 41. $\frac{t}{9t}$ 42. $\frac{q}{7q}$ 43. $\frac{40p}{15p}$ 44. $\frac{25x}{75x}$

218 Chapter 8

REVIEW CAPSULE FOR SECTION 8-2

Solve each equation. (Pages 200–202)

1. $5x = 75$
2. $13n = 130$
3. $546 = 21p$
4. $90 = 8w$
5. $4p = 50$
6. $4k = 45$
7. $105 = 8t$
8. $80 = 25q$

The answers are on page 249.

8-2 Proportions

A **proportion** is an equation that shows equivalent ratios.

$$\frac{r}{7} = \frac{15}{21} \quad \text{Proportion} \qquad \frac{3}{4} = \frac{n}{8} \quad \text{Proportion}$$

A ratio has two terms. A proportion has four terms.

First term ⟶ $\frac{3}{4} = \frac{15}{20}$ ⟵ Third term
Second term ⟶ ⟵ fourth term

Solving a proportion is similar to solving a multiplication equation.

Property of Proportions

If $\frac{a}{b} = \frac{c}{d}$, then $ad = cb$, $b \neq 0$, $d \neq 0$.

If $ad = cb$, then $\frac{a}{b} = \frac{c}{d}$, $b \neq 0$, $d \neq 0$.

PROCEDURE

To solve a proportion:

1. Write the cross-products.
2. Solve the resulting equation for the variable.

EXAMPLE 1 Solve and check: $\frac{t}{30} = \frac{5}{6}$

Solution: $\frac{t}{30} = \frac{5}{6}$

1. $t \cdot 6 = 30 \cdot 5$
   $6t = 150$
2. $\frac{6t}{6} = \frac{150}{6}$
   $t = 25$

Check: $\frac{25}{30} \stackrel{?}{=} \frac{5}{6}$

$25 \cdot 6$	$30 \cdot 5$
150	150

Ratio, Proportion, and Percent   219

Margin Notes

SELF-CHECK
1. r = 1   2. p = 1
3. w = 13   4. p = $2\frac{3}{5}$

ADDITIONAL EXAMPLES
Example 2
Solve and check.
1. $\frac{7}{x} = \frac{6}{12}$ Ans: x = 14
2. $\frac{30}{48} = \frac{c}{4}$ Ans: c = $2\frac{1}{2}$

SELF-CHECK
5. w = $2\frac{2}{5}$ 6. q = 4
7. p = 8   8. t = 16

CLASSROOM EXERCISES
1. 16n = 80
2. 18p = 18   3. 45 = 7x
4. 40 = 6y   5. 24 = 4m
6. 3r = 18   7. 3   8. 4
9. 8   10. 21   11. 2
12. 48

ASSIGNMENT GUIDE
BASIC
pp. 220-221: 1-29 odd
AVERAGE
pp. 220-221: 1-33 odd
ABOVE AVERAGE
pp. 220-221: 1-35 odd

220

Main Content

Self-Check *Solve and check.*

1. $\frac{r}{8} = \frac{3}{24}$ 2. $\frac{p}{7} = \frac{6}{42}$ 3. $\frac{w}{20} = \frac{65}{100}$ 4. $\frac{p}{13} = \frac{1}{5}$

A variable can appear in any term of the proportion.

EXAMPLE 2 Solve and check: $\frac{10}{3} = \frac{x}{2}$

Solution:

$\frac{10}{3} = \frac{x}{2}$

[1] $10 \cdot 2 = 3 \cdot x$

$20 = 3x$

[2] $\frac{20}{3} = \frac{3x}{3}$

$6\frac{2}{3} = x$

Check: $\frac{10}{3} = \frac{6\frac{2}{3}}{2}$

$10 \cdot 2 \;\bigg|\; 3 \cdot 6\frac{2}{3}$

$20 \;\bigg|\; \cancel{3} \cdot \frac{20}{\cancel{3}}$

$20 \;\bigg|\; 20$

Self-Check *Solve and check.*

5. $\frac{3}{10} = \frac{w}{8}$ 6. $\frac{14}{35} = \frac{q}{10}$ 7. $\frac{5}{p} = \frac{15}{24}$ 8. $\frac{10}{t} = \frac{5}{8}$

CLASSROOM EXERCISES

*Use cross-products to write an equation. Do **not** solve the equation.* (Examples 1 and 2, Step 1)

1. $\frac{n}{8} = \frac{10}{16}$ 2. $\frac{p}{3} = \frac{6}{18}$ 3. $\frac{15}{x} = \frac{7}{3}$ 4. $\frac{8}{6} = \frac{y}{5}$ 5. $\frac{3}{4} = \frac{m}{8}$ 6. $\frac{3}{2} = \frac{9}{r}$

Solve and check each proportion. (Examples 1 and 2)

7. $\frac{x}{6} = \frac{1}{2}$ 8. $\frac{n}{7} = \frac{20}{35}$ 9. $\frac{5}{p} = \frac{15}{24}$ 10. $\frac{9}{s} = \frac{3}{7}$ 11. $\frac{6}{27} = \frac{k}{9}$ 12. $\frac{6}{7} = \frac{m}{56}$

WRITTEN EXERCISES

Goal: To solve proportions

Solve and check each proportion. (Examples 1 and 2)

1. $\frac{x}{30} = \frac{3}{5}$ 2. $\frac{y}{24} = \frac{5}{6}$ 3. $\frac{n}{8} = \frac{3}{24}$ 4. $\frac{p}{9} = \frac{24}{54}$ 5. $\frac{m}{3} = \frac{10}{15}$ 6. $\frac{r}{84} = \frac{3}{14}$

Chapter 8

7. $\frac{6}{x} = \frac{18}{21}$ 8. $\frac{10}{t} = \frac{20}{30}$ 9. $\frac{12}{s} = \frac{4}{9}$ 10. $\frac{24}{k} = \frac{6}{5}$ 11. $\frac{12}{a} = \frac{9}{12}$ 12. $\frac{n}{3} = \frac{10}{15}$

13. $\frac{9}{7} = \frac{p}{42}$ 14. $\frac{2}{3} = \frac{w}{21}$ 15. $\frac{35}{90} = \frac{b}{18}$ 16. $\frac{25}{75} = \frac{c}{15}$ 17. $\frac{2}{5} = \frac{m}{45}$ 18. $\frac{8}{70} = \frac{d}{35}$

19. $\frac{3}{12} = \frac{15}{n}$ 20. $\frac{3}{5} = \frac{21}{x}$ 21. $\frac{15}{33} = \frac{5}{y}$ 22. $\frac{30}{70} = \frac{6}{q}$ 23. $\frac{7}{9} = \frac{28}{m}$ 24. $\frac{30}{50} = \frac{3}{t}$

25. $\frac{c}{10} = \frac{8}{25}$ 26. $\frac{7}{y} = \frac{8}{15}$ 27. $\frac{9}{16} = \frac{x}{8}$ 28. $\frac{4}{5} = \frac{10}{p}$ 29. $\frac{3}{4} = \frac{k}{15}$ 30. $\frac{7}{5} = \frac{6}{n}$

APPLICATIONS: USING PROPORTIONS

Solve each problem.

31. The scale on a map compares distance on the map to actual distance. The map of California on the right uses 1 inch to represent 145 miles. How many inches represent 290 miles?

32. On the map of California, how many miles do 3 inches represent?

33. On a map of California, 3 centimeters represent 100 kilometers. How many kilometers do 12 centimeters represent?

34. On a map of California, 2.5 centimeters represent 40 kilometers. How many centimeters represent 200 kilometers?

35. On a map, $\frac{1}{2}$ inch represents 200 miles. How many inches would represent 500 miles?

REVIEW CAPSULE FOR SECTION 8-3

Write each ratio in lowest terms. (Pages 216–218)

1. 12 to 30   2. 6 to 58   3. 20 : 18   4. 21 : 35   5. $\frac{7}{91}$ 6. $\frac{42}{15}$

Solve each equation. (Pages 200–202)

7. $6x = 24$ 8. $12p = 108$ 9. $350 = 14m$ 10. $73 = 8n$

The answers are on page 249.

Ratio, Proportion, and Percent   **221**

WRITTEN EXERCISES

1. 18   2. 20   3. 1
4. 4   5. 2   6. 18
7. 7   8. 15   9. 27
10. 20   11. 16   12. 2
13. 54   14. 14   15. 7
16. 5   17. 18   18. 4
19. 60   20. 35   21. 11
22. 14   23. 36   24. 5
25. $3\frac{1}{5}$ 26. $13\frac{1}{8}$
27. $4\frac{1}{2}$ 28. $12\frac{1}{2}$
29. $11\frac{1}{4}$ 30. $4\frac{2}{7}$ 31. 2
32. 435   33. 400
34. 12.5   35. $1\frac{1}{4}$

WRITTEN EXERCISES p. 218

19. No   20. No   21. $\frac{5}{6}$
22. $\frac{3}{2}$ 23. $\frac{5r}{2}$ 24. $\frac{2}{5k}$
25. $\frac{10}{7}$ 26. $\frac{2}{5p}$ 27. $\frac{3}{7}$
28. $\frac{8c}{3}$ 29. $\frac{5}{12}$ 30. $\frac{4c}{1}$
31. $\frac{15a}{8}$ 32. $\frac{3}{10}$ 33. $\frac{1}{1}$
34. $\frac{1}{1}$ 35. $\frac{4}{3}$ 36. $\frac{2}{7}$
37. $\frac{r}{2}$ 38. $\frac{4y}{1}$ 39. $\frac{r}{1}$
40. $\frac{r}{1}$ 41. $\frac{1}{9}$ 42. $\frac{1}{7}$
43. $\frac{8}{3}$ 44. $\frac{1}{3}$

Teaching Suggestions
p. M-21

QUICK QUIZ
Write each ratio in lowest terms.
<u>1</u>. 10 to 40   Ans: $\frac{1}{4}$
<u>2</u>. 18 to 10   Ans: $\frac{9}{5}$
Solve and check.
<u>3</u>. 5a = 175   Ans: a = 35
<u>4</u>. 204 = 17b
   Ans: b = 12
<u>5</u>. 8 x = 44   Ans: x = $5\frac{1}{2}$

ADDITIONAL EXAMPLE
Example 1
In Example 1, Anna mixed yellow and blue paint in the ratio 5 : 2. She plans to use 8 quarts of blue paint. How much yellow paint will she need? Ans: 20 quarts

SELF-CHECK
<u>1</u>. $21\frac{1}{3}$ quarts

PROBLEM SOLVING AND APPLICATIONS

8-3 Using Proportions

Proportions are useful in solving many kinds of problems.

> **PROCEDURE**
> **To use proportions to solve word problems:**
> 1. Choose a variable to represent the unknown.
> 2. Write two ratios. Compare the quantities <u>in the same order</u> in both ratios.
> 3. Use the ratios to write a proportion.
> 4. Solve the proportion.

EXAMPLE 1 To obtain a certain shade of green paint, Anna mixed yellow and blue paint in the ratio 4 : 3. She plans to use 12 quarts of yellow paint. How much blue paint will be needed?

Solution: 1. Let x = the number of quarts of blue paint.

2.

	One Ratio	Ratio with the Unknown
	$\frac{4}{3}$ ← Yellow paint ← Blue paint	$\frac{12}{x}$ ← Quarts of yellow paint ← Quarts of blue paint

3. $\frac{4}{3} = \frac{12}{x}$ ◄ Use the table to write a proportion.

4. $4 \cdot x = 3 \cdot 12$
   $4x = 36$
   $x = 9$ Thus, **9 quarts** of blue paint will be needed.

Self-Check 1. In Example 1, suppose that Anna plans to use 16 quarts of blue paint. How much yellow paint will she need?

222 Chapter 8

EXAMPLE 2 In checking the heartbeat of a patient, a nurse counted 19 beats in 15 seconds. Find the patient's heartbeat for 60 seconds.

Solution: ① Let $h =$ the heartbeat for 60 seconds.

②
One Ratio	Ratio with the Unknown
$\frac{19}{15}$ ← Number of beats / Number of seconds	$\frac{h}{60}$ ← Number of beats / Number of seconds

③ $\quad \dfrac{19}{15} = \dfrac{h}{60}$

④ $\quad 19 \cdot 60 = 15 \cdot h$

$\quad\quad 1140 = 15h$

$\quad\quad \dfrac{1140}{15} = \dfrac{15h}{15}$

$\quad\quad 76 = h$

There will be **76 beats** in 60 seconds.

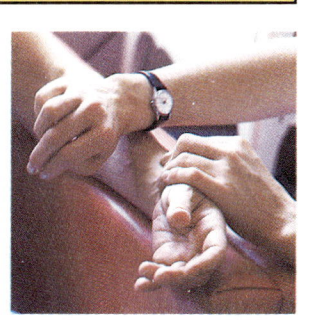

Self-Check 2. In Example 2, find the patient's heartbeat for 75 seconds.

CLASSROOM EXERCISES

Complete each table. (Examples 1 and 2)

1. 2 pounds of bananas at 96¢
   5 pounds at x cents

One Ratio	Ratio with the Unknown
$\frac{2}{96}$ ← pounds / cents	?

2. 20 pounds of dog food at $8
   35 pounds at x dollars

One Ratio	Ratio with the Unknown
$\frac{20}{8}$ ← pounds / dollars	?

3. 64 feet of rope weigh 20 pounds
   90 feet weigh x pounds

One Ratio	Ratio with the Unknown
$\frac{64}{20}$ ← feet / pounds	?

4. 2 meters of cable weigh 5 kg.
   12 meters weigh x kg

One Ratio	Ratio with the Unknown
$\frac{2}{5}$ ← meters / kilograms	?

Ratio, Proportion, and Percent

ASSIGNMENT GUIDE
BASIC
p. 224: 1–10
AVERAGE
pp. 224–225: 1–14
ABOVE AVERAGE
pp. 224–225: 1–14

WRITTEN EXERCISES

<u>1</u>. 585 km   <u>2</u>. $12\frac{4}{5}$ in
<u>3</u>. 18   <u>4</u>. 10   <u>5</u>. $77\frac{7}{9}$
<u>6</u>. 84 beats   <u>7</u>. 10.5 in
<u>8</u>. 106.25 mm   <u>9</u>. 30 in

WRITTEN EXERCISES

Goal: To use proportions to solve word problems

1. On a map, the distance from San Francisco to Pasadena is 9 centimeters. On the map, 2 centimeters represent 130 kilometers. Find the actual distance between the cities.

2. On a blue print for a house, 2 inches represent 5 feet. If the actual length of the family room is 32 feet, what is its length on the blueprint?

3. A painter mixed yellow paint with blue paint in the ratio of 2:3 to obtain a shade of green. How many gallons of blue paint are needed to mix with 12 gallons of yellow paint to obtain the same shade of green?

4. A certain shade of paint can be obtained by mixing yellow with red in the ratio of 1:2. How many gallons of red paint are needed to mix with 5 gallons of yellow to obtain the same shade of paint?

5. The Santini family drove 280 miles on a tank of gasoline. The tank held about 18 gallons. How many miles could they drive on 5 gallons?

6. In checking a patient's heartbeat, a nurse counted 28 beats in 20 seconds. Find the patient's heartbeat for 60 seconds.

7. A photograph has a length of 7 inches and a width of 5 inches. The photograph is enlarged so that the width is 7.5 inches. What is the length of the enlarged photograph?

8. A photograph measuring 200 millimeters wide and 250 millimeters long must be reduced to a width of 85 millimeters. What will be its reduced length?

9. Alan is building a radio controlled model airplane. One inch on the model represents 12 inches on the actual plane. If the actual plane is 360 inches long, what is the length of the model airplane?

10. Virginia is building a model of an old sailing ship. The scale of the model is in the ratio of 1:15. If the length of the actual ship is 22.5 meters, what is the length of the model?

Chapter 8

11. A 20-acre field yields 610 bushels of soybeans. About how many bushels would a 25-acre field yield under similar conditions?

12. A farmer had a yield of 228 bushels of corn for every 3 acres. What yield could he expect for 11 acres under similar conditions?

13. There are about 60 calories in 4 ounces of apple juice. How many calories are there in 20 ounces of apple juice?

14. Jogging for 15 minutes uses about 105 calories. How many minutes of jogging would be needed to use 525 calories?

REVIEW: SECTIONS 8-1–8-3

Write a fraction in lowest terms for each ratio. (Section 8-1)

1. 12 to 15
2. 42 to 28
3. 16 : 10
4. 5 : 25
5. $\frac{18}{30}$
6. $\frac{65}{25}$
7. $\frac{18t}{75t}, t \neq 0$
8. $\frac{16k}{24}$

Determine whether each pair of ratios is equivalent. Answer Yes or No. (Section 8-1)

9. $\frac{2}{3}$ and $\frac{16}{24}$
10. $\frac{6}{13}$ and $\frac{18}{39}$
11. $\frac{4}{5}$ and $\frac{8}{20}$
12. $\frac{35}{59}$ and $\frac{5}{7}$

Solve each proportion. (Section 8-2)

13. $\frac{4}{12} = \frac{5}{d}$
14. $\frac{15}{x} = \frac{40}{16}$
15. $\frac{9}{12} = \frac{b}{8}$
16. $\frac{a}{21} = \frac{4}{5}$

Solve by using a proportion. (Section 8-3)

17. An 8 ounce glass of whole milk contains about 160 calories. How many calories are there in 24 ounces of whole milk?

18. Bicycling for 10 minutes uses about 54 calories of energy. How many minutes of bicycling would be needed to use 270 calories?

REVIEW CAPSULE FOR SECTION 8-4

Solve each proportion. (Pages 219–221)

1. $\frac{2}{5} = \frac{x}{100}$
2. $\frac{3}{10} = \frac{p}{100}$
3. $\frac{4}{25} = \frac{a}{100}$
4. $\frac{23}{50} = \frac{b}{100}$
5. $\frac{3}{4} = \frac{c}{100}$
6. $\frac{9}{20} = \frac{t}{100}$
7. $\frac{23}{10} = \frac{r}{100}$
8. $\frac{12}{5} = \frac{y}{100}$
9. $\frac{x}{100} = \frac{3}{8}$
10. $\frac{a}{100} = \frac{7}{12}$
11. $\frac{b}{100} = \frac{7}{8}$
12. $\frac{d}{100} = \frac{1}{3}$

The answers are on page 249.

Ratio, Proportion, and Percent

WRITTEN EXERCISES
10. 1.5 m 11. $762\frac{1}{2}$ bu
12. 836 bu 13. 300
14. 75

QUIZ: SECTIONS 8-1-8-3
After completing this Review, you may wish to administer a quiz covering the same sections. A Quiz is provided in the Teacher's Edition: Part II.

REVIEW: SECTIONS 8-1-8-3
1. $\frac{4}{5}$ 2. $\frac{3}{2}$ 3. $\frac{8}{5}$
4. $\frac{1}{5}$ 5. $\frac{3}{5}$ 6. $\frac{13}{5}$
7. $\frac{6}{25}$ 8. $\frac{2k}{3}$ 9. Yes
10. Yes 11. No
12. No 13. 15 14. 6
15. 6 16. $16\frac{4}{5}$
17. 480 18. 50

COMPUTER APPLICATIONS

This feature illustrates the use of a computer program for solving word problems involving ratios, as introduced in Chapter 7 and related to percent in Section 8-4 (pages 227-230). Two fundamental problem solving steps, identifying the input/output variables and writing the correct formula, are logically extended to writing and checking the BASIC program. The computer solution and printout provides verification and reinforcement of the student's problem solving skills.

1. H = number of hours
   R = pay per hour
   P = pay
   ```
   10  INPUT H, R
   20  LET P = H * R
   30  PRINT P
   40  END
   ```

2. M = miles traveled
   G = gallons used
   A = miles-per-gallon
   ```
   10  INPUT M, G
   20  LET A = M/G
   30  PRINT A
   40  END
   ```

COMPUTER APPLICATIONS Problem Solving

In baseball, a player's **batting average** is defined as a ratio of the **number of hits** to the **number of times-at-bat**. This ratio is then expressed as a decimal.

PROBLEM Write a program which accepts a baseball player's times-at-bat and the number of hits. Then have the program print the player's batting average.

Solution: There are four logical phases for writing a program to solve a problem.

Phase 1 Identify the input variables and the output variables.

B = number of times-at-bat ← INPUT variable
H = number of hits ← INPUT variable
A = batting average ← OUTPUT variable

Phase 2 Write a formula for calculating the value of the output variable.

Batting Average (A) = Number of Hits (H) ÷ Times at Bat (B)

Phase 3 Write a program.
```
10  PRINT "ENTER TIMES-AT-BAT."
20  INPUT B
30  PRINT "ENTER NUMBER OF HITS."
40  INPUT H
50  LET A = H / B
60  PRINT "BATTING AVERAGE = "; A
70  END
```

Phase 4 Check the program by using test values. In the following output, B = 50 and H = 15.

```
ENTER TIMES-AT-BAT.
? 50
ENTER NUMBER OF HITS.
? 15
BATTING AVERAGE = .3
```
← **Final zeros after the decimal point are not usually printed.**

Exercises Write a program for each problem.

1. Given the number of hours worked and the pay per hour, compute the worker's pay.

2. Given the number of miles a car has traveled and the number of gallons of gasoline used, compute the average miles-per-gallon.

Computer Application

8-4 Meaning of Percent

The ratio 77 out of 100 or $\frac{77}{100}$ can be read as 77 percent. It is written as 77%.

77 out of 100 persons PREFER CLEAN-UP TOOTHPASTE

> A percent is a ratio in which the second term is 100.
>
> $\frac{3}{100} = 3\%$
> $\frac{y}{100} = y\%$

If the second term of a ratio is not 100, you can use a proportion to write a percent for the ratio.

PROCEDURE

To write a percent for a ratio whose denominator is not 100:

1. Choose a variable for the first term of the percent ratio. Using 100 as the second term, write the percent ratio.
2. Write a proportion.
3. Solve the proportion.

EXAMPLE 1 Write a percent for $\frac{17}{20}$.

Solution:

1. Let p = the first term of the percent ratio.

   Percent ratio: $\frac{p}{100}$ ◀ The second term is 100.

2. Solve for p.

   $\frac{p}{100} = \frac{17}{20}$

3. $20p = (100)(17)$

   $20p = 1700$

   $\frac{20p}{20} = \frac{1700}{20}$

   $p = 85$ Thus, $\frac{17}{20} = \frac{85}{100} = 85\%$

Self-Check *Write a percent for each ratio.*

1. $\frac{1}{4}$ 2. $\frac{3}{5}$ 3. $\frac{7}{10}$ 4. $\frac{0}{3}$ 5. $\frac{9}{20}$ 6. $\frac{41}{50}$

Ratio, Proportion, and Percent

Teaching Suggestions p. M-21

QUICK QUIZ

Solve and check.

1. $\frac{18}{12} = \frac{x}{42}$ Ans: x = 63
2. $\frac{7}{5} = \frac{p}{100}$ Ans: p = 140
3. $\frac{a}{100} = \frac{13}{20}$ Ans: a = 65
4. $\frac{q}{28} = \frac{25}{100}$ Ans: q = 7
5. $\frac{9}{t} = \frac{36}{100}$ Ans: t = 25

ADDITIONAL EXAMPLES

Example 1

Write a percent for each ratio.

1. $\frac{6}{5}$ Ans: 120%
2. $\frac{3}{4}$ Ans: 75%
3. $\frac{7}{35}$ Ans: 20%

SELF-CHECK

1. 25%   2. 60%   3. 70%
4. 0%   5. 45%   6. 82%

ADDITIONAL EXAMPLES

Example 2
Write a percent for each ratio.
1. $\frac{3}{8}$ Ans: $37\frac{1}{2}$% or 37.5%
2. $\frac{3}{11}$ Ans: $27\frac{3}{11}$% or 27.27%

SELF-CHECK
7. $33\frac{1}{3}$%   8. $83\frac{1}{3}$%
9. $42\frac{6}{7}$%   10. $22\frac{2}{9}$%
11. $41\frac{2}{3}$%   12. $56\frac{1}{4}$%

Example 3
Write a percent for each ratio.
1. $\frac{60}{40}$ Ans: 150%
2. $\frac{27}{6}$ Ans: 450%

SELF-CHECK
13. 100%   14. 100%
15. 350%   16. 625%
17. $216\frac{2}{3}$%   18. 325%

EXAMPLE 2 *Write a percent for $\frac{1}{6}$.*

Solution: ① Let m = the first term of the percent ratio.

Percent ratio: $\frac{m}{100}$

② $\frac{m}{100} = \frac{1}{6}$

③ $6m = 100$

$m = 16\frac{4}{6} = 16\frac{2}{3}$

Thus, $\frac{1}{6} = 16\frac{2}{3}$%.

Self-Check *Write a percent for each ratio.*

7. $\frac{1}{3}$ 8. $\frac{5}{6}$ 9. $\frac{3}{7}$ 10. $\frac{2}{9}$ 11. $\frac{5}{12}$ 12. $\frac{9}{16}$

If the first and second terms of a ratio are the same, the ratio *equals* 100%.

$\frac{50}{50} = 100\%$ $\frac{7.3}{7.3} = 100\%$ $\frac{2\frac{1}{2}}{2\frac{1}{2}} = 100\%$

If the first term of a ratio is greater than the second term, the ratio is greater than 100%.

EXAMPLE 3 *Write a percent for $\frac{11}{8}$.*

Solution: ① Let q = the first term of the percent ratio.

Percent ratio: $\frac{q}{100}$

② $\frac{q}{100} = \frac{11}{8}$

③ $8q = 1100$

$q = 137\frac{4}{8} = 137\frac{1}{2}$

Thus, $\frac{11}{8} = \frac{137\frac{1}{2}}{100} = 137\frac{1}{2}$%.

Self-Check *Write a percent for each ratio.*

13. $\frac{100}{100}$ 14. $\frac{1.5}{1.5}$ 15. $\frac{7}{2}$ 16. $\frac{25}{4}$ 17. $\frac{13}{6}$ 18. $\frac{39}{12}$

Chapter 8

CLASSROOM EXERCISES

Write a percent for each ratio. (Definition)

1. $\dfrac{9}{100}$ 2. $\dfrac{13}{100}$ 3. $\dfrac{7.6}{100}$ 4. $\dfrac{113.6}{100}$ 5. $\dfrac{16\frac{2}{3}}{100}$ 6. $\dfrac{\frac{1}{2}}{100}$

Write the proportion you would use in writing a percent for each ratio. Use n for the variable. (Examples 1–3, Steps 1 and 2)

7. $\dfrac{3}{4}$ 8. $\dfrac{3}{10}$ 9. $\dfrac{7}{20}$ 10. $\dfrac{2}{5}$ 11. $\dfrac{1}{2}$ 12. $\dfrac{3}{8}$

13. $\dfrac{2}{3}$ 14. $\dfrac{5}{12}$ 15. $\dfrac{4}{15}$ 16. $\dfrac{4}{9}$ 17. $\dfrac{1}{6}$ 18. $\dfrac{3}{7}$

19. $\dfrac{16}{5}$ 20. $\dfrac{21}{10}$ 21. $\dfrac{20}{7}$ 22. $\dfrac{11}{7}$ 23. $\dfrac{20}{19}$ 24. $\dfrac{24}{11}$

WRITTEN EXERCISES

Goal: To write a percent for a ratio

For Exercises 1–54, write a percent for each ratio. (Example 1)

1. $\dfrac{1}{4}$ 2. $\dfrac{3}{10}$ 3. $\dfrac{1}{5}$ 4. $\dfrac{1}{20}$ 5. $\dfrac{2}{25}$ 6. $\dfrac{3}{50}$

7. $\dfrac{4}{5}$ 8. $\dfrac{29}{50}$ 9. $\dfrac{4}{25}$ 10. $\dfrac{11}{20}$ 11. $\dfrac{17}{20}$ 12. $\dfrac{37}{50}$

(Example 2)

13. $\dfrac{5}{8}$ 14. $\dfrac{1}{12}$ 15. $\dfrac{1}{40}$ 16. $\dfrac{3}{16}$ 17. $\dfrac{5}{7}$ 18. $\dfrac{8}{9}$

19. $\dfrac{1}{50}$ 20. $\dfrac{9}{11}$ 21. $\dfrac{7}{11}$ 22. $\dfrac{5}{9}$ 23. $\dfrac{13}{40}$ 24. $\dfrac{9}{16}$

(Example 3)

25. $\dfrac{12}{10}$ 26. $\dfrac{14}{10}$ 27. $\dfrac{30}{20}$ 28. $\dfrac{15}{8}$ 29. $\dfrac{21}{16}$ 30. $\dfrac{17}{6}$

31. $\dfrac{51}{50}$ 32. $\dfrac{7}{4}$ 33. $\dfrac{11}{5}$ 34. $\dfrac{76}{25}$ 35. $\dfrac{17}{5}$ 36. $\dfrac{167}{20}$

MIXED PRACTICE

37. $\dfrac{4}{10}$ 38. $\dfrac{18}{5}$ 39. $\dfrac{7}{9}$ 40. $\dfrac{1}{8}$ 41. $\dfrac{22}{15}$ 42. $\dfrac{9}{2}$

43. $\dfrac{9}{50}$ 44. $\dfrac{15}{4}$ 45. $\dfrac{7}{11}$ 46. $\dfrac{3}{40}$ 47. $\dfrac{1}{12}$ 48. $\dfrac{16}{40}$

49. $\dfrac{61}{10}$ 50. $\dfrac{8}{9}$ 51. $\dfrac{119}{10}$ 52. $\dfrac{11}{12}$ 53. $\dfrac{7}{10}$ 54. $\dfrac{25}{4}$

Ratio, Proportion, and Percent

CLASSROOM EXERCISES

1. 9%   2. 13%   3. 7.6%
4. 113.6%   5. $16\frac{2}{3}$%
6. $\frac{1}{2}$%   7. $\dfrac{n}{100} = \dfrac{3}{4}$

8. $\dfrac{n}{100} = \dfrac{3}{10}$

9. $\dfrac{n}{100} = \dfrac{7}{20}$

10. $\dfrac{n}{100} = \dfrac{2}{5}$

11. $\dfrac{n}{100} = \dfrac{1}{2}$

12. $\dfrac{n}{100} = \dfrac{3}{8}$

13. $\dfrac{n}{100} = \dfrac{2}{3}$

14. $\dfrac{n}{100} = \dfrac{5}{12}$

15. $\dfrac{n}{100} = \dfrac{4}{15}$

16. $\dfrac{n}{100} = \dfrac{4}{9}$

17. $\dfrac{n}{100} = \dfrac{1}{6}$

18. $\dfrac{n}{100} = \dfrac{3}{7}$

19. $\dfrac{n}{100} = \dfrac{16}{5}$

20. $\dfrac{n}{100} = \dfrac{21}{10}$

21. $\dfrac{n}{100} = \dfrac{20}{7}$

22. $\dfrac{n}{100} = \dfrac{11}{7}$

23. $\dfrac{n}{100} = \dfrac{20}{19}$

24. $\dfrac{n}{100} = \dfrac{24}{11}$

ASSIGNMENT GUIDE
BASIC
p. 229: 1-53 odd
AVERAGE
pp. 229-230: 1-57 odd
ABOVE AVERAGE
pp. 229-230: 1-53 odd, 55-61

WRITTEN EXERCISES

1. 25% 2. 30% 3. 20%
4. 5% 5. 8% 6. 6%
7. 80% 8. 58% 9. 16%
10. 55% 11. 85%
12. 74% 13. $62\frac{1}{2}$%
14. $8\frac{1}{3}$% 15. $2\frac{1}{2}$%
16. $18\frac{3}{4}$% 17. $71\frac{3}{7}$%
18. $88\frac{8}{9}$% 19. 2%
20. $81\frac{9}{11}$% 21. $63\frac{7}{11}$%
22. $55\frac{5}{9}$% 23. $32\frac{1}{2}$%
24. $56\frac{1}{4}$% 25. 120%
26. 140% 27. 150%
28. $187\frac{1}{2}$% 29. $131\frac{1}{4}$%
30. $283\frac{1}{3}$% 31. 102%
32. 175% 33. 220%
34. 304% 35. 340%
36. 835% 37. 40%
38. 360% 39. $77\frac{7}{9}$%
40. $12\frac{1}{2}$% 41. $146\frac{2}{3}$%
42. 450% 43. 18%
44. 375% 45. $63\frac{7}{11}$%
46. $7\frac{1}{2}$% 47. $8\frac{1}{3}$%
48. 40% 49. 610%
50. $88\frac{8}{9}$% 51. 1190%
52. $91\frac{2}{3}$% 53. 70%
54. 625% 55. 40%
56. 32% 57. 88%
58. 38% 59. 68%
60. 12% 61. 62%

APPLICATIONS: FROM RATIOS TO PERCENTS

Write a percent for each ratio.

55. In 1970, $\frac{2}{5}$ of the population of the United States was under age 21.

56. By 1990, $\frac{8}{25}$ of the population of the United States will be under age 21.

57. By 1990, $\frac{22}{25}$ of the population of the United States will be under age 65.

58. By 1990, $\frac{19}{50}$ of the population of the United States will be 40 years old or older.

59. Refer to your answer to Exercise 56 to determine what percent of the population will be 21 years old or older by 1990.

60. Refer to your answer to Exercise 57 to determine what percent of the population will be 65 years old or older by 1990.

61. Refer to your answer to Exercise 58 to determine what percent of the population will be under 40 years old by 1990.

REVIEW CAPSULE FOR SECTION 8-5

Multiply each number by 100. (Pages 38-41)

| 1. 0.35 | 2. 0.71 | 3. 0.08 | 4. 0.05 | 5. 0.875 | 6. 0.125 |
| 7. 1.45 | 8. 1.97 | 9. 0.005 | 10. 0.001 | 11. 0.305 | 12. 0.109 |

Divide each number by 100. (Pages 44-46, 173-175)

| 13. 18 | 14. 26 | 15. 4 | 16. 7 | 17. 19.5 | 18. 50.9 |
| 19. 119 | 20. 256 | 21. 103 | 22. 405 | 23. 1.5 | 24. 47.6 |

The answers are on page 249.

230 Chapter 8

8-5 Percents and Decimals

A manufacturer found that, on the average, 2 out of every 100 TV's produced by a plant were defective. That is, 0.02 of the total number produced were defective. Recall that

$$\frac{2}{100} = 0.02 = 2\%.$$

Thus, **2% of the TV's produced** by the plant were defective.

To write a percent for a decimal, first write the decimal as a fraction with 100 as the denominator.

TABLE 1

Decimal	Fraction	Percent
0.16	$\frac{16}{100}$	16%
0.9	$\frac{90}{100}$	90%
0.005	$\frac{0.5}{100}$	0.5%
3.8	$\frac{380}{100}$	380%

Table 1 suggests this procedure.

PROCEDURE

To write a percent for a decimal:

[1] Move the decimal point two places to the right. Annex zeros as needed.

[2] Write the percent symbol.

EXAMPLE 1 Write a percent for each decimal.

    a. 0.35   **b.** 0.3   **c.** 9   **d.** 0.006

Solutions:

	a.	b.	c.	d.
[1]	0.35 = 0.35	0.3 = 0.30	9 = 9.00	0.006 = 0.006
[2]	= 35%	= 30%	= 900%	= 0.6%

Ratio, Proportion, and Percent

Teaching Suggestions p. M-21

QUICK QUIZ

Multiply each number by 100.

1. 27 Ans: 2700
2. 3.6 Ans: 360
3. 0.8 Ans: 80
4. 1.85 Ans: 185
5. 0.072 Ans: 7.2

Divide each number by 100.

6. 384 Ans: 3.84
7. 5.27 Ans: 0.0527
8. 63 Ans: 0.63
9. 1.9 Ans: 0.019
10. 0.06 Ans: 0.0006

ADDITIONAL EXAMPLES

Example 1

Write a percent.

1. 6.2 Ans: 620%
2. 0.057 Ans: 5.7%

SELF-CHECK
1. 8%   2. 63%
3. 50.1%   4. 940%
5. 358%

Self-Check *Write a percent for each decimal.*

  1. 0.08 **2.** 0.63 **3.** 0.501 **4.** 9.4 **5.** 3.58

To write a decimal for a percent, first write the percent ratio.

TABLE 2

Percent	Percent Ratio	Decimal
22%	$\frac{22}{100}$	0.22
8%	$\frac{8}{100}$	0.08
0.6%	$\frac{0.6}{100}$	0.006
100%	$\frac{100}{100}$	1.00
225%	$\frac{225}{100}$	2.25

Table 2 suggests this procedure.

PROCEDURE

To write a decimal for a percent:

1. Move the decimal point two places to the left. Insert zeros as needed.
2. Drop the percent symbol.

ADDITIONAL EXAMPLES
Example 2
Write a decimal for each percent.
1. 8%   Ans: 0.08
2. 63.5%   Ans: 0.635

SELF-CHECK
6. 0.15   7. 0.06
8. 0.215   9. 0.004
10. 1.13

EXAMPLE 2 Write a decimal for each percent.

  a. 17% **b.** 2% **c.** 0.1% **d.** 150%

Solutions: **a.** **b.** **c.** **d.**

1 17% = 17% 2% = 02% 0.1% = 00.1% 150% = 150%
2   = 0.17 = 0.02 = 0.001 = 1.50

Self-Check *Write a decimal for each per cent.*

  6. 15% **7.** 6% **8.** 21.5% **9.** 0.4% **10.** 113%

CLASSROOM EXERCISES

Write a fraction with 100 as the denominator for each decimal. (Table 1)

1. 0.41
2. 0.8
3. 0.97
4. 0.08
5. 7.9
6. 0.223

Write a decimal for each ratio. (Table 2)

7. $\frac{21}{100}$
8. $\frac{1}{100}$
9. $\frac{0.7}{100}$
10. $\frac{110}{100}$
11. $\frac{6.3}{100}$
12. $\frac{15.76}{100}$
13. $\frac{200}{100}$
14. $\frac{900}{100}$
15. $\frac{1.5}{100}$
16. $\frac{8.9}{100}$
17. $\frac{0.05}{100}$
18. $\frac{0.01}{100}$

WRITTEN EXERCISES

Goals: To write a percent for a decimal
To write a decimal for a percent

Write a percent for each decimal. (Example 1)

1. 0.4
2. 3.6
3. 9.4
4. 0.15
5. 0.203
6. 4.57
7. 1.5
8. 0.9
9. 0.26
10. 5.1
11. 0.1
12. 2.04
13. 8.6
14. 0.005
15. 0.04
16. 0.03
17. 3.62
18. 0.8

Write a decimal for each percent. (Example 2)

19. 18%
20. 91%
21. 66.1%
22. 17.8%
23. 3%
24. 10%
25. 101%
26. 507%
27. 0.5%
28. 0.1%
29. 20%
30. 80%
31. 0.25%
32. 0.66%
33. 400%
34. 900%
35. 4.5%
36. 9.8%

MIXED PRACTICE

Complete the tables below. Write all fractions in lowest terms. (Tables 1 and 2, Examples 1 and 2)

	Percent	Decimal	Fraction
37.	?	0.14	?
39.	33%	?	?
41.	?	?	$\frac{17}{20}$
43.	60%	?	?
45.	?	?	$\frac{9}{10}$

	Percent	Decimal	Fraction
38.	?	0.05	?
40.	118%	?	?
42.	?	?	$\frac{16}{25}$
44.	?	4.08	?
46.	51%	?	?

Ratio, Proportion, and Percent   **233**

CLASSROOM EXERCISES
1. $\frac{41}{100}$ 2. $\frac{80}{100}$ 3. $\frac{97}{100}$
4. $\frac{8}{100}$ 5. $\frac{790}{100}$
6. $\frac{22.3}{100}$ 7. 0.21
8. 0.01 9. 0.007
10. 1.10 11. 0.063
12. 0.1576 13. 2.00
14. 9.00 15. 0.015
16. 0.089 17. 0.0005
18. 0.0001

ASSIGNMENT GUIDE
BASIC
p. 233: 1–45
AVERAGE
pp. 233–234: 1–57 odd
ABOVE AVERAGE
pp. 233–234: 1–61 odd

WRITTEN EXERCISES
1. 40% 2. 360%
3. 940% 4. 15%
5. 20.3% 6. 457%
7. 150% 8. 90%
9. 26% 10. 510%
11. 10% 12. 204%
13. 860% 14. 0.5%
15. 4% 16. 3%
17. 362% 18. 80%
19. 0.18 20. 0.91
21. 0.661 22. 0.178
23. 0.03 24. 0.10
25. 1.01 26. 5.07
27. 0.005 28. 0.001
29. 0.20 30. 0.80
31. 0.0025 32. 0.0066
33. 4.00 34. 9.00
35. 0.045

WRITTEN EXERCISES

36. 0.098 37. 14%; $\frac{7}{50}$
38. 5%; $\frac{1}{20}$
39. 0.33; $\frac{33}{100}$
40. 1.18; $\frac{59}{50}$
41. 85%; 0.85
42. 64%; 0.64
43. 0.60; $\frac{3}{5}$
44. 408%; $\frac{102}{25}$
45. 90%; 0.90
46. 0.51; $\frac{51}{100}$
47. 0.47   48. 0.28
49. 0.05   50. 0.024
51. 0.025 52. 0.036
53. 0.025 54. 0.081
55. 0.52   56. 0.75
57. 0.131 58. 0.061

APPLICATIONS: FROM PERCENT TO DECIMALS

This circle graph below shows the percent (by weight) which each element forms of the earth's surface.

Write a decimal for each percent.

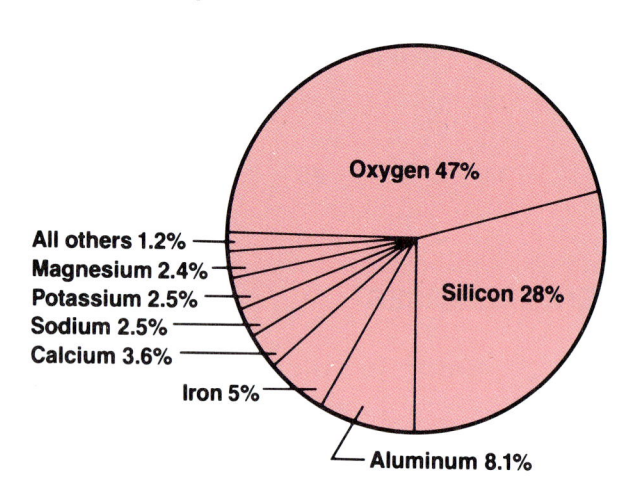

47. oxygen
48. silicon
49. iron
50. magnesium
51. potassium
52. calcium
53. sodium
54. aluminum
55. oxygen and iron
56. silicon and oxygen
57. iron and aluminum
58. calcium and sodium

Oxygen 47%
Silicon 28%
Aluminum 8.1%
Iron 5%
Calcium 3.6%
Sodium 2.5%
Potassium 2.5%
Magnesium 2.4%
All others 1.2%

REVIEW CAPSULE FOR SECTION 8-6

Evaluate each expression. (Pages 18–21)

1. (1 ÷ 4) × 100
2. (2 ÷ 5) × 100
3. (1 ÷ 2) × 100
4. (3 ÷ 10) × 100
5. (3 ÷ 5) × 100
6. (7 ÷ 20) × 100
7. (8 ÷ 25) × 100
8. (3 ÷ 4) × 100

Multiply. (Pages 38–41)

9. 800(0.15)
10. 912(0.38)
11. 461(0.03)
12. 53(0.21)
13. 2644(0.45)
14. 8678(0.15)
15. 6990(0.12)
16. 9003(0.48)

(Pages 129–131)

17. $920 \cdot \frac{7}{8}$
18. $618 \cdot \frac{2}{3}$
19. $804 \cdot \frac{5}{6}$
20. $275 \cdot \frac{4}{15}$
21. $680 \cdot \frac{3}{50}$
22. $400 \cdot \frac{19}{20}$
23. $760 \cdot \frac{11}{25}$
24. $405 \cdot \frac{3}{5}$
25. $927 \cdot \frac{2}{3}$
26. $3072 \cdot \frac{1}{6}$
27. $3072 \cdot \frac{5}{8}$
28. $3128 \cdot \frac{1}{8}$

The answers are on page 249.

234 Chapter 8

8-6 Finding a Percent of a Number

You can use an equation to find a percent of a number.

The **variable** in the equation represents the unknown (what you are trying to find).

Inflation Rises 1.5%
11.37% YIELD ON NEW BONDS!
JUNIORS FALL SALE 30%–40% off

PROCEDURE
To find a percent of a number:
1. Write an equation for the problem.
2. Solve the equation.

EXAMPLE 1 What number is 5% of 120?

Solution:
1. $n = 0.05 \cdot 120$
2. $n = 6$

Replace "is" with "=".
Replace "of" with "·" (times).

Thus, **6** is 5% of 120.

Self-Check 1. What number is 1% of 30?   2. What number is 12% of 20?

Sometimes it is easier to write a fraction for a percent. This table can help you. You should memorize the equivalent fractions and percents in the table.

TABLE

Equivalent Fractions and Percents			
$\frac{1}{4} = 25\%$	$\frac{1}{2} = 50\%$	$\frac{3}{4} = 75\%$	
$\frac{1}{5} = 20\%$	$\frac{2}{5} = 40\%$	$\frac{3}{5} = 60\%$	$\frac{4}{5} = 80\%$
$\frac{1}{6} = 16\frac{2}{3}\%$	$\frac{1}{3} = 33\frac{1}{3}\%$	$\frac{2}{3} = 66\frac{2}{3}\%$	$\frac{5}{6} = 83\frac{1}{3}\%$
$\frac{1}{8} = 12\frac{1}{2}\%$	$\frac{3}{8} = 37\frac{1}{2}\%$	$\frac{5}{8} = 62\frac{1}{2}\%$	$\frac{7}{8} = 87\frac{1}{2}\%$

Ratio, Proportion, and Percent   **235**

Teaching Suggestions
p. M-21

QUICK QUIZ
Multiply.
1. 78 (0.05)   Ans: 3.9
2. 127 (0.20) Ans: 25.4
3. 280 (0.65) Ans: 182
4. 795 ($\frac{2}{3}$)   Ans: 530
5. 505 ($\frac{3}{5}$)   Ans: 303

ADDITIONAL EXAMPLES
Example 1
1. What number is 70% of 86? Ans: 60.2
2. What number is 3% of 1200? Ans: 36

SELF-CHECK
1. 0.3   2. 2.4

ADDITIONAL EXAMPLES

Example 2

1. What number is $37\frac{1}{2}\%$ of 288? Ans: 108
2. What number is $33\frac{1}{3}\%$ of 627? Ans: 209

SELF-CHECK

3. 2   4. 60

Example 3

A salesperson receives a 7% commission on sales. What is the commission on sales of $51,500? Ans: $3605

SELF-CHECK

5. $3900

EXAMPLE 2   What number is $87\frac{1}{2}\%$ of 1600?

Solution:
  [1] $n = \frac{7}{8} \cdot 1600$ $\frac{7}{\overset{1}{8}} \cdot \overset{200}{1600} = \frac{7 \cdot 200}{1}$
  [2] $n = 7 \cdot 200$
  $n = 1400$ **1400** is $87\frac{1}{2}\%$ of 1600.

Self-Check
3. What number is 25% of 8?
4. What number is $66\frac{2}{3}\%$ of 90?

Follow a similar procedure to solve word problems involving percent.

> **PROCEDURE**
> To solve word problems involving percent:
> [1] Identify what is given and what you are to find.
> [2] Use the given information to restate the problem.
> [3] Write an equation for the problem.
> [4] Solve the equation.

Example 3 shows how to find commission. **Commission** is earnings based on a percent of sales.

EXAMPLE 3   In addition to her regular salary at Central Auto Sales, Louise Wong receives a commission of 6% on total sales. Her total sales last week amounted to $32,000. How much commission did she receive?

Solution:
  [1] Given: Rate of commission = 6%   Total sales = $32,000
      Find: Amount of commission, or 6% of $32,000
  [2] Restate the problem.   What number is 6% of 32,000?
  [3] Write an equation. $n = 0.06 \cdot 32{,}000$
  [4] Solve the equation. $n = 1920$
  Louise's commission was **$1920.00.**

Self-Check
5. A real estate agent receives a 6% commission of sales. What is the agent's commission on the sale of a house for $65,000?

236   Chapter 8

CLASSROOM EXERCISES

Write an equation for each percent problem.
(Examples 1 and 2, step 1)

1. n is 7% of 121.
2. q is 1% of 30.
3. r is 15% of 144.
4. t is 30% of 612.
5. w is 60% of 75.
6. a is 105% of 60.
7. $r = 33\frac{1}{3}\%$ of 96.
8. $p = 2\frac{1}{2}\%$ of 16.
9. c is 75% of 4.

Write an equation for each word problem. Do NOT solve.
(Example 3, steps 1, 2, and 3)

10. What is the commission on sales of $4,600 if the rate of commission is 1%?
11. A baseball team won 75% of the 30 games played. How many games were won?
12. A state has a 4% sales tax. What is the sales tax on a suit that sells for $118.50?

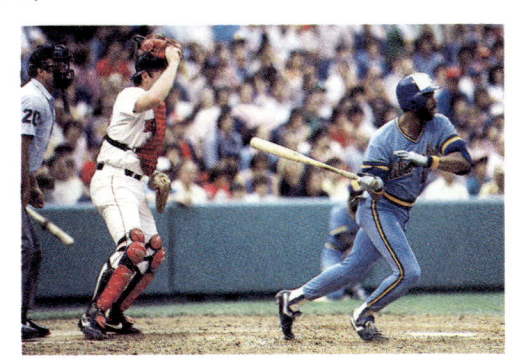

WRITTEN EXERCISES

Goals: To find a percent of a number
To apply the skill of finding a percent of a number to solving word problems

For Exercises 1–6, write an equation for each problem.
(Examples 1 and 2, step 1)

1. What is 35% of 400?
2. What is 9% of 800?
3. What is 83% of 100?
4. What is 16% of 900?
5. Find $37\frac{1}{2}\%$ of 16.
6. Find $66\frac{2}{3}\%$ of 60.

Solve. (Example 1)

7. What is 10% of 516?
8. What is 30% of 505?
9. What is 45% of 120?
10. What is 18% of 150?
11. 4% of 7.8 is n. Find n.
12. 3% of 24 is w. Find w.
13. Find 12% of 900.
14. Find 1% of 69.

Ratio, Proportion, and Percent   **237**

CLASSROOM EXERCISES

1. $n = 0.07 \cdot 121$
2. $q = 0.01 \cdot 30$
3. $r = 0.15 \cdot 144$
4. $t = 0.3 \cdot 612$
5. $w = \frac{3}{5} \cdot 75$
6. $a = 1.05 \cdot 60$
7. $r = \frac{1}{3} \cdot 96$
8. $p = 0.025 \cdot 16$
9. $c = \frac{3}{4} \cdot 4$
10. $n = 0.01 \cdot 4600$
11. $n = \frac{3}{4} \cdot 30$
12. $n = 0.04 \cdot 118.5$

ASSIGNMENT GUIDE
BASIC
Day 1 p. 237: 1–14
Day 2 p. 238: 15–26
AVERAGE
Day 1 pp. 237–238: 1–25 odd
Day 2 pp. 237–238: 8–32 even
ABOVE AVERAGE
pp. 237–238: 7–25 odd, 27–38

WRITTEN EXERCISES
1. $n = 0.35 \cdot 400$
2. $n = 0.09 \cdot 800$
3. $n = 0.83 \cdot 100$
4. $n = 0.16 \cdot 900$
5. $n = \frac{3}{8} \cdot 16$

WRITTEN EXERCISES

6. $n = \frac{2}{3} \cdot 60$ 7. 51.6
8. 151.5   9. 54
10. 27   11. 0.312
12. 0.72   13. 108
14. 0.69   15. 20
16. 44   17. 48
18. 3.8   19. 93   20. 2
21. 8   22. 30.6
23. 34   24. 56
25. 120   26. 90
27. $7500   28. $5600
29. $300   30. 24
31. 15   32. $1428

(Example 2)

15. What is 25% of 80?
16. What is 50% of 88?
17. What is 75% of 64?
18. What is 10% of 38?
19. 60% of 155 is *q*. Find *q*.
20. What is $16\frac{2}{3}$% of 12?
21. What is $66\frac{2}{3}$% of 12?
22. 30% of 102 is *r*. Find *r*.
23. What is $33\frac{1}{3}$% of 102?
24. What is $87\frac{1}{2}$% of 64?
25. Find $83\frac{1}{3}$% of 144.
26. Find $62\frac{1}{2}$% of 144.

APPLICATIONS: FINDING A PERCENT OF A NUMBER

Solve each problem.   (Example 3)

27. An auctioneer sold goods amounting to $125,000. She received a commission of 6% on the total sales. How much was her commission?

28. A real estate agent received a 5% on the sales of a house. The house sold for $112,000. What was the agent's commission?

29. After a meal in a restaurant, Eileen gave the server a tip which was 15% of the total bill. The total bill amounted to $20.00. How much was the tip?

30. Of 120 students in a high school band, 20% play woodwind instruments. How many play woodwind instruments?

31. A basketball team played 25 games this year. The team won 60% of the games played. How many games did the team win?

32. The **depreciation** (loss in value) of a car was 14% of its original value of $10,200. Find the amount of depreciation.

Chapter 8

33. A family budgets 28% of its take home pay for food. The family's monthly take home pay is $912. How much is their monthly food budget?

34. An automobile supply store found that 2% of the batteries it buys are defective. The store recently ordered 250 batteries. How many would the store expect to be defective?

33. $255.36   34. 5
35. 12   36. $93.08
37. $5.45   38. $360

MORE CHALLENGING EXERCISES

Solve each problem.

35. There are 15 girls on the tennis team at Jansen Senior High. Of these, 20% are seniors. How many are not seniors?

36. Ceanne Brophy bought a suit for $89.50. She also paid a sales tax which was 4% of the purchase price. Find the total cost of the suit.

37. A worker who earns $5.00 per hour receives a 9% raise. What is the new hourly wage?

38. Luis Hernandez earns a base salary of $225 per week. He also earns a 3% commission on all sales. What is his total salary for a week in which his total sales amount to $4500?

REVIEW CAPSULE FOR SECTION 8-7

Write a decimal for each percent. (Pages 231–234)

1. 6%   **2.** 9%   **3.** 10%   **4.** 60%   **5.** 12%   **6.** 14%

Write as a fraction in lowest terms. (Pages 124–125)

7. $\frac{6}{12}$ **8.** $\frac{18}{12}$ **9.** $\frac{38}{12}$ **10.** $\frac{9}{12}$ **11.** $\frac{46}{12}$ **12.** $\frac{10}{12}$

Multiply. (Pages 38–41, 129–131)

13. $900 \times 0.04 \times \frac{2}{3}$ **14.** $2500 \times 0.06 \times \frac{3}{4}$ **15.** $700 \times 0.13 \times \frac{1}{2}$

16. $(4500)(0.085)\frac{20}{3}$ **17.** $(9400)(0.12)\frac{9}{4}$ **18.** $(6000)(0.10)\frac{5}{2}$

The answers are on page 249.

Ratio, Proportion, and Percent   **239**

Sidebar (left margin)

Teaching Suggestions
p. M-21

QUICK QUIZ

Write a decimal for each percent.

1. 8%   Ans: 0.08
2. 11%   Ans: 0.11
3. 1%   Ans: 0.01

Write in lowest terms.

4. $\frac{8}{12}$ Ans: $\frac{2}{3}$
5. $\frac{21}{12}$ Ans: $\frac{7}{4}$

Multiply.

6. 300 × 0.06 × 7
   Ans: 126
7. 11,000 (0.12)($\frac{1}{2}$)
   Ans: 660
8. (120,000)(0.08)($\frac{2}{3}$)
   Ans: 6400

ADDITIONAL EXAMPLE

Example 1
Find the interest on $8000 borrowed for 5 years at a rate of 9%.
Ans: $3600

SELF-CHECK

1. $240.00

PROBLEM SOLVING AND APPLICATIONS

8-7 Formulas: Simple Interest

An important use of percent is in computing interest.

> **Interest** is an amount of money charged for the use of borrowed money. The amount of money borrowed is called the **principal**.
>
> **Interest** is also paid on money deposited in certain bank accounts.

Interest is usually expressed as a percent of the principal. This percent is called the **interest rate**.

Use this formula to compute simple interest, i.

$i = prt$

  p = principal
  r = rate (expressed as a decimal or a fraction)
  t = time (expressed in years)

PROCEDURE

To compute simple interest:

1. Write the formula.
2. Identify known values.
3. Substitute and solve for the unknown value.

EXAMPLE 1 Find the interest on $400 for one year at a yearly rate of 13%.

Solution:
1. $i = prt$
2. $p = 400$; $r = 13\% = 0.13$; $t = 1$
3. $i = 400\,(0.13)(1)$
   $i = 52$

The interest on $400 for one year at a yearly rate of 13% is **$52.00**.

Self-Check 1. Find the interest on $1000 borrowed for 2 years at a rate of 12%.

240   Chapter 8

When using the interest formula, you must express the time given in months in terms of years.

Expressed in Months	Expressed in Years
6	$\frac{6}{12}$, or $\frac{1}{2}$
11	$\frac{11}{12}$
40	$\frac{40}{12}$, or $\frac{10}{3}$

EXAMPLE 2 The Velez family opened a savings account with $2400. The yearly interest rate is 8.5%. How much will the account earn in 4 months?

Solution:
1. $i = prt$
2. $p = 2400$; $r = 8.5\% = 0.085$; $t = \frac{4}{12} = \frac{1}{3}$
3. $i = (2400)(0.085)\left(\frac{1}{3}\right)$
   $i = (204)\left(\frac{1}{3}\right)$
   $i = 68$ The account will earn **$68** in 4 months.

Self-Check 2. Find the interest on $1000 at $5\frac{1}{2}\%$ for 6 months.

EXAMPLE 3 Carla borrowed $45,000 from her credit union to open a custom furniture shop. The yearly interest rate for the loan is $9\frac{1}{2}\%$ and the loan is to be paid back in 40 months. Find the total interest.

Credit Union 9½% INTEREST

Solution:
1. $i = prt$
2. $p = 45{,}000$; $r = 9\frac{1}{2}\% = 0.09\frac{1}{2} = 0.095$; $t = \frac{40}{12} = \frac{10}{3}$ years
3. $i = (45{,}000)(0.095)\frac{10}{3}$
   $i = (4275)\frac{10}{3}$
   $i = 14{,}250$ Carla will pay **$14,250** in interest.

Self-Check 3. Find the interest on $2500 at $12\frac{1}{2}\%$ for 30 months.

Ratio, Proportion, and Percent **241**

ADDITIONAL EXAMPLES
Example 2
Find the interest on $5300 at 12% for 9 months.   Ans: $477

SELF-CHECK
2. $27.50

Example 3
Find the interest on $2520 at 8% for 31 months.   Ans: $520.80

SELF-CHECK
3. $781.25

CLASSROOM EXERCISES

1. i = (950)(0.08)(1)
2. i = (10,000)(0.12)(1)
3. i = (1956)(0.05)($\frac{1}{2}$)
4. i = (864)(0.09)($\frac{1}{4}$)
5. i = (2575)(0.125)(2)
6. i = (900)(0.13)($\frac{3}{2}$)
7. $5.00   8. $6.00
9. $16.00   10. $9.00
11. $3.00   12. $15.00
13. $8.00   14. $264.00
15. $100.00   16. $32.50
17. $19.00   18. $312.50

ASSIGNMENT GUIDE
BASIC
pp. 242-243: 1-18
AVERAGE
pp. 242-243: 1-17 odd, 18-22
ABOVE AVERAGE
pp. 242-243: 1-17 odd, 18-22

WRITTEN EXERCISES
1. $40.00   2. $60.00
3. $12.00   4. $26.00
5. 6.00   6. $27.00
7. $22.50   8. $18.00
9. $30.00   10. $200.00
11. $75.00   12. $16.20

CLASSROOM EXERCISES

Replace the variables in $i = prt$ *with the known values.*
(Examples 1, 2, and 3, step 3)

1. Interest on $950 at 8% for one year
2. Interest on $10,000 for one year at 12%
3. Interest on $1956 at a yearly rate of 5% for 6 months
4. Interest on $864 at a yearly interest rate of 9% for 3 months
5. Interest on $2575 borrowed for 2 years at a yearly rate of $12\frac{1}{2}$%
6. Interest on $900 for 18 months at a yearly interest rate of 13%

For Exercises 7–18, find the simple interest. (Example 1)

7. $100 borrowed for one year at 5%
8. $100 borrowed for one year at 6%
9. $200 borrowed for one year at 8%
10. $100 borrowed for one year at 9%

(Example 2)

11. $100 borrowed for 3 months at 12%
12. $200 borrowed for 9 months at 10%
13. $100 borrowed for 8 months at 12%
14. $800 borrowed for 3 years at 11%

(Example 3)

15. $500 borrowed for 2 years at 10%
16. $300 borrowed for 10 months at 13%
17. $400 borrowed for 6 months at $9\frac{1}{2}$%
18. $1000 borrowed for 30 months at $12\frac{1}{2}$%

WRITTEN EXERCISES

Goals: To use the formula $i = prt$ to compute simple interest
To apply computing simple interest to solving word problems

For Exercises 1–18, find the simple interest. (Example 1)

1. $500 borrowed for one year at 8%
2. $600 borrowed for one year at 10%
3. $100 borrowed for one year at 12%
4. $200 borrowed for one year at 13%
5. $100 borrowed for one year at 6%
6. $300 borrowed for one year at 9%

(Example 2)

7. $500 borrowed for 6 months at 9%
8. $900 borrowed for 3 months at 8%
9. $1200 borrowed for 3 months at 10%
10. $5000 borrowed for 8 months at 6%
11. $1800 borrowed for 4 months at 12.5%
12. $720 borrowed for 2 months at 13.5%

(Example 3)

13. $600 borrowed for 2 years at 12%
14. $1700 borrowed for 30 months at 13%
15. $7500 borrowed for 3 years at $12\frac{1}{2}$%
16. $9000 borrowed for 3 years at $11\frac{1}{2}$%
17. $15,000 borrowed for 4 months at $14\frac{1}{2}$%
18. $3200 borrowed for 3 months at $13\frac{1}{2}$%

Solve each problem. (Examples 1–3)

19. George and Clara Lightfoot borrow $20,000 to start a business. The yearly interest rate is $11\frac{1}{2}$%. How much interest is due on the loan in 3 months?

20. Miriam Taylor borrowed $950 to operate a stand at the state fair. The yearly interest rate is 18%. How much is the interest for one month?

21. The Nishifue family obtained a loan of $6500 to buy a car. The yearly interest rate is 12%. Find the interest paid on the loan at the end of 4 years.

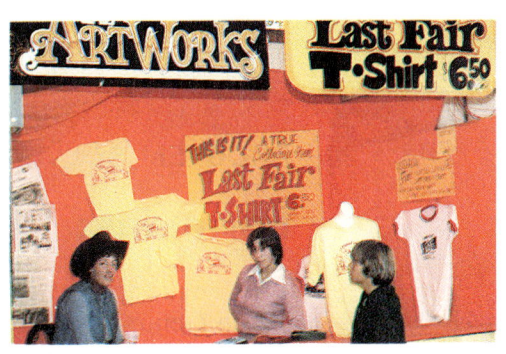

22. A drama club has $3500 in a savings account. The account pays a yearly interest rate of 6.5%. How much interest will the account earn in 2 years?

WRITTEN EXERCISES
13. $144.00
14. $552.50
15. $2812.50
16. $3105.00
17. $725.00
18. $108.00
19. $575.00
20. $14.25
21. $3120.00
22. $455.00

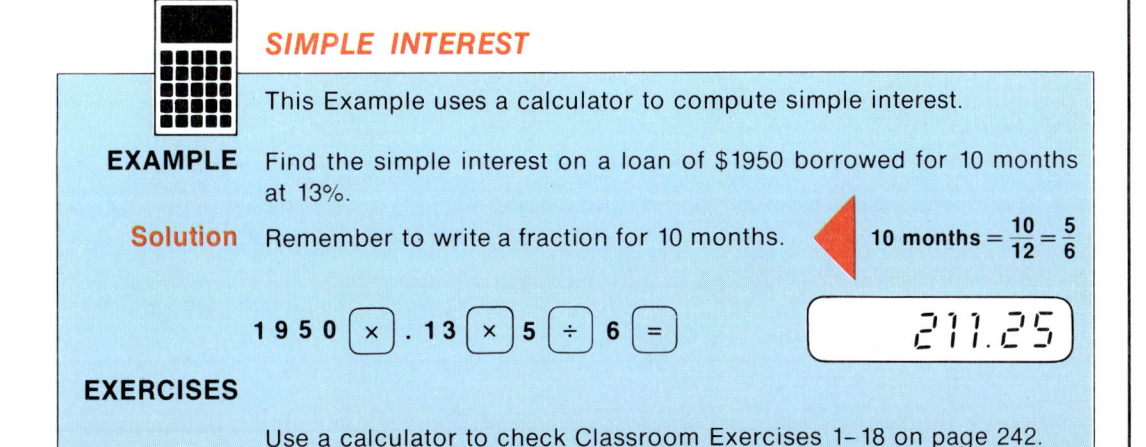

SIMPLE INTEREST

This Example uses a calculator to compute simple interest.

EXAMPLE Find the simple interest on a loan of $1950 borrowed for 10 months at 13%.

Solution Remember to write a fraction for 10 months.

10 months = $\frac{10}{12} = \frac{5}{6}$

1 9 5 0 [×] . 1 3 [×] 5 [÷] 6 [=]   *211.25*

EXERCISES

Use a calculator to check Classroom Exercises 1–18 on page 242.

CALCULATOR EXERCISES
1. $76.00   2. $1200.00
3. $48.90   4. $19.44
5. $643.75 6. $175.50
7. $5.00   8. $6.00
9. $16.00   10. $9.00
11. $3.00   12. $15.00
13. $8.00   14. $264.00
15. $100.00 16. $32.50
17. $19.00 18. $312.50

Ratio, Proportion, and Percent   **243**

Teaching Suggestions
p. M-21

QUICK QUIZ
Write a decimal for each percent.
1. 7%   Ans: 0.07
2. 5.9% Ans: 0.059
Multiply.
3. (700)(0.09) Ans: 63
4. (629)(0.30)
   Ans: 188.7
Subtract.
5. 310.60
   − 62.01
   Ans: 248.59
6. 500 − 87.25
   Ans: 412.75

ADDITIONAL EXAMPLE
Example 1
An item with a list price of $83.90 is marked for a 10% discount. Find the amount of discount.
Ans: $8.39

REVIEW CAPSULE FOR SECTION 8-8

Write a decimal for each percent. (Pages 231–234)

1. 3%   2. 2%   3. 9%   4. 10%   5. 15%   6. $4\frac{1}{2}\%$

Multiply. (Pages 38–41)

7. (55.80)(0.33)   8. (151.80)(0.40)   9. (41.90)(0.10)

Subtract. (Pages 35-37)

10. 18.91 − 2.48   11. 10.05 − 1.96   12. 93.80 − 11.72

The answers are on page 249.

PROBLEM SOLVING AND APPLICATIONS

8-8 Formulas: Discount

Phil Maquire works as a salesperson in Central Department Store. As an employee, he receives a 20% discount on all items he purchases.

A **discount** is the amount that an article of merchandise is reduced in price. Discounts are often expressed as a percent of the list price. The **list price** is the regular price of an article.

Use this formula to find the amount of discount.

$$d = rp$$

d = amount of discount
r = rate of discount
p = list price

EXAMPLE 1 Phil decided to purchase a bicycle he saw at Central Department Store. The list price was $117.50 and his employee discount was 20%. How much was the discount?

Solution: $d = rp$ $p = 117.50;$ $r = 20\% = 0.20$

$d = (0.20)(117.50)$

$d = \$23.50$

The amount of discount was **$23.50**.

244 Chapter 8

Self-Check

1. An item with a list price of $50 is marked for a 25% discount. Find the amount of discount.

The price of an item after the discount is deducted is the **net price** or **sale price**.

Use this formula to find the sale price.

$$s = p - d$$

p = list price
d = discount
s = sale price

To find the sale price when the list price and rate of discount are given, you must first answer the question.

What is the amount of discount?

This is the **hidden question** in the problem. Then you use subtraction to find the sale price.

EXAMPLE 2 A calculator with a list price of $26.50 is on sale at a 40% discount. Find the sale price.

Solution: ⬜1 Find the amount of discount.

$$d = rp$$

p = $26.50;
r = 40% = 0.40

$$d = (26.50)(0.40)$$

$$d = 10.60$$

The amount of discount is $10.60.

⬜2 Find the sale price.

$$s = p - d$$

p = $26.50;
d = $10.60

$$s = 26.50 - 10.60$$

$$s = 15.90$$

The sale price is **$15.90**.

$26.50
ON SALE
40% DISCOUNT

Self-Check *Find the sale price.*

2. List price: $52.90; discount: 20%
3. List price: $17.32; discount: 25%
4. List price: $149.00; discount: 33%

SELF-CHECK

1. $12.50

ADDITIONAL EXAMPLES

Example 2
Find the sale price.
1. List price: $280.90; discount: 30%
   Ans: $196.63
2. List price: $3250.00; discount: 15%
   Ans: $2762.50

SELF-CHECK

2. $42.32
3. $12.99
4. $99.83

Ratio, Proportion, and Percent

CLASSROOM EXERCISES
1. $6.00   2. $1.85
3. $27.30   4. $66.00
5. $15.80   6. $13.26
7. $25.13   8. $8095.50

ASSIGNMENT GUIDE
BASIC
pp. 246-247: 1-13
AVERAGE
pp. 246-247: 1-13
ABOVE AVERAGE
pp. 246-247: 1-9 odd, 10-13

WRITTEN EXERCISES
1. $57.00   2. $15.58
3. $19.50   4. $23.40
5. $52.80   6. $13.77
7. $323.00 8. $211.20
9. $13.00   10. $20.00
11. $2144.00
12. $21.93
13. $504.00

CLASSROOM EXERCISES

Find the amount of discount on each item. (Example 1)

1. A baseball glove listed at $24 with a discount rate of 25%
2. A hockey stick listed at $18.50 with a discount rate of 10%
3. A $78 dress on sale at 35% discount
4. A washer listed at $198 with a $33\frac{1}{3}$% discount

Find the sale price of each item. (Example 2)

5. A shirt listed at $19.75 with a 20% discount
6. A book listed at $15.60 with a 15% discount
7. Running shoes listed at $35.90 with a 30% discount
8. An automobile listed at $8995 with a 10% discount

WRITTEN EXERCISES

Goals: To find the amount of discount

To apply the skill of finding the amount of discount and the technique of the "hidden question" to solving word problems

Find the amount of discount on each item. (Example 1)

1. SALE! 15% off REGULARLY $380
2. 40% off regularly $38.95
3. CLOSE-OUT warm-up suits ORIGINALLY $32.50 NOW 60% OFF
4. 14 KARAT GOLD CHAINS 30% off REGULARLY $78
5. 3-Day Sale 20% off REGULAR PRICE $264.00
6. JEANS SALE 30% OFF REGULAR PRICE $45.90

246 Chapter 8

Find the sale price of each item. Refer to the advertisements in Exercise 1-6. (Example 2)

7. Typewriter **8.** Golf Clubs **9.** Warm-up Suit

For Exercises 10–15, find the sale price. (Example 2)

10. A theater ticket purchased for $25 is sold at a 20% discount.

11. A sailboat with a list price of $3200 is offered on sale at a 33% discount.

12. Senior citizens receive a special discount of 15% off at Wilson's Pharmacy. The regular price of a certain prescription is $25.80.

13. A motorbike listed at $576 is on sale at a $12\frac{1}{2}$% discount.

REVIEW: SECTIONS 8-4 — 8-8

Write a percent for each ratio. (Section 8-4)

1. $\frac{1}{2}$ **2.** $\frac{3}{20}$ **3.** $\frac{27}{50}$ **4.** $\frac{21}{25}$ **5.** $\frac{49}{40}$ **6.** $\frac{31}{5}$

Write a percent for each decimal. (Section 8-5)

7. 0.9 **8.** 3.8 **9.** 0.17 **10.** 1.04 **11.** 0.1 **12.** 0.01

Write a decimal for each percent. (Section 8-5)

13. 12% **14.** 2% **15.** 85% **16.** 70% **17.** 9.5% **18.** 15.3%

Solve. (Section 8-6)

19. What number is 15% of 44?
20. What is 2% of 64?
21. A real estate agent receives a 5% commission on sales. Find the commission on $82,600.

Solve. (Section 8-7)

22. $p = \$550$; $r = 6\%$, $t = 2$ years
23. $p = \$8000$; $r = 12\%$, $t = 6$ months
24. Clare borrowed $2000 at an interest rate of 8% for 9 months. How much interest will she pay?

(Section 8-8)

25. Compute the discount on a radio if the rate of discount is 22% and the list price is $65.50.

26. Jim Barnes bought slacks on sale at 24% off. The list price was $27. How much did Jim pay for the slacks?

Ratio, Proportion, and Percent **247**

QUIZ: SECTIONS 8-4-8-8
After completing this Review, you may wish to administer a quiz covering the same sections. A Quiz is provided in the *Teacher's Edition: Part II.*

REVIEW: SECTIONS 8-4-8-8
1. 50% 2. 15% 3. 54%
4. 84% 5. $122\frac{1}{2}$%
6. 620% 7. 90%
8. 380% 9. 17%
10. 104% 11. 10%
12. 1% 13. 0.12
14. 0.02 15. 0.85
16. 0.70 17. 0.095
18. 0.153 19. 6.6
20. 1.28 21. $4130
22. $66.00 23. $480.00
24. $120.00 25. $14.41
26. $20.52

CHAPTER REVIEW

PART 1: VOCABULARY

For Exercises 1–10, choose from the box at the right the word(s) that best corresponds to each description.

1. Earnings based on a percent of total sales __?__
2. Uses division to compare two numbers __?__
3. The amount that an article of merchandise is reduced in price __?__
4. A ratio in which the second term is 100 __?__
5. An equation that shows equivalent ratios __?__
6. An amount charged for the use of borrowed money __?__
7. In Exercise 6, the amount of money borrowed __?__
8. The regular price of an article of merchandise __?__
9. The price of an item after the discount is subtracted __?__
10. The amount paid for borrowing money expressed as a percent of the principal __?__

> ratio
> proportion
> percent
> commission
> interest
> interest rate
> principal
> discount
> list price
> net price

PART 2: SKILLS

Write each ratio in lowest terms. No variable equals zero. (Section 8-1)

11. 12 to 20
12. 6 to 8
13. 15 to 10
14. 9:21
15. 24:14
16. 7:28
17. $\dfrac{15}{30}$
18. $\dfrac{21}{35}$
19. $\dfrac{18}{42}$
20. $\dfrac{6}{15p}$
21. $\dfrac{30a}{50a}$
22. $\dfrac{36c}{24c}$

Determine whether each pair of ratios is equivalent. Answer Yes or No. (Section 8-1)

23. $\dfrac{1}{2}$ and $\dfrac{4}{6}$
24. $\dfrac{2}{3}$ and $\dfrac{16}{24}$
25. $\dfrac{8}{20}$ and $\dfrac{4}{5}$
26. $\dfrac{6}{14}$ and $\dfrac{2}{7}$
27. $\dfrac{5}{12}$ and $\dfrac{15}{36}$
28. $\dfrac{4}{5}$ and $\dfrac{23}{30}$
29. $\dfrac{24}{9}$ and $\dfrac{8}{3}$
30. $\dfrac{49}{35}$ and $\dfrac{7}{5}$

Solve and check each proportion. (Section 8-2)

31. $\dfrac{n}{14} = \dfrac{2}{7}$
32. $\dfrac{x}{12} = \dfrac{5}{6}$
33. $\dfrac{6}{8} = \dfrac{a}{4}$
34. $\dfrac{10}{15} = \dfrac{p}{18}$
35. $\dfrac{21}{28} = \dfrac{12}{c}$
36. $\dfrac{4}{8} = \dfrac{5}{y}$
37. $\dfrac{8}{n} = \dfrac{10}{15}$
38. $\dfrac{24}{p} = \dfrac{16}{18}$
39. $\dfrac{3}{5} = \dfrac{m}{8}$
40. $\dfrac{9}{10} = \dfrac{4}{b}$
41. $\dfrac{d}{15} = \dfrac{4}{9}$
42. $\dfrac{7}{g} = \dfrac{12}{5}$

248   Chapter 8

Complete. (Sections 8-4, 8-5)

	Ratio	Decimal	Percent
43.	?	0.05	?
45.	?	?	2.5%

	Ratio	Decimal	Percent
44.	$\frac{1}{100}$	?	?
46.	?	1.05	?

Complete. (Section 8-6)

47. What number is 15% of 300?
48. What is 1% of 82?
49. What number is 9% of 812?

	Amount of Sales	Rate of Commission	Commission
50.	$80,000	7%	?
51.	$92,000	2%	?

Find the interest. (Section 8-7)

	Principal	Rate	Time
52.	$2000	9%	3 years
53.	$5000	6%	4 months

Find the discount. (Section 8-8)

	List Price	Rate of Discount
54.	$55.00	20%
55.	$29.80	30%

PART 3: APPLICATIONS

56. A car travels 225 kilometers on 15 liters of fuel. At this rate, how many liters of fuel are needed to travel 375 kilometers? (Section 8-3)

57. Juan obtained a loan of $5400. The yearly interest rate is 12%. Find the interest paid on the loan at the end of 4 months. (Section 8-7)

58. A real estate agent receives a 6% commission on sales. How much commission does the agent receive on an acre of land that sells for $2400. (Section 8-6)

59. Senior citizens receive a 20% discount on bus fares during off-peak hours. If a one-way fare is 75¢, find how much a senior citizen pays during off-peak hours. (Section 8-8)

ANSWERS TO REVIEW CAPSULES

Page 219 **1.** 15 **2.** 10 **3.** 26 **4.** $11\frac{1}{4}$ **5.** $12\frac{1}{2}$ **6.** $11\frac{1}{4}$ **7.** $13\frac{1}{8}$ **8.** $3\frac{1}{5}$

Page 221 **1.** $\frac{2}{5}$ **2.** $\frac{3}{29}$ **3.** $\frac{10}{9}$ **4.** $\frac{3}{5}$ **5.** $\frac{1}{13}$ **6.** $\frac{14}{5}$ **7.** 4 **8.** 9 **9.** 25 **10.** $9\frac{1}{8}$

Page 225 **1.** 40 **2.** 30 **3.** 16 **4.** 46 **5.** 75 **6.** 45 **7.** 230 **8.** 240 **9.** $37\frac{1}{2}$ **10.** $58\frac{1}{3}$ **11.** $87\frac{1}{2}$ **12.** $33\frac{1}{3}$

Page 230 **1.** 35 **2.** 71 **3.** 8 **4.** 5 **5.** 87.5 **6.** 12.5 **7.** 145 **8.** 197 **9.** 0.5 **10.** 0.1 **11.** 30.5 **12.** 10.9 **13.** 0.18 **14.** 0.26 **15.** 0.04 **16.** 0.07 **17.** 0.195 **18.** 0.509 **19.** 1.19 **20.** 2.56 **21.** 1.03 **22.** 4.05 **23.** 0.015 **24.** 0.476

Page 234 **1.** 25 **2.** 40 **3.** 50 **4.** 30 **5.** 60 **6.** 35 **7.** 32 **8.** 75 **9.** 120 **10.** 346.56 **11.** 13.83 **12.** 11.13 **13.** 1189.8 **14.** 1301.7 **15.** 838.8 **16.** 4321.44 **17.** 805 **18.** 412 **19.** 670 **20.** $73\frac{1}{3}$ **21.** $40\frac{4}{5}$ **22.** 380 **23.** $334\frac{2}{5}$ **24.** 243 **25.** 618 **26.** 512 **27.** 1920 **28.** 391

Page 239 **1.** 0.06 **2.** 0.09 **3.** 0.10 **4.** 0.60 **5.** 0.12 **6.** 0.14 **7.** $\frac{1}{2}$ **8.** $\frac{3}{2}$ **9.** $\frac{19}{6}$ **10.** $\frac{3}{4}$ **11.** $\frac{23}{6}$ **12.** $\frac{5}{6}$ **13.** 24 **14.** 112.5 **15.** 45.5 **16.** 2550 **17.** 2538 **18.** 1500

Page 244 **1.** 0.03 **2.** 0.02 **3.** 0.09 **4.** 0.10 **5.** 0.15 **6.** 0.045 **7.** 18.414 **8.** 60.72 **9.** 4.19 **10.** 16.43 **11.** 8.09 **12.** 82.08

CHAPTER REVIEW

23. No   24. Yes
25. No   26. No
27. Yes   28. No
29. Yes   30. Yes
31. 4   32. 10   33. 3
34. 12   35. 16   36. 10
37. 12   38. 27   39. $4\frac{4}{5}$
40. $4\frac{4}{9}$ 41. $6\frac{2}{3}$
42. $2\frac{11}{12}$ 43. $\frac{1}{20}$; 5%
44. .01; 1%
45. $\frac{1}{40}$; .025
46. $\frac{21}{20}$; 105%   47. 45
48. 0.82   49. 73.08
50. $5600   51. $1840
52. $540   53. $100
54. $11.00   55. $8.94
56. 25   57. $216
58. $144   59. 60¢

Ratio, Proportion, and Percent

CHAPTER TEST

Write the cross-product to determine whether each pair of ratios is equivalent. Answer Yes or No.

1. $\frac{2}{3}$ and $\frac{14}{21}$
2. $\frac{2}{7}$ and $\frac{21}{63}$
3. $\frac{8}{3}$ and $\frac{18}{48}$
4. $\frac{39}{26}$ and $\frac{3}{2}$

Solve and check each proportion.

5. $\frac{5}{8} = \frac{15}{x}$
6. $\frac{9}{21} = \frac{b}{7}$
7. $\frac{y}{48} = \frac{7}{8}$
8. $\frac{16}{c} = \frac{2}{9}$

Write a percent for each of the following.

9. $\frac{3}{5}$
10. $\frac{11}{5}$
11. 0.7
12. 0.07

Write a decimal for each percent.

13. 18%
14. 3%
15. 8.5%
16. 140%

17. What number is 28% of 75?
18. What number is 36% of 25?
19. What is 3% of 27?
20. What is 90% of 82?
21. On a blueprint for a house, 2 inches represent 6 feet. If the actual length of the living room is 24 feet, what is its length on the blueprint?
22. Elizabeth borrowed $1200 at a yearly interest rate of 12%. How much interest will she owe at the end of 4 months?
23. A tent is on sale at a 30% discount. The list price is $425. How much is the discount?
24. A stereo system with a list price of $620.50 is on sale at a 20% discount. Find the sale price.
25. After a meal in a restaurant, David gave the server a tip which was 15% of the total bill. The total bill amounted to $25.00. How much was the tip?

ENRICHMENT

Compound Interest

Compound interest is computed on the principal plus the interest previously earned. The compound interest formula shown below is used to calculate the compounded amount (principal plus interest).

$$A = P(1 + r)^n$$

- A: compounded amount
- P: principal
- r: yearly interest rate
- n: number of years

EXAMPLE Four years ago, Erica Caulo put $400 into a savings account. The account earns 9% interest compounded yearly. Find the compounded amount.

Solution: $A = P(1 + r)^n$ P = $400; r = 9% = 0.09; n = 4

$A = 400(1 + 0.09)^4$

$A = 400(1.09)^4$ 1 + 0.09 = 1.09

$A = 400 \times 1.09 \times 1.09 \times 1.09 \times 1.09$ Use a calculator.

`4 0 0 [×] 1 . 0 9 [=][=][=][=]`   `564.63264`

The compounded amount after 4 years is **$564.63** (rounded <u>down</u> to the nearest cent).

EXERCISES

Use the compound interest formula to find the compounded amount. The interest is compounded yearly. Round down to the nearest cent.

	Principal	Rate	Years			Principal	Rate	Years
1.	$900	7%	2		2.	$8000	9%	2
3.	$750	8%	3		4.	$4000	7%	3
5	$1000	10%	4		6.	$12,000	12%	4

7. Five years ago, Ed Fong put $3000 in a savings account at 7% interest compounded yearly. What is the compounded amount?

Ratio, Proportion, and Percent

ENRICHMENT

You may wish to use this lesson for students who performed well on the formal Chapter Test.

1. $1030.41
2. $9504.80
3. $944.78
4. $4900.17
5. $1464.10
6. $18,882.23
7. $4207.66

ADDITIONAL PRACTICE

You may wish to use all or some of these exercises, depending on how well students performed on the formal Chapter Test.

1. $\frac{1}{5}$ 2. $\frac{3}{5}$ 3. $\frac{4}{9}$
4. $\frac{1}{2}$ 5. $\frac{4}{1}$ 6. $\frac{7}{9}$
7. No 8. Yes 9. Yes
10. No 11. $\frac{2}{5}$ 12. $\frac{4x}{5}$
13. $\frac{5t}{1}$ 14. $\frac{1}{3}$ 15. $\frac{6}{25y}$
16. $\frac{1}{8t}$ 17. $11\frac{2}{3}$
18. $24\frac{3}{4}$ 19. 36
20. 49 21. $8\frac{3}{4}$
22. $y = 60$ 23. $36\frac{4}{11}\%$
24. $58\frac{1}{3}\%$ 25. 275%
26. $55\frac{5}{9}\%$ 27. 14%
28. 115% 29. 25%; $\frac{1}{4}$
30. 15%; $\frac{3}{20}$
31. 0.42; $\frac{21}{50}$
32. 60%; 0.60
33. 85%; 0.85
34. 0.12; $\frac{3}{25}$ 35. 52.91
36. 320 37. $24
38. $700 39. 132 beats
40. $21.45

ADDITIONAL PRACTICE

SKILLS *Write each ratio in lowest terms.* (Pages 216-218)

1. $\frac{8}{40}$ 2. 21:35 3. 12 to 27 4. $\frac{9}{18}$ 5. 84 to 21 6. 70:90

Determine whether each pair of ratios is equivalent. Answer Yes or No. (Pages 216-218)

7. $\frac{2}{3}$ and $\frac{11}{16}$ 8. $\frac{5}{6}$ and $\frac{60}{72}$ 9. $\frac{5}{7}$ and $\frac{30}{42}$ 10. $\frac{3}{7}$ and $\frac{8}{19}$

Write each ratio in lowest terms. No variable equals zero. (Pages 216-218)

11. $\frac{10x}{25x}$ 12. $\frac{8x}{10}$ 13. $\frac{40t}{8}$ 14. $\frac{15y}{45y}$ 15. $\frac{12}{50y}$ 16. $\frac{7}{56t}$

Solve and check each proportion. (Pages 219-221)

17. $\frac{7}{y} = \frac{3}{5}$ 18. $\frac{11}{c} = \frac{4}{9}$ 19. $\frac{x}{99} = \frac{4}{11}$ 20. $\frac{x}{91} = \frac{7}{13}$ 21. $\frac{4}{7} = \frac{5}{y}$ 22. $\frac{132}{y} = \frac{11}{5}$

Write a percent for each ratio. (Pages 227-230)

23. $\frac{4}{11}$ 24. $\frac{7}{12}$ 25. $\frac{11}{4}$ 26. $\frac{5}{9}$ 27. $\frac{7}{50}$ 28. $\frac{23}{20}$

Complete the tables. Write all fractions in lowest terms. (Pages 231-234)

	Percent	Decimal	Fraction
29.	?	0.25	?
31.	42%	?	?
33.	?	?	$\frac{17}{20}$

	Percent	Decimal	Fraction
30.	?	0.15	?
32.	?	?	$\frac{3}{5}$
34.	12%	?	?

Solve. (Pages 235-239)

35. What is 37% of 143?
36. What is $66\frac{2}{3}\%$ of 480?

APPLICATIONS *Find the simple interest.* (Pages 240-243)

37. $400 borrowed for one year at 6%
38. $800 borrowed for 7 years at $12\frac{1}{2}\%$
39. After running a 100 meter race, a runner had a heartbeat of 22 beats in 10 seconds. Find the heartbeat for one minute. (Pages 222-225)
40. Jeans are on sale at a 25% discount. The regular price of the jeans is $28.60. Find the sale price of the jeans. (Pages 244-247)

252 Chapter 8

CHAPTER 9: Percent and Applications

SECTIONS
- **9-1** Finding What Percent One Number Is of Another
- **9-2** Problem Solving and Applications: Percent of Increase and Decrease
- **9-3** Finding a Number Given a Percent
- **9-4** Using Proportions to Solve Percent Problems
- **9-5** Bar Graphs
- **9-6** Circle Graphs
- **9-7** Problem Solving and Applications: Using Estimation

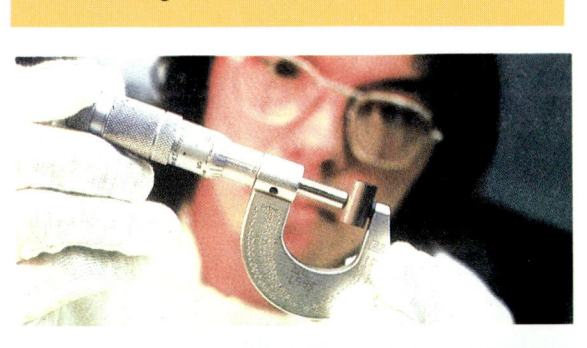

FEATURES
Calculator Application: *Finding Discount*
Consumer Application: *Nutrition*
Enrichment: *Sequences*
Common Errors

Teaching Suggestions
p. M-22

QUICK QUIZ
Write a percent for each fraction.

1. $\frac{3}{4}$ Ans: 75%
2. $\frac{78}{120}$ Ans: 65%
3. $\frac{15}{10}$ Ans: 150%

Solve and check.

4. $51a = 17$ Ans: $\frac{1}{3}$
5. $24z = 60$ Ans: $2\frac{1}{2}$

ADDITIONAL EXAMPLES
Example 1
1. What percent of 90 is 18?   Ans: 20%
2. What percent of 320 is 144?   Ans: 45%

SELF-CHECK
1. 20%   2. 40%

9-1 Finding What Percent One Number Is of Another

Uranium is a radioactive element producing enormous amounts of energy. One pellet of uranium (worth $7 in a recent year) has the energy of one ton of coal (worth $28 in a recent year).

You can use an equation to find what percent 7 is of 28.

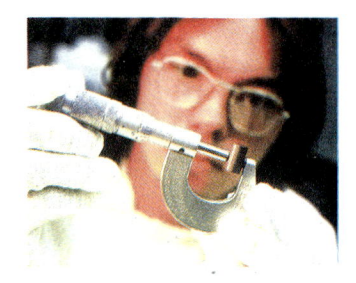

PROCEDURE

To find what percent one number is of another:

1. Write an equation for the problem.
2. Solve the equation for the variable.
3. Write a percent for the value of the variable.

EXAMPLE 1 What percent of 28 is 7?

Solution:

1. Write an equation.

   What percent of 28 is 7?

   $p \cdot 28 = 7$

   $28p = 7$

2. Solve the equation.

   $\frac{28p}{28} = \frac{7}{28}$

   $p = \frac{1}{4}$

3. Write a percent for $\frac{1}{4}$.

   $\frac{1}{4} = \frac{25}{100} = 25\%$

Thus, 7 is **25%** of 28.

Self-Check   1. What percent of 85 is 17?   2. What percent of 90 is 36?

Some word problems involve finding a percent. Follow the procedure for solving word problems involving percent given on page 236. Remember to write a percent for the value of the variable.

254 Chapter 9

EXAMPLE 2 In a recent year, the United States produced 9 million barrels of oil. That same year, the United States imported 6 million barrels of oil. What percent of the number of barrels imported is the number of barrels produced?

Solution:

1. Barrels produced: 9 million   Barrels imported: 6 million
Find: What percent of the number of barrels imported is the number of barrels produced?

2. Restate the problem.   What percent of 6 is 9?

3. Write an equation. $p \cdot 6 = 9$
$$6p = 9$$

4. Solve the equation. $\dfrac{6p}{6} = \dfrac{9}{6}$

$$p = \dfrac{9}{6} = \dfrac{3}{2}$$

5. Write a percent for $\dfrac{3}{2}$. $p = 1.50 = 150\%$

The number of barrels produced is **150%** of the number of barrels imported.

Self-Check

3. Karen borrowed $300 from Fred. He charged her $24 interest. What percent of the amount borrowed is the interest?

CLASSROOM EXERCISES

Write an equation for each percent problem. (Example 1, step 1)

1. What percent of 36 is 20?
2. What percent of 50 is 5?
3. What percent of 108 is 81?
4. What percent of 32 is 8?
5. 120 is what percent of 12?
6. 12 is what percent of 8?

Write an equation for each word problem. Do not solve. (Example 2, steps 1, 2, and 3)

7. Mike spent $4000 on his car last year. Of this amount, $200 was spent on tune-ups and oil change. What percent was spent on tune-ups and oil changes?

8. In May, the depth of water in a reservoir was 30 feet. In August, the depth had fallen 12 feet. What percent of the depth in May was the amount of decrease?

Percent and Applications   **255**

ADDITIONAL EXAMPLES
Example 2
1. Sue scored 114 points out of a possible 120. What percent did she score? Ans: 95%
2. Kirk had sales of $3500. If his commission was $560, what percent commission does he receive? Ans: 16%

SELF-CHECK
3. 8%

CLASSROOM EXERCISES
1. p · 36 = 20
2. p · 50 = 5
3. p · 108 = 81
4. p · 32 = 8
5. 120 = p · 12
6. 12 = p · 8
7. p · 4000 = 200
8. p · 30 = 12

ASSIGNMENT GUIDE
BASIC
Day 1 p. 256: 1-14
Day 2 p. 256: 15-26
AVERAGE
p. 256: 1-19 odd, 21-26
ABOVE AVERAGE
p. 256: 7-19 odd, 21-26

WRITTEN EXERCISES
1. p · 20 = 15
2. p · 12 = 18
3. p · 15 = 9
4. p · 4 = 1
5. 56 = p · 64
6. 64 = p · 56
7. $37\frac{1}{2}$%   8. 250%
9. 250%   10. 80%
11. $33\frac{1}{3}$%   12. 50%
13. 20%   14. 75%
15. 30%   16. 10%
17. $83\frac{1}{3}$%   18. 75%
19. $66\frac{2}{3}$%   20. $33\frac{1}{3}$%
21. 60%   22. 12%
23. 80%   24. 4.5%
25. 16%   26. 20%

REVIEW CAPSULE
This Review Capsule reviews prior-taught skills used in Section 9-2. The reference is to the pages where the skills were taught.

WRITTEN EXERCISES

Goal: To determine what percent one number is of another

Write an equation for each percent problem. (Example 1, step 1)

1. What percent of 20 is 15?
2. What percent of 12 is 18?
3. What percent of 15 is 9?
4. What percent of 4 is 1?
5. 56 is what percent of 64?
6. 64 is what percent of 56?

Solve. (Example 1)

7. What percent of 40 is 15?
8. What percent of 100 is 250?
9. What percent of 30 is 75?
10. What percent of 75 is 60?
11. 16 is what % of 48?
12. 36 is what percent of 72?
13. What percent of 70 is 14?
14. What percent of 96 is 72?
15. What percent of 80 is 24?
16. What percent of 50 is 5?
17. 25 is what percent of 30?
18. 3 is what percent of 4?
19. 6 is what percent of 9?
20. 4 is what percent of 12?

Solve each problem. (Example 2)

21. In a survey, 45 out of 75 persons chose Daisy vacuum cleaners. What percent was this?
22. The fee for selling a $40,000 house is $4800. What percent of the sales price is the fee?
23. A concert by the Supersounds had an audience of 24,000 people. The next night, a concert by the same group drew a crowd of 30,000 people. Find what percent the smaller crowd was of the larger.
24. A salesperson received commission of $324 during one month. The sales for the month were $7200. Find what percent commission the salesperson receives.
25. The population of a city of 2,500,000 dropped by 400,000 in ten years. Find what percent of its original population the city lost.
26. The discount on a table listed at $95 is $19. What is the rate of discount?

REVIEW CAPSULE FOR SECTION 9-2

Write a percent for each fraction. (Pages 227-230)

1. $\frac{1}{4}$ 2. $\frac{1}{5}$ 3. $\frac{7}{10}$ 4. $\frac{1}{3}$ 5. $\frac{7}{8}$ 6. $\frac{5}{6}$

Round each decimal to the nearest whole number. (Pages 47-49)

7. 6.05   8. 11.9   9. 7.28   10. 25.1   11. 18.55   12. 9.05

The answers are on page 286.

Chapter 9

PROBLEM SOLVING AND APPLICATIONS

9-2 Percent of Increase and Decrease

You can save on the cost of energy during the warmer months of the year by keeping the thermostat of your air conditioner set at about 80°F.

The figure at the right shows the approximate percent of increase in energy costs when the thermostat of the air conditioner is set at 79°F, 78°F, 77°F, 76°F, and 75°F.

80°	
79°	10% more
78°	19% more
77°	29% more
76°	40% more
75°	52% more
74°	

The **percent of increase** is the ratio of the amount of increase to the original amount.

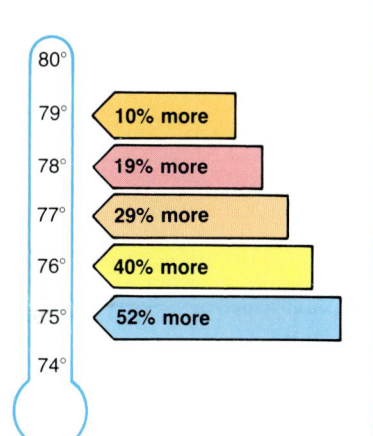

PROCEDURE

To find the percent of increase:

1. Find the amount of increase.
2. Write this ratio in lowest terms: $\dfrac{\text{Amount of Increase}}{\text{Original Amount}}$
3. Write a percent for the ratio.

EXAMPLE 1 The Bassett family paid $120 in cooling costs for the month of June. During August, they lowered the thermostat setting by 1°. Cooling costs for August amounted to $132. Find the percent of increase.

Solution:
1. Amount of increase: $132 − 120 = $12
2. $\dfrac{\text{Amount of increase}}{\text{Original amount}}$: $\dfrac{12}{120}$, or $\dfrac{1}{10}$ ◀ Lowest terms
3. Percent of increase: Let n = the first term.

$$\frac{n}{100} = \frac{1}{10}$$

$$10n = 100$$

$$n = 10 \quad \text{The increase is } \mathbf{10\%}.$$

Self-Check 1. The Wilson's paid $140 in cooling costs last month. Their bill increased to $196 this month. Find the percent of increase.

Percent and Applications 257

Teaching Suggestions
p. M-22

QUICK QUIZ

Write a percent for each fraction.

1. $\dfrac{15}{45}$ Ans: $33\tfrac{1}{3}\%$
2. $\dfrac{26}{65}$ Ans: 40%
3. $\dfrac{36}{96}$ Ans: $37\tfrac{1}{2}\%$

Round each to the nearest whole number.

4. 18.49   Ans: 18
5. 0.81   Ans: 1

ADDITIONAL EXAMPLES
Example 1

1. The Harpers paid $125 in cooling costs last month. Their bill increased to $170 this month. Find the percent of increase.
   Ans: 36%

2. The Wedmans spent $80 for groceries last week. They spent $88 this week. Find the percent of increase.
   Ans: 10%

SELF-CHECK
1. 40%

You can save on heating costs during the cooler months of the year by lowering the thermostat setting to about 65°F.

The **percent of decrease** is the ratio of the amount of decrease to the original amount.

> **PROCEDURE**
>
> **To find the percent of decrease:**
>
> 1. Find the amount of decrease.
> 2. Write this ratio in lowest terms: $\dfrac{\text{Amount of Decrease}}{\text{Original Amount}}$
> 3. Write a percent for the ratio.

EXAMPLE 2 The Gadbois family paid $90 in heating costs last month. This month's bill was reduced by lowering the thermostat setting 3°. Heating costs for this month were $82. Find the percent of decrease (nearest whole percent).

Solution:
1. Amount of decrease: $90 − 82 = $8
2. $\dfrac{\text{Amount of decrease}}{\text{Original amount}}$: $\dfrac{8}{90}$, or $\dfrac{4}{45}$ ◀ **Lowest terms**
3. Percent of decrease: Let p = the first term.

$$\frac{p}{100} = \frac{4}{45}$$

$$45p = 400$$

$$p = 8.8 \quad \blacktriangleleft \text{ Decrease: 9\%}$$

Self-Check   2. The Lopez family paid $123 in heating costs during December. Their bill for January was $105. Find the percent of decrease (nearest whole percent).

CLASSROOM EXERCISES

Complete. (Example 1)

	Original Cost	Present Cost	Amount of Increase	$\dfrac{\text{Amount of Increase}}{\text{Original Amount}}$	Percent of Increase
1.	$100	$119	?	?	?
2.	$150	$210	?	?	?
3.	$ 80	$ 96	?	?	?
4.	$120	$156	?	?	?

258   Chapter 9

(Example 2)

	Original Cost	Present Cost	Amount of Decrease	Amount of Decrease / Original Amount	Percent of Decrease
5.	$80	$68	?	?	?
6.	$100	$88	?	?	?
7.	$150	$141	?	?	?

WRITTEN EXERCISES

Goals: To find the percent of increase and decrease

To apply this skill to solving word problems

For Exercises 1–8, find the percent of increase. Round each answer to the nearest whole percent. (Example 1)

	Original Cost	Present Cost	Percent of Increase			Original Cost	Present Cost	Percent of Increase
1.	$110	$121	?		2.	$125	$175	?
3.	$135	$150	?		4.	$90	$137	?
5.	$140	$160	?		6.	$180	$216	?

7. The Donaldson family paid $150 in cooling costs last month. This month, their bill was $179.

8. The Haas family paid $121 in cooling costs for the month of June. Cooling costs for the month of July amounted to $169.

For Exercises 9–16, find the percent of decrease. Round each answer to the nearest whole percent. (Example 2)

	Original Cost	Present Cost	Percent of Decrease			Original Cost	Present Cost	Percent of Decrease
9.	$100	$91	?		10.	$140	$119	?
11.	$80	$75	?		12.	$120	$106	?
13.	$106	$87	?		14.	$200	$164	?

15. The Emanuel family paid $150 in heating costs one month. The next month, their heating costs amounted to $118.

16. The Caron family paid $89 in heating costs during January. Their bill for February was $78.

Percent and Applications **259**

---

5. $12; $\frac{12}{80}$; 15%
6. $12; $\frac{12}{100}$; 12%
7. $9; $\frac{9}{150}$; 6%

ASSIGNMENT GUIDE
BASIC
p. 259: 1–16
AVERAGE
pp. 259–260: 1–19
ABOVE AVERAGE
pp. 259–260: 1–15 odd, 17–23

WRITTEN EXERCISES
1. 10%   2. 40%   3. 11%
4. 52%   5. 14%   6. 20%
7. 19%   8. 40%   9. 9%
10. 15% 11. 6%
12. 12% 13. 18%
14. 18% 15. 21%
16. 12%

WRITTEN EXERCISES

17. 12%   18. 30%
19. 12%   20. $71.69
21. 26%   22. 8%
23. 20%

MIXED PRACTICE

Solve. Round each answer to the nearest whole percent.

17. The Stafford Company paid $330 in heating costs one month. The next month, after lowering the thermostat setting, their bill decreased to $290. Find the percent of decrease.

18. Beverly Kindell paid $104 in cooling costs for the month of July. For the month of August, her bill increased to $135. What was the percent of increase?

19. Don Tran paid $95 for cooling this month. Last month his thermostat was set at a higher temperature and his bill was $85. Find the percent of increase.

MORE CHALLENGING EXERCISES

The rate schedule at the right shows the cost for electricity in a certain area based on the number of kilowatt-hours used. Use this rate schedule and the information in the table at the right below to solve each problem.

Rate Schedule	
First 10 kwh or less	$2.79
Next 90 kwh	0.13 per kwh
Next 600 kwh	0.11 per kwh

Andros Family's Heating Costs		
Month	Kilowatt-hours Used	Cost
December	620	?
January	580	?
February	420	?

20. Find the Andros family's heating costs for the month of December.

21. Find the percent of decrease between the heating costs for January and the heating costs for February. Round the answer to the nearest whole percent.

22. The table shows the number of kilowatt-hours used this December. Last December, the Andros family used 570 kilowatt-hours. Using the same rate schedule, find the percent of increase between the heating costs for last December and the heating costs for this December. Round the answer to the nearest whole percent.

23. During March, the Andros family used their new fireplace. They spent $25 for wood and used 104 kilowatt-hours of electricity for the month. Find the percent of decrease between the heating cost for February and the total heating cost for March. Round the answer to the nearest whole percent.

Chapter 9

REVIEW CAPSULE FOR SECTION 9-3

Write a decimal for each percent. (Pages 231–234)

1. 8%   2. 1%   3. 10%   4. 80%   5. 21%   6. 19%

Divide. (Pages 44–46)

7. $26 \div 0.20$ 8. $84 \div 0.56$ 9. $26 \div 0.05$ 10. $45 \div 1.2$

Solve each equation. (Pages 200–202)

11. $0.5r = 90$ 12. $0.7n = 140$ 13. $0.75t = 18$ 14. $1.15q = 34.5$

The answers are on page 286.

9-3 Finding a Number Given a Percent

It is estimated that by 1991, Texas will have 2,600,000 more residents than it had in 1981. This number, 2,600,000, is 20% of the 1981 population. To find the population in 1981, follow this procedure.

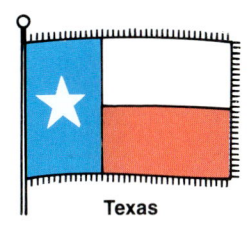

Texas

PROCEDURE

To find a number when a percent of it is known:

1 Write an equation.
2 Solve the equation.

EXAMPLE 1   2,600,000 is 20% of what number?

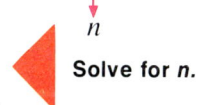

Solution: 1 $2,600,000 = 20\% \cdot n$

$2,600,000 = 0.20n$ Solve for *n*.

2   $\dfrac{2,600,000}{0.20} = \dfrac{0.20n}{0.20}$

$13,000,000 = n$ Thus, 2,600,000 is 20% of **13,000,000**.

Self-Check   1. 3 is 8% of what number?   2. 45 is 12% of what number?

The **wholesale price** is the price that a store pays for an item it will sell to the public. The public pays the **retail price** for these items.

Percent and Applications   **261**

ADDITIONAL EXAMPLES

Example 2

Find the retail price to the nearest dollar.

1. The wholesale price of $129.00 is 86% of the retail price.
Ans: $150

2. A salesperson makes a 7.5% commission. One week the commission was $900. Find the amount of sales.
Ans: $12,000

SELF-CHECK

3. $53

CLASSROOM EXERCISES

1. $50 = 0.20n$
2. $4 = 0.80q$
3. $9 = 0.15t$
4. $10 = 0.15a$
5. $26 = 0.15p$
6. $45 = 1.20r$
7. $18 = 0.45n$
8. $1800 = 0.10n$
9. $6 = 0.30n$
10. $24 = 0.32n$
11. 40; $18,000; 20; 75

EXAMPLE 2 The wholesale price of a watch is $84.00. This is 58% of the retail price. Find the retail price. Round your answer to the nearest dollar.

Solution:

⬜1 Wholesale price: $84.00; $84.00 is 58% of the retail price.
Find: Retail price

⬜2 Restate the problem. 84 is 58% of what number?

⬜3 Write an equation. $84 = 58\% \cdot n$, or
$84 = 0.58n$

⬜4 Solve the equation. $\dfrac{84}{0.58} = \dfrac{0.58n}{0.58}$

$144.827 = n$ ◀ **Round to the nearest dollar.**

The retail price is about **$145.00**.

Self-Check

3. The wholesale price of a pair of running shoes is $32.80. This is 62% of the retail price. What is the retail price? Round your answer to the nearest dollar.

CLASSROOM EXERCISES

Write an equation for each problem. Do not solve the equation.
(Example 1, step 1)

1. 50 is 20% of *n*.
2. 4 is 80% of *q*.
3. 9 is 15% of *t*.
4. 10 is 15% of *a*.
5. 26 is 15% of *p*.
6. 45 is 120% of *r*.

Write an equation for each word problem. Do not solve.
(Example 2, steps 1, 2, and 3)

7. Kozo won 18 chess games. This was 45% of the games he played. How many games did he play?

8. Natalie received a raise of $1800. This is 10% of her present yearly salary. What is her present salary?

9. Cora Whitefeather spent 6 vacation days at the beach. This was 30% of her total number of vacation days. Find the number of vacation days.

10. A baseball player got a hit 32% of the times at bat. The player's record showed 24 hits. Find the number of times at bat.

11. Solve the equations you wrote for Exercises 7–10.

262 Chapter 9

WRITTEN EXERCISES

Goal: To find a number when a percent of it is known.

Solve. (Example 1)

1. 20% of r is 15.
2. 30% of x is 21.
3. 42 is 21% of a.
4. 19 is 50% of n.
5. 42 is 25% of what number?
6. 6 is 15% of what number?
7. 9 is 2% of what number?
8. 93 is 10% of what number?
9. 16 is 80% of what number?
10. 27% of what number is 81?
11. 20% of what number is 19.5?
12. 16.8 is 15% of what number?
13. 54 is 45% of what number?
14. 204 is 85% of what number?
15. 65% of what number is 195?
16. 52% of what number is 234?

Solve each problem. (Example 2)

17. The wholesale price of a camera is $60. This is 55% of the retail price. Find the retail price. Round your answer to the nearest dollar.

18. A typewriter has a wholesale price of $355. This is 48% of the retail price. Find the retail price. Round your answer to the nearest dollar.

19. A carpenter spends 18% of his monthly take-home pay for rent. If he spends $350 a month for rent, what is his monthly take-home pay (to the nearest dollar)?

20. In the 1984 presidential elections, 53% of the population of voting age voted. About 93,000,000 people voted. Find the population of voting age that year. Round the answer to the nearest ten million.

MORE CHALLENGING EXERCISES

Solve Exercises 21–24. Write a fraction for each percent.

21. 7 is $16\frac{2}{3}$% of what number?
22. 50 is $33\frac{1}{3}$% of what number?
23. 80 is $12\frac{1}{2}$% of what number?
24. 12 is $66\frac{2}{3}$% of what number?
25. It takes $4\frac{1}{5}$ minutes to play the longest song on a record. This is 20% of the total time it takes to play the entire record. What is the total playing time of the record?
26. By 1991, it is projected that the population of Vermont will be 60,000 more than it was in 1981. The amount of increase is $12\frac{1}{2}$%. Find the 1981 population.

ASSIGNMENT GUIDE
BASIC
p. 263: 1–20
AVERAGE
p. 263: 1–20
ABOVE AVERAGE
p. 263: 1–21 odd, 22–26

WRITTEN EXERCISES
1. 75   2. 70   3. 200
4. 38   5. 168   6. 40
7. 450   8. 930   9. 20
10. 300   11. 97.5
12. 112   13. 120
14. 240   15. 300
16. 450   17. $109
18. $740   19. $1944
20. 180,000,000
21. 42; $\frac{1}{6}$
22. 150; $\frac{1}{3}$
23. 640; $\frac{1}{8}$
24. 18; $\frac{2}{3}$
25. 21 min; $\frac{1}{5}$
26. 480,000; $\frac{1}{8}$

Percent and Applications **263**

Teaching Suggestions
p. M-22

QUICK QUIZ

1. Change 6% to a decimal. Ans: 0.06

Change each to a percent.

2. 0.8 Ans: 80%
3. $\frac{3}{5}$ Ans: 60%
4. $\frac{19}{76}$ Ans: 25%

5. Multiply:
(320)(0.51)
Ans: 163.2

6. Divide:
720 ÷ 0.30
Ans: 2400

REVIEW CAPSULE FOR SECTION 9-4

Multiply. (Pages 38–41)

1. 39 · 100
2. 68 · 100
3. 10.8 · 100
4. 0.09 · 100
5. 45.5 · 100

Find each answer. (Pages 18–21)

6. 35 × 84 ÷ 100
7. 27 × 4 ÷ 36
8. 15 × 32 ÷ 16
9. 3 × 28 ÷ 21

Solve each proportion. (Pages 219-221)

10. $\frac{x}{30} = \frac{3}{5}$
11. $\frac{p}{10} = \frac{27}{15}$
12. $\frac{1}{3} = \frac{p}{15}$
13. $\frac{1}{4} = \frac{q}{28}$

The answers are on page 286.

9-4 Using Proportions to Solve Percent Problems

The three basic types of percent problems can also be solved by using proportions. To help you write the proportion, use this statement.

100% of a number is the number.

When writing a proportion, you will also find it helpful to write each percent problem in the same general form.

Percent Problem	General Form
1. A number is 5% of 120.	1. 5% of 120 is some number.
2. What percent of 28 is 42?	2. Some % of 28 is 42.
3. 45.5 is 91% of what number?	3. 91% of some number is 45.5.

To solve percent problems by the proportion method, follow this procedure.

PROCEDURE

To solve percent problems by using the proportion method:

1 Write the problem in general form.
2 Use a variable to represent the unknown.
3 Write a statement using 100%.
4 Use the statements in steps 2 and 3 to write a proportion.
5 Solve the proportion.

Chapter 9

EXAMPLE 1 A number is 5% of 120. Find the number.

Solution:

[1] Write the problem in general form.   5% of 120 is some number.

[2] Use a variable to represent the unknown.   5% of 120 is n.

[3] Write a statement using 100%.   100% of 120 is 120.

[4] Use the statements in steps 2 and 3 to write a proportion.
$$\frac{5}{100} = \frac{n}{120}$$
▶ n corresponds to 5; 120 corresponds to 100.

[5] Solve the proportion.
$$(5)(120) = 100n$$
$$600 = 100n$$
$$6 = n$$

Thus, 5% of 120 is **6**.

Self-Check 1. Use the proportion method to find 8% of 120.

Example 2 shows how to use the proportion method to solve the other two types of percent problems.

EXAMPLE 2 a. What percent of 28 is 42?   b. 45.5 is 91% of what number?

Solutions:

[1] a. Some % of 28 is 42.   b. 91% of some number is 45.5.

[2] p % of 28 is 42.   91% of t is 45.5.

[3] 100 % of 28 is 28.   100% of t is t.

[4] $\dfrac{p}{100} = \dfrac{42}{28}$ $\dfrac{91}{100} = \dfrac{45.5}{t}$

[5] $28p = (100)(42)$ $91t = (100)(45.5)$
$28p = 4200$ $91t = 4550$
$p = 150$ $t = 50$

42 is **150**% of 28.   45.5 is 91% of **50**.

Self-Check *Use the proportion method to solve.*

2. What percent of 14 is 49?   3. 16 is 80% of what number?

Percent and Applications **265**

ADDITIONAL EXAMPLES

Example 1

<u>1</u>. What number is 3% of 825?   Ans: 24.75

<u>2</u>. What number is 12% of 275?   Ans: 33

<u>3</u>. What number is 18% of 521?   Ans: 93.78

SELF-CHECK

<u>1</u>. 9.6

Example 2

<u>1</u>. What percent of 18 is 24?   Ans: 133.$\overline{3}$%

<u>2</u>. 10.40 is 65% of what number?   Ans: 16

<u>3</u>. 23.7 is 27% of what number?   Ans: 87.778

SELF-CHECK

<u>2</u>. 350%   <u>3</u>. 20

CLASSROOM EXERCISES

1. 8% of 21 is n
2. 20% of n is 4
3. n% of 85 is 17
4. n% of 60 is 20
5. 15% of n is 4
6. 30% of 612 is n
7. 100% of k is k; $\frac{100}{300} = \frac{k}{86}$
8. 100% of 4 is 4; $\frac{w}{100} = \frac{1}{4}$
9. 100% of t is t; $\frac{100}{128} = \frac{t}{45}$
10. 100% of q is q; $\frac{100}{5} = \frac{q}{1.55}$
11. 100% of 126 is 126; $\frac{3}{100} = \frac{h}{126}$
12. 100% of 90 is 90; $\frac{8}{100} = \frac{g}{90}$
13. 100% of a is a; $\frac{100}{15} = \frac{a}{10}$
14. 100% of 156 is 156; $\frac{z}{100} = \frac{23}{156}$

ASSIGNMENT GUIDE
BASIC
p. 266: 1–18
AVERAGE
p. 266: 1–19 odd, 20–26
ABOVE AVERAGE
pp. 266–267: 1–19 odd, 20–34

CLASSROOM EXERCISES

Write each percent statement in general form. Use the variable n to represent the unknown. (Examples 1 and 2, Steps 1 and 2)

1. What number is 8% of 21?
2. 4 is 20% of what number?
3. 17 is what percent of 85?
4. What percent of 60 is 20?
5. 4 is 15% of what number?
6. Find 30% of 612.

For Exercises 7–14, first write a statement using 100%. Then write a proportion. Do NOT *solve.* (Examples 1 and 2, Steps 3 and 4)

7. 300% of k is 86.
8. w% of 4 is 1.
9. 128% of t is 45.
10. 5% of q is 1.55.
11. 3% of 126 is h.
12. 8% of 90 is g.
13. 15% of a is 10.
14. z% of 156 is 23.

WRITTEN EXERCISES

Goal: To solve percent problems by using proportions

For Exercise 1–32, use the proportion method to solve. (Example 1)

1. 25% of 28 is x.
2. A number is 10% of 261.
3. 12% of 90 is a certain number.
4. p is 8% of 75.
5. 21% of 500 = r.
6. Find 23% of 500.
7. What number is 1% of 512?
8. 13% of 651 is some number.

(Example 2)

9. r% of 92 is 69.
10. 30% of b is 21.
11. 12% of t is 3.
12. 30 is z% of 48.
13. 20 is 80% of b.
14. 14 is 40% of p.
15. 7 is 16% of what number?
16. What percent of 65 is 39?
17. What percent of 52 is 78?
18. 91.84 is 82% of what number?

MIXED PRACTICE

19. 42 is 84% of what number?
20. What number is 28% of 30?
21. What percent of 36 is 9?
22. A number is 2% of 72.
23. 28 is 5% of what number?
24. What percent of 102 is 56.1?
25. 102 is 30% of what number?
26. Find 42% of 510.

266 Chapter 9

APPLICATIONS: SOLVING PERCENT PROBLEMS

Solve each problem.

27. In three hours, there were 80 landings at an air terminal. 68 of the planes landing were jets. What percent of planes landing were jets?

28. At 7 years of a age, a girl is about 74% of adult height. If a 7 year old girl is 50 inches tall, how tall can she expect to be as an adult? Round the answer to the nearest inch.

29. A certain cereal contains 12% protein. How many ounces of protein are there in a 26-ounce package of the cereal?

30. One season, a softball player got a hit 30% of her times at bat. She had 30 hits in all. How many times was she at bat?

31. The Blouin family spends 15% of total income on food each month. Last month, they spent $312 on food. What was their total income?

32. An appliance salesperson receives a 10% commission on sales. What is the commission on sales of $1500?

33. An architect charged a fee of $2250 for designing a house. The cost of the house was $90,000. What percent of the cost was the fee?

34. Lisa earns a 5% commission on sales. Last week, her commission was $35.50. What were her sales?

REVIEW: SECTIONS 9-1 — 9-4

Solve. (Section 9-1)

1. What percent of 30 is 27?

2. What percent of 75 is 90?

3. 25 is what percent of 60?

4. 42 is what percent of 30?

5. The cruising range of a car is 432 miles for city driving and 675 miles for highway driving. What percent of the highway driving range is the city driving range?

6. Warren spelled 30 words correctly on a 40-word spelling test. What percent of the words on the test did Warren correctly spell?

Percent and Applications **267**

WRITTEN EXERCISES

1. 7   2. 26.1
3. 10.8   4. 6   5. 105
6. 115   7. 5.12
8. 84.63   9. 75%
10. 70   11. 25
12. $62\frac{1}{2}$%   13. 25
14. 35   15. 43.75
16. 60%   17. 150%
18. 112   19. 50
20. 8.4   21. 25%
22. 1.44   23. 560
24. 55%   25. 340
26. 214.2   27. 85%
28. 68 in   29. 3.12 oz
30. 100 times
31. $2080   32. $150
33. $2\frac{1}{2}$%   34. $710

QUIZ: SECTIONS 9-1-9-4
After completing this Review, you may wish to administer a quiz covering the same sections. A Quiz is provided in the *Teacher's Edition: Part II.*

REVIEW: SECTIONS 9-1-9-4
1. 90%   2. 120%
3. $41\frac{2}{3}$%   4. 140%
5. 64%   6. 75%

REVIEW: SECTIONS 9-1–9-4
7. 6%   8. 52%   9. 30%
10. 9%   11. 90
12. 30   13. 80
14. 200   15. 60
16. 62.5   17. 6%
18. $300

Find the percent of increase or decrease as indicated. Round each answer to the nearest whole percent. (Section 9-2)

7. Original Cost: $98
   Present Cost: $92
   Percent of Decrease: __?__

8. Original Cost: $85
   Present Cost: $129
   Percent of increase: __?__

Solve. Round each answer to the nearest whole percent. (Section 9-2)

9. The Ramirez family paid $115 last month for cooling. This month their cooling bill was $150. Find the percent of increase.

10. The Tobin family's heating cost for January was $92. After lowering the thermostat setting, their bill for February was $84. Find the percent of decrease.

Solve. (Section 9-3)

11. 18 is 20% of what number?
12. 18 is 60% of what number?
13. 60 is 75% of what number?
14. 4 is 2% of what number?
15. 25% of what number is 15?
16. 40% of what number is 25?

Solve. (Section 9-4)

17. A salesperson received a commission of $450 during one month. The sales for the month were $7500. Find what percent commission the salesperson receives.

18. A family budgets 25% of their take-home pay for food. The family's monthly take-home pay is $1200. How much is their monthly food budget?

REVIEW CAPSULE FOR SECTION 9-5

Complete. Write a percent for each fraction. (Pages 227–230)

1. $\frac{3}{4} =$ __?__   2. $\frac{2}{5} =$ __?__   3. $\frac{5}{6} =$ __?__   4. $\frac{1}{8} =$ __?__   5. $\frac{2}{3} =$ __?__

(Pages 227–230)

6. $\frac{85}{170} =$ __?__   7. $\frac{900}{3000} =$ __?__   8. $\frac{150}{400} =$ __?__   9. $\frac{355}{568} =$ __?__   10. $\frac{1800}{8000} =$ __?__

The answers are on page 286.

Chapter 9

CONSUMER APPLICATION
Nutrition

The **RDA** (recommended daily allowance) indicates the total amount of vitamins, minerals, and protein that should be included in a daily diet. The table below shows the adult **RDA** for certain vitamins and minerals.

Item	Recommended Daily Allowance
Vitamin A	5000 International Units
Vitamin C	45 milligrams
Vitamin D	400 International Units
Calcium	800 milligrams
Phosphorus	800 milligrams

The four basic food groups are shown: protein products (upper left), carbohydrates (upper right), dairy products (lower left) and fruits and vegetables (lower right).

Exercises

Solve each problem. Refer to the table above for the adult RDA.

1. Find the amount of Vitamin A in 100 grams of milk when 100 grams of milk contain 3% of the RDA for Vitamin A.

2. Find the amount of calcium in 100 grams of milk when 100 grams of milk contain 15% of the total RDA for calcium.

3. One large apple contains 20% of the total RDA for Vitamin C. Find the amount of Vitamin C in an apple.

4. One tablespoon of butter contains 1¼% of the total RDA for Vitamin D. Find the amount of Vitamin D in one tablespoon of butter.

5. One large banana contains 16 milligrams of calcium. What per cent of the total RDA for calcium is this?

6. One cup of cooked green beans contains 0.9 milligrams of iron. This is 5% of the total RDA for iron. Find the total RDA.

7. One slice of enriched white bread contains 0.5 milligrams or 2.5% of the total RDA for niacin. Find the total RDA for niacin.

8. One cup of shredded carrots contains 12,000 International Units of Vitamin A. What per cent of the total RDA for Vitamin A is this?

CONSUMER APPLICATIONS
This feature applies the skills for solving word problems involving percent to comparing the vitamin or mineral content of selected foods to the adult RDA. Therefore, it can be used after Section 9-4 (pages 264-267) is taught.

1. 150 I.U.
2. 120 mg
3. 9 mg
4. 5 I.U.
5. 2%
6. 18 mg
7. 20 mg
8. 240%

Teaching Suggestions
p. M-22

QUICK QUIZ
Write each fraction as a percent.

1. $\frac{1}{4}$ Ans: 25%
2. $\frac{5}{8}$ Ans: 62.5%
3. $\frac{1}{3}$ Ans: $33\frac{1}{3}$%
4. $\frac{90}{1200}$ Ans: 7.5%
5. $\frac{3078}{8550}$ Ans: 36%

ADDITIONAL EXAMPLES
Example 1
Use the given bar graph.

1. What percent of the cars were sold by companies Y and Z together?
   Ans: 56.6%
2. By what percent did the sales of company Y exceed the sales of company X?
   Ans: $16\frac{2}{3}$%

SELF-CHECK
1. 40%

9-5 Bar Graphs

A **bar graph** has two axes, a horizontal axis and a vertical axis. On a bar graph, the length of each bar represents a number. The **scale** on one of the axes tells you how to find the number.

A graph is one way to present information or **data**. This **bar graph** shows the number of cars sold during the month of July by four automobile manufacturers.

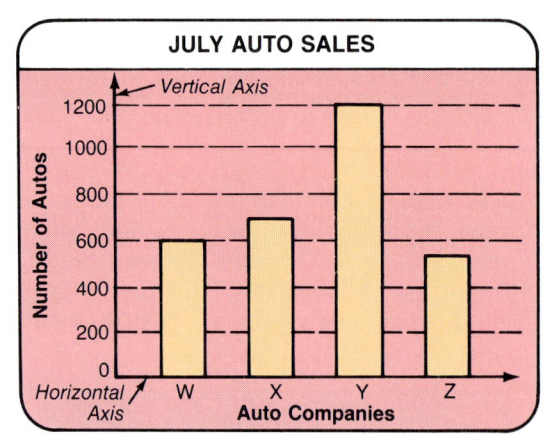

JULY AUTO SALES (bar graph: W=600, X=700, Y=1200, Z=500; Vertical Axis: Number of Autos; Horizontal Axis: Auto Companies)

EXAMPLE 1 Use the bar graph above to answer each question.

Question	Solutions	
a. What was the total number of cars sold in July?	a. $600 + 700 + 1200 + 500 = 3000$	From the graph
b. What percent of the cars was sold by company W?	b. $\frac{600}{3000} = \frac{1}{5} = 20\%$	
c. What percent of the cars was sold by company Z?	c. $\frac{500}{3000} = \frac{1}{6} = 16\frac{2}{3}\%$	See the table on page 235
d. By what percent did the sales of company W exceed the sales of company Z?	d. $20\% - 16\frac{2}{3}\% = 3\frac{1}{3}\%$	

Self-Check

1. Use the bar graph above to find what percent of the cars was sold by company Y?

Since the graph shown above has vertical bars, it is called a **vertical bar graph**. Bar graphs may also be **horizontal**.

PROCEDURE
To construct a bar graph:

1. Draw the vertical and horizontal axes. Mark off a convenient scale at equal intervals on each axis. Label the axes.
2. Draw vertical or horizontal bars to represent the data. (The bars should be the same width and distance apart.)

270 Chapter 9

EXAMPLE 2 Use the given data to construct a horizontal bar graph.

BUSINESS SCHOOL GRADUATES

Fields of Study	Number
Managerial	600
Technical	450
Business	250
Medical	400

Business School Graduates (horizontal bar graph showing Managerial ~600, Technical ~450, Business ~250, Medical ~400; x-axis "Number of Graduates" from 0 to 600)

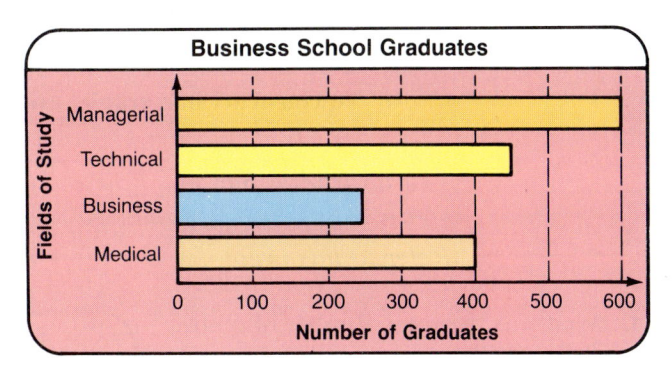

Self-Check

2. Use the data in the table below to construct a horizontal bar graph.

Average July Temperatures

City	Phoenix	Chicago	Boston	Los Angeles	Cheyenne
Average Temperature (Degrees Celsius)	34°	24°	22°	22°	20°

CLASSROOM EXERCISES

For Exercises 1–4, refer to the bar graph below. (Example 1)

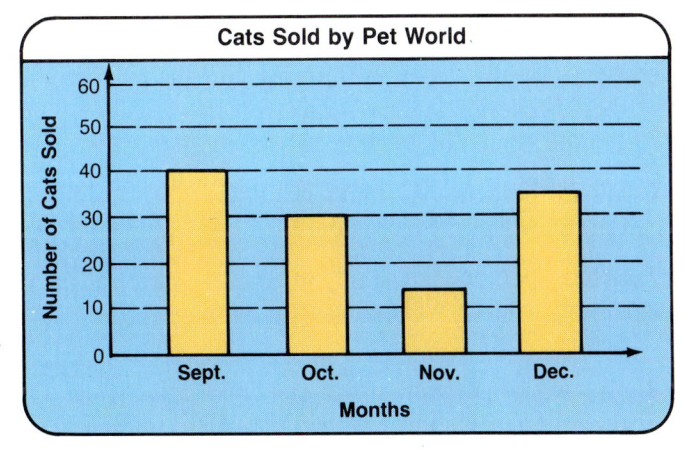

Cats Sold by Pet World (bar graph: Sept. 40, Oct. 30, Nov. ~14, Dec. ~35)

1. What was the total number of cats sold?

2. What percent of the cats was sold in November?

3. By what percent did the September sales exceed the October sales?

4. The percent of profit each month was in the same ratio as the percent of sales. The total profit for the four months was $3600. What was the amount of profit for the month of September?

Example 2

Use the data below to construct a horizontal bar graph.

Pen Sales

Pen color	Number sold
Blue	14
Red	10
Black	16
Green	5

Ans:

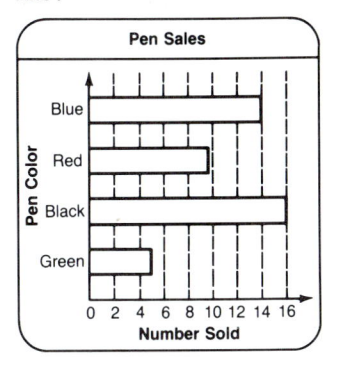

(Horizontal bar graph "Pen Sales" showing Blue 14, Red 10, Black 16, Green 5)

SELF-CHECK

2.

(Horizontal bar graph "July Temperatures for Five Cities" showing Phoenix 34, Chicago 24, Boston 22, Los Angeles 22, Cheyenne 20)

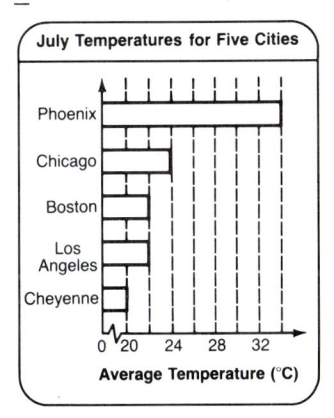

CLASSROOM EXERCISES

1. 120   2. $12\frac{1}{2}$%

3. $33\frac{1}{3}$%   4. $1200

Percent and Applications 271

ASSIGNMENT GUIDE
BASIC
pp. 272-273: 1-10
AVERAGE
pp. 272-273: 1-12
ABOVE AVERAGE
pp. 272-273: 1-12

WRITTEN EXERCISES
1. 185,000   2. 31%
3. 14%   4. 43%
5. 86   6. 27%
7. 52%   8. 340%

WRITTEN EXERCISES

Goals: To read bar graphs

To construct bar graphs using data from tables

The horizontal bar graph at the right below shows the number of metric tons of wheat produced by 6 countries in a recent year.
Refer to this graph for Exercises 1-4. (Example 1)

1. What was the total number of metric tons of wheat produced by the six countries?

2. What percent of the wheat was produced by the United States?

3. What percent of the wheat was produced by Canada?

4. By what percent did the production of wheat by Canada exceed the production of wheat by Turkey?

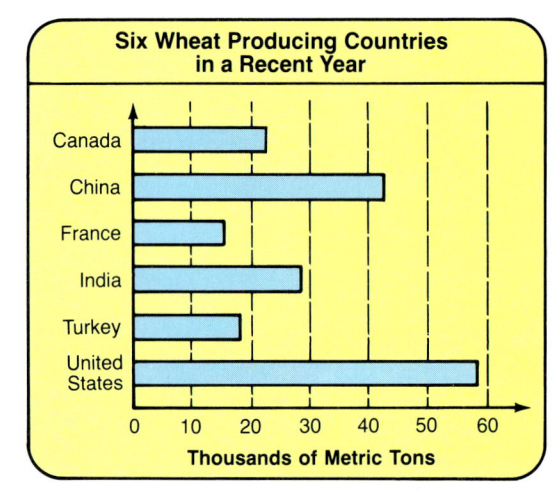

Six Wheat Producing Countries in a Recent Year

This bar graph shows predicted energy sources for the United States in 1985.
Refer to this graph for Exercises 5-8.

5. What is the total number of predicted energy source units for the United States in 1990?

6. What percent of our predicted energy will come from coal?

7. What percent of our predicted energy source will come from coal and gas?

8. By what percent does the predicted energy source of gas exceed the predicted energy source of hydropower?

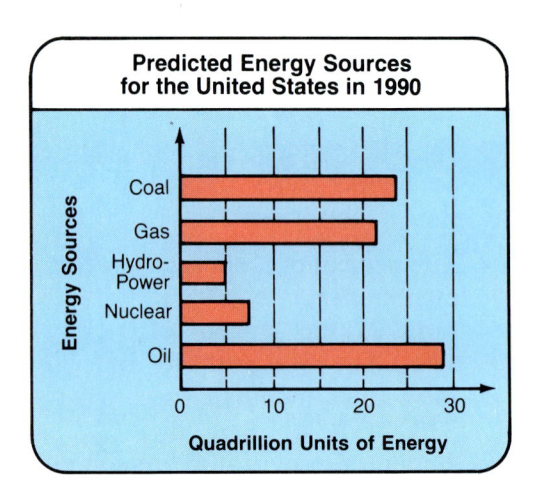

Predicted Energy Sources for the United States in 1990

272 Chapter 9

For Exercises 9–12. use the given data to construct a bar graph.
(Example 2)

9. Book Sales

Month	Books Sold
January	150
February	200
March	300
April	400
May	175
June	500

10. Land Use in a Town

Use	Per Cent
Roads	25%
Railroad	3%
Public Use	12%
Farm	8%
Housing	52%

11. Approximate Heights in Meters of Six Waterfalls

Waterfall	Height
Yosemite Lower Falls	100
Yosemite Upper Falls	435
Nevada Falls	180
Great Falls	255
Ribbon Falls	490
Silver Strand Falls	360

12. Approximate Yearly Amount of Food Consumed by the Average Person

Food	Pounds
Beef	75
Poultry	100
Fruit	130
Vegetables	150
Grains	130
Dairy	285

WRITTEN EXERCISES

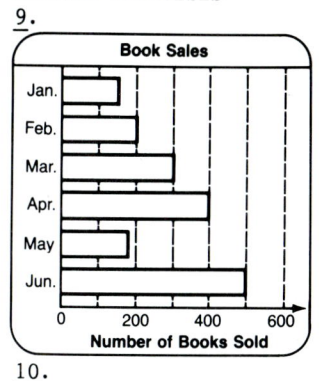

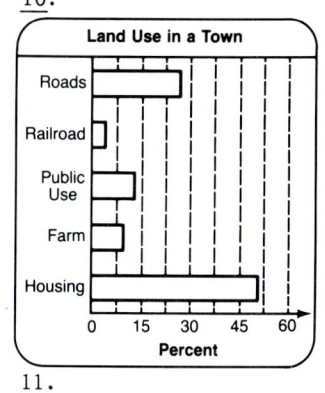

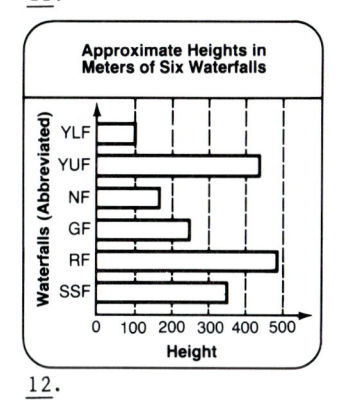

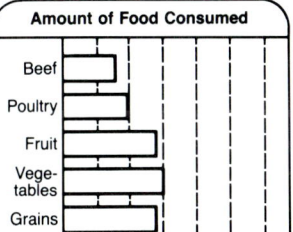

9. Book Sales bar graph (Jan.–Jun., Number of Books Sold 0–600)

10. Land Use in a Town bar graph (Roads, Railroad, Public Use, Farm, Housing; Percent 0–60)

11. Approximate Heights in Meters of Six Waterfalls bar graph (YLF, YUF, NF, GF, RF, SSF; Height 0–500)

12. Amount of Food Consumed bar graph (Beef, Poultry, Fruit, Vegetables, Grains, Dairy; Pounds 0–300)

REVIEW CAPSULE FOR SECTION 9-6

For Exercises 1–10, write a decimal for each percent.
(Pages 231–234)

1. 40% 2. 65% 3. 8% 4. 10% 5. $17\frac{1}{2}\%$
6. 7% 7. 81% 8. $12\frac{3}{4}\%$ 9. 92% 10. $9\frac{1}{2}\%$

For Exercise 11–16:
a. Write an equation for the problem.
b. Solve the equation. (Pages 235–239)

11. What is 10% of 516?
12. Find 45% of 200.
13. What is $66\frac{2}{3}\%$ of 90?
14. Find 60% of 75.
15. What is $37\frac{1}{2}\%$ of 400?
16. Find $83\frac{1}{3}\%$ of 144.

The answers are on page 286.

Percent and Applications

Teaching Suggestions
p. M-22

QUICK QUIZ

1. What is 8% of 75?
   Ans: 6
2. What is $62\frac{1}{2}$% of 800?
   Ans: 500
3. What is 47% of 650?
   Ans: 305.5
4. Find 7.3% of 600.
   Ans: 43.8
5. Find $33\frac{1}{3}$% of 1287.
   Ans: 429

ADDITIONAL EXAMPLES
Example 1
Use the given circle graph.
1. How many vehicles were cars?
   Ans: 225
2. How many vehicles were not trucks or pickups?
   Ans: 365

SELF-CHECK
1. 25

9-6 Circle Graphs

Circle graphs are often used to show data given as percents. The entire circle represents 100%.

The circle graph below shows what percent of each type of vehicle was serviced at a station during a certain day.

Vehicles Serviced in a Day

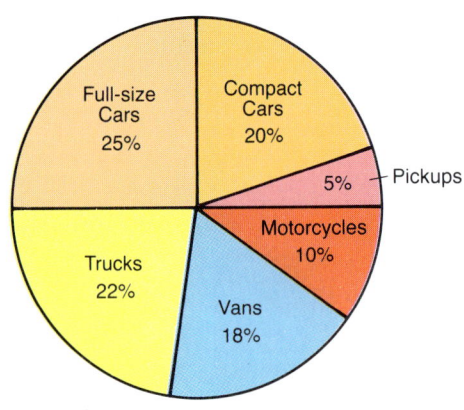

- Full-size Cars 25%
- Compact Cars 20%
- Pickups 5%
- Motorcycles 10%
- Vans 18%
- Trucks 22%

EXAMPLE 1 The circle graph above represents a total of 500 vehicles. Find how many of these vehicles were compact cars.

Solution:
① Percent of compact cars: 20%
② Write an equation. What is 20% of 500?
   $n = 0.20 \ (500)$
③ Solve the equation. $n = 100$

Thus, **100** of the **vehicles** serviced were compact cars.

Self-Check 1. Refer to the circle graph above to find how many of the 500 vehicles serviced were pickups.

To construct a circle graph, you use a protractor.

PROCEDURE

To construct a circle graph:

① Write a decimal for each percent.
② Multiply the decimal by 360°. Round the answer to the nearest degree.
③ Use a protractor to construct the graph.

274 Chapter 9

To construct a circle graph, you use this fact from geometry.

There are 360° in a circle.

EXAMPLE 2 This table shows the percent of each of four types of tissues in the human body. Construct a circle graph to show the data.

Body Tissue

Type	Percent
Muscle	47%
Supporting	33%
Blood	7%
Surface	13%

Solution:

1 Write a decimal for each percent.

$47\% = 0.47$ $33\% = 0.33$
$13\% = 0.13$ $7\% = 0.07$

2 Multiply each decimal by 360°. Round to the nearest degree.

$0.47 \times 360° = 169.2$, or $169°$ $0.33 \times 360° = 118.8$, or $119°$
$0.13 \times 360° = 46.8$, or $47°$ $0.07 \times 360° = 25.2$, or $25°$

3 Construct the graph.

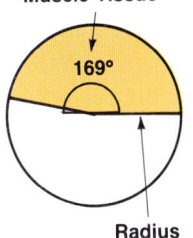

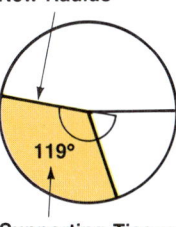

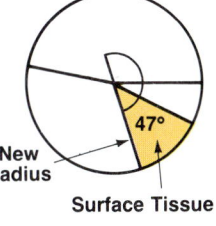

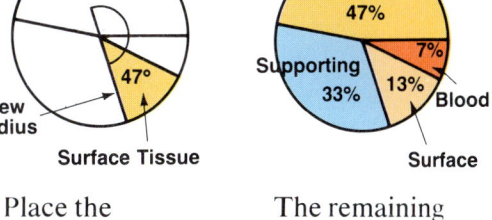

Muscle Tissue (169°) — Draw a radius. Place the protractor on the radius. Draw the angle for the muscle tissue.

Supporting Tissue (New Radius, 119°) — Place the protractor on the "new" radius. Draw the angle for supporting tissue.

Surface Tissue (New Radius, 47°) — Place the protractor on the "new" radius. Draw the angle for surface tissue.

Body Tissue — Muscle 47%, Supporting 33%, Blood 7%, Surface 13%. The remaining angle represents the blood tissue.

Self-Check 2. Construct a circle graph for the following data. Round each angle measure to the nearest degree.

Popular Vacation Cities

City	New York	Los Angeles	San Francisco	Washington, D.C.	Orlando
Percent of Total	37%	19%	17%	14%	13%

Percent and Applications   **275**

Example 2

1. Use the data below to construct a circle graph.

Bowl Game Viewers

Category	Percent of Audience
Men	52%
Women	32%
Teens	8%
Children	8%

Ans:

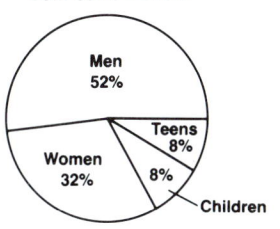

Bowl Game Viewers — Men 52%, Women 32%, Teens 8%, Children 8%

SELF-CHECK

2.

Popular Vacation Cities — New York 37%, Los Angeles 19%, San Francisco 17%, Washington D.C. 14%, Orlando 13%

CLASSROOM EXERCISES

1. $378   2. $162
3. n = 0.20(900); 180
4. 5%; n = 0.05(900); $45
5. 15%; n = 0.15(900); $135
6. 108°   7. 90°
8. 0.42; 151°
9. 0.18; 65°
10. 0.08; 29°
11. 0.21; 76°
12.

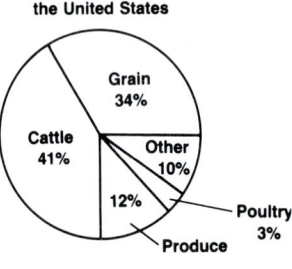

Kinds of Farms in the United States

(Circle graph: Grain 34%, Cattle 41%, Produce 12%, Poultry 3%, Other 10%)

CLASSROOM EXERCISES

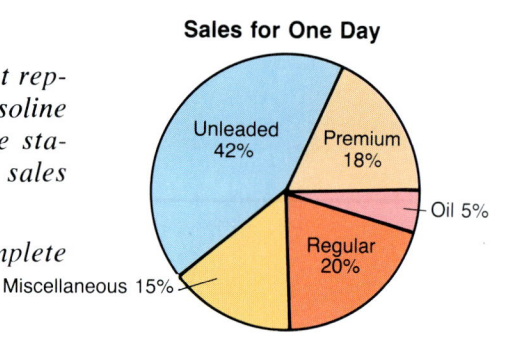

The circle graph at the right represents the total sales of gasoline and other items at a service station for one day. The total sales were $900.

Refer to the graph to complete Exercises 1-5. (Example 1)

Sales for One Day

(Circle graph: Unleaded 42%, Premium 18%, Oil 5%, Regular 20%, Miscellaneous 15%)

	Item	Percent of Total Sales	Related Equation	Dollar Amount Sold
1.	Unleaded Gas	42%	n = 0.42(900)	?
2.	Premium Gas	18%	n = 0.18(900)	?
3.	Regular Gas	20%	?	?
4.	Oil	?	?	?
5.	Miscellaneous	?	?	?

Find the decimal equivalent of the percent and the number of degrees needed to represent each percent on a circle graph. Round each answer to the nearest degree. (Example 2, Steps 1 and 2)

	Percent	Decimal	Degrees
6.	30%	0.3	?
8.	42%	?	?
10.	8%	?	?

	Percent	Decimal	Degrees
7.	25%	0.25	?
9.	18%	?	?
11.	21%	?	?

12. Construct a circle graph for the following data. (Example 2)

Kinds of Farms in the United States

Farm	Grain	Cattle	Produce	Poultry	Other
Percent	34%	41%	12%	3%	10%

276   Chapter 9

WRITTEN EXERCISES

Goals: To read and interpret circle graphs

To construct circle graphs

The circle graph at the right shows how a school club raised $3000 for a special project. Refer to this graph for Exercises 1-4. (Example 1)

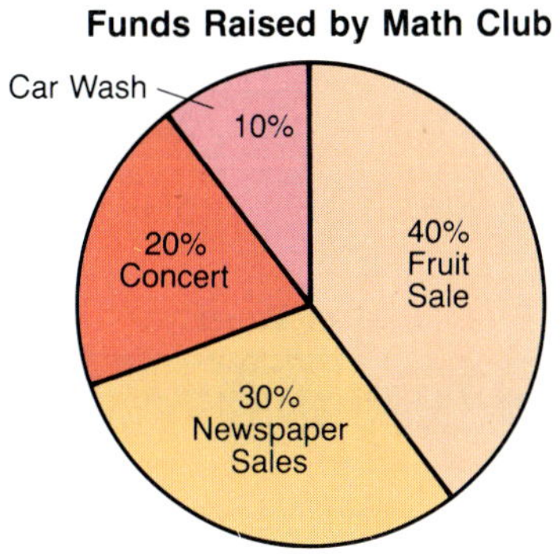

Funds Raised by Math Club

- Car Wash 10%
- 20% Concert
- 40% Fruit Sale
- 30% Newspaper Sales

1. How much did the club earn selling newspapers?
2. How much did the club earn from the concert?
3. How much did the club earn washing cars and selling newspapers?
4. How much greater was the income from the fruit sale than the income from the car wash?

The circle graph at the right shows the percent of each type of book found in a public library. The collection consists of 4000 books. Refer to this graph for Exercises 5-8. (Example 1)

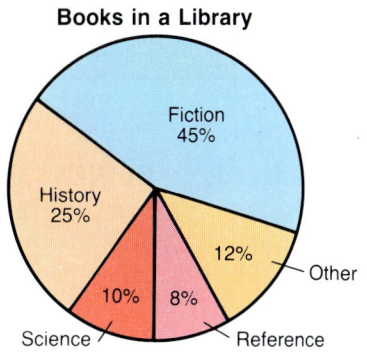

Books in a Library

- Fiction 45%
- History 25%
- Science 10%
- Reference 8%
- Other 12%

5. How many are history books?
6. How many are reference books?
7. How many are either science or fiction books?
8. What is the difference between the number of fiction and the number of history books?

Use the given data to construct a circle graph. Round each angle measure to the nearest degree. (Example 2)

9.

City Budget

Item	Schools	Public Works	Health and Safety	Other
Percent	35%	15%	40%	10%

ASSIGNMENT GUIDE

BASIC
pp. 277-278: 1-11

AVERAGE
pp. 277-278: 1-13

ABOVE AVERAGE
pp. 277-278: 1-13

WRITTEN EXERCISES

1. $900   2. $600
3. $1200 4. $900
5. 1000   6. 320
7. 2200   8. 800
9.

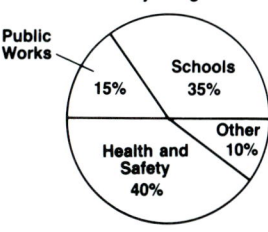

City Budget

- Public Works 15%
- Schools 35%
- Other 10%
- Health and Safety 40%

Percent and Applications 277

277

WRITTEN EXERCISES

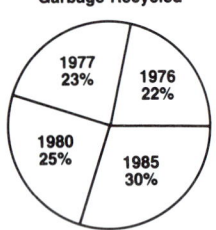

10. United States Energy Sources

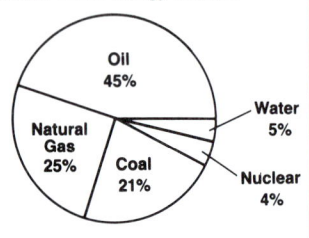

Source	Percent
Oil	45%
Natural Gas	25%
Coal	21%
Nuclear	4%
Water	5%

11. Causes of Forest Fires

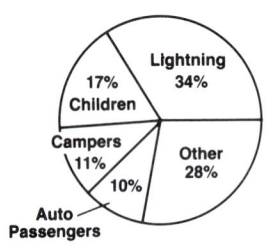

Cause	Percent
Lightning	34%
Children	17%
Campers	11%
Auto Passengers	10%
Other	28%

12. Uses of Paper

Use	Percent
Packaging	48%
Writing Paper	29%
Tissues	7%
Other	16%

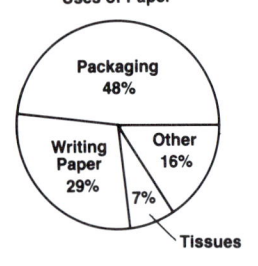

13. Garbage Recycled

Year	Percent
1976	22%
1977	23%
1980	25%
1985	30%

REVIEW CAPSULE FOR SECTION 9-7

Round to the nearest whole number. (Pages 47-49)

1. 58.75 **2.** 16.50 **3.** 11.21 **4.** 32.95 **5.** 70.55 **6.** 0.085
7. 4.38 **8.** 5.499 **9.** 6.3 **10.** 19.6 **11.** 0.927 **12.** 69.5

Round to the nearest ten. (Pages 47-49)

13. 58.75 **14.** 16.50 **15.** 11.21 **16.** 32.95 **17.** 70.55 **18.** 83.05
19. 39.5 **20.** 46.81 **21.** 50.05 **22.** 86.91 **23.** 70.62 **24.** 78.62

Multiply. (Pages 129-131)

25. $\frac{1}{10} \times 900$ **26.** $\frac{1}{8} \times 512$ **27.** $\frac{1}{6} \times 426$ **28.** $\frac{1}{12} \times 1560$ **29.** $\frac{2}{5} \times 75$
30. $\frac{2}{3} \times 960$ **31.** $\frac{1}{4} \times 188$ **32.** $\frac{5}{8} \times 200$ **33.** $\frac{3}{10} \times 900$ **34.** $\frac{5}{6} \times 900$

Write a fraction for each percent. (Pages 235-239)

35. 75% **36.** 40% **37.** $12\frac{1}{2}$% **38.** $66\frac{2}{3}$% **39.** $16\frac{2}{3}$%

The answers are on page 286.

PROBLEM SOLVING AND APPLICATIONS

9-7 Using Estimation

To estimate a percent of a number, you can sometimes round the number only.

EXAMPLE 1 A car radio with a list price of $78.69 is on sale at a 25% discount. Estimate the amount of discount.

Solution:
1. $25\% = \frac{1}{4}$
2. Round to a <u>convenient</u> price; that is, to a price that is close to $78.69 <u>and</u> is easy to multiply by $\frac{1}{4}$.
3. Estimated discount: $\frac{1}{4} \cdot 80 = 20$

The amount of discount is about $20.

Self-Check 1. A sport coat with a list price of $62.25 is on sale at a 20% discount. Estimate the amount of discount.

Sometimes you can round to a convenient percent.

EXAMPLE 2 The Athlete's Shoe Shop has running shoes on sale at a 38% discount. Estimate the amount of discount on the running shoes if the list price is $45.

Solution:
1. Round to a <u>convenient</u> percent; that is, to a percent close to 38% and easy to multiply by 45.

   38% is about 40%. $40\% = \frac{2}{5}$

2. Estimated discount: $\frac{2}{\cancel{5}} \cdot \cancel{45}^{9} = 18$

The amount of discount is about $18.

Self-Check 2. A canoe with a list price of $200 is on sale at a 18% discount. Estimate the amount of discount on the canoe.

It is sometimes useful to estimate what percent a number is of another.

Percent and Applications **279**

Teaching Suggestions p. M-22

QUICK QUIZ
Round to the nearest ten.
1. 31.8   Ans: 30
2. 1287   Ans: 1290
3. 632.86   Ans: 630

Multiply.
4. $\frac{2}{3}(1200)$ Ans: 800
5. $\frac{3}{8}(3240)$ Ans: 1215

ADDITIONAL EXAMPLES
Example 1
Estimate the discount.
1. List price: $29.95; discount: 40%
   Ans: $12
2. List price: $1289; discount: 50%
   Ans: $650

SELF-CHECK
1. $12

Example 2
Estimate the discount.
1. List price: $250; discount: 21%
   Ans: $50
2. List price: $400; discount: 14%
   Ans: $60

SELF-CHECK
2. $40

ADDITIONAL EXAMPLES

Example 3

Estimate the percent.

1. What percent of 124 is 25?   Ans: 20%

2. 72 is what percent of 79.95?   Ans: 90%

SELF-CHECK

3. 20%

CLASSROOM EXERCISES

1. c   2. c   3. b
4. a   5. c   6. c
7. b   8. b   9. a
10. c 11. b 12. c
13. b 14. c 15. b
16. a 17. b

EXAMPLE 3 The total cost of repairs on Mario's car was $58. The cost of parts alone was $15. Estimate what percent of the total cost was the cost of parts.

Solution:
[1] Restate the problem.   What percent of 58 is 15?

[2] Think: 15 is about $\frac{1}{4}$ of 58.

58 is about 60 and 15 is $\frac{1}{4}$ of 60.

[3] Therefore, 15 is about **25%** of 58.

Self-Check 3. Of the 150 students in a school band, 29 play woodwind instruments. Estimate what percent play woodwind instruments.

CLASSROOM EXERCISES

Choose the best estimate. Choose a, b, or c.   (Example 1)

1. 25% of $58.56   a. 25% of $50   b. 25% of $55   c. 25% of $60
2. 10% of $29.95   a. 10% of $20   b. 10% of $25   c. 10% of $30
3. 50% of $91.30   a. 50% of $80   b. 50% of $90   c. 50% of $100
4. 20% of $34.75   a. 20% of $35   b. 20% of $30   c. 20% of $40
5. 30% of $10.06   a. 30% of $20   b. 30% of $15   c. 30% of $10

(Example 2)

6. 21% of $150   a. 30% of $150   b. 25% of $150   c. 20% of $150
7. 9.5% of $700   a. 5% of $700   b. 10% of $700   c. 20% of $700
8. 26% of $400   a. 20% of $400   b. 25% of $400   c. 30% of $400
9. 33% of $90   a. $33\frac{1}{3}$% of $90   b. 30% of $90   c. 40% of $90
10. 12% of $24   a. 20% of $24   b. $16\frac{2}{3}$% of $24   c. $12\frac{1}{2}$% of $24

(Example 3)

11. What percent of 73 is 12?   a. 10%   b. $16\frac{2}{3}$%   c. 25%
12. What percent of 41 is 8?   a. 10%   b. 15%   c. 20%
13. What percent of 14 is 3?   a. 10%   b. 20%   c. 30%
14. What percent of 68 is 23?   a. 50%   b. 40%   c. $33\frac{1}{3}$%
15. What percent of 95 is 46?   a. 40%   b. 50%   c. 60%
16. 58 is what percent of 79?   a. $66\frac{2}{3}$%   b. 75%   c. 80%
17. 39 is what percent of 60?   a. 40%   b. 60%   c. 50%

Chapter 9

WRITTEN EXERCISES

Goal: To use estimation to find the percent of a number

To use estimation to find what percent a number is of another

Choose the best estimate. Choose a, b, or c. (Example 1)

1. 50% of $121.90   **a.** $60   **b.** $65   **c.** $70
2. 20% of $98.50   **a.** $15   **b.** $20   **c.** $25
3. 25% of $20.98   **a.** $5   **b.** $6   **c.** $7
4. 75% of $101.17   **a.** $65   **b.** $85   **c.** $75
5. $33\frac{1}{3}$% of $134.90   **a.** $33   **b.** $50   **c.** $45

(Example 2)

6. 19% of $80   **a.** $8   **b.** $16   **c.** $20
7. 8.5% of $40   **a.** $4   **b.** $2   **c.** $8
8. 41% of $1600   **a.** $800   **b.** $480   **c.** $640
9. 49% of $500   **a.** $200   **b.** $250   **c.** $300
10. 24.2% of $60   **a.** $15   **b.** $18   **c.** $12

Estimate each percent. (Example 3)

11. What percent of 83 is 20?
12. What percent of 50 is 9?
13. What percent of 73 is 15?
14. What percent of 88 is 9?
15. What percent of 130 is 12?
16. What percent of 60 is 5.5?

Choose the best estimate for each discount. Choose a, b, or c.

17. SALE! 20% off LAWN CHAIR REGULARLY—$19.50
   a. $4   **b.** $15   **c.** $20

18. Sale! 40% off Tennis Racquet REGULARLY $49.95
   a. $18   **b.** $24   **c.** $20

19. SALE! Snow Tires 9% Off REGULARLY $39.95
   a. $4   **b.** $5   **c.** $6

Percent and Applications **281**

ASSIGNMENT GUIDE

BASIC

p. 281: 1-19

AVERAGE

pp. 281-282: 1-25

ABOVE AVERAGE

pp. 281-282: 1-25

WRITTEN EXERCISES

1. a   2. b   3. a
4. c   5. c   6. b
7. a   8. c   9. b
10. a   11. 25%
12. 20%   13. 20%
14. 10%   15. 10%
16. 10%   17. a
18. c   19. a

WRITTEN EXERCISES

20. $33\frac{1}{3}$%   21. 25%
22. $18   23. $16
24. $16\frac{2}{3}$%   25. 50%

CALCULATOR EXERCISES

1. $200; $208.00
2. $80; $80.30
3. $22; $21.98
4. $300; $299.06
5. $2; $2.13

Solve each problem.

20. A quarterback completed 10 of 28 passes in a football game. Estimate what percent of the passes he completed.

21. A reporter spends $300 a month for rent. Her monthly salary is $1235. Estimate what percent of her monthly salary she spends on rent.

22. A rain coat with a list price of $91.20 is on sale at a 20% discount. Estimate the amount of discount.

23. A tape deck with a list price of $47.88 is on sale at a $33\frac{1}{3}$% discount. Estimate the amount of discount.

24. In a class of 32 students, 5 students were absent. Estimate what percent of the class was absent.

25. A tennis player made 32 of 61 first serves. Estimate what percent of first serves were made.

FINDING DISCOUNT

Before using a calculator to find the amount of discount:

1. Estimate the answer.
2. Use the calculator to find the exact answer.
3. Compare the exact answer with the estimate to see whether the exact answer is reasonable.

EXAMPLE: Regular Price: $300.00   Rate of Discount: 9.8%   Discount: __?__

Solution: Estimated Discount: $\frac{1}{10} \times \$300 =$ **$30**   ◄ 9.8% is about 10%.

Exact Discount: 300 × 9.8%

[3 0 0] [×] [9 . 8] [%] [=]   **29.4**

Since the estimate is $30, **$29.40** is a reasonable answer.

EXERCISES

First estimate the discount. Then find the exact discount.

	Regular Price	Rate of Discount	Estimated Discount	Exact Discount
1.	$800.00	26%	?	?
2.	$401.50	20%	?	?
3.	$ 87.90	25%	?	?
4.	$996.85	30%	?	?
5.	$ 25.00	8.5%	?	?

282   Chapter 9

REVIEW: SECTIONS 9-5—9-7

1. Use the data below to construct a horizontal bar graph. (Section 9-5)

Renell High School Enrollment

Year	1976	1977	1978	1979	1980
Number of Students	1030	1020	1095	1090	1075

The circle graph at the right shows how a library club raised $2000. Refer to this graph for Exercises 2-4. (Section 9-6)

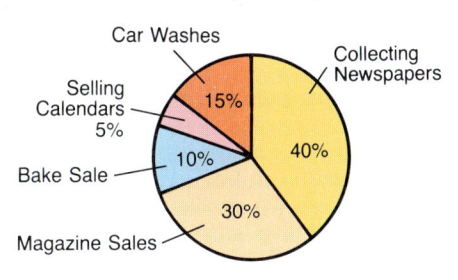

Funds Raised by Library Club
- Car Washes 15%
- Collecting Newspapers 40%
- Selling Calendars 5%
- Bake Sale 10%
- Magazine Sales 30%

2. How much did the library club earn by selling calendars?
3. How much did the library club earn by washing cars?
4. By how much did the amount received by collecting newspapers exceed the amount made by selling magazines?
5. Use the data below to construct a circle graph. (Section 9-6)

Population of a City by Age

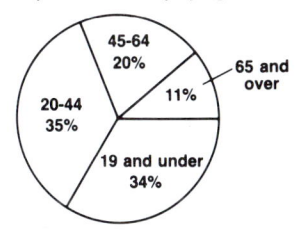

Age	65 and over	45-64	20-44	19 and under
Percent	11%	20%	35%	34%

Choose the best estimate. Choose a, b, or c. (Section 9-7)

6. 20% of $61.01   a. $12   b. $10   c. $14
7. 40% of $148.71   a. $80   b. $70   c. $60
8. 58% of $91.20   a. $48   b. $54   c. $60
9. 19% of $50   a. $1   b. $5   c. $10

Estimate what percent a number is of another. (Section 9-7)

10. What percent of 75 is 24?
11. What percent of 49 is 8?
12. The regular price of a television set is $510.00. The set is marked at 25% off. Estimate the amount of discount. (Section 9-7)
13. A warm-up suit with a list price of $61.50 is on sale at a discount of 40%. Estimate the amount of discount. (Section 9-7)

Percent and Applications **283**

CHAPTER REVIEW

PART 1: VOCABULARY

For Exercises 1–5, choose from the box at the right the word(s) that best corresponds to each description.

1. The ratio of the amount of decrease to the original amount __?__
2. The amount that the public pays for an item __?__
3. A ratio in which the second term is 100 __?__
4. The amount that a store pays for an item it sells to the public __?__
5. The ratio of the amount of increase to the original amount __?__

> percent
> retail price
> percent of increase
> percent of decrease
> wholesale price

PART 2: SKILLS

Solve. (Section 9-1)

6. What percent of 16 is 10?
7. What percent of 10 is 6?
8. What percent of 16 is 24?
9. What percent of 35 is 30?
10. What percent of 30 is 12?
11. What percent of 72 is 9?

Find the percent of increase or decrease. Identify each answer as a percent of increase or decrease. Round each answer to the nearest whole percent. (Section 9-2)

	Original Cost	Present Cost		Original Cost	Present Cost
12.	$100	$120	13.	$ 75	$ 89
14.	$130	$104	15.	$112	$109
16.	$250	$322	17.	$ 97	$ 74
18.	$174	$156	19.	$150	$212

Solve. (Section 9-3)

20. 14 is 40% of what number?
21. 42 is 35% of what number?
22. 17 is 85% of what number?
23. 9 is 15% of what number?
24. 80% of what number is 56?
25. 36% of what number is 18?

284 Chapter 9

CHAPTER REVIEW

1. percent of decrease
2. retail price
3. percent
4. wholesale price
5. percent of increase
6. $62\frac{1}{2}\%$ 7. 60%
8. 150%   9. $85\frac{5}{7}\%$
10. 40%   11. $12\frac{1}{2}\%$
12. 20% increase
13. 19% increase
14. 20% decrease
15. 3% decrease
16. 29% increase
17. 24% decrease
18. 10% decrease
19. 41% increase
20. 35   21. 120
22. 20   23. 60   24. 70

(Section 9-4)

26. 35 is what percent of 50?
27. What number is 60% of 315?
28. 75 is 30% of what number?
29. 20 is what percent of 200?
30. What number is 11% of 100?
31. 64 is 25% of what number?

For Exercises 32-34, refer to the graph below.
(Section 9-5)

32. What was the total number of yearbooks sold?

33. What percent was sold by the eighth grade?

34. By what percent did the seventh grade yearbook sales exceed the eighth grade yearbook sales?

(Section 9-6)

35. Use this data to construct a circle graph.

Area of World's Oceans

Ocean	Pacific	Atlantic	Indian	Arctic	Other
Per Cent	46%	23%	20%	4%	7%

Choose the best estimate. Choose a, b, or c. (Section 9-7)

36. 50% of $23.85 a. $10 b. $12 c. $15
37. 25% of $198.50 a. $50 b. $40 c. $30
38. 21% of $50 a. $8 b. $10 c. $12
39. 32% of $3300 a. $700 b. $900 c. $1100

Estimate each percent. (Section 9-7)

40. What percent of 98 is 24?
41. What percent of 149 is 50?
42. What percent of 62 is 30?
43. What percent of 80 is 7?

PART 3: APPLICATIONS

44. The Pelletier family budgets $480 per month for food. The family's monthly income is $1920. What percent of the monthly income is the amount budgeted for food? (Section 9-1)

45. The Haber Family paid $109 for cooling in June. After lowering the thermostat setting, their bill increased to $119.90 in July. Find the percent of increase. (Section 9-2)

Percent and Applications

CHAPTER REVIEW

25. 50   26. 70%
27. 189   28. 250
29. 10%   30. 11
31. 256   32. 1200
33. 25%   34. $66\frac{2}{3}$%   35.

Area of World's Oceans

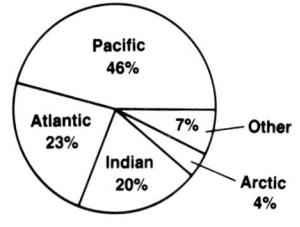

36. b   37. a   38. b
39. c   40. 25%
41. $33\frac{1}{3}$%   42. 50%
43. 10%   44. 25%
45. 10%

CHAPTER REVIEW

46. 11% 47. $250
48. $200

46. The Las-Tech Corporation's heating costs last month were $526. After lowering the thermostat setting, the heating costs this month are $468.14. Find the percent of decrease. (Section 9-2)

47. The wholesale price of a gold watch is $150. This is 60% of the retail price. What is the retail price? (Sections 9-3 and 9-4)

48. A stereo with a list price of $795 is on sale at a 25% discount. Estimate the amount of discount. (Section 9-7)

ANSWERS TO REVIEW CAPSULES

Page 256   **1.** 25%   **2.** 20%   **3.** 70%   **4.** $33\frac{1}{3}$%   **5.** $87\frac{1}{2}$%   **6.** $83\frac{1}{3}$%   **7.** 6   **8.** 12   **9.** 7   **10.** 25   **11.** 19   **12.** 9

Page 261   **1.** 0.08   **2.** 0.01   **3.** 0.10   **4.** 0.80   **5.** 0.21   **6.** 0.19   **7.** 130   **8.** 150   **9.** 520   **10.** 37.5   **11.** 180   **12.** 200   **13.** 24   **14.** 30

Page 264   **1.** 3900   **2.** 6800   **3.** 1080   **4.** 9   **5.** 4550   **6.** 29.4   **7.** 3   **8.** 30   **9.** 4   **10.** 18   **11.** 18   **12.** 5   **13.** 7

Page 268   **1.** 75%   **2.** 40%   **3.** $83\frac{1}{3}$%   **4.** $12\frac{1}{2}$%   **5.** $66\frac{2}{3}$%   **6.** 50%   **7.** 30%   **8.** $37\frac{1}{2}$%   **9.** $62\frac{1}{2}$%   **10.** $22\frac{1}{2}$%

Page 273   **1.** 0.40   **2.** 0.65   **3.** 0.08   **4.** 0.10   **5.** 0.175   **6.** 0.07   **7.** 0.81   **8.** 0.1275   **9.** 0.92   **10.** 0.095   **11.** $n = 0.10 \cdot 516$; 51.6   **12.** $n = 0.45 \cdot 200$; 90   **13.** $n = \frac{2}{3} \cdot 90$; 60   **14.** $n = 0.60 \cdot 75$; 45   **15.** $n = \frac{3}{8} \cdot 400$; 150   **16.** $n = \frac{5}{6} \cdot 144$; 120

Page 278   **1.** 59   **2.** 17   **3.** 11   **4.** 33   **5.** 71   **6.** 0   **7.** 4   **8.** 5   **9.** 6   **10.** 20   **11.** 1   **12.** 70   **13.** 60   **14.** 20   **15.** 10   **16.** 30   **17.** 70   **18.** 80   **19.** 40   **20.** 50   **21.** 50   **22.** 90   **23.** 70   **24.** 80   **25.** 90   **26.** 64   **27.** 71   **28.** 130   **29.** 30   **30.** 640   **31.** 47   **32.** 125   **33.** 270   **34.** 750   **35.** $\frac{3}{4}$ **36.** $\frac{2}{5}$ **37.** $\frac{1}{8}$ **38.** $\frac{2}{3}$ **39.** $\frac{1}{6}$

CHAPTER TEST

Solve.

1. What number is 32% of 25?
2. What is 45% of 780?
3. 16 is 80% of what number?
4. 27 is what percent of 90?
5. 7 is 50% of what number?
6. What percent of 84 is 14?
7. 64 is 25% of what number?
8. 35 is what percent of 50?

Complete. Round each answer to the nearest whole percent.

9. Original Cost: $150
   Present Cost: $179
   Percent of Increase: ___?___

10. Original Cost: $104
    Present Cost: $94
    Percent of Decrease: ___?___

11. In last week's football game, Philip threw 20 passes. He completed 6 of them. What percent did he complete?

12. Theresa bought a new car at a 12% discount. The regular price was $9600. How much money did she save on the purchase?

13. Luis earned $30 on Monday. This was 20% of his weekly earnings. What does he earn per week?

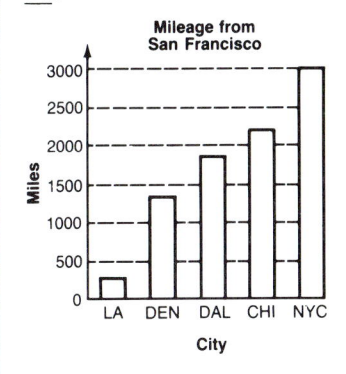

14. The table at the right shows the mileage from San Francisco to five cities in the United States. Use the data from the table to construct a vertical bar graph.

15. Company A has an advertising budget of $8000. Refer to the data below to construct a circle graph.

Mileage from San Francisco

City	Miles
To Los Angeles	250
To Denver	1300
To Dallas	1800
To Chicago	2200
To New York City	3000

Advertising Budget

Item	Newspaper	TV	Radio	Other
Percent	15%	55%	20%	10%

Percent and Applications **287**

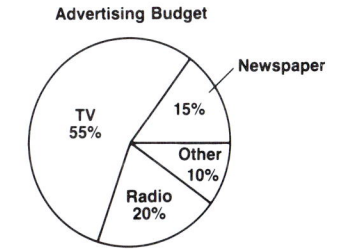

ENRICHMENT

You may wish to use this lesson for students who performed well on the formal Chapter Test.

1. 9, 11, 13
2. 25, 30, 35
3. 512, 2048, 8192
4. 7500, 37,500, 187,500
5. 16, 8, 4
6. 27, 9, 3
7. 21, 34, 55
8. 32, 52, 84
9. 35, 57, 92

ENRICHMENT

Sequences

Suppose a new sports car is worth $25,600. This table shows how the value of the car depreciates (lessens) for each of four years.

New	After One Year	After Two Years	After Three Years	After Four Years	After Five Years
$25,600	$19,200	$14,400	$10,800	$8100	?

The pattern of numbers shown in the table is a **sequence**. Each number in the sequence is three-fourths of the preceeding number. Thus, to find the value of the car after five years, multiply the value after four years by $\frac{3}{4}$.

$$8100 \times \frac{3}{4} = 6075$$

◀ Value after 5 years: $6075

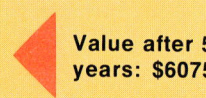

EXAMPLES Find the next three numbers in each sequence.

a. 1, 4, 7, 10, 13, · · ·   b. 1, 3, 9, 27, · · ·

Solutions: a. Think: Each number is <u>3 more than</u> the preceeding number.

1, 4, 7, 10, 13, **16, 19, 22,** · · ·

b. Think: Each number is <u>3 times</u> the preceeding number.

1, 3, 9, 27, **81, 243, 729,** · · ·

EXERCISES

Write the next three numbers in each sequence.

1. 1, 3, 5, 7, · · ·   2. 5, 10, 15, 20, · · ·   3. 2, 8, 32, 128, · · ·
4. 12, 60, 300, 1500, · · ·   5. 256, 128, 64, 32, · · ·   6. 2187, 729, 243, 81, · · ·

Each number in this **Fibonacci sequence** is the sum of the two preceeding numbers.

1, 1, 2, 3, 5, 8, 13, · · ·

Write the next three numbers in each sequence.

7. 2, 3, 5, 8, 13, · · ·   8. 4, 4, 8, 12, 20, · · ·   9. 1, 4, 9, 13, 22, · · ·

288 Chapter 9

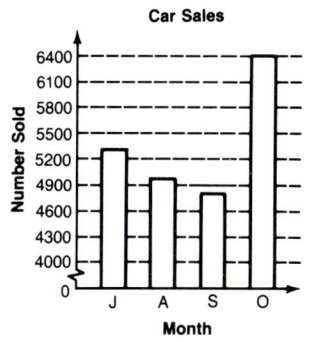

ADDITIONAL PRACTICE

SKILLS

Solve. (Pages 254–256)

1. What percent of 40 is 20?
2. What percent of 72 is 12?

Complete the table. Round each answer to the nearest whole percent. (Pages 257–260)

	Original Cost	Present Cost	Percent of Increase
3.	$40	$60	?
5.	$57	$76	?

	Original Cost	Present Cost	Percent of Decrease
4.	$12	$ 6	?
6.	$72	$60	?

Solve. (Pages 261–263)

7. 12 is 25% of x.
8. 6 is 12% of n.
9. 108.8 is 16% of y.

Use the given data to construct a bar graph. (Pages 270–273)

10. **Car Sales**

Month	Number Sold
July	5285
August	4950
September	4800
October	6350

Use the given data to construct a circle graph. (Pages 274–278)

11. **Boutique Sales**

Item	Percent
Suits	40%
Dresses	35%
Skirts	15%
Shoes	10%

Choose the best estimate. Choose a, b, or c. (Pages 279–281)

12. 24% of 160   a. 40   b. 20   c. 60
13. What percent of 41 is 8?   a. 50%   b. 20%   c. 15%
14. 49% of 281   a. 150   b. 130   c. 140

APPLICATIONS

15. Ted scored 180 points during last year's basketball season. This is 15% of the total number of points his team scored. How many points did Ted's team score? (Pages 261–263)

16. René was earning $3.80 per hour as a cook in a restaurant. After working for one year, her wage was increased to $4.75 per hour. Find the percent of increase. (Pages 257–260)

Percent and Applications **289**

COMMON ERRORS

In preparation for the Cumulative Review, these exercises focus the student's attention on the most common errors to be avoided.

1. 11   2. 25   3. 2
4. $\dfrac{2}{68} = \dfrac{6}{x}$
5. t = 16.$\overline{6}$
6. 66$\dfrac{2}{3}$%   7. n = 6
8. 300%   9. 300
10. 25%   11. 25%

COMMON ERRORS

Each of these problems contains a common error.

a Find the correct answer.
b. Find the error.

Solve each equation.

1. $x + 7 = 18$
   $x + 7 - 7 = 18 + 7$
   $x = 25$

2. $y - 5 = 20$
   $y - 5 + 5 = 20$
   $y = 20$

3. $3x = 6$
   $\dfrac{3x}{6} = \dfrac{6}{6}$
   $x = 1$

Solve each problem.

4. Write a proportion.
   2 apples for 68¢
   6 apples for x cents
   $\dfrac{2}{68} = \dfrac{x}{6}$

5. $\dfrac{t}{20} = \dfrac{5}{6}$; $t =$ ___?___
   $\dfrac{t}{20} = \dfrac{5}{6}$
   $5t = 120$
   $t = $ **24**

6. Write a percent for $\dfrac{2}{3}$.
   $\dfrac{2}{3} = \dfrac{100}{n}$
   $2n = 300$
   $n = 150$ ◀ $\dfrac{2}{3}$ = **150%**

7. What number is 5% of 120?
   $n = 0.5(120)$
   $n = $ **60**

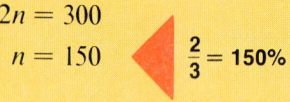

8. What percent of 16 is 48?
   $p \cdot 48 = 16$
   $48p = 16$
   $p = \dfrac{1}{3}$, or **33$\dfrac{1}{3}$%**

9. 36 is 12% of what number?
   $36 = 12 \cdot p$
   $36 = 12p$
   $3 = p$

10. Original cost: $80
    Present cost: $60
    Percent of decrease: ___?___
    $\dfrac{n}{100} = \dfrac{60}{80}$
    $80n = 6000$
    $n = 75$ ◀ Percent of Decrease: **75%**

11. Original cost: $160
    Present cost: $200
    Percent of increase: ___?___
    $\dfrac{p}{100} = \dfrac{40}{200}$
    $200p = 4000$
    $p = 20$ ◀ Percent of Increase: **20%**

290 Chapter 9

CUMULATIVE REVIEW: CHAPTERS 7–9

Choose the correct answer. Choose a, b, c, or d.

1. Solve for c: $13.7 = c + 8.9$
   - a. 21.6
   - b. 22.6
   - c. 5.2
   - d. 4.8

2. Solve for y: $y - 24 = 33$
   - a. 9
   - b. 57
   - c. 65
   - d. 14

3. Last month, Frank's gross earnings were $1158.45. The total deductions were $283.14. Use the formula $n + d = g$ to find the net pay.
   - a. $1341.59
   - b. $975.31
   - c. $875.31
   - d. $935.31

4. A plane is flying into a head wind of 35 kilometers per hour. The ground speed is 477 kilometers per hour.
   Use the formula $a - h = g$ to find the air speed.
   - a. 442
   - b. 502
   - c. 412
   - d. 512

5. Solve for t: $11t = 132$
   - a. 121
   - b. 127
   - c. 12
   - d. 143

6. Solve for n: $\frac{n}{2} = 1\frac{2}{3}$
   - a. $\frac{5}{12}$
   - b. $3\frac{1}{3}$
   - c. $\frac{1}{3}$
   - d. $3\frac{2}{3}$

7. A plane flies the 3000 miles from New York to San Francisco in $3\frac{3}{4}$ hours. Find the average speed.
   - a. 495
   - b. 925
   - c. 750
   - d. 800

8. Write this fraction in lowest terms: $\frac{16x}{50x}$
   - a. $\frac{8x}{25}$
   - b. $\frac{8}{25}$
   - c. $\frac{4x}{25}$
   - d. $\frac{4}{25}$

9. Solve for m: $\frac{9}{27} = \frac{m}{6}$
   - a. 3
   - b. 2
   - c. 4
   - d. 5

10. On a map of California, 1 centimeter equals 40 kilometers. How many kilometers do 4.2 centimeters represent?
    - a. 16.8
    - b. 1.68
    - c. 168
    - d. 1680

11. The Evans family drove 368 miles on 16 gallons of gas. How many miles could they drive on 7 gallons?
    - a. 181
    - b. 161
    - c. 171
    - d. 191

CUMULATIVE REVIEW

A cumulative test is provided on copying masters in the *Teacher's Edition: Part II.*

1. d 2. b 3. c
4. d 5. c 6. b
7. d 8. b 9. b
10. c 11. b

Cumulative Review: Chapters 7–9

CUMULATIVE REVIEW

12. a   13. b   14. d
15. a   16. b   17. d
18. d   19. a   20. c
21. a   22. a

12. Write a percent for $\frac{9}{8}$.
    a. $112\frac{1}{2}\%$
    b. 108%
    c. 109%
    d. $110\frac{1}{2}\%$

13. Write a decimal for 63%.
    a. 0.063
    b. 0.63
    c. 6.3
    d. 630

14. On a math test, Rene answered 75% of the 80 questions asked. How many questions did Rene answer?
    a. 65
    b. 55
    c. 70
    d. 60

15. The Garbey family invested $1400 at a yearly interest rate of 10.5%. How much interest will the account earn in 6 months?
    a. $73.50
    b. $147.00
    c. $95.75
    d. $355.00

16. A camera sells at a regular price of $60. It is on sale at a 20% discount. Find the sale price.
    a. $48.80
    b. $48
    c. $45
    d. $58.80

17. What percent of 20 is 4?
    a. 25%
    b. 15%
    c. 40%
    d. 20%

18. 27 is 30% of what number?
    a. 8.1
    b. 81
    c. 900
    d. 90

19. Thirty of forty club members attended a meeting. What percent of the members attended the meeting?
    a. 75%
    b. 65%
    c. 70%
    d. 80%

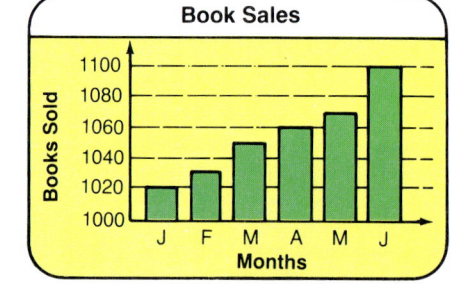

20. Use the bar graph at the right to estimate the percent of increase between the books sold in January and the books sold in April.
    a. 40%
    b. 45%
    c. 3.9%
    d. 4.5%

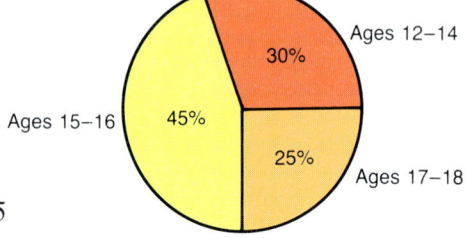

21. In the circle graph at the right, how many degrees represent the section "Ages 12–14"?
    a. 108°
    b. 30°
    c. 100°
    d. 106°

22. Choose the best estimate: 24.7% of 1019.5
    a. 250
    b. 260
    c. 350
    d. 200

292   Cumulative Review: Chapters 7–9

CHAPTER

10 Rational Numbers: Addition/Subtraction

SECTIONS
10-1 Integers
10-2 Rational Numbers
10-3 Addition on the Number Line
10-4 Adding Rational Numbers
10-5 Subtracting Rational Numbers
10-6 Properties of Addition
10-7 Adding Polynomials

FEATURES
Calculator Application:
Adding Rational Numbers
Career Application:
Travel Agent
Consumer Application:
Driving Safety
Enrichment: *Clock Arithmetic*

Teaching Suggestions
p. M-23

QUICK QUIZ

Replace each __?__ with >, <, or =.

1. 13 __?__ 7   Ans: >
2. 0 __?__ 5   Ans: <
3. $4\frac{3}{8}$ __?__ $4\frac{1}{2}$ Ans: <
4. 8.01 __?__ 8.1   Ans: <
5. 7.2 __?__ 7.22   Ans: <

ADDITIONAL EXAMPLES

Example 1

Write a positive or negative number to represent each description.

1. A temperature of 17° above zero.
   Ans: 17 or +17
2. A drop in unemployment of 0.5%
   Ans: ⁻0.5%

SELF-CHECK

1. +200, or 200
2. ⁻120

Example 2

Graph 0; ⁻4; 4; 2; ⁻3

Write the integers in order from least to greatest.

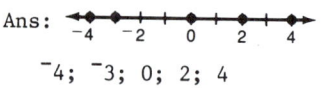

Ans:

⁻4; ⁻3; 0; 2; 4

294

10-1 Integers

Positive and negative numbers are used to represent temperatures above and below zero. Positive and negative numbers can also be used to describe other situations.

EXAMPLE 1 Write a positive or negative number to represent each word description.

Word Description	Number
a. 54 meters below sea level	a. ⁻54
b. 3 minutes before rocket lift-off	b. ⁻3
c. A gain of 25 yards	c. +25, or 25

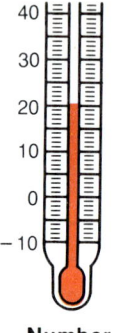

Self-Check *Write a positive or negative number to represent each description.*

1. A profit of $200
2. A loss in altitude of 120 meters

Numbers such as ⁻54, ⁻3, 25, and 0 are called <u>integers</u>.

> The set of **integers** is made up of the positive integers, zero, and the negative integers.
>
> Integers: {··· ⁻3, ⁻2, ⁻1, 0, 1, 2, 3, ···}

The 3 dots mean the numbers continue indefinitely.

To compare integers, it is often helpful to graph them. Numbers become larger as you move <u>from left to right</u> on the number line.

> **PROCEDURE**
>
> **To graph on a number line:**
>
> ① Draw and label a number line.
> ② Draw a "dot" at the point that corresponds to the integer being graphed.

EXAMPLE 2 a. Graph: 5; ⁻2; 0; 3; ⁻7

b. Write the integers in order from least to greatest.

Solutions: a.

Each dot is the graph of an integer.

b. Write the integers in order from left to right: ⁻7, ⁻2, 0, 3, 5

294 Chapter 10

Self-Check

*Graph the integers.
Then write the integers in order from least to greatest.*

3. ⁻6; ⁻7; 4; ⁻5; 3   4. 10; 8; ⁻9; ⁻3; 0

Statements that use > or < are **inequalities**.

">" means **"is greater than."**   "<" means **"is less than."**

EXAMPLE 3 Replace each ? with > or <. Refer to the number line in Example 2.

Problem	Think	Solution
a. ⁻6 ? 0	⁻6 is to the left of 0.	⁻6 < 0
b. 5 ? ⁻9	5 is to the right of ⁻9.	5 > ⁻9
c. ⁻2 ? ⁻8	⁻2 is to the right of ⁻8.	⁻2 > ⁻8

Self-Check

Replace each ? with > or <. Refer to the number line in Example 2.

5. 0 ? ⁻4   6. 7 ? ⁻8   7. ⁻10 ? ⁻2   8. ⁻5 ? 1

NOTE: The inequality symbol always points toward the smaller number.

CLASSROOM EXERCISES

Write a positive or negative number to represent each description. (Example 1)

1. A loss of 8 yards
2. A gain of 15 yards
3. A temperature of 20°F below zero
4. A drop of 6 degrees in temperature

For each of Exercises 5–10:
  a. *Graph the integers on a number line.*
  b. *Write the integers in order from least to greatest.* (Example 2)

5. ⁻3, 2, 7, 1, 3
6. 4, 8, ⁻7, 3, 0
7. 2, 0, 11, ⁻8, ⁻1
8. 2, ⁻5, 3, 0, ⁻2
9. 6, 12, ⁻12, 0, 13
10. 11, 7, ⁻9, 8, ⁻14

Replace each ? with > or <. (Example 3)

11. ⁻4 ? 8   12. 0 ? ⁻1   13. 11 ? ⁻1   14. ⁻1 ? 1

Rational Numbers: Addition/Subtraction   **295**

ASSIGNMENT GUIDE
BASIC
p. 296: 1-28
AVERAGE
p. 296: 1-28
ABOVE AVERAGE
p. 296: 1-28

WRITTEN EXERCISES
1. 10   2. 65   3. ⁻2000
4. 156   5. ⁻2   6. 9
7. ⁻3; 0; 3; 6
8. ⁻6; ⁻4; 2; 4
9. ⁻2; 0; 1; 3
10. ⁻2; 1; 2; 3
11. ⁻4; ⁻2; 1; 3; 8
12. ⁻2; ⁻1; 0; 1; 5
13. <   14. >   15. >
16. <   17. <   18. >
19. <   20. >   21. >
22. >   23. <   24. >
25. >   26. <   27. <
28. >

WRITTEN EXERCISES

Goal: To graph and compare integers

Write a positive or negative number to represent each description. (Example 1)

1. A rise of 10° in temperature
2. A pay raise of 65¢ per hour
3. A loss of 2000 feet in altitude
4. 156 meters above sea level
5. Two years ago
6. Nine years from now

For each of Exercises 7-12:
 a. *Graph the integers on a number line.*
 b. *Write the integers in order from least to greatest.* (Example 2)

7. 0; 3; ⁻3; 6
8. ⁻6; 2; 4; ⁻4
9. 0; 3; ⁻2; 1
10. 3; 1; ⁻2; 2
11. ⁻4; 8; 1; 3; ⁻2
12. 5; ⁻2; 1; ⁻1; 0

For Exercises 13-28, replace each __?__ with > or <. (Example 3)

13. 0 __?__ 4
14. 0 __?__ ⁻4
15. 3 __?__ ⁻3
16. ⁻4 __?__ 2
17. ⁻6 __?__ ⁻5
18. 4 __?__ 0
19. ⁻7 __?__ 7
20. 7 __?__ ⁻7
21. ⁻1 __?__ ⁻2
22. 1 __?__ ⁻2
23. ⁻1 __?__ 2
24. 0 __?__ ⁻8

APPLICATIONS: USING INTEGERS

25. The lowest temperature ever recorded in Florida was ⁻2°F. The lowest temperature ever recorded in Georgia was ⁻17°F.

    Complete: ⁻2° __?__ ⁻17°

26. The lowest temperature ever recorded in Alaska was ⁻80°F. The lowest temperature ever recorded in Hawaii was 12°F.

    Complete: ⁻80° __?__ 12°

27. The lowest temperature ever recorded in Vermont was ⁻50°F. The lowest temperature ever recorded in Maryland was ⁻40°F.

    Complete: ⁻50° __?__ ⁻40°

28. The lowest temperature ever recorded in Illinois was ⁻35°F. The lowest temperature ever recorded in Iowa was ⁻47°F.

    Complete: ⁻35° __?__ ⁻47°

296   Chapter 10

REVIEW CAPSULE FOR SECTION 10-2

Write a fraction for each mixed numeral. (Pages 135-137)

1. $2\frac{1}{5}$
2. $5\frac{2}{3}$
3. $8\frac{7}{9}$
4. $3\frac{1}{6}$
5. $10\frac{1}{2}$
6. $1\frac{11}{12}$
7. $6\frac{1}{4}$
8. $1\frac{2}{5}$
9. $3\frac{1}{10}$
10. $6\frac{3}{10}$
11. $4\frac{1}{7}$
12. $3\frac{2}{9}$

The answers are on page 320.

10-2 Rational Numbers

Numbers such as these are **positive rational numbers.**

$$\frac{2}{3} \quad 1 \quad \frac{5}{4} \quad 2\frac{2}{5} \quad 5.6 \quad 27$$

Numbers such as these are **negative rational numbers.**

$$\frac{^-1}{4} \quad ^-2 \quad \frac{^-8}{5} \quad ^-4\frac{2}{3} \quad ^-3.4 \quad ^-32$$

Any <u>rational number</u> can be written as a fraction.

EXAMPLE 1 Write a fraction for each rational number.

a. 6 b. $^-3\frac{1}{3}$ c. 0 d. $^-4.2$

Solutions: a. $6 = \frac{6}{1}$ b. $^-3\frac{1}{3} = \frac{^-10}{3}$ c. $0 = \frac{0}{1}$ d. $^-4.2 = ^-4\frac{2}{10} = \frac{^-42}{10}$

Self-Check *Write a fraction for each rational number.*

1. 4
2. $^-4\frac{1}{2}$
3. 3.8
4. $^-5$
5. $2\frac{1}{4}$

NOTE: In Example 1c, any number except 0 could have been written in the denominator. Thus,

$$0 = \frac{0}{4} = \frac{0}{10} = \frac{0}{21} = \frac{0}{100}, \text{ and so on.}$$

The rational numbers 4 and $^-4$ are the same distance from 0.

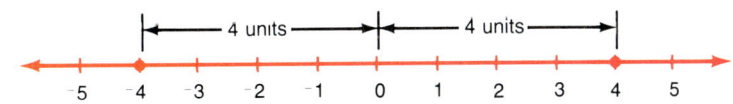

Therefore, 4 and $^-4$ are <u>opposites</u>. The number 0 is its own opposite.

> Two numbers that are the same distance from 0 and are in opposite directions from 0 are **opposites.**

Rational Numbers: Addition/Subtraction

ADDITIONAL EXAMPLES

Example 2

Write the opposite.

1. ⁻3.6   Ans: 3.6
2. 5   Ans: ⁻5

SELF-CHECK

6. 15   7. $\frac{8}{3}$ 8. 0
9. 2.7 10. $\frac{-4}{5}$

CLASSROOM EXERCISES

1. $\frac{4}{1}$ 2. $\frac{12}{1}$ 3. $-\frac{10}{1}$
4. $-\frac{9}{1}$ 5. $\frac{32}{10}$ 6. $\frac{45}{10}$
7. $-\frac{11}{5}$ 8. $-\frac{43}{6}$ 9. $\frac{9}{4}$
10. $\frac{4}{3}$ 11. $-\frac{18}{10}$
12. $-\frac{4}{10}$ 13. -12
14. -9 15. 6   16. 15
17. -3.2 18. -6.8
19. 1.5   20. 12.4
21. $-\frac{1}{4}$ 22. $-2\frac{1}{2}$
23. $1\frac{4}{5}$ 24. $\frac{3}{8}$

ASSIGNMENT GUIDE

BASIC
pp. 298–299: 1–48

AVERAGE
pp. 298–299: 1–48

ABOVE AVERAGE
pp. 298–299: 1–49 odd, 50–56

EXAMPLE 2 Write the opposite of each number.

   a. 7   b. $-2\frac{1}{2}$ c. 3.5   d. $\frac{-3}{4}$

Solutions: a. ⁻7   b. $2\frac{1}{2}$ c. ⁻3.5   d. $\frac{3}{4}$

Self-Check Write the opposite of each number.

6. ⁻15   7. $\frac{-8}{3}$ 8. 0   9. ⁻2.7   10. $\frac{4}{5}$

A dash, —, is used to represent "the opposite of." For example,

$$^-4 = -4$$

Read: "Negative 4 equals the opposite of 4."

> Since ⁻4 = −4, we will use the symbol for "the opposite of" to indicate a negative number. Thus, −4 can be read "negative 4."

CLASSROOM EXERCISES

Write a fraction for each rational number. (Example 1)

1. 4   2. 12   3. −10   4. −9   5. 3.2   6. 4.5
7. $-2\frac{1}{5}$ 8. $-7\frac{1}{6}$ 9. $2\frac{1}{4}$ 10. $1\frac{1}{3}$ 11. −1.8   12. −0.4

Write the opposite of each number. (Example 2)

13. 12   14. 9   15. −6   16. −15   17. 3.2   18. 6.8
19. −1.5 20. −12.4   21. $\frac{1}{4}$ 22. $2\frac{1}{2}$ 23. $-1\frac{4}{5}$ 24. $-\frac{3}{8}$

WRITTEN EXERCISES

Goals: To write a fraction for a rational number
To write the opposite of a rational number

Write a fraction for each rational number. (Example 1)

1. 17   2. 8   3. −7   4. −5   5. 3.8   6. 6.2
7. $-3\frac{1}{3}$ 8. $-2\frac{4}{5}$ 9. $3\frac{1}{4}$ 10. $3\frac{1}{10}$ 11. $5\frac{2}{5}$ 12. $-1\frac{1}{6}$
13. 0   14. −1.7   15. −9.3   16. −1   17. 7.5   18. −0.8

Chapter 10

Write the opposite of each number. (Example 2)

19. 12
20. 15
21. −20
22. −3
23. $2\frac{1}{3}$
24. $6\frac{2}{5}$
25. $-\frac{3}{4}$
26. $-\frac{9}{4}$
27. $\frac{2}{5}$
28. $\frac{5}{6}$
29. $-8\frac{1}{2}$
30. $-9\frac{2}{3}$
31. 4.8
32. 7.25
33. −8.9
34. −12.6
35. 0
36. 0.3

APPLICATIONS: USING RATIONAL NUMBERS

Write a rational number to represent each word description.

37. A growth of $2\frac{1}{2}$ inches
38. A loss of $8.60
39. 32.6 meters below sea level
40. A gain of $3\frac{1}{2}$ pounds
41. An increase of $4.35
42. 7.6° below 0
43. 132.5 meters above sea level
44. A fall of $6\frac{3}{4}$ feet
45. A loss of 5.5 kilograms
46. A deduction of $4.75
47. 12.5° above 0
48. An increase of 3.6 centimeters

MORE CHALLENGING EXERCISES

Arrange the numbers in order from least to greatest.

49. 1; $\frac{1}{4}$; $-2\frac{1}{5}$; $3\frac{1}{3}$; $-5\frac{1}{5}$; $-\frac{1}{4}$
50. −8; 0; $4\frac{1}{3}$; $-7\frac{1}{2}$; $-\frac{2}{5}$; 2
51. 2; $-\frac{3}{4}$; 1; $-\frac{2}{3}$; $\frac{4}{5}$; $-1\frac{1}{2}$
52. 0; $-\frac{2}{3}$; $1\frac{1}{2}$; $-2\frac{3}{4}$; $\frac{2}{3}$; $-3\frac{1}{5}$
53. 2.25; −1.83; −1.875; 1.5; 0; −0.687
54. −0.875; 1; −2; 0.937; −1; 0.88
55. −1.125; −3; −2.5; −1; −0.25; −2.875
56. 0.2; 0; −0.25; 1.5; −2.7; −5

REVIEW CAPSULE FOR SECTION 10-3

For Exercises 1-4:
a. Graph the integers on a number line.
b. Write the integers in order from least to greatest. (Pages 294-296)

1. 3; 0; 1; −1
2. −5; 1; 2; −3
3. 0; 4; −2; 6
4. −4; 5; −6; 7

Replace each __?__ with > or <. (Pages 294-296)

5. 0 __?__ −4
6. −3 __?__ −5
7. −4 __?__ 4
8. 9 __?__ −13
9. −6 __?__ 5
10. 0 __?__ 7
11. 3 __?__ −5
12. −8 __?__ −10

The answers are on page 320.

Rational Numbers: Addition/Subtraction **299**

WRITTEN EXERCISES
1. $\frac{17}{1}$ 2. $\frac{8}{1}$ 3. $-\frac{7}{1}$
4. $-\frac{5}{1}$ 5. $\frac{38}{10}$ 6. $\frac{62}{10}$
7. $-\frac{10}{3}$ 8. $-\frac{14}{5}$
9. $\frac{13}{4}$ 10. $\frac{31}{10}$ 11. $\frac{27}{5}$
12. $-\frac{7}{6}$
13. Answers will vary.
14. $-\frac{17}{10}$ 15. $-\frac{93}{10}$
16. $-\frac{1}{1}$ 17. $\frac{75}{10}$
18. $-\frac{8}{10}$ 19. −12
20. −15 21. 20 22. 3
23. $-2\frac{1}{3}$ 24. $-6\frac{2}{5}$
25. $\frac{3}{4}$ 26. $\frac{9}{4}$ 27. $-\frac{2}{5}$
28. $-\frac{5}{6}$ 29. $8\frac{1}{2}$
30. $9\frac{2}{3}$ 31. −4.8
32. −7.25 33. 8.9
34. 12.6 35. 0
36. −0.3 37. $\frac{5}{2}$
38. $-\frac{86}{10}$ 39. $-\frac{326}{10}$
40. $\frac{7}{2}$ 41. $\frac{435}{100}$
42. $-\frac{76}{10}$ 43. $\frac{1325}{10}$
44. $-\frac{27}{4}$ 45. $-\frac{55}{10}$
46. $-\frac{475}{100}$ 47. $\frac{125}{10}$
48. $\frac{36}{10}$
49. $-5\frac{1}{5}$; $-2\frac{1}{5}$; $-\frac{1}{4}$; $\frac{1}{4}$; 1; $3\frac{1}{3}$
50. −8; $-7\frac{1}{2}$; $-\frac{2}{5}$; 0; 2; $4\frac{1}{3}$

See page 303 for the answers to Exercises 51-56.

299

Teaching Suggestions
p. M-23

QUICK QUIZ
Graph on a number line.
1. -3; 3; 0
   Ans: 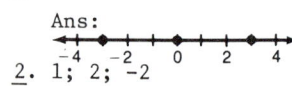
2. 1; 2; -2
   Ans:
Replace each ? with > or <.
3. 0 ? -6   Ans: >
4. -8 ? -10   Ans: >

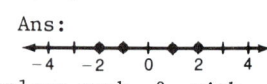

ADDITIONAL EXAMPLES
Example 1
Use a number line to add.
1. -6 + (-4)   Ans: -10
2. -1 + (-5)   Ans: -6

SELF-CHECK
1. -4   2. -5   3. -9
4. -14

Example 2
Use a number line to add.
1. (-3) + 8   Ans: 5
2. (-5) + 1   Ans: -4

SELF-CHECK
5. 1   6. -3   7. 0
8. -7

300

10-3 Addition on the Number Line

You can use a number line to find the sum of two rational numbers.

PROCEDURE
To add on the number line:
1 Graph the first number.
2 Draw an arrow to represent the second number.
   a. Move to the right if the second number is positive.
   b. Move to the left if the second number is negative.
3 Read the answer at the tip of the arrow.

EXAMPLE 1   Use a number line to add $-1 + (-4)$.
Solution:
1 Graph the first number, -1.
2 From this point, draw an arrow 4 units to the left.
3 Read the answer at the tip of the arrow.

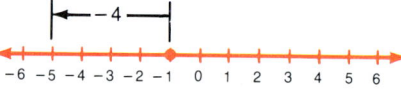

$$-1 + (-4) = -5$$

Self-Check   Use a number line to add.
1. $-1 + (-3)$ 2. $-3 + (-2)$ 3. $-4 + (-5)$ 4. $-8 + (-6)$

Examples 2 and 3 show how to add a positive and a negative integer on the number line.

EXAMPLE 2   Use a number line to add $-4 + 9$.
Solution:
1 Graph the first number, -4.
2 From this point, draw an arrow 9 units to the right.
3 Read the answer at the tip of the arrow.

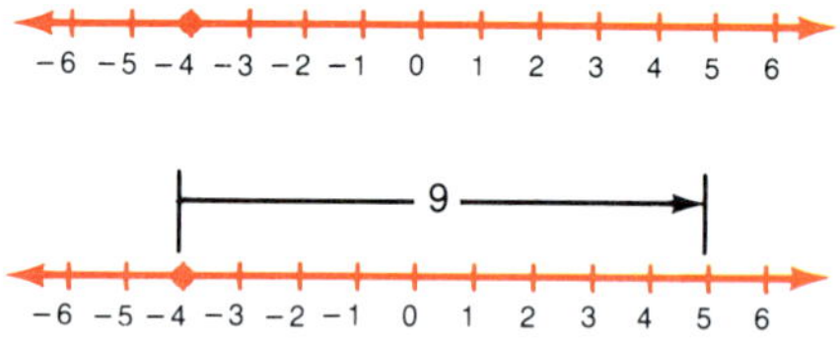

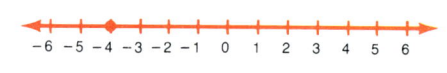

$$-4 + 9 = 5$$

Self-Check   Use a number line to add.
5. $-2 + 3$ 6. $-5 + 2$ 7. $-7 + 7$ 8. $-12 + 5$

300   Chapter 10

EXAMPLE 3 Use a number line to add $5 + (-8)$.

Solution: ①and ②

③ $5 + (-8) = -3$

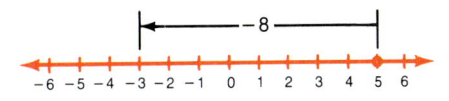

Self-Check Use a number line to add.

9. $2 + (-5)$ 10. $6 + (-6)$ 11. $8 + (-5)$ 12. $9 + (-4)$

CLASSROOM EXERCISES

For Exercises 1–24, use a number line to add. (Example 1)

1. $-2 + (-4)$ 2. $-7 + (-4)$ 3. $-5 + (-5)$ 4. $-2 + (-7)$
5. $-6 + (-4)$ 6. $-8 + (-3)$ 7. $-5 + (-2)$ 8. $-3 + (-4)$

(Example 2)

9. $-7 + 10$ 10. $-3 + 8$ 11. $-1 + 4$ 12. $-7 + 5$
13. $-6 + 2$ 14. $-4 + 1$ 15. $-2 + 5$ 16. $-8 + 4$

(Example 3)

17. $2 + (-7)$ 18. $5 + (-6)$ 19. $1 + (-4)$ 20. $8 + (-10)$
21. $3 + (-2)$ 22. $6 + (-5)$ 23. $7 + (-2)$ 24. $4 + (-4)$

WRITTEN EXERCISES

Goal: To use a number line to add integers

For Exercises 1–48, use a number line to add. (Example 1)

1. $-5 + (-4)$ 2. $-6 + (-5)$ 3. $-1 + (-6)$ 4. $-4 + (-1)$
5. $-3 + (-7)$ 6. $-8 + (-4)$ 7. $-2 + (-2)$ 8. $-3 + (-3)$
9. $-7 + (-8)$ 10. $-1 + (-5)$ 11. $-3 + (-2)$ 12. $-4 + (-6)$

(Example 2)

13. $-4 + 7$ 14. $-3 + 6$ 15. $-5 + 10$ 16. $-6 + 14$
17. $-2 + 8$ 18. $-1 + 5$ 19. $-3 + 7$ 20. $-6 + 8$
21. $-8 + 4$ 22. $-5 + 1$ 23. $-4 + 4$ 24. $-7 + 5$

Rational Numbers: Addition/Subtraction

WRITTEN EXERCISES

16. 8   17. 6   18. 4
19. 4   20. 2   21. -4
22. -4 23. 0   24. -2
25. -2 26. -3 27. 2
28. 4   29. -5 30. -4
31. 5   32. 3   33. -6
34. -1 35. 7   36. 4
37. -3 38. -7 39. -3
40. 6   41. -5 42. 5
43. 6   44. 0   45. -6
46. -6 47. 3   48. -9
49. -5 50. 14 lb
51. loss of 4 yd

QUIZ: SECTIONS 10-1–10-3
After completing this Review, you may wish to administer a quiz covering the same sections. A Quiz is provided in the Teacher's Edition: Part II.

REVIEW: SECTIONS 10-1–10-3

1. <--•--•--•--•--•-->
   -4 -2 0 2 4 6
   -3; -1; 0; 2; 5

2. <--•--•--•--•--•-->
   -8 -4 0 4 8 12
   -8; -4; 1; 7; 9

3. <--•--•--•--•--•-->
   -4 -2 0 2 4 6
   -4; -2; 0; 1; 4

4. $\frac{15}{1}$ 5. $-\frac{8}{1}$ 6. $-\frac{58}{10}$

See page 304 for the answers to Exercises 6-18.

(Example 3)

25. $3 + (-5)$	26. $4 + (-7)$	27. $5 + (-3)$	28. $6 + (-2)$
29. $1 + (-6)$	30. $3 + (-7)$	31. $7 + (-2)$	32. $4 + (-1)$
33. $2 + (-8)$	34. $6 + (-7)$	35. $9 + (-2)$	36. $10 + (-6)$

MIXED PRACTICE

37. $-7 + 4$	38. $-3 + (-4)$	39. $6 + (-9)$	40. $-1 + 7$
41. $-2 + (-3)$	42. $10 + (-5)$	43. $9 + (-3)$	44. $-8 + 8$
45. $-1 + (-5)$	46. $-8 + 2$	47. $6 + (-3)$	48. $-4 + (-5)$

APPLICATIONS: USING ADDITION OF INTEGERS

Use a number line to solve each problem.

49. Ian scored 10 points by answering his first quiz-bowl question correctly. He missed the second question and lost 15 points. What was his score after two questions?

50. Andrea lost 8 pounds last month. This month she lost 6 pounds. How much did she lose in all?

51. A football team lost 12 yards on one play and gained 8 yards on the next play. Find the total gain or loss for the two plays.

REVIEW: SECTIONS 10-1–10-3

Graph the integers on a number line. Then write the integers in order from least to greatest. (Section 10-1)

1. $5; -3; 0; 2; -1$ 2. $-8; 9; 1; 7; -4$ 3. $4; 0; -2; 1; -4$

Write a fraction for each rational number. (Section 10-2)

4. 15   5. -8 6. -5.8 7. $2\frac{1}{8}$ 8. -2.2 9. 0

For Exercises 10–18, use a number line to solve. (Section 10-3)

10. $-2 + (-1)$ 11. $-5 + (-8)$ 12. $-3 + 11$ 13. $-10 + 2$
14. $4 + (-15)$ 15. $9 + (-7)$ 16. $10 + (-10)$ 17. $-25 + 25$

18. The temperature was $-5°C$ and rose 7 degrees. Find the new temperature.

302   Chapter 10

CONSUMER APPLICATION Driving Safety

The ability of a motorist to stop quickly in an emergency depends greatly on the vehicle's speed, brakes, and the condition of the road.

This formula can be used to estimate the stopping distance of a vehicle.

$$s = 1.1r + \frac{r^2}{30F}$$

s = stopping distance (ft)
r = rate in miles per hour
F = driving surface factor

The table at the right shows several values for the driving surface factor.

Type of Surface	Driving Surface Factor Dry Road	Wet Road
Asphalt	0.85	0.65
Concrete	0.90	0.60
Gravel	0.65	0.65
Packed Snow	0.45	0.45

EXAMPLE Carlos Dorman was driving 36 miles per hour on a dry concrete road when he saw a tree branch lying in the road. Estimate the stopping distance to the nearest foot.

SOLUTION $s = 1.1r + \frac{r^2}{30F}$ ← r = 36, F = 0.9 (from the table)

$s = 1.1(36) + \frac{(36)^2}{30(0.9)}$ ← $(36)^2$ means 36 · 36.

Using a calculator, first compute 1.1r. Store the value in memory.

1 . 1 [×] 3 6 [=] [M+]   M 39.6

Compute $\frac{r^2}{30F}$. Add the value stored in memory.

36 [×] 36 [÷] 30 [÷] .9 [=] [+] [MR] [=]   87.6

The stopping distance is about **88 feet**.

Exercises

Complete the table. Round each answer to the nearest foot.

Rate	Road Condition	Stopping Distance		Rate	Road Condition	Stopping Distance
1. 39 mi/hr	wet asphalt	?		2. 27 mi/hr	packed snow	?
3. 45 mi/hr	wet concrete	?		4. 54 mi/hr	dry concrete	?
5. 35 mi/hr	dry gravel	?		6. 39 mi/hr	dry asphalt	?

Career Application

CONSUMER APPLICATIONS
This feature applies the skills for using formulas to finding the stopping distance of a vehicle when vehicle speed and the road conditions are known. Thus it can be used anyplace in Chapter 10.

1. 121 ft
2. 84 ft
3. 162 ft
4. 167 ft
5. 101 ft
6. 103 ft

Teaching Suggestions
p. M-23

QUICK QUIZ

Replace the ? with > or <.

1. 12 ? 0 Ans: >
2. 5 ? $5\frac{1}{3}$ Ans: <
3. 3.2 ? 3.22 Ans: <

Write the opposite.

4. −8 Ans: 8
5. 16.3 Ans: −16.3

WRITTEN EXERCISES p. 299

51. $-1\frac{1}{2}$; $-\frac{3}{4}$; $-\frac{2}{3}$; $\frac{4}{5}$; 1; 2
52. $-3\frac{1}{5}$; $-2\frac{3}{4}$; $-\frac{2}{3}$; 0; $\frac{2}{3}$; $1\frac{1}{2}$
53. −1.875; −1.83; −0.687; 0; 1.5; 2.25
54. −2; −1; −0.875; 0.88; 0.937; 1
55. −3; −2.875; −2.5; −1.125; −1; −0.25
56. −5; −2.7; −0.25; 0; 0.2; 1.5

REVIEW EXERCISES p. 302

6. $-\frac{58}{10}$ 7. $\frac{17}{8}$
8. $-\frac{22}{10}$
9. Answers will vary.
10. −3   11. −13   12. 8
13. −8   14. −11   15. 2
16. 0   17. 0   18. 2°C

REVIEW CAPSULE FOR SECTION 10-4

Replace the ? with > or <. (Pages 294-296)

1. 14 ? 8
2. 16 ? 2
3. 0 ? 5
4. 24 ? 25
5. 37 ? 29
6. 101 ? 110
7. 3.5 ? 3.45
8. 9.06 ? 9.6

Write the opposite of each number. (Pages 297-299)

9. 6
10. 15
11. −8
12. 3
13. 0
14. −7

The answers are on page 320.

10-4 Adding Rational Numbers

You use **absolute value notation** when you want to indicate distance, but not direction, from 0. The symbol for absolute value is | |.

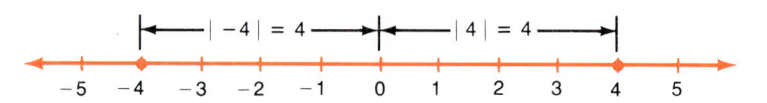

$$\begin{array}{c} \longleftarrow |-4| = 4 \longrightarrow \quad \longleftarrow |4| = 4 \longrightarrow \\ -5 \; -4 \; -3 \; -2 \; -1 \; 0 \; 1 \; 2 \; 3 \; 4 \; 5 \end{array}$$

> The **absolute value** of a number is the distance of the number from 0.

Some examples are given in the table below.

TABLE

Absolute Value	Meaning	Solution
$|5|$	Distance of 5 from 0	$|5| = \mathbf{5}$
$|-6.7|$	Distance of −6.7 from 0	$|-6.7| = \mathbf{6.7}$
$|0|$	Distance of 0 from 0	$|0| = \mathbf{0}$
$\left|-\frac{1}{2}\right|$	Distance of $-\frac{1}{2}$ from 0	$\left|-\frac{1}{2}\right| = \mathbf{\frac{1}{2}}$

You can use absolute value to add two negative numbers.

> **PROCEDURE**
> **To add two negative numbers:**
> ① Add the absolute values.
> ② Write the opposite of the result.

304 Chapter 10

EXAMPLE 1 Add: $-6 + (-9)$

Solution:
[1] Add the absolute values. $|-6| + |-9| = 6 + 9$
$= 15$

[2] Write the opposite of the result. ⟶ -15

Thus, $-6 + (-9) = \mathbf{-15}$

Self-Check Add: **1.** $-12 + (-6)$ **2.** $-7 + (-15)$ **3.** $-4 + (-9)$

Follow this procedure to add a positive and a negative number.

> **PROCEDURE**
>
> **To add a positive and a negative number:**
>
> [1] Subtract the absolute values.
>
> [2] If the positive number has the greater absolute value, write the difference.
>
> If the negative number has the greater absolute value, write the opposite of the difference.

EXAMPLE 2 Add: $8 + (-3)$

Solution:
[1] Subtract the absolute values. $|8| - |-3| = 8 - 3$
[2] Since $|8| > |-3|$, write the difference. $= 5$

$8 + (-3) = \mathbf{5}$ ◀ The positive number has the greater absolute value.

Self-Check Add: **4.** $12 + (-6)$ **5.** $-13 + 21$ **6.** $-8 + 15$

EXAMPLE 3 Add: $-20 + 8$

Solution:
[1] Subtract the absolute values. $|-20| - |8| = 20 - 8$
$= 12$
[2] Since $|-20| > |8|$, write the opposite of the difference. ⟶ -12

$-20 + 8 = \mathbf{-12}$ ◀ The negative number has the greater absolute value.

Self-Check Add: **7.** $8 + (-9)$ **8.** $-18 + 12$ **9.** $-16 + 10$

Rational Numbers: Addition/Subtraction   **305**

ADDITIONAL EXAMPLES
Example 1
Add.
1. (-8) + (-17)
   Ans: -25
2. -6 + (-15)   Ans: -21

SELF-CHECK
1. -18   2. -22   3. -13

Example 2
Add.
1. 21 + (-3)   Ans: 18
2. 47 + (-28)   Ans: 19

SELF-CHECK
4. 6   5. 8   6. 7

Example 3
Add.
1. (-31) + 12   Ans: -19
2. 19 + (-50)   Ans: -31

SELF-CHECK
7. -1   8. -6   9. -6

CLASSROOM EXERCISES
<u>1</u>. 8   <u>2</u>. 8   <u>3</u>. 6
<u>4</u>. 10   <u>5</u>. 15   <u>6</u>. 24
<u>7</u>. -18   <u>8</u>. -9   <u>9</u>. -22
<u>10</u>. -13   <u>11</u>. 29
<u>12</u>. 25   <u>13</u>. -22
<u>14</u>. -20   <u>15</u>. 2   <u>16</u>. 6
<u>17</u>. 11   <u>18</u>. 16   <u>19</u>. 4
<u>20</u>. 11   <u>21</u>. 12   <u>22</u>. 17
<u>23</u>. -3   <u>24</u>. -9   <u>25</u>. -8
<u>26</u>. -8   <u>27</u>. -4
<u>28</u>. -20   <u>29</u>. -15
<u>30</u>. -12

ASSIGNMENT GUIDE
BASIC
Day 1 p. 306: 1-34
Day 2 p. 307: 35-62
AVERAGE
Day 1 pp. 306-307: 1-49 odd
Day 2 p. 307: 51-82
ABOVE AVERAGE
pp. 306-307: 3, 6, 9, ..., 84, 87-89

WRITTEN EXERCISES
<u>1</u>. 7   <u>2</u>. 18   <u>3</u>. 5
<u>4</u>. 21   <u>5</u>. 15   <u>6</u>. 36
<u>7</u>. 8   <u>8</u>. 53   <u>9</u>. 32
<u>10</u>. 16   <u>11</u>. 0   <u>12</u>. 1
<u>13</u>. 1.3   <u>14</u>. 3.4
<u>15</u>. 6.7   <u>16</u>. 0.9
<u>17</u>. $\frac{1}{3}$ <u>18</u>. $\frac{3}{5}$ <u>19</u>. -16
<u>20</u>. -12   <u>21</u>. 28
<u>22</u>. 29   <u>23</u>. -36
<u>24</u>. -18   <u>25</u>. -18

CLASSROOM EXERCISES

Evaluate. (Table)

1. $|8|$ **2.** $|-8|$ **3.** $|-6|$ **4.** $|10|$ **5.** $|-15|$ **6.** $|24|$

For Exercises 7-30, add the given numbers. (Example 1)

7. $-4 + (-14)$ **8.** $-6 + (-3)$ **9.** $-7 + (-15)$ **10.** $-9 + (-4)$
11. $16 + 13$ **12.** $7 + 18$ **13.** $-8 + (-14)$ **14.** $-5 + (-15)$

(Example 2)

15. $6 + (-4)$ **16.** $7 + (-1)$ **17.** $-4 + 15$ **18.** $-5 + 21$
19. $20 + (-16)$ **20.** $-10 + 21$ **21.** $30 + (-18)$ **22.** $-7 + 24$

(Example 3)

23. $6 + (-9)$ **24.** $-16 + 7$ **25.** $-22 + 14$ **26.** $12 + (-20)$
27. $9 + (-13)$ **28.** $-31 + 11$ **29.** $-18 + 3$ **30.** $7 + (-19)$

WRITTEN EXERCISES

Goals: To write the absolute value of a number
To add two rational numbers
To apply the skill of adding rational numbers to solving word problems

Evaluate. (Table)

1. $|7|$ **2.** $|18|$ **3.** $|-5|$ **4.** $|-21|$ **5.** $|15|$ **6.** $|36|$
7. $|-8|$ **8.** $|-53|$ **9.** $|32|$ **10.** $|-16|$ **11.** $|0|$ **12.** $|1|$
13. $|1.3|$ **14.** $|-3.4|$ **15.** $|-6.7|$ **16.** $|0.9|$ **17.** $|\frac{1}{3}|$ **18.** $|-\frac{3}{5}|$

For Exercises 19-34, add the given numbers. (Example 1)

19. $-11 + (-5)$ **20.** $-9 + (-3)$ **21.** $6 + 22$ **22.** $15 + 14$
23. $-22 + (-14)$ **24.** $-8 + (-10)$ **25.** $-15 + (-3)$ **26.** $-32 + (-21)$
27. $-1.3 + (-4.2)$ **28.** $-1.6 + (-1.2)$ **29.** $-3.4 + (-1.8)$ **30.** $-6.3 + (-0.4)$
31. $-\frac{1}{3} + \left(-\frac{1}{3}\right)$ **32.** $-\frac{2}{7} + \left(-\frac{3}{7}\right)$ **33.** $-\frac{1}{2} + \left(-\frac{1}{4}\right)$ **34.** $-\frac{1}{3} + \left(-\frac{1}{5}\right)$

306   Chapter 10

For Exercises 35–86, add the given numbers. (Example 2)

35. 21 + (−17)	**36.** 8 + (−7)	**37.** −4 + 12	**38.** −1 + 21
39. −16 + 25	**40.** 13 + (−4)	**41.** 8 + (−3)	**42.** −12 + 7
43. 2.5 + (−1.8)	**44.** −1.7 + 3.2	**45.** 8.7 + (−4.8)	**46.** −0.8 + 3.5
47. $-\frac{1}{5} + \frac{2}{5}$	**48.** $\frac{3}{4} + \left(-\frac{1}{4}\right)$	**49.** $\frac{1}{2} + \left(-\frac{1}{3}\right)$	**50.** $-\frac{1}{8} + \frac{1}{2}$

(Example 3)

51. −12 + 7	**52.** −21 + 14	**53.** 8 + (−20)	**54.** 4 + (−19)
55. −27 + 6	**56.** 7 + (−8)	**57.** 22 + (−30)	**58.** −15 + 10
59. 1.3 + (−3.2)	**60.** −2.5 + 1.2	**61.** 1.1 + (−2.2)	**62.** −4.6 + 0.8
63. $-\frac{3}{8} + \frac{1}{8}$	**64.** $\frac{1}{3} + \left(-\frac{2}{3}\right)$	**65.** $\frac{1}{2} + \left(-\frac{5}{8}\right)$	**66.** $-\frac{4}{5} + \frac{1}{4}$

MIXED PRACTICE

67. −16 + 21	**68.** −17 + 30	**69.** −24 + (−10)	**70.** −2 + (−17)
71. −13 + 8	**72.** −18 + 7	**73.** 27 + (−12)	**74.** 16 + 32
75. −4 + (−21)	**76.** −14 + 12	**77.** 30 + (−24)	**78.** −9 + (−22)
79. −8.4 + 2.7	**80.** −7.3 + (−6.2)	**81.** 3.5 + 0.4	**82.** −4.6 + 1.4
83. $-\frac{1}{5} + \left(-\frac{2}{5}\right)$	**84.** $\frac{3}{8} + \left(-\frac{5}{8}\right)$	**85.** $-\frac{1}{4} + \frac{1}{2}$	**86.** $\frac{3}{10} + \left(-\frac{3}{4}\right)$

APPLICATIONS: USING ADDITION OF RATIONAL NUMBERS

87. A submarine at a depth of 40 meters below the ocean's surface fired a missile which rose 225 meters. How far above the ocean's surface did the missile rise?

88. A plane was flying at an altitude of 4000 feet. Then it dropped 320 feet. What was the plane's new altitude?

89. The temperature was −2°C and dropped 6 degrees. Find the new temperature.

WRITTEN EXERCISES

26. -53 **27.** -5.5
28. -2.8 **29.** -5.2
30. -6.7 **31.** $-\frac{2}{3}$
32. $-\frac{5}{7}$ **33.** $-\frac{3}{4}$
34. $-\frac{8}{15}$ **35.** 4 **36.** 1
37. 8 **38.** 20 **39.** 9
40. 9 **41.** 5 **42.** -5
43. 0.7 **44.** 1.5
45. 3.9 **46.** 2.7
47. $\frac{1}{5}$ **48.** $\frac{1}{2}$ **49.** $\frac{1}{6}$
50. $\frac{3}{8}$ **51.** -5 **52.** -7
53. -12 **54.** -15
55. -21 **56.** -1
57. -8 **58.** -5
59. -1.9 **60.** -1.3
61. -1.1 **62.** -3.8
63. $-\frac{1}{4}$ **64.** $-\frac{1}{3}$
65. $-\frac{1}{8}$ **66.** $-\frac{11}{20}$
67. 5 **68.** 13
69. -34 **70.** -19
71. -5 **72.** -11
73. 15 **74.** 48
75. -25 **76.** -2 **77.** 6
78. -31 **79.** -5.7
80. -13.5 **81.** 3.9
82. -3.2 **83.** $-\frac{3}{5}$
84. $-\frac{1}{4}$ **85.** $\frac{1}{4}$
86. $-\frac{9}{20}$ **87.** 185 m
88. 3680 ft **89.** -8°C

REVIEW CAPSULE FOR SECTION 10-5

Write the opposite of each number. (Pages 297–299)

1. 12 **2.** 4 **3.** −28 **4.** −15 **5.** 3.7 **6.** $-1\frac{1}{3}$

The answers are on page 320.

Rational Numbers: Addition/Subtraction

Teaching Suggestions
p. M-23

QUICK QUIZ

Write the opposite.
1. 12   Ans: -12
2. -8   Ans: 8

Subtract.
3. 6.2 - 1.8   Ans: 4.4
4. $7\frac{2}{3} - 1\frac{1}{2}$ Ans: $6\frac{1}{6}$
5. $5\frac{3}{4} - 2\frac{7}{8}$ Ans: $2\frac{7}{8}$

ADDITIONAL EXAMPLES

Example 1
Write an addition problem for each subtraction problem.
1. 16 - 4
   Ans: 16 + (-4)
2. -21 - 17
   Ans: (-21) + (-17)

SELF-CHECK
1. 4 = (-9)
2. -10 + (-18)
3. -1 + 12   4. 17 + 21

Example 2
Subtract.
1. -28 - 13   Ans: -41
2. -14.3 - 8.6
   Ans: -22.9

SELF-CHECK
5. -6   6. -85   7. -2.6
8. -5.9

308

10-5 Subtracting Rational Numbers

Notice that the answers to the two problems below are the same.

Subtraction	Related Addition
$7 - 4 = 3$	$7 + (-4) = 3$

Subtracting an integer is the same as adding its opposite.

> **PROCEDURE**
>
> To write an addition problem for a subtraction problem, <u>add</u> the <u>opposite</u> of the number you are subtracting.

EXAMPLE 1 Write an addition problem for each subtraction problem.

Subtraction Problem	Think	Addition Problem
a. $9 - 4$	The opposite of 4 is -4.	$9 + (-4)$
b. $5 - (-2)$	The opposite of -2 is 2.	$5 + 2$
c. $-6 - 3$	The opposite of 3 is -3.	$-6 + (-3)$
d. $-7 - (-1)$	The opposite of -1 is 1.	$-7 + 1$

Self-Check *Write an addition problem for each subtraction problem.*

1. $4 - 9$ 2. $-10 - 18$ 3. $-1 - (-12)$ 4. $17 - (-21)$

> **PROCEDURE**
>
> To subtract two rational numbers:
> 1. Write the related addition problem.
> 2. Follow the rules for addition.

EXAMPLE 2 Subtract: $-5 - 6$

Solution:
1. Write the related addition problem. $-5 - 6 = -5 + (-6)$
2. Add. $= -11$

Self-Check *Subtract.*

5. $-2 - 4$ 6. $-47 - 38$ 7. $-0.5 - 2.1$ 8. $-4.5 - 1.4$

308   Chapter 10

Example 3 shows how to subtract a negative number.

EXAMPLE 3 Subtract: **a.** $8 - (-9)$ **b.** $-6 - (-4)$

Solutions: ① **a.** $8 - (-9) = 8 + 9$ **b.** $-6 - (-4) = -6 + 4$
② $\qquad\qquad\quad = 17 \qquad\qquad\qquad\qquad\quad = -2$

Self-Check Subtract.

9. $12 - (-17)$ **10.** $24 - (-39)$ **11.** $-10 - (-25)$ **12.** $-9 - (-16)$

CLASSROOM EXERCISES

Complete. (Example 1)

1. $14 - 5 = 14 + \underline{\ ?\ }$ **2.** $2 - 9 = 2 + \underline{\ ?\ }$ **3.** $5 - (-14) = 5 + \underline{\ ?\ }$
4. $-6 - 8 = -6 + \underline{\ ?\ }$ **5.** $0 - 17 = 0 + \underline{\ ?\ }$ **6.** $-1 - (-10) = (-1) + \underline{\ ?\ }$

For Exercises 7-22, subtract. (Example 2)

7. $5 - 8$ **8.** $3 - 13$ **9.** $-5 - 17$ **10.** $-6 - 10$
11. $6 - 9$ **12.** $8 - 17$ **13.** $-4 - 21$ **14.** $-12 - 30$

(Example 3)

15. $4 - (-5)$ **16.** $12 - (-4)$ **17.** $-6 - (-3)$ **18.** $-4 - (-10)$
19. $10 - (-11)$ **20.** $15 - (-32)$ **21.** $-13 - (-19)$ **22.** $-21 - (-43)$

WRITTEN EXERCISES

Goal: To subtract two rational numbers

Write an addition problem for each subtraction problem. (Example 1)

1. $11 - 3$ **2.** $14 - 2$ **3.** $14 - (-6)$ **4.** $12 - (-9)$
5. $-6 - 1$ **6.** $-9 - 7$ **7.** $-8 - (-5)$ **8.** $-12 - (-1)$
9. $-7 - 7$ **10.** $-8 - 8$ **11.** $4.7 - 8.6$ **12.** $-1\frac{1}{2} - \left(-2\frac{1}{2}\right)$

For Exercises 13-52, subtract. (Example 2)

13. $11 - 15$ **14.** $17 - 22$ **15.** $-19 - 22$ **16.** $-16 - 24$
17. $0 - 24$ **18.** $18 - 43$ **19.** $-37 - 84$ **20.** $-10 - 26$
21. $2.7 - 3.6$ **22.** $1.8 - 5.1$ **23.** $4\frac{3}{4} - 6\frac{1}{4}$ **24.** $3\frac{1}{2} - 5\frac{1}{4}$

Rational Numbers: Addition/Subtraction

ADDITIONAL EXAMPLES
Example 3
Subtract.
1. $21 - (-12)$ Ans: 33
2. $-29 - (-102)$ Ans: 73

SELF-CHECK
9. 29 10. 63 11. 15
12. 7

CLASSROOM EXERCISES
1. (-5) 2. (-9)
3. 14 4. (-8)
5. (-17) 6. 10 7. -3
8. -10 9. -22
10. -16 11. -3
12. -9 13. -25
14. -42 15. 9 16. 16
17. -3 18. 6 19. 21
20. 47 21. 6 22. 22

ASSIGNMENT GUIDE
BASIC
Day 1 p. 309: 1-24
Day 2 p. 310: 25-48
AVERAGE
Day 1 pp. 309-310: 1-35 odd
Day 2 p. 310: 37-55
ABOVE AVERAGE
pp. 309-310: 3,6,9,
..., 51, 53-55

WRITTEN EXERCISES
1. 11 + (-3)
2. 14 + (-2)
3. 14 + 6
4. 12 + 9

WRITTEN EXERCISES

5. −6 + (−1)
6. −9 + (−7)
7. −8 + 5
8. −12 + 1
9. −7 + (−7)
10. −8 + (−8)
11. 4.7 + (−8.6)
12. $-1\frac{1}{2} + 2\frac{1}{2}$
13. −4   14. −5
15. −41   16. −40
17. −24   18. −25
19. −121 20. −36
21. −0.9 22. −3.3
23. $-1\frac{1}{2}$ 24. $-1\frac{3}{4}$
25. 4   26. 7   27. 8
28. 42   29. 25
30. 53   31. 57
32. 81   33. 23.4
34. 8.4 35. $5\frac{1}{2}$
36. $1\frac{1}{3}$ 37. 0
38. 4   39. 7
40. −171 41. −188
42. −85   43. −351
44. 0   45. 12
46. 42   47. −81
48. −67   49. 11.5
50. −8.3 51. $2\frac{7}{8}$
52. $-3\frac{2}{3}$ 53. 56°C
54. 41°F
55. 14,776 ft

(Example 3)

25. −3 − (−7)
26. −8 − (−15)
27. 6 − (−2)
28. 25 − (−17)
29. 18 − (−7)
30. 24 − (−29)
31. 9 − (−48)
32. 5 − (−76)
33. 9.5 − (−13.9)
34. −8.9 − (−17.3)
35. $3\frac{3}{4} - \left(-1\frac{3}{4}\right)$
36. $\frac{1}{2} - \left(-\frac{5}{6}\right)$

MIXED PRACTICE

37. −127 − (−127)
38. −4 − (−8)
39. 0 − (−7)
40. −128 − 43
41. 48 − 236
42. 39 − 124
43. −254 − 97
44. −64 − (−64)
45. −24 − (−36)
46. 27 − (−15)
47. −36 − 45
48. 15 − 82
49. 4.8 − (−6.7)
50. −5.4 − 2.9
51. $2\frac{3}{8} - 5\frac{1}{4}$
52. $-1\frac{1}{6} - 2\frac{1}{2}$

APPLICATIONS: USING SUBTRACTION OF RATIONAL NUMBERS

53. The greatest recorded temperature change in one day in North America took place in Montana. The change was from 7°C to −49°C. Find the difference in temperatures. (Hint: 7 − (−49) = __?__)

54. The actual temperature is 32°F. The wind chill temperature is −9°F. How much colder than the actual temperature is the wind chill temperature? (Hint: 32 − (−9) = __?__)

55. The highest point in California is 14,494 feet above sea level (+). The lowest point is 282 feet below sea level (−). Find the difference in altitude between the highest and lowest point.

REVIEW CAPSULE FOR SECTION 10-6

Evaluate each expression. (Pages 18–21)

1. 2 + 17 − 9
2. 16 + 32 + 25
3. 32 + 57 − 65 + 10
4. 23 − 16 + 10 + 52
5. 81 + 125 + 51
6. 13 + 42 + 180 − 57

Add. (Pages 300–302)

7. −5 + (−12)
8. 16 + (−20)
9. −32 + 55
10. −48 + 19
11. 25 + (−17)
12. −19 + (−13)
13. 18 + (−18)
14. −54 + 54

The answers are on page 320.

310 Chapter 10

CAREER APPLICATIONS
Travel Agent

Travel agents must be able to determine the difference in time between any two cities in the world. The longitude of the cities can be used to find this difference.

The figure at the right shows the longitude of New York City to be 74°W. This means that New York City is 74° west of the prime meridian. A point east of the prime meridian has <u>east longitude</u>.

This table shows the relation between the longitude of a point and its time zone.

One hour of time = 15° in longitude

A **zone description (ZD)** of +5 (west longitude) means that the time is 5 hours earlier than the time at Zone 0. A ZD of −3 (east longitude) means that the time is 3 hours later than the time at Zone 0. To find the ZD of a point, first divide its longitude by 15°. Then round to the nearest whole number.

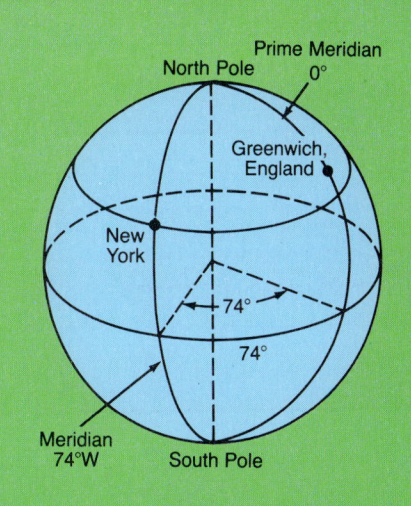

Zone Description Table

37½°W	22½°W	7½°W	7½°E	22½°E	37½°E
Zone +2	Zone +1	Zone 0	Zone −1	Zone −2

30°W   15°W   0°   15°E   30°E

EXAMPLE At point A with longitude 0°, the time is 10:45 A.M. Find the time at point B with longitude 74°W.

SOLUTION Find the ZD of point B.

$$\frac{74}{15} = 4.9\overline{3} \quad ZD = 5$$

▶ **West longitude is positive. East longitude is negative.**

The time at point B is 5 hours <u>earlier</u> than the time at point A.

10:45 − 5 = **5:45 A.M.**

Exercises

Find the zone description (ZD) of each city. Then find its time when it is 6:05 P.M. in London (longitude 0°).

1. Chicago (87.6°W)
2. Seattle (122.3°W)
3. Houston (95.4°W)
4. Moscow (37.7°E)
5. New Delhi (77.2°E)
6. Rome (12.5°E)

Consumer Application

Teaching Suggestions
p. M-23

QUICK QUIZ
Evaluate.
1. |23| Ans: 23
2. |-7.6| Ans: 7.6
3. 10 + 12 - 10 Ans: 12
4. 27 - 15 + 32 Ans: 44
5. 102 + 87 - 99
   Ans: 90
6. -10 + 8 Ans: -2
7. (-21) + (-37)
   Ans: -58
8. 23 + (-23) Ans: 0
9. -211 + (-211)
   Ans: -422
10. 100 + (-81) Ans: 19

ADDITIONAL EXAMPLES
Example 1
Add.
1. -13 + (-27) + 13
   Ans: -27
2. 592 + (-28) + (-272)
   Ans: 292

SELF-CHECK
1. -10   2. 12

10-6 Properties of Addition

The properties of addition also apply to addition with rational numbers.

Addition Property of Zero
For any rational number a,
$$a + 0 = a \text{ and } 0 + a = a$$
$4 + 0 = 4$
$0 + (-2.7) = -2.7$

Addition Property of Opposites
(Additive Inverse Property)
For any rational number a,
$$a + (-a) = 0 \text{ and } -a + a = 0$$
$35 + (-35) = 0$
$-2\frac{1}{4} + 2\frac{1}{4} = 0$

Commutative Property of Addition
For any rational numbers a and b,
$$a + b = b + a$$
$17 + (-21) = -21 + 17$
$-\frac{1}{2} + \frac{2}{5} = \frac{2}{5} + (-\frac{1}{2})$

Associative Property of Addition
For any rational numbers a, b, and c,
$$(a + b) + c = a + (b + c)$$
$(5 + 3) + (-2) = 5 + (3 + (-2))$
$(1.2 + 5.8) + 4 = 1.2 + (5.8 + 4)$

The Commutative and Associative Properties can be used when adding three or more numbers.

EXAMPLE 1 Add: $(-3) + 8 + (-6)$

Solution:

Method 1
$-3 + 8 + (-6)$
$5 + (-6)$
-1
◀ Add from left to right.

Method 2
$-3 + 8 + (-6)$
$-3 + (-6) + 8$
$-9 + 8$
-1
◀ First add the negative numbers. Then add 8 to the sum.

Self-Check Add.

1. $3 + (-6) + (-7)$
2. $9 + (-4) + 7$

Example 2 shows how the Commutative and Associative Properties can be used to add four rational numbers.

312 Chapter 10

EXAMPLE 2 Add: $5 + (-1) + 8 + (-3)$

Solution:

Method 1
$$5 + (-1) + 8 + (-3)$$
$$4 \quad + \quad 5$$
$$9$$

Method 2
$$5 + (-1) + 8 + (-3)$$
$$5 + 8 + (-1) + (-3)$$
$$13 \quad + \quad (-4)$$
$$9$$

◂ Group the positive and negative numbers. Then add.

Self-Check *Add.*

3. $-4 + 9 + 6 + (-5)$ **4.** $-3 + 4 + (-5) + 8$

CLASSROOM EXERCISES

For Exercises 1–6, complete. (Example 1)

1. $12 + (-3) + 2 = 12 + \underline{\ ?\ } = \underline{\ ?\ }$ **2.** $9 + (-6) + 3 = \underline{\ ?\ } + (-6) = \underline{\ ?\ }$
3. $-6 + 4 + (-8) = -14 + \underline{\ ?\ } = \underline{\ ?\ }$ **4.** $-8 + 3 + (-1) = \underline{\ ?\ } + 3 = \underline{\ ?\ }$
5. $4 + (-1) + (-2) = 3 + \underline{\ ?\ } = \underline{\ ?\ }$ **6.** $3 + (-13) + 5 = \underline{\ ?\ } + (-13) = \underline{\ ?\ }$

Add. (Example 2)

7. $1 + (-3) + (-2) + 4$ **8.** $-15 + 4 + 7 + (-6)$ **9.** $-10 + 9 + 4 + (-6)$
10. $-1 + 1 + (-1) + 9$ **11.** $-12 + 7 + 13 + (-8)$ **12.** $-6 + (-12) + 9 + 3$

WRITTEN EXERCISES

Goal: To add more than two rational numbers

For Exercises 1–36, add. (Example 1)

1. $-8 + 6 + (-3)$ **2.** $-12 + 5 + (-2)$ **3.** $7 + (-13) + 8$
4. $5 + (-12) + 13$ **5.** $-6 + (-9) + (-14)$ **6.** $(-3) + (-9) + (-15)$
7. $-10 + 18 + (-32)$ **8.** $-9 + 21 + (-21)$ **9.** $16 + (-8) + 8$
10. $-31 + (-19) + (-7)$ **11.** $-51 + 27 + (-40)$ **12.** $16 + (-29) + 54$

(Example 2)

13. $-5 + (-2) + (-7) + 4$ **14.** $-6 + 2 + (-11) + 8$ **15.** $3 + (-9) + 16 + 3$
16. $42 + (-6) + 6 + (-9)$ **17.** $-8 + (-3) + 9 + 7$ **18.** $9 + (-7) + 13 + (-6)$

Rational Numbers: Addition/Subtraction   **313**

ADDITIONAL EXAMPLES
Example 2
<u>1</u>. $-16 + (-24) + 81 + (-10)$ Ans: 31
<u>2</u>. $200 + (-36) + (-64) + 84$ Ans: 184

SELF-CHECK
<u>3</u>. 6 <u>4</u>. 4

CLASSROOM EXERCISES
<u>1</u>. (-1); 11 <u>2</u>. 12; 6
<u>3</u>. 4; -10 <u>4</u>. -9; -6
<u>5</u>. (-2); 1 <u>6</u>. 8; -5
<u>7</u>. 0 <u>8</u>. -10 <u>9</u>. -3
<u>10</u>. 8 <u>11</u>. 0 <u>12</u>. -6

ASSIGNMENT GUIDE
BASIC
Day 1 pp. 313–314:
1–29 odd
Day 2 pp. 313–314:
2–36 even
AVERAGE
pp. 313–314: 1–23 odd, 25–36
ABOVE AVERAGE
pp. 313–314: 1–35 odd, 37–44

WRITTEN EXERCISES
<u>1</u>. -5 <u>2</u>. -9 <u>3</u>. 2
<u>4</u>. 6 <u>5</u>. -29 <u>6</u>. -27
<u>7</u>. -24 <u>8</u>. -9 <u>9</u>. 16
<u>10</u>. -57 <u>11</u>. -64
<u>12</u>. 41 <u>13</u>. -10
<u>14</u>. -7 <u>15</u>. 13
<u>16</u>. 33 <u>17</u>. 5 <u>18</u>. 9

WRITTEN EXERCISES

19. 0   20. -16   21. -6
22. 8   23. 8   24. -10
25. -9   26. -4   27. -9
28. -4   29. 26
30. -25 31. -4
32. -50 33. -12
34. -14 35. 2   36. -3
37. -10.7 38. -5.97
39. $-\frac{1}{4}$ 40. $-\frac{5}{8}$
41. -0.7 42. 0.14
43. $1\frac{1}{4}$ 44. $-\frac{11}{12}$

CALCULATOR EXERCISES

1. -97   2. -101
3. -0.67 4. 0.33
5. -6.99 6. 26

---

19. $-9 + 12 + 4 + (-7)$ 20. $-8 + 8 + (-12) + (-4)$ 21. $-7 + 13 + (-16) + 4$
22. $-7 + 14 + 6 + (-5)$ 23. $13 + (-3) + (-9) + 7$ 24. $-10 + 2 + (-6) + 4$

MIXED PRACTICE

25. $-14 + 3 + (-5) + 7$ 26. $-14 + 3 + (-5) + 12$ 27. $3 + (-6) + 4 + (-10)$
28. $-12 + 14 + (-6)$ 29. $5 + 9 + (-5) + 17$ 30. $-35 + 25 + (-15)$
31. $-8 + 9 + (-5)$ 32. $-40 + 15 + (-25)$ 33. $-8 + 3 + (-11) + 4$
34. $-5 + 28 + (-37)$ 35. $2 + (-5) + 6 + (-1)$ 36. $5 + (-12) + 4$

MORE CHALLENGING EXERCISES

Add.

37. $0.3 + (-5.8) + 4.2 + (-9.4)$ 38. $-6.8 + 3.02 + (-9.46) + 14.5 + (-7.23)$
39. $-\frac{1}{4} + \frac{3}{8} + \frac{1}{4} + (-\frac{5}{8})$ 40. $-\frac{3}{8} + 1\frac{1}{4} + (-1) + (-\frac{1}{2})$
41. $0.8 + (-1.5) + 2.3 + 1.2 + (-3.5)$ 42. $-0.3 + 0.47 + 0.28 + (-0.31)$
43. $\frac{1}{3} + (-\frac{1}{2}) + 1\frac{2}{3} + (-\frac{1}{4})$ 44. $-\frac{3}{8} + (-1\frac{1}{2}) + (-\frac{1}{6}) + 1\frac{1}{8}$

ADDING RATIONAL NUMBERS

A calculator with a "+/−" or "sign change" key (sometimes indicated as "SC") allows you to operate with negative numbers. This key enables you to add rational numbers from left to right without grouping.

EXAMPLE Add: $-6.9 + (-2.3) + 4.01 + (-7.2)$

Solution Press the sign change key for any negative number <u>after</u> you enter the value of the number.

6 . 9 [+/−] [+] 2 . 3 [+/−] [+]

4 . 0 1 [+] 7 . 2 [+/−] [=]   ▢ −12.39

EXERCISES *Add.*

1. $-17 + 52 + (-94) + (-38)$ 2. $-52 + (-16) + 31 + (-64)$
3. $-6.0 + 3.23 + (-5.2) + 7.3$ 4. $-0.2 + 0.41 + 0.37 + (-0.25)$
5. $-6.8 + 3.02 + (-9.46) + (-7.23) + 14.5 + 6.9 + (-7.92)$
6. $-104 + (-602) + (-523) + 344 + 681 + (-712) + 942$

314 Chapter 10

REVIEW CAPSULE FOR SECTION 10-7

For Exercises 1–8, combine like terms. (Pages 76–80)

1. $4a + 3a$
2. $5m - 2m$
3. $8x^2 - x^2$
4. $10b^2 + 4b^2$
5. $2a + 4a + 5$
6. $3x + 5x + 9$
7. $3t^2 + 12 + 2t^2$
8. $5n^2 + 6 + 7n^2$

Add. (Pages 304–307)

9. $-5 + 12$
10. $-4 + (-3)$
11. $9 + (-6)$
12. $10 + (-1)$
13. $1 + (-5)$
14. $-7 + (-3)$
15. $-8 + 10$
16. $7 + (-11)$

The answers are on page 320.

10-7 Adding Polynomials

Each of the following is an example of a **monomial**.

a. $5y^3$ **b.** $-q$ **c.** $\frac{1}{3}r$ **d.** -9.6 **e.** g^2

A **binomial** is the sum or difference of two monomials.

a. $t + 5$ **b.** $2g - 1$ **c.** $5x^2 - 4$ **d.** $a + b$

A **trinomial** is the sum or difference of three monomials.

a. $x^2 + 5x - 9$ **b.** $2t - t^2 - 8$ **c.** $5p^3 - 1 + p^2$

Monomials and binomials belong to a larger class of expressions called polynomials.

> A **polynomial** is a monomial or the sum or difference of two or more nonomials.
>
> $-5v$
> $8t - 9$
> $x^2 - 2x + 3$

Combining like terms is similar to adding polynomials.

EXAMPLE 1 Add: $(-3x^2 + 2x) + (5x^2 - 3x)$

Solution:
1 Write without parentheses and combine like terms.

$-3x^2 + 2x + 5x^2 - 3x$
$-3x^2 + 5x^2 + 2x - 3x$

2 Arrange terms in order with the greatest exponent first.

$2x^2 - x$

Self-Check 1. $(3x + 7) + (x - 4) = \underline{\ ?\ }$ 2. $(-2p^2 + 5p) + (-3p^2 - 8p) = \underline{\ ?\ }$

Rational Numbers: Addition/Subtraction **315**

Teaching Suggestions
p. M-23

QUICK QUIZ
Combine like terms.
1. 6x + 7x   Ans: 13x
2. 9n - n   Ans: 8n
3. 2y^2 + 6y^2 + 3
   Ans: 8y + 3
Add.
4. -5 + (-4) Ans: -9
5. -8 + (13) Ans: 5

ADDITIONAL EXAMPLES
Example 1
Add.
1. (4m^2 - 3m) +
   (-m^2 + 5m)
   Ans: 3m^2 + 2m
2. (-7a + 6) + (-2a - 4)
   Ans: -9a + 2

SELF-CHECK
1. 4x + 3   2. -5p^2 - 3p

ADDITIONAL EXAMPLES

Example 2

Add.

1. $(3h^2 - h + 1) + (-4h^2 + 5h - 7)$
   Ans: $-h^2 + 4h - 6$
2. $(-8t^2 + 7t - 5) + (8t^2 - 3t + 4)$
   Ans: $4t - 1$

SELF-CHECK

3. $-3y - 2$
4. $5n^2 + 3n + 7$

CLASSROOM EXERCISES

1. monomial
2. trinomial
3. binomial
4. trinomial
5. 2   6. 3   7. 3
8. 4

ASSIGNMENT GUIDE

BASIC
pp. 316-317: Omit

AVERAGE
pp. 316-317: Omit

ABOVE AVERAGE
pp. 316-317: 1-23 odd

WRITTEN EXERCISES

1. $7p + 3$ 2. $-t^2 - 2$
3. $6c^2 - 6c$ 4. $-2x + 5$
5. $7d - 2$
6. $-2y^2 + 5y + 3$
7. $5n^2 - 6n + 5$
8. $-7y^2 + y - 1$
9. $3x^2 - 2x - 2$

It is sometimes helpful to arrange polynomials in vertical order.

EXAMPLE 2 Add: $(2x^2 - 5x + 6) + (-2x^2 + 3x - 10)$

Solution: Write like terms in the same column.

$$\begin{array}{r} 2x^2 - 5x + 6 \\ -2x^2 + 3x - 10 \\ \hline -2x - 4 \end{array}$$

◀ Add the columns.

Self-Check 3. $(3y - 5) + (-6y + 3)$ 4. $(4n^2 + 6n + 3) + (n^2 - 3n + 4)$

The **degree of a polynomial** is the highest degree of its monomials.

Polynomial	Degree
$3x^2 - 5x + 9$	2
$5y + 7$	1
$a^3 + 3a + 8$	3

◀ Read the degree from the exponent of the monomial.

CLASSROOM EXERCISES

Classify each of the following as a monomial, binomial, or trinomial. (Definition)

1. $-5h$ 2. $3x^2 + 5x - 4$ 3. $3p - 4$ 4. $7t^2 - 5t + 21$

Write the degree of each polynomial.

5. $t^2 - 5$ 6. $-9p^3$ 7. $q^3 + q^2$ 8. $-x + x^4 - 21$

WRITTEN EXERCISES

Goal: To add polynomials

Add. (Examples 1 and 2)

1. $(3p - 2) + (4p + 5)$
2. $(-5t^2 + 1) + (4t^2 - 3)$
3. $(2c^2 - c) + (4c^2 - 5c)$
4. $(3x + 7) + (-5x - 2)$
5. $(4d - 3) + (3d + 1)$
6. $(-2y^2 - 7) + (5y + 10)$
7. $(4n^2 - 3n + 1) + (n^2 - 3n + 4)$
8. $(-5y^2 + 4y - 7) + (-2y^2 - 3y + 6)$
9. $(x^2 - 3x + 1) + (2x^2 + x - 3)$
10. $(4g^2 - g + 5) + (-g^2 + 6g - 3)$
11. $(5r^2 - 2r + 3) + (-r^2 + 6r - 5)$
12. $(3p^2 - 2p + 10) + (2p^2 + 4p - 6)$

316 Chapter 10

MORE CHALLENGING EXERCISES

Subtract.

EXAMPLE: $5t - (-6t)$

Solution: $5t - (-6t) = 5t + 6t$
$= 11t$

▶ Add the opposite of the number you are subtracting.

13. $8x - (-3x)$
14. $7y - (-5y)$
15. $-4p^2 - 5p^2$
16. $-3b^2 - 7b^2$
17. $-9y - (-3y)$
18. $-2n^2 - (-7n^2)$
19. $6t^2 - 15t^2$
20. $4s - 10s$
21. $3x^2 - (-5x^2)$
22. $-4ab - 6ab$
23. $-6y - (-4y)$
24. $b - 8b$

REVIEW: SECTIONS 10-4 — 10-7

Add. (Section 10-4)

1. $-6 + 12$
2. $-14 + (-7)$
3. $-18 + 13$
4. $-11 + 22$
5. $-3 + 14$
6. $16 + 34$
7. $16 + (-30)$
8. $-31 + (-8)$

Subtract. (Section 10-5)

9. $25 - 18$
10. $23 - 47$
11. $6 - (-18)$
12. $-3 - 14$
13. $18 - (-21)$
14. $-4 - 12$
15. $-5 - (-11)$
16. $-15 - (-3)$

Solve.

17. On a certain day, the temperature at noon was 8°F. By midnight it had dropped 17°. What was the temperature at midnight? (Section 10-4)

18. The average daytime temperature on Mars is about −30°C. On Earth it is about 20°C. About how much warmer is the average daytime temperature on Earth? (Section 10-5)

Add. (Section 10-6)

19. $-10 + 4 + (-12)$
20. $-5 + (-12) + (-9)$
21. $12 + (-10) + (-32) + 16$
22. $-4 + 15 + 16 + (-24)$
23. $-6 + 18 + (-34) + 25$
24. $-12 + (-25) + 40$

Add. (Section 10-7)

25. $(x + 3) + (-2x + 5)$
26. $(-2y^2 + 4y - 5) + (-4y^2 - 5y + 2)$
27. $(12t - 3) + (8t + 4)$
28. $(2p^2 - p) + (p^2 + 5p)$
29. $(y^2 + 2y - 3) + (4y^2 - 6y + 5)$
30. $(3n - 2n^2 - 4) + (5n^2 + n - 3)$

Rational Numbers: Addition/Subtraction   **317**

WRITTEN EXERCISES

10. $3g^2 + 5g + 2$
11. $4r^2 + 4r - 2$
12. $5p^2 + 2p + 4$
13. $11x$ 14. $12y$
15. $-9p^2$ 16. $-10b^2$
17. $-6y$ 18. $5n^2$
19. $-9t^2$ 20. $-6s$
21. $8x^2$ 22. $-10ab$
23. $-2y$ 24. $-7b$

QUIZ: SECTIONS 10-4–10-7
After completing this Review, you may wish to administer a quiz covering the same sections. A Quiz is provided in the *Teacher's Edition: Part II.*

REVIEW: SECTIONS 10-4–10-7
1. 6   2. -21   3. -5
4. 11   5. 11   6. 50
7. -14   8. -39   9. 7
10. -24   11. 24
12. -17   13. 39
14. -16   15. 6
16. -12   17. -9°F
18. 50°C   19. -18
20. -26   21. -14
22. 3   23. 3   24. 3
25. -x + 8
26. $-6y^2 - y - 3$
27. $20t + 1$
28. $3p^2 + 4p$
29. $5y^2 - 4y + 2$
30. $3n^2 + 4n - 7$

CHAPTER REVIEW

PART 1: VOCABULARY

For Exercises 1–11, choose from the box at the right below, the word(s) or numerical expression(s) that best corresponds to each description.

1. Commutative Property of Addition ?
2. Distance of a number from zero ?
3. Numbers that are the same distance from zero and are in opposite directions from zero ?
4. Addition Property of Zero ?
5. A monomial or the sum or difference of two or more monomials ?
6. Associative Property of Addition ?
7. Integers less than zero ?
8. Any number that can be written as a fraction ?
9. Addition Property of Opposites ?
10. Set of positive integers, zero, and negative integers ?
11. Integers to the right of zero ?

> integers
> opposites
> $5 + 0 = 5$
> $-2 + (5 + 3) = (-2 + 5) + 3$
> $3 + (-3) = 0$
> $4 + (-7) = -7 + 4$
> absolute value
> positive integers
> rational number
> negative integers
> polynomial

PART 2: SKILLS

Graph the integers on a number line. (Section 10-1)

12. 0; −1; 3; −2   **13.** 4; 2; −5; 1   **14.** −3; −4; 6; 0   **15.** 5; −1; 0; −3

Replace each ? with < or >. (Section 10-1)

16. 4 ? −4   **17.** 6 ? −2   **18.** −4 ? 3   **19.** 0 ? 4
20. 0 ? −1   **21.** −3 ? 3   **22.** −7 ? 6   **23.** −2 ? −4

Write a fraction for each rational number. (Section 10-2)

24. 3   **25.** −5   **26.** $2\frac{1}{2}$ **27.** $-3\frac{1}{3}$ **28.** 4.5   **29.** −2.3

Write the opposite of each number. (Section 10-2)

30. 6   **31.** −6   **32.** 0   **33.** 8   **34.** 12   **35.** −7

318 Chapter 10

Add. (Sections 10-3 and 10-4)

36. $9 + (-2)$
37. $-4 + 12$
38. $16 + (-20)$
39. $-10 + 5$
40. $-4 + (-8)$
41. $-13 + (-9)$
42. $9 + 17$
43. $-8 + 24$
44. $-15 + 7$
45. $-11 + (-22)$
46. $-1 + 6$
47. $-4 + 4$

Subtract. (Section 10-5)

48. $6 - 8$
49. $-3 - 14$
50. $-9 - (-12)$
51. $13 - 8$
52. $15 - (-9)$
53. $3 - 15$
54. $-4 - (-3)$
55. $8 - (-8)$
56. $19 - 32$
57. $-7 - (-20)$
58. $8 - (-5)$
59. $-3 - 14$

Add. (Section 10-6)

60. $3 + (-8) + 12$
61. $-9 + (-4) + 6$
62. $-12 + 15 + (-3)$
63. $-7 + (-11) + 8 + (-13)$
64. $12 + 20 + (-18) + 30$
65. $12 + 16 + 22 + (-14)$

(Section 10-7)

66. $(4r - 7) + (-r + 3)$
67. $(-6x^2 + 5x) + (2x^2 - 8x)$
68. $(-2y^2 - 5y + 7) + (5y^2 + 11y + 3)$
69. $(8b^2 - 5b + 2) + (-2b^2 + 6b - 3)$

PART 3: APPLICATIONS

Write a rational number to represent each word description. (Sections 10-1 and 10-2)

70. 4° below zero
71. $32\frac{1}{2}$ feet below sea level
72. A gain of 2.5 kilograms
73. An increase of $245

Solve each problem.

74. The lowest temperature ever recorded in Los Angeles, California, was 28°F. The lowest temperature ever recorded in Atlanta, Georgia, was −3°F. Replace the ? with < or >: 28° ? −3°. (Section 10-1)

75. A football team lost 8 yards on one play and gained 10 yards on the next play. Use a number line to find the total yardage on the two plays. (Section 10-3)

76. A submarine dove to a level of 320 feet below the ocean's surface. It later descended 150 feet. What was the new depth? (Section 10-4)

77. The lowest temperature ever recorded in Louisiana was −16°F. The highest temperature ever recorded was 114°F. Find the difference. (Section 10-5)

CHAPTER REVIEW

36. 7   37. 8   38. -4
39. -5   40. -12
41. -22   42. 26
43. 16   44. -8
45. -33   46. 5   47. 0
48. -2   49. -17   50. 3
51. 5   52. 24   53. -12
54. -1   55. 16
56. -13   57. 13
58. 13   59. -17
60. 7   61. -7   62. 0
63. -23   64. 44
65. 36   66. $3r - 4$
67. $-4x^2 - 3x$
68. $3y^2 + 6y + 10$
69. $6b^2 + b - 1$
70. $-\frac{4}{1}$ 71. $-\frac{65}{2}$
72. $\frac{5}{2}$ 73. $\frac{245}{1}$ 74. >
75. gain of 2 yd
76. 470 ft
77. 130°F

Rational Numbers: Addition/Subtraction

CHAPTER TEST

Two forms of a chapter test, Form A and Form B, are provided on copying masters in the *Teacher's Edition: Part II.*

1. > 2. > 3. >
4. < 5. 7 6. -7
7. -72 8. 12 9. -9
10. 7 11. -28
12. -29 13. -25
14. 18 15. -13
16. 17 17. -15
18. 22 19. -20
20. -25 21. -13
22. 0 23. 21
24. gain of 3 yd
25. 34°F

CHAPTER TEST

Replace each __?__ with < or >.

1. 5 __?__ −2
2. 2 __?__ −2
3. 0 __?__ −6
4. −9 __?__ −3

Add.

5. 21 + (−14)
6. −16 + 9
7. −43 + (−29)
8. −48 + 60
9. 9 + (−18)
10. −28 + 35
11. −11 + (−17)
12. 27 + (−56)

Subtract.

13. −4 − 21
14. −14 − (−32)
15. 9 − 22
16. 5 − (−12)
17. 20 − 35
18. 14 − (−8)
19. −8 − 12
20. −43 − (−18)

For Exercises 21–23, add.

21. −5 + 4 + (−12)
22. 8 + (−3) + 2 + (−7)
23. 12 + (−9) + 18

24. A football team lost 4 yards on one play and gained 7 yards on the next play. Find the total yardage for the two plays.

25. The actual temperature is 23°F. The wind chill temperature is −11°F. How much colder than the actual temperature is the wind chill temperature?

ANSWERS TO REVIEW CAPSULES

Page 297 1. $\frac{11}{5}$ 2. $\frac{17}{3}$ 3. $\frac{79}{9}$ 4. $\frac{19}{6}$ 5. $\frac{21}{2}$ 6. $\frac{23}{12}$ 7. $\frac{25}{4}$ 8. $\frac{7}{5}$ 9. $\frac{31}{10}$ 10. $\frac{63}{10}$ 11. $\frac{29}{7}$ 12. $\frac{29}{9}$

Page 299 1. The points on the number line from left to right in the following order: −1; 0; 1; 3 2. The points on the number line from left to right in the following order: −5; −3; 1; 2 3. The points on the number line from left to right in the following order: −2; 0; 4; 6 4. The points on the number line from left to right in the following order: −5; −4; 5; 7 5. > 6. > 7. < 8. > 9. < 10. < 11. > 12. >

Page 303 1. > 2. > 3. < 4. < 5. > 6. < 7. > 8. < 9. −6 10. −15 11. 8 12. −3 13. 0 14. 7

Page 306 1. −12 2. −4 3. 28 4. 15 5. −3.7 6. $1\frac{1}{3}$

Page 309 1. 10 2. 73 3. 34 4. 69 5. 257 6. 178 7. −17 8. −4 9. 23 10. −29 11. 8 12. −32 13. 0 14. 0

Page 315 1. $7a$ 2. $3m$ 3. $7x^2$ 4. $14b^2$ 5. $6a + 5$ 6. $8x + 9$ 7. $5t^2 + 12$ 8. $12n^2 + 6$ 9. 7 10. −7 11. 3 12. 9 13. −4 14. −10 15. 2 16. −4

320   Chapter 10

ENRICHMENT

Clock Arithmetic

Think of a 7-hour clock such as that shown at the right. Addition on this clock is the same as on a 12-hour clock. That is,

$4 + 5 = 2$ ◀ Start at 4. Move 5 units in a clockwise direction.

You learned in algebra that the sum of opposites is 0.

In Algebra: $7 + (-7) = 0$ $(-1) + 1 = 0$ $0 + 0 = 0$

On this 7-hour clock, the sum of opposites is also 0.

$1 + 6 = 0$ $2 + 5 = 0$ $3 + 4 = 0$ $0 + 0 = 0$ ◀ **Opposites:** 1 and 6; 2 and 5; 3 and 4; 0 and 0

NOTE: On the 7-hour clock, opposites are <u>directly opposite</u> each other.

To subtract in algebra, you add the opposite of the number you are subtracting. To subtract on a 7-hour clock, you follow a similar procedure.

EXAMPLES Evaluate on the 7-hour clock: **a.** $6 + 6$ **b.** $1 - 3$

Solutions: **a.** Start at 6. Move 6 units in a clockwise direction. $6 + 6 = 5$

b. $1 - 3 = 1 + 4$ ◀ Add 4, the opposite of 3.
      $= 5$

EXERCISES

Evaluate. Use a 7-hour clock.

1. $5 + 3$ 2. $5 + 5$ 3. $4 + 2$ 4. $6 + 1$ 5. $2 + 0$ 6. $6 + 5$
7. $1 - 5$ 8. $3 - 6$ 9. $3 - 4$ 10. $2 - 2$ 11. $0 - 1$ 12. $6 - 4$

Multiplication on a 7-hour clock can be defined as "repeated addition." That is, $4 \times 3 = 4 + 4 + 4 = 1 + 4 = 5$. Multiply.

13. 4×2 14. 3×6 15. 3×5 16. 4×5 17. 6×6 18. 3×3

Rational Numbers: Addition/Subtraction

ENRICHMENT

You may wish to use this lesson for students who performed well on the formal Chapter Test.

1. 1   2. 3   3. 6
4. 0   5. 2   6. 4
7. 3   8. 4   9. 6
10. 0   11. 6   12. 2
13. 1   14. 4   15. 1
16. 6   17. 1   18. 2

ADDITIONAL PRACTICE

You may wish to use all or some of these exercises, depending on how well students performed on the formal Chapter Test.

1. ●—●—●—● −6; −4; 0; 2
2. ●—●—●—● −5; −3; 1; 3
3. ●—●—●—● −1; 2; 3; 4; 5
4. >   5. >   6. <
7. <   8. $-\frac{4}{1}$ 9. $\frac{6}{10}$
10. $\frac{72}{10}$ 11. $-\frac{35}{10}$
12. $-\frac{5}{2}$ 13. $\frac{10}{3}$
14. 6.1   15. −4
16. 0   17. 7
18. $-3\frac{1}{3}$ 19. 2.5
20. 8   21. −4
22. −4   23. −6
24. 6   25. 3   26. 17
27. 32   28. 1   29. 0
30. 4   31. 9
32. −48   33. −78
34. 10.5   35. −7.9
36. 1   37. 2   38. −9
39. −14   40. 11
41. 9   42. $-\frac{1}{2}$
43. $\frac{1}{8}$ 44. $1\frac{1}{4}$ 45. $\frac{1}{2}$
46. −3   47. 0   48. −1
49. −1   50. 14°C
51. gain of 2 yd

ADDITIONAL PRACTICE

SKILLS   For Exercises 1–3:
 a. *Graph the integers on a number line.*
 b. *Write the integers in order from least to greatest.*
    (Pages 294–296)

1. 0; −4; −6; 2   2. −5; 1; 3; −3   3. 4; 5; 3; 2; −1

Replace each __?__ with > or <. (Pages 294–296)

4. 0 __?__ −6   5. 3 __?__ −1   6. −10 __?__ −2   7. −13 __?__ −1

Write a fraction for each rational number. (Pages 297–299)

8. −4   9. 0.6   10. 7.2   11. −3.5   12. $-2\frac{1}{2}$ 13. $3\frac{1}{3}$

Write the opposite of each number. (Pages 297–299)

14. −6.1   15. 4   16. 0   17. −7   18. $3\frac{1}{3}$ 19. −2.5

Use a number line to add. (Pages 300–302)

20. 5 + 3   21. −8 + 4   22. 4 + (−8)   23. −4 + (−2)

Evaluate. (Pages 304–306)

24. |6| 25. |−3| 26. |−17| 27. |32| 28. |1| 29. |0|

Perform the indicated operations. (Pages 304–314)

30. −27 + 31   31. −54 + 63   32. −12 + (−36)   33. −22 + (−56)
34. 18.4 + (−7.9)   35. 19.2 + (−27.1)   36. −3 − (−4)   37. −4 − (−6)
38. −2 − 7   39. −5 − 9   40. 7 − (−4)   41. 5 − (−4)
42. $-\frac{2}{3} + \frac{1}{6}$ 43. $-\frac{1}{4} + \frac{3}{8}$ 44. $\frac{1}{2} - \left(-\frac{3}{4}\right)$ 45. $\frac{1}{3} - \left(-\frac{1}{6}\right)$
46. (−5) + (−3) + 6 + (−1)   47. 8 + (−3) + 4 + (−9)
48. −6 + 17 + (−19) + 7   49. (−1) + (−6) + (−3) + 9

APPLICATIONS

50. The high temperature for the day was 8°C and the low temperature was −6°C. How much warmer was the high temperature than the low temperature? (Pages 308–310)

51. A football team lost 7 yards on first down, gained 13 yards on second down, and lost 4 yards on third down. Find the total yardage for the three downs. (Pages 312–314)

322   Chapter 10

CHAPTER

11 Equations: Addition/Subtraction

SECTIONS
11-1 Equations: Addition/Subtraction
11-2 Like Terms/More Than One Operation
11-3 Variable on Both Sides
11-4 Problem Solving and Applications: Words to Symbols: Addition and Subtraction
11-5 Inequalities on the Number Line
11-6 Using Addition and Subtraction to Solve Inequalities

FEATURES
Calculator Application: *Checking Equations*
Calculator Application: *Checking Inequalities*
Consumer Application: *Banking*
Computer Application: *Adding Fractions*
Enrichment: *Compound Inequalities*

Teaching Suggestions
p. M-24

QUICK QUIZ

Add or subtract.

1. −17 + 23   Ans: 6
2. −21 + (−8)   Ans: −29
3. 26 + (−32)   Ans: −6
4. −23 −12   Ans: −35
5. −40 (−10)
    Ans: 30

ADDITIONAL EXAMPLES

Example 1
Solve and check.
1. x −82 = 36   Ans: 118
2. −2.9 = y −1.7
    Ans: −1.2

SELF-CHECK
1. −4   2. 21   3. −3

Example 2
Solve and check.
1. a + 43 = −3.2
    Ans: −46.2
2. b + $5\frac{1}{3}$ = $8\frac{2}{3}$
    Ans: $3\frac{1}{3}$

SELF-CHECK
4. −15   5. −4   6. −17

11-1 Equations: Addition/Subtraction

You used the following properties when you solved equations in Chapter 7.

> **Addition Property for Equations**
> Adding the same number to each side of an equation forms an equivalent equation.
>
> $p - 8 = -10$
> $p - 8 + 8 = -10 + 8$
> $p = -2$

EXAMPLE 1 Solve and check: $x - 6 = -3$

Solution: To get x alone, add 6 to each side.

$x - 6 = -3$
$x - 6 + 6 = -3 + 6$
$x = 3$

Check: $x - 6 = -3$
$3 - 6$
$3 + (-6)$
-3

Self-Check *Solve and check.*

1. $t - 4 = -8$ 2. $a - 13 = 8$ 3. $-15 = b - 12$

> **Subtraction Property for Equations**
> Subtracting the same number from each side of an equation forms an equivalent equation.
>
> $t + 4 = -15$
> $t + 4 - 4 = -15 - 4$
> $t = -15 + (-4)$
> $t = -19$

EXAMPLE 2 Solve and check: $y + 5 = -8$

Solution:
$y + 5 = -8$
$y + 5 - 5 = -8 - 5$ ◀ −8 − 5 means −8 + (−5)
$y = -8 + (-5)$
$y = -13$

Check: $y + 5 = -8$
$-13 + 5$
-8

Self-Check *Solve and check.*

4. $p + 6 = -9$ 5. $m + 8 = 4$ 6. $-12 = r + 5$

324 Chapter 11

CLASSROOM EXERCISES

Write the equivalent equations formed as the first step in solving each equation. (Examples 1 and 2)

1. $y + 18 = 7$
2. $x + 24 = 20$
3. $a - 5 = 22$
4. $q - 18 = -30$
5. $-15 = 23 + b$
6. $-28 = y + 39$

For Exercises 7–18 solve and check. (Example 1)

7. $a - 4 = -5$
8. $r - 8 = -20$
9. $q - 9 = 20$
10. $b - 4 = 10$
11. $-8 = y - 6$
12. $20 = m - 14$

(Example 2)

13. $p + 4 = -8$
14. $c + 5 = -19$
15. $n + 1 = 4$
16. $x + 3 = 8$
17. $0 = d + 7$
18. $-6 = m + 22$

WRITTEN EXERCISES

Goal: To solve equations by using the Addition and Subtraction Properties for Equations

For Exercises 1–39, solve and check each equation. (Example 1)

1. $x - 10 = 16$
2. $y - 42 = 42$
3. $a - 6 = -4$
4. $c - 10 = -15$
5. $d - 3 = -16$
6. $b - 8 = -11$
7. $x - 14 = 25$
8. $-15 = d - 24$
9. $13 = k - 18$
10. $p - 12 = -15$
11. $-6.2 = q - 4.4$
12. $t - 5.5 = 9.8$

(Example 2)

13. $h + 2 = 8$
14. $k + 8 = 2$
15. $n + 5 = -4$
16. $b + 15 = -32$
17. $t + 9 = -12$
18. $w + 8 = 3$
19. $p + 29 = 16$
20. $t + 8 = -19$
21. $-23 = m + 12$
22. $27 = q + 35$
23. $b + 13.7 = 19.6$
24. $5.1 = c + 9.4$

MIXED PRACTICE

25. $a + 48 = 23$
26. $p + 38 = -5$
27. $x - 47 = -29$
28. $-10 = g + 4$
29. $7 = t - 15$
30. $12 = y + 14$
31. $12 + a = -90$
32. $t - 15 = -50$
33. $t - 16 = -16$
34. $-19.5 = 8.7 + c$
35. $n + 22.9 = 15.4$
36. $a - 11.9 = -27.4$
37. $33.4 = k + 22.9$
38. $y - 5\frac{3}{4} = -3\frac{1}{4}$
39. $m + 2\frac{3}{5} = 1\frac{1}{5}$

Equations: Addition/Subtraction

CLASSROOM EXERCISES

1. y + 18 −18 = 7 −18
2. x + 24 −24 = 20 −24
3. a −5 + 5 = 22 + 5
4. q −18 + 18 = −30 + 18
5. -15 −23 = 23 + b −23
6. -28 −39 = y + 39 −39
7. -1   8. -12   9. 29
10. 14   11. -2   12. 34
13. -12   14. -24
15. 3   16. 5   17. -7
18. -28

ASSIGNMENT GUIDE
BASIC
Day 1 p. 325: 1–33 odd
Day 2 p. 325: 2–34 even
AVERAGE
p. 325: 1–39 odd
ABOVE AVERAGE
p. 325: 2–38 even

WRITTEN EXERCISES
1. 26   2. 84   3. 2
4. -5   5. -13   6. -3
7. 39   8. 9   9. 31
10. -3   11. -1.8
12. 15.3   13. 6
14. -6   15. -9
16. -47   17. -21
18. -5   19. -13
20. -27   21. -35
22. -8   23. 5.9
24. -4.3   25. -25
26. -43   27. 18

See page 328 for the answers to Exercises 28–39.

Teaching Suggestions
p. M-24

REVIEW CAPSULE
This Review Capsule reviews prior-taught skills used in Section 11-2. The reference is to the pages where the skills were taught.

QUICK QUIZ
Combine like terms.
1. 3a + 9a   Ans: 12a
2. 7b - b   Ans: 6b
3. -7y + y   Ans: -6y
4. -5a - 6a   Ans: -11a
5. x - 4x + 3x Ans: 0

ADDITIONAL EXAMPLES
Example 1
Combine like terms.
1. 8x - 7 - x
   Ans: 7x - 7
2. -y + 13 + 9y
   Ans: 8y + 13

SELF-CHECK
1. 4   2. 5

Example 2
Solve and check.
1. -6x - 15 + 7x = 23 - 37   Ans: x = 1
2. 17.9 - 10.4 = 9.1y + 3.3 - 8.1y
   Ans: y = 4.2

REVIEW CAPSULE FOR SECTION 11-2

Combine like terms. (Pages 76–78)   The answers are on page 346.

1. $3x + 2x$
2. $7p + 6p$
3. $8q - 6q$
4. $4t - 2t$
5. $8x + x$
6. $9b + b$
7. $2m - m$
8. $5q - 4q$

Solve each equation. (Pages 200–202)

9. $8t = 352$
10. $15 = 60z$
11. $1.2q = 1.32$
12. $16 = 0.04r$

11-2 Like Terms/More Than One Operation

When solving an equation, combine like terms on each side of the equation before applying the Addition and Subtraction Properties for Equations. Then divide each side by the same number (if necessary) as you did in Chapter 7 (see pages 200–202).

EXAMPLE 1 Solve and check. $5x + 3x = -8 + 24$

Solution: ① Combine like terms. $8x = 16$

② Divide each side by 8. $\frac{8x}{8} = \frac{16}{8}, \; x = 2$

Check: $5x + 3x = -8 + 24$
$5(2) + 3(2) = 16$
$10 + 6$
16

Self-Check *Solve and check.*

1. $2x + 4x = -5 + 29$ 2. $-x + 4x = 8 + 7$

EXAMPLE 2 Solve and check: $2x - 4 + 3x = -3 + 14$

Solution: ① Combine like terms. $2x - 4 + 3x = -3 + 14$
$5x - 4 = 11$

② To get 5x alone, add 4 to each side. $5x - 4 + 4 = 11 + 4$
$5x = 15$

③ Divide each side by 5. $\frac{5x}{5} = \frac{15}{5}, \; x = 3$ ◀ The check is left for you

326   Chapter 11

Self-Check Solve and check.

3. $x - 3 + 2x = -2 + 17$
4. $5x + 6 - x = -7 + 25$

CLASSROOM EXERCISES

Combine like terms on both sides of each equation. Do <u>not</u> solve the equation. (Example 1, step 1)

1. $6x + 2x = -2 + 34$
2. $y + 2y = -7 + 10$
3. $-c + 2c = 25 - 6$
4. $4n - n = -4 + 13$

(Example 2, step 1)

5. $3x - 5 + 4x = -3 + 19$
6. $p - 2 + 3p = -4 + 18$
7. $-a + 4 + 3a = -5 + 11$
8. $4g + 3 - g = -6 + 36$

Solve and check. (Example 1)

9. $6x + 3x = -4 + 58$
10. $y + 2y = -12 + 21$
11. $5h - h = 34 - 2$
12. $8k - k = 54 - 5$

(Example 2)

13. $4x - 7 + 2x = -2 + 37$
14. $q - 5 + 4q = -11 + 26$
15. $-b + 3 + 2b = -1 + 8$
16. $7a + 8 - a = -9 + 35$

WRITTEN EXERCISES

Goal: To solve equations containing like terms

Solve and check. (Example 1)

1. $4a + 5a = 37 + 44$
2. $8c + 9c = -12 + 80$
3. $8e + 12e = -15 + 95$
4. $2x + 11x = 144 + 25$
5. $5p + 6p = 110 - 11$
6. $2y + y = -5 + 32$
7. $11m - m = -4 + 34$
8. $9v - v = -25 + 89$
9. $-n + 8n = -7 + 168$
10. $-q + 2q = -9 + 36$
11. $-b + 3b = -17 + 21$
12. $5.2g + 6.3g = -2 + 71$
13. $2\frac{3}{4}f - \frac{1}{4}f = -6 + 16$
14. $4\frac{1}{2}d - 1\frac{1}{2}d = 107 - 17$

Equations: Addition/Subtraction **327**

SELF-CHECK
<u>3</u>. 6   <u>4</u>. 3

CLASSROOM EXERCISES
<u>1</u>. 8x = 32   <u>2</u>. 3y = 3
<u>3</u>. c = 19   <u>4</u>. 3n = 9
<u>5</u>. 7x - 5 = 16
<u>6</u>. 4p - 2 = 14
<u>7</u>. 2a + 4 = 6
<u>8</u>. 3g + 3 = 30   <u>9</u>. 6
<u>10</u>. 3   <u>11</u>. 8   <u>12</u>. 7
<u>13</u>. 7   <u>14</u>. 4   <u>15</u>. 4
<u>16</u>. 3

ASSIGNMENT GUIDE
BASIC
Day 1 pp. 327-328:
1-35 odd
Day 2 pp. 327-328:
2-36 even
AVERAGE
pp. 327-328: 1-29 odd,
30-36
ABOVE AVERAGE
pp. 327-328: 3, 6, 9,
..., 48

WRITTEN EXERCISES
<u>1</u>. 9   <u>2</u>. 4   <u>3</u>. 4
<u>4</u>. 13   <u>5</u>. 9   <u>6</u>. 9
<u>7</u>. 3   <u>8</u>. 8   <u>9</u>. 23
<u>10</u>. 27   <u>11</u>. 2   <u>12</u>. 6
<u>13</u>. 4   <u>14</u>. 30

WRITTEN EXERCISES

15. 6   16. 2   17. 2
18. 7   19. 5   20. 7
21. 6   22. 12   23. 19
24. 11   25. 33   26. 7
27. $5\frac{3}{5}$ 28. $2\frac{3}{8}$ 29. $3\frac{1}{2}$
30. $12\frac{3}{4}$ 31. 13   32. $\frac{1}{3}$
33. $6\frac{1}{2}$ 34. 8   35. 6
36. 6   37. 8   38. $1\frac{3}{4}$
39. 10   40. 7   41. 5
42. 3   43. 10   44. 4
45. 4   46. 3   47. 5.5
48. 11.25

WRITTEN EXERCISES p. 325

28. −14   29. 22
30. −2   31. −102
32. −35   33. 0
34. −28.2   35. −7.5
36. −15.5   37. 10.5
38. $2\frac{1}{2}$ 39. $-1\frac{2}{5}$

Solve and check. (Example 2)

15. $3x - 8 + x = -3 + 19$
16. $w - 5 + 4w = -10 + 15$
17. $2h + 8 + 3h = -9 + 27$
18. $k + 6 + 6k = -13 + 68$
19. $4z - 12 - z = -6 + 9$
20. $13y - 9 - y = -10 + 85$
21. $7m - 7 + 4m = -15 + 74$
22. $3n - 14 + 5n = -23 + 105$
23. $-p + 13 + 3p = -7 + 58$
24. $-c + 9 + 10c = -17 + 125$
25. $-k - 21 + 6k = -11 + 155$
26. $12a - 52 + 13a = -12 + 135$
27. $3\frac{1}{2}g - 8 + 1\frac{1}{2}g = -3\frac{3}{5} + 23\frac{3}{5}$
28. $2\frac{1}{3}x - 5 + 1\frac{2}{3}x = -3\frac{1}{4} + 7\frac{3}{4}$

MIXED PRACTICE

29. $-q + 3q = -8 + 15$
30. $3y - 8 + y = -7 + 50$
31. $3n - 14 + 5n = -23 + 113$
32. $4x + 2x = -5 + 43$
33. $4\frac{1}{2}a + 5\frac{1}{2}a = -6 + 71$
34. $\frac{3}{4}k - 8 + 6\frac{1}{4}k = -12 + 60$
35. $3x + 5x = 17 + 31$
36. $t + 3t = 18 + 6$
37. $11\frac{1}{8}m - 5\frac{1}{8} - 3\frac{1}{8}m = -5\frac{1}{8} + 64$
38. $-b + 3b = -17\frac{1}{2} + 21$
39. $8.5v - 2.1v = -25 + 89$
40. $12c - 52.5 + 13c = -11.5 + 134$
41. $4m - 7 + 5m = 29 + 9$
42. $5z - 3 + 2z = 11 + 7$
43. $6.5x - 9.5 - 1.3x = 32 + 10.5$
44. $-2.3y + 7.2 + 6.8y = 25.2$
45. $-13.1 + 3.2n + 2.8n = 3.2 + 7.7$
46. $-11.4 - 3.7v + 6.7v = 3.1 - 5.5$
47. $y + 3 + 3y = 14 + 11$
48. $5w - 11 - w = 26 + 8$

REVIEW CAPSULE FOR SECTION 11-3

Combine like terms. (Pages 79–80)

1. $4d + 2d + 6$
2. $3p - 2p + 8$
3. $4x - 6 - 2x$
4. $9 + 6x - 2x$
5. $4 + 6 + 2r + 3r$
6. $7 + 2b - 5 + 3b$
7. $2 + 3 + 5y - y$
8. $8c + 4 - 3c + 9$
9. $-x + 3 + 4x - 5$
10. $2x + 3 - 5x + 8$
11. $n - 7 - 14n + 2$
12. $3y + 8 + 9y - 5$

Solve. (Pages 324–325)

13. $x + 3 = 12$
14. $y - 4 = 12$
15. $a - 4 = -6$
16. $b + 7 = -9$
17. $m + 6 = 3$
18. $n + 15 = 4$
19. $p - 5 = -1$
20. $q - 9 = -3$
21. $8 + x = 12$
22. $7 + 2y = 19$
23. $-5 + 3g = -20$
24. $-12 + 2t = -18$
25. $3\frac{1}{2} + 2d = 7\frac{1}{2}$
26. $5\frac{1}{5} + 3x = 11\frac{1}{5}$
27. $n - 4\frac{2}{3} = -5\frac{2}{3}$

The answers are on page 346.

328   Chapter 11

11-3 Variable on Both Sides

Some equations have a variable on both sides. Use the Addition and Subtraction Properties for Equations to get the variable <u>alone on one side</u>.

EXAMPLE 1 Solve and check: $y = 14 - y$

Solution:
$\boxed{1}$ To get rid of $-y$ on the right side, add y to both sides.

$y = 14 - y$
$y + y = 14 - y + y$
$2y = 14$

$\boxed{2}$ To get y alone, divide both sides by 2.

$\dfrac{2y}{2} = \dfrac{14}{2}$
$y = 7$

Check: $y = 14 - y$
  7 | $14 - 7$
  7 | 7

Self-Check *Solve and check.*

1. $n = 10 - n$ **2.** $2m = 9 - m$ **3.** $4b = 60 - b$

Combine like terms on each side of an equation before applying the Addition and Subtraction Properties for Equations.

EXAMPLE 2 Solve and check: $2x + 9 = 38 + x$

Solution:
$\boxed{1}$ To get rid of x on the right side, subtract x from both sides.

$2x + 9 = 38 + x$
$2x - x + 9 = 38 + x - x$
$x + 9 = 38$

$\boxed{2}$ To get x alone, subtract 9 from both sides.

$x + 9 - 9 = 38 - 9$
$x = \mathbf{29}$

Check: $2x + 9 = 38 + x$
  $2(29) + 9$ | $38 + 29$
  $58 + 9$ | $38 + 29$
     67 | 67

Self-Check *Solve and check:* **4.** $3e + 7 = 11 + 2e$ **5.** $6 + 2k = k + 10$

Equations: Addition/Subtraction **329**

Teaching Suggestions p. M-24

QUICK QUIZ

Combine like terms.
1. $3x - 8 + 7x$
   Ans: $10x - 8$
2. $12y - 15y + 8$
   Ans: $-3y + 8$

Solve and check.
3. $-13 = y - 10$
   Ans: -3
4. $-7 = a + 2$ Ans: -9
5. $b + 3 = 3$ Ans: 0

ADDITIONAL EXAMPLES

Example 1
Solve and check.
1. $15x - 4 = 14x + 7$
   Ans: 11
2. $\dfrac{3}{8}x - 6 = -\dfrac{5}{8}x + 10$
   Ans: 16

SELF-CHECK
1. 5   2. 3   3. 12

Example
Solve and check.
1. $5x - x - 7 = 3x + 8 - 5$ Ans: 10
2. $3a + 4\dfrac{1}{4} - a = a + 7\dfrac{1}{2} - 3$ Ans: $\dfrac{1}{4}$

SELF-CHECK
4. 4   5. 4

329

CLASSROOM EXERCISES
1. v; v; 2v; 3
2. a; a; 4a; 5
3. d; d; 6d; 12
4. 3g; 3g; g; 3
5. 5h; 5h; h; 12; 7

ASSIGNMENT GUIDE
BASIC
Day 1 p. 330: 1-31 odd
Day 2 p. 330: 2-32 even
AVERAGE
Day 1 p. 330: 1-20
Day 2 p. 330: 21-32
ABOVE AVERAGE
p. 330: 1-31 odd

WRITTEN EXERCISES
1. 10 2. 14 3. 7
4. 9 5. 16 6. 9
7. $7\frac{1}{2}$ 8. $11\frac{1}{2}$ 9. $3\frac{1}{2}$
10. $9\frac{2}{3}$ 11. 50
12. 27 13. 7 14. 2
15. 10 16. 12 17. 8
18. 5 19. 7 20. 9
21. 8 22. 4 23. 5
24. 5 25. 3 26. 6
27. 2 28. 11 29. 30
30. 14 31. 4 32. 2

CLASSROOM EXERCISES

Complete. (Example 1)

1. $v = 6 - v$
$v + \underline{?} = 6 - v + \underline{?}$
$\underline{?} = 6$
$v = \underline{?}$

2. $3a = 20 - a$
$3a + \underline{?} = 20 - a + \underline{?}$
$\underline{?} = 20$
$a = \underline{?}$

3. $5d = 72 - d$
$5d + \underline{?} = 72 - d + \underline{?}$
$\underline{?} = 72$
$d = \underline{?}$

(Example 2)

4. $4g + 8 = 11 + 3g$
$4g - \underline{?} + 8 = 11 + 3g - \underline{?}$
$\underline{?} + 8 = 11$
$g = \underline{?}$

5. $6h + 5 = 12 + 5h$
$6h - \underline{?} + 5 = 12 + 5h - \underline{?}$
$\underline{?} + 5 = \underline{?}$
$h = \underline{?}$

WRITTEN EXERCISES

Goal: To solve an equation with the variable on both sides

For Exercises 1–32, solve and check. (Example 1)

1. $a = 20 - a$
2. $b = 28 - b$
3. $2c = 21 - c$
4. $3d = 36 - d$
5. $4e = 80 - e$
6. $5f = 54 - f$
7. $g = 15 - g$
8. $h = 23 - h$
9. $7k = 28 - k$
10. $2m = 29 - m$
11. $3.6n = 100 + 1.6n$
12. $6.3p = 81 + 3.3p$

(Example 2)

13. $5q + 5 = 12 + 4q$
14. $3r + 3 = 5 + 2r$
15. $2s - 3 = 7 + s$
16. $7t - 7 = 5 + 6t$
17. $4u - 12 = -4 + 3u$
18. $10v - 15 = -10 + 9v$
19. $6.5w + 10 = 17 + 5.5w$
20. $8.3x - 21 = -12 + 7.3x$
21. $3y - 5\frac{1}{2} = 2\frac{1}{2} + 2y$
22. $15z + 3\frac{1}{8} = 7\frac{1}{8} + 14z$
23. $5\frac{1}{2}a + 14\frac{1}{4} = 19\frac{1}{4} + 4\frac{1}{2}a$
24. $\frac{13}{3}b - \frac{11}{3} = \frac{4}{3} + \frac{10}{3}b$

MIXED PRACTICE

25. $15c = 48 - c$
26. $12e + 10 = 16 + 11e$
27. $3f + 6 = 8 + 2f$
28. $11d = 132 - d$
29. $4.7g = 90 + 1.7g$
30. $3.8h = 28 + 1.8h$
31. $13k - 4\frac{1}{3} = 3\frac{2}{3} + 11k$
32. $6m - 5\frac{1}{4} = 2\frac{3}{4} + 2m$

330 Chapter 11

CHECKING EQUATIONS

You can use a calculator with a $\boxed{^+/_-}$ key to check solutions of equations. First, use the solution to evaluate the left side of the equation. Store this value in memory. Then use the solution to evaluate the right side of the equation. Compare with the value in memory.

EXAMPLE Check the solution $x = -16$ for the equation $8 - 2x = 24 - x$.

Left side: 8 [−] 2 [×] 1 6 [+/−] [=] [M+]   → 40.

Right side: 2 4 [−] 1 6 [+/−] [=]   → 40.

EXERCISES Check the given solution to each equation.

1. $-14m - 35 = 46 - 13m$; $m = -3$
2. $-12v + 40 = -13v + 112$; $v = 72$
3. $4.7x - 1.1 = 5.7x - 2.5$; $x = 1.5$
4. $19t - 80 = -16t - 1025$; $t = 27$

REVIEW: SECTIONS 11-1 — 11-3

Solve and check. (Section 11-1)

1. $n + 5 = 12$
2. $p - 7 = 3$
3. $q + 9 = -15$
4. $r - 3 = -7$
5. $s - 6 = -4$
6. $t + 10 = 3$
7. $a - 12 = -15$
8. $d - 9 = -12$

Solve and check. (Section 11-2)

9. $3p + 8 + 2p = 5 + 8$
10. $8a - 14 + 7a = -4 + 65$
11. $22 - 5 = -7 - y + 5y$
12. $-c + 14 + 7c = -8 + 40$

Solve and check. (Section 11-3)

13. $p = 6 - p$
14. $2a = 27 - a$
15. $5v + 9 = 17 + 4v$
16. $3r - 15 = -10 + 2r$

REVIEW CAPSULE FOR SECTION 11-4

Solve and check. (Pages 324–325)

1. $n + 6 = 18$
2. $t + 9 = 15$
3. $b + 14 = 27$
4. $r + 21 = 55$
5. $m - 9 = 3$
6. $n - 4 = 6$
7. $a - 17 = 8$
8. $c - 32 = 15$

The answers are on page 346.

Equations: Addition/Subtraction **331**

CONSUMER APPLICATIONS

This feature illustrates the use of a calculator to finding the yearly percentage rate of interest of a bank loan. Because it applies the skills for evaluating an expression by more than one operation, it can be used with Section 11-2 (pages 326-328).

1. 14.7%
2. 11.3%
3. 17.2%
4. 12.7%
5. 15.3%

CONSUMER APPLICATION Banking

When you borrow money from a bank, the bank charges you interest for the use of the money.

You can estimate the yearly percentage rate of interest by using this formula.

$$r = \frac{2Mi}{B(n+1)}$$

r: yearly interest rate
i: interest charge
B: amount of loan
M: number of payment periods per year
n: number of payments

The formula can only be used when the same number of payments are made each year.

EXAMPLE A $5000 loan was repaid in 36 monthly payments. The interest charge was $800. Estimate to the nearest tenth the yearly percentage rate of interest.

SOLUTION $r = \dfrac{2Mi}{B(n+1)}$ $M = 12; i = \$800;$
$B = \$5000; n = 36$

$r = \dfrac{2(12)(800)}{5000(36+1)}$ Use a calculator.

2 × 12 × 800 ÷ 5000 ÷ 37 =   `0.1037837`

The annual percentage rate of interest is about **10.4%**.

Exercises

Complete the table. Round each answer to the nearest tenth.

	Loan Amount	Interest Charge	Payment Periods per year	Number of Payments	Yearly Percentage Rate of Interest
1.	$ 700	$ 30	12	6	?
2.	$ 900	$ 55	12	12	?
3.	$1100	$150	12	18	?
4.	$2500	$410	12	30	?
5.	$3400	$800	12	36	?

Consumer Application

COMPUTER APPLICATIONS Adding Fractions

You can find the sum of two fractions by using the following program.

Program:
```
100 INPUT A, B, C, D
110 IF A*B*C*D <> 0 THEN 140
120 PRINT "NO ZERO VALUES.  TRY AGAIN."
130 GOTO 100
140 LET N = A*D + B*C
150 LET D1 = B*D
160 IF N = 0 THEN 290
170 LET N1 = ABS(N)
180 LET D2 = ABS(D1)
190 IF N1 > D2 THEN 220
200 LET S = N1
210 GOTO 230
220 LET S = D2
230 FOR Y = S TO 1 STEP -1
240 IF N1/Y <> INT(N1/Y) THEN 260
250 IF D2/Y = INT(D2/Y) THEN 270
260 NEXT Y
270 LET N = N/Y
280 LET D1 = D1/Y
290 PRINT "(";A;"/";B") + (";C;"/";D;") =;
300 IF ABS(D1) = 1 THEN 360
310 IF N * D1 < 0 THEN 340
320 PRINT ABS(N);"/";ABS(D1)
330 GOTO 370
340 PRINT "-";ABS(N);"/";ABS(D1)
350 GOTO 370
360 PRINT N * D1
370 PRINT
380 END
```

This program is based on the following.

In Algebra

$$\frac{a}{b} + \frac{c}{d} = \frac{ad}{bd} + \frac{bc}{bd} = \frac{ad + bc}{bd}$$

In Arithmetic

$$\frac{2}{3} + \frac{5}{7} = \frac{2 \cdot 7}{3 \cdot 7} + \frac{5 \cdot 3}{3 \cdot 7} = \frac{(2 \cdot 7) + (5 \cdot 3)}{3 \cdot 7}$$

Exercises Use the program to find each sum.

1. $\frac{1}{2} + \frac{1}{5}$
2. $\frac{2}{3} + \frac{7}{8}$
3. $\frac{4}{9} + \left(-\frac{1}{3}\right)$
4. $\frac{5}{16} + \frac{4}{5}$
5. $\left(-\frac{5}{6}\right) + \frac{1}{3}$
6. $\frac{3}{8} + \left(-\frac{1}{10}\right)$
7. $\left(-\frac{9}{5}\right) + \left(-\frac{3}{5}\right)$
8. $\left(-\frac{7}{2}\right) + \left(-\frac{12}{15}\right)$

Computer Application

COMPUTER APPLICATIONS
This feature illustrates the use of a computer program for finding the sum of two fractions. Since it applies the skill of adding rational numbers, it can be used after Section 10-4 (pages 304-307) is taught.

1. $\frac{7}{10}$
2. $\frac{37}{24}$
3. $\frac{1}{9}$
4. $\frac{89}{80}$
5. $-\frac{3}{6}$
6. $\frac{11}{40}$
7. $-\frac{12}{5}$
8. $-\frac{129}{30}$

Teaching Suggestions
p. M-24

QUICK QUIZ

Evaluate.

<u>1</u>. x + 12 when x = 8
   Ans: 20

<u>2</u>. -17 + y when y = 10
   Ans: -7

<u>3</u>. a - 13 when a = -13
   Ans: -26

<u>4</u>. b + 27 when b = -40
   Ans: -13

Solve and check.

<u>5</u>. h + 7 = -8
   Ans: -15

<u>6</u>. 21 = k - 8   Ans: 29

<u>7</u>. a - 3 = -9
   Ans: -6

<u>8</u>. 16 = b - 20
   Ans: 36

<u>9</u>. x + 7.6 = 1.8
   Ans: 5.8

<u>10</u>. y + $2\frac{1}{3}$ = $7\frac{1}{2}$
    Ans: $5\frac{1}{6}$

PROBLEM SOLVING AND APPLICATIONS

11-4 Words to Symbols: Addition and Subtraction

Jenna's basketball coach told her that she had scored 12 more points this season than last season. Jenna thought:

Points last season: p

12 more points than
last season: $p + 12$

The following table shows some <u>word expressions</u> with the corresponding <u>algebraic expressions</u>.

TABLE

Operation	Word Expression	Algebraic Expression
Addition	Twelve **more than** the number of points, p, scored	$p + 12$
	The **sum** of the distance, d, and 20	$d + 20$
	Carol's score, s, **plus** 8	$s + 8$
	The number of problems, p, **increased by** 10	$p + 10$
	The number of hours, h, **added to** 40	$40 + h$
	The **total** of 16 and $-r$	$16 + (-r)$
Subtraction	The **difference** between Juan's age, a, and 18	$a - 18$
	The number of days, d, **minus** 6	$d - 6$
	Thirty **decreased by** the number of years, y	$30 - y$
	Fifteen **less than** the number of meters, m	$m - 15$
	t **decreased** by -4	$t - (-4)$

Jenna knows that she scored 240 points this season. That is 12 points more than she scored last season. She can use an equation to find how many points she scored last season.

334   Chapter 11

> **PROCEDURE**
> **To use an equation to solve word problems:**
> 1. Choose a variable to represent the unknown.
> 2. Write an equation for the problem.
> 3. Solve the equation.
> 4. Check the solution with the statements in the problem. Answer the question.

EXAMPLE 1 Twelve more than the number of points Jenna scored last season is 240. Find how many points she scored last season.

Solution:
1. Choose a variable.   Let p = the number of points.
2. Write an equation.   12 more than the number of points is 240.

$$p + 12 = 240$$

3. Solve the equation. $p + 12 - 12 = 240 - 12$
$$p = 288$$

4. Check: Does $228 + 12 = 240$?   Yes ✓

Jenna scored **288 points**.

Self-Check 1. The original cost of a radio increased by $15 is $68. Find the original cost of the radio.

EXAMPLE 2 Juan's weekly salary decreased by $24 in deductions is $198. What is his weekly salary?

Solution:
1. Let x = Juan's weekly salary.
2. The weekly salary decreased by $24 is $198.

$$x - 24 = 198$$

3. $x - 24 + 24 = 198 + 24$
$$x = 222$$

4. Check: Does $222 - 24 = 198$?   Yes ✓

Juan's weekly salary is **$222**.

Self-Check 2. Twelve pounds less than Laura's weight is 92 pounds. Find Laura's weight.

Equations: Addition/Subtraction

ADDITIONAL EXAMPLES
Example 1
1. The sum of her age and 17 is 29. Find her age.
   Ans: a + 17 = 29; 12
2. Thirty-six added to the temperature is 20. Find the temperature.
   Ans: t + 36 = 20; -16

SELF-CHECK
1. $53

Example 2
1. The cost minus $19 is $113. Find the cost.
   Ans: c - 19 = 113; $132
2. Twenty-seven less than the wages is 100. Find the wages.
   Ans: w - 27 = 100; $127

SELF-CHECK
2. 104 lb

CLASSROOM EXERCISES
1. y − 6   2. 6 + y
3. y + 6   4. 6 + y
5. 6 + y; y + 6
6. y − 6
7. 6 + y; y + 6
8. y + 6   9. y − 6
10. 6 − y   11. c + 12
12. m + 5
13. 60 − d = 24
14. n + 40 = 98
15. k − 100 = 425
16. 9 = s − 45

ASSIGNMENT GUIDE
BASIC
Day 1 pp. 336–337: 1–20
Day 2 p. 337: 21–36
AVERAGE
Day 1 pp. 336–337: 1–20
Day 2 p. 337: 21–36
ABOVE AVERAGE
pp. 336–337: 1–35 odd

WRITTEN EXERCISES
1. 25 − n   2. p + 16
3. x + 5, or 5 + x
4. y − 17   5. e − 6
6. n + 11   7. 32 + b
8. 45 − w
9. 85 + c, or c + 85
10. t − 25
11. d − 14.5
12. a + 0.6
13. y + 25
14. r − 2.6
15. 42 − (−p)
16. 5 + (−t)

CLASSROOM EXERCISES

For Exercises 1–10, choose from the box at the right the algebraic expression that represents each word expression. More than one answer may be possible in some cases. (Table)

1. y minus 6
2. 6 plus y
3. y increased by 6
4. the total of 6 and y
5. y more than 6
6. 6 less than y
7. the sum of 6 and y
8. y plus 6
9. y decreased by 6
10. 6 decreased by y

| $y + 6$ | $6 + y$ |
| $y - 6$ | $6 - y$ |

Complete. (Examples 1 and 2, step 2)

Word Sentence	Equation
11. Twelve dollars more than the cost, c, is $27	$\underline{\ ?\ } = 27$
12. The number of club members, m, increased by 5 is 24	$\underline{\ ?\ } = 24$
13. Sixty workdays decreased by the number of vacation days, d, is 24	?
14. The number of desks, n, increased by 40 is 98	?
15. The number of kilometers, k, on the odometer minus 100 is 425	?
16. Nine is the difference between Tom's score, s, and 45	?

WRITTEN EXERCISES

Goals: To write an algebraic equation for a word sentence
To apply the skill of writing algebraic equations to solving word problems

Write an algebraic expression for each word expression. (Table)

1. The difference of 25 and n
2. 16 more than p
3. The total of x and 5
4. y decreased by 17
5. Evelyn's age, e, minus 6
6. Eleven added to n
7. 32 plus b
8. 45 decreased by w
9. The sum 85 and c
10. t minus 25
11. 14.5 less than d
12. a increased by 0.6
13. 25 years more than y
14. r minus 2.6
15. 42 decreased by $(-p)$
16. $(-t)$ added to 5

336   Chapter 11

For Exercises 17–24, write an equation for each word sentence. Use t for the variable in each exercise. Do not solve the equation.
(Examples 1 and 2, step 2)

17. Nineteen more than a certain number is 42.
18. Twelve dollars more than the cost of a jacket is $57.
19. The number of tickets sold increased by 32 is 78.
20. The total of Rick's age and 15 is 31.
21. The difference of 24 and a certain number is 15.
22. Fourteen less than the number of sandwiches is 264.
23. The number of apples picked minus 12 is 52.
24. The length of a room decreased by 3 meters is 15 meters.

For Exercises 25–36, solve each problem. (Example 1)

25. Thirty-nine kilometers more than the distance traveled by the Blair family is 721 kilometers. How far did the Blair family travel?
26. Last Tuesday's average temperature plus 7° is 88°. What was last Tuesday's average temperature?
27. The amount of the Ohira family's electric bill increased by $8.50 is $45.25. Find the amount.

(Example 2)

29. The original cost of a pair of skis decreased by a discount of $30 is $135. Find the original cost of the skis.
30. The number of words Raymond types per minute decreased by 13 is 75. How many words does Raymond type per minute?
31. Sixty minutes minus the number of minutes for commercials is 52. How many minutes are used for commercials?
32. Thirteen dollars less than the original cost of a hair dryer is $18. Find the original cost of the hair dryer.

28. The number of passengers increased by a flight crew of 7 is 192. Find the number of passengers.

MIXED PRACTICE

33. The amount of snow plus 7 centimeters of rain is 16 centimeters. Find the amount of snow.
34. The number of library books minus 43 science books is 325. Find the number of library books.
35. The number of students in a karate class decreased by 15 is 19. Find the number of students.
36. Thirty-one more than the number of marathon runners is 124. Find the number of marathon runners.

WRITTEN EXERCISES

17. $t + 19 = 42$
18. $t + 12 = 57$
19. $t + 32 = 78$
20. $t + 15 = 31$
21. $24 - t = 15$
22. $t - 14 = 264$
23. $t - 12 = 52$
24. $t - 3 = 15$
25. 682 km
26. 81°
27. $36.75
28. 185   29. $165
30. 88   31. 8
32. $31   33. 9 cm
34. 368   35. 34
36. 93

Equations: Addition/Subtraction

Teaching Suggestions
p. M-24

QUICK QUIZ

Replace the ? with > or <.

1. 0 ? 21   Ans: <
2. -12 ? 17   Ans: <
3. -8 ? -10   Ans: >
4. 9 ? -2   Ans: >
5. -120 ? -118 Ans: <

ADDITIONAL EXAMPLES
Example 1
Graph each inequality.
1. a > -2
Ans: [number line with open circle at -2, shaded right]
2. b < 3
Ans: [number line with open circle at 3, shaded left]

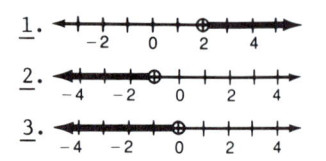

SELF-CHECK
1. [number line with open circle at 2, shaded right]
2. [number line with open circle at -1, shaded left]
3. [number line with open circle at 0, shaded left]

REVIEW CAPSULE FOR SECTION 11-5

Complete. Replace each ? with < or >. (Pages 294–296)

1. 4 ? 6
2. 4 ? -6
3. -3 ? -5
4. 6 ? 2
5. 0 ? 5
6. -1 ? -9
7. 0 ? -3
8. 15 ? -10
9. -9 ? -8
10. 7 ? 70

The answers are on page 346.

11-5 Inequalities on the Number Line

The fuel gauge on a plane shows that it has enough fuel for almost 4 hours of flying time. Thus, the flying time of the plane can be expressed by the inequality $t < 4$, where t is the number of hours.

The symbols $<$ or $>$ are used to write inequalities.

"$<$" means "is less than."   "$>$" means "is greater than."

You can graph inequalities on a number line.

EXAMPLE 1 Graph each inequality: **a.** $x > -6$ **b.** $x < 4$

Solutions:
a. [number line from -7 to 7 with open circle at -6, heavy line and arrow pointing right]

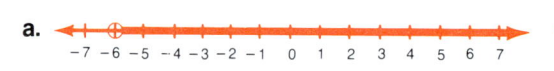

▶ The open circle at -6 means -6 is *not* on the graph.

The heavy line and arrow pointing to the right indicate that the graph includes all numbers greater than -6. Thus, **all numbers greater than -6 are** solutions of the inequality $x > -6$.

b.  [number line from -7 to 7 with open circle at 4, heavy line and arrow pointing left]

The heavy line and arrow pointing to the left indicate that the graph includes all numbers less than 4. Thus, **all numbers less than 4 are** solutions of the inequality $x < 4$.

Self-Check *Graph each inequality:* **1.** $y > 2$ **2.** $y < -1$ **3.** $x < 0$

The symbol "$\leq$" means "is less than or equal to."

The symbol "$\geq$" means "is greater than or equal to."

338 Chapter 11

EXAMPLE 2 Graph each inequality.

a. $n \leq -3$ b. $n \geq 1$

Solutions:
a. [number line with solid dot at -3, shaded left, from -8 to 6]
b. [number line with solid dot at 1, shaded right, from -4 to 10]

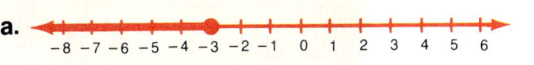

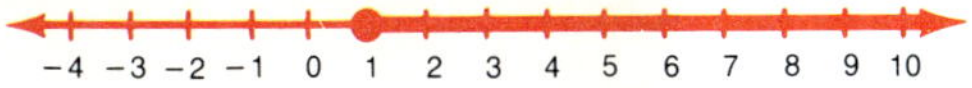

The large dot at −3 means −3 *is* a solution.

Self-Check Graph each inequality: **4.** $n \leq 3$ **5.** $n \geq -5$ **6.** $y \geq 0$

CLASSROOM EXERCISES

Which of these are solutions for $d > -3$? Write <u>Yes</u> or <u>No</u>.
(Example 1)

1. -4 No
2. 3 Yes
3. 5 Yes
4. -3 No
5. 0 Yes
6. -8 No
7. -1.5 Yes
8. $-\frac{2}{3}$ Yes
9. -3.01 No
10. $-10\frac{1}{4}$ No
11. 6.17 Yes
12. $\frac{1}{2}$ Yes

Which of these are solutions for $x \leq 3$? Write <u>Yes</u> or <u>No</u>.
(Example 2)

13. 2 Yes
14. 4 No
15. 8.7 No
16. 3 Yes
17. -1.9 Yes
18. 0 Yes
19. -3 Yes
20. $2\frac{1}{2}$ Yes
21. $3\frac{1}{9}$ No
22. 6.1 No
23. $\frac{4}{3}$ Yes
24. -1 Yes

WRITTEN EXERCISES

Goal: To graph inequalities on a number line

For Exercises 1–35, graph each inequality. (Example 1)

1. $a > 2$
2. $n < 3$
3. $s < -1$
4. $y > -4$
5. $m < 4$
6. $e > 5$
7. $x > 0$
8. $p < -6$
9. $r > 1$
10. $t < 6$

(Example 2)

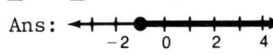

11. $y \geq 1$
12. $l \leq 6$
13. $c \geq -5$
14. $x \geq 0$
15. $z \leq 0$
16. $s \geq 3$
17. $t \leq -2$
18. $m \leq 5$
19. $n \geq -2$
20. $p \leq 1$

MIXED PRACTICE

21. $x > 3$
22. $y \leq 2$
23. $a \geq -3$
24. $c < -4$
25. $s \leq -1$
26. $t > -1$
27. $b \geq 5$
28. $d < 7$
29. $e > -5$
30. $g \leq 7$
31. $h < -3$
32. $j \geq -7$
33. $k \leq 4$
34. $n > 6$
35. $m \geq -4$

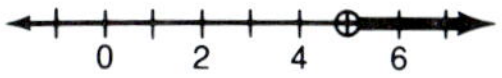

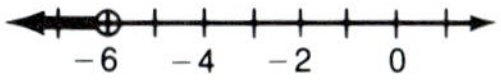

Equations: Addition/Subtraction

Teaching Suggestions
p. M-24

QUICK QUIZ

Solve and check.

1. x - 8 = 20
   Ans: 28
2. a + 3 = -10
   Ans: -13
3. 21 = b + 7
   Ans: 14
4. b + 121 = -87
   Ans: -208
5. x - 12.3 = -19
   Ans: -6.7

ADDITIONAL EXAMPLES

Example 1

Solve and graph.

1. x - 16 < -18
   Ans: x < -2;
2. 1 < y - 2
   Ans: y > 3;

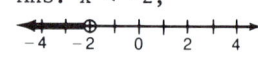

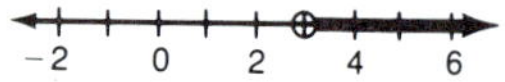

SELF-CHECK

1. p > 7; 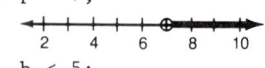
2. b < 5;
3. t < 1;

REVIEW CAPSULE FOR SECTION 11-6

Solve and check. (Pages 324-325)

1. $x - 5 = 7$
2. $u - 2 = 9$
3. $16 = y - 8$
4. $235 = r - 36$
5. $b - 64 = -7$
6. $t - 15 = -4$
7. $x + 4 = 8$
8. $a + 12 = 17$
9. $w + 5 = 4$
10. $n + 11 = 3$
11. $d + 6 = 7$
12. $c + 15 = -1$
13. $y - 3 = -5$
14. $d + 3 = -3$
15. $y + 2 = -4$
16. $t - 5 = -7$

The answers are on page 346.

11-6 Using Addition and Subtraction to Solve Inequalities

The Addition Property for solving inequalities is similar to the Addition Property for solving equations.

> **Addition Property for Inequalities**
>
> Adding the same number to each side of an inequality forms an equivalent inequality.
>
> $k - 5 > 7$
> $k - 5 + 5 > 7 + 5$
> $k > 12$

EXAMPLE 1 Solve and graph: $x - 3 > 2$

Solution:
1. To get x alone, add 3 to each side.
   $x - 3 + 3 > 2 + 3$
   $x > 5$
2. Graph the inequality.

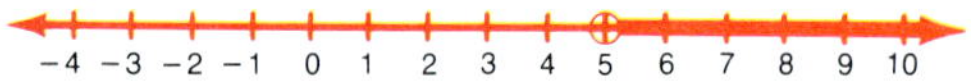

The solution is **all numbers greater than 5.**

Self-Check Solve and graph: 1. $p - 3 > 4$ 2. $b - 7 < -2$ 3. $0 > t - 1$

There is also a Subtraction Property for Inequalities.

> **Subtraction Property for Inequalities**
>
> Subtracting the same number from each side of an inequality forms an equivalent inequality.
>
> $q + 6 < 8$
> $q + 6 - 6 < 8 - 6$
> $q < 2$

340 Chapter 11

EXAMPLE 2 Solve and graph: $c + 2 \leq 6$

Solution:
1. To get c alone, subtract 2 from each side.
$$c + 2 - 2 \leq 6 - 2$$
$$c \leq 4$$
2. Graph the inequality.

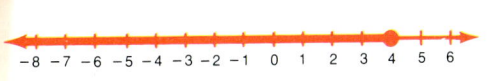

The solution is **all numbers less than 4**.

Self-Check *Solve and graph.*

4. $x + 7 < 10$
5. $c + 5 > -1$
6. $a + 7 > 2$

CLASSROOM EXERCISES

Write an equivalent inequality as in Step 1 of Example 1. Do <u>not</u> solve the inequality. (Step 1, Examples 1 and 2)

1. $a - 5 > 3$
2. $t - 7 > -5$
3. $w - 4 < 8$
4. $p - 11 < -9$
5. $y - 2 > 6$
6. $h - 4 < -7$
7. $m + 12 \leq 3$
8. $x + 14 \leq -2$
9. $t + 15 \geq -5$
10. $b + 3 \geq -4$
11. $c + 5 < 12$
12. $d + 4 \geq -8$

Solve and graph. (Example 1)

13. $g - 5 > 7$
14. $t - 8 \geq 11$
15. $w - 1 \leq -4$
16. $a - 4 < -9$
17. $-3 > b - 8$
18. $6 < c - 12$

(Example 2)

19. $p + 3 < 9$
20. $x + 5 \leq -4$
21. $y + 8 \geq 12$
22. $m + 7 \geq -15$
23. $14 < b + 4$
24. $-15 > t + 6$

WRITTEN EXERCISES

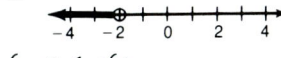

Goal: To solve and graph inequalities by using the Addition and Subtraction Properties

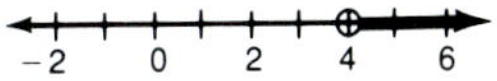

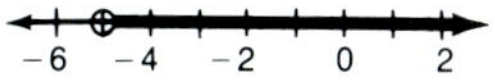

For Exercises 1–9, solve and graph. (Example 1)

1. $t - 3 > 2$
2. $r - 7 > -13$
3. $p - 5 < -3$
4. $x - 2 < -4$
5. $x - 12 > 4$
6. $-8 > q - 2$
7. $10 < c - 7$
8. $x - 9 \leq -9$
9. $b - 3 \leq -9$

Equations: Addition/Subtraction

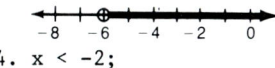

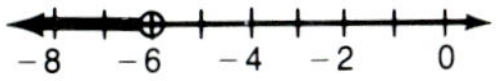

8. $x \leq 0$;
10. $x < 5$;
12. $y > 2$;
14. $r < -6$;
16. $t > -4$;
18. $m \geq 8$;
20. $p \geq -3$;
22. $t < -3$;
24. $y > 12$;
26. $r < -7$;
28. $x > -2$;
30. $m \leq -4$;
32. $t < 10$;
34. $n < 7$;
36. $x \leq 10$;

See page 347 for the answers to Exercises 38–48 even.

CALCULATOR EXERCISES

1. Checks
2. Checks
3. Does not check
4. Checks
5. Does not check
6. Checks

342

For Exercises 10–48, solve and graph. (Example 2)

10. $x + 3 < 8$
11. $z + 5 < 6$
12. $y + 5 > 7$
13. $n + 6 < 10$
14. $r + 4 < -2$
15. $k + 5 > 3$
16. $t + 7 > 3$
17. $w + 14 < 8$
18. $m + 24 \geq 32$
19. $79 \geq 68 + d$
20. $15 \leq 18 + p$
21. $18 + b \geq 3$

MIXED PRACTICE

22. $t + 4 < 1$
23. $n + 3 < 4$
24. $y - 5 > 7$
25. $x - 15 \geq -22$
26. $r + 16 < 9$
27. $a - 8 > 8$
28. $x - 3 > -5$
29. $t - 15 \leq 35$
30. $m + 1 \leq -3$
31. $x + 7 \geq -2$
32. $4 > t - 6$
33. $19 < y + 20$
34. $2n + 5 \leq 19$
35. $5x - 1 < 19$
36. $23 \geq 2x + 3$

MORE CHALLENGING EXERCISES

37. $5n - 13 - 4n < -17$
38. $-8t + 14 + 9t > 25$
39. $3r - 6 - 2r < -4$
40. $-23 - 2y + 3y > -15$
41. $12w - 15 - 11w < 28$
42. $-18b - 42 + 19b > -56$
43. $4p - 3 \leq 5p + 6$
44. $-4y - 2 \geq 8 - 3y$
45. $3x + 5 > 2x - 2$
46. $4 + 6d < 5d - 3$
47. $-3c - 1 \leq 8 - 2c$
48. $6p + 2p + 4 \geq 7p - 12$

CHECKING INEQUALITIES

You can use a calculator to check whether a given number is a solution of a given inequality.

EXAMPLE Is -117 a solution of $t + 49 < -78$?

Solution Substitute -117 in the left side of the inequality.

1 1 7 [+/−] [+] 4 9 [=]   −68.

Since $-68 > -78$, -117 is **not** a solution.

EXERCISES Check whether the given number is a solution for the inequality.

1. $-47 > x - 94$; 46
2. $-133 < y - 51$; -81
3. $s - 163 < -85$; 79
4. $72 < x + 37$; 100
5. $y + 71 > -84$; -155
6. $m - 13.4 < -47.7$; -61.1

Chapter 11

REVIEW: SECTIONS 11-4 — 11-6

Write an algebraic expression for each word expression. (Section 11-4)

1. Some number, n, increased by 5
2. The number of days, d, minus 7
3. Five pounds less than the weight, w
4. The number of books, b, plus 31
5. The price of a coat, c, decreased by $5
6. The number, y, reduced by 27
7. The quantity, x, increased by 14
8. A number, n, increased by 3

Write an equation for each word sentence. (Section 11-4)

9. Eight dollars more than the cost, c, is $33.
10. The sum of 12 and the number of points, p, is 56.
11. Robert's age, a, decreased by 3 is 13.
12. The distance, d, minus 10 meters is 36 meters.
13. The number, x, increased by 11 is 34.
14. The difference of y and 5 is 19.
15. Twenty reduced by n is 13.
16. The price of a suit, s, decreased by $28 is $103.

Solve. (Section 11-4)

17. The amount of Gina's gas bill minus $13 is $52. Find the amount of her gas bill.
18. Acme Appliances sold washers for $103. The regular price was reduced $38 for the sale. Find the regular price. 
19. James grew $2\frac{1}{2}$ inches since his birthday last year. He is now 51 inches tall. What was his height a year ago?
20. Twenty-two kilometers more than the distance traveled on Tuesday is 630 kilometers. Find the distance traveled on Tuesday.

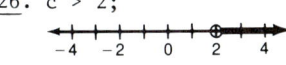

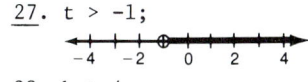

For Exercises 21–24, graph each inequality. (Section 11-5)

21. $x > -8$
22. $m < 5$
23. $p \leq -3$
24. $n \geq 6$

For Exercises 25–28, solve and graph. (Section 11-6)

25. $y - 5 < 4$
26. $c - 8 > -6$
27. $t + 1 > 0$
28. $5 \leq k + 1$

Equations: Addition/Subtraction

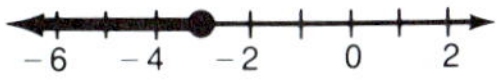

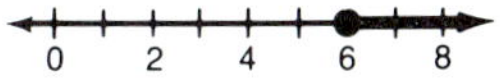

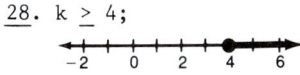

CHAPTER REVIEW

PART 1: VOCABULARY

For Exercises 1-5, choose from the box at the right the word(s) that best correspond to each description.

1. Statements that use $<, >, \leq$, and $\geq$ are called __?__
2. The words difference, minus, less than, and decreased by suggest the operation of __?__.
3. "Eight points more than 45" is an example of a __?__ expression.
4. The words sum, more than, plus, and increased by suggest the suggest the operation of __?__.
5. "$c - 6$" is an example of an __?__ expression.

| addition |
| inequalities |
| algebraic |
| word |
| subtraction |

PART 2: SKILLS

Solve and check. (Section 11-1)

6. $t - 8 = 2$
7. $p - 13 = -9$
8. $n - 47 = -28$
9. $a + 29 = -15$
10. $-18 = y + 36$
11. $4 = b + 8$
12. $x + 25 = 13$
13. $y - 9 = -12$

Solve and check. (Section 11-2)

14. $13m - 14 + 12m = -13 + 149$
15. $-5 - 18 = -c - 95 + 9c$
16. $3d - 8 - d = 18 - 12$
17. $-p + 6p - 3 = -4 + 31$
18. $19 - 33 = 4q - 91 + 3q$
19. $29 - x + 11x = -7 + 66$

Solve and check. (Section 11-3)

20. $11s = 27 + 2s$
21. $6d = 28 - d$
22. $17x = 90 - x$
23. $y = 18 - y$
24. $15 + 10m = 24 + 9m$
25. $39 + 36p = 44 + 35p$
26. $8 + 5\frac{1}{3}y = 17 + 2\frac{1}{3}y$
27. $9 + 4\frac{1}{2}t = 10 + 3\frac{1}{2}t$

Write an algebraic expression for each word expression. (Section 11-4)

28. The difference of 4 and p
29. The sum of 35 apples and t
30. y increased by 32
31. The cost, c, decreased by $6
32. Twelve feet less than w
33. The distance, d, plus 5

344 Chapter 11

Write an equation for each word sentence. Do not solve.
(Section 11-4)

34. Three dollars more than the original cost, c, is $19.
35. The number of runners, r, at the starting line decreased by 15 is 42.
36. The difference between $10 and the price, p, of a ticket is $4.
37. The width, w, plus 16 centimeters is 84 centimeters.

Graph each inequality. (Section 11-5)

38. $n > -2$
39. $r < 5$
40. $t < -3$
41. $y > -4$
42. $p \leq 2$
43. $t \leq -4$
44. $a \geq -5$
45. $q \geq 0$

Solve and graph. (Section 11-6)

46. $x + 2 < 5$
47. $-15 > t - 19$
48. $-37 + m > -41$
49. $y + 4 \geq 1$
50. $n - 18 \leq -23$
51. $-24 > t - 19$
52. $-19 + r > -25$
53. $-27 < k - 24$

PART 3: APPLICATIONS

Solve each problem. (Section 11-4)

54. Eighty-five calories plus the number of calories in a fruit salad totals 164. How many calories are there in the fruit salad?
55. Twelve dollars more than the cost of a pair of jeans is $28. Find the cost of the jeans.
56. A house painter's weekly salary minus $72 in deductions is $208. What is the painter's weekly salary?
57. Twenty dollars less than the amount of the Lopez family's gas bill is $68. Find the amount of the gas bill.

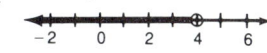

CHAPTER REVIEW

34. $c + 3 = 19$
35. $r - 15 = 42$
36. $10 - p = 4$
37. $w + 16 = 84$
38. [number line -4 to 4, open circle at -2, shaded right]
39. [number line -2 to 6, open circle at 5, shaded left]
40. [number line -6 to 2, open circle at -3, shaded left]
41. [number line -6 to 2, open circle at -4, shaded right]
42. [number line -4 to 4, closed circle at 2, shaded left]
43. [number line -6 to 2, closed circle at -4, shaded left]
44. [number line -6 to 2, closed circle at -5, shaded right]
45. [number line -4 to 4, closed circle at 0, shaded right]
46. $x < 3$; [number line -2 to 6]
47. $t < 4$; [number line -2 to 6]
48. $m > -4$; [number line -6 to 2]
49. $y \geq -3$; [number line -6 to 2]
50. $n \leq -5$; [number line -8 to 0]
51. $t < -5$; [number line -8 to 0]
52. $r > -6$; [number line -8 to 0]
53. $k > -3$; [number line -6 to 2]
54. 79
55. $16
56. $280
57. $88

Equations: Addition/Subtraction

CHAPTER TEST
Two forms of a chapter test, Form A and Form B, are provided on copying masters in the Teacher's Edition: Part II.

1. 27 2. 8 3. -15
4. -4.7 5. 2 6. 2
7. 3 8. 3

CLASSROOM EXERCISES p.341
2. t - 7 + 7 > -5 + 7
4. p - 11 + 11 < -9 + 11
6. h - 4 + 4 < -7 + 4
8. x + 14 - 14 ≤ -2 - 14
10. b + 3 - 3 ≥ -4 - 3
12. d + 4 - 4 ≥ -8 - 4
13. g > 12;
14. t ≥ 19;
15. w ≤ -3;
16. a < -5;
17. b < 5;
18. c > 18;
19. p < 6;
20. x ≤ -9;
21. y ≥ 4;
22. m ≥ -22;
23. b > 10;
24. t < -21;

CHAPTER TEST

Solve and check.

1. $x - 12 = 15$
2. $d - 9 = -1$
3. $t + 10 = -5$
4. $n + 1.8 = -2.9$
5. $2x + 5x = 5 + 9$
6. $x + 2 + 3x = 10$
7. $3x - 9 = 3 - x$
8. $7 + 3k = 19 - k$
9. $5\frac{3}{4}w = 1\frac{3}{4}w + 16$

Write an algebraic expression for each word expression.

10. Nineteen weeks increased by x weeks
11. Forty days decreased by d days.
12. Fifteen less than v
13. Eleven more than n apples

Solve.

14. The amount of Ted's grocery bill minus a refund of $5 is $32. Find the amount of his bill.
15. James bought a sweater and a tie for $37. The sweater cost $12 more than the tie. Find the cost of the sweater.
16. The number of people in an elevator increased by 6 is 14. Find the number of people in the elevator.
17. When $105 is subtracted from the amount in Jill's checking account, the result is $48.20. How much is in the account?

Solve and graph.

18. $b \geq -5$
19. $z - 13 > -5$
20. $p + 8 < 3$

ANSWERS TO REVIEW CAPSULES

Page 326 1. $5x$ 2. $13p$ 3. $2q$ 4. $2t$ 5. $9x$ 6. $10b$ 7. m 8. q 9. 44 10. $\frac{1}{4}$ 11. 1.1 12. 400

Page 328 1. $6d + 6$ 2. $p + 8$ 3. $2x - 6$ 4. $9 + 4x$ 5. $10 + 5r$ 6. $2 + 5b$ 7. $5 + 4y$ 8. $5c + 13$ 9. $3x - 2$ 10. $-3x + 11$ 11. $-13n - 5$ 12. $12y + 3$ 13. 9 14. 16 15. -2 16. -16 17. -3 18. -11 19. 4 20. 6 21. 4 22. 6 23. -5 24. -3 25. 2 26. 2 27. -1

Page 331 1. 12 2. 6 3. 13 4. 34 5. 12 6. 10 7. 25 8. 47

Page 338 1. < 2. > 3. > 4. > 5. < 6. > 7. > 8. > 9. < 10. <

Page 340 1. 12 2. 11 3. 24 4. 271 5. 57 6. 11 7. 4 8. 5 9. -1 10. -8 11. 1 12. -16 13. -2 14. -6 15. -6 16. -2

Chapter 11

ENRICHMENT

Compound Inequalities

In a high school wrestling weight class, wrestlers must weigh more than 167 pounds but no more than 185 pounds. If w represents the weight of a wrestler, you can write an inequality for this guideline.

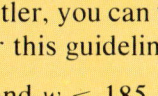

$$w > 167 \text{ and } w \leq 185$$

A sentence containing two inequalities with the connecting words <u>and</u> or <u>or</u> is a **compound inequality**.

EXAMPLES Graph each compound inequality.

a. $w > 167$ <u>and</u> $w \leq 185$ **b.** $x \leq -2$ <u>or</u> $x > 1$

Solutions: **a.** When the connective <u>and</u> is used, the solutions are those numbers that make <u>both</u> inequalities true.

$w > 167$ and $w \leq 185$

b. When the connective <u>or</u> is used, the solutions are those numbers that make either inequality true.

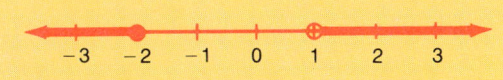

$x \leq -2$ or $x > 1$

EXERCISES

Graph the solution for each compound inequality.

1. $x > 3$ <u>and</u> $x < 4$
2. $x \geq -3$ <u>and</u> $x < 1$
3. $x \geq -2$ <u>and</u> $x < 3$
4. $x > 1$ <u>or</u> $x < -1$
5. $x > 0$ <u>or</u> $x \leq -3$
6. $x > 2$ <u>or</u> $x \leq 0$

The weight of each player on a high school football team is greater than 150 pounds and no greater than 220 pounds.

7. Represent this weight as a compound inequality.
8. Graph the solution of the compound inequality.

Equations: Addition/Subtraction **347**

ENRICHMENT

You may wish to use this lesson for students who performed well on the formal Chapter Test.

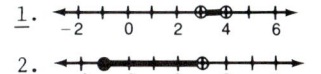

1. [number line graph from -2 to 6]
2. [number line graph from -4 to 4]
3. [number line graph from -4 to 4]
4. [number line graph from -4 to 4]
5. [number line graph from -4 to 4]
6. [number line graph from -4 to 4]
7. x > 150 and x ≤ 220
8. [number line graph from 0 to 400]

WRITTEN EXERCISES

p. 342

38. t > 11; [graph 8 to 16]
40. y > 8; [graph 2 to 10]
42. b > -14; [graph -18 to -10]
44. y ≤ -10; [graph -14 to -6]
46. d < -7; [graph -10 to -2]
48. p ≥ 16; [graph -20 to -12]

ADDITIONAL PRACTICE

You may wish to use all or some of these exercises, depending on how well students performed on the formal Chapter Test.

1. 23   2. -58   3. -93
4. -26   5. -41   6. 36
7. -44.6   8. -8.1
9. -40.7   10. 4
11. 3   12. -8
13. -13   14. 8   15. 3
16. 18   17. 7   18. 9
19. 7   20. 9   21. 50
22. 20   23. 3   24. 2
25. 19   26. 7   27. 33
28. 10
29. [number line graph]
30. [number line graph]
31. [number line graph]
32. [number line graph]
33. $n < 3$; [number line graph]
34. $m < -9$; [number line graph]
35. $a > 14$; [number line graph]
36. $y < 7$; [number line graph]
37. $3 + n = 84$
38. $y + 16 = 43$
39. $c - 24 = 36$
40. $x - 16.1 = 24.3$
41. $c - 55 = 326$; $381
42. $w + 13 = 171$; 158 lb

ADDITIONAL PRACTICE

SKILLS

Solve and check. (Pages 324–328)

1. $x + 24 = 27$
2. $p + 41 = -17$
3. $c + 12 = -81$
4. $t + 13 = -13$
5. $a - 13 = -54$
6. $t - 42 = -6$
7. $a + 11.5 = -33.1$
8. $n + 22.7 = 14.6$
9. $x + 27.8 = -12.9$
10. $3a + 2a = 9 + 11$
11. $5c + 4c = 18 + 9$
12. $12x - 11x = 14 - 22$
13. $17b - 16b = 8 - 21$
14. $10y - 2y = -25 + 89$
15. $12m - 2m = -4 + 34$
16. $6x - 12 - 5x = -12 + 18$
17. $14r - 9 - 2r = -10 + 85$
18. $4n - 8 + 6n = -23 + 105$

(Pages 329–330)

19. $c = 21 - 2c$
20. $x = 36 - 3x$
21. $4.6n = 100 + 2.6n$
22. $7.3p = 80 + 3.3p$
23. $6q + 7 = 13 + 4q$
24. $4y + 3 = 5 + 3y$
25. $6x - 6 = 13 + 5x$
26. $8t - 9 = 5 + 6t$
27. $9.3a - 21 = 12 + 8.3a$
28. $4.3w - 9 = 21 + 1.3w$

Graph each inequality on a number line. (Pages 338–339)

29. $x > 5$
30. $m < 7$
31. $c \geq -4$
32. $p \leq -2$

Solve and graph. (Pages 340–342)

33. $n + 6 < 9$
34. $m + 3 < -6$
35. $a - 7 > 7$
36. $y - 5 < 2$

APPLICATIONS

Write an equation for each word sentence. (Pages 334–337)

37. The sum of 3 and n is 84.
38. 16 more than y is 43.
39. The difference of c and 24 is 36.
40. 16.1 less than x is 24.3.
41. The original cost of a television decreased by $55 is $326. Find the original cost. (Pages 334–337)
42. Thirteen pounds more than Carl's weight is 171 pounds. Find Carl's weight. (Pages 334–337)

348   Chapter 11

CHAPTER 12
Rational Numbers: Multiplication/Division

SECTIONS
- **12-1** Multiplication: Unlike Signs
- **12-2** Multiplication: Like Signs
- **12-3** Properties of Multiplication
- **12-4** Dividing Rational Numbers
- **12-5** Factoring

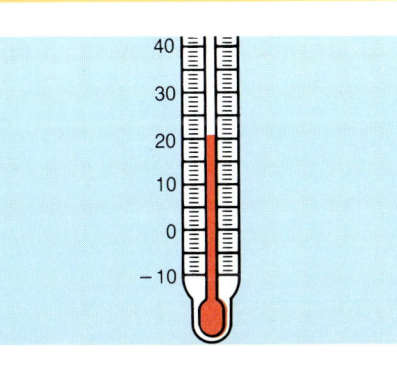

FEATURES
Calculator Application:
Multiplying Rational Numbers
Calculator Application:
Distributive Property
Consumer Application:
Conserving Energy
Enrichment:
Multiplying Binomials

12-1 Multiplication: Unlike Signs

Study the patterns in each column.

Column 1

$2 \times 2 = 4$
$2 \times 1 = 2$
$2 \times 0 = 0$
$2 \times (-1) = -2$
$2 \times (-2) = -4$
$2 \times (-3) = -6$

The products are decreasing by 2.

Column 2

$2 \times 3 = 6$
$1 \times 3 = 3$
$0 \times 3 = 0$
$(-1) \times 3 = -3$
$(-2) \times 3 = -6$
$(-3) \times 3 = -9$

The products are decreasing by 3.

These patterns suggest that the product of two numbers having unlike signs is negative.

> **PROCEDURE**
>
> To multiply two numbers with <u>unlike</u> signs:
>
> [1] Multiply as with whole numbers.
> [2] Write the opposite of the result.

EXAMPLE 1   Multiply: $(12)(-8)$

Solution: [1] Multiply. $12 \cdot 8 = 96$
   [2] Write the opposite. -96

Thus, $(12)(-8) = \mathbf{-96}$

Self-Check   *Multiply:*   **1.** $(14)(-5)$ **2.** $(9)(-25)$ **3.** $(6)(-16)$

Example 2 shows the product of a negative number and a positive number.

EXAMPLE 2   Multiply: $(-8)(11)$

Solution: [1] Multiply. $8 \cdot 11 = 88$
   [2] Write the opposite. -88

Thus, $(-8)(11) = \mathbf{-88}$

Self-Check   *Multiply:*   **4.** $(-4)(9)$ **5.** $-12(6)$ **6.** $(-18)(3)$

Chapter 12

Teaching Suggestions
p. M-25

QUICK QUIZ
Multiply.
<u>1.</u> (23)(71)   Ans: 1633
<u>2.</u> 102(78)   Ans: 7956
<u>3.</u> (0.7)(11.8)
   Ans: 8.26
<u>4.</u> $\frac{2}{3} \cdot \frac{6}{7}$ Ans $\frac{4}{7}$
<u>5.</u> $(\frac{3}{8})(\frac{12}{21})$ Ans: $\frac{3}{14}$

ADDITIONAL EXAMPLES
Example 1
Multiply:
(25)(-18)   Ans: -450

SELF-CHECK
<u>1.</u> -70   <u>2.</u> -225
<u>3.</u> -96

SELF-CHECK
<u>4.</u> -36   <u>5.</u> -72
<u>6.</u> -54

The following properties that apply to whole numbers are also true for rational numbers.

> **Multiplication Property of Zero**
> For any rational number a, $0 \cdot (-3) = 0$
> $a \cdot 0 = 0$ and $0 \cdot a = 0$ $\left(-1\frac{3}{4}\right) \cdot 0 = 0$
>
> **Multiplication Property of One**
> For any rational number a, $(-10)(1) = -10$
> $a \cdot 1 = a$ and $1 \cdot a = a$ $(1)(-2.51) = -2.51$

CLASSROOM EXERCISES

For Exercises 1–4, complete the pattern. (Columns 1 and 2)

1. $8 \times 1 = 8$	**2.** $2 \times 3 = 6$	**3.** $0 \times 5 = 0$	**4.** $16 \times 1 = 16$
$8 \times 0 = 0$	$1 \times 3 = 3$	$-1 \times 5 = -5$	$16 \times 0 = 0$
$8 \times (-1) = -8$	$0 \times 3 = 0$	$-2 \times 5 = \underline{\ ?\ }$	$16 \times (-1) = \underline{\ ?\ }$
$8 \times (-2) = \underline{\ ?\ }$	$-1 \times 3 = \underline{\ ?\ }$	$-3 \times 5 = \underline{\ ?\ }$	$16 \times (-2) = \underline{\ ?\ }$

For Exercises 5–24, multiply the given numbers. (Example 1)

5. $0(-10)$ **6.** $1(-15)$ **7.** $5(-13)$ **8.** $7(-18)$
9. $11(-21)$ **10.** $19(-32)$ **11.** $27(-128)$ **12.** $349(-2)$

(Example 2)

13. $-1(6)$ **14.** $-3(2)$ **15.** $-4(5)$ **16.** $-7(8)$
17. $-5(0)$ **18.** $-15(10)$ **19.** $(-32)(12)$ **20.** $(-20)(8)$
21. $(-25)(17)$ **22.** $(-16)(13)$ **23.** $(-41)(27)$ **24.** $(-15)(52)$

WRITTEN EXERCISES

Goal: To multiply rational numbers having unlike signs

For Exercises 1–36, multiply the given numbers. (Example 1)

1. $6(-5)$ **2.** $9(-3)$ **3.** $15(-12)$ **4.** $11(-17)$
5. $19(-35)$ **6.** $34(-21)$ **7.** $42(-50)$ **8.** $52(-48)$
9. $2.1(-1.5)$ **10.** $0.7(-3.2)$ **11.** $\frac{1}{2}(-6)$ **12.** $\left(1\frac{1}{4}\right)\left(-1\frac{2}{5}\right)$

Rational Numbers: Multiplication/Division   **351**

CLASSROOM EXERCISES
1. -16   2. -3
3. -10; -15
4. -16; -32
5. 0   6. -15   7. -65
8. -126   9. -231
10. -608   11. 3456
12. -698   13. -6
14. -6   15. -20
16. -56   17. 0
18. -150   19. -384
20. -160   21. -425
22. -208   23. -1107
24. -780

ASSIGNMENT GUIDE
BASIC
Day 1 pp. 351–352: 1–24
Day 2 p. 352: 25–44
AVERAGE
Day 1 pp. 351–352: 1–24
Day 2 p. 352: 25–50
ABOVE AVERAGE
pp. 351–352: 3, 6, 9, ..., 36, 37–42

WRITTEN EXERCISES
1. -30   2. -27
3. -180   4. -187
5. -665   6. -714
7. -2100   8. -2496
9. -3.15   10. -2.24
11. -3   12. $-1\frac{3}{4}$
13. 0   14. -5
15. -81   16. -42
17. -156   18. -360
19. -840   20. -2520

WRITTEN EXERCISES

21. -32.8   22. -0.195
23. $-\frac{1}{12}$ 24. $-\frac{9}{14}$
25. 0   26. -81
27. -2016   28. -2961
29. -6272   30. -2808
31. -4402   32. -9546
33. -86.8   34. -1.44
35. $-\frac{4}{5}$ 36. $-\frac{7}{20}$
37. -3(5); -15
38. 5(-100); -500
39. 6(-15); -90
40. 2(-200); -400
41. 4(-6); -24
42. 3(-4); -12
43. -2(78); -156
44. 2(-3); -6
45. -145   46. -28
47. -102   48. -95
49. -27   50. -10
51. -50y   52. -48x
53. -8ab   54. -18pq
55. $-56a^2$ 56. $-6x^2$
57. $-6m^3$ 58. $6rt^3$
59. $-2v^5$ 60. $-12c^7$
61. $-30n^3$ 62. $-3e^4$

(Example 2)

13. $-8(0)$ 14. $-5(1)$ 15. $-9(9)$ 16. $-7(6)$
17. $-12(13)$ 18. $-18(20)$ 19. $-35(24)$ 20. $-42(60)$
21. $-8.2(4)$ 22. $-0.13(1.5)$ 23. $\left(-\frac{1}{3}\right)\left(\frac{1}{4}\right)$ 24. $-\frac{3}{4}\left(\frac{6}{7}\right)$

MIXED PRACTICE

25. $-75(0)$ 26. $-81(1)$ 27. $36(-56)$ 28. $47(-63)$
29. $224(-28)$ 30. $156(-18)$ 31. $-31(142)$ 32. $-258(37)$
33. $12.4(-7)$ 34. $(-0.4)(3.6)$ 35. $\left(1\frac{1}{5}\right)\left(-\frac{2}{3}\right)$ 36. $\left(-\frac{1}{8}\right)\left(2\frac{4}{5}\right)$

APPLICATIONS: USING MULTIPLICATION WITH RATIONAL NUMBERS

Use positive and negative numbers to represent each situation by a multiplication problem. Then find the product.

37. Three penalties of 5 yards each
38. Five times a debt of $100
39. Six times a depth of 15 meters below sea level
40. Twice a loss of 200 feet in altitude
41. Four times a temperature drop of 6 degrees
42. Three times a loss of 4 yards in rushing
43. Two debts of $78 each
44. Twice a drop in temperature of 3 degrees

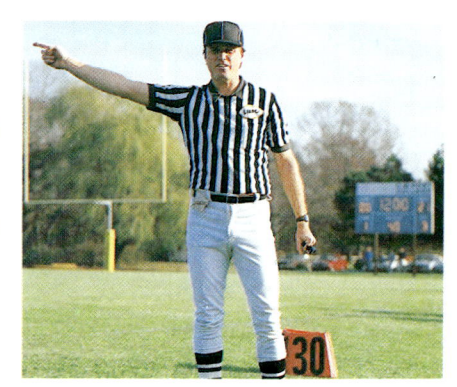

MORE CHALLENGING EXERCISES

Evaluate each expression.

45. $4(-29) + 8(-3) + (-1)(5)$ 46. $2(-1) + 7(-2) + 3(-4)$
47. $9(-2) + 3(-22) + (6)(-3)$ 48. $(-7)(10) + (-5)(3) + (-2)(5)$
49. $12(-21 + 19) + (-3)$ 50. $8 + (-3) \cdot (-6 + 12)$

Multiply the monomials.

51. $-10(5y)$ 52. $4(-12x)$ 53. $(-2a)(4b)$ 54. $(6p)(-3q)$
55. $(-8a)(7a)$ 56. $(2x)(-3x)$ 57. $(-3m^2)(2m)$ 58. $(2rt)(-3t^2)$
59. $2v^2(-v^3)$ 60. $(-3c^3)(4c^4)$ 61. $(5n)(-6n^2)$ 62. $(-e^2)(3e^2)$

352   Chapter 12

REVIEW CAPSULE FOR SECTION 12-2

Evaluate each expression. (Pages 22-24)

1. $a + a$ when $a = -3$
2. $b + b$ when $b = -6$
3. $x + x + x$ when $x = -5$
4. $m + m + m$ when $m = -1$
5. $p + p + p + p$ when $p = 12$
6. $t + t + t + t$ when $t = -10$
7. rs when $r = 4$ and $s = 12$
8. xy when $x = 13$ and $y = 11$

The answers are on page 370.

12-2 Multiplication: Like Signs

Study the pattern in each column.

Column 1

$2 \times (-4) = -8$
$1 \times (-4) = -4$
$0 \times (-4) = 0$
$-1 \times (-4) = 4$
$-2 \times (-4) = 8$
$-3 \times (-4) = 12$

The products are increasing by 4.

Column 2

$-6 \times 2 = -12$
$-6 \times 1 = -6$
$-6 \times 0 = 0$
$-6 \times (-1) = 6$
$-6 \times (-2) = 12$
$-6 \times (-3) = 18$

The products are increasing by 6.

These patterns suggest that the product of two rational numbers having like signs is positive.

> **PROCEDURE**
>
> **To multiply two rational numbers with <u>like</u> signs:**
>
> 1. Multiply as with whole numbers.
> 2. Write the result.

EXAMPLE Multiply: $(-24)(-3)$

Solution: 1. Multiply. $24 \cdot 3 = 72$
2. Result. $(-24)(-3) = \mathbf{72}$

Self-Check *Multiply.*

1. $-2(-7)$
2. $-10(-13)$
3. $(-12)(-25)$
4. $(-7)(-11)$

Rational Numbers: Multiplication/Division 353

REVIEW CAPSULE
This Review Capsule reviews prior-taught skills used in Section 12-2. The reference is to the pages where the skills were taught.

Teaching Suggestions p. M-25

QUICK QUIZ
Evaluate.
1. $(-7) + (-7) + (-7) + (-7)$
   Ans: -28
2. $(-23) + (-23) + (-23)$ Ans: -69
3. $17(108)$ Ans: 1836
4. $(0.0071)(200)$
   Ans: 1.42
5. $(3\frac{2}{5})(\frac{5}{8})$ Ans: $2\frac{1}{8}$

ADDITIONAL EXAMPLES
Example
Multiply.
1. $(-23)(-306)$
   Ans: 7038
2. $(-5.06)(-12)$
   Ans: 60.72

SELF-CHECK
1. 14   2. 130
3. 300   4. 77

CLASSROOM EXERCISES
1. 12   2. 16
3. 9; 18   4. 7; 14

ASSIGNMENT GUIDE
BASIC
Day 1 p. 354: 1-31 odd
Day 2 p. 354: 2-32 even
AVERAGE
p. 354: 1-32
ABOVE AVERAGE
p. 354: 3, 6, 9, ...,
30, 33-48

WRITTEN EXERCISES
1. 35   2. 32   3. 9
4. 18   5. 735   6. 896
7. 810   8. 770
9. 625   10. 1989
11. 2992   12. 3206
13. 1498   14. 4650
15. 7812   16. 21,424
17. 6.6   18. 2.8
19. 8.28   20. 4.5
21. 5.7   22. 0.48
23. 21.294   24. 58.077
25. $\frac{1}{2}$ 26. $\frac{3}{10}$ 27. $10\frac{1}{5}$
28. $1\frac{1}{2}$ 29. $\frac{1}{8}$ 30. $\frac{3}{10}$
31. $4\frac{1}{6}$ 32. $3\frac{3}{10}$
33. 13   34. 1.21
35. 36   36. -35
37. 52   38. 58   39. 1
40. -250   41. 18a
42. 18b   43. 15x²
44. 40c²   45. 10a²b²
46. 48p³   47. 12x³y
48. 14p²t²

354

CLASSROOM EXERCISES

For Exercises 1-4, complete the pattern. (Columns 1 and 2)

1. $2 \times (-6) = -12$
   $1 \times (-6) = -6$
   $0 \times (-6) = 0$
   $-1 \times (-6) = 6$
   $-2 \times (-6) = \underline{\ ?\ }$

2. $-8 \times 2 = -16$
   $-8 \times 1 = -8$
   $-8 \times 0 = 0$
   $-8 \times (-1) = 8$
   $-8 \times (-2) = \underline{\ ?\ }$

3. $-9 \times 2 = -18$
   $-9 \times 1 = -9$
   $-9 \times 0 = 0$
   $-9 \times (-1) = \underline{\ ?\ }$
   $-9 \times (-2) = \underline{\ ?\ }$

4. $2 \times (-7) = -14$
   $1 \times (-7) = -7$
   $0 \times (-7) = 0$
   $-1 \times (-7) = \underline{\ ?\ }$
   $-2 \times (-7) = \underline{\ ?\ }$

WRITTEN EXERCISES

Goal: To multiply two negative rational numbers

Multiply. (Example)

1. $(-7)(-5)$
2. $(-4)(-8)$
3. $(-1)(-9)$
4. $(-3)(-6)$
5. $(-21)(-35)$
6. $(-32)(-28)$
7. $(-45)(-18)$
8. $(-14)(-55)$
9. $(-5)(-125)$
10. $(-221)(-9)$
11. $(-8)(-374)$
12. $(-458)(-7)$
13. $(-14)(-107)$
14. $(-186)(-25)$
15. $(-36)(-217)$
16. $(-412)(-52)$
17. $(-3)(-2.2)$
18. $(-1.4)(-2)$
19. $(-2.07)(-4)$
20. $(-0.9)(-5)$
21. $(-1.5)(-3.8)$
22. $(-0.6)(-0.8)$
23. $(-5.07)(-4.2)$
24. $(-2.7)(-21.51)$
25. $\left(-\frac{2}{3}\right)\left(-\frac{3}{4}\right)$
26. $\left(-\frac{3}{8}\right)\left(-\frac{4}{5}\right)$
27. $\left(-4\frac{1}{4}\right)\left(-2\frac{2}{5}\right)$
28. $\left(-3\frac{3}{10}\right)\left(-\frac{5}{11}\right)$
29. $\left(-\frac{1}{2}\right)\left(-\frac{1}{4}\right)$
30. $\left(-\frac{2}{5}\right)\left(-\frac{3}{4}\right)$
31. $\left(-1\frac{1}{4}\right)\left(-3\frac{1}{3}\right)$
32. $\left(-2\frac{2}{5}\right)\left(-1\frac{3}{8}\right)$

MORE CHALLENGING EXERCISES

Evaluate each expression.

33. $(-2)^2 + (-3)^2$
34. $(-1.1)^2 + (-0.2)^2$
35. $(-9 + 6)(-12)$
36. $(-1)^2 + 12(-13 + 10)$
37. $-8(-8 + 2) + (-2)^2$
38. $(-21 + 7)(-3) + (-4)^2$
39. $(-1)^2 \cdot (-1) \cdot (-1)$
40. $(-5)(-5)(-5)(-1)(-2)$

Multiply the monomials.

41. $(-6)(-3a)$
42. $(-2b)(-9)$
43. $(-3x)(-5x)$
44. $(-10c)(-4c)$
45. $(-5ab)(-2ab)$
46. $(-6p^2)(-8p)$
47. $(-3xy)(-4x^2)$
48. $(-2pt)(-7pt)$

354 Chapter 12

MULTIPLYING RATIONAL NUMBERS

A calculator with a [%] or [sc] key enables you to multiply two or more rational numbers.

EXAMPLE 1 Multiply: $(-928)(57)(-1239)$

Solution 9 2 8 [+/-] [×] 5 7

[×] 1 2 3 9 [+/-] [=]   `65538144.`

NOTE: To multiply without the "sign change" key, observe that the answer is:
a. <u>positive</u>, when the number of multiplied negative numbers is <u>even</u>,
b. <u>negative</u>, when the number of multiplied negative numbers is <u>odd</u>.

EXAMPLE 2 Multiply: $(-3.6)(-4)(-6.5)$ Do not use the "sign change" key.

Solution 3 . 6 [×] 4 [×] 6 . 5 [=]   `93.6`

Since <u>three</u> negative numbers are multiplied, the answer is -93.6.

EXERCISES Multiply

1. $(-128)(209)$
2. $(436)(-157)$
3. $(-89.6)(-56.3)$
4. $(-37.59)(-92.47)$
5. $(-0.33829)(0.48)$
6. $(-0.5286)(-0.8025)$
7. $(-48)(-29)(37)$
8. $(-65)(44)(-23)$
9. $(-28.6)(-42.3)(-38.7)$

REVIEW CAPSULE FOR SECTION 12-3

Evaluate each expression. (Pages 350-352)

1. ab when $a = 16$ and $b = -3$
2. xy when $x = -4$ and $y = 12$
3. cd when $c = -15$ and $d = 20$
4. rt when $r = 14$ and $t = -11$
5. pq when $p = 7$ and $q = 12$
6. fg when $f = 24$ and $g = -8$

Multiply. (Pages 67–69)

7. $10(5y)$
8. $4(12t)$
9. $(9r)9$
10. $(4k)7$
11. $(8a)(7a)$
12. $(2x)(x)$
13. $(3z)(6z)$
14. $(5p)(3p)$

The answers are on page 370.

Rational Numbers: Multiplication/Division **355**

CALCULATOR EXERCISES
1. $-26,752$
2. $-68,452$
3. 5044.48
4. 3475.9473
5. -0.1623792
6. 0.42420150
7. $51,504$
8. $65,780$
9. $46,818.486$

WRITTEN EXERCISES p. 366
27. $t(t + 2)$
28. $p(5 - p)$
29. $13(y + 1)$
30. $5(p - 7)$
31. $7(n + 6)$
32. $2(1 - y)$
33. $3(t + 4)$
34. $d(d + 3)$
35. $5(a - 4b)$
36. $3(12 - x)$
37. $x(x - 12)$
38. $3(7n + p)$
39. $y(9 - y)$
40. $8(t - 1)$
41. $3(x^2 + 3x + 5)$
42. $2(2y^2 + 5y + 4)$
43. $5(a^2 + 2a + 3)$
44. $6(c^2 + 2c + 4)$
45. $7(d^2 + 2d + 4)$
46. $4(2n^2 + 5n + 8)$
47. $3(6m^2 + 5m + 9)$
48. $5(2x^2 + 6x + 3)$
49. $2(x^2 - x - 1)$

355

12-3 Properties of Multiplication

The following properties that apply to whole numbers are also true for rational numbers.

> **Commutative Property of Multiplication**
>
> For any rational numbers a and b,
> $$ab = ba. \qquad (-7)(4) = 4(-7)$$
>
> **Associative Property of Multiplication**
>
> For any rational numbers a and b,
> $$(ab)c = a(bc). \qquad (-3 \cdot 5)6 = -3(5 \cdot 6)$$
>
> **Distributive Property of Multiplication over Addition**
>
> For any rational numbers a, b, and c,
> 1. $a(b + c) = ab + ac$, and $\qquad 2(-4 + 8) = 2(-4) + 2(8)$
> 2. $(b + c)a = ba + ca.$ $\qquad (x + 0.5)3 = x(3) + 0.5(3)$
>
> **Distributive Property of Multiplication over Subtraction**
>
> For any rational numbers a, b, and c,
> 1. $a(b - c) = ab - ac$, and $\qquad 3(4 - 1) = 3(4) - 3(1)$
> 2. $(b - c)a = ba - ca.$ $\qquad (x - 5)2 = x(2) - 5(2)$

The Distributive Property can be used to write products as sums or differences.

EXAMPLE 1 Multiply: $4(2x + 3)$

Solution: [1] Use the Distributive Property. $\quad 4(2x + 3) = 4(2x) + 4(3)$

[2] Multiply. Write in simplest form. $\qquad\qquad\qquad = 8x + 12 \quad \blacktriangleleft$ Sum

Self-Check *Multiply:* **1.** $2(5n + 4)$ **2.** $3(-8y + 6)$ **3.** $(p + 7)(4)$

EXAMPLE 2 Multiply: $-2(-6x + 5)$

Solution: [1] $-2(-6x + 5) = (-2)(-6x) + (-2)(5)$
[2] $\qquad\qquad\qquad = 12x + (-10)$
$\qquad\qquad\qquad\quad = 12x - 10$

356 Chapter 12

Self-Check Multiply: **4.** $(4p + 7)(-3)$ **5.** $(x + 12)(-4)$ **6.** $-2(4x + 1)$

The answers are on page 356.

This example shows multiplication by a term containing a variable.

EXAMPLE 3 Multiply: $(4y - 1)3y$

Solution:
1. $(4y - 1)3y = (4y)(3y) - (1)(3y)$
2. $\qquad\qquad\quad = 12y^2 - 3y$ ◀ Difference

Self-Check Multiply: **7.** $(4x - 2)6x$ **8.** $(5t - 2)(3t)$ **9.** $-2r(3r + 1)$

CLASSROOM EXERCISES

For Exercises 1-12, complete each problem. (Example 1)

1. $2(x + 5) = 2(x) + 2(5)$
$\qquad\quad = \underline{\;?\;} + 10$

2. $(3g + 2)5 = \underline{\;?\;} + \underline{\;?\;}$
$\qquad\qquad = 15g + 10$

(Example 2)

3. $-2(y + 4) = -2(y) + (-2)(4)$
$\qquad\qquad = -2y + \underline{\;?\;}$
$\qquad\qquad = -2y - 8$

4. $-6(p + 5) = \underline{\;?\;} + (-6)(5)$
$\qquad\qquad = -6p + (-30)$
$\qquad\qquad = -6p - 30$

5. $(2x + 6)(-3) = 2x(-3) + \underline{\;?\;}$
$\qquad\qquad\;\; = \underline{\;?\;} + (-18)$
$\qquad\qquad\;\; = -6x - 18$

6. $(-3c + 2)(-4) = \underline{\;?\;} + 2(-4)$
$\qquad\qquad\quad = 12c + \underline{\;?\;}$
$\qquad\qquad\quad = 12c - 8$

(Example 3)

7. $2n(n - 5) = (2n)(n) - (2n)(5)$
$\qquad\quad = \underline{\;?\;} - 10n$

8. $(5b - 6)4b = (5b)(4b) - \underline{\;?\;}$
$\qquad\qquad = 20b^2 - 24b$

WRITTEN EXERCISES

Goal: To use the Distributive Property to express a product as a sum or difference

For Exercises 1-39, multiply. Show all work. (Example 1)

1. $5(3s + 10)$
2. $4(m + 9)$
3. $2(4x + 7)$
4. $3(5p + 6)$
5. $(2n + 3)7$
6. $(y + 4)8$
7. $6(4q + 1)$
8. $8(-a + 9)$
9. $9(-3c + 2)$

Rational Numbers: Multiplication/Division   357

WRITTEN EXERCISES

28. $-7p - 7$
29. $8x + 20$
30. $2y^2 - 3y$
31. $2t + 18$
32. $2m^2 - 2m$
33. $-3p - 15$
34. $3a^2 - 18a$
35. $-12n - 4$
36. $16c + 48$
37. $40b + 70$
38. $2q - 7$
39. $16d^2 - 8d$
40. $6x + 20$
41. $4n + 12$
42. $-17a + 16$
43. $-b - 6$
44. $12c - 10$
45. $-14p + 12$
46. $3a^2 + a$
47. $m^2 + 3md$
48. $3z^2 - 3z$

QUIZ: SECTIONS 12-1—12-3

After completing this Review, you may wish to administer a quiz covering the same sections. A Quiz is provided in the Teacher's Edition: Part II.

(Example 2)

10. $-1(t + 4)$ $-t - 4$
11. $-2(r + 7)$ $-2r - 14$
12. $-4(x + 2)$ $-4x - 8$
13. $-3(-2p + 1)$ $6p - 3$
14. $-2(-3y + 5)$ $6y - 10$
15. $(-4c + 3)(-1)$ $4c - 3$
16. $(-2m + 6)(-4)$ $8m - 24$
17. $-5(z + (-2))$ $-5z + 10$
18. $-6(2d + (-5))$ $-12d + 30$

(Example 3)

19. $(a - 1)a$ $a^2 - a$
20. $(r - 3)r$ $r^2 - 3r$
21. $(q - 2)q$ $q^2 - 2q$
22. $2x(x - 3)$ $2x^2 - 6x$
23. $3y(y - 4)$ $3y^2 - 12y$
24. $(n - 6)(5n)$ $5n^2 - 30n$
25. $(p - 2)(6p)$ $6p^2 - 12p$
26. $2b(2b - 1)$ $4b^2 - 2b$
27. $4x(3x - 7)$ $12x^2 - 28x$

MIXED PRACTICE

28. $-7(p + 1)$
29. $4(2x + 5)$
30. $y(2y - 3)$
31. $2(t + 9)$
32. $2m(m - 1)$
33. $(p + 5)(-3)$
34. $(a - 6)(3a)$
35. $-4(3n + 1)$
36. $8(2c + 6)$
37. $10(4b + 7)$
38. $(-1)(-2q + 7)$
39. $4d(4d - 2)$

MORE CHALLENGING EXERCISES

Multiply, then combine like terms.

40. $2(x + 10) + 4x$
41. $3(2n + 4) - 2n$
42. $-4(3a - 4) + -5a$
43. $-4b + 3(b - 2)$
44. $6c + 2(3c - 5)$
45. $-9p - 4(2p + -3) + 3p$
46. $a(3a - 1) + 2a$
47. $m(m + d) + 2md$
48. $-5z - z(4 - 3z) + 6z$

REVIEW: SECTIONS 12-1—12-3

For Exercises 1-16, multiply. (Section 12-1)

1. $8(-5)$ -40
2. $10(-6)$ -60
3. $(-15)9$ -135
4. $(-17)(21)$ -357
5. $-42(16)$ -672
6. $35(-24)$ -840
7. $(-126)(25)$ -3150
8. $(57)(-206)$ $-11,742$

(Section 12-2)

9. $(-8)(-4)$ 32
10. $(-5)(-3)$ 15
11. $(-9)(-7)$ 63
12. $(-16)(-32)$ 512
13. $(-25)(-40)$ 1000
14. $(-38)(-118)$ 4484
15. $(-12)(-308)$ 3696
16. $(-429)(-67)$ $28,743$

Multiply. Show all work. (Section 12-3)

17. $2(5a + 3)$ $10a + b$
18. $(b + 8)6$ $6b + 48$
19. $-4(x + 2)$ $-4x - 8$
20. $(3t + 1)(-5)$ $-15t - 5$
21. $y(y - 5)$ $y^2 - 5y$
22. $(2r - 1)4r$ $8r^2 - 4r$
23. $(-5c)(4c + 3)$ $-20c^2 - 15c$
24. $6v(7 - 3v)$ $-18v^2 + 42v$

CONSUMER APPLICATION
Conserving Energy

In warm weather, outside temperatures are usually higher than inside temperatures. The following formula is used to estimate the rate at which heat passes from the outside to the inside of a building through the walls and windows. This **heat transfer** is measured in British Thermal Units (BTU's) per hour.

Heat transfer $= A \cdot U(i - o)$

A = surface area in ft^2
U = heat transfer factor
i = inside temperature in °F
o = outside temperature in °F

The variable U in the formula represents a number that varies depending on the type of surface through which the heat passes.

Surface	Value of U
Concrete, 6 inches thick	0.58
Glass, single pane	1.13
Brick, 8 inches thick	0.41
Wood, 2 inches thick	0.43

EXAMPLE Estimate the rate of heat transfer through a 200 foot wall of 8-inch brick. The inside temperature is 74°F and the outside temperature is 88°F.

SOLUTION Heat transfer $= A \cdot U(i - o)$ $A = 200; U = 0.41; i = 74; o = 88$

Heat transfer $= (200)(0.41)(74 - 88)$
Heat transfer $= 82(-14)$
Heat transfer $= -1148$ ← The negative number means that heat is gained.

Thus, heat is being gained at a rate of **1148 BTU's per hour**.

Exercises

Find the rate of heat transfer.

1. Surface: glass
   $A = 50$ ft^2; $i = 78$°F; $o = 90$°F

2. Surface: concrete
   $A = 300$ ft^2; $i = 75$°F; $o = 95$°F

3. Surface: wood
   $A = 400$ ft^2; $i = 72$°F; $o = 85$°F

4. Surface: brick
   $A = 100$ ft^2; $i = 70$°F; $o = 84$°F

CONSUMER APPLICATIONS
This feature applies the skills for multiplying rational numbers to calculating the rate of heat transfer through a surface. Because this feature requires use of the Distributive Property to express products as a sum or a difference, it can be used after Section 12-3 (pages 356-358) is taught.

1. -678 BTU's
2. -3480 BTU's
3. -2236 BTU's
4. -574 BTU's

Teaching Suggestions
p. M-25

QUICK QUIZ
Evaluate.
1. $\frac{136}{x}$ when x = 17
   Ans: 8
2. y ÷ 3 when y = 2724
   Ans: 908
3. a ÷ b when a = 216 and b = 27   Ans: 8
4. x ÷ y when x = 96 and y = 1.2   Ans: 80
5. $\frac{m}{n}$ when m = 10.2 and n = 0.17   Ans: 60

REVIEW: SECTIONS
12-4–12-5
29. x(x − 2)
30. x(x + 12)
31. 2(x + 7)
32. x(x − 3)
33. 7(x − 6)
34. 3(5 + x)
35. 6(3 + x)
36. 4(3 + x)
37. x(16 − x)
38. x(3 + x)
39. x(7 + x)
40. x(13 − x)

360

REVIEW CAPSULE FOR SECTION 12-4

Evaluate each expression.   (Pages 124-125, 148-151)

1. $m \div p$ when $m = 81$ and $p = 9$
2. $\frac{c}{d}$ when $c = 186$ and $d = 3$
3. $\frac{375}{t}$ when $t = 15$
4. $a \div b$ when $a = 6$ and $b = \frac{2}{3}$
5. $\frac{p}{q}$ when $p = \frac{3}{4}$ and $q = 9$
6. $\frac{t}{r}$ when $t = \frac{1}{3}$ and $r = 6$

The answers are on page 370.

12-4 Dividing Rational Numbers

A multiplication problem has two related division problems.

TABLE 1

Multiplication	Related Division
4 · 8 = 32	32 ÷ 4 = 8 and 32 ÷ 8 = 4

You can use this fact to find the pattern for dividing a positive number by a negative number.

TABLE 2

Multiplication	Related Division
(−3)(−8) = 24	24 ÷ (−3) = −8 and 24 ÷ (−8) = −3
(−5)(−7) = 35	35 ÷ (−5) = −7 and 35 ÷ (−7) = −5

Conclusion: The quotient of a positive number and a negative number is negative.

Now study these patterns.

TABLE 3

Multiplication	Related Division
−5(3) = −15	−15 ÷ (−5) = 3 and −15 ÷ 3 = −5
−9(6) = −54	−54 ÷ (−9) = 6 and −54 ÷ 6 = −9

Conclusion: The quotient of two negative numbers is positive.

Conclusion: The quotient of a negative number and a positive number is negative.

You can use these patterns to divide integers.

360   Chapter 12

PROCEDURE
To divide two rational numbers:

[1] Divide.

[2] **a.** When the two numbers have like signs, the quotient is positive.

 b. When the two numbers have unlike signs, the quotient is negative.

EXAMPLE 1 Divide: **a.** $-36 \div (-4)$ **b.** $-27 \div 9$ **c.** $12 \div (-2)$

Solutions: **a.** **b.** **c.**

[1] $36 \div 4 = 9$ [1] $27 \div 9 = 3$ [1] $12 \div 2 = 6$

[2a] $-36 \div (-4) = \mathbf{9}$ [2b] $-27 \div 9 = \mathbf{-3}$ [2b] $12 \div (-2) = \mathbf{-6}$

Self-Check Divide: **1.** $-18 \div (-2)$ **2.** $-28 \div 4$ **3.** $\frac{-36}{6}$ **4.** $\frac{24}{-3}$

Recall that two numbers are reciprocals if their product is 1.

$$\tfrac{2}{1} \cdot \tfrac{1}{2} = 1 \qquad \left(-\tfrac{3}{8}\right)\left(-\tfrac{8}{3}\right) = 1 \qquad \left(-\tfrac{5}{4}\right)\left(-\tfrac{4}{5}\right) = 1$$

Reciprocals: 2 and $\tfrac{1}{2}$ $-\tfrac{3}{8}$ and $-\tfrac{8}{3}$ $-\tfrac{5}{4}$ and $-\tfrac{4}{5}$

Dividing by a number is the same as multiplying by its reciprocal.

EXAMPLE 2 Divide: **a.** $-30 \div \tfrac{1}{6}$ **b.** $-16 \div \left(-\tfrac{4}{3}\right)$

Solutions: **a.** $-30 \div \tfrac{1}{6} = -30 \cdot 6$
$= -180$

b. $-16 \div \left(-\tfrac{4}{3}\right) = -\tfrac{\cancel{16}^{4}}{1} \cdot \left(-\tfrac{3}{\cancel{4}_{1}}\right)$

$= \left(-\tfrac{4}{1}\right)\left(-\tfrac{3}{1}\right)$

$= \tfrac{12}{1}$, or **12**

Self-Check Divide: **5.** $-4 \div \tfrac{1}{3}$ **6.** $\tfrac{4}{5} \div (-8)$ **7.** $\tfrac{7}{8} \div \left(-\tfrac{1}{4}\right)$

In Algebra	In Arithmetic
$a \div b = a \cdot \tfrac{1}{b} (b \neq 0)$	$8 \div (-3) = 8 \cdot \left(-\tfrac{1}{3}\right)$

Rational Numbers: Multiplication/Division **361**

ADDITIONAL EXAMPLES
Example 1
Divide.
1. 1206 ÷ (-6)
   Ans: -201
2. -198 ÷ (-2)
   Ans: 9

SELF-CHECK
1. 9   2. -7
3. -6   4. -8

Example 2
Divide.
1. (-102) ÷ (-$\tfrac{1}{2}$)
   Ans: 240
2. (-$\tfrac{2}{3}$) ÷ ($\tfrac{7}{6}$)
   Ans: $\tfrac{4}{7}$

SELF-CHECK
5. -12   6. $\tfrac{1}{10}$
7. -3$\tfrac{1}{2}$

CLASSROOM EXERCISES
1. 8; 6   2. -6; -4
3. 7; -8 4. -4; 5
5. 6   6. 3   7. -9
8. -9 9. -10 10. -7
11. 11 12. -10
13. -6 14. 12
15. -4 16. -19
17. -12 18. -15
19. 10 20. 10
21. $-\frac{1}{5}$ 22. $-\frac{2}{3}$
23. $1\frac{1}{2}$ 24. $1\frac{2}{7}$

ASSIGNMENT GUIDE
BASIC
pp. 362-363: 1-35 odd
AVERAGE
p. 362: 1-36
ABOVE AVERAGE
pp. 362-363: 9-31 odd, 37-56

WRITTEN EXERCISES
1. 9   2. 5   3. 7
4. 7   5. -3 6. -9
7. -7 8. -4 9. -3
10. -9 11. -18
12. -7 13. 5 14. 3
15. -15 16. -4
17. -8 18. -5
19. -23 20. 7
21. -5 22. 14 23. -7
24. 25 25. -48
26. -30 27. $\frac{1}{27}$
28. $\frac{5}{96}$ 29. -6
30. $-\frac{2}{3}$ 31. $-1\frac{1}{2}$
32. $-2\frac{1}{4}$ 33. $1\frac{2}{9}$

CLASSROOM EXERCISES

Complete. (Tables 1-3)

1. $6 \cdot 8 = 48$
   $48 \div 6 = \underline{\ ?\ }$
   $48 \div 8 = \underline{\ ?\ }$
2. $-4(-6) = 24$
   $24 \div (-4) = \underline{\ ?\ }$
   $24 \div (-6) = \underline{\ ?\ }$
3. $-8(7) = -56$
   $-56 \div (-8) = \underline{\ ?\ }$
   $-56 \div 7 = \underline{\ ?\ }$
4. $5(-4) = -20$
   $-20 \div 5 = \underline{\ ?\ }$
   $-20 \div (-4) = \underline{\ ?\ }$

Divide. (Example 1)

5. $(-54) \div (-9)$
6. $-24 \div (-8)$
7. $63 \div (-7)$
8. $81 \div (-9)$
9. $\frac{-100}{10}$
10. $\frac{42}{-6}$
11. $\frac{-99}{-9}$
12. $\frac{-20}{2}$
13. $30 \div (-5)$
14. $-72 \div (-6)$
15. $-44 \div 11$
16. $19 \div (-1)$

(Example 2)

17. $-3 \div \frac{1}{4}$
18. $-5 \div \frac{1}{3}$
19. $-6 \div \left(-\frac{3}{5}\right)$
20. $-8 \div \left(-\frac{4}{5}\right)$
21. $-\frac{1}{4} \div \frac{5}{4}$
22. $\frac{8}{9} \div \left(-\frac{4}{3}\right)$
23. $-\frac{12}{5} \div \left(-\frac{8}{5}\right)$
24. $-\frac{6}{7} \div \left(-\frac{2}{3}\right)$

WRITTEN EXERCISES

Goal: To divide two rational numbers

Divide. (Example 1)

1. $-36 \div (-4)$
2. $-30 \div (-6)$
3. $-63 \div (-9)$
4. $-35 \div (-5)$
5. $21 \div (-7)$
6. $81 \div (-9)$
7. $49 \div (-7)$
8. $36 \div (-9)$
9. $-39 \div 13$
10. $-45 \div 5$
11. $-72 \div 4$
12. $-14 \div 2$
13. $\frac{-15}{-3}$
14. $\frac{-24}{-8}$
15. $\frac{-75}{5}$
16. $\frac{-48}{12}$
17. $\frac{64}{-8}$
18. $\frac{100}{-20}$
19. $\frac{-138}{6}$
20. $\frac{-98}{-14}$
21. $200 \div (-40)$
22. $-98 \div (-7)$
23. $133 \div (-19)$
24. $-200 \div (-8)$

(Example 2)

25. $-16 \div \frac{1}{3}$
26. $-25 \div \frac{5}{6}$
27. $-\frac{2}{3} \div (-18)$
28. $-\frac{5}{8} \div (-12)$
29. $\frac{4}{3} \div \left(-\frac{2}{9}\right)$
30. $\frac{1}{2} \div \left(-\frac{3}{4}\right)$
31. $-\frac{3}{8} \div \frac{1}{4}$
32. $-\frac{9}{10} \div \frac{2}{5}$
33. $-\frac{11}{12} \div \left(-\frac{3}{4}\right)$
34. $-\frac{7}{10} \div \left(-\frac{3}{4}\right)$
35. $4 \div \left(-\frac{2}{5}\right)$
36. $9 \div \left(-\frac{3}{7}\right)$

MORE CHALLENGING EXERCISES

EXAMPLE: $\dfrac{w}{-3} \div \dfrac{2w}{9} = \dfrac{w}{-3} \cdot \dfrac{9}{2w}$ ◄ The reciprocal of $\dfrac{9}{2w}$ is $\dfrac{2w}{9}$.

$$= \dfrac{\cancel{w}^{\,1}}{\cancel{-3}_{-1}} \cdot \dfrac{\cancel{9}^{\,3}}{\cancel{2w}_{2}}$$

$$= \dfrac{1 \cdot 3}{-1 \cdot 2} = \dfrac{3}{-2}, \text{ or } -1\tfrac{1}{2}$$

37. $\dfrac{a}{-4} \div 2$
38. $\dfrac{2a}{5} \div a$
39. $\dfrac{3e}{-7} \div \dfrac{-9}{14}$
40. $\dfrac{4}{9} \div \dfrac{-22}{27g}$

41. $\dfrac{b}{-6} \div (-3)$
42. $\dfrac{5v}{-7} \div v$
43. $\dfrac{6h}{-5} \div \dfrac{12}{25}$
44. $\dfrac{-8}{13} \div \dfrac{-4}{39k}$

45. $\dfrac{7}{x} \div (-7)$
46. $\dfrac{z}{-3} \div \dfrac{1}{3}$
47. $\dfrac{-1}{9} \div \dfrac{1}{-9}$
48. $\dfrac{11}{-3m} \div \dfrac{22}{3}$

49. $\dfrac{y}{4} \div \dfrac{y}{c}$
50. $(-d) \div \dfrac{1}{d}$
51. $(-10) \div \dfrac{f}{10}$
52. $(-6) \div \dfrac{18}{-5s}$

The *Closure Property for Addition* states that the sum of two rational numbers is a rational number. The *Closure Property for Multiplication* states that the product of two rational numbers is a rational number.

53. Does the Closure Property hold for subtraction of whole numbers; that is, is the difference of two whole numbers always a whole number? Give a reason for your answer.

54. Does the Closure Property hold for division of natural numbers; that is, is the quotient of two natural numbers always a natural number? Give a reason for your answer.

55. Does the Closure Property hold for division of rational numbers?

56. Does the Closure Property hold for multiplication of integers?

REVIEW CAPSULE FOR SECTION 12-5

Multiply. (Pages 356-358) The answers are on page 370.

1. $2(a + 4)$
2. $3(b + 1)$
3. $6(x - 3)$
4. $5(y - 5)$
5. $4(a - b)$
6. $7(x + 2)$
7. $x(x + 4)$
8. $d(d - 3)$
9. $(5 + m)m$
10. $(6 - n)n$
11. $(r + 7)r$
12. $(t - 4)8$

Find the GCF of each pair of numbers. (Pages 115-117)

13. 4, 8
14. 6, 12
15. 8, 12
16. 4, 10
17. 6, 9
18. 12, 15

Rational Numbers: Multiplication/Division   **363**

---

34. $\dfrac{14}{15}$ 35. -10
36. -21
37. $-\dfrac{a}{8}$, or $-\dfrac{1}{8}a$ 38. $\dfrac{2}{5}$
39. $\dfrac{2e}{3}$, or $\dfrac{2}{3}e$
40. $-\dfrac{6}{11g}$
41. $\dfrac{b}{18}$, or $\dfrac{1}{18}b$ 42. $-\dfrac{5}{7}$
43. $-\dfrac{5h}{2}$, or $-\dfrac{5}{2}h$
44. 6k   45. $-\dfrac{1}{x}$
46. -z   47. 1
48. $-\dfrac{1}{2m}$ 49. $\dfrac{c}{4}$, or $\dfrac{1}{4}c$
50. $-d^2$ 51. $-\dfrac{100}{f}$
52. $\dfrac{5s}{3}$, or $\dfrac{5}{3}s$
53. No; difference may be a negative integer.
54. No; quotient may be a fraction.
55. Yes   56. Yes

ADDITIONAL PRACTICE
p. 372
48. -180   49. $\dfrac{2}{3}$ 50. $\dfrac{2}{7}$
51. -10   52. $-\dfrac{1}{4}$
53. 14(x + 1)
54. 6(m + 6)
55. 7(3m + 1)
56. 7(v + 1)
57. 7(p - 5)
58. 5(t - 3)
59. 4(c - 1)
60. 4(a - 5)
61. 4(9 - v)
62. 7(7 - y)
63. x(x - 6)
64. x(x - 8)
65. 4(-15); -60
66. 5(-3); -15

363

12-5 Factoring

Teaching Suggestions
p. M-25

QUICK QUIZ

Find the GCF of each pair of numbers.

<u>1</u>. 20, 12   Ans: 4

<u>2</u>. 36, 6   Ans: 6

Multiply.

<u>3</u>. 3(x + 8)
      Ans: 3x + 24

<u>4</u>. c(c - 8)
      Ans: c^2 - 8c

<u>5</u>. -4(3a - 2)
      Ans: -12a + 8

You have used the Distributive Property to express a product as a sum or as a difference.

Product	Sum	Product	Difference
3(4 + 6)	3(4) + 3(6)	2(8 − 5)	2(8) − 2(5)
2(x + 4)	2x + 8	3(y − 2)	3y − 6

The Distributive Property is also used to express a sum or a difference as a product. This is called **factoring**.

Sum or Difference		Product
2(5) + 2(6)	→	2(5 + 6)
4(8) − 4(3)	→	4(8 − 3)
3a + 6	→	3(a + 2)
5b − 10	→	5(b − 2)

To factor a sum or a difference, find all the number factors and variable factors shared by each term. These **common factors** become one of the factors of the product.

PROCEDURE

To factor using the Distributive Property:

[1] Write each term of the sum or difference as the product of factors.

[2] Use the common factor(s) to write a product.

ADDITIONAL EXAMPLES

Example 1

Factor.

<u>1</u>. 7x + 14
    Ans: 7(x + 2)

<u>2</u>. 11 + 22y
    Ans: 11(1 + 2y)

EXAMPLE 1 Factor: 5y + 5

Solution: [1] Write the factors of each term. $5y + 5 = 5 \cdot y + 5 \cdot 1$

◄ 5 is the common factor.

[2] Use the common factor, 5, to write a product. $= 5(y + 1)$

◄ By the Distributive Property

SELF-CHECK

<u>1</u>. 3(a + 1)

<u>2</u>. 7(x + 1)

<u>3</u>. 2(2 + p)

<u>4</u>. 5(3 + n)

Self-Check *Factor.*

1. 3a + 3   2. 7x + 7   3. 4 + 2p   4. 15 + 5n

364 Chapter 12

Example 2 shows how to write a difference as a product.

EXAMPLE 2 Factor: $3p - 6$
① $3p - 6 = 3 \cdot p - 3 \cdot 2$ ◀ **3 is the common factor.**
② $ = 3(p - 2)$

Self-Check Factor: **5.** $3x - 6$ **6.** $5m - 10$ **7.** $12n - 3$

Sometimes the common factor is a variable such as the variable x in Example 3.

EXAMPLE 3 Factor: $x^2 + 5x$
① $x^2 + 5x = x \cdot x + 5 \cdot x$ ◀ **x is the common factor.**
$ = x(x + 5)$

Self-Check Factor: **8.** $t^2 + 3t$ **9.** $p^2 + 5p$ **10.** $4y - y^2$

CLASSROOM EXERCISES

Identify the common factor of each pair of terms.
(Examples 1-3, Step 2)

1. $2x, 2$ **2.** $5m, 5$ **3.** $2t, 6r$ **4.** $7n, 21p$
5. $15, 3y$ **6.** $t^2, 3t$ **7.** $m^2, 6m$ **8.** $6a, 3$

For Exercises 9–32, one factor is given for each product. Write the other factor. (Example 1)

9. $3x + 3 = \underline{}(x + 1)$
10. $5p + 5 = \underline{}(p + 1)$
11. $2m + 10 = \underline{}(m + 5)$
12. $3q + 18 = \underline{}(q + 6)$
13. $5r + 35 = 5(\underline{})$
14. $7p + 49 = 7(\underline{})$
15. $18 + 2t = 2(\underline{})$
16. $21 + 3r = 3(\underline{})$

(Example 2)

17. $3p - 6 = \underline{}(p - 2)$
18. $5r - 10 = \underline{}(r - 2)$
19. $12m - 3 = \underline{}(4m - 1)$
20. $7x - 42 = \underline{}(x - 6)$
21. $2r - 2 = 2(\underline{})$
22. $9m - 3 = 3(\underline{})$
23. $4x - 2 = 2(\underline{})$
24. $15a - 3 = 3(\underline{})$

Rational Numbers: Multiplication/Division   **365**

ADDITIONAL EXAMPLES
Example 2
Factor.
1. 4 - 4x
   Ans: 4(1 - x)
2. 15a - 10
   Ans: 5(3a - 2)

SELF-CHECK
5. 3(x - 2) 6. 5(m - 2)
7. 3(4n - 1)

Example 3
Factor.
1. 6y + xy
   Ans: y(6 + x)
2. 8a - a²
   Ans: a(8 - a)

SELF-CHECK
8. t(t + 3)   9. p(p + 5)
10. y(4 - y)

CLASSROOM EXERCISES
1. 2   2. 5   3. 2
4. 7   5. 3   6. t
7. m   8. 3   9. 3
10. 5   11. 2   12. 3
13. r + 7   14. p + 7
15. 9 + t   16. 7 + r
17. 3   18. 5   19. 3
20. 7   21. r - 1
22. 3m - 1   23. 2x - 1
24. 5a - 1   25. n
26. a   27. t + 10
28. y + 100   29. x
30. t   31. 12 - c
32. 9 - m

ASSIGNMENT GUIDE
BASIC
p. 366: Omit
AVERAGE
p. 366: 1-39 odd
ABOVE AVERAGE
p. 366: 1-49 odd

WRITTEN EXERCISES
1. $2(r + 1)$
2. $7(y + 1)$
3. $3(r + 1)$
4. $5(t + 1)$
5. $3(y + 5)$
6. $5(3 + m)$
7. $7(3 + c)$
8. $3(6 + d)$
9. $7(n - 2)$
10. $13(r - 2)$
11. $5(3a - 1)$
12. $7(5x - 1)$
13. $7(n - 4)$
14. $11(2t - 1)$
15. $3(p - 4)$
16. $2(c - 1)$
17. $y(y + 5)$
18. $p(p + 9)$
19. $c(c - 4)$
20. $t(t - 5)$
21. $x(13 + x)$
22. $b(15 - b)$
23. $r(r + 1)$
24. $q(1 + q)$
25. $r(r - 1)$
26. $q(1 - q)$

See page 355 for the answers to Exercises 27-49.

366

(Example 3)

25. $n^2 + 2n = \underline{}(n + 2)$
26. $a^2 + 5a = \underline{}(a + 5)$
27. $t^2 + 10t = t(\underline{})$
28. $y^2 + 100y = y(\underline{})$
29. $5x - x^2 = \underline{}(5 - x)$
30. $8t - t^2 = \underline{}(8 - t)$
31. $12c - c^2 = c(\underline{})$
32. $9m - m^2 = m(\underline{})$

WRITTEN EXERCISES

Goal: To factor a sum or difference of a binomial by applying the Distributive Property

For Exercises 1-49, factor. (Example 1)

1. $2n + 2$
2. $7y + 7$
3. $3r + 3$
4. $5t + 5$
5. $3y + 15$
6. $15 + 5m$
7. $21 + 7c$
8. $18 + 3d$

(Example 2)

9. $7n - 14$
10. $13r - 26$
11. $15a - 5$
12. $35x - 7$
13. $7n - 28$
14. $22t - 11$
15. $3p - 12$
16. $2c - 2$

(Example 3)

17. $y^2 + 5y$
18. $p^2 + 9p$
19. $c^2 - 4c$
20. $t^2 - 5t$
21. $13x + x^2$
22. $15b - b^2$
23. $r^2 + r$
24. $q + q^2$
25. $r^2 - r$
26. $q - q^2$
27. $t^2 + 2t$
28. $5p - p^2$

MIXED PRACTICE

29. $13y + 13$
30. $5p - 35$
31. $7n + 42$
32. $2 - 2y$
33. $3t + 12$
34. $d^2 + 3d$
35. $5a - 20b$
36. $36 - 3x$
37. $x^2 - 12x$
38. $21n + 3p$
39. $9y - y^2$
40. $8t - 8$

MORE CHALLENGING EXERCISES

EXAMPLE: $2x^2 + 14x + 10 = 2(x^2 + 7x + 5)$ ▶ The common factor is 2.

41. $3x^2 + 9x + 15$
42. $4y^2 + 10y + 8$
43. $5a^2 + 10a + 15$
44. $6c^2 + 12c + 24$
45. $7d^2 + 14d + 28$
46. $8n^2 + 20n + 32$
47. $18m^2 + 15m + 27$
48. $10x^2 + 30x + 15$
49. $2x^2 - 2x - 2$

366 Chapter 12

DISTRIBUTIVE PROPERTY

When using a calculator to evaluate an expression of the form $(b+c)a$ or $(b-c)a$, it is quicker to use the product form than the sum/difference form.

EXAMPLE Evaluate: $(876 + 958)0.75$

Solution First find the sum. Then find the product.

$$8\ 7\ 6\ \boxed{+}\ 9\ 5\ 8\ \boxed{=}\ \boxed{\times}\ .7\ 5\ \boxed{=}\quad\boxed{1375.5}$$

EXERCISES Evaluate.

1. $(4.5 - 3.9)52$
2. $(94 + 133)0.5$
3. $(6.72 + 8.86)76$
4. $(1258 + 9355)2.5$
5. $(24{,}856 - 19{,}072)0.9$
6. $(12{,}804 + 9496)12.9$
7. $(1092.8 - 948.9)109.2$
8. $(6294 - 4807)0.8$
9. $(8.31 - 5.85)29$
10. $(6255 - 4075)1.4$
11. $(12{,}806.3 - 8948.8)436.8$
12. $(12.064 + 6.066)0.25$

REVIEW: SECTIONS 12-4 — 12-5

For Exercises 1–24, divide. (Section 12-4)

1. $-24 \div (-3)$
2. $-35 \div (-5)$
3. $36 \div (-2)$
4. $81 \div (-3)$
5. $\frac{-45}{15}$
6. $\frac{-56}{7}$
7. $\frac{-64}{-4}$
8. $\frac{-65}{-13}$
9. $\frac{-124}{4}$
10. $\frac{-200}{40}$
11. $\frac{125}{-25}$
12. $\frac{84}{-7}$
13. $300 \div (-30)$
14. $-256 \div 16$
15. $-120 \div (-24)$
16. $350 \div (-50)$
17. $-4 \div \frac{1}{4}$
18. $-5 \div \frac{1}{3}$
19. $8 \div \left(-\frac{4}{5}\right)$
20. $10 \div \left(-\frac{2}{5}\right)$
21. $-\frac{12}{5} \div \left(-\frac{8}{5}\right)$
22. $-\frac{8}{9} \div \left(-\frac{4}{3}\right)$
23. $\frac{7}{8} \div \left(-\frac{7}{8}\right)$
24. $-\frac{1}{4} \div \frac{5}{4}$

Factor each expression. (Section 12-5)

25. $2x + 2$
26. $5x - 10$
27. $3x - 3$
28. $5x + 15$
29. $x^2 - 2x$
30. $x^2 + 12x$
31. $2x + 14$
32. $x^2 - 3x$
33. $7x - 42$
34. $15 + 3x$
35. $18 + 6x$
36. $12 + 4x$
37. $16x - x^2$
38. $3x + x^2$
39. $7x + x^2$
40. $13x - x^2$

Rational Numbers: Multiplication/Division **367**

CALCULATOR EXERCISES
1. 31.2   2. 113.5
3. 1184.08   4. 26,532.5
5. 5205.6   6. 287,670
7. 15,713.88   8. 1189.6
9. 71.34   10. 3052
11. 1,684,956
12. 4.5325

QUIZ: SECTIONS
12-4–12-5
After completing this
Review, you may wish to
administer a quiz
covering the same
sections. A quiz is
provided in the
Teacher's Edition:
Part II.

REVIEW: SECTIONS
12-4–12-5
1. 8   2. 7   3. -18
4. -27   5. -3   6. -8
7. 16   8. 5   9. -31
10. -5   11. -5
12. -12   13. -10
14. -16   15. 5   16. -7
17. -16   18. -15
19. -10   20. -25
21. $1\frac{1}{2}$ 22. $\frac{2}{3}$ 23. -1
24. $-\frac{1}{5}$ 25. $2(x + 1)$
26. $5(x - 2)$
27. $3(x - 1)$
28. $5(x + 3)$
See page 360 for the
answers to Exercises
29–40.

CHAPTER REVIEW

1. (6)(−3) = (−3)(6)
2. 5(6−3) =
    5(6) − 5(3)
3. 5 · 0 = 0
4. common factors
5. (−2 · 5) 4 =
    −2(5 · 4)
6. −7 · 1 = −7
7. factoring
8. 2(1 + 4) =
    2(1) + 2(4)

9. −108   10. −135
11. −84   12. −75
13. −391   14. −627
15. −960   16. −2240
17. −2394   18. −18,620
19. −4536   20. −17,784
21. −10.44   22. −7.276
23. −$\frac{1}{24}$ 24. −$\frac{1}{4}$
25. 24   26. 27
27. 144   28. 330
29. 1598   30. 1260
31. 3654   32. 6715
33. 1526   34. 2934
35. 2856   36. 15,736
37. 8.64   38. 1.290
39. 3$\frac{3}{4}$ 40. 18$\frac{9}{20}$
41. 2x + 6
42. 3a + 21
43. −4b − 20
44. 6d − 3

CHAPTER REVIEW

PART 1: VOCABULARY

For Exercises 1–8, choose from the box at the right the word(s) or numerical expression(s) that best corresponds to each description.

1. Commutative Property of Multiplication __?__
2. Distributive Property of Multiplication over Subtraction __?__
3. Multiplication Property of Zero __?__
4. All number and variable factors shared by each term __?__
5. Associative Property of Multiplication __?__
6. Multiplication Property of One __?__
7. To write a sum or a difference as a product __?__
8. Distributive Property of Multiplication over Addition __?__

$5 \cdot 0 = 0$
$-7 \cdot 1 = -7$
$(6)(-3) = (-3)(6)$
$(-2 \cdot 5)4 = -2(5 \cdot 4)$
$2(1 + 4) = 2(1) + 2(4)$
$5(6 - 3) = 5(6) - 5(3)$
common factors
factoring

PART 2: SKILLS

For Exercises 9–52, multiply. (Section 12-1)

9. 18(−6)
10. 15(−9)
11. −7(12)
12. −5(15)
13. 17(−23)
14. 11(−57)
15. −20(48)
16. −35(64)
17. 126(−19)
18. 532(−35)
19. −42(108)
20. −78(228)
21. 3.6(−2.9)
22. 1.07(−6.8)
23. $\left(-\frac{1}{4}\right)\left(\frac{1}{6}\right)$
24. $\left(-\frac{2}{5}\right)\left(\frac{5}{8}\right)$

(Section 12-2)

25. (−6)(−4)
26. (−3)(−9)
27. (−12)(−12)
28. (−15)(−22)
29. (−34)(−47)
30. (−28)(−45)
31. (−58)(−63)
32. (−85)(−79)
33. (−7)(−218)
34. (−326)(−9)
35. (−24)(−119)
36. (−281)(−56)
37. (−2.7)(−3.2)
38. (−0.6)(−2.15)
39. $\left(-1\frac{2}{3}\right)\left(-2\frac{1}{4}\right)$
40. $\left(-3\frac{3}{5}\right)\left(-5\frac{1}{8}\right)$

(Section 12-3)

41. 2(x + 3)
42. (a + 7)3
43. −4(b + 5)
44. (−2d + 1)(−3)
45. r(r − 6)
46. 2t(t − 9)
47. 4(−2n + 4)
48. (−3m + 1)5
49. −1(−2p + 3)
50. (−4q + 5)(−2)
51. 3y(2y − 2)
52. (6c − 5)2c

368   Chapter 12

Divide. (Section 12-4)

53. $-50 \div (-5)$
54. $-42 \div (-6)$
55. $-18 \div (-6)$
56. $-85 \div (-5)$
57. $\frac{56}{-14}$
58. $\frac{81}{-3}$
59. $\frac{32}{-8}$
60. $\frac{138}{-6}$
61. $\frac{-51}{3}$
62. $\frac{-132}{22}$
63. $\frac{-65}{5}$
64. $\frac{-212}{4}$
65. $-8 \div \frac{1}{2}$
66. $-16 \div \frac{2}{3}$
67. $-\frac{1}{2} \div \left(-\frac{1}{3}\right)$
68. $-\frac{5}{2} \div \left(-\frac{10}{3}\right)$
69. $\frac{1}{4} \div \left(-\frac{3}{8}\right)$
70. $\frac{3}{5} \div \left(-\frac{3}{5}\right)$
71. $-\frac{1}{4} \div (-5)$
72. $\frac{5}{6} \div (-10)$

Factor each expression. (Section 12-5)

73. $2x + 2$
74. $3b + 12$
75. $k^2 + k$
76. $x - x^2$
77. $39r - 13$
78. $7y + 7$
79. $2 + 2h$
80. $t^2 + 4t$
81. $14 - 7f$
82. $15 - 3n$
83. $13y - y^2$
84. $12 - 3d$

PART 3: APPLICATIONS

Use positive and negative numbers to represent each situation by a multiplication problem. Then find the product. (Section 12-1)

85. Two penalties of 10 yards each
86. Three debts of $42 each
87. Twice a drop in temperature of 6 degrees
88. Four times a depth of 18 meters below sea level
89. Five times a debt of $12
90. Twice a loss of 125 feet in altitude

---

45. $r^2 - 6r$
46. $2t^2 - 18t$
47. $-8n + 16$
48. $-15m + 5$
49. $2p - 3$
50. $8q - 10$
51. $6y^2 - 6y$
52. $12c^2 - 10c$ 53. 10
54. 7   55. 3   56. 17
57. -4 58. -27
59. -4 60. -23
61. -17 62. -6
63. -13 64. -53
65. -16 66. -24
67. $1\frac{1}{2}$ 68. $\frac{3}{4}$ 69. $-\frac{2}{3}$
70. -1 71. $\frac{1}{20}$
72. $-\frac{1}{12}$ 73. $2(x + 1)$
74. $3(b + 4)$
75. $k(k + 1)$
76. $x(1 - x)$
77. $13(3r - 1)$
78. $7(y + 1)$
79. $2(1 + h)$
80. $t(t + 4)$
81. $7(2 - f)$
82. $3(5 - n)$
83. $y(13 - y)$
84. $3(4 - d)$
85. $2(-10); -20$
86. $3(-42); -126$
87. $2(-6); -12$
88. $4(-18); -7$
89. $5(-12); -60$
90. $2(-125); -250$

Rational Numbers: Multiplication/Division

CHAPTER TEST
==========

Two forms of a chapter test, Form A and Form B, are provided on copying masters in the *Teacher's Edition: Part II*.

1. -384   2. 40
3. -306   4. -6.48
5. -7.75   6. -2632
7. 900   8. -1876
9. 2a + 2 10. -3c - 9
11. $4y^2$ - 8y
12. 5t + 20
13. -2r - 12
14. $18x^2$ - 6x   15. 9
16. -4   17. -8
18. -16   19. $-\frac{1}{3}$
20. 15   21. -5
22. $2\frac{1}{2}$ 23. 5(x + 1)
24. 3(p + 4)
25. 2(y - 8)
26. 7(2 - a)
27. m(m + 2)
28. x(1 - x)
29. 5(y - 4)
30. 2(x + 4)

CHAPTER TEST

For Exercises 1–8, multiply.

1. $-16(24)$
2. $(-8)(-5)$
3. $9(-34)$
4. $-1.2(5.4)$
5. $(3.1)(-2.5)$
6. $-47(56)$
7. $(-12)(-75)$
8. $28(-67)$

For Exercises 9–14, multiply. Show all work.

9. $2(a + 1)$
10. $-3(c + 3)$
11. $4y(y - 2)$
12. $(t + 4)5$
13. $(r + 6)(-2)$
14. $(3x - 1)(6x)$

Divide.

15. $-36 \div (-4)$
16. $68 \div (-17)$
17. $\frac{-24}{3}$
18. $\frac{80}{-5}$
19. $\frac{3}{4} \div \left(-\frac{9}{4}\right)$
20. $-120 \div (-8)$
21. $125 \div (-25)$
22. $-\frac{1}{2} \div \left(-\frac{1}{5}\right)$

Factor each expression.

23. $5x + 5$
24. $3p + 12$
25. $2y - 16$
26. $14 - 7a$
27. $m^2 + 2m$
28. $x - x^2$
29. $5y - 20$
30. $2x + 8$

ANSWERS TO REVIEW CAPSULES

Page 353   1. -6 2. -12 3. -15 4. -3 5. 48 6. -40 7. 48 8. 143

Page 355   1. -48 2. -48 3. -300 4. -154 5. 84 6. -192 7. $50y$ 8. $48t$ 9. $81r$ 10. $28k$ 11. $5a^2$ 12. $2x^2$ 13. $18z^2$ 14. $15p^2$

Page 360   1. 9 2. 62 3. 25 4. 9 5. $\frac{1}{12}$ 6. $\frac{1}{18}$

Page 363   1. $2a + 8$ 2. $3b + 3$ 3. $6x - 18$ 4. $5y - 25$ 5. $4a - 4b$ 6. $7x + 14$ 7. $x^2 + 4x$ 8. $d^2 - 3d$ 9. $5m + m^2$ 10. $6n - n^2$ 11. $r^2 + 7r$ 12. $8t - 32$ 13. 4 14. 6 15. 4 16. 2 17. 3 18. 3

370 Chapter 12

ENRICHMENT

Multiplying Binomials

You can use the distributive property to multiply two binomials.

$$(x + 2)(x + 4) = x(x + 4) + 2(x + 4)$$
$$= x \cdot x + x \cdot 4 + 2 \cdot x + 2 \cdot 4$$
$$= x^2 + 4x + 2x + 8$$
$$= x^2 + 6x + 8$$

Here is a short way to multiply binomials. This is called the **FOIL** method.

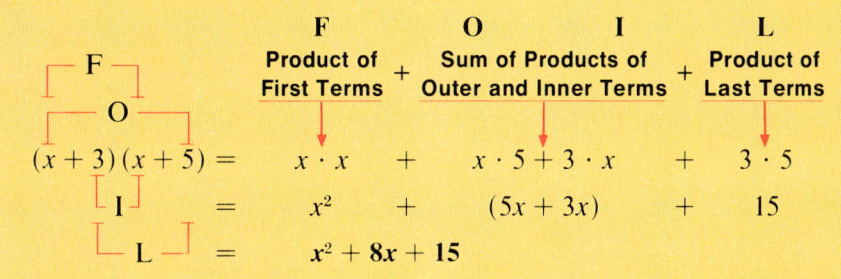

	F Product of First Terms	+	O I Sum of Products of Outer and Inner Terms	+	L Product of Last Terms
$(x + 3)(x + 5) =$	$x \cdot x$	+	$x \cdot 5 + 3 \cdot x$	+	$3 \cdot 5$
=	x^2	+	$(5x + 3x)$	+	15
=	$x^2 + 8x + 15$				

EXERCISES

Multiply.

1. $(x + 2)(x + 3)$
2. $(a + 1)(a + 2)$
3. $(w + 5)(w + 4)$
4. $(y + 9)(y + 4)$
5. $(m + 8)(m + 7)$
6. $(x + 6)(x + 5)$
7. $(a - 1)(a - 4)$
8. $(n - 2)(n - 3)$
9. $(p - 1)(p - 3)$
10. $(y - 2)(y - 8)$
11. $(w - 7)(w - 4)$
12. $(d - 9)(d - 11)$
13. $(c + 7)(c - 3)$
14. $(t + 2)(t - 6)$
15. $(b + 5)(b - 4)$
16. $(b - 7)(b + 3)$
17. $(y - 4)(y + 2)$
18. $(p - 4)(p + 2)$

19. A rectangular garden has a length of $(t + 12)$ units and a width of $(t + 7)$ units. Find the area.

20. Each side of a square lot has a length of $(x + 4)$ units. Find the area of the lot.

Rational Numbers: Multiplication/Division

ENRICHMENT

You may wish to use this lesson for students who performed well on the formal Chapter Test.

1. $x^2 + 5x + 6$
2. $a^2 + 3a + 2$
3. $w^2 + 9w + 20$
4. $y^2 + 13y + 36$
5. $m^2 + 15m + 56$
6. $x^2 + 11x + 30$
7. $a^2 - 5a + 4$
8. $n^2 - 5n + 6$
9. $p^2 - 4p + 3$
10. $y^2 - 10y + 16$
11. $w^2 - 11w + 28$
12. $d^2 - 20d + 99$
13. $c^2 + 4c - 21$
14. $t^2 - 4t - 12$
15. $b^2 + b - 20$
16. $b^2 - 4b - 21$
17. $y^2 - 2y - 8$
18. $p^2 - 2p - 8$
19. $t^2 + 19t + 84$
20. $x^2 + 8x + 16$

ADDITIONAL PRACTICE

You may wish to use all or some of these exercises, depending on how well students performed on the formal Chapter Test.

1. -156   2. -735
3. -189   4. -1914
5. 0   6. -71   7. -2520
8. -204   9. -8.40
10. -7.84   11. $-\frac{5}{8}$
12. $-\frac{5}{6}$ 13. 24
14. 63   15. 66   16. 65
17. 1404   18. 8862
19. 6405   20. 14,784
21. 1.12   22. 24.647
23. 2   24. $1\frac{1}{2}$
25. 15x + 20
26. 20s + 40
27. -30b - 50
28. -6m - 6
29. -12n - 12
30. -25q - 35
31. $4y^2 - 2y$
32. $25d^2 - 15d$
33. $28p^2 - 35p$
34. $5c^2 - 25c$
35. $12f^2 - 24f$
36. $18a^2 - 12a$ 37. 10
38. 7   39. -4   40. -7
41. -4   42. -9   43. -3
44. -16   45. -4
46. -5   47. -6

See page 363 for the answers to Exercises 48-66.

ADDITIONAL PRACTICE

SKILLS

Multiply. (Pages 350-352)

1. 13(-12)
2. 21(-35)
3. 21(-9)
4. 66(-29)
5. -63(0)
6. -71(1)
7. -42(60)
8. (-17)(12)
9. -3.5(2.4)
10. (-4.9)(1.6)
11. $\left(-\frac{1}{4}\right)\left(2\frac{1}{2}\right)$
12. $\left(-\frac{2}{3}\right)\left(1\frac{1}{4}\right)$

(Pages 353-355)

13. (-6)(-4)
14. (-9)(-7)
15. (-11)(-6)
16. (-13)(-5)
17. (-13)(-108)
18. (-42)(-211)
19. (-183)(-35)
20. (-308)(-48)
21. (-1.4)(-0.8)
22. (-5.03)(-4.9)
23. $\left(-1\frac{1}{2}\right)\left(-1\frac{1}{3}\right)$
24. $\left(-2\frac{3}{4}\right)\left(-\frac{6}{11}\right)$

(Pages 356-358)

25. 5(3x + 4)
26. 4(5s + 10)
27. -10(3b + 5)
28. -6(m + 1)
29. -4(3n + 3)
30. -5(5q + 7)
31. y(4y - 2)
32. 5d(5d - 3)
33. 7p(4p - 5)
34. 5c(c - 5)
35. (3f - 6)(4f)
36. (6a - 4)(3a)

Divide. (Pages 360-363)

37. -50 ÷ (-5)
38. -49 ÷ (-7)
39. -44 ÷ 11
40. -56 ÷ 8
41. $\frac{36}{-9}$
42. $\frac{45}{-5}$
43. $\frac{-18}{6}$
44. $\frac{-96}{6}$
45. $3 \div \left(-\frac{3}{4}\right)$
46. $4 \div \left(-\frac{4}{5}\right)$
47. $-3 \div \frac{1}{2}$
48. $-30 \div \frac{1}{6}$
49. $-\frac{1}{2} \div \left(-\frac{3}{4}\right)$
50. $-\frac{1}{4} \div \left(-\frac{7}{8}\right)$
51. $-\frac{5}{6} \div \frac{1}{12}$
52. $-\frac{7}{10} \div \frac{14}{5}$

Factor. (Pages 364-366)

53. 14x + 14
54. 6m + 36
55. 21m + 7
56. 7v + 7
57. 7p - 35
58. 5t - 15
59. 4c - 4
60. 4a - 20
61. 36 - 4v
62. 49 - 7y
63. $x^2 - 6x$
64. $x^2 - 8x$

APPLICATIONS

Use positive and negative numbers to represent each situation by a multiplication problem. Then find the product. (Pages 350-352)

65. Four fines of $15 each
66. A temperature drop of 3°C per hour for five hours

372   Chapter 12

CHAPTER 13 Equations: Multiplication/Division

SECTIONS
- **13-1** Equations: Multiplication/Division
- **13-2** Problem Solving and Applications: Words to Symbols: Multiplication and Division
- **13-3** Combined Operations
- **13-4** Equations with Parentheses
- **13-5** Problem Solving and Applications: More Than One Unknown: Angles in a Triangle
- **13-6** Problem Solving and Applications: More Than One Unknown: Consecutive Numbers
- **13-7** Inequalities: Multiplication/Division
- **13-8** Problem Solving and Applications: Using Inequalities

FEATURES
Computer Application: *Solving Equations*
Enrichment: *Sets*
Common Errors

Teaching Suggestions
p. M-26

QUICK QUIZ
Solve and check.
1. $3a = 129$ Ans: 43
2. $5.2y = 312$ Ans: 60
3. $\frac{a}{0.5} = 12$ Ans: 6
4. $0.6 = \frac{h}{1.3}$ Ans: 0.78
5. $\frac{b}{9} = \frac{2}{3}$ Ans: 6

ADDITIONAL EXAMPLES
Example 1
Solve and check.
1. $\frac{y}{-10} = -21$ Ans: 210
2. $\frac{x}{-12} = \frac{5}{6}$ Ans: -10

SELF-CHECK
1. 42 2. -10
3. -112 4. 20

Example 2
Solve and check.
1. $-8a = 1616$ Ans: -202
2. $-10.8 = 1.8x$ Ans: 6

SELF-CHECK
5. -9   6. -7

13-1 Equations: Multiplication/Division

The following properties were used to solve division and multiplication equations in Chapter 7.

> **Multiplication Property for Equations**
> Multiplying each side of an equation by the same nonzero number forms an equivalent equation.
>
> $\frac{m}{2} = -3$
> $\frac{m}{2}(2) = -3(2)$
> $m = -6$

EXAMPLE 1   Solve and check: $\frac{x}{-12} = 6$

Solution: To get x alone, multiply each side by -12.

$\frac{x}{-12} = 6$ Check: $\frac{x}{-12} = 6$

$\frac{x}{-12}(-12) = 6(-12)$ $\frac{-72}{-12}$

$x = -72$ 6

Self-Check   Solve and check.

1. $\frac{a}{-7} = -6$ 2. $\frac{c}{5} = -2$ 3. $14 = \frac{y}{-8}$ 4. $-5 = \frac{b}{-4}$

> **Division Property for Equations**
> Dividing each side of an equation by the same nonzero number forms an equivalent equation.
>
> $-3w = 27$
> $\frac{-3w}{-3} = \frac{27}{-3}$
> $w = -9$

EXAMPLE 2   Solve and check: $-4p = 16$

Solution: To get p alone, divide each side by -4.

$-4p = 16$ Check: $-4p = 16$

$\frac{-4p}{-4} = \frac{16}{-4}$ $-4(-4)$

$p = -4$ 16

Self-Check   Solve and check: 5. $-2y = 18$ 6. $21 = -3q$

374   Chapter 13

NOTE: By the Multiplication Property of One,
$$x = 1 \cdot x, \text{ or } \mathbf{1}x.$$
Thus, $\quad -x = -1 \cdot x, \text{ or } \mathbf{-1}x.$

Therefore, to solve an equation such as $-x = 8$, divide each side of the equation by -1.

$$-x = 8 \quad\blacktriangleleft\quad -x = -1 \cdot x$$
$$\frac{-1 \cdot x}{-1} = \frac{8}{-1}$$

Since $\frac{-1}{-1} = 1,\qquad x = \frac{8}{-1}.$

$$x = \mathbf{-8}$$

CLASSROOM EXERCISES

For Exercises 1-8, by what number would you multiply each side of the equation in order to solve for x? (Example 1)

1. $\frac{x}{2} = -9$
2. $\frac{x}{5} = -2$
3. $\frac{x}{-7} = 1$
4. $\frac{x}{-3} = 8$
5. $\frac{x}{-3} = -5$
6. $\frac{x}{-6} = -8$
7. $8 = \frac{x}{12}$
8. $-3 = \frac{x}{-15}$

For Exercises 9–16, by what number would you divide each side of the equation in order to solve for n? (Example 2)

9. $-4n = 52$
10. $-6n = 36$
11. $7n = -21$
12. $5n = -15$
13. $-3n = -24$
14. $-10n = -40$
15. $36 = -2n$
16. $-48 = -12n$

For Exercises 17-36, solve and check. (Example 1)

17. $\frac{w}{4} = -9$
18. $\frac{p}{6} = -12$
19. $\frac{x}{-9} = -7$
20. $\frac{y}{-3} = -8$
21. $\frac{a}{-2} = 14$
22. $\frac{d}{-5} = 10$
23. $-11 = \frac{b}{4}$
24. $-15 = \frac{c}{7}$

(Example 2)

25. $-2q = 6$
26. $-8t = 64$
27. $10r = -90$
28. $4a = -144$
29. $-12r = -132$
30. $-7p = -49$
31. $-54 = 9c$
32. $-24 = 4m$
33. $-x = 3$
34. $-x = -5$
35. $-x = -3$
36. $-x = 0$

CLASSROOM EXERCISES
1. −18 2. −10 3. −7
4. −24 5. 15   6. 48
7. 96   8. 45   9. −4
10. −6 11. 7   12. 5
13. −3 14. −10
15. −2 16. −12
17. −36 18. −72
19. 63 20. 24
21. −28 22. −50
23. −44 24. −105
25. −3 26. −8 27. −9
28. −36 29. 11 30. 7
31. −6 32. −6 33. −3
34. 5   35. 3   36. 0

Equations: Multiplication/Division

ASSIGNMENT GUIDE
BASIC
Day 1 p. 376: 1-39 odd
Day 2 p. 376: 2-40 even
AVERAGE
p. 376: 1-39 odd
ABOVE AVERAGE
p. 376: 1-39 odd

WRITTEN EXERCISES
1. -96   2. -21   3. -60
4. -64   5. 24   6. 20
7. -72   8. -125
9. -114 10. -36
11. -516 12. -208
13. -4   14. -3
15. -11 16. -7
17. 24   18. 4   19. -3
20. -3   21. $-72\frac{1}{2}$
22. -1   23. $16\frac{2}{3}$
24. $-4\frac{3}{4}$ 25. -28
26. 28   27. 29
28. -29 29. 175
30. 260 31. -140
32. 320 33. -8   34. 0
35. 0   36. -2
37. -1.4 38. 16.8
39. 12.5 40. 6

REVIEW CAPSULE
This Review Capsule reviews prior-taught skills used in Section 13-2. The reference is to the pages where the skills were taught.

WRITTEN EXERCISES

Goal: To solve equations by using the Multiplication and Division Properties for Equations

For Exercises 1-40, solve and check.   (Example 1)

1. $\dfrac{p}{12} = -8$
2. $\dfrac{q}{3} = -7$
3. $\dfrac{x}{-5} = 12$
4. $\dfrac{y}{-4} = 16$
5. $\dfrac{x}{-3} = -8$
6. $\dfrac{y}{-5} = -4$
7. $\dfrac{a}{4} = -18$
8. $\dfrac{b}{-25} = -5$
9. $\dfrac{c}{6} = -19$
10. $\dfrac{m}{9} = -4$
11. $12 = \dfrac{d}{-43}$
12. $52 = \dfrac{c}{-4}$

(Example 2)

13. $-4c = 16$
14. $-6y = 18$
15. $12x = -132$
16. $15g = -105$
17. $-3t = -72$
18. $-7x = -28$
19. $15 = -5p$
20. $21 = -7r$
21. $2w = -145$
22. $18w = -18$
23. $-50 = -3x$
24. $-19 = 4m$
25. $-x = 28$
26. $-x = -28$
27. $-c = -29$
28. $-c = 29$

MIXED PRACTICE

29. $\dfrac{x}{-5} = -35$
30. $\dfrac{c}{-2} = -130$
31. $\dfrac{n}{-2} = 70$
32. $\dfrac{t}{-4} = -80$
33. $9a = -72$
34. $-5y = 0$
35. $-7b = 0$
36. $13p = -26$
37. $3x = -4.2$
38. $8 = \dfrac{y}{2.1}$
39. $\dfrac{b}{-2.5} = -5$
40. $-7.2 = -1.2p$

REVIEW CAPSULE FOR SECTION 13-2

Determine whether the solution for the given equation is 5. Answer <u>Yes</u> or <u>No</u>.

1. $4y = 20$
2. $3x = 24$
3. $\dfrac{b}{2} = 10$
4. $\dfrac{c}{5} = 1$

Solve and check.

5. $7y = 21$
6. $2p = 14$
7. $12w = 60$
8. $9r = 81$
9. $\dfrac{b}{6} = 8$
10. $\dfrac{c}{10} = 11$
11. $\dfrac{w}{5} = 10$
12. $\dfrac{t}{13} = 7$

The answers are on page 401.

376 Chapter 13

PROBLEM SOLVING AND APPLICATIONS

13-2 Words to Symbols: Multiplication and Division

Albert and Carmen picked apples last Saturday. Carmen stated that Albert had picked four times the number of apples she had picked.

Albert thought:

Number of apples Carmen picked: a

4 times the number Carmen picked: $4a$

The following table shows some word expressions and the corresponding algebraic expressions.

TABLE

Operation	Word Expression	Algebraic Expression
Multiplication	4 times the number of apples, a	$4a$
	The product of 18 and the number of tables, t	$18t$
	Twice the number of swimmers, s	$2s$
	Double Bill's weight, w	$2w$
	Triple the cost, c	$3c$
	Eight multiplied by the number of gallons, g	$8g$
Division	The quotient of the length, l, and 10	$\dfrac{l}{10}$
	The number of days, d, divided by 3	$\dfrac{d}{3}$
	84 divided by the number of kilometers, k	$\dfrac{84}{k}$

Albert knows that he picked 360 apples. He can use an equation to find how many apples Carmen picked.

Equations: Multiplication/Division   **377**

Teaching Suggestions p. M-26

QUICK QUIZ
Evaluate.
1. $4x$ when $x = 309$
   Ans: 1236
2. $7y$ when $y = 10.3$
   Ans: 72.1
3. $\dfrac{a}{12}$ when $a = 1.2$
   Ans: 0.1
4. $\dfrac{h}{7}$ when $h = 602$
   Ans: 86

Solve and check.
5. $3x = 51$ Ans: 17
6. $6 = 18a$ Ans: $\dfrac{1}{3}$
7. $\dfrac{y}{8} = 31$ Ans: 248
8. $1.8 = \dfrac{c}{3}$ Ans: 5.4

377

ADDITIONAL EXAMPLES

Example 1

1. Triple Jim's age will be 174. Find Jim's age. Ans: 58

2. The product of Alice's salary and 23 is $3588. Find Alice's salary. Ans: $156

SELF-CHECK

1. 6

Example 2

1. The number of turtles divided by 9 is 414. Find the number of turtles. Ans: 46

2. The quotient of h and 18 is 1440. Find the value of h. Ans: 80

SELF-CHECK

2. 120

CLASSROOM EXERCISES

1. $5t$ 2. $\frac{t}{5}$ 3. $\frac{q}{4}$
4. $\frac{x}{3}$ 5. $4q$ 6. $3x$
7. $2t$ 8. $\frac{t}{2}$
9. $7m$ 10. $2s$
11. $\frac{c}{12} = 25$
12. $\frac{p}{6} = 42$
13. $\frac{h}{7} = 6$

EXAMPLE 1 Four times the number of apples Carmen picked is 360. Find how many apples Carmen picked.

Solution:
1. Choose a variable. Let $a =$ the number of apples Carmen picked.
2. Write an equation. 4 times the number of apples Carmen picked is 360.
$$4a = 360$$
3. Solve the equation.
$$\frac{4a}{4} = \frac{360}{4}$$
$$a = 90$$
4. Check: Does $4(90) = 360$? Yes ✔ Carmen picked **90 apples.**

Self-Check 1. The product of 22 and the number of chairs per table is 132. Find how many chairs there are per table.

EXAMPLE 2 The number of kilometers traveled by the Phillips family divided by 3 hours is 62. Find the distance traveled.

Solution:
1. Let $d =$ the number of kilometers.
2. Number of kilometers divided by 3 is 62.
$$\frac{d}{3} = 62$$
3.
$$\frac{d}{3}(3) = 62(3)$$
$$d = 186$$
4. Check: Does $\frac{186}{3} = 62$? Yes ✔ Distance: **186 km**

Self-Check 2. The quotient of t and 8 is 15. Find t.

CLASSROOM EXERCISES

Choose from the box at the right the algebraic expression that represents each word expression. (Table)

1. The product of t and 5
2. The quotient of t and 5
3. q divided by 4
4. The quotient of x and 3
5. 4 multiplied by q
6. x tripled
7. t doubled
8. t divided by 2

$4q$	$5t$
$\frac{t}{2}$	$\frac{t}{5}$
$\frac{q}{4}$	$\frac{x}{3}$
$2t$	$3x$

Chapter 13

Complete. (Examples 1 and 2, step 2)

Word Sentence	Equation
9. Seven times the number of miles, m, is 140.	__?__ = 140
10. Twice the number of club members, s, is 36.	__?__ = 36
11. The quotient of the total cost, c, and 12 payments is $25.	__?__
12. The number of passengers, p, divided by 6 buses is 42.	__?__
13. The quotient of the number of hours, h, and 7 is 6.	__?__

WRITTEN EXERCISES

Goals: To write an algebraic equation for a word sentence
To apply the skill of writing algebraic equations to solving word problems

Represent each word expression by an algebraic expression. (Table)

1. The quotient of x and 9
2. q multiplied by 7
3. The product of 2 and q
4. Twice p
5. Three times r
6. t divided by 212
7. a multiplied by (-4)
8. a divided by -4
9. 3 times p
10. p times 5
11. w tripled
12. The quotient of y and 12
13. t divided by 14
14. x doubled

For Exercises 15–22, write an equation for each word sentence. Do not solve the equation. (Example 1, step 2)

15. Four times the number of sailboats, s, is 28.
16. Twice the number of millimeters of rain, m, is 82 millimeters.
17. The product of the number of hours, h, and $6 is $132.
18. Twenty books multiplied by the number of shelves, n, is 240.

(Example 2, Step 2)

19. The number of kilometers, k, divided by 78 is 4.
20. The quotient of the total cost, c, and 24 payments is $30.
21. The quotient of the number of club members, m, and 7 is 10.
22. The number of points, p, divided by 12 questions is 13.

Equations: Multiplication/Division

ASSIGNMENT GUIDE
BASIC
Day 1 p. 379: 1-20
Day 2 pp. 379-380: 21-30
AVERAGE
Day 1 p. 379: 1-20
Day 2 pp. 379-380: 21-34
ABOVE AVERAGE
pp. 379-380: 1-29 odd
31-34

WRITTEN EXERCISES
1. $\frac{x}{9}$ 2. $7q$ 3. $2q$
4. $2p$ 5. $3r$ 6. $\frac{t}{212}$
7. $-4a$ 8. $\frac{a}{-4}$ 9. $3p$
10. $5p$ 11. $3w$ 12. $\frac{y}{12}$
13. $\frac{t}{14}$ 14. $2x$
15. $4s = 28$
16. $2m = 82$
17. $6h = 132$
18. $20n = 240$
19. $\frac{k}{78} = 4$
20. $\frac{c}{24} = 30$
21. $\frac{m}{7} = 10$ 22. $\frac{p}{12} = 13$

WRITTEN EXERCISES

23. 3   24. 12 cm
25. 38   26. 5   27. 400
28. 735 29. $675
30. 432 31. 71
32. 384 33. 270
34. $12

For Exercises 23-34, solve each problem.   (Example 1)

23. The number of hours times 52 miles per hour is 156 miles. Find the number of hours.

24. Four times the width of a rectangle is 48 centimeters. Find the width of the rectangle.

25. Double the number of points Pedro scored is 76. How many points did Pedro score?

26. The product of 12 players and the number of teams is 60. Find the number of teams.

(Example 2)

27. The number of dimes saved divided by 50 is 8. Find the number of dimes.

28. The number of cars washed divided by 7 is 105. Find the number of cars.

29. The quotient of the amount of a loan and 15 payments is $45. Find the amount of the loan.

30. The quotient of the number of names in a phone directory and 6 is 72. Find the number of names.

MIXED PRACTICE

31. Twice the number of words Marilyn can type per minute is 142. How many words can she type per minute?

32. The number of balcony seats in a theater divided by 24 rows is 16. Find the number of balcony seats.

33. The quotient of the number of 9th grade students and 30 is 9. Find the number of 9th grade students.

34. Eight times the cost of a concert ticket is $96. Find the cost of a concert ticket.

REVIEW CAPSULE FOR SECTION 13-3

Simplify each expression.   (Pages 64-66)

1. $3x + 5 - 5$
2. $4n + 6 - 6$
3. $7b - 1 + 1$
4. $5d - 7 + 7$
5. $\frac{x}{3} - 1 + 1$
6. $\frac{d}{4} - 2 + 2$
7. $\frac{h}{3} + 8 - 8$
8. $\frac{r}{6} + 2 - 2$

Solve and check.   (Pages 190-195)

9. $x + 5 = 7$
10. $y + 6 = -4$
11. $b - 3 = 4$
12. $a - 9 = -6$
13. $p - 5 = -10$
14. $t + 7 = -10$
15. $r + 12 = 3$
16. $s - 6 = 14$

The answers are on page 401.

380   Chapter 13

13-3 Combined Operations

This procedure will help you to solve equations that involve more than one operation.

> **PROCEDURE**
>
> **To solve equations with more than one operation:**
> 1. In general, use the Addition or Subtraction Property first.
> 2. Then use the Multiplication or Division Property.

EXAMPLE 1 Solve and check: $3x + 11 = 5$

Solution:

$3x + 11 = 5$

1. Subtract 11 from each side. $3x + 11 - 11 = 5 - 11$

    $3x = -6$

2. Divide each side by 3. $\dfrac{3x}{3} = \dfrac{-6}{3}$

    $x = \mathbf{-2}$

Check: $3x + 11 = 5$

$3(-2) + 11$

$-6 + 11$

5

Self-Check *Solve and check.*

1. $7n + 20 = 13$
2. $2y - 5 = 7$
3. $81 = -5a + 6$

Be sure to check your answer in the <u>original equation</u>.

EXAMPLE 2 Solve and check: $\dfrac{n}{3} - 7 = -8$

Solution:

$\dfrac{n}{3} - 7 = -8$

1. Add 7 to each side. $\dfrac{n}{3} - 7 + 7 = -8 + 7$

    $\dfrac{n}{3} = -1$

2. Multiply each side by 3. $(3)\left(\dfrac{n}{3}\right) = -1(3)$

    $n = \mathbf{-3}$

Check: $\dfrac{n}{3} - 7 = -8$

$\dfrac{-3}{3} - 7$

$-1 - 7$

-8

Self-Check *Solve and check.*

4. $\dfrac{a}{3} - 6 = -14$
5. $\dfrac{x}{-9} + 5 = 3$
6. $1 = \dfrac{t}{5} + 7$

Equations: Multiplication/Division

Teaching Suggestions
p. M-26

QUICK QUIZ
Solve and check.
1. x + 8 = 15
   Ans: x = 7
2. a - 29 = 29   Ans: 58
3. 3x = 171   Ans: 57
4. 8 = 16y   Ans: $\frac{1}{2}$
5. $\frac{c}{7}$ = 10   Ans: 70

ADDITIONAL EXAMPLES
Example 1
Solve and check.
1. 8 - 4a = -12   Ans: 5
2. 0 = 18 + 9c   Ans: -2

SELF-CHECK
1. -1   2. 6   3. -15

Example 2
Solve and check.
1. $\frac{a}{3}$ + 5 = 14
   Ans: a = 27
2. 0 = 6 - $\frac{t}{4}$
   Ans: t = 24

SELF-CHECK
4. -24   5. 18   6. -30

CLASSROOM EXERCISES

CLASSROOM EXERCISES
1. $3k + 5 - 5 = 2 - 5$
2. $4n + 6 - 6 = 14 - 6$
3. $7b + 1 - 1 = 15 - 1$
4. $5d - 7 + 7 = 13 + 7$
5. $12 - 8 = -2y + 8 - 8$
6. $15 - 3 = -3t + 3 - 3$
7. $4x + 7 - 7 = -13 - 7$
8. $5m + 1 - 1 = -29 - 1$
9. $\frac{x}{3} - 1 + 1 = -2 + 1$
10. $\frac{d}{4} - 2 + 2 = -5 + 2$
11. $\frac{v}{-2} + 3 - 3 = 7 - 3$
12. $\frac{f}{-5} + 6 - 6 = 12 - 6$
13. $2 - 4 = \frac{y}{2} + 4 - 4$
14. $6 - 13 = \frac{a}{3} + 13 - 13$
15. $\frac{c}{-2} - 3 + 3 = 4 + 3$
16. $\frac{g}{-5} - 4 + 4 = 2 + 4$

ASSIGNMENT GUIDE
BASIC
Day 1 p. 382: 1-12
Day 2 p. 382: 13-32
AVERAGE
Day 1 p. 382: 1-39 odd
Day 2 p. 382: 2-40 even
ABOVE AVERAGE
p. 382: 1-39 odd

WRITTEN EXERCISES
1. -2   2. -1   3. 2
4. 3   5. -4   6. -1
7. -4   8. -3   9. -5
10. -6   11. -8   12. -5
See page 384 for the answers to Exercises 13-40.

For Exercises 1–16, write an equivalent equation as in step 1 of Examples 1 and 2. Do <u>not</u> solve the equation. (Example 1, step 1)

1. $3k + 5 = 2$
2. $4n + 6 = 14$
3. $7b + 1 = 15$
4. $5d - 7 = 13$
5. $12 = -2y + 8$
6. $15 = -3t + 3$
7. $4x + 7 = -13$
8. $5m + 1 = -29$

(Example 2, step 1)

9. $\frac{x}{3} - 1 = -2$
10. $\frac{d}{4} - 2 = -5$
11. $\frac{v}{-2} + 3 = 7$
12. $\frac{f}{-5} + 6 = 12$
13. $2 = \frac{y}{2} + 4$
14. $6 = \frac{a}{3} + 13$
15. $\frac{c}{-2} - 3 = 4$
16. $\frac{g}{-5} - 4 = 2$

WRITTEN EXERCISES

Goal: To solve an equation with more than one operation

For Exercises 1–40, solve and check. (Example 1)

1. $2x + 7 = 3$
2. $3y + 9 = 6$
3. $5a - 1 = 9$
4. $6p - 2 = 16$
5. $13 = -2b + 5$
6. $7 = -4g + 3$
7. $8z + 10 = -22$
8. $12d + 7 = -29$
9. $5h - 9 = -34$
10. $7w - 5 = -47$
11. $22 = -4f - 10$
12. $44 = -13k - 21$

(Example 2)

13. $\frac{x}{2} - 1 = -4$
14. $\frac{t}{4} - 3 = -5$
15. $\frac{n}{-2} + 1 = 5$
16. $\frac{a}{-3} + 2 = 7$
17. $10 = \frac{y}{5} + 6$
18. $12 = \frac{w}{4} + 5$
19. $\frac{c}{-6} - 5 = 6$
20. $\frac{g}{-5} - 7 = 9$
21. $\frac{h}{-2} + 8 = -2$
22. $\frac{r}{-3} + 7 = -4$
23. $-11 = \frac{q}{8} + 9$
24. $-10 = \frac{z}{7} + 6$

MIXED PRACTICE

25. $6x + 15 = 3$
26. $4p + 21 = 1$
27. $10a - 23 = 27$
28. $9c - 15 = 30$
29. $\frac{y}{5} - 6 = -13$
30. $\frac{b}{7} - 8 = -14$
31. $\frac{m}{-4} + 3 = 6$
32. $\frac{t}{-7} + 5 = 11$
33. $17 = \frac{u}{6} + 7$
34. $22 = \frac{v}{8} + 14$
35. $\frac{e}{-3} - 8 = -15$
36. $\frac{k}{-6} - 9 = -20$
37. $54 = -12x + 18$
38. $79 = -9s - 11$
39. $-4w - 6 = -18$
40. $-5d + 9 = -16$

Chapter 13

REVIEW CAPSULE FOR SECTION 13-4

For Exercises 1-8, multiply. (Pages 356-358)

1. $3(4n + 6)$
2. $4(-7y + 3)$
3. $(p + 7)(5)$
4. $(2x + 3)(-4)$
5. $(z + 11)(-5)$
6. $-7(b + 6)$
7. $(4a - 3)5a$
8. $(2t)(4t - 4)$

Solve and check. (Pages 326-330)

9. $-x + 3 + 2x = 33$
10. $6y - 7 - 5y = 83$
11. $5 + 3p + 7 = 44 + 2p$
12. $4 + 10d - 13 = 7d + 8 + 2d$

The answers are on page 401.

13-4 Equations With Parentheses

Equations sometimes contain parentheses. To remove the parentheses, use the Distributive Property.

> **PROCEDURE**
> **To solve equations containing parentheses:**
> 1 Use the Distributive Property to express the product as a sum or difference.
> 2 Solve the equation.

EXAMPLE 1 Solve and check: $4(y + 2) = 28$

Solution:
1 Use the Distributive Property. $\quad 4(y + 2) = 28$

$$(4 \cdot y) + (4 \cdot 2) = 28$$

2 Solve the equation.
$$4y + 8 = 28$$
$$4y + 8 - 8 = 28 - 8$$
$$4y = 20$$
$$\frac{4y}{4} = \frac{20}{4}$$
$$y = 5$$

◀ The check is left for you.

Self-Check *Solve and check.*

1. $5(n + 3) = 25$
2. $6(q - 1) = -12$
3. $2(x + 4) = -4$

Equations: Multiplication/Division **383**

Teaching Suggestions
p. M-26

QUICK QUIZ
Multiply.
1. $3(2x - 7)$
   Ans: $6x - 21$
2. $(4x + 2)(-5)$
   Ans: $-20x - 10$
3. $-6(x - 8)$
   Ans: $-6x + 48$
4. $2x(x + 3)$
   Ans: $2x^2 + 6x$
5. $(x - 6)(-4x)$
   Ans: $-4x^2 + 24x$
6. $\frac{2}{3}(9x + 15)$
   Ans: $6x + 10$

ADDITIONAL EXAMPLES
Example 1
Solve and check.
1. $-4(16 + x) = 40$
   Ans: -26
2. $\frac{1}{3}(6x + 21) = 9$
   Ans: 1

SELF-CHECK
1. 2   2. -1   3. -6

ADDITIONAL EXAMPLES

Example 2

Solve and check.

1. 2(x + 5) = 7(3x − 4)

    Ans: x = 2

2. −6(x − 8) =
    2(5x + 48)

    Ans: x = −3

SELF-CHECK

4. 2   5. −6

CLASSROOM EXERCISES

1. 21   2. 14
3. 9e; 72   4. 2q   5. −
6. −   7. 35; −3z
8. 2p; −   9. 12; −
10. −; −3k; −   11. Yes
12. Yes   13. No
14. No   15. Yes
16. No   17. Yes
18. No

WRITTEN EXERCISES p. 382

13. −6   14. −8   15. −8
16. −15   17. 20
18. 28   19. −66
20. −80   21. 20
22. 33   23. −160
24. −112   25. −2
26. −5   27. 5   28. 5
29. −35   30. −42
31. −12   32. −42
33. 60   34. 64   35. 21
36. 66   37. −3
38. −10   39. 3   40. 5

384

EXAMPLE 2 Solve and check: $4(a - 3) = -2(a + 3)$

Solution: ① $4(a - 3) = -2(a + 3)$

$(4 \cdot a) + (4)(-3) = (-2 \cdot a) + (-2 \cdot 3)$

② $\qquad 4a + (-12) = -2a - 6$

$4a + 2a - 12 = -2a + 2a - 6$

$6a - 12 = -6$

$6a - 12 + 12 = -6 + 12 \qquad$ Check: $4(a - 3) \mid -2(a + 3)$

$6a = 6 \qquad\qquad\qquad\qquad\quad 4(1 - 3) \mid -2(1 + 3)$

$\dfrac{6a}{6} = \dfrac{6}{6} \qquad\qquad\qquad\qquad\quad 4(-2) \mid -2(4)$

$\qquad\qquad\qquad\qquad\qquad\qquad\qquad\quad -8 \mid -8$

$a = 1$

Self-Check Solve and check.

4. $5(v - 4) = -2(v + 3)$ 5. $3(b + 6) = -2(b + 6)$

CLASSROOM EXERCISES

For Exercises 1–6, complete.   (Example 1, step 1)

1. $3(c + 7) = 32$ 2. $7(2 + d) = 63$ 3. $9(e + 8) = 100$
   $3c + \underline{\ ?\ } = 32$ $\underline{\ ?\ } + 7d = 63$ $\underline{\ ?\ } + \underline{\ ?\ } = 100$
4. $2(-4 - q) = 7$ 5. $6(m - 5) = 11$ 6. $10(n - 5) = 81$
   $-8 - \underline{\ ?\ } = 7$ $6m \underline{\ ?\ } 30 = 11$ $10n \underline{\ ?\ } 50 = 81$

For Exercises 7–10, complete.   (Example 2, step 1)

7. $5(z + 7) = -3(z + 7)$ 8. $2(p + 3) = -6(p + 7)$
   $5z + \underline{\ ?\ } = \underline{\ ?\ } - 21$ $\underline{\ ?\ } + 6 = -6p \underline{\ ?\ } 42$
9. $4(m - 3) = 6(m - 5)$ 10. $-8(k + 7) = -3(k + 2)$
   $4m - \underline{\ ?\ } = 6m \underline{\ ?\ } 30$ $-8k \underline{\ ?\ } 56 = \underline{\ ?\ } \underline{\ ?\ } 6$

Determine whether the given value of the variable is a solution of each equation. Answer <u>Yes</u> or <u>No</u>.   (Examples 1 and 2, Check)

11. $7(v + 4) = 119$ when $v = 13$ 12. $6(r + 2) = 66$ when $r = 9$
13. $3(t - 5) = 15$ when $t = 11$ 14. $2(n - 4) = -4$ when $n = 3$
15. $-4(b + 2) = 3(b - 5)$ when $b = 1$ 16. $5(d - 2) = -6(d + 2)$ when $d = 3$
17. $8(c + 4.5) = 68$ when $c = 4$ 18. $3.5(e - 4) = -2.5(e + 20)$ when $e = 6$

384   Chapter 13

WRITTEN EXERCISES

Goal: To solve equations containing parentheses

For Exercises 1–40, solve and check. (Example 1)

1. $8(x + 2) = 48$
2. $2(y - 6) = 10$
3. $6(z - 3) = 4$
4. $11(f + 3) = 88$
5. $5(u - 2) = -25$
6. $4(r - 5) = -16$
7. $3(a + 3) = -39$
8. $7(c + 5) = -77$
9. $\frac{1}{2}(6t + 8) = 37$
10. $\frac{1}{8}(24m + 40) = 26$
11. $8(g + 1.5) = 92$
12. $4(2.5v + 7.5) = 100$

(Example 2)

13. $5(p - 6) = -3(p + 2)$
14. $8(n - 2) = -2(n + 3)$
15. $7(m - 7) = -5(m + 5)$
16. $6(q - 9) = -4(q + 1)$
17. $-1(c + 40) = -3(c + 10)$
18. $-2(d + 83) = -1(d + 15)$
19. $6(r - 4) = -2(r + 10)$
20. $3(s - 4) = -1(s + 11)$
21. $\frac{1}{3}(6e - 12) = \frac{1}{4}(12e - 4)$
22. $\frac{1}{2}(10f - 8) = \frac{1}{8}(24f + 96)$
23. $6(r - 4.5) = -2(r + 22.5)$
24. $3.5(2v + 4) = 1.5(4v - 10)$

MIXED PRACTICE

25. $3(a + 5) = 60$
26. $10(c - 5) = -3(c - 5)$
27. $12(d + 3) = -2(d + 38)$
28. $7(b + 2) = 77$
29. $4(e - 10) = -6(e + 15)$
30. $2(y - 8) = -44$
31. $5(x - 12) = -100$
32. $9(h + 3) = 3(h - 1)$
33. $10(1.5q + 2.5) = 175$
34. $6(0.5k + 6.5) = -4(1.5k + 10.5)$

MORE CHALLENGING EXERCISES

EXAMPLE: $-(-5 - 3x) = -7$ ◀ $-(-5 - 3x) = -1(-5 - 3x)$
$(-1)(-5) - (-1)(3x) = -7$
$5 + 3x = -7$
$3x = -12$
$x = -4$ ◀ The check is left for you.

35. $-(-10 - 2x) = 16$
36. $-(-2y - 4) = -6$
37. $-3(-6 - 4a) = -6$
38. $-2(-6b - 4) = 44$
39. $-3(x - 4) = -(-2x + 8)$
40. $-2(x - 5) = -4(2x + 2)$

Equations: Multiplication/Division

ASSIGNMENT GUIDE
BASIC
Day 1 p. 385: 1–16
Day 2 p. 385: 17–30
AVERAGE
Day 1 p. 385: 1–33 odd
Day 2 p. 385: 2–34 even
ABOVE AVERAGE
p. 385: 3, 6, 9, ..., 33, 35–40

WRITTEN EXERCISES

1. 4 2. 11 3. $3\frac{2}{3}$
4. 5 5. -3 6. 1
7. -16 8. -16 9. 11
10. 7 11. 10 12. 7
13. 3 14. 1 15. 2
16. 5 17. 5 18. -151
19. $\frac{1}{2}$ 20. $\frac{1}{4}$ 21. -3
22. 8 23. $-2\frac{1}{4}$
24. -29 25. 15
26. -5 27. -8 28. 9
29. -5 30. -14
31. -8 32. -5 33. 10
34. -9 35. 3 36. -5
37. -2 38. 3 39. 4
40. -3

QUIZ: SECTIONS 13-1-13-4

After completing this Review, you may wish to administer a quiz covering the same sections. A Quiz is provided in the *Teacher's Edition: Part II.*

REVIEW: SECTIONS 13-1-13-4

1. -30 2. -56
3. -108 4. -45 5. -7
6. -5 7. 6 8. 3
9. 128 10. 15 11. 4
12. 336 13. 1 14. -2
15. -3 16. -2 17. -8
18. -6 19. -30
20. 18 21. 4 22. 1
23. -20 24. -5
25. -6 26. -10

REVIEW: SECTIONS 13-1—13-4

For Exercises 1-8, solve and check. (Section 13-1)

1. $\frac{n}{6} = -5$
2. $\frac{x}{8} = -7$
3. $36 = \frac{c}{-3}$
4. $9 = \frac{a}{-5}$
5. $-6p = 42$
6. $-10x = 50$
7. $-24 = -4y$
8. $-21 = -7z$

For Exercises 9-12, solve each problem. (Section 13-2)

9. The quotient of the number of fish caught and 4 is 32. Find the number of fish caught.
10. Five times the number of acres is 75. Find the number of acres.
11. The product of the number of weeks and $320 is $1280. Find the number of weeks.
12. The number of miles divided by 42 is 8. Find the number of miles.

For Exercises 13-26, solve and check. (Section 13-3)

13. $3x + 5 = 8$
14. $2a + 7 = 3$
15. $5w - 4 = -19$
16. $4t - 5 = -13$
17. $\frac{y}{2} - 2 = -6$
18. $\frac{b}{6} - 3 = -4$
19. $-4 = \frac{w}{5} + 2$
20. $10 = \frac{p}{6} + 7$

(Section 13-4)

21. $3(a + 6) = 30$
22. $5(y - 7) = -30$
23. $2(x - 4) = 4(x + 8)$
24. $7(b + 1) = -4(b + 12)$
25. $-3(d + 4) = 2(9 + d)$
26. $7(c + 5) = 5(c + 3)$

REVIEW CAPSULE FOR SECTION 13-5

Combine like terms. (Pages 76-80)

1. $7x + 9x + 2$
2. $8y + y - 9$
3. $4a + 2a - a$
4. $8a + 3a - 9$
5. $6t - 5t + 1$
6. $6y - y + 9y$
7. $4r - 9r + 32$
8. $10b - 7 + 15b$
9. $3c + 3c + 10$
10. $5g - 12 - 3g$
11. $6p + 3p + 4p$
12. $9s + 4s + 2s$

The answers are on page 401.

386 Chapter 13

COMPUTER APPLICATIONS
Solving Equations

The following program will solve equations of the form **Ax + B = Cx + D** where $A \neq C$.

To use this program, you must first write the equation in the given form. Then you identify A, B, C, and D.

EXAMPLE Write these equations in the form $Ax + B = Cx + D$. Then identify A, B, C, and D.

a. $-2 + 7x = 5x - 14$
b. $8 - 3x - 14 = 3x$

Solutions:
a. $7x - 2 = 5x - 14$
 $A = 7, b = -2, C = 5, D = -14$
b. $-3x + 8 = 3x + 14$
 $A = -3, B = 8, C = 3, D = 14$

Program:

```
100 PRINT
110 PRINT "FOR THE EQUATION AX + B = CX + D,"
120 PRINT "WHAT ARE A, B, C, AND D";
130 INPUT A, B, C, D
140 LET X = (D - B) / (A - C)
150 PRINT "X = "; X
160 PRINT
170 PRINT "ANY MORE EQUATIONS (1 = YES, 0 = NO)";
180 INPUT Z
190 IF Z = 1 THEN 100
200 PRINT
210 END
```

NOTE: Solving $Ax + B = Cx + D$ for x yields statement 140.

$$x = \frac{D - B}{A - C}$$

Exercises

Write each equation in the form $Ax + B = Cx + D$. Then use the program to solve the equation.

1. $8x - 13 = 2x + 56$
2. $8x - 20 = -32 - 7x$
3. $11 - 2x - 16 = 8x$
4. $4x + 12 = 3 + 7x$
5. $-16 - 3x = 5x + 16$
6. $-5x + 2 = -9x - 16$
7. $8(x + 2) = 4x - 10$
8. $5x - 13 = 6x - 18$
9. $-3(x - 8) = 2x - 1$

COMPUTER APPLICATIONS
This feature illustrates the use of a computer program for solving equations. It can be used after Section 13-4 (pages 383-386) is taught.

1. x = 11.5
2. x = -0.8
3. x = -0.5
4. x = 3
5. x = -4
6. x = -4.5
7. x = -6.5
8. x = 5
9. x = 5

Teaching Suggestions
p. M-26

QUICK QUIZ
Combine like terms.
1. 3x + 7 + 2x
   Ans: 5x + 7
2. 5x + x + 9 + 7x
   Ans: 13x + 9
3. 4a − 2 + 5 − a
   Ans: 3a + 3

Write an algebraic expression for each.
4. 17 less than x
   Ans: x − 17
5. the product of 27 and y   Ans: 27y

ADDITIONAL EXAMPLE
Example
In triangle DEF, the measure of angle D is 7 times the measure of angle E. The measure of angle F is 76°. Find the measure of angle D. Ans: 91°.

SELF-CHECK
15°, 75°, 90°

CLASSROOM EXERCISES
1. 30; 60; 90
2. 32; 64; 84
3. 60; 100; 0

PROBLEM SOLVING AND APPLICATIONS

13-5 More Than One Unknown: Angles in a Triangle

Many problems that relate to geometry can be solved by writing an equation.

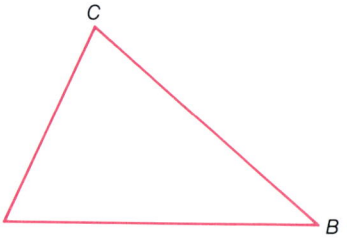

Word Rule: The sum of the measures of the angles of a triangle is 180°.

Formula: $A + B + C = 180°$,

where A, B, and C represent the measures of the angles of a triangle.

EXAMPLE Find the measure of each angle of this triangle.

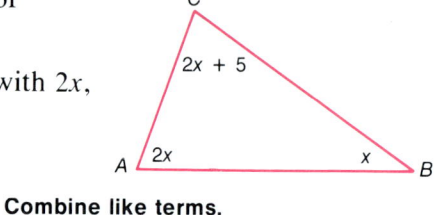

Solution:

[1] Use the formula. Replace A with $2x$, B with x, and C with $2x + 5$.

$A\ +\ B\ +\ \ \ C\ = 180°$
$2x + x + 2x + 5 = 180$ ◀ **Combine like terms.**

[2] Solve the equation. $5x + 5 = 180$
$5x = 175$
$x = 35$ ◀ **Find 2x and 2x + 5.**

$2x = 2(35) = 70$ $2x + 5 = 2(35) + 5 = 75$

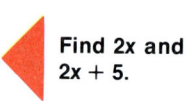

[3] Check: $35 + 70 + 75 = 180$?   Yes ✓
The angle measures are **35°, 70°,** and **75°**.

Self-Check 1. The measures of the angles of a triangle can be represented by a, $5a$, and $6a$. Find the measure of each angle.

CLASSROOM EXERCISES

Complete. (Example)

1. $x + 2x + 3x = 180$; $x =$ __?__ ; $2x =$ __?__ ; $3x =$ __?__
2. $t + 2t + 2t + 20 = 180$; $t =$ __?__ ; $2t =$ __?__ ; $2t + 20 =$ __?__
3. $3s + 5s + 2s - 20 = 180$; $3s =$ __?__ ; $5s =$ __?__ ; $2s - 20 =$ __?__

388 Chapter 13

WRITTEN EXERCISES

Find the measure of each angle of the triangle. (Example)

1. Triangle ABC with angles x at A, $x+20$ at B, $2x$ at C.

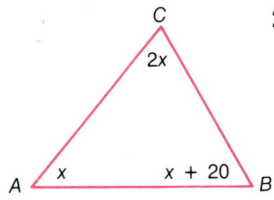

2. Triangle DGF with angles $5s$ at D, $5s$ at G, $5s$ at F.

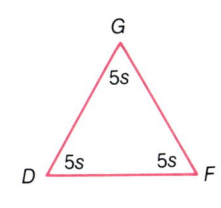

3. Triangle RTS with angles $2y$ at R, y at T, y at S.

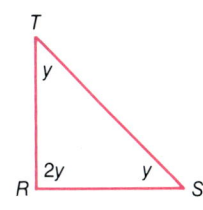

4. Triangle PNQ with angles $d+30$ at P, d at N, d at Q.

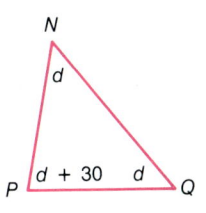

5. Triangle ACB with angles $y+15$ at A, y at C, $y+15$ at B.

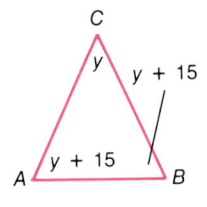

6. Triangle KLM with angles $8t$ at L, $3t$ at K, t at M.

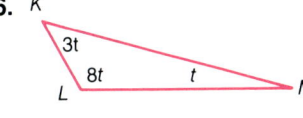

Solve.

7. The measures of the angles of a triangle can be represented by d, $d - 15$, and $d - 30$. Find the measure of each angle of the triangle.

8. The measures of the angles of a triangle can be represented by s, $2s$, and $6s$. Find the measure of each angle of the triangle.

9. The measures of the angles of a triangle can be represented by x, $x + 10$, and $x + 20$. Find the measure of each angle of the triangle.

10. The measures of the angles of a triangle can be represented by c, $5c + 20$, and $10c$. Find the measure of each angle of the triangle.

REVIEW CAPSULE FOR SECTION 13-6

Combine like terms. (Pages 76-80)

1. $m + m + 1$
2. $x + x + 1$
3. $p + p + 1 + p + 2$
4. $y + y + 1 + y + 2$
5. $2t + 2t + 1 + 2t + 2$
6. $3g + 3g + 1 + 3g + 2$

Solve and check. (Pages 326-328)

7. $n + n + 1 = 13$
8. $p + p + 1 = 19$
9. $s + s + 1 + s + 2 = 27$
10. $q + q + 1 + q + 2 = 45$
11. $99 = 2y + 2y + 1 + 2y + 2$
12. $57 = 3b + 3b + 1 + 3b + 2$

The answers are on page 401.

Equations: Multiplication/Division

ASSIGNMENT GUIDE
BASIC
p. 389: Omit
AVERAGE
p. 389: 1-10
ABOVE AVERAGE
p. 389: 1-10

WRITTEN EXERCISES
1. $x = 40°$; $2x = 80°$; $x + 20° = 60°$
2. $5s = 60°$
3. $y = 45°$; $2y = 90°$
4. $d = 50°$; $d + 30 = 80°$
5. $y = 50°$; $y + 15 = 65°$
6. $t = 15°$; $3t = 45°$; $8t = 120°$
7. $d = 75°$; $d - 15 = 60°$; $d - 30 = 45°$
8. $s = 20°$; $2s = 40°$; $6s = 120°$
9. $x = 50°$; $x + 10 = 60°$; $x + 20 = 70°$
10. $c = 10°$; $5c + 20 = 70°$; $10c = 100°$

Teaching Suggestions
p. M-26

QUICK QUIZ
Solve and check.
1. $2y + 1 = 17$ Ans: 8
2. $3a + 3 = 45$ Ans: 14
3. $x + x + 1 = 131$
   Ans: 65
4. $y + y + 1 + y + 2 = 252$ Ans: 83
5. $4b + 4b + 4 = 60$
   Ans: 7

ADDITIONAL EXAMPLE
Example 1
The sum of two consecutive whole numbers is 283. Find the numbers.
Ans: 141; 142

SELF-CHECK
1. 7; 8

PROBLEM SOLVING AND APPLICATIONS

13-6 More Than One Unknown: Consecutive Numbers

Michael is one year younger than Janet. Thus, their ages can be written as *consecutive whole numbers*.

Consecutive whole numbers are numbers such as 3, 4, 5, 6, 7, and so on. When you find consecutive whole numbers, list them in order from least to greatest.

EXAMPLE 1 The sum of Michael's and Janet's present ages is 31. Their ages are consecutive whole numbers. Find their present ages.

Solution:

[1] Choose a variable. Represent the two unknowns.

Let n = the age of Michael.
Then $n + 1$ = the age of Janet. ◂ Janet is one year older.

Michael's age + Janet's age = 31

[2] Write an equation. $n + n + 1 = 31$

[3] Solve the equation.
$2n + 1 = 31$
$2n = 30$
$n = 15$
$n + 1 = 16$ ◂ Don't forget to find $n + 1$.

[4] Check: a. Are the two numbers consecutive?
Does $15 + 1 = 16$? Yes ✓
b. Is the sum of their ages 31?
Does $15 + 16 = 31$? Yes ✓

Michael is **15 years old** and Janet is **16 years old**.

Self-Check 1. The sum of two consecutive whole numbers is 15. Find the two numbers.

390 Chapter 13

You can find three consecutive whole numbers when given their sum.

EXAMPLE 2 The sum of Lorraine's three diving scores is 24. The scores are consecutive whole numbers. Find each score.

Solution: ⓵ Choose a variable. Represent the three unknowns.

Let q = the smallest score.
Then $q + 1$ = the second score.
And $q + 2$ = the third score.

⓶ Write an equation.

Smallest Score + Second Score + Third Score = 24
q + $q + 1$ + $q + 2$ = 24

⓷ Solve the equation.
$3q + 3 = 24$
$3q = 21$
$q = 7$ *Don't forget to find $q + 1$ and $q + 2$.*
$q + 1 = 8$
$q + 2 = 9$

⓸ Check: **a.** Are the numbers consecutive?
Does $7 + 1 = 8$ and $8 + 1 = 9$?   Yes ✔
b. Is the sum of the numbers 24?
Does $7 + 8 + 9 = 24$?   Yes ✔   **Scores: 7, 8, 9**

Self-Check 2. The sum of three consecutive whole numbers is 27. Find the numbers.

CLASSROOM EXERCISES

For Exercises 1–4, determine whether the numbers listed are consecutive whole numbers. Write Yes or No. (Examples 1 and 2, step 4)

1. 86, 87   2. 56, 57, 58, 60   3. 5, 10, 11, 15   4. 21, 22, 23, 24

Write an equation for each sentence. Let h represent the smallest whole number. (Examples 1 and 2, Step 2)

5. The sum of two consecutive whole numbers is 5.
6. The sum of two consecutive whole numbers is 13.
7. The sum of three consecutive whole numbers is 21.
8. The sum of three consecutive whole numbers is 39.

Equations: Multiplication/Division   **391**

ADDITIONAL EXAMPLE
Example 2
The sum of three consecutive whole numbers is 264. Find the numbers.
Ans: 87; 88; 89

SELF-CHECK
2. 8; 9; 10

CLASSROOM EXERCISES
1. Yes   2. No   3. No
4. Yes
5. $h + h + 1 = 5$
6. $h + h + 1 = 13$
7. $h + h + 1 + h + 2 = 21$
8. $h + h + 1 + h + 2 = 39$

ASSIGNMENT GUIDE

BASIC

p. 392: Omit

AVERAGE

p. 392: 1–14

ABOVE AVERAGE

p. 392: 1–14

WRITTEN EXERCISES

1. Louise: 14; Reynaldo: 15
2. Kyoshi: 19; Rosa: 18
3. William: 85; Juan: 84
4. Julia: 93; Roberta: 94
5. 104; 105; 106
6. Jason: 11; Amy: 10; Tasha: 9
7. Morris: 21; Eric: 20; Maria: 22
8. 78; 79; 80
9. Hans: 17; Rosaline: 16
10. 86; 87
11. 6; 7; 8
12. Stan: 36; Mark: 37; Wayne: 38
13. 64; 65
14. 24; 25; 26

WRITTEN EXERCISES

Goal: To solve word problems involving consecutive whole numbers

For Exercises 1–14, solve and check. (Example 1)

1. Louise is one year younger than Reynaldo. The sum of their ages is 29. Find their ages.

2. Kyoshi is one year older than Rosa. The sum of their ages is 37. Find their ages.

3. William's score on a spelling test was one point higher than Juan's score. The sum of their scores was 169. Find their scores.

4. Julia's test score was one point lower than Roberta's score. The sum of their scores was 187. Find their scores.

(Example 2)

5. Carol Soong's scores for 3 games of bowling were consecutive whole numbers. The sum of her scores was 315. Find her scores.

6. The ages of Jason, Amy, and Tasha are consecutive whole numbers. Jason is the oldest and Tasha is the youngest. The sum of their ages is 30. Find their ages.

7. The ages of Morris, Eric, and Maria are consecutive whole numbers. Maria is the oldest and Eric is the youngest. The sum of their ages is 63. Find their ages.

8. William Blackfeather's three math scores are consecutive whole numbers. The sum of his scores is 237. Find his scores.

MIXED PRACTICE

9. Hans is one year older than Rosaline. The sum of their ages is 33. Find their ages.

10. Regina's two scores were consecutive whole numbers. The sum of her scores was 173. Find her scores.

11. The sum of Winnie's three diving scores was 21. The scores were consecutive whole numbers. Find her scores.

12. Stan's, Mark's and Wayne's ages are consecutive whole numbers. Stan is the youngest and Wayne is the oldest. The sum of their ages is 111. Find their ages.

13. The sum of two consecutive whole numbers is 129. Find the numbers.

14. The sum of three consecutive whole numbers is 75. Find the numbers.

Chapter 13

REVIEW CAPSULE FOR SECTION 13-7

Replace each __?__ with < or >. (Pages 294–296)

1. -2 __?__ 4
2. 2 __?__ -4
3. 0 __?__ 5
4. 0 __?__ -5
5. -6 __?__ 10

Solve each equation. (Pages 374–376)

6. $\frac{x}{3} = 8$
7. $\frac{t}{-4} = 2$
8. $6m = -42$
9. $-5p = 30$
10. $-7y = -56$

The answers are on page 401.

13-7 Inequalities: Multiplication/Division

Solving an inequality is similar to solving an equation except when you multiply or divide an inequality by a negative number.

Multiplication Property for Inequalities

1. The direction of the inequality symbol <u>remains unchanged</u> when each side of the inequality is multiplied by the <u>same positive number</u>.

   $\frac{n}{2} < 8$
   $(2)\frac{n}{2} < 8(2)$
   $n < 16$

2. The direction of the inequality symbol is <u>reversed</u> when each side of the inequality is multiplied by the <u>same negative number</u>.

   $3 < \frac{x}{-4}$
   $(-4)3 > \frac{x}{-4}(-4)$
   $-12 > x$, or $x < -12$

EXAMPLE 1 Solve and graph: **a.** $\frac{x}{2} > 3$ **b.** $\frac{y}{-3} < 6$

Solutions:

a. $\frac{x}{2} > 3$

$(2)\frac{x}{2} > 3(2)$

$x > 6$

The solution is **all numbers greater than 6.**

b. $\frac{y}{-3} < 6$

$(-3)\frac{y}{-3} > 6(-3)$ ◀ Reverse the inequality symbol.

$y > -18$

The solution is **all numbers greater than -18.**

Equations: Multiplication/Division **393**

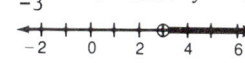

SELF-CHECK

1. c < 8;
2. x > −3;
3. d > −18;
4. r < 20;

ADDITIONAL EXAMPLES

Example 2

Solve and graph.

1. −2x > −20

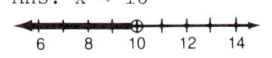

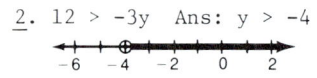

    Ans: x < 10

2. 12 > −3y   Ans: y > −4

SELF-CHECK

5. m > −5;
6. b < −6;
7. q ≤ −5

CLASSROOM EXERCISES

1. 5	2. 3	3. −2
4. −4	5. 9	6. 6
7. 7	8. −7	9. −8
10. −5	11. −3	12. −4
13. −6	14. −3	15. 5
16. −2	17. −7	18. −11

Self-Check Solve and graph: 1. $\frac{c}{4} < 2$ 2. $\frac{x}{3} > -1$ 3. $\frac{d}{-6} < 3$ 4. $-4 < \frac{r}{-5}$

REMEMBER: The open circle on a graph means that the number circled is <u>not</u> one of the solutions of the inequality.

> **Division Property for Inequalities**
> 1. The direction of the inequality symbol <u>remains unchanged</u> when each side of the inequality is divided by the <u>same positive number</u>.
>
> $5y < 15$
> $\frac{5y}{5} < \frac{15}{5}$
> $y < 3$
>
> 2. The direction of the inequality symbol is <u>reversed</u> when each side of the inequality is divided by the <u>same negative number</u>.
>
> $-3a > 9$
> $\frac{-3a}{-3} < \frac{9}{-3}$
> $a < -3$

EXAMPLE 2 Solve and graph: **a.** $-5y < 40$ **b.** $-24 > 8p$

Solutions: **a.** $-5y < 40$

$\frac{-5y}{-5} > \frac{40}{-5}$ ◀ Reverse the inequality symbol.

$y > -8$

b. $-24 > 8p$

$\frac{-24}{8} > \frac{8p}{8}$

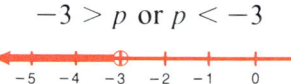

$-3 > p$ or $p < -3$

The solution is **all numbers greater than −8**.

The solution is **all numbers less than −3**.

Self-Check Solve and graph: **5.** $-3m < 15$ **6.** $-54 > 9b$ **7.** $25 \leq -5q$

CLASSROOM EXERCISES

For Exercises 1–10, by what number would you multiply each side of the inequality in order to solve for x? (Example 1)

1. $\frac{x}{5} > 4$
2. $\frac{x}{3} < 6$
3. $\frac{x}{-2} < 8$
4. $\frac{x}{-4} > 1$
5. $\frac{x}{9} < -5$
6. $\frac{x}{6} \geq -10$
7. $\frac{x}{7} \leq -5$
8. $-2 < \frac{m}{-7}$
9. $-9 > \frac{h}{-8}$
10. $\frac{x}{-5} > 12$

394 Chapter 13

For Exercises 11–18, by what number would you divide each side of the inequality in order to solve for p? (Example 2)

11. $-3p > 12$
12. $-4p < 24$
13. $-6p < -48$
14. $-3p > -15$
15. $-20 \geq 5p$
16. $-18 \leq -2p$
17. $21 \leq -7p$
18. $22 \geq -11p$

WRITTEN EXERCISES

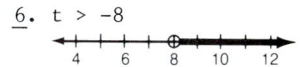

Goal: To solve and graph inequalities using the Multiplication and Division Properties for Inequalities

For Exercises 1–32, solve and graph each inequality. (Example 1)

1. $\frac{y}{3} < 3$
2. $\frac{x}{5} > 1$
3. $\frac{a}{7} > -10$
4. $\frac{g}{6} \leq -8$
5. $\frac{w}{-2} \geq 3$
6. $\frac{t}{-4} < 2$
7. $-5 < \frac{k}{-10}$
8. $-11 > \frac{n}{-8}$

(Example 2)

9. $-5n < 35$
10. $-3y > 27$
11. $-10b < -40$
12. $-8a > -24$
13. $-20 > 2m$
14. $-30 < 6x$
15. $20 \leq -4t$
16. $12 \geq -3z$

MIXED PRACTICE

17. $\frac{x}{4} > 24$
18. $\frac{k}{3} > 3$
19. $\frac{a}{5} < -8$
20. $\frac{t}{6} > -6$
21. $-12j < -120$
22. $-9m > -108$
23. $-8n < 16$
24. $-4y > 32$
25. $\frac{b}{-5} > 3$
26. $\frac{c}{-7} < 3$
27. $-8 < \frac{d}{-4}$
28. $-9 > \frac{e}{-5}$
29. $22 \leq -2p$
30. $14 \geq -14w$
31. $-32 \leq 2n$
32. $-12 \geq 12x$

REVIEW CAPSULE FOR SECTION 13-8

Write an algebraic expression for each word expression. (Pages 334–337)

1. Five more than a certain number, y
2. Nine less than a certain number, p
3. Eight less than a certain number, t
4. Ten more than a certain number, b

Find the average score. (Pages 10–12)

5. Team A: 16, 19, 7, 14, 14
6. Team B: 33, 25, 29, 25

The answers are on page 401.

Equations: Multiplication/Division **395**

Teaching Suggestions
p. M-26

QUICK QUIZ
Solve and check.
1. x + 8 = 21
   Ans: x = 13
2. 3 = 2x - 7
   Ans: x = 5
3. a + 29 = 84
   Ans: a = 55

Find the average.
4. 21, 36, 27, 32
   Ans: 29
5. 102, 86; 113, 91, 103
   Ans: 99

ADDITIONAL EXAMPLES
Example 1
The number of people increased by 17 is less than 81. Find the largest number of people possible. Ans: 63

SELF-CHECK
1. $27.99

Example 2
Given quiz scores of 19, 16, 20, 16, and 19, find the lowest score for the 6th quiz to make the average greater than 18. Ans: 19

PROBLEM SOLVING AND APPLICATIONS

13-8 Using Inequalities

Marla plans to spend less than $50 for a shirt and sweater. She buys a shirt for $21 and wants to know the greatest amount she can spend for a sweater. Marla can use an inequality to find the amount.

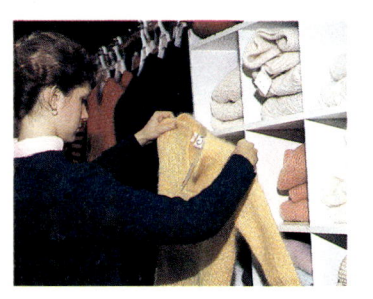

The cost of a sweater plus $21 is less than $50.

$$c + 21 \quad < \quad 50$$

The procedure for solving word problems involving inequalities is similar to the procedure for solving word problems involving equations.

EXAMPLE 1 Find the greatest amount Marla can spend for a sweater.

Solution: [1] Choose a variable. Let c = the cost of the sweater.

The cost of a sweater and $21 is less than $50.

[2] Write an inequality. $\quad c + 21 \quad < \quad 50$

[3] Solve the inequality. $\quad c + 21 - 21 < 50 - 21$
$$c < 29$$

The greatest amount Marla can spend for a sweater is **$28.99.**

Self-Check 1. Ted plans to spend less than $60 for a baseball glove and bat. He spends $32 for a glove. Find the greatest amount he can spend for a bat.

Recall that the **average** of two or more scores is the sum of the scores divided by the number of scores.

EXAMPLE 2 Raoul's scores on 4 tests were 80, 77, 81, and 84. He wants his average on **five** tests to be greater than 80. Find the lowest score Raoul can get on the fifth test.

Solution: [1] Let x = Raoul's score on the fifth test.

[2] $\dfrac{80 + 77 + 81 + 84 + x}{5} > 80 \qquad$ Average = $\dfrac{\text{Sum of scores}}{5}$.

396 Chapter 13

3. $$5\left(\frac{80+77+81+84+x}{5}\right) > 80(5)$$
$$80+77+81+84+x > 400$$
$$322+x > 400$$
$$322+x-322 > 400-322$$
$$x > 78$$

Raoul's score must be **at least 79.**

Self-Check 2. Lisa's scores on 4 tests were 90, 92, 89, and 85. She wants her average on five tests to be greater than 90. Find the lowest score Lisa can get on the fifth test.

CLASSROOM EXERCISES

Write an inequality for each word sentence. Use b as the variable. Do not solve the inequality. (Example 1, Step 2)

1. The cost of a lamp and $8 is less than $45.
2. The number of persons increased by 5 is less than 290.
3. The sum of the cost of a shirt and $15 is less than $32.
4. The number of hats increased by 12 is less than 72.

Write an inequality for each word sentence. Do not solve the inequality. (Example 2, Step 2)

5. The average of 87, 92, and y is greater than 91.
6. The average of 85, 87, 84, and p is greater than 86.
7. The average of 65, 70, 68 and x is greater than 66.
8. The average of 68, 70, 72, and c is greater than 71.

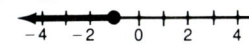

WRITTEN EXERCISES

Goal: To solve word problems involving inequalities

For Exercises 1–8, solve (Example 1)

1. Bill plans to buy a jacket and sweater for less than $70. He spends $42 for a jacket. Find the greatest amount he can spend for the sweater.
2. A family plans to spend less than $1100 for a new sofa and chair. They spend $800 for a sofa. Find the greatest amount they can spend for a chair.

Equations: Multiplications/Divisions **397**

SELF-CHECK
2. 95

CLASSROOM EXERCISES
1. b + 8 < 45
2. b + 5 < 290
3. b + 15 < 32
4. b + 12 < 72
5. $\frac{87 + 92 + y}{3} > 91$
6. $\frac{85 + 87 + 84 + p}{4} > 86$
7. $\frac{65 + 70 + 68 + x}{4} > 66$
8. $\frac{68 + 70 + 72 + c}{4} > 71$

ASSIGNMENT GUIDE
BASIC
pp. 397-398: Omit
AVERAGE
pp. 397-398: Omit
ABOVE AVERAGE
pp. 397-398: 1-8

WRITTEN EXERCISES
1. $27.99   2. $299.99

WRITTEN EXERCISES p. 395
24. y < -8
26. c > -21
28. e > 45
30. w ≥ -1
32. x ≤ -1

WRITTEN EXERCISES
3. $44.99   4. $219.99
5. 89   6. 93   7. 101
8. 94

QUIZ: SECTIONS 13-5–13-8
After completing this Review, you may wish to administer a quiz covering the same sections. A Quiz is provided in the Teacher's Edition: Part II.

REVIEW: SECTIONS 13-5–13-8
1. 8, 9   2. 12, 13, 14
3. $x < -8$
   [number line from -12 to -4, open circle at -8]
4. $c < -6$
   [number line from -10 to -2, open circle at -6]
5. $n < -2$
   [number line from -4 to 4, open circle at -2]
6. $y < 4$
   [number line from 0 to 8, open circle at 4]
7. $x = 25°$; $2x = 50°$; $4x + 5 = 105°$
8. $349.99

---

3. Carl plans to spend less than $70 for a tennis racket and a pair of shoes. He spends $25 for a pair of shoes. Find the greatest amount he can spend for a tennis racket.

4. The Chin family plans to spend less than $650 a month for rent and food. They spend $430 a month for rent. What is the most they can spend for food a month?

(Example 2)

5. Ava's three math quiz scores were 78, 81, and 93. She wants her average on four quizzes to be greater than 85. Find the lowest score she can get on the fourth quiz.

6. Ronald's grades on four tests are 68, 82, 78, and 80. He would like to have an average of more than 80 on five tests. What is the lowest grade he can get on the fifth test?

7. Rocco's scores for three games of bowling were 76, 103, and 121. In order to have an average greater than 100 for four games, what is the lowest score he can bowl on the fourth game?

8. Lillian's scores for four games of bowling were 76, 85, 104 and 92. She wants her average for five games to be greater than 90. Find the lowest score she can bowl on the fifth game.

REVIEW: SECTIONS 13-5 — 13-8

Solve and check. (Section 13-6)

1. The sum of two consecutive whole numbers is 17. Find the numbers.

2. The sum of three consecutive whole numbers is 39. Find the numbers.

Solve and graph. (Section 13-7)

3. $\dfrac{x}{4} < -2$

4. $\dfrac{c}{-2} > 3$

5. $5n < -10$

6. $-8y > -32$

Solve. (Section 13-8)

7. The measures of the angles of a triangle can be represented by x, $2x$, and $4x + 5$. Find the measure of each angle of the triangle. (Section 13-5)

8. The Luna family plans to spend less than $750 a month for rent and food. They spend $400 each month for rent. What is the most they can spend each month for food? (Section 13-8)

398   Chapter 13

CHAPTER REVIEW

PART 1: VOCABULARY

For Exercises 1–6, choose from the box at the right below the word(s) that best corresponds to each description.

1. Words such as product, times, twice, and tripled suggest the operation of __?__.
2. The sum of the measures of the angles of a __?__ is 180°.
3. Numbers such as 12, 13, 14, 15 and so on are __?__ whole numbers.
4. The words quotient and divided by suggest the operation of __?__.
5. The direction of the symbol of inequality is __?__ when each side is multiplied by the same negative number.
6. The direction of the symbol of inequality is __?__ when each side is multiplied by the same positive number.

> triangle
> unchanged
> multiplication
> reversed
> division
> consecutive

PART 2: SKILLS

Solve and check. (Section 13-1)

7. $\dfrac{x}{5} = -8$
8. $\dfrac{y}{-3} = -12$
9. $\dfrac{b}{4} = 9$
10. $\dfrac{c}{-8} = 22$
11. $3p = 24$
12. $-5q = -45$
13. $-8d = 64$
14. $12t = -60$

Write an equation for each word sentence. (Section 13-2)

15. Three times the number of club members, m, is 42.
16. The number of letters, l, divided by 6 is 14.
17. The number of peanuts, p, divided by 16 is 8.
18. Twice the length, l of a rectangle is 56 centimeters.

Solve and check. (Section 13-3)

19. $3m + 6 = -3$
20. $2p + 4 = -6$
21. $-4t - 8 = 16$
22. $-2y - 9 = -13$
23. $\dfrac{p}{7} + 4 = 5$
24. $\dfrac{x}{3} + 5 = 2$
25. $\dfrac{a}{-4} - 2 = 7$
26. $\dfrac{b}{-5} - 6 = -4$

(Section 13-4)

27. $3(c + 9) = 54$
28. $8(m - 2) = 40$
29. $-4(2n + 3) = -2(3n + 18)$
30. $-5(t + 4) = -7(t + 12)$

Equations: Multiplication/Division

CHAPTER REVIEW
1. multiplication
2. triangle
3. consecutive
4. division
5. reversed
6. unchanged
7. -40   8. 36   9. 36
10. -176   11. 8   12. 9
13. -8   14. -5
15. $3m = 42$ 16. $\dfrac{1}{6} = 14$
17. $\dfrac{p}{16} = 8$ 18. $2l = 56$
19. -3   20. -5   21. -6
22. 2   23. 7   24. -9
25. -36   26. -10
27. 9   28. 7   29. 12
30. -32

CHAPTER REVIEW

31. x = 60°
32. m = 35°; 3m = 105°; m + 5 = 40°
33. t = 30°; 4t = 120°
34. 6; 7   35. 17; 18
36. 7; 8; 9
37. 12; 13; 14
38. x > −24;
 −28 −26 −24 −22 −20
39. p < −135;
 −138   −134   −130
40. a > 45;
 42 44 46 48 50
41. b < 48;
 44 46 48 50 52
42. y > −4;
 −8 −6 −4 −2 0
43. t > 11;
 8 10 12 14 16
44. c > −5;
 −6 −4 −2 0 2
45. q > −2;
 −4 −2 0 2 4
46. 7   47. 90
48. b = 35°; 2b + 20 = 90°; b + 20 = 55°
49. Kermit: 18; Vera: 19
50. $64.99

Use the formula A + B + C = 180° to find the measure of each angle of the triangle. (Section 13-5)

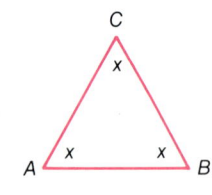

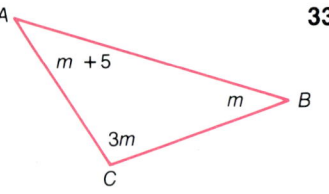

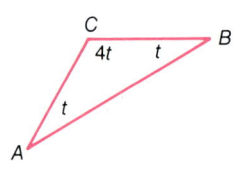

31. Triangle ABC with angles x, x, x
32. Triangle ABC with angles m + 5, 3m, m
33. Triangle ABC with angles 4t, t, t

Solve each equation. Then complete. (Section 13-6)

34. $n + n + 1 = 13$; $n = $ _?_ ; $n + 1 = $ _?_
35. $p + p + 1 = 35$; $p = $ _?_ ; $p + 1 = $ _?_
36. $x + x + 1 + x + 2 = 24$; $x = $ _?_ ; $x + 1 = $ _?_ ; $x + 2 = $ _?_
37. $y + y + 1 + y + 2 = 39$; $y = $ _?_ ; $y + 1 = $ _?_ ; $y + 2 = $ _?_

Solve and graph. (Section 13-7)

38. $\frac{x}{3} > -8$
39. $\frac{p}{-9} > 15$
40. $\frac{a}{-5} < -9$
41. $\frac{b}{4} < 12$
42. $6y > -24$
43. $-4t < -44$
44. $-5c < 25$
45. $8q > -16$

PART 3:   APPLICATIONS

Solve each problem.

46. Twice the number of fish Steve caught is 14. How many fish did he catch? (Section 13-2)

47. The number of bicycles divided by 6 is 15. Find the number of bicycles. (Section 13-2).

48. The measures of the angles of a triangle can be represented by b, $2b + 20$, and $b + 20$. Find the measure of each angle of the triangle. (Section 13-5)

49. Kermit is one year younger than Vera. Find their present ages if the sum of their ages is 37. (Section 13-6)

50. Glenda plans to spend less than $100 for a lamp and a table. She buys a lamp for $35. Find the greatest amount she can spend on a table. (Section 13-8)

CHAPTER TEST

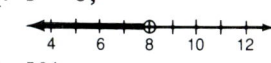

For Exercises 1–10, solve and check.

1. $\dfrac{x}{15} = -4$
2. $\dfrac{c}{-7} = 3$
3. $-6y = 48$
4. $-15z = -45$
5. $6x + 10 = 52$
6. $3a - 8 = -114$
7. $\dfrac{n}{-5} - 2 = 13$
8. $\dfrac{y}{4} + 7 = -5$
9. $10(e + 2) = 50$
10. $6(x + 5) = -3(x + 2)$

11. Solve and graph: $\dfrac{b}{4} < 2$

Solve each problem.

12. The number of seats in an auditorium divided by 36 rows is 14. Find the number of seats.

13. The measures of the angles of a triangle can be represented by x, $3x$, and $6x$. Find the measure of each angle of the triangle.

14. Kurt and Melissa were born in consecutive years. Kurt is 1 year younger than Melissa. Find their present ages if the sum of their ages is 35.

15. Bert plans to buy a shirt and a pair of slacks for less than $53. He spends $21 for a shirt. Find the greatest amount he can spend for a pair of slacks.

ANSWERS TO REVIEW CAPSULES

Page 376 1. Yes 2. No 3. No 4. Yes 5. 3 6. 7 7. 5 8. 9 9. 48 10. 110 11. 50 12. 91

Page 380 1. $3x$ 2. $4n$ 3. $7b$ 4. $5d$ 5. $\dfrac{x}{3}$ 6. $\dfrac{d}{4}$ 7. $\dfrac{h}{3}$ 8. $\dfrac{r}{6}$ 9. 2 10. -10 11. 7 12. 3 13. -5 14. -17 15. -9 16. 20

Page 383 1. $12n + 18$ 2. $-28y + 12$ 3. $5p + 35$ 4. $-8x - 12$ 5. $-5z - 55$ 6. $-7b - 42$ 7. $20a^2 - 15a$ 8. $8t^2 - 8t$ 9. 30 10. 90 11. 32 12. 17

Page 386 1. $16x + 2$ 2. $9y - 9$ 3. $5a$ 4. $11a - 9$ 5. $t + 1$ 6. $14y$ 7. $-5r + 32$ 8. $25b - 7$ 9. $6c + 10$ 10. $2g - 12$ 11. $13p$ 12. $15s$

Page 389 1. $2m + 1$ 2. $2x + 1$ 3. $3p + 3$ 4. $3y + 3$ 5. $6t + 3$ 6. $9g + 3$ 7. 6 8. 9 9. 8 10. 14 11. 16 12. 6

Page 393 1. < 2. > 3. < 4. > 5. < 6. 24 7. -8 8. -7 9. -6 10. 8

Page 395 1. $y + 5$ 2. $p - 9$ 3. $t - 8$ 4. $b + 10$ 5. 14 6. 28

Equations: Multiplications/Division **401**

ENRICHMENT

You may wish to use this lesson for students who performed well on the formal Chapter Test.

1. Yes 2. No 3. No
4. No 5. Yes 6. Yes

ENRICHMENT

Sets

A **set** is a collection of objects called **elements**.

Here are two ways to describe a set.

Rule Method	Roster Method
D = The set of natural numbers less than 6	D = {1, 2, 3, 4, 5}
W = The set of whole numbers	W = {0, 1, 2, 3, $\cdots$}

The numbers continue without end.

Symbol or Term	Explanation	Example
$\in$	Is an element of	$6 \in \{2, 4, 6, 8\}$
$\notin$	Is not an element of	$3 \notin \{6, 7, 8\}$
$=$	Is equal to; has the <u>same</u> elements	$\{1, 4, 6\} = \{6, 4, 1\}$
$\neq$	Is not equal to	$\{2, 3, 5\} \neq \{4, 6, 8\}$
Finite set	The elements can be listed.	$\{1, 2, 5, 6\}$
Infinite set	The elements can <u>not</u> be completely listed.	$\{2, 4, 6, 8, \cdots\}$
ϕ	Empty set; has no elements	The set of whole numbers less than $0 = \phi$.
Equivalent sets	Sets that have the <u>same number</u> of elements	$\{1, 2, 3, 4\}$ $\{6, 7, 8, 9\}$ — One-to-one correspondence

EXERCISES

Determine whether each of the following is a true statement. Answer <u>Yes</u> or <u>No</u>

1. $6 \in \{1, 2, 3, 4, \cdots\}$
2. $15 \notin \{3, 6, 9, \cdots\}$
3. $\{1, 2, 3\} = \{1, 2, 3, \cdots\}$
4. $\{6, 7, 8, 9\} \neq \{9, 8, 7, 6\}$
5. $\{0\}$ is a finite set.
6. $\{2, 4, 6, \cdots\}$ is an infinite set.

402 Chapter 13

ADDITIONAL PRACTICE

SKILLS

Solve and check. (Pages 374–376)

1. $\dfrac{t}{-5} = -30$
2. $\dfrac{x}{-6} = -36$
3. $\dfrac{y}{3} = -12$
4. $\dfrac{r}{4} = -52$
5. $8c = -64$
6. $4w = -56$
7. $-3x = -9.6$
8. $-3v = -8.1$

(Pages 381–382)

9. $5x + 1 = 51$
10. $10v + 26 = 6$
11. $36 = -6g + 6$
12. $14 = -8g + 6$
13. $\dfrac{y}{5} - 5 = -12$
14. $\dfrac{k}{3} - 7 = -35$
15. $20 = \dfrac{x}{8} + 14$
16. $10 = \dfrac{w}{3} + 4$

(Pages 383–385)

17. $7(x + 3) = 42$
18. $8(a + 10) = 32$
19. $2(4w - 18) = -28$
20. $4(b - 3) = -48$
21. $7(y + 7) = 5(y + 5)$
22. $9(g + 3) = 3(g + 1)$

Solve and graph each inequality. (Pages 393–395)

23. $\dfrac{x}{2} < 4$
24. $\dfrac{n}{-3} > 3$
25. $6x > -24$
26. $-2v < -6$

APPLICATIONS *Represent each word expression by an algebraic equation.* (Pages 377–380)

27. The product of 3 and x is 33.
28. The product of 12 and w is 144.
29. The quotient of w and 4 is 9.
30. The quotient of v and 11 is 4.

Solve each problem.

31. The Hearth family plans to spend less than $1200 for a new table and chairs. The chairs cost $360. Find the greatest amount they can spend for the table. (Pages 396–398)

32. Jeff's three test scores were 79, 84, and 94. He wants his average on four tests to be greater than 86. Find the lowest score he can get on the fourth test. (Pages 396–398)

33. Laura's scores for three games of bowling were consecutive whole numbers. The sum of her scores was 327. Find Laura's scores for each game. (Pages 390–392)

34. The measures of the angles of a triangle can be represented by x, $x + 20$, and $6x$. Find the measure of each angle of the triangle. (Pages 388–389)

Equations: Multiplication/Division **403**

COMMON ERRORS

In preparation for the Cumulative Review, these exercises focus the student's attention on the most common errors to be avoided.

1. -13   2. -3   3. -16
4. 9   5. 3   6. -2
7. 7   8. q < 3   9. -18
10. 6   11. $-16\frac{1}{2}$
12. w < 5   13. 4m + 36
14. $p^2 + 5p$

COMMON ERRORS

Each of these problems contains a common error.
  a. Find the correct answer.
  b. Find the error.

1. $-9 + (-4) = \underline{\ ?\ }$
$|-9| + |-4| = 9 + 4$
$ = 13$

Thus, $-9 + (-4) = \mathbf{13}$

3. $-9 - 7 = -9 + 7$
$ = \mathbf{-2}$

2. $4 + (-7) = \underline{\ ?\ }$
$|4| + |-7| = -4 + 7$
$ = 3$

Thus, $4 + (-7) = \mathbf{3}$

4. $6 - (-3) = 6 - 3$
$ = \mathbf{3}$

Solve.

5. $x - 6 = -3$
$x - 6 = -3 - 6$
$x = \mathbf{-9}$

7. $2c = 21 - c$
$2c - c = 21 - c + c$
$c = \mathbf{21}$

9. $\dfrac{x}{-3} = 6$
$(-3)\dfrac{x}{-3} = 6(-3)$
$x = \mathbf{18}$

11. $2x + 9 = -24$
$\dfrac{2x}{2} + 9 = \dfrac{-24}{2}$
$x + 9 = -12$
$x = \mathbf{-21}$

6. $y + 5 = 3$
$y + 5 - 5 = 3 - 5$
$y = \mathbf{2}$

8. $q + 6 < 9$
$q + 6 - 6 < 9 - 6$
$q = \mathbf{3}$

10. $-4y = -24$
$\dfrac{-4y}{-4} = \dfrac{-24}{-4}$
$y = \mathbf{-6}$

12. $-6w > -30$
$\dfrac{-6w}{-6} > \dfrac{-30}{-6}$
$w > \mathbf{5}$

Multiply.

13. $4(m + 9) = 4(m) + 9$
$ = \mathbf{4m + 9}$

14. $p(p + 5) = p(p) + p(5)$
$ = 2p + 5p = \mathbf{7p}$

404   Chapter 13

CUMULATIVE REVIEW: CHAPTERS 10–13

Choose the correct answer. Choose a, b, c, or d.

1. Which integers are arranged in order from least to greatest?
   - **a.** 4, 7, 13, -44
   - **b.** $-4, -6, -9, -13$
   - **c.** $-15, -11, 0, 8$
   - **d.** $-17, -12, -13, 0$

2. Which integer is the opposite of -2?
   - **a.** 2
   - **b.** -1
   - **c.** -3
   - **d.** 3

3. Add: $-93 + 112$
   - **a.** -21
   - **b.** 21
   - **c.** 19
   - **d.** -19

4. Subtract: $-12 - (-36)$
   - **a.** 48
   - **b.** -48
   - **c.** 24
   - **d.** -24

5. The highest temperature on the planet Mars is about 35° F above zero and the lowest temperature is about 120° F below zero. About how much warmer is the highest temperature than the lowest temperature?
   - **a.** 155° F
   - **b.** 85° F
   - **c.** 95° F
   - **d.** 105° F

6. Add: $-35 + 47 + (-74) + 81$
   - **a.** 21
   - **b.** 19
   - **c.** -19
   - **d.** -21

7. Solve for y: $y - 41 = -135$
   - **a.** -176
   - **b.** 176
   - **c.** 94
   - **d.** -94

8. Solve for m: $-13 = m + 49$
   - **a.** 36
   - **b.** 62
   - **c.** -62
   - **d.** -36

9. Solve for t: $-3t + 8 + 4t = -7 + 12$
   - **a.** -3
   - **b.** 13
   - **c.** 3
   - **d.** -13

10. Solve for p: $4p - 17 = 3p + 4$
    - **a.** -13
    - **b.** 21
    - **c.** 13
    - **d.** -21

11. Write an algebraic expression: Three more than the sum of x and y.
    - **a.** $3 > x + y$
    - **b.** $xy + 3$
    - **c.** $(x + y) + 3$
    - **d.** $3 > xy$

12. Write an algebraic expression: Seven less than some number is 12.
    - **a.** $n - 7 = 12$
    - **b.** $n - 12 = 7$
    - **c.** $7 - n = 12$
    - **d.** $12 - n = 7$

13. Last Wednesday's average temperature plus 2° equals $-7°$. Find last Wednesday's average temperature.
    - **a.** $-14°$
    - **b.** $-5°$
    - **c.** $-9°$
    - **d.** $-11°$

CUMULATIVE REVIEW

14. b
15. d
16. a
17. b
18. c
19. c
20. c
21. b
22. b
23. a
24. c
25. d
26. c
27. b

Cumulative Review: Chapters 10–13 **405**

CUMULATIVE REVIEW

A cumulative test is provided on copying masters in the *Teacher's Edition: Part II.*

1. c
2. a
3. c
4. c
5. a
6. b
7. d
8. c
9. a
10. b
11. c
12. a
13. c

14. Which graph below shows the solution to the inequality $-3 > x + 1$?

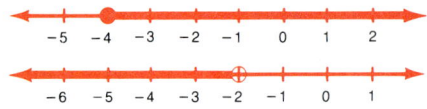

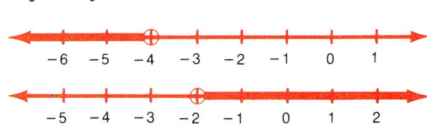

15. Multiply: $20(-35)$
    a. 700 b. -55 c. 55 d. -700

16. Multiply: $(-43)(-35)$
    a. 1505 b. -1505 c. -78 d. 78

17. Multiply: $8(2x - 10)$
    a. $16x + 10$ b. $16x + (-80)$ c. $-10x + 2$ d. $10x + (-80)$

18. Divide: $-36 \div 4$
    a. $-\frac{1}{9}$ b. 9 c. -9 d. $\frac{1}{9}$

19. Divide: $-\frac{3}{8} \div (-12)$
    a. $-\frac{1}{32}$ b. $4\frac{1}{2}$ c. $\frac{1}{32}$ d. $-4\frac{1}{2}$

20. Factor: $9x - 54$
    a. $9(x - 5)$ b. $9(x + 6)$ c. $9(x - 6)$ d. $9(x + 5)$

21. Factor: $y^2 - 4y$
    a. $4y(y - 1)$ b. $y(y - 4)$ c. $y(y + 4)$ d. $y(3y - 2)$

22. Solve for a: $\frac{a}{-4} = -16$
    a. -64 b. 64 c. -4 d. 4

23. Solve for w: $11w = -286$
    a. -26 b. -297 c. -275 d. -24

24. Write an algebraic expression for the quotient of 12 and y.
    a. $12 - y$ b. $y \div 12$ c. $12 \div y$ d. $12y$

25. Solve for h: $7(h + 5) = -77$
    a. 16 b. -112 c. 112 d. -16

26. One angle of a triangle is 50°. The measures of the other two angles are equal. Find the measures of the equal angles.
    a. 60° b. 130° c. 65° d. 70°

27. The sum of three consecutive whole numbers is 84. Find the three numbers.
    a. 26, 27, 28 b. 27, 28, 29 c. 26, 28, 30 d. 25, 27, 29

406 *Cumulative Review: Chapters 10-13*

CHAPTER 14 Graphing and Equations

SECTIONS
14-1 Graphing Ordered Pairs
14-2 Graphing Equations
14-3 Problem Solving and Applications: Broken Line Graphs
14-4 Slope of a Line
14-5 Intercepts
14-6 Direct and Indirect Variation
14-7 Solving a System of Equations by Graphing
14-8 Graphing Linear Inequalities

FEATURES
Calculator Application: *Computing y Values for Equations*
Computer Application: *Slope of a Line*
Enrichment: *Venn Diagrams*

Teaching Suggestions
p. M-27

QUICK QUIZ
Graph and label each point on a number line.
1. A = 2
   Ans: 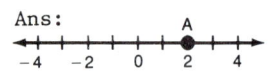
2. B = 0
   Ans:

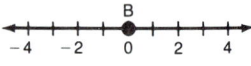

Graph each inequality.
3. x > -2
   Ans: 
4. x ≥ 2
   Ans:

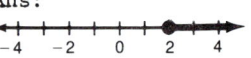

ADDITIONAL EXAMPLES
Example 1
Use the graph below to give the ordered pair for each point.

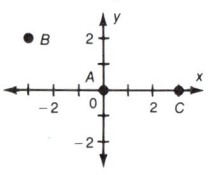

1. A   Ans: (0, 0)
2. B   Ans: (-3, 2)
3. C   Ans: (3, 0)

SELF-CHECK
1. (-2, 4)   2. (-5, -5)
3. (2, -2)   4. (2, 3)

14-1 Graphing Ordered Pairs

In the figure at the right, the two intersecting number lines are called **axes**. The horizontal number line is the *x* **axis**. The vertical number line is the *y* **axis**. The two axes separate the **coordinate plane** into four **quadrants**.

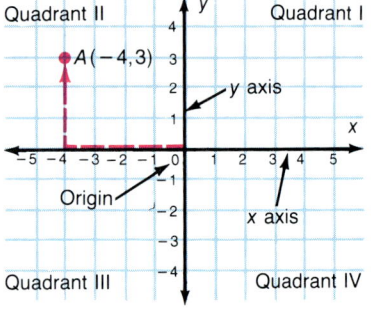

Note that point *A* is 4 units to the left of the *y* axis. It is also 3 units above the *x* axis.

The ordered pair (−4, 3) gives the location of point *A* in the coordinate plane. The *first* number, −4, is the *x* **coordinate**; the *second* number, 3, is the *y* **coordinate**.

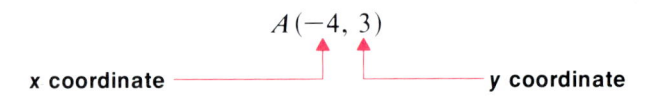

$$A(-4, 3)$$

x coordinate ⟶   ⟵ y coordinate

The point with coordinates (0, 0) is called the **origin**. It is the point where the *x* axis and the *y* axis intersect.

EXAMPLE 1 Use the graph at the right to give the ordered pair for each point.

Point	Think	Ordered Pair
a. Q	Start at the origin. Move 3 units to the right. Then move 4 units down.	(3, −4)
b. S	Start at the origin. Move 4 units to the left. Then move 3 units up.	(−4, 3)

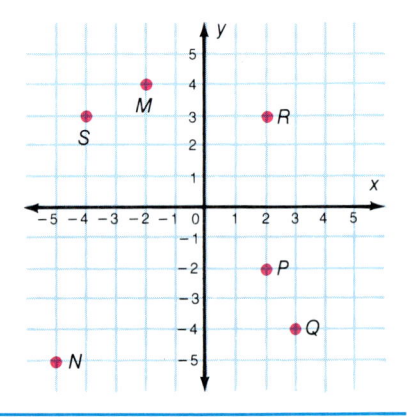

Self-Check *Refer to the graph above to give the ordered pair for each point.*
1. M   2. N   3. P   4. R

Note that the points S(−4, 3) and Q(3, −4) were graphed using the same two numbers, −4 and 3. However, they are not the same point. Thus, the order of the numbers is important.

408 Chapter 14

EXAMPLE 2   Graph: **a.** $B(2, 1)$ **b.** $C(-3, 0)$

Solutions: **a.** Start at the origin. Move 2 units to the right. Then move 1 unit up. Label the point B.

b. Start at the origin. Move 3 units to the left. Move neither up nor down. Label the point C.

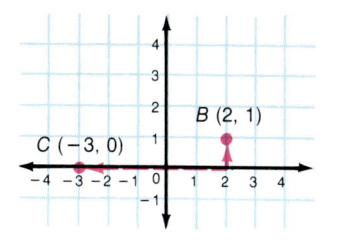

Self-Check Graph each point: **5.** $D(-1, -4)$ **6.** $F(4, 0)$ **7.** $G(0, -2)$

CLASSROOM EXERCISES

For Exercises 1-12, name the point on the graph that corresponds to each ordered pair. (Example 1)

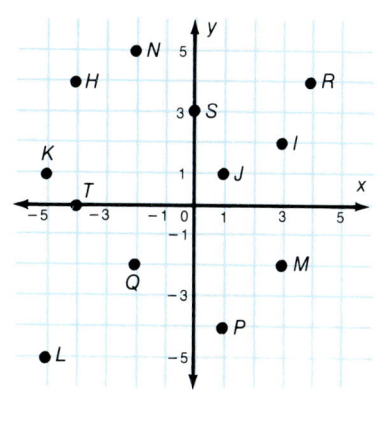

1. $(-2, 5)$ 2. $(3, 2)$
3. $(1, -4)$ 4. $(-5, 1)$
5. $(-2, -2)$ 6. $(4, 4)$
7. $(1, 1)$ 8. $(3, -2)$
9. $(-5, -5)$ 10. $(-4, 0)$
11. $(0, 3)$ 12. $(-4, 4)$

WRITTEN EXERCISES

Goals: To write the ordered pair for a point on a graph

To graph a point in a coordinate plane

Write the ordered pair for each point graphed below. (Example 1)

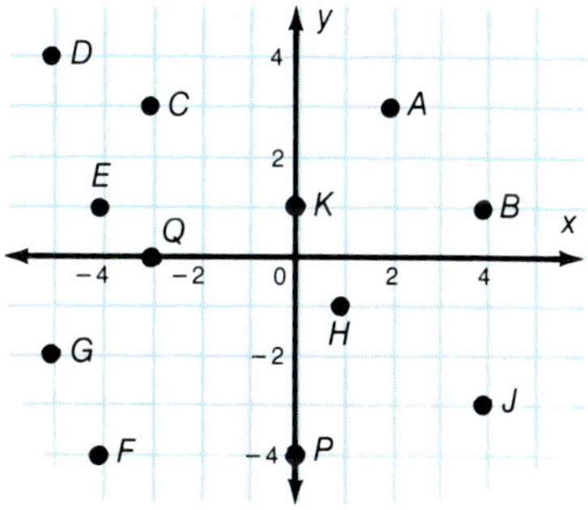

1. A 2. B 3. C
4. D 5. E 6. F
7. G 8. H 9. J
10. K 11. P 12. Q

Graph the point for each ordered pair. Label each point. (Example 2)

13. $A(4, 2)$ 14. $C(-3, -5)$ 15. $D(-2, -4)$ 16. $E(4, -5)$ 17. $F(6, -3)$
18. $G(-5, 4)$ 19. $H(-2, 6)$ 20. $J(6, 0)$ 21. $K(-4, 0)$ 22. $L(0, -5)$

Graphing and Equations   **409**

ADDITIONAL EXAMPLES

Example 2

Graph each point on the same set of axes.

1. $A(-3, -2)$
2. $B(4, 0)$

Ans:

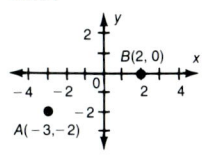

SELF-CHECK

5.-7.

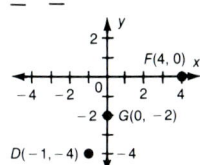

CLASSROOM EXERCISES

1. N   2. I   3. P
4. K   5. Q   6. R
7. J   8. M   9. L
10. T   11. S   12. H

ASSIGNMENT GUIDE

BASIC
p. 410: Omit
AVERAGE
p. 410: 1-22
ABOVE AVERAGE
p. 410: 1-22

WRITTEN EXERCISES

1. (2, 3)   2. (4, 1)
3. (-3, 3)   4. (-5, 4)

See page 414 for the answers to Exercises 5-24.

Sidebar

REVIEW CAPSULE

This Review Capsule reviews prior-taught skills used in Section 14-2. The reference is to the pages where the skills were taught.

Teaching Suggestions p. M-27

QUICK QUIZ

Evaluate.

1. $x - 5$ when $x = 2$
   Ans: -3
2. $y + 6$ when $y = 0$
   Ans: 6
3. $x + 1$ when $x = -3$
   Ans: -2
4. $2x + 1$ when $x = -8$
   Ans: -15

ADDITIONAL EXAMPLE

Example 1
Make a table of ordered pairs for $d = 500t$. Use $t = 0$, $t = 3$, and $t = 6$.

Ans:

t (sec)	d (ft)
0	0
3	1500
6	3000

SELF-CHECK

1.

t (seconds)	d (feet)
1	1100
3	3300
5	5500

410

REVIEW CAPSULE FOR SECTION 14-2

Evaluate each expression. (Pages 22–24)

1. $y + 3$ when $y = 4$
2. $n + 3$ when $n = 10$
3. $z - 5$ when $z = 8$
4. $3r - 5$ when $r = 4$
5. $2t + 5$ when $t = -4$
6. $4q - 3$ when $q = -2$

The answers are on page 435.

14-2 Graphing Equations

Jay was watching an approaching thunderstorm. Four seconds after he saw a flash of lightning, he heard the thunder. He wondered how far away the lightning was.

You can use this equation to find out how far away a flash of lightning is.

$d = 1100t$

d = distance in feet
t = time in seconds

EXAMPLE 1 Make a table of ordered pairs for the equation $d = 1100t$.

Solution:

[1] Choose at least 3 values for t.

Let $t = 0$. → $d = 1100 \cdot 0 = 0$
Let $t = 2$. → $d = 1100 \cdot 2 = 2200$
Let $t = 4$. → $d = 1100 \cdot 4 = 4400$

[2] Find d.

$d = 1100t$

[3] Make a table.

t (seconds)	d (feet)
0	0
2	2200
4	4400

Self-Check

1. Make a table of ordered pairs for the equation $d = 1100t$. Use $t = 1$, $t = 3$, and $t = 5$.

You can use the table of ordered pairs to graph a <u>linear equation</u>. A **linear equation** is an equation whose graph is a straight line.

PROCEDURE

To graph a linear equation:

[1] Make a table of ordered pairs.

[2] Graph the ordered pairs. Connect them with a straight line.

Chapter 14

EXAMPLE 2 Graph the equation $d = 1100t$. Use the table of ordered pairs from Example 1.

Solution: ⟨1⟩

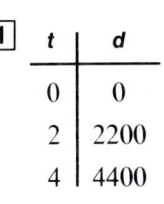

t	d
0	0
2	2200
4	4400

⟨2⟩ Graph with points (0, 0), (2, 2200), (4, 4400).

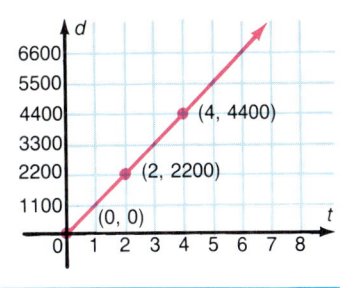

Self-Check 2. Graph the equation $d = 1100t$. Use the table of ordered pairs from Self-Check 1.

In Example 2, different scales were used on the horizontal and vertical axes for convenience in graphing. In Example 3, the same scale is used on each axis.

You can graph linear equations such as $y = 2x + 1$ in the coordinate plane.

EXAMPLE 3 Graph the linear equation $y = 2x + 1$.

Solution: ⟨1⟩ Make a table.

x	2x + 1
−2	2(−2) + 1 = −3
0	2(0) + 1 = 1
1	2(1) + 1 = 3

⟨2⟩ $y = 2x + 1$

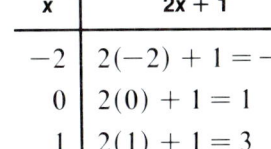

x	y
−2	−3
0	1
1	3

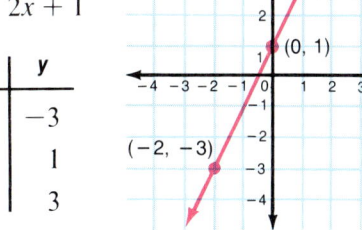

Graph with points (−2, −3), (0, 1), (1, 3).

Self-Check 3. Graph the equation $y = x - 1$.

CLASSROOM EXERCISES

Make a table of ordered pairs for each equation. Use the three given values for t. (Examples 1 and 3)

1. $d = 80t$
   Let $t = 0$.
   Let $t = 2$.
   Let $t = 4$.

2. $d = 60t$
   Let $t = 1$.
   Let $t = 2$.
   Let $t = 3$.

3. $d = 500t$
   Let $t = 0$.
   Let $t = 1$.
   Let $t = 2$.

4. $d = 1000t$
   Let $t = 1$.
   Let $t = 3$.
   Let $t = 5$.

Graphing and Equations **411**

ADDITIONAL EXAMPLES
Example 2
Graph the equation $d = 500t$. Use your table of ordered pairs from Additional Example 1.
Ans:

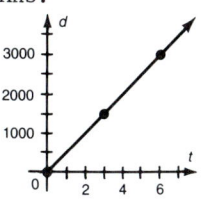

SELF-CHECK
2.

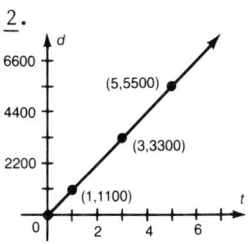

Example 3
Graph the equation $y = -2x + 2$
Ans:

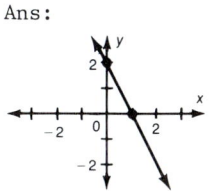

SELF-CHECK
3.

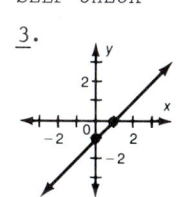

CLASSROOM EXERCISES
See page 416 for the answers to Exercises 1-4.

411

ASSIGNMENT GUIDE
BASIC
p. 412: Omit
AVERAGE
Day 1 p. 412: 1-10
Day 2 p. 412: 11-26
ABOVE AVERAGE
p. 412: 1-9 odd, 11-26

WRITTEN EXERCISES
7. 8. 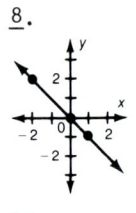
9. 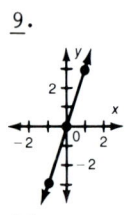 10. 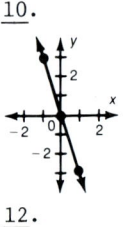
11. 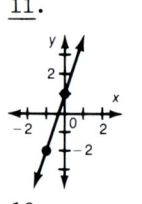 12. 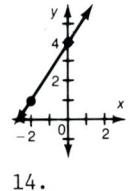
13. 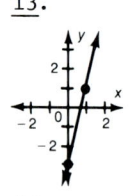 14. 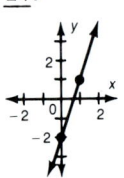
15. 16.

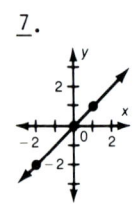

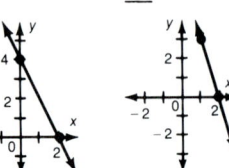

WRITTEN EXERCISES

Goal: To graph linear equations

For Exercises 1–2, copy the table and the coordinate axes. Then complete the table and draw the graph. (Examples 1 and 2)

1. The equation $d = 50t$ relates distance in miles and time in hours for a car traveling at a rate of 50 miles per hour.

   $d = 50t$

t	d
0	? 0
2	? 100
4	? 200

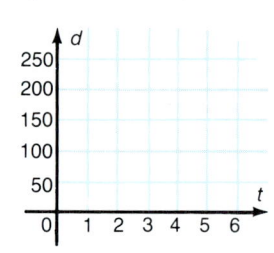

2. The equation $C = \frac{5F - 160}{9}$ relates temperature in degrees Fahrenheit to temperature in degrees Celsius.

   $C = \frac{5F - 160}{9}$

°F	°C
−40	? −40
32	? 0
68	? 20

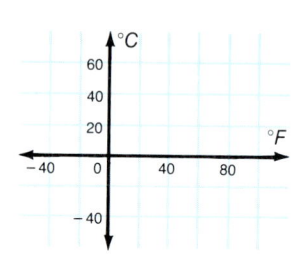

Copy and complete the table of ordered pairs for each equation. (Example 1 and Example 3, Step 1)

3. $y = x$

x	y
1	? 1
0	? 0
−2	? −2

4. $y = -x$

x	y
1	? −1
0	? 0
−2	? 2

5. $y = 3x$

x	y
2	? 6
0	? 0
−2	? −6

6. $y = -3x$

x	y
1	? −3
0	? 0
−1	? 3

Graph each equation. Use the corresponding tables from Exercises 3–6. (Examples 2 and 3, Step 2)

7. $y = x$
8. $y = -x$
9. $y = 3x$
10. $y = -3x$

Graph each equation. (Example 3)

11. $y = 3x + 1$
12. $y = 2x + 5$
13. $y = 4x - 3$
14. $y = 3x - 2$
15. $y = -2x + 4$
16. $y = -3x + 6$
17. $y = 2x - 3$
18. $y = 3x - 4$
19. $y = -x + 1$
20. $y = -x - 1$
21. $y = 2x$
22. $y = -2x$
23. $y = -2x - 1$
24. $y = -2x + 1$
25. $y = \frac{1}{2}x$
26. $y = -\frac{1}{2}x - 1$

412 Chapter 14

COMPUTING Y VALUES FOR EQUATIONS

You can use a calculator to compute y values for an equation.

EXAMPLE Complete the table of values. $y = 3x + 12$

x	y
2	?
0	?
-2	?

SOLUTION Substitute x values from the table in the equation.

$$y = 3x + 12 \longrightarrow y = 3(2) + 12$$

`3` `×` `2` `+` `1` `2` `=`   → 18.

Repeat the procedure for $x = 0$ and $x = -2$.
Then $y = 12$ and $y = 6$.

EXERCISES

1. $y = 3x + 1$

x	y
0	?
-2	?
-3	?

2. $y = -2x + 2$

x	y
4	?
-4	?
-8	?

3. $y = 4x - 3$

x	y
9	?
-3	?
-6	?

REVIEW CAPSULE FOR SECTION 14-3

Add. (Pages 35–37)

1. $22 + 8$ **2.** $17 + 4$ **3.** $35 + 19$ **4.** $4.7 + 2.4$ **5.** $12.8 + 3.2$

Subtract. (Pages 35–37)

6. $20 - 9$ **7.** $27 - 8$ **8.** $9.2 - 1.8$ **9.** $7.0 - 1.1$ **10.** $87 - 38$

Find the average. (Pages 10–12)

11. 1; 7; 13; 15   **12.** 2.4; 1.5; 3.6   **13.** 10; 12; 9; 15; 14   **14.** 22.1; 19.5; 20.5

The answers are on page 435.

Graphing and Equations   **413**

WRITTEN EXERCISES

17. 18. 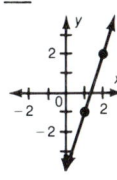
19. 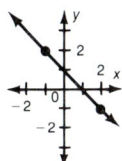 20. 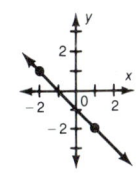
21. 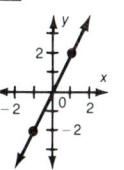 22. 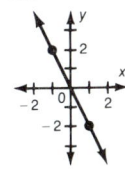
23. 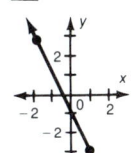 24. 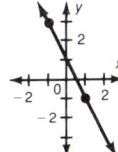
25. 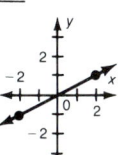 26.

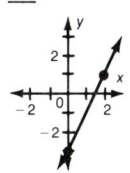

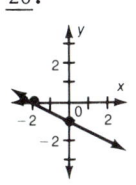

CALCULATOR EXERCISES

<u>1</u>. 1; -5; -8
<u>2</u>. -6; 10; 18
<u>3</u>. 33; -15; -27

Teaching Suggestions
p. M-27

QUICK QUIZ
Add or subtract.

<u>1</u>. 27 + 80   Ans: 107
<u>2</u>. 91 - 17   Ans: 74

Find the average.

<u>3</u>. 12, 17, 16   Ans: 15
<u>4</u>. 3, 9, 8, 8, 7, 2, 5
    Ans: 6

ADDITIONAL EXAMPLE
Example 1
Use the broken line graph on page 414 to find the total rainfall for January through June.   Ans: 25 in

SELF-CHECK
<u>1</u>. 2 in

WRITTEN EXERCISES p. 409
<u>5</u>. (-4, 1) <u>6</u>. (-4, -4)
<u>7</u>. (-5, -2) <u>8</u>. (1, -1)
<u>9</u>. (4, -3) <u>10</u>. (0, 1)
<u>11</u>. (0, -4) <u>12</u>. (-3, 0)
<u>13</u>. - <u>24</u>.

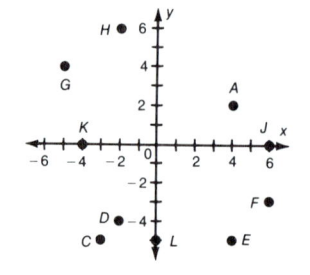

PROBLEM SOLVING AND APPLICATIONS

14-3 Broken Line Graphs

Broken line graphs show the amount of change over a certain period of time. This broken line graph shows the amount of rainfall in New Orleans for each month of the year.

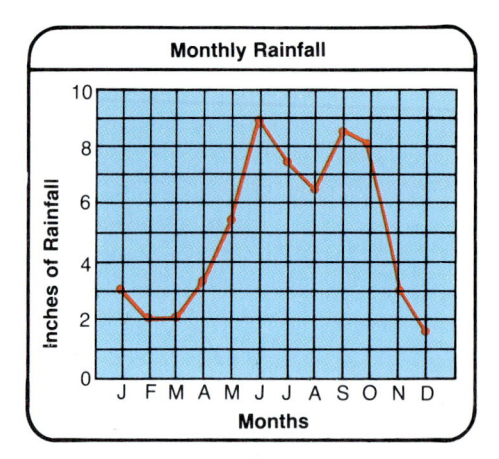

A broken line graph has a horizontal axis and a vertical axis. Thus, each point on the graph can be represented by this ordered pair.

(month, inches of rainfall)

Monthly Rainfall

EXAMPLE 1   Use the broken line graph above to answer each question.

  a. What is the difference in the amount of rainfall for June and March?
  b. What is the total amount of rainfall for the year?
  c. What is the average amount of rainfall per month?

Solutions: a. Rainfall for June: 9 inches   Rainfall for March: 7 inches
        Difference: 9 inches − 2 inches = **7 inches**

  b. Total: 3 + 2 + 2 + 3.5 + 5.5 + 9 + 7.5 + 6.5 + 8.5 + 8 + 3 + 1.5
     = 60   The total rainfall is **60 inches.**

  c. Average = $\frac{\text{Total Yearly Rainfall}}{\text{Number of Months}} = \frac{60}{12} =$ **5 inches**

Self-Check   1. Use the broken line graph above to find the difference in the amount of rainfall for the months of July and May.

The procedure for constructing a broken line graph is similar to the procedure for graphing equations.

> **PROCEDURE**
> **To construct a broken line graph:**
> **1** Use the data table as a table of ordered pairs.
> **2** Graph the ordered pairs. Connect the points in order.

414   Chapter 14

EXAMPLE 2 Construct a broken line graph for the following data.

Solution:

PUBLIC SCHOOL ATTENDANCE

Year	Pupils (in millions)	Ordered Pair
1940	25	(1940, 25)
1950	25	(1950, 25)
1960	36	(1960, 36)
1970	45	(1970, 45)
1980	41	(1980, 41)

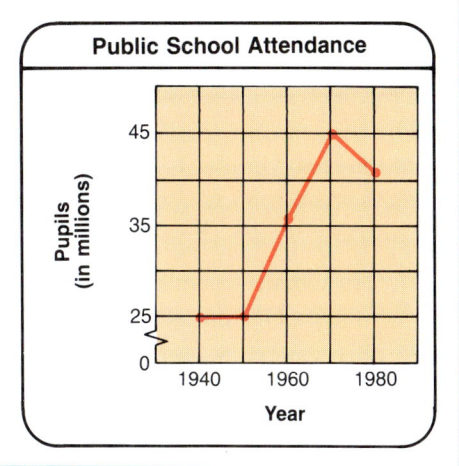

Self-Check 2. Construct a broken line graph for the following data.

PRICE OF A SHARE OF STOCK

Month	JAN	FEB	MAR	APR	MAY	JUN
Price	$10	$8	$9	$16	$20	$18

CLASSROOM EXERCISES

For Exercises 1–4, refer to the broken line graph at the right to answer each question. (Example 1)

1. How many minutes of daylight are there on the first day of June?
2. How many minutes of daylight are there on the first day of January?
3. How many more minutes of daylight are there on the first day of June than on the first day of October?
4. Find the total minutes of daylight for April 1, May 1, June 1, and July 1.

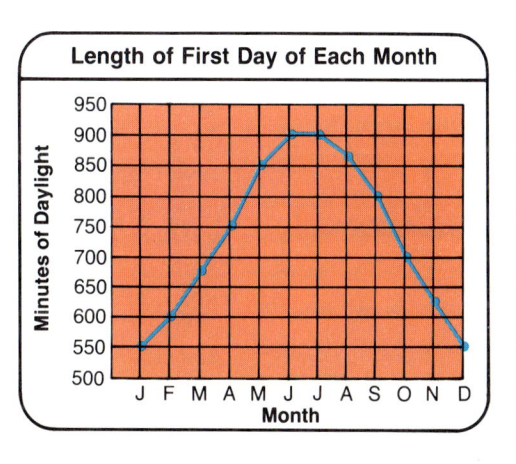

Graphing and Equations **415**

ADDITIONAL EXAMPLE

Example 2

Construct a broken line graph from the following data.

Chicago Temperatures

time	temp.	time	temp.
1 am	69°F	1 pm	84°F
5 am	65°F	5 pm	92°F
9 am	78°F	9 pm	80°F

Ans:

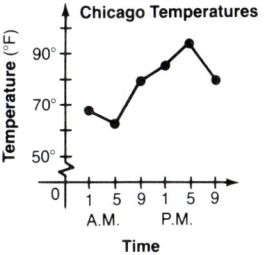

SELF-CHECK

CLASSROOM EXERCISES

1. 900   2. 550
3. 200   4. 3400

ASSIGNMENT GUIDE
BASIC
p. 416: Omit
AVERAGE
p. 416: 1-6
ABOVE AVERAGE
p. 416: 1-6

WRITTEN EXERCISES
1. 49   2. 12
3. $12\frac{1}{2}$%   4. 34.7
5.

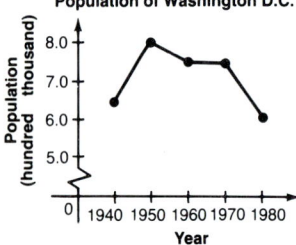

Population of Washington D.C.

6.

Morning Temperatures

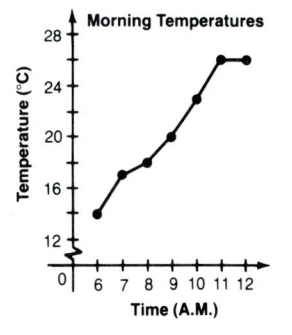

CLASSROOM EXERCISES
p. 411

1.
t	d
0	0
2	160
4	320

2.
t	d
1	60
2	120
3	180

3.
t	d
0	0
1	500
2	1000

4.
t	d
1	1000
3	3000
5	5000

416

WRITTEN EXERCISES

Goals: To read broken line graphs
To construct broken line graphs

The broken line graph at the right shows the number of countries that participated in the Winter Olympics. Refer to this graph for Exercises 1–4. (Example 1)

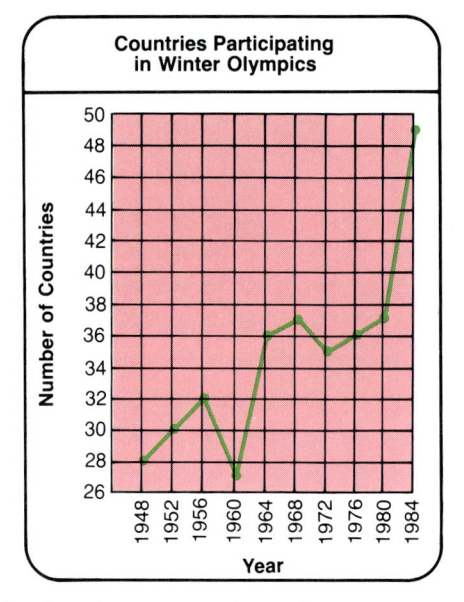

1. How many countries participated in the Winter Olympics in 1984?

2. How many more countries participated in the Winter Olympics in 1984 than in 1980?

3. Find the percent of increase for the number of countries that participated in 1956 and the number that participated in 1976.

4. Find the average number of countries that participated in the Winter Olympics for the years shown.

Countries Participating in Winter Olympics

For Exercises 5–6, construct a broken line graph for the data. (Example 2)

5. POPULATION OF WASHINGTON, D.C.

Year	Population (in hundred thousands)
1940	6.6
1950	8.0
1960	7.6
1970	7.6
1980	6.4

6. MORNING TEMPERATURES

Time	Temperature
6 A.M.	14
7 A.M.	17
8 A.M.	18
9 A.M.	20
10 A.M.	23
11 A.M.	26

REVIEW CAPSULE FOR SECTION 14-4

Subtract. (Pages 308–309)

1. $3 - 4$
2. $5 - 12$
3. $8 - 10$
4. $0 - 9$
5. $2 - (-3)$
6. $5 - (-6)$
7. $4 - (-4)$
8. $0 - (-2)$
9. $-1 - 6$
10. $-2 - 4$
11. $-8 - 3$
12. $-2 - (-2)$
13. $-1 - (-3)$
14. $-4 - (-4)$
15. $-3 - (-8)$

The answers are on page 435.

Chapter 14

14-4 Slope of a Line

When you talk about the grade or steepness of the road shown at the right, you say that it rises a vertical distance of 100 feet over a horizontal distance of 400 feet. That is,

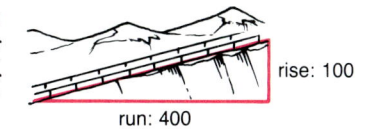

rise: 100
run: 400

$$\text{steepness} = \frac{\text{rise}}{\text{run}} = \frac{\text{vertical change}}{\text{horizontal change}} = \frac{100}{400} = \frac{1}{4}.$$

The measure of the steepness or slant of a line is called <u>slope</u>.

> The **slope** of a line is the ratio of the vertical change to the horizontal change between any two points on the line.
>
> $$\text{slope} = \frac{\text{vertical change}}{\text{horizontal change}}, \text{ (horizontal change} \neq 0)$$

When you know the coordinates of any two points on a line, you can determine the slope of a line.

PROCEDURE

To find the slope of a line using two points on a line:

1. Subtract the *y* coordinates to find the vertical change.
2. Subtract the *x* coordinates in the same order as the *y* coordinates to find the horizontal change.
3. Write this ratio in lowest terms: $\dfrac{\text{vertical change}}{\text{horizontal change}}$

EXAMPLE 1   Find the slope of the line containing the points (1, 3) and (6, 7).

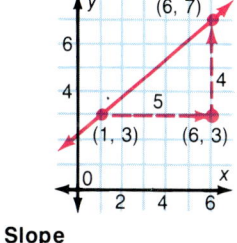

Solution:
1. $7 - 3 = 4$ ◄ **Vertical change**
2. $6 - 1 = 5$ ◄ **Horizontal change**
3. Write the ratio: $\dfrac{\text{Vertical change}}{\text{Horizontal change}} = \dfrac{4}{5}$ ◄ **Slope**

Self-Check   *Find the slope of the line containing the given points.*

1. $A(4, 2)$; $B(6, 4)$
2. $C(5, 8)$; $D(2, 3)$

Graphing and Equations

ADDITIONAL EXAMPLES

Example 2

Find the slope of the line containing the given points.

1. $(6, -1); (0, 3)$
    Ans: $-\frac{2}{3}$
2. $(7, 1); (7, -3)$
    Ans: undefined

SELF-CHECK

3. $-\frac{1}{2}$ 4. $\frac{2}{3}$

CLASSROOM EXERCISES

1. negative
2. undefined
3. zero
4. positive

The slope of a line can also be negative or zero.

EXAMPLE 2 Find the slope of the line containing the given points.

   a. $F(1, 4); G(4, 2)$ b. $M(2, 2); N(-2, 2)$

Solutions:

a.
1. $4 - 2 = 2$
2. $1 - 4 = 1 + (-4) = -3$
3. Slope: $\frac{2}{-3}$ or $-\frac{2}{3}$

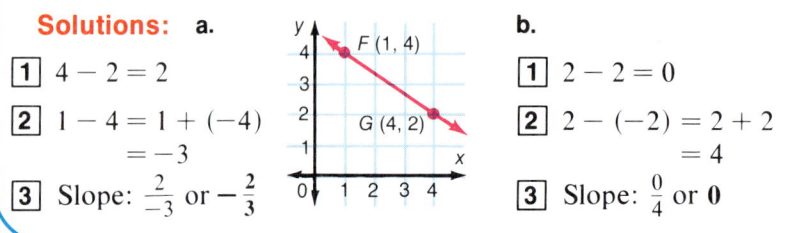

b.
1. $2 - 2 = 0$
2. $2 - (-2) = 2 + 2 = 4$
3. Slope: $\frac{0}{4}$ or 0

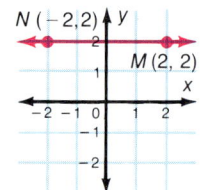

Self-Check *Find the slope of the line containing the given points.*

3. $R(-1, 4); S(3, 2)$ 4. $Y(-2, -1); Z(4, 3)$

When computing slope, always be careful to subtract the x coordinates in the <u>same order</u> as you subtracted the y coordinates.

The table below shows how the slope of a line determines its slant.

TABLE

Slope: Positive	Slope: Negative	Slope: 0	Slope: Undefined
The line rises from left to right.	The line falls from left to right.	A horizontal line	A vertical line

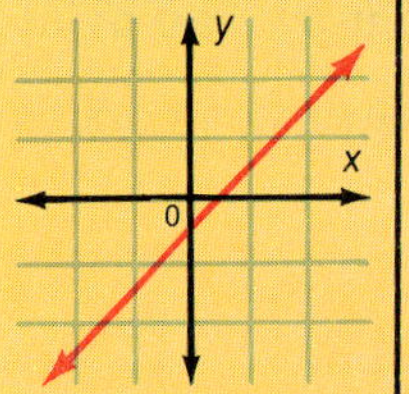

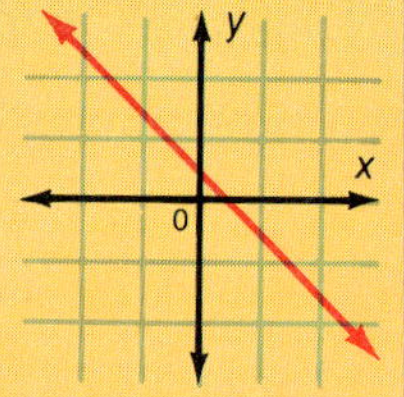

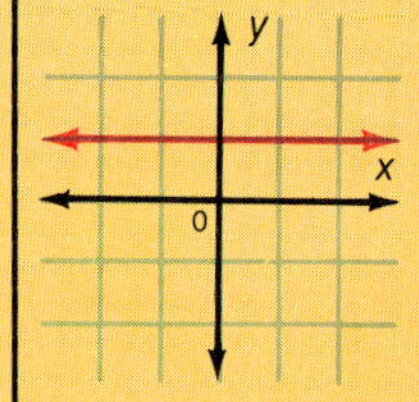

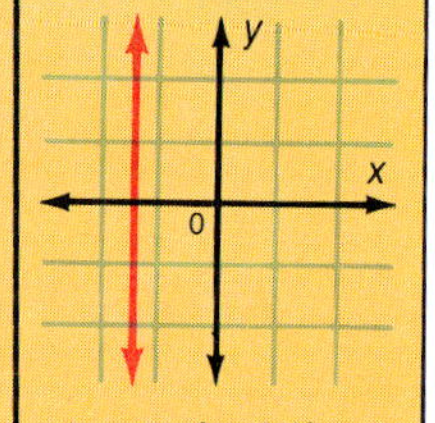

CLASSROOM EXERCISES

Classify the slope of each line as positive, negative, equal to zero, or undefined. (Table)

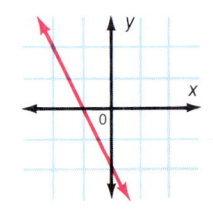

1. 2. 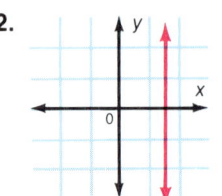 3. 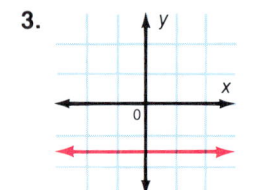 4.

418 Chapter 14

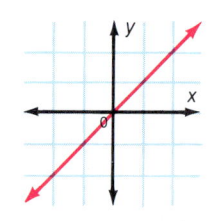

For Exercises 5–11, complete the table.
(Examples 1 and 2)

	Coordinates	Vertical Change	Horizontal Change	Slope
5.	$A(5, 4); B(1, 3)$	$4 - 3 = 1$	$5 - 1 = 4$	?
6.	$T(3, 1); V(2, -2)$	?	$1 - (-2) = 3$	?
7.	$E(-1, -5); W(-3, -8)$	$-5 - (-8) = 3$	?	?
8.	$G(-2, -3); H(-4, -6)$	?	?	?
9.	$L(4, 2); M(3, 5)$	$2 - 5 = -3$	$4 - 3 = 1$	?
10.	$S(-2, -1); T(3, -4)$	?	$-2 - 3 = -5$	?
11.	$D(-2, 4); G(-6, 4)$	?	?	?

WRITTEN EXERCISES

Goal: To find the slope of a line

For Exercises 1–36, find the slope of the line containing the given points. (Example 1)

1. $A(6, 8); B(3, 2)$
2. $C(7, 3); D(3, 1)$
3. $E(4, 2); F(-1, -3)$
4. $G(1, 4); H(-5, -2)$
5. $J(-1, -3); K(-5, -4)$
6. $L(-3, -2); M(-6, -9)$
7. $N(2, -3); P(0, -4)$
8. $Q(-1, 5); R(-7, 0)$
9. $S(2, 3); T(7, 4)$
10. $V(-1, -2); W(6, 3)$
11. $A(2, -4); C(3, -1)$
12. $D(5, 3); F(9, 10)$

(Example 2)

13. $B(3, 4); C(6, 2)$
14. $G(1, 4); H(4, 3)$
15. $K(7, 2); L(3, 5)$
16. $M(4, 3); P(1, 7)$
17. $R(-1, -2); S(-5, 3)$
18. $Q(-3, -7); T(-7, 2)$
19. $V(-4, -1); A(-3, -1)$
20. $E(8, 2); F(-4, 2)$
21. $J(-6, -4); L(-2, -6)$
22. $M(-1, -3); N(4, -5)$
23. $K(-2, 5); P(0, 5)$
24. $A(0, -4); D(-3, -4)$

MIXED PRACTICE

25. $C(4, 8); D(3, 1)$
26. $S(7, 1); T(5, 9)$
27. $A(-4, 3); B(-8, 6)$
28. $Q(5, -2); P(-7, 4)$
29. $G(-5, -1); W(-6, -4)$
30. $T(-6, 3); R(-2, -9)$
31. $H(2, 6); T(-2, -3)$
32. $J(3, -5); K(-3, 9)$
33. $B(-8, 2); C(10, -6)$
34. $P(0, 4); S(-6, 4)$
35. $D(2, -8); G(-1, -8)$
36. $E(-6, -2); F(0, 9)$

Graphing and Equations

QUIZ: SECTIONS 14-1–14-4

After completing this Review, you may wish to administer a quiz covering the same sections. A Quiz is provided in the *Teacher's Edition: Part II.*

REVIEW: SECTIONS 14-1–14-4

1. (-4, 4)   2. (1, 3)
3. (0, -1)   4. (3, -2)
5. -8.

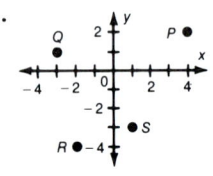

9. 0; 80; 160
10. -1; 1; 3
11. 2; 4; 6
12. 0; 4; 8
13.   14. 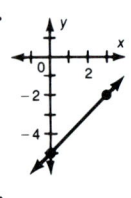
15. 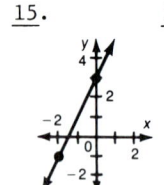   16.

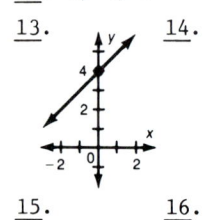

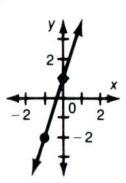

See page 424 for the answers to Exercises 17–20.

420

REVIEW: SECTIONS 14-1–14-4

Write the ordered pair for each point shown on the graph. (Section 14-1)

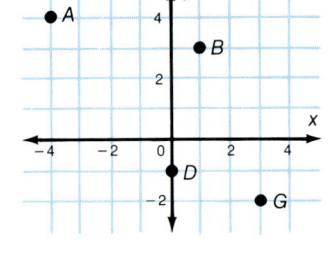

1. A   2. B   3. D   4. G

Draw and label a pair of axes. Then graph the point for each ordered pair. Label each point. (Section 14-1)

5. $P(4, 2)$ 6. $Q(-3, 1)$
7. $R(-2, -4)$ 8. $S(1, -3)$

For Exercises 9–12, evaluate each equation for the given values of t. Use t = 0, t = 2, and t = 4. (Section 14-2)

9. $d = 40t$ 10. $y = t - 1$ 11. $y = t + 2$ 12. $y = 2t$

Graph each equation. (Section 14-2)

13. $y = x + 4$ 14. $y = x - 5$ 15. $y = 2x + 3$ 16. $y = 3x + 1$

17. Use the table to construct a broken line graph. (Section 14-3)

CALORIES BURNED IN ONE WEEK

Day	Mon	Tue	Wed	Thur	Fri	Sat	Sun
Calories	2000	2500	1500	2400	2100	3000	1200

Find the slope of the line containing the given points. (Section 14-4)

18. $A(4, 2)$; $B(1, 1)$ 19. $C(6, 2)$; $D(1, 4)$ 20. $F(4, -3)$; $C(-1, -2)$

REVIEW CAPSULE FOR SECTION 14-5

Find the value of y when x = 0. (Pages 64–66)

1. $y = x + 4$ 2. $y = x - 6$ 3. $y = 2x + 3$ 4. $y = 3x - 4$

Find the value of x when y = 0. (Pages 190–195)

5. $y = x + 2$ 6. $y = x - 5$ 7. $y = 2x + 2$ 8. $y = 3x - 3$

The answers are on page 435.

420 Chapter 14

COMPUTER APPLICATIONS: Slope of a Line

Given the coordinates of two points on a line, the following program will compute and print the slope of the line. This program includes the possibility that the line is vertical. In this case, the slope is undefined.

Program:

```
100 PRINT "WHAT ARE THE COORDINATES OF POINT 1";
110 INPUT X1, Y1
120 PRINT "WHAT ARE THE COORDINATES OF POINT 2";
130 INPUT X2, Y2
140 IF X1 = X2 THEN 180
150 LET M = (Y2 - Y1) / (X2 - X1)
160 PRINT "SLOPE = ";M
170 GOTO 190
180 PRINT "SLOPE IS UNDEFINED"
190 PRINT
200 PRINT "ANY MORE TO COMPUTE (1 = YES, 0 = NO)";
210 INPUT Z
220 IF Z = 1 THEN 100
230 PRINT
240 END
```

Statement 160 contains the following formula for slope.

$$\text{Slope} = \frac{\text{vertical change}}{\text{horizontal change}}, \text{ or } m = \frac{y_2 - y_1}{x_2 - x_1}$$

where (x_1, y_1) are the coordinates of one point and (x_2, y_2) are the coordinates of a second point.

Since the formula cannot be used if $x_2 - x_1 = 0$, statement 150 first tests whether $x_1 = x_2$. If $x_1 = x_2$, the computer is sent to statement 190 and prints SLOPE IS UNDEFINED

Exercises

Use the program to find the slope of the line containing the given points.

1. $(-4,3), (6,-9)$
2. $(-1,-3), (2,-3)$
3. $(4,5), (4,9)$
4. $(1,2), (-3,5)$
5. $(-5,-1), (-2,-4)$
6. $(-3,-11), (2,-7)$
7. $(4,-2), (3,2)$
8. $(3,-2), (0,0)$
9. $(6,-9), (-1,-1)$

Computer Application 421

Teaching Suggestions
p. M-27

QUICK QUIZ
Find the value of y when x = 0.
1. y = 7x   Ans: y = 0
2. y = 3x - 2
   Ans: y = -2

Find the value of x when y = 0.
3. y = x + 4
   Ans: x = -4
4. y = 4x + 3
   Ans: x = -$\frac{3}{4}$

ADDITIONAL EXAMPLES
Example 1
Find the y intercept.
1. y = x + 3
   Ans: y = 3
2. y = 7x   Ans: y = 0

SELF-CHECK
1. 4   2. -1

Example 2
Find the x intercept.
1. y = 5x   Ans: x = 0
2. y = 4x + 12
   Ans: x = -3

SELF-CHECK
3. 2   4. 3

14-5 Intercepts

Point A in the figure at the right is on the y axis. Therefore, its x value is 0. Its y value, -3, is called the **y intercept** of the line. This means that the graph of the line crosses the y axis at the point $(0, -3)$.

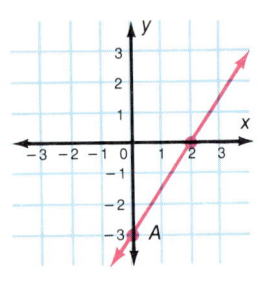

> **PROCEDURE**
>
> **To find the y intercept of a linear equation:**
> [1] Replace x with 0.   [2] Solve for y.

EXAMPLE 1   Find the y intercept of $y = 2x + 6$.

[1] Replace x with 0. ⟶ $y = 2x + 6$
$y = 2(0) + 6$

[2] Solve for y. ⟶ $y = 0 + 6$
$y = 6$ ⟵ y intercept

Self-Check   Find the y intercept: **1.** $y = 3x + 4$ **2.** $y = -2x - 1$

Point B in the figure above is on the x axis. Therefore, its y value is 0. Its x value, 2, is called the **x intercept** of the line. This means that the graph of the line crosses the x axis at the point $(2, 0)$.

> **PROCEDURE**
>
> **To find the x intercept of a linear equation:**
> [1] Replace y with 0.   [2] Solve for x.

EXAMPLE 2   Find the x intercept of $y = x - 4$.

Solution: [1] Replace y with 0.
$y = x - 4$
$0 = x - 4$

[2] Solve for x.
$0 + 4 = x - 4 + 4$
$4 = x$ ⟵ x intercept

Self-Check   Find the x intercept: **3.** $y = x - 2$ **4.** $y = 2x - 6$

422   Chapter 14

EXAMPLE 3 Graph $y = 3x + 6$. Use the x and y intercepts.

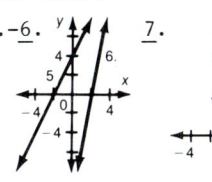

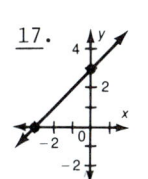

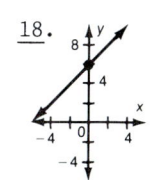

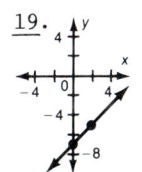

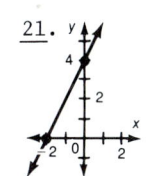

Solution:
1. Find the y intercept. $y = 6$ See Example 1.
2. Find the x intercept. $x = -2$ See Example 2.

Since the y intercept is 6 and the x intercept is -2, graph the points $(0, 6)$ and $(-2, 0)$.

Draw a line through the two points.

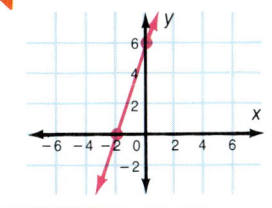

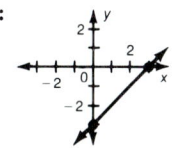

Self-Check Graph each linear equation. Use the x and y intercepts.

5. $y = 2x + 4$
6. $y = 5x - 10$
7. $y = -3x + 9$

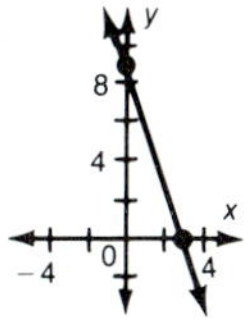

CLASSROOM EXERCISES

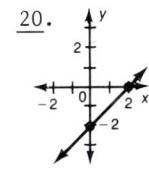

Find the x and y intercept of each equation. (Examples 1 and 2)

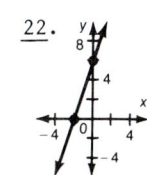

1. $y = 4x - 8$ 2; -8
2. $y = 3x + 9$ -3; 9
3. $y = -x + 1$ 1; 1
4. $y = -2x + 6$ 3; 6

WRITTEN EXERCISES

Goals: To find the x and y intercepts of a linear equation
To graph a linear equation using the x and y intercepts

Find the y intercept of each equation. (Example 1)

1. $y = 3x + 2$ 2
2. $y = 6x + 1$ 1
3. $y = 4x + 7$ 7
4. $y = 4x + 3$ 3
5. $y = x - 3$ -3
6. $y = x - 6$ -6
7. $y = -x + 4$ 4
8. $y = 3x$ 0

Find the x intercept of each equation. (Example 2)

9. $y = x - 3$ 3
10. $y = x - 10$ 10
11. $y = x + 6$ -6
12. $y = x + 8$ -8
13. $y = 3x - 9$ 3
14. $y = 5x - 10$ 2
15. $y = 4x + 12$ -3
16. $y = 6x + 24$ -4

Graph each equation. Use the x and y intercepts. (Example 3)

17. $y = x + 3$
18. $y = x + 5$
19. $y = x - 7$
20. $y = x - 2$
21. $y = 2x + 4$
22. $y = 3x + 6$
23. $y = 4x - 4$
24. $y = 2x - 4$
25. $y = -2x + 8$
26. $y = -5x + 10$
27. $y = -3x - 6$
28. $y = -4x - 8$

Graphing and Equations

Teaching Suggestions
p. M-27

QUICK QUIZ
Multiply.
1. $15(5\frac{2}{3})$ Ans: 85
2. $10(7\frac{1}{2})$ Ans: 75

Solve each equation.
3. $7x = 147$
   Ans: $x = 21$
4. $12x = 132$
   Ans: $x = 11$
5. $8x = 42$
   Ans: $x = 5\frac{1}{4}$ or 5.25

ADDITIONAL EXAMPLE
Example 1
Use April's table of values to find how much she earns in 32 hours.
Ans: $320

REVIEW: SECTIONS
14-1–14-4

17.

Calories Burned in One Week
[graph showing calories vs. days M T W Th F S S, values ranging 1200-3000]

18. $\frac{1}{3}$ 19. $-\frac{2}{5}$
20. $-\frac{1}{5}$

REVIEW CAPSULE FOR SECTION 14-6

Multiply. (Pages 135-137)

1. $8\left(13\frac{1}{2}\right)$ 2. $6\left(10\frac{2}{3}\right)$ 3. $12\left(20\frac{1}{4}\right)$ 4. $24\left(15\frac{1}{6}\right)$ 5. $30\left(15\frac{1}{2}\right)$ 6. $25\left(25\frac{1}{5}\right)$

Solve each equation. (Pages 200–202)

7. $8t = 48$ 8. $12x = 132$ 9. $9p = 126$ 10. $24w = 180$

Round to the nearest whole number. (Pages 47–49)

11. 12.35   12. 27.91   13. 9.55   14. $10.\overline{33}$ 15. $52.\overline{45}$ 16. $16.\overline{66}$

The answers are on page 435.

14-6 Direct and Indirect Variation

April Connors works during the summer to help pay her college expenses. She earns $10 per hour painting houses. April made a table and a graph to show the total amount she earns.

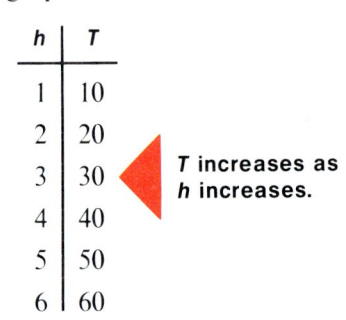

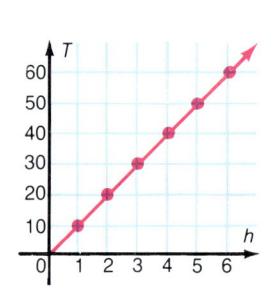

h	T
1	10
2	20
3	30
4	40
5	50
6	60

▶ *T increases as h increases.*

[graph of T vs. h, linear through origin]

From the table: $\frac{T}{h} = \frac{10}{1} = \frac{20}{2} = \frac{30}{3} = \frac{40}{4} = \frac{50}{5} = \frac{60}{6} = 10$

The variables *T* and *h* are in **direct variation** because the ratio $\frac{T}{h}$ is always the same. We say that *T* **varies directly** as *h*.

EXAMPLE 1 a. Use April's table of values to write an equation for the variation.
b. Use the equation to find how much April earns in $22\frac{1}{2}$ hours.

Solution: a. **Think:** *T* equals 10 times *h*.
   Equation: $T = 10h$

b. $T = 10h$
   $T = 10\left(22\frac{1}{2}\right)$
   $T = 225$ April earns **$225** in $22\frac{1}{2}$ hours.

424   Chapter 14

Self-Check 1. Refer to Example 1 to find how much April earns in 47 hours.

Two variables may also be related by indirect variation. We say that two variables, x and y, are indirectly or inversely related, when their **product is always the same.** That is, when x increases, y decreases. When x decreases, y increases.

EXAMPLE 2 The table below shows that for a trip of 480 kilometers, the rate, r, and the time, t, vary inversely.

r	80	60	40	30	20	10
t	6	8	12	16	24	48

As r decreases, t increases.

a. Refer to the table to write an equation for the variation.

b. Use the equation to find t when $r = 88$ kilometers per hour. Round the answer to the nearest hour.

Solution: a. **Think:** For **every ordered pair** in the table, $r \cdot t = 480$.
Equation: $rt = 480$

b. $rt = 480$ Replace r with 88.
$88t = 480$
$t = 5.\overline{45}$ Thus, $t =$ **5 hours.**

Self-Check 2. Use the inverse variation equation in Example 2 to find r when $t = 7$ hours. Round the answer to the nearest tenth.

CLASSROOM EXERCISES

Write an equation for the direct variation. (Example 1)

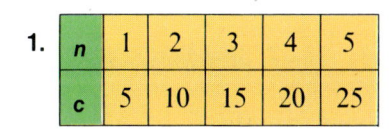

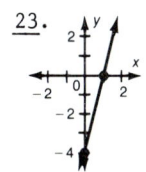

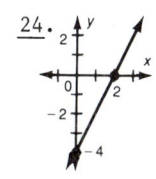

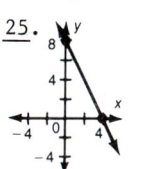

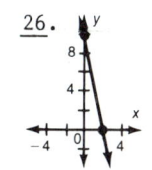

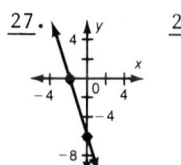

1.
n	1	2	3	4	5
c	5	10	15	20	25

2.
x	1	2	3	4	5
y	3	6	9	12	15

Write an equation for the inverse variation. (Example 2)

3.
r	10	15	20	25	30
t	30	20	15	12	10

4.
l	2	5	8	10	20	40
w	20	8	5	4	2	1

Graphing and Equations **425**

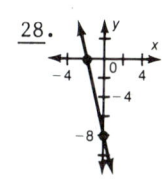

ASSIGNMENT GUIDE

BASIC
p. 426: Omit

AVERAGE
p. 426: 1-4

ABOVE AVERAGE
p. 426: 1-4

WRITTEN EXERCISES
1. $c = 2n$
2. $V = 3t$; 45 L
3. $lw = 400$; $3\frac{1}{5}$
4. $nd = 24$; 6.9

ADDITIONAL PRACTICE
p. 438

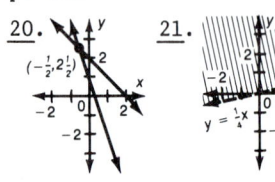

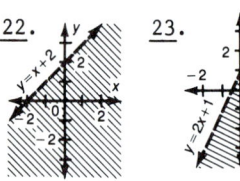

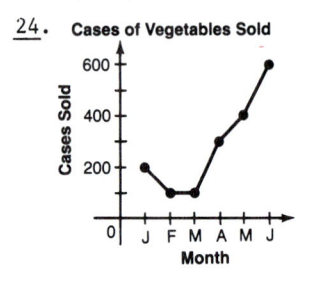

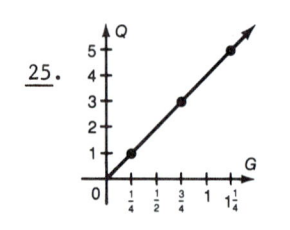

20. [graph with $(-\frac{1}{2}, 2\frac{1}{2})$]
21. [graph $y = \frac{1}{4}x$]
22. [graph $y = x + 2$]
23. [graph $y = 2x + 1$]
24. Cases of Vegetables Sold [graph]
25. [graph with points Q, G]

WRITTEN EXERCISES

Solve each problem. (Example 1)

1. The cost, c, in cents, of operating a TV varies directly as the number of hours, n, it is in operation.
   a. Refer to the table below to write an equation for the operation.

n	1	2	5	6	7
c	2	4	10	12	14

   b. Use the equation to find the cost in dollars of operating the TV 5 hours a day for 30 days.

2. The volume, V, in liters, of water in a tank varies directly as the time, t, in minutes, that an inlet valve is open.
   a. Refer to the table below to write an equation for the variation.

t	1	2	3	4
V	3	6	9	12

   b. Use the equation to find how many liters are in the tanks after 15 minutes.

(Example 2)

3. The length, l, of a rectangle varies inversely as the width, w.
   a. Refer to the table below to write an equation for the variation.

l	5	20	50	100	200
w	80	40	8	4	2

   b. Use the equation to find the length when $w = 125$.

4. The number of days, d, needed to complete a job varies inversely as the number of persons working, n.
   a. Refer to the table to write an equation for the variation.

n	1	2	3	4	5
d	24	12	8	6	4.8

   b. Use the equation to find how many workers would be needed to complete the job in $3\frac{1}{2}$ days.

REVIEW CAPSULE FOR SECTION 14-7

Find the value of y for each given value of x. (Pages 22–24)

1. $y = x + 4$; $x = -6$
2. $y = -4x + 1$; $x = -3$
3. $y = x - 2$; $x = -1$

Write <u>Yes</u> *or* <u>No</u> *to indicate whether the given ordered pair is a solution of the given equation.* (Pages 410-413)

4. $y = x + 5$; (1, 6)
5. $y = 2x - 3$; (4, 0)
6. $y = -3x + 4$; (−1, 7)

The answers are on page 435.

14-7 Solving a System of Equations by Graphing

The Modern Microchip Company plans to produce a new pocket calculator. When the income from sales equals the cost of producing the calculator, the company is said to **break-even**. Any further sales represent a profit.

A pair of linear equations is a **system of equations**. To solve a system of equations means to find the coordinates of the point where the graphs of the equations intersect.

PROCEDURE

To solve a system of equations:

1. Make a table of ordered pairs for each equation.
2. Graph each equation on the same coordinate plane.
3. Write the solution as an ordered pair.

EXAMPLE 1 Find the number of pocket calculators, x, that must be sold in order for the Modern Microchip Company to break-even. Each calculator will sell for $40. Use the following system of equations.

$$\begin{cases} y = 20x + 2000 & \longleftarrow \text{ Cost of producing } x \text{ calculators} \\ y = 40x & \longleftarrow \text{ Amount made for selling } x \text{ calculators} \end{cases}$$

1 $y = 20x + 2000$

x	20x + 2000
0	20(0) + 2000 = 2000
50	20(50) + 2000 = 3000
100	20(100) + 2000 = 4000

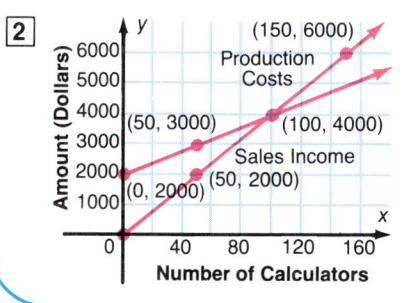

x	y
0	2000
50	3000
100	4000

$y = 40x$

x	40x
0	40(0) = 0
50	40(50) = 2000
150	40(150) = 6000

x	y
0	0
50	2000
150	6000

2 [Graph showing Production Costs line and Sales Income line intersecting at (100, 4000), with points (0, 2000), (50, 3000), (150, 6000) on Production Costs and (0, 0), (50, 2000), (100, 4000) on Sales Income. X-axis: Number of Calculators (0 to 160). Y-axis: Amount (Dollars) (1000 to 6000).]

3 The graphs meet at **(100, 4000)**.

One hundred calculators must be sold for the company to break-even.

Check: The check is left for you.

Teaching Suggestions
p. M-27

QUICK QUIZ

Find the value of y for each given value of x.

<u>1</u>. y = x - 7; x = -1
   Ans: y = -8

<u>2</u>. y = 7x + 2; x = 3
   Ans: y = 23

Write <u>Yes</u> or <u>No</u> to indicate whether the given ordered pair is a solution of the given equation.

<u>3</u>. y = x - 3; (-2, -5)
   Ans: Yes

<u>4</u>. y = 4x + 1; (3, 13)
   Ans: Yes

<u>5</u>. y = 2x - 5; (3, -1)
   Ans: No

ADDITIONAL EXAMPLE
Example 1
Find the number of calculators, x, that must be sold the Modern Microchip Company to break-even. Each calculator will sell for $20. Use the equations
y = 20x and y = 10x + 1500.
Ans: 150

Graphing and Equations **427**

SELF-CHECK

1. x = 50

ADDITIONAL EXAMPLE

Example 2

Graph $\begin{cases} y = -x + 4 \\ y = 2x + 1 \end{cases}$

in the coordinate plane. Determine the point of intersection.

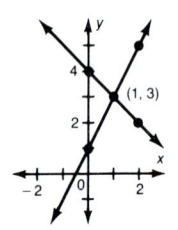

SELF-CHECK

2. (-3, -4)
3. (-1, 2)
4. (0, 0)

CLASSROOM EXERCISES

1. Yes 2. No 3. Yes
4. No   5. Yes 6. Yes

Self-Check

1. Find the number of calculators, x, that must be sold for the Modern Microchip Company to break-even. Each calculator will sell for $60. Use the equations $y = 20x + 2000$ and $y = 60x$.

NOTE: As shown in Example 1, you check the solution of a system of equations by substituting the ordered pair in both equations of the system.

EXAMPLE 2 Graph this system of equations in the coordinate plane. Determine the point of intersection.

$$\begin{cases} y = 2x + 3 \\ y = x + 2 \end{cases}$$

Solution: ① $y = 2x + 3$ $y = x + 2$

x	y
-2	-1
-1	1
0	3

x	y
-1	1
0	2
2	4

②

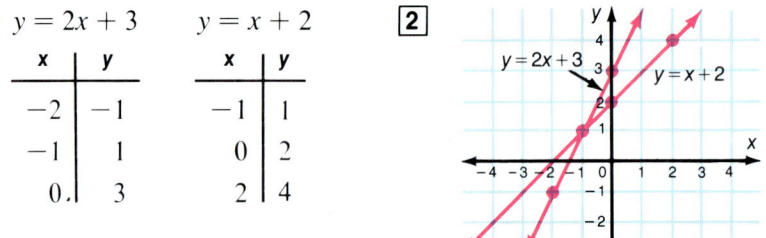

③ The graphs meet at (−1, 1).

Check: $y = 2x + 3$ $y = x + 2$
         $1 = 2(-1) + 3$ $1 = -1 + 2$
             $-2 + 3$ 1
                1

Self-Check Graph each system of equations. Determine the point of intersection.

2. $\begin{cases} y = 2x + 2 \\ y = x - 1 \end{cases}$ 3. $\begin{cases} y = -x + 1 \\ y = x + 3 \end{cases}$ 4. $\begin{cases} y = -2x \\ y = x \end{cases}$

CLASSROOM EXERCISES

Determine whether the given ordered pair is a solution of the system of equations. Answer Yes or No. (Example 2, step 3)

1. $\begin{cases} y = 2x - 3 \\ y = x - 4 \end{cases}$; (−1, −5)   2. $\begin{cases} y = 3x - 1 \\ y = x - 1 \end{cases}$; (1, 0)   3. $\begin{cases} y = -x + 2 \\ y = x - 6 \end{cases}$; (4, −2)

4. $\begin{cases} y = -x - 1 \\ y = x + 1 \end{cases}$; (0, −1)   5. $\begin{cases} y = -4x \\ y = x \end{cases}$; (0, 0)   6. $\begin{cases} y = -x \\ y = 5x \end{cases}$; (0, 0)

428 Chapter 14

WRITTEN EXERCISES

Goal: To solve a system of equations by graphing

For Exercises 1–2:

a. *Complete the tables. Then copy the coordinate axes on a piece of graph paper and draw the graph.*

b. *Write the solution.* (Example 1)

1. Find the number of office chairs, x, that must be sold in order for the Break-Time Office Supplies Company to break-even. Each office chair sells for $75.

   $y = 25x + 2000$ $y = 75x$

x	y
0	? 2000
20	? 2500
80	? 4000

x	y
0	? 0
20	? 1500
60	? 4500

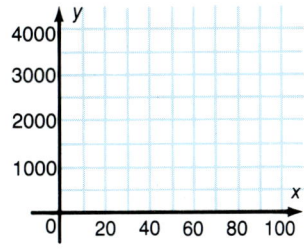

2. Find the number of camera cases, x, that must be sold in order for the Cogswell Camera Company to break-even. Each camera case sells for $10.

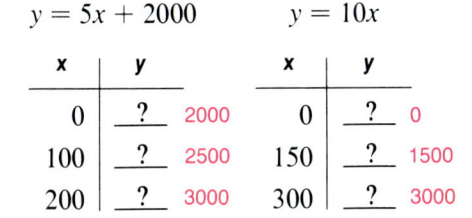

   $y = 5x + 2000$ $y = 10x$

x	y
0	? 2000
100	? 2500
200	? 3000

x	y
0	? 0
150	? 1500
300	? 3000

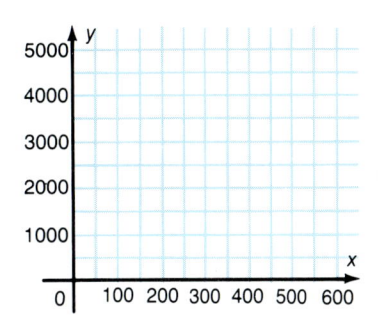

Solve by graphing. (Example 2)

3. $\begin{cases} y = 2x + 3 \\ y = x + 1 \end{cases}$
4. $\begin{cases} y = 2x + 1 \\ y = x + 4 \end{cases}$
5. $\begin{cases} y = 3x + 1 \\ y = x - 1 \end{cases}$
6. $\begin{cases} y = 3x + 2 \\ y = x - 2 \end{cases}$
7. $\begin{cases} y = -x + 3 \\ y = x + 3 \end{cases}$
8. $\begin{cases} y = -x + 6 \\ y = x + 2 \end{cases}$
9. $\begin{cases} y = -x - 2 \\ y = x - 4 \end{cases}$
10. $\begin{cases} y = -x - 1 \\ y = x - 3 \end{cases}$

REVIEW CAPSULE FOR SECTION 14-8

Tell whether each of the following is a solution for $x - 1 < 4$.
Write Yes or No. (Pages 338–341)   The answers are on page 435.

1. 1   2. 5   3. 0   4. −4   5. −3   6. 6   7. −1   8. 8

Graphing and Equations   **429**

ASSIGNMENT GUIDE

BASIC
p. 429: Omit

AVERAGE
p. 429: 1-11

ABOVE AVERAGE
p. 429: 1-11

WRITTEN EXERCISES

3. (−2, 1)
4. (3, 7) 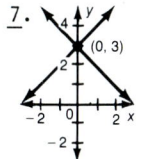
5. (−1, −2)
6. (−2, 4)
7. (0, 3)
8. (2, 4)
9. (1, −3)
10. (1, −2)

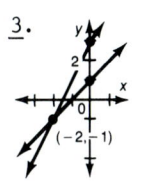

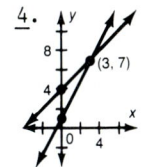

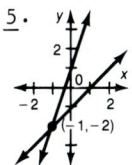

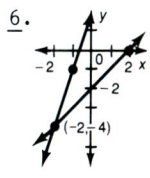

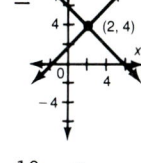

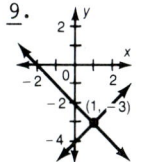

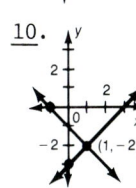

Teaching Suggestions
p. M-27

QUICK QUIZ

Tell whether each of the following is a solution for $x + 3 > -2$.
Write Yes or No.
1. 5   Ans: Yes
2. -6   Ans: No
3. 6   Ans: Yes
4. -3   Ans: Yes
5. -10 Ans: No

ADDITIONAL EXAMPLES

Example
1. Graph $y > x - 3$
   Ans:

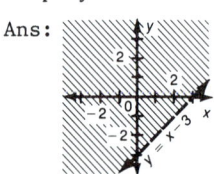

2. Graph $y > -2x + 1$
   Ans:

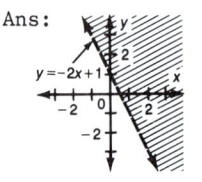

SELF-CHECK
1. 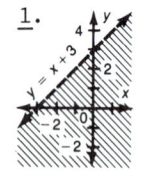
2.

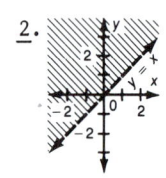

430

14-8 Graphing Linear Inequalities

Mathematical sentences such as $y > \frac{1}{3}x$ and $y < \frac{1}{3}x$ are called **linear inequalities**. To graph a linear inequality, first draw the graph of the related linear equation as a dashed line.

PROCEDURE

To graph a linear inequality:

1. Graph the related linear equation.
2. Test an ordered pair from each side of the graph of the equation to determine which point is a solution of the inequality.
3. Shade the part of the coordinate plane containing the solution.

EXAMPLE Graph $y < x + 1$

Solution:
1. $y = x + 1$ *Replace < with =.*

   Draw the graph of $y = x + 1$ as a dashed line.

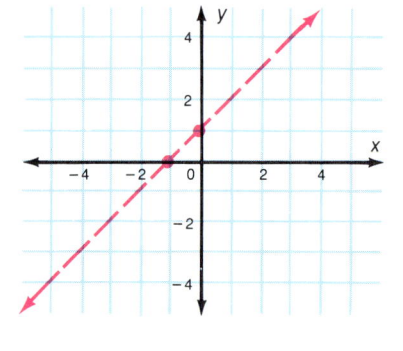

2. Choose two ordered pairs.
   Choose one ordered pair on one side of the line.
   Choose another ordered pair on the other side of the line.

   a. Test $(0, 2)$.   b. Test $(3, -2)$.
      $y < x + 1$ $y < x + 1$
      $2 \stackrel{?}{<} 0 + 1$ $-2 \stackrel{?}{<} 3 + 1$
      $2 \stackrel{?}{<} 1$ No $-2 \stackrel{?}{<} 4$ Yes

   Thus, $(0, 2)$ is not   Thus, $(3, -2)$ is a
   a solution.   solution.

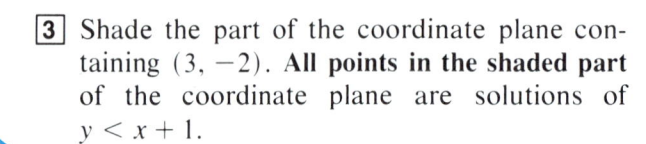

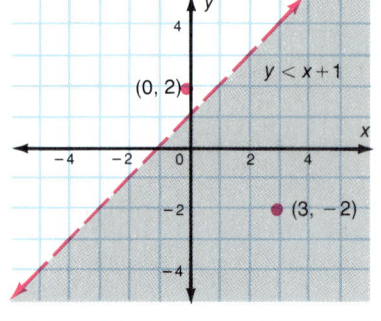

3. Shade the part of the coordinate plane containing $(3, -2)$. **All points in the shaded part of the coordinate plane are solutions of $y < x + 1$.**

Self-Check Graph each inequality: **1.** $y < x + 3$ **2.** $y > x$

430 Chapter 14

NOTE: The points on the dashed line, $y = x + 1$, are **not** solutions of $y < x + 1$ or $y > x + 1$.

CLASSROOM EXERCISES

Determine whether the given point is a solution of the given inequality. Write Yes or No. (Example, step 2)

1. $y > x + 1$; $(-3, 2)$
2. $y < 4x$; $(1, 1)$
3. $y > x$; $(0, 0)$
4. $y < 3x - 4$; $(2, 3)$
5. $y > -x + 1$; $(3, 4)$
6. $y < -2x - 1$; $(1, -5)$

For Exercises 7–10, choose a graph from a–d that matches each inequality. (Example, step 2)

7. $y > \frac{1}{2}x$
8. $y < \frac{1}{2}x$
9. $y > 4x$
10. $y < 4x$

a. b. 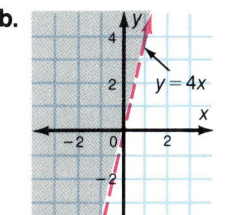 c. 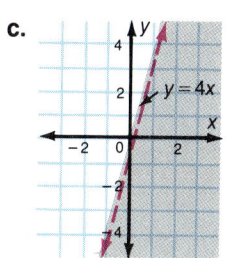 d.

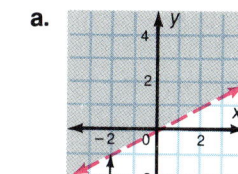

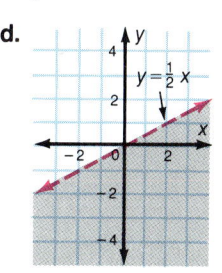

WRITTEN EXERCISES

Goal: To graph linear inequalities in the coordinate plane

Determine whether each point is a solution of $y > x - 2$. Write Yes or No. (Example)

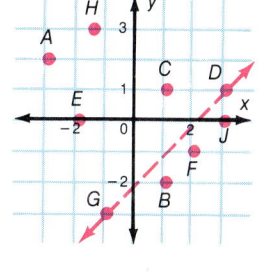

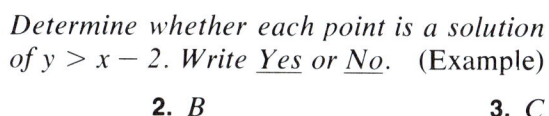

1. A
2. B
3. C
4. D
5. E
6. F
7. G
8. H
9. J

For Exercises 10–21, graph each inequality in the coordinate plane. (Example)

10. $y > x + 2$
11. $y < x - 1$
12. $y < 2x$
13. $y > 3x$
14. $y > 2x - 1$
15. $y < 3x - 1$
16. $y > -x$
17. $y < -3x$
18. $y < -x + 5$
19. $y > -x - 2$
20. $y < -2x - 4$
21. $y > -4x + 3$

Graphing and Equations

CLASSROOM EXERCISES
1. Yes 2. Yes 3. No
4. No   5. Yes 6. Yes
7. a   8. d   9. b
10. c

ASSIGNMENT GUIDE
BASIC
p. 431: Omit
AVERAGE
p. 431: Omit
ABOVE AVERAGE
p. 431: 1–21

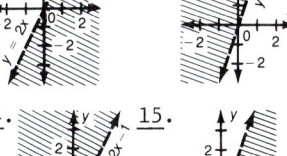

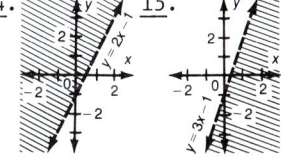

WRITTEN EXERCISES
1. Yes 2. No   3. Yes
4. No   5. Yes 6. No
7. No   8. Yes 9. No

10. 11. 12. 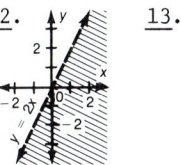 13. 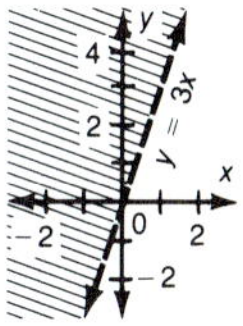 14. 15. 16. 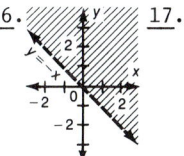 17.

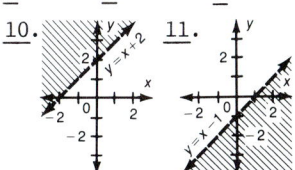

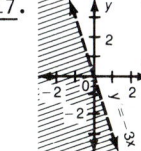

See Page 435 for the answers to Exercises 18–21.

QUIZ: SECTIONS 14-5–14-8

After completing this Review, you may wish to administer a quiz covering the same sections. A Quiz is provided in the Teacher's Edition: Part II.

REVIEW: SECTIONS 14-5-14-8

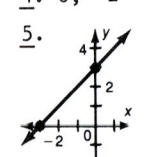

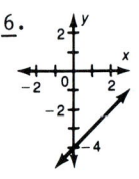

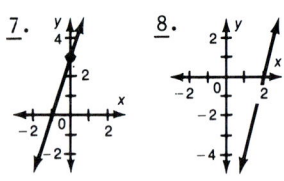

1. 2; -2
2. -5; 5
3. -4; 2
4. 6; -2
5. [graph]
6. [graph]
7. [graph]
8. [graph]
9. D = 60m; 660
10. wd = 24; 6
11. [graph (1, 3)]
12. [graph (1, 0)]
13. [graph (1, 1)]
14. [graph (0, 0)]

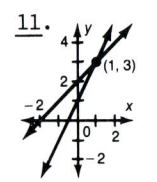

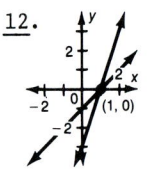

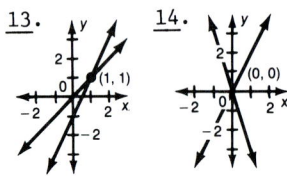

See page 434 for the answers to the Exercises 15-18.

432

REVIEW: SECTIONS 14-5–14-8

For Exercises 1–4:
  a. *Find the y intercept of each linear equation.*
  b. *Find the x intercept of each linear equation.* (Section 14-5)

1. $y = x + 2$
2. $y = x - 5$
3. $y = 2x - 4$
4. $y = 3x + 6$

For Exercises 5–8:
  a. *Find the x and y intercepts.*
  b. *Use the x and y intercepts to graph the equation.* (Section 14-5)

5. $y = x + 3$
6. $y = x - 4$
7. $y = 3x + 3$
8. $y = 4x - 8$

Solve each problem. (Section 14-6)

9. The actual distance, *D*, in miles between two towns varies directly as the distance, *m*, in inches between the towns on a map.

  a. Refer to the table below to write an equation for the variation.

m	2	3	6	8
D	120	180	360	480

  b. Use the equation to find the actual distance when *m* equals 11 inches.

10. The number of days, *d*, needed to repair a sidewalk varies inversely as the number of persons, *w*, working. Assume that all persons work at the same rate.

  a. Refer to the table below to write an equation for the variation.

w	2	3	4	12
d	12	8	6	2

  b. Use the equation to find how many workers are needed to complete the job in 4 days.

Graph each system of equations in a coordinate plane. Determine the point of intersection. (Section 14-7)

11. $\begin{cases} y = x + 2 \\ y = 2x + 1 \end{cases}$
12. $\begin{cases} y = x - 1 \\ y = 3x - 3 \end{cases}$
13. $\begin{cases} y = x \\ y = 2x - 1 \end{cases}$
14. $\begin{cases} y = 2x \\ y = -3x \end{cases}$

Graph each inequality in the coordinate plane. (Section 14-8)

15. $y < x + 3$
16. $y > x - 4$
17. $y > -3x$
18. $y < 2x - 4$

Chapter 14

CHAPTER REVIEW

PART 1: VOCABULARY

For Exercises 1–15, choose from the box at the right the word(s) that best corresponds to each description.

1. The point where the *x* axis and *y* axis cross __?__
2. For a line, the *y* value of its point on the *y* axis __?__
3. The vertical number line in the coordinate plane __?__
4. The second number in an ordered pair __?__
5. A graph that shows the amount of change over a certain period of time __?__
6. A pair of linear equations __?__
7. When the income from sales equals the cost of production, a company is said to __?__ .
8. The ratio of two variables is always the same __?__
9. The ratio of the vertical change to the horizontal change between any two points on a line __?__
10. An equation whose graph is a straight line __?__
11. The product of two variables is always the same __?__
12. The horizontal number line in the coordinate plane __?__
13. The first number in an ordered pair __?__
14. For a line, the *x* value of its point on the *x* axis __?__
15. The solution of a system of equations __?__

> *x* axis
> *y* axis
> origin
> *x* coordinate
> *y* coordinate
> slope
> *x* intercept
> *y* intercept
> linear equation
> line graph
> direct variation
> indirect variation
> system of equations
> point of intersection
> break-even

PART 2: SKILLS

For Exercises 16–27, give the ordered pair for the points graphed at the right. (Section 14-1)

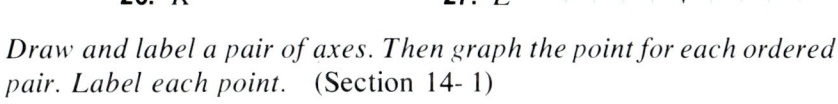

16. *A*
17. *B*
18. *C*
19. *D*
20. *E*
21. *F*
22. *G*
23. *H*
24. *I*
25. *J*
26. *K*
27. *L*

Draw and label a pair of axes. Then graph the point for each ordered pair. Label each point. (Section 14-1)

28. $P(-3, 2)$
29. $Q(4, -4)$
30. $R(1, 2)$
31. $S(3, -2)$
32. $T(0, -3)$
33. $V(5, 0)$
34. $U(-3, -4)$
35. $W(5, 4)$

CHAPTER REVIEW

1. origin
2. y intercept
3. y axis
4. y coordinate
5. line graph
6. system of equations
7. break-even
8. direct variation
9. slope
10. linear equation
11. indirect variation
12. x axis
13. x coordinate
14. x intercept
15. point of intersection
16. (0, 4)
17. (1, -1)
18. (3, -2)
19. (-4, 2)
20. (-1, -3)
21. (4, 3)
22. (-3, 4)
23. (2, 1)
24. (-2, 0)
25. (-4, -2)
26. (-1, 3)
27. (1, 2)
28.–35.

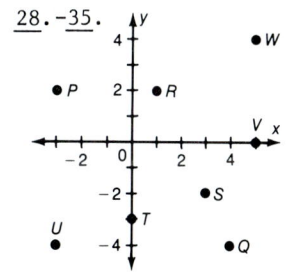

Graphing and Equations **433**

CHAPTER REVIEW

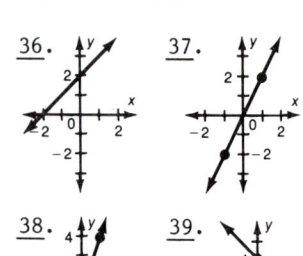

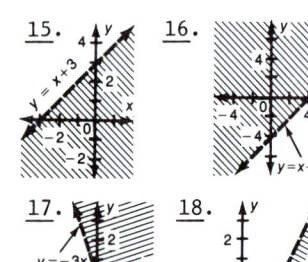

Graph each equation. (Section 14-2)

36. $y = x + 2$
37. $y = 2x$
38. $y = 3x + 1$
39. $y = -x + 4$

This line graph shows the number of car sales for one year. Use this graph for Exercises 40–42. (Section 14-3)

CAR SALES FOR ONE YEAR

40. How many cars were sold in June?
41. How many more cars were sold in June than in January?
42. What is the difference between the number of cars sold in September and the number sold in November?

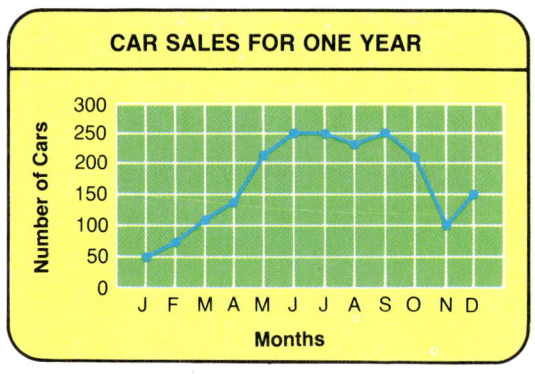

Find the slope of the line containing the given points. (Section 14-4)

43. $A(3, 5); B(1, 4)$
44. $F(5, -1); G(3, 4)$
45. $H(-2, -3); J(-5, 2)$
46. $K(7, -2); L(3, -2)$
47. $P(-3, 6); Q(4, 0)$
48. $R(0, -1); S(-3, 6)$

Find the y intercept of each linear equation. (Section 14-5)

49. $y = x + 7$
50. $y = x - 12$
51. $y = 2x + 3$
52. $y = 3x - 5$

Find the x intercept of each linear equation. (Section 14-5)

53. $y = x + 6$
54. $y = -x + 3$
55. $y = 2x + 8$
56. $y = 4x - 12$

Solve each problem. (Section 14-6)

57. The toll, T, on a bridge varies directly with the number of axles, a, of a vehicle.

   a. Refer to the table below to write an equation for the variation.

a	2	4	8
T	1.50	3.00	6.00

   b. Use the equation to find the amount of the toll for a 6-axle truck.

58. The number of tomato plants, p, in a row in a garden varies inversely as the space, s, between them.

   a. Refer to the table below to write an equation for the variation.

s	15	20	30
p	60	45	30

   b. Use the equation to find how many plants can fit in a row if he places them 25 centimeters apart.

434 Chapter 14

Graph each system of equations in a coordinate plane. Determine the point of intersection. (Section 14-7)

59. $\begin{cases} y = x - 1 \\ y = 2x - 2 \end{cases}$
60 $\begin{cases} y = x + 2 \\ y = -x + 4 \end{cases}$
61. $\begin{cases} y = x \\ y = 2x + 1 \end{cases}$
62 $\begin{cases} y = 2x \\ y = 3x - 1 \end{cases}$

Graph each inequality in the coordinate plane. (Section 14-8)

63 $y < -x + 3$ **64** $y > x - 2$ **65** $y > 2x + 1$ **66** $y < 3x - 1$

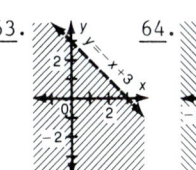

PART 3: APPLICATIONS

67. Use the information in the table below to construct a broken line graph. (Section 14-3)

CALORIE INTAKE FOR ONE WEEK

Day	MON.	TUES.	WED.	THURS.	FRI.	SAT.	SUN.
Calories	2500	2400	3000	2200	1900	2500	2000

68. Find the number of phones that must be sold in order for the Dial-A-Tone Company to break-even. Each phone sells for $50. (HINT: Copy the tables and the coordinate axes. Then complete the table and draw the graph.) (Section 14-8)

$y = 25x + 1000$ $y = 50x$

x	y
0	? 1000
40	? 2000
80	? 3000

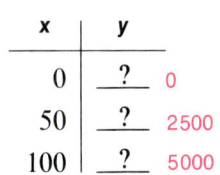

x	y
0	? 0
50	? 2500
100	? 5000

(40, 2000)

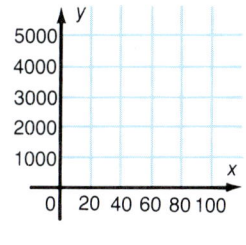

ANSWERS TO REVIEW CAPSULES

Page 410   **1.** 7   **2.** 13   **3.** 3   **4.** 7   **5.** −3   **6.** −11

Page 413   **1.** 30   **2.** 21   **3.** 54   **4.** 7.1   **5.** 16.0   **6.** 11   **7.** 19   **8.** 7.4   **9.** 5.9   **10.** 49   **11.** 9   **12.** 2.5   **13.** 12   **14.** 20.7

Page 416   **1.** −1   **2.** −7   **3.** −2   **4.** −9   **5.** 5   **6.** 11   **7.** 8   **8.** 2   **9.** −7   **10.** −6   **11.** −11   **12.** 0   **13.** 2   **14.** 0   **15.** 5

Page 420   **1.** 4   **2.** −6   **3.** 3   **4.** −4   **5.** −2   **6.** 5   **7.** −1   **8.** 1

Page 424   **1.** 108   **2.** 64   **3.** 243   **4.** 364   **5.** 465   **6.** 630   **7.** 6   **8.** 11   **9.** 14   **10.** $7\frac{1}{2}$ **11.** 12   **12.** 28   **13.** 10   **14.** 10   **15.** 52   **16.** 17

Page 426   **1.** −2   **2.** 13   **3.** −3   **4.** Yes   **5.** No   **6.** Yes

Page 429   **1.** Yes   **2.** No   **3.** Yes   **4.** Yes   **5.** Yes   **6.** No   **7.** Yes   **8.** No

Graphing and Equations   **435**

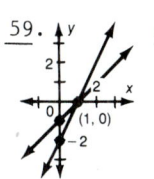

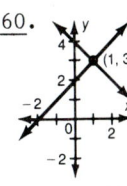

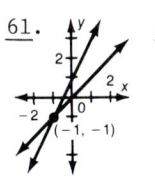

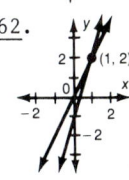

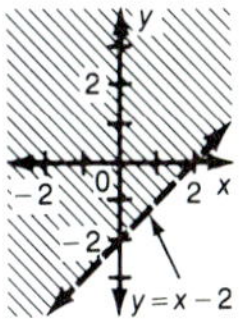

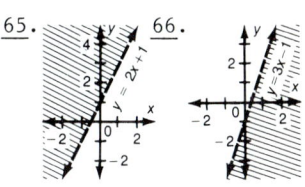

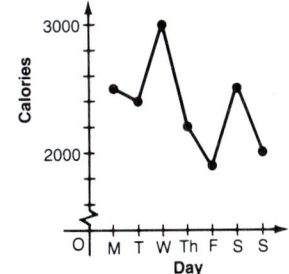

WRITTEN EXERCISES
p. 431

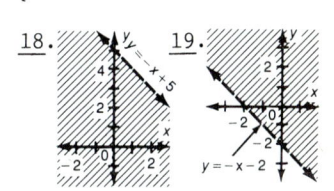

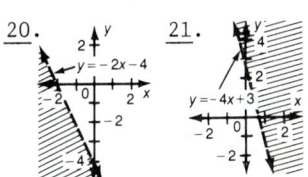

CHAPTER TEST

Two forms of a chapter test, Form A and Form B, are provided on copying masters in the Teacher's Edition: Part II.

1. (4, 0) 2. (-2, 1)
3. (2, 4) 4. (-2, -2)

5. [graph]
6. [graph]

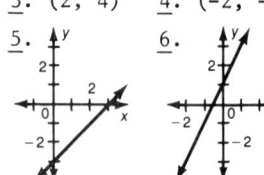

7. $\frac{3}{2}$ 8. -2

9. [graph]
10. [graph]

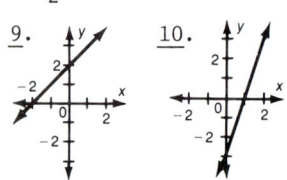

11. [graph]
12. [graph]

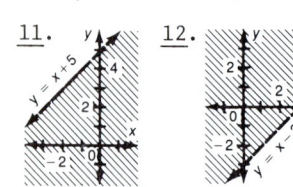

13. [graph with (3, 9)]

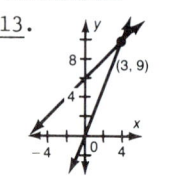

14.

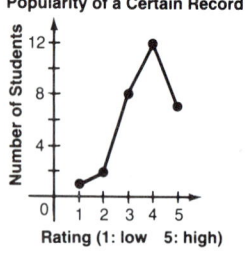

Popularity of a Certain Record
[broken line graph]

15. $v = 9w$; 45

436

CHAPTER TEST

For Exercises 1–4, give the ordered pair for each point graphed at the right.

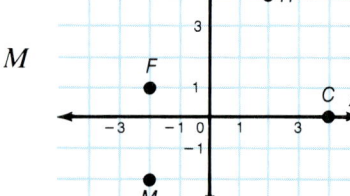

1. C 2. F 3. H 4. M

Graph each equation.

5. $y = x - 3$
6. $y = 2x + 1$

Find the slope of the line containing the given points.

7. $A(3, 4)$; $B(1, 1)$
8. $C(2, -1)$; $D(4, -5)$

Find the x and y intercepts of each equation. Then graph the equation.

9. $y = x + 2$
10. $y = 3x - 3$

Graph each inequality in the coordinate plane.

11. $y < x + 5$
12. $y > x - 3$

13. Solve by graphing: $\begin{cases} y = 3x \\ y = x + 6 \end{cases}$

14. Use the table below to construct a broken line graph.

POPULARITY OF A CERTAIN RECORD

Rating (1: low – 5: high)	Number of Students
1	1
2	2
3	8
4	12
5	7

15. The speed, v, in miles per hour of the blade tips of a windmill varies directly with the speed, w, of the wind.

   a. Refer to the table below to write an equation for the variation.

w	20	10	8	2
v	180	90	72	18

   b. Use the equation to find the speed of the blade tips in a 5 mile per hour wind.

436 Chapter 14

ENRICHMENT

Venn Diagrams

You can use **Venn diagrams** to show how sets are related. The Venn diagram at the right shows that every element in set B is also in set A. Thus, B is a **subset** of A. Every element of A, however, is *not* in B. Thus, A is not a subset of B.

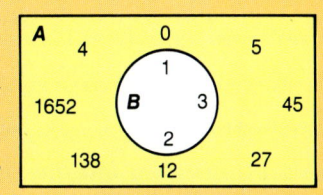

A = {whole numbers}

B ⊂ A ◀ Read: "B is a subset of A."

A ⊄ B ◀ Read: "A is not a subset of B."

The union of sets B and C (in symbols: B ∪ C) is shown by the shaded portion of the Venn diagram at the right. The **union** of two sets is the set that contains all the elements in both sets. Thus,

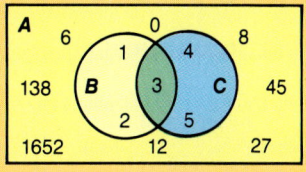

A = {whole numbers}

B ∪ C = {1, 2, 3} ∪ {3, 4, 5} = {1, 2, 3, 4, 5} ◀ The "3" is not repeated.

The intersection of sets B and C (in symbols: B ∩ C) is shown by the shaded portion of the Venn diagram at the right. The **intersection** of two sets is the set that contains all the elements common to both sets. Thus,

B ∩ C = {1, 2, 3} ∩ {3, 4, 5} = {3}

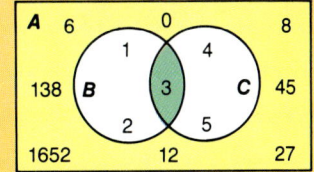

A = {whole numbers}

EXERCISES

Write the elements of each set.

1. B ∪ D
2. A ∩ B
3. D ∪ F
4. B ∪ F
5. B ∩ D
6. A ∩ F

Replace each ? with ⊂ or ⊄.

7. A __?__ D
8. F __?__ D
9. F __?__ A
10. B __?__ A
11. {3} __?__ A
12. {6} __?__ B

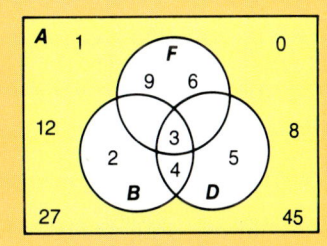

A = {whole numbers}

ENRICHMENT
You may wish to use this lesson for students who performed well on the formal Chapter Test.

1. {2, 3, 4, 5}
2. {2, 3, 4}
3. {3, 4, 5, 6, 9}
4. {2, 3, 4, 6, 9}
5. {3, 4}
6. {3, 6, 9}
7. ⊄
8. ⊄
9. ⊂
10. ⊂
11. ⊂
12. ⊄

Graphing and Equations

ADDITIONAL PRACTICE

SKILLS *Graph each ordered pair in the same coordinate plane.* (Pages 408–409)

1. $A(1, 1)$
2. $B(-1, 4)$
3. $F(-2, -2)$
4. $E(3, -2)$

Graph each equation in the coordinate plane. (Pages 410–412)

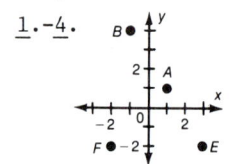

5. $y = x$
6. $y = 2x$
7. $y = x + 2$
8. $y = 3x + 1$

Find the slope of the line containing the given points. (Pages 417–419)

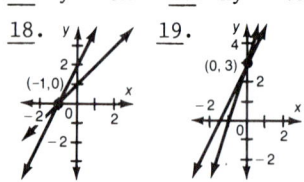

9. $A(1, 3); B(4, 7)$
10. $C(-1, 2); D(4, -3)$
11. $G(-4, -3); H(5, 3)$

Find the x and y intercepts. (Pages 422–423)

12. $y = x + 2$
13. $y = 3x + 6$
14. $y = -2x + 4$
15. $y = -3x + 9$

16. Write an equation for the direct variation. (Pages 426–428)

x	2	5	7	10	20
y	6	15	21	30	60

17. Write an equation for the indirect variation. (Pages 426–428)

x	2	4	5	8	40
y	20	10	8	5	1

Solve each system of equations. (Pages 427–429)

18. $\begin{cases} y = x + 1 \\ y = 2x + 2 \end{cases}$
19. $\begin{cases} y = 2x + 3 \\ y = 3x + 3 \end{cases}$
20. $\begin{cases} y = -x - 5 \\ y = -3x + 1 \end{cases}$

Graph each inequality in the coordinate plane. (Pages 430–431)

21. $y > \frac{1}{4}x$
22. $y < x + 2$
23. $y < 2x + 1$

APPLICATIONS

24. Use the information at the right to construct a broken line graph. (Pages 414–416)

CASES OF VEGETABLES SOLD

Month	JAN	FEB	MAR	APR	MAY	JUN
Cases Sold	200	100	100	300	400	600

25. There are four quarts in a gallon. The equation $Q = 4G$ describes this relation. Complete the table. Then draw the graph of the equation. (Pages 410–412)

G	Q
1	?
3	?
5	?

438 Chapter 14

CHAPTER

15 Special Triangles

FEATURES
Calculator Application: *Tangent Values*

Career Application: *Meteorologist*

Calculator/Computer Application: *Area of Triangles*

Enrichment: *Arithmetic Method for Finding $\sqrt{N}$*

Common Errors

SECTIONS
15-1 Meaning of Square Root
15-2 Irrational Numbers
15-3 Table of Squares and Square Roots
15-4 Problem Solving and Applications: Formulas: Using Square Roots
15-5 Pythagorean Theorem and Applications
15-6 Tangent Ratio and Applications

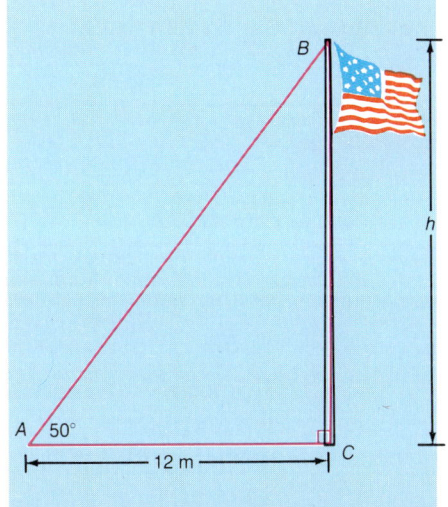

Teaching Suggestions
p. M-28

QUICK QUIZ
Divide.
1. 64 ÷ 8   Ans: 8
2. 121 ÷ 11   Ans: 11
3. 289 ÷ 17   Ans: 17

Evaluate.
4. 7^2 Ans: 49
5. 9^2 Ans: 81

ADDITIONAL EXAMPLES
Example 1
Square each number.
1. 27   Ans: 729
2. 16   Ans: 256

SELF-CHECK
1. 9   2. 169
3. 324   4. 961

15-1 Meaning of Square Root

You learned in Chapter 3 that the area of a square equals the product of the lengths of any two sides. For the figure at the right,

$$A = 12 \cdot 12 = 144 \text{ ft}^2.$$

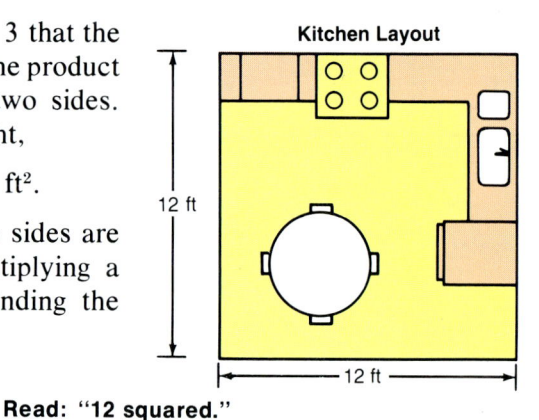

Since the lengths of the sides are the same, you are multiplying a number by itself, or finding the **square** of the number.

Kitchen Layout
12 ft
12 ft

$A = 12^2$ ◀ Read: "12 squared."
$ = 12 \cdot 12$
$ = 144$ ◀ 144 is the square of 12.

PROCEDURE
To square a number, multiply the number by itself.

EXAMPLE 1 Square each number.
  a. 8   b. 15   c. 24

Solutions: a. $8^2 = 8 \cdot 8$ b. $15^2 = 15 \cdot 15$ c. $24^2 = 24 \cdot 24$
              $= 64$ $= 225$ $= 576$

Self-Check *Square each number.*
  1. 3   **2.** 13   **3.** 18   **4.** 31

The inverse operation of squaring a number is taking its **square root**.

Since $8^2 = 64$, $\sqrt{64} = 8$ ◀ Read: "The square root of 64 is 8."

Since $8^2 = 64$ and $(-8)^2 = 64$, the number 64 has two square roots, 8 and -8.

To avoid confusion, we define

$\sqrt{64} = 8$ ◀ Principal square root
and $-\sqrt{64} = -8$.

440   Chapter 15

EXAMPLE 2 Find each answer.

    a. $\sqrt{25}$ b. $\sqrt{121}$ c. $-\sqrt{36}$

Solutions: a. Think: $25 = 5 \cdot 5$.   So $\sqrt{25} = 5$.
             b. Think: $121 = 11 \cdot 11$.   So $\sqrt{121} = 11$.
             c. Think: $36 = 6 \cdot 6$.   So $-\sqrt{36} = -6$.

Self-Check   Find each answer.

5. $\sqrt{49}$ 6. $\sqrt{9}$ 7. $-\sqrt{16}$ 8. $-\sqrt{81}$

CLASSROOM EXERCISES

Complete. (Example 1)

1. $4^2 = 4 \cdot 4$
   = ___?___

2. $1^2 = $ ___?___ $\cdot$ ___?___
   = ___?___

3. $9^2 = 9 \cdot $ ___?___
   = ___?___

4. $11^2 = $ ___?___ $\cdot$ ___?___
   = ___?___

(Example 2)

5. $16 = 4 \cdot 4$.   So $\sqrt{16} = $ ___?___ .
6. $9 = 3 \cdot 3$.   So $\sqrt{9} = $ ___?___ .
7. $49 = $ ___?___ $\cdot 7$.   So $\sqrt{49} = $ ___?___ .
8. $81 = $ ___?___ $\cdot 9$.   So $\sqrt{81} = $ ___?___ .

WRITTEN EXERCISES

Goals: To square a number
       To find the square root of a number

Square each number. (Example 1)

1. 2   2. 3   3. 6   4. 9   5. 10   6. 1
7. 4   8. 11   9. 15   10. 25   11. 30   12. 40
13. 21   14. 51   15. 13   16. 19   17. 18   18. 14
19. 60   20. 80   21. 24   22. 36   23. 100   24. 200

Find each answer. (Example 2)

25. $\sqrt{4}$ 26. $\sqrt{1}$ 27. $\sqrt{49}$ 28. $\sqrt{64}$ 29. $\sqrt{9}$ 30. $\sqrt{81}$
31. $\sqrt{36}$ 32. $\sqrt{16}$ 33. $\sqrt{100}$ 34. $\sqrt{400}$ 35. $\sqrt{169}$ 36. $\sqrt{225}$
37. $-\sqrt{9}$ 38. $-\sqrt{1}$ 39. $-\sqrt{49}$ 40. $-\sqrt{121}$ 41. $-\sqrt{36}$ 42. $-\sqrt{100}$
43. $\sqrt{196}$ 44. $\sqrt{484}$ 45. $\sqrt{1600}$ 46. $\sqrt{2500}$ 47. $\sqrt{3600}$ 48. $\sqrt{4900}$

Special Triangles   **441**

ADDITIONAL EXAMPLES
Example 2
Find each answer.
1. $-\sqrt{625}$ Ans: -25
2. $-\sqrt{400}$ Ans: -20

SELF-CHECK
5. 7   6. 3
7. -4 8. -9

CLASSROOM EXERCISES
1. 16   2. 1; 1; 1
3. 9; 81
4. 11; 11; 121
5. 4   6. 3   7. 7; 7
8. 9; 9

ASSIGNMENT GUIDE
BASIC
p. 441: Omit
AVERAGE
p. 441: 1–48
ABOVE AVERAGE
p. 441: 1–47 odd

WRITTEN EXERCISES
1. 4   2. 3   3. 36
4. 81   5. 100   6. 1
7. 4   8. 121   9. 225
10. 625   11. 900
12. 1600   13. 441
14. 2601   15. 169
16. 361   17. 324
18. 196   19. 3600
20. 6400   21. 576
22. 1296   23. 10,000
24. 40,000   25. 2
See page 445 for the answers to Exercises 26–48.

REVIEW CAPSULE

This Review Capsule reviews prior-taught skills used in Section 15-2. The reference is to the pages where the skills were taught.

Teaching Suggestions p. M-28

QUICK QUIZ

Divide. Round each answer to the nearest hundredth.

1. $6 \div 8$ Ans: 0.75
2. $10 \div 3$ Ans: 3.33
3. $4 \div 5$ Ans: 0.20

Write each fraction as a decimal.

4. $\frac{5}{9}$ Ans: $0.\overline{5}$
5. $\frac{3}{8}$ Ans: 0.375

ADDITIONAL EXAMPLES
Example 1
Identify each as rational or irrational.
1. $\sqrt{125}$ Ans: Irrational
2. $\sqrt{225}$ Ans: Rational

SELF-CHECK
1. Rational
2. Irrational
3. Irrational
4. Rational

442

REVIEW CAPSULE FOR SECTION 15-2

Divide. Round each answer to the nearest hundredth. (Pages 32–34)

1. $2\overline{)1}$ 2. $3\overline{)2}$ 3. $5\overline{)3}$ 4. $4\overline{)1}$ 5. $6\overline{)7}$ 6. $4\overline{)9}$

Write a decimal for each fraction. Use a bar to indicate any digit or digits that repeat. (Pages 173–175)

7. $\frac{1}{5}$ 8. $\frac{3}{4}$ 9. $\frac{1}{3}$ 10. $\frac{5}{6}$ 11. $\frac{3}{20}$ 12. $\frac{5}{8}$
13. $\frac{7}{8}$ 14. $\frac{9}{20}$ 15. $\frac{10}{9}$ 16. $\frac{19}{5}$ 17. $\frac{7}{4}$ 18. $\frac{11}{3}$

The answers are on page 464.

15-2 Irrational Numbers

Rational numbers such as the following are <u>perfect</u> <u>squares</u>.

$$16 \qquad 144 \qquad \frac{1}{4} \qquad 0.81$$

> A rational number is a **perfect square** if it is the square of a rational number.
>
> $16 = 4 \cdot 4$
> $144 = 12 \cdot 12$
> $\frac{1}{4} = \frac{1}{2} \cdot \frac{1}{2}$
> $0.81 = (0.9)(0.9)$

A rational number that is not a perfect square has an **irrational square root**. Thus, the following numbers are irrational.

$$\sqrt{2} \qquad \sqrt{5} \qquad \sqrt{17} \qquad \sqrt{31}$$

EXAMPLE 1 Identify each number as rational or irrational. Give a reason.
  a. $\sqrt{15}$ b. $\sqrt{49}$ c. $\sqrt{3}$

Solutions: a. $\sqrt{15}$ is **irrational** since 15 is <u>not</u> a perfect square.
  b. $\sqrt{49}$ is **rational** since 49 is a perfect square.
  c. $\sqrt{3}$ is **irrational** since 3 is <u>not</u> a perfect square.

Self-Check Identify each number as rational or irrational.

1. $\sqrt{64}$ 2. $\sqrt{33}$ 3. $\sqrt{114}$ 4. $\sqrt{100}$

442 Chapter 15

Every rational number can be written either as a **terminating** or as a **repeating decimal**.

$$2\tfrac{1}{2} = \tfrac{5}{2} = 2.5 \quad \blacktriangleleft \text{ Terminating decimal}$$

$$\tfrac{2}{3} = 0.6666\cdots = 0.\overline{6} \quad \blacktriangleleft \text{ Repeating decimal}$$

The decimal equivalents of irrational numbers are **non-terminating** and **non-repeating**.

$$\sqrt{2} = 1.141421\cdots \quad \sqrt{3} = 1.73205\cdots \quad \sqrt{5} = 2.23606\cdots$$
$$\sqrt{6} = 2.44948\cdots \quad \sqrt{7} = 2.64575\cdots \quad \sqrt{8} = 2.82842\cdots$$

◀ There is no repeating pattern.

EXAMPLE 2 Identify each number as rational or irrational. Give a reason for each answer.

   a. 4.2   b. 3.02$\overline{16}$ c. 5.292292229···

Solutions:

Number	Rational or Irrational?	Reason
a. 4.2	Rational	It is a terminating decimal.
b. 3.02$\overline{16}$	Rational	It is a repeating decimal.
c. 5.292292229···	Irrational	It is <u>neither</u> a terminating <u>nor</u> a repeating decimal.

Self-Check *Identify each number as rational or irrational.*

5. 5.4772···   6. 7.$\overline{6}$ 7. 3.4215   8. 7.07106···

CLASSROOM EXERCISES

Tell whether each of the following is a perfect square. Write <u>Yes</u> or <u>No</u>. (Definition)

1. 12   2. 18   3. 16   4. 25   5. 30   6. 36
7. 49   8. 80   9. 75   10. 81   11. 125   12. 144

Identify each number as rational or irrational (Example 1)

13. $\sqrt{10}$ 14. $\sqrt{4}$ 15. $\sqrt{64}$ 16. $\sqrt{23}$ 17. $\sqrt{5}$ 18. $\sqrt{49}$
19. $\sqrt{81}$ 20. $\sqrt{7}$ 21. $\sqrt{9}$ 22. $\sqrt{21}$ 23. $\sqrt{15}$ 24. $\sqrt{144}$

Special Triangles   **443**

ADDITIONAL EXAMPLES
Example 2
Identify each as rational or irrational.
1. 3.$\overline{16}$ Ans: Rational
2. 5.01732...
   Ans: Irrational

SELF-CHECK
5. Irrational
6. Rational
7. Rational
8. Irrational

CLASSROOM EXERCISES
1. No   2. No   3. Yes
4. Yes   5. No   6. Yes
7. Yes   8. No   9. No
10. Yes   11. No
12. Yes   13. Irrational
14. Rational
15. Rational
16. Irrational
17. Irrational
18. Rational
19. Rational
20. Irrational
21. Rational
22. Irrational
23. Irrational
24. Rational
25. Rational
26. Irrational
27. Irrational
28. Rational
29. Rational
30. Rational
31. Irrational
32. Rational

ASSIGNMENT GUIDE

BASIC
p. 444: Omit

AVERAGE
p. 444: 1–42

ABOVE AVERAGE
p. 444: 1–48

WRITTEN EXERCISES

1. R; perfect square
2. I; not a perfect square
3. R: perfect square
4. R; perfect square
5. I; not a perfect square
6. I; not a perfect square
7. I; not a perfect square
8. R; perfect square
9. I; not a perfect square
10. R; perfect square
11. R; perfect square
12. I; not a perfect square
13. I; not a perfect square
14. I; not a perfect square
15. R; perfect square
16. R; perfect square
17. R; perfect square
18. R; perfect square
19. R; terminating

See page 457 for the answers to Exercises 20–48.

Identify each number as rational or irrational: (Example 2)

25. 3.5
26. 2.828423···
27. 3.31662···
28. $2.\overline{142857}$
29. 6.4512
30. $9.\overline{1}$
31. 5.41621···
32. $9.02\overline{6}$

WRITTEN EXERCISES

Goal: To identify a number as rational or irrational

For Exercises 1–42, identify each number as rational or irrational. Give a reason for each answer. (Example 1)

1. $\sqrt{25}$
2. $\sqrt{3}$
3. $\sqrt{100}$
4. $\sqrt{400}$
5. $\sqrt{18}$
6. $\sqrt{96}$
7. $\sqrt{22}$
8. $\sqrt{121}$
9. $\sqrt{7}$
10. $\sqrt{1}$
11. $\sqrt{36}$
12. $\sqrt{50}$
13. $\sqrt{112}$
14. $\sqrt{500}$
15. $\sqrt{1600}$
16. $\sqrt{2500}$
17. $\sqrt{1000}$
18. $\sqrt{900}$

(Example 2)

19. 0.52
20. 9.05
21. 2.6457···
22. 3.8729···
23. $1.\overline{378}$
24. $4.\overline{6}$
25. $1.\overline{857142}$
26. 2.052
27. 3.06006···
28. $9.\overline{16}$
29. 3.506
30. 1.414114···

MIXED PRACTICE

31. $\sqrt{19}$
32. $\sqrt{100}$
33. 2.95
34. $3.8\overline{3}$
35. 4.5825···
36. $8.\overline{6}$
37. $3.\overline{142857}$
38. $\sqrt{32}$
39. $\sqrt{61}$
40. $\sqrt{169}$
41. 14.001
42. 9.11043···

MORE CHALLENGING EXERCISES

Choose one or more letters below to describe each number.
a. whole number **b.** integer **c.** rational number **d.** irrational number

43. −6
44. $\sqrt{3}$
45. $\frac{1}{8}$
46. $3.\overline{3}$
47. 2.23606···
48. $-\sqrt{21}$

REVIEW CAPSULE FOR SECTION 15-3

Evaluate. Round each answer to the nearest whole number. (Pages 32–34)

1. 6^2
2. 15^2
3. 1.414^2
4. 10.1^2
5. 42^2
6. 2.2^2
7. 7.5^2
8. 58^2
9. 3.6^2
10. 9.4^2
11. 90^2
12. 118^2

The answers are on page 464.

444 Chapter 15

15-3 Table of Squares and Square Roots

You can use the table on page 447 to find the squares and square roots of numbers from 1 to 150.

> **PROCEDURE**
> **To use a table of squares and square roots:**
> 1 Find the number in the **NUMBER** column.
> 2 Look directly to the right.
>   a. Read the square of the number in the **SQUARE** column, or
>   b. Read the square root of the number in the **SQUARE ROOT** column.

EXAMPLE 1 Find each answer.

a. 24^2 b. $\sqrt{23}$

Solutions: a. $24^2 = 576$ b. $\sqrt{23} = 4.796$

To check your answer in **b**, use a calculator to multiply 4.796 by 4.796.

Number	Square	Square Root
21	441	4.583
22	484	4.690
23	529	4.796
24	576	4.899
25	625	5.000

4 . 7 9 6 × 4 . 7 9 6 =

23.001616

Thus, $\sqrt{23}$ is about **4.796**.

Self-Check *Find each answer.*

1. 35^2 2. 91^2 3. $\sqrt{21}$ 4. $\sqrt{84}$

NOTE: Most numbers in the square root column in the table are approximations (correct to thousandths) for irrational numbers. Thus, when you find the square root of a number in the table, you may be asked to round your answer to the nearest tenth, or to the nearest hundredth, and so on.

You can use the table to find the square roots of some numbers greater than 150. First, find the number in the **SQUARE** column. Then read the square root in the **NUMBER** column.

Special Triangles **445**

Teaching Suggestions p. M-28

QUICK QUIZ
Evaluate. Round each to the nearest whole number.
1. 7^2 Ans: 49
2. 1.4^2 Ans: 2
3. 2.646^2 Ans: 7
4. 25^2 Ans: 625
5. 26.5^2 Ans: 702

ADDITIONAL EXAMPLES
Example 1
Use the table to find each.
1. 58^2 Ans: 3364
2. $\sqrt{134}$ Ans: 11.576

SELF-CHECK
1. 1225   2. 8281
3. 4.583   4. 9.165

WRITTEN EXERCISES p. 441
26. 1   27. 7   28. 8
29. 3   30. 9   31. 6
32. 4   33. 10   34. 20
35. 13   36. 15   37. -3
38. -1   39. -7
40. -11 41. -6
42. -10 43. 14
44. 22   45. 40   46. 50
47. 60   48. 70

445

ADDITIONAL EXAMPLES
Example 2
Use the Table to find each.
1. $\sqrt{4761}$ Ans: 69
2. $\sqrt{15,625}$ Ans: 125

SELF-CHECK
5. 29   6. 85
7. 20   8. 99

CLASSROOM EXERCISES
1. 3.742   2. 19; 4.359
3. 400; 4.472
4. 80; 8.944
5. 324; 4.243
6. 35; 5.916
7. 3025; 7.416
8. 23; 4.796

ASSIGNMENT GUIDE
BASIC
p. 446: Omit
AVERAGE
p. 446: 1–36
ABOVE AVERAGE
p. 446: 1–35 odd

WRITTEN EXERCISES
1. 169   2. 289   3. 441
4. 729   5. 1024
6. 2025   7. 6561
8. 10,201   9. 14,400
10. 13,689   11. 16,900
12. 22,500   13. 2,828
See page 453 for the answers to Exercises 14–36.

EXAMPLE 2 $\sqrt{4761} = \underline{\ ?\ }$

Solution: ① Find 4761 in the SQUARE column.
② Look directly to the left.
$\sqrt{4761} = 69$

Number	Square
66	4356
67	4489
68	4624
69	← 4761
70	4900

Self-Check   *Refer to the table on page 447 to find each answer.*

5. $\sqrt{841}$ 6. $\sqrt{7225}$ 7. $\sqrt{400}$ 8. $\sqrt{9801}$

CLASSROOM EXERCISES

Complete. Refer to the table on page 447.   (Example 1)

	Number	Square	Square Root
1.	14	196	?
3.	20	?	?
5.	18	?	?
7.	55	?	?

	Number	Square	Square Root
2.	?	361	?
4.	?	6400	?
6.	?	1225	?
8.	?	529	?

WRITTEN EXERCISES

Goal: To use a table of squares and square roots

Refer to the table on page 447 to find each answer.   (Example 1)

1. 13^2 2. 17^2 3. 21^2 4. 27^2 5. 32^2 6. 45^2
7. 81^2 8. 101^2 9. 120^2 10. 117^2 11. 130^2 12. 150^2
13. $\sqrt{8}$ 14. $\sqrt{15}$ 15. $\sqrt{21}$ 16. $\sqrt{32}$ 17. $\sqrt{56}$ 18. $\sqrt{41}$
19. $\sqrt{121}$ 20. $\sqrt{144}$ 21. $\sqrt{45}$ 22. $\sqrt{27}$ 23. $\sqrt{149}$ 24. $\sqrt{130}$

(Example 2)

25. $\sqrt{225}$ 26. $\sqrt{256}$ 27. $\sqrt{676}$ 28. $\sqrt{841}$ 29. $\sqrt{1521}$ 30. $\sqrt{1024}$
31. $\sqrt{6561}$ 32. $\sqrt{7225}$ 33. $\sqrt{3600}$ 34. $\sqrt{9025}$ 35. $\sqrt{3025}$ 36. $\sqrt{17,956}$

446   Chapter 15

Table of Squares and Square Roots

Number	Square	Square Root	Number	Square	Square Root	Number	Square	Square Root
1	1	1.000	51	2601	7.141	101	10,201	10.050
2	4	1.414	52	2704	7.211	102	10,404	10.100
3	9	1.732	53	2809	7.280	103	10,609	10.149
4	16	2.000	54	2916	7.348	104	10,816	10.198
5	25	2.236	55	3025	7.416	105	11,025	10.247
6	36	2.449	56	3136	7.483	106	11,236	10.296
7	49	2.646	57	3249	7.550	107	11,449	10.344
8	64	2.828	58	3364	7.616	108	11,664	10.392
9	81	3.000	59	3481	7.681	109	11,881	10.440
10	100	3.162	60	3600	7.746	110	12,100	10.488
11	121	3.317	61	3721	7.810	111	12,321	10.536
12	144	3.464	62	3844	7.874	112	12,544	10.583
13	169	3.606	63	3969	7.937	113	12,769	10.630
14	196	3.742	64	4096	8.000	114	12,996	10.677
15	225	3.873	65	4225	8.062	115	13,225	10.724
16	256	4.000	66	4356	8.124	116	13,456	10.770
17	289	4.123	67	4489	8.185	117	13,689	10.817
18	324	4.243	68	4624	8.246	118	13,924	10.863
19	361	4.359	69	4761	8.307	119	14,161	10.909
20	400	4.472	70	4900	8.367	120	14,400	10.954
21	441	4.583	71	5041	8.426	121	14,641	11.000
22	484	4.690	72	5184	8.485	122	14,884	11.045
23	529	4.796	73	5329	8.544	123	15,129	11.091
24	576	4.899	74	5476	8.602	124	15,376	11.136
25	625	5.000	75	5625	8.660	125	15,625	11.180
26	676	5.099	76	5776	8.718	126	15,876	11.225
27	729	5.196	77	5929	8.775	127	16,129	11.269
28	784	5.292	78	6084	8.832	128	16,384	11.314
29	841	5.385	79	6241	8.888	129	16,641	11.358
30	900	5.477	80	6400	8.944	130	16,900	11.402
31	961	5.568	81	6561	9.000	131	17,161	11.446
32	1024	5.657	82	6724	9.055	132	17,424	11.489
33	1089	5.745	83	6889	9.110	133	17,689	11.533
34	1156	5.831	84	7056	9.165	134	17,956	11.576
35	1225	5.916	85	7225	9.220	135	18,225	11.619
36	1296	6.000	86	7396	9.274	136	18,496	11.662
37	1369	6.083	87	7569	9.327	137	18,769	11.705
38	1444	6.164	88	7744	9.381	138	19,044	11.747
39	1521	6.245	89	7921	9.434	139	19,321	11.790
40	1600	6.325	90	8100	9.487	140	19,600	11.832
41	1681	6.403	91	8281	9.539	141	19,881	11.874
42	1764	6.481	92	8464	9.592	142	20,164	11.916
43	1849	6.557	93	8649	9.644	143	20,449	11.958
44	1936	6.633	94	8836	9.695	144	20,736	12.000
45	2025	6.708	95	9025	9.747	145	21,025	12.042
46	2116	6.782	96	9216	9.798	146	21,316	12.083
47	2209	6.856	97	9409	9.849	147	21,609	12.124
48	2304	6.928	98	9604	9.899	148	21,904	12.166
49	2401	7.000	99	9801	9.950	149	22,201	12.207
50	2500	7.071	100	10,000	10.000	150	22,500	12.247

CAREER APPLICATIONS

This feature applies the skills for problem solving, using square roots, to calculating the wind chill temperature and illustrates the use of a calculator to evaluate a formula. Thus, it can be used after Section 15-4 (pages 451-453) is taught.

1. −11.9°C
2. −29.9°C
3. −13.7°C
4. 1.3°C

CAREER APPLICATIONS
Meteorologist

Meteorology is the study of the atmosphere. A meteorologist must be able to observe, analyze, and forecast the weather.

One statistic that is often given in winter weather reports in cold climates is **wind chill temperature**. Meteorologists can obtain the wind chill temperature from tables or they can use a formula such as the one below.

$$C = 33 - \frac{(10\sqrt{r} + 10.45 - r)(33 - t)}{22.1}$$

C = wind chill temperature in degrees Celsius.
t = air temperature in degrees Celsius.
r = wind speed in meters per second.

EXAMPLE On a winter day, the air temperature is 2°C. The wind speed is 6 meters per second. Find the wind chill temperature to the nearest tenth.

SOLUTION

$$C = 33 - \frac{(10\sqrt{r} + 10.45 - r)(33 - t)}{22.1} \quad r = 6, \ t = 2$$

$$C = 33 - \frac{(10\sqrt{6} + 10.45 - 6)(33 - 2)}{22.1}$$

$$C = 33 - \frac{(10\sqrt{6} + 10.45 - 6)(31)}{22.1} \quad \text{Use a calculator.}$$

6 [√] [×] 1 0 [+] 1 0 . 4 5 [−]

6 [=] [×] 3 1 [÷] 2 2 . 1 [=]

Display: 40.60144

Round 40.60144 to the nearest tenth: 40.6
Thus, $C = 33 - 40.6 = $ **−7.6°C**.

Exercises

Find each wind chill temperature to the nearest tenth. The air temperature is given in degrees Celsius (°C) and the wind speed is given in meters per second (m/s).

1. Air temperature: 4°C
   Wind speed: 15 m/s

2. Air temperature: −15°C
   Wind speed: 6 m/s

3. Air Temperature: −5°C
   Wind speed: 4.5 m/s

4. Air temperature: 7.2°C
   Wind speed: 4.5 m/s

CALCULATOR/COMPUTER APPLICATIONS
Area of Triangles

If you know the lengths of three sides of a triangle, you can use **Hero's Formula** to find the area.

$$A = \sqrt{s(s-a)(s-b)(s-c)}, \text{ where } s = \frac{a+b+c}{2}$$

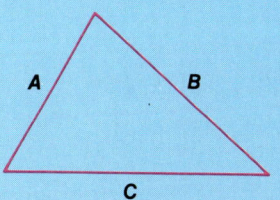

Calculator: The sides of a triangle are 212, 341, and 285 centimeters long. Find the area to the nearest whole number.

Solution: First find s, $s-a$, $s-b$, and $s-c$.

$$s = 212 \;[+]\; 341 \;[+]\; 285 \;[\div]\; 2 \;[=] \quad \boxed{419.}$$

$s - a = 419 - 212 = 207 \qquad s - b = 419 - 341 = 78 \qquad s - c = 419 - 285 = 134$

$$A = \sqrt{419(207)(79)(139)}$$

$$A = 419 \;[\times]\; 207 \;[\times]\; 79 \;[\times]\; 139 \;[=]\; [\sqrt{}] \quad \boxed{30861.22281}$$

The area is **30,861 cm²** to the nearest whole number.

Computer: The following program also uses Hero's Formula.

Program:
```
10 PRINT "ENTER THE LENGTHS OF THE SIDES."
20 INPUT A, B, C
30 LET S = .5 * (A + B + C)
40 LET X = SQR(S * (S - A) * (S - B) * (S - C))
50 PRINT "AREA = "; X
60 PRINT "ANY MORE TRIANGLES (1 = YES, 0 = NO)";
70 INPUT Y
80 IF Y = 1 THEN 10
90 END
```

Exercises

Find the area of each triangle by using a calculator or the program. Round each answer to the nearest whole number.

1. $a = 6$; $b = 25$; $c = 29$
2. $a = 19$; $b = 23$; $c = 14$
3. $a = 62$; $b = 48$; $c = 36$
4. $a = 24$; $b = 36$; $c = 18$
5. $a = 17$; $b = 18$; $c = 19$
6. $a = 16$; $b = 21$; $c = 33$

CALCULATOR/COMPUTER APPLICATIONS

This feature illustrates the use of a calculator or a computer program for finding the area of a triangle using Hero's Formula. It can be used after Section 15-3 (pages 445-447) is taught, as it applies the skill of finding the square root of a number.

1. 60
2. 133
3. 862
4. 192
5. 139
6. 136

Computer Application 449

QUIZ: SECTIONS 15-1–15-3

After completing this Review, you may wish to administer a quiz covering the same sections. A quiz is provided in the Teacher's Edition: Part II.

REVIEW: SECTIONS 15-1–15-3

1. 16 2. 121 3. 361
4. 64 5. 576 6. 900
7. 7 8. 2 9. 10
10. 1 11. -6 12. -9
13. -4 14. 5 15. -8
16. -12 17. 3 18. 11
19. Irrational
20. Rational
21. Rational
22. Irrational
23. Rational
24. Irrational
25. Irrational
26. Rational
27. Rational
28. Rational
29. Rational
30. Irrational
31. 324 32. 484
33. 12,544 34. 4225
35. 5.657 36. 8.660
37. 11.180 38. 11.790
39. 31 40. 97
41. 130 42. 62
43. 29 44. 96
45. 146 46. 71
47. 135 48. 149

450

REVIEW: SECTIONS 15-1—15-3

Square each number. (Section 15-1)

1. 4   2. 11   3. 19   4. 8   5. 24   6. 30

Find each answer. (Section 15-1)

7. $\sqrt{49}$ 8. $\sqrt{4}$ 9. $\sqrt{100}$ 10. $\sqrt{1}$ 11. $-\sqrt{36}$ 12. $-\sqrt{81}$
13. $-\sqrt{16}$ 14. $\sqrt{25}$ 15. $-\sqrt{64}$ 16. $-\sqrt{144}$ 17. $\sqrt{9}$ 18. $\sqrt{121}$

Identify each number as rational or irrational. (Section 15-2)

19. $\sqrt{21}$ 20. $\sqrt{36}$ 21. 4.84   22. 2.3132···   23. $2.\overline{6}$ 24. $\sqrt{50}$
25. $\sqrt{14}$ 26. $\sqrt{64}$ 27. 3.75   28. $12.\overline{326}$ 29. $-\sqrt{9}$ 30. $\sqrt{55}$

Refer to the table on page 447 to find each answer. (Section 15-3)

31. 18^2 32. 22^2 33. 112^2 34. 65^2 35. $\sqrt{32}$ 36. $\sqrt{75}$
37. $\sqrt{125}$ 38. $\sqrt{139}$ 39. $\sqrt{961}$ 40. $\sqrt{9409}$ 41. $\sqrt{16,900}$ 42. $\sqrt{3844}$
43. $\sqrt{841}$ 44. $\sqrt{9216}$ 45. $\sqrt{21,316}$ 46. $\sqrt{5041}$ 47. $\sqrt{18,225}$ 48. $\sqrt{22,201}$

REVIEW CAPSULE FOR SECTION 15-4

Round each number to the nearest tenth. (Pages 32–34)

1. 43.24   2. 16.71   3. 27.83   4. 14.01   5. 82.87   6. 93.14
7. 12.521   8. 39.765   9. 55.543   10. 68.331   11. 10.897   12. 50.112

Evaluate each expression. (Pages 38–41)

13. $3.5p$ when $p = 1.214$ 14. $4.2x$ when $x = 4.561$
15. $2.5t$ when $t = 3.125$ 16. $1.6y$ when $y = 2.496$

Round each of the following to the nearest whole number. (Pages 32–34)

17. 38.3   18. 27.6   19. 42.1   20. 16.9   21. 72.84   22. 24.62
23. 6.27   24. 18.39   25. 9.064   26. 2.307   27. 8.099   28. 123.724

The answers are on page 464.

Chapter 15

PROBLEM SOLVING AND APPLICATIONS

15-4 Formulas: Using Square Roots

The area of the square park at the right is 2 square kilometers. You can use the formula $A = s^2$ and the Table of Squares and Square Roots on page 447 to find the length of one side of the park.

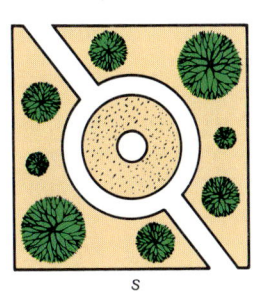

EXAMPLE 1 Find the length of one side of the park shown above. Round the answer to the nearest tenth.

Solution:
1. Formula: $A = s^2$ ← Known value: $A = 2$
2. $2 = s^2$, or $s^2 = 2$
   $s = \sqrt{2}$ ← Since $s^2 = 2$, $s = \sqrt{2}$.
   $s = 1.414$ ← From the Table on page 447.

The length of one side of the park is about **1.4 kilometers**.

Self-Check 1. The area of a square is 144 square feet. Find the length of one side.

You can use the formula below to find the distance in miles to the horizon from an aircraft above the ground.

Formula $D = 1.2\sqrt{A}$

D: distance in mile
A: number of feet from the ground

EXAMPLE 2 Find the distance to the horizon from a plane flying at a height of 4096 feet. Round the answer to the nearest mile.

Solution:
1. Formula: $D = 1.2\sqrt{A}$ ← Known value: $A = 4096$
2. $D = 1.2\sqrt{4096}$
   $D = 1.2(64)$
   $D = 76.8$ The distance is about **77 miles**.

Self-Check 2. Find the distance to the horizon from a plane flying at a height of 3136 feet. Round the answer to the nearest mile.

Special Triangles **451**

Teaching Suggestions
p. M-28

QUICK QUIZ
Round each to the
nearest
1. tenth: 2.236
   Ans: 2.2
2. whole number: 11.533
   Ans: 12
Evaluate.
3. 1.2x when x = 2.36
   Ans: 2.832
4. 20.3y when y = 1.732
   Ans: 35.1596

ADDITIONAL EXAMPLES
Example 1
Given the area of a
square, find the length
of a side to the
nearest tenth.
1. A = 121 m^2 Ans: 11 m
2. A = 75 ft^2
   Ans: 8.7 ft

SELF-CHECK
1. 12 ft

Example 2
Find the distance to the
horizon, to the nearest
mile, given the height.
1. H = 100 ft Ans: 12 mi
2. H = 5184 ft
   Ans: 86 mi

SELF-CHECK
2. 67 mi

451

CLASSROOM EXERCISES

1. 3.7 m   2. 5.5 mm
3. 1.7 km 4. 4.9 cm
5. 6 in   6. 9 ft
7. 12 yd   8. 2 mi
9. 60 mi   10. 84 mi
11. 71; 85 mi
12. 47; 56 mi
13. 98; 118 mi
14. 78; 94 mi
15. 114; 137 mi
16. 130; 156 mi

ASSIGNMENT GUIDE
BASIC
pp. 452-453: Omit
AVERAGE
pp. 452-453: 1-22
ABOVE AVERAGE
pp. 452-453: 1-22

WRITTEN EXERCISES

1. 4.8 m   2. 2.8 km
3. 5.9 cm 4. 10.5 mm
5. 11.7 cm 6. 3.2 km
7. 6.5 m   8. 8.8 mm
9. 6 ft   10. 10 mi
11. 11 in 12. 4 yd

452

CLASSROOM EXERCISES

For Exercises 1–8, use the formula $A = s^2$ to find the length, s, of one side of a square with the given area, A. For Exercises 1–4, round each answer to the nearest tenth.   (Example 1)

1. $A = 14$ m²   2. $A = 30$ mm²   3. $A = 3$ km²   4. $A = 24$ cm²
   $s = \underline{\ ?\ }$ $s = \underline{\ ?\ }$ $s = \underline{\ ?\ }$ $s = \underline{\ ?\ }$

5. $A = 36$ in² 6. $A = 81$ ft²   7. $A = 144$ yd² 8. $A = 4$ mi²
   $s = \underline{\ ?\ }$ $s = \underline{\ ?\ }$ $s = \underline{\ ?\ }$ $s = \underline{\ ?\ }$

For Exercises 9–16, use the formula $D = 1.2\sqrt{A}$ to find the distance, D, (in miles) to the horizon from an aircraft that is the given number of feet, A, from the ground. Round each answer to the nearest mile. (Example 2)

	A	$\sqrt{A}$	D		A	$\sqrt{A}$	D
9.	2500 ft	50	?	10.	4900 ft	70	?
11.	5041 ft	?	?	12.	2209 ft	?	?
13.	9604 ft	?	?	14.	6084 ft	?	?
15.	12,996 ft	?	?	16.	16,900 ft	?	?

WRITTEN EXERCISES

Goal: To use a formula to solve word problems involving square roots

For Exercises 1–12, use the formula $A = s^2$ to find the length, s, of one side of a square with the given area, A. Round each answer to the nearest tenth.   (Example 1)

	A	s		A	s		A	s
1.	23 m²	?	2.	8 km²	?	3.	35 cm²	?
4.	110 mm²	?	5.	136 cm²	?	6.	10 km²	?
7.	42 m²	?	8.	78 mm²	?	9.	36 ft²	?
10.	100 mi²	?	11.	121 in²	?	12.	16 yd²	?

452 Chapter 15

Solve. Round each answer to the nearest tenth. (Example 1)

13. A square swimming pool has an area of 324 square feet. Find the length of one side of the pool.

14. The area of the top of a square table is 130 square centimeters. Find the length of one side of the table.

For Exercises 15–18, use the formula $D = 1.2\sqrt{A}$ to find the distance, D, in miles to the horizon from an aircraft that is the given number of feet, A, from the ground. Round each answer to the nearest mile. (Example 2)

	Feet from the Ground	Distance to the Horizon			Feet from the Ground	Distance to the Horizon
15.	3025	?		16.	7225	?
17.	14,884	?		18.	10,816	?

Solve. Round each answer to the nearest mile. (Example 2)

19. The height of a hot-air balloon above the ground is 3844 feet. Find the distance to the horizon.

20. A plane is flying at a height of 19,321 feet above the ground. Find the distance to the horizon.

21. A plane is flying at a height of 10,404 feet. Find the distance to the horizon.

22. The height of a plane above the ground is 20,736 feet. Find the distance to the horizon.

REVIEW CAPSULE FOR SECTION 15-5

Evaluate. (Pages 440–441)

1. 3^2
2. 8^2
3. 10^2
4. 20^2
5. 36^2
6. 19^2

Solve and check. (Pages 190–192)

7. $n + 6 = 10$
8. $p + 17 = 30$
9. $x + 3 = 12$
10. $b + 12 = 25$
11. $t + 15 = 40$
12. $y + 30 = 75$
13. $a + 4 = 21$
14. $c + 9 = 27$

Round each number to the nearest tenth. (Pages 32–34)

15. 12.35
16. 21.72
17. 18.09
18. 32.15
19. 49.53
20. 62.43

The answers are on page 464.

WRITTEN EXERCISES

13. 18 ft
14. 11.4 cm
15. 66 mi
16. 102 mi
17. 146 mi
18. 125 mi
19. 74 mi
20. 167 mi
21. 122 mi
22. 173 mi

WRITTEN EXERCISES p. 446

14. 3.873
15. 4.583
16. 5.657
17. 7.483
18. 6.403
19. 11
20. 12
21. 6.708
22. 5.196
23. 12.207
24. 11.402
25. 15
26. 16
27. 26
28. 29
29. 39
30. 32
31. 81
32. 85
33. 60
34. 95
35. 55
36. 134

Special Triangles 453

Teaching Suggestions
p. M-28

QUICK QUIZ

Evaluate.

1. 26^2 Ans: 676
2. $\sqrt{90}$ Ans: 9.487

Solve and check.

3. $y + 25 = 49$ Ans: 24
4. $64 + a = 100$ Ans: 36

Round to the nearest tenth.

5. 61.37   Ans: 61.4
6. 3.648   Ans: 3.6

ADDITIONAL EXAMPLES

Example 1

Find the length of the hypotenuse, c, to the nearest tenth.

1. $a = 3$ ft; $b = 4$ ft
   Ans: 5 ft
2. $a = 17$ in; $b = 20$ in
   Ans: 26.2 in

15-5 Pythagorean Theorem and Applications

A **right triangle** is a triangle with one right angle (90°).

When you know the lengths of two sides of a right triangle, you can use the **Pythagorean Theorem** to find the length of the other side.

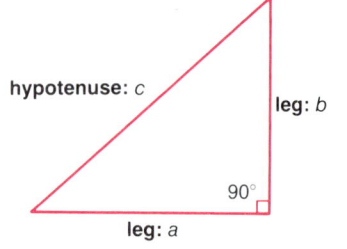

> **Pythagorean Theorem:** (hypotenuse)² = (leg)² + (leg)²
> $$c^2 = a^2 + b^2$$

NOTE: The **hypotenuse** of a right triangle is the side opposite the right angle. The hypotenuse is the longest side of the right triangle.

PROCEDURE

To find the measure of a side of a right triangle:

1. Draw the triangle. Label the sides with the given measures.
2. Write the Pythagorean Theorem. Replace two letters in the formula by the corresponding given measures.
3. Solve for the third measure.

EXAMPLE 1   The sail on this boat is 5 yards high and 3 yards wide. Find the length of the third side. Round your answer to the nearest tenth.

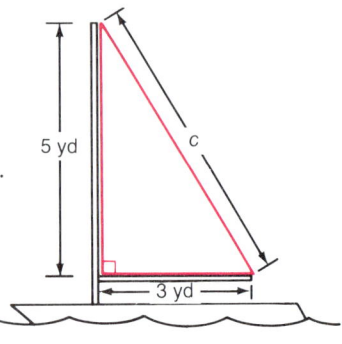

Solution: Use the Pythagorean Theorem to find c.

2   $c^2 = a^2 + b^2$ **Replace *a* with 5 and *b* with 3.**
   $c^2 = 5^2 + 3^2$
3   $c^2 = 25 + 9$
   $c^2 = 34$
   $c = \sqrt{34}$ **Use the table on page 447.**
   $c = 5.831$ The length is about **5.8 yards.**

454   Chapter 15

Self-Check *Find the length of the hypotenuse, c, of each right triangle with legs a and b.*

1. $a = 5$ in; $b = 7$ in **2.** $a = 7$ m; $b = 8$ m

When you know the lengths of the hypotenuse and one leg of a right triangle, you can use the Pythagorean Theorem to find the length of the other leg.

EXAMPLE 2 A 7-meter ladder is leaning against a shed. If the base of the ladder is 3 meters from the wall, how high up wall does the ladder reach? Round your answer to the nearest tenth.

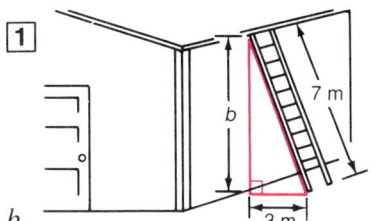

Solution: Use the Pythagorean Theorem to find b.

⬛2
$$c^2 = a^2 + b^2$$
$$7^2 = 3^2 + b^2$$

⬛3
$$49 = 9 + b^2$$
$$49 - 9 = 9 - 9 + b^2$$
$$40 = b^2$$
$$\sqrt{40} = b$$
$$6.325 = b$$

◀ Replace c with 7 and a with 3.

◀ Use the table on page 447.

The ladder reaches about **6.3 meters** above the ground.

Self-Check *Refer to this right triangle with hypotenuse c and legs a and b. Find each unknown length.*

3. $b = 4$ m; $c = 5$ m; $a = $ __?__
4. $a = 5$ ft; $c = 13$ ft; $b = $ __?__

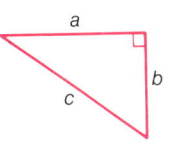

CLASSROOM EXERCISES

In Exercises 1–2, find the length of the hypotenuse, c, of each right triangle. Use the Table of Squares and Square Roots on page 447. Round each answer to the nearest tenth. (Example 1)

1. $a = 6$, $b = 8$, $c = $ __?__ **2.** $a = 8$, $b = 8$, $c = $ __?__

Special Triangles

SELF-CHECK
1. 8.6 in 2. 10.6 m

ADDITIONAL EXAMPLES
Example 2
Given the hypotenuse, c, find the missing leg to the nearest tenth.
1. c = 26 yd; a = 10 yd
   Ans: b = 24 yd
2. b = 25 mi; c = 35 mi
   Ans: a = 24.5 mi

SELF-CHECK
3. 3 m 4. 7.5 ft

CLASSROOM EXERCISES
1. 10 2. 11.3

CLASSROOM EXERCISES

<u>3.</u> 16   <u>4.</u> 28   <u>5.</u> 35

ASSIGNMENT GUIDE
BASIC
pp. 456-457: Omit
AVERAGE
pp. 456-457: 1-6
ABOVE AVERAGE
pp. 456-457: 1-6

WRITTEN EXERCISES
<u>1.</u> 9 m   <u>2.</u> 15 yd

For Exercises 3–5, the length of the hypotenuse, c, and the length of one leg of the right triangle are given. Find the unknown length. Refer to the Table of Squares and Square Roots on page 447. Round each answer to the nearest tenth. (Example 2)

3. $a = 30$, $c = 34$, $b = ?$

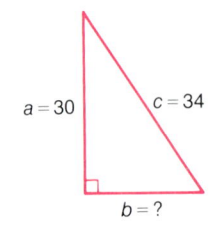

4. $a = ?$, $c = 53$, $b = 45$

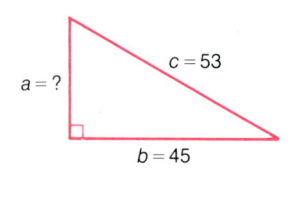

5. $c = 91$, $a = 84$, $b = ?$

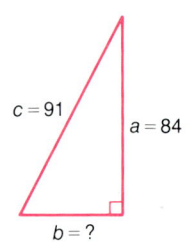

WRITTEN EXERCISES

Goals: To use the Pythagorean Theorem to find the measure of the hypotenuse of a right triangle

To use the Pythagorean Theorem to find the measure of one leg of a right triangle

To apply the skill of finding the measure of a side of a right triangle to solving word problems

In Exercises 1–6, use the Pythagorean Theorem to solve each problem. Use the Table of Squares and Square Roots on page 447. (Example 1)

1. In the figure below, a wire supports a pole that is 5 meters high. The distance from the foot of the pole to the point where the wire is fastened to a stake in the ground is 7 meters. How long is the wire? Round the answer to the nearest meter.

2. The figure below shows the escalator between two floors of a department store. How many yards are you carried as you travel on the escalator from one floor to the next? Round the answer to the nearest yard.

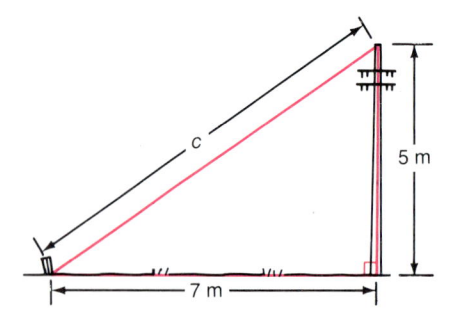

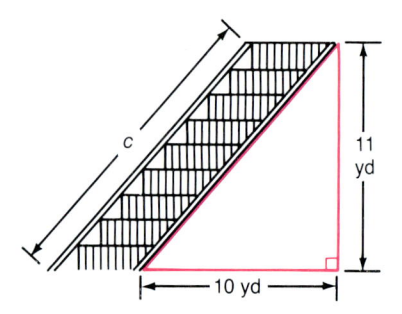

456 *Chapter 15*

3. A ladder is placed against the side of the house as shown in the figure at the right. Find the length of the ladder.

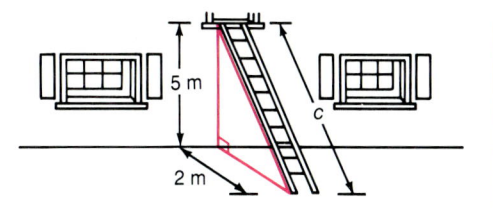

(Example 2)

4. A 18-meter wire staked to the ground supports a pole that is 15 meters high. How far from the pole is the stake?

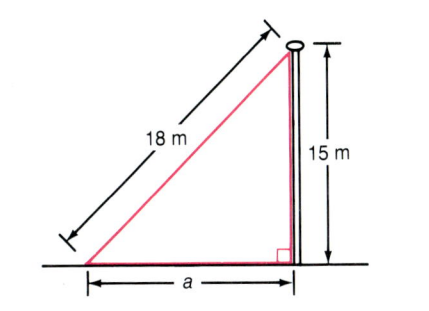

5. The front wall of an A-frame house is as shown in the figure at the right. How tall is the front of the house?

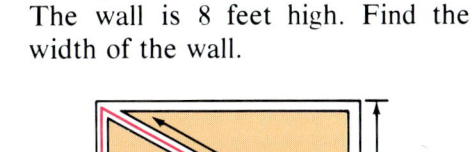

6. A diagonal wall brace is 14 feet long. The wall is 8 feet high. Find the width of the wall.

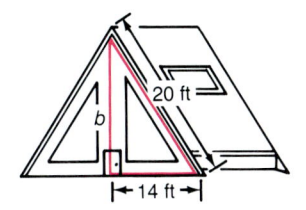

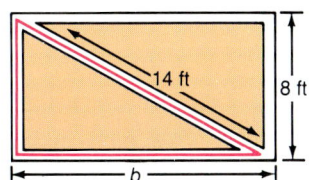

REVIEW CAPSULE FOR SECTION 15-6

Write a two-place decimal for each fraction. (Pages 173-175)

1. $\frac{1}{2}$ 2. $\frac{9}{10}$ 3. $\frac{5}{2}$ 4. $\frac{6}{5}$ 5. $\frac{9}{4}$ 6. $\frac{7}{20}$

Round to the nearest whole number. (Pages 32-34)

7. 12.1 8. 18.7 9. 37.32 10. 18.08 11. 100.49 12. 5.98

Solve and check. (Pages 203-205)

13. $\frac{n}{10} = 7$ 14. $\frac{x}{8} = 4$ 15. $15 = \frac{b}{100}$ 16. $9 = \frac{y}{9}$ 17. $\frac{c}{50} = 2$

18. $\frac{y}{25} = 5$ 19. $\frac{p}{10} = 1.5$ 20. $\frac{t}{7} = 6.8$ 21. $0.2 = \frac{y}{9}$ 22. $0.9 = \frac{x}{4}$

The answers are on page 464.

Special Triangles **457**

WRITTEN EXERCISES

3. 5.4 m   4. 9.9 m
5. 14.3 ft 6. 11.5 ft

WRITTEN EXERCISES p. 444

20. R; terminating
21. I; non-terminating; non-repeating
22. I; non-terminating; non-repeating
23. R; repeating
24. R; repeating
25. R; repeating
26. R; terminating
27. I; non-terminating; non-repeating
28. R; repeating
29. R; terminating
30. I; non-terminating; non-repeating
31. I; not a perfect square
32. R; perfect square
33. R; terminating
34. R; repeating
35. I; non-terminating; non-repeating
36. R; repeating
37. R; repeating
38. I; not a perfect square
39. I; not a perfect square
40. R; perfect square
41. R; terminating
42. I; non-terminating; non-repeating
43. b; c 44. d 45. c
46. c 47. d 48. d

Teaching Suggestions
p. M-28

QUICK QUIZ

Change to a decimal to the nearest hundredth.

1. $\frac{5}{12}$ Ans: 0.42
2. $\frac{8}{17}$ Ans: 0.47

Round to the nearest whole number.

3. 9.83   Ans: 10

Solve and check.

4. $\frac{y}{12} = 17$ Ans: 204
5. $\frac{a}{20} = 1.375$ Ans: 27.5

ADDITIONAL EXAMPLES

Example 1

Find tan A.

1. 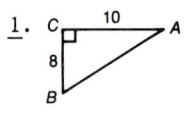 [triangle with C, 10, A on top; 8 on side; B at bottom]
   Ans: 0.8

2. 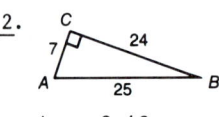 [triangle with C at top; 7, 24; A, 25, B]
   Ans: 3.43

SELF-CHECK

1. 1.5   2. 0.42

Example 2

Solve each to the nearest whole number.

1. $\tan 75° = \frac{a}{12}$ Ans: 45
2. $\tan 15° = \frac{b}{17}$ Ans: 5

458

15-6 Tangent Ratio and Applications

Another way to find the lengths of the legs of a right triangle is to use the **tangent ratio**.

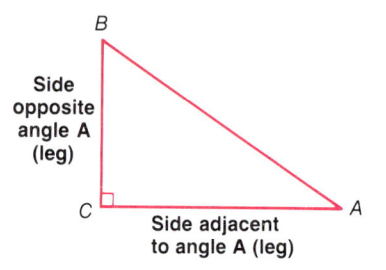

The abbreviation for the "tangent of angle A" is "tan A."

$$\tan A = \frac{\text{length of the side opposite angle } A}{\text{length of the side adjacent to angle } A}$$

EXAMPLE 1 Find tan A (nearest hundredth).

Solution: $\tan A = \frac{\text{length of side opposite angle } A}{\text{length of side adjacent to angle } A}$

$\tan A = \frac{3}{4}$

$\tan A = \mathbf{0.75}$

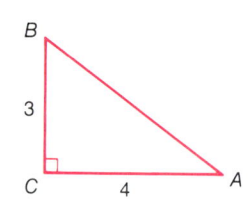

Self-Check Find tan A: 1. 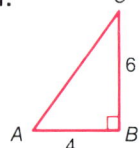 [triangle with C at top, 6 on side, A-4-B]   2. 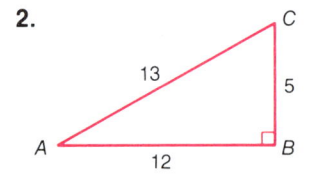 [triangle with C at top right, 13, 5, A-12-B]

When you know the measure of an acute angle of a right triangle such as angle A, you can use a table (see page 459) to find tan A.

EXAMPLE 2 Find the height of this flagpole. Round your answer to the nearest meter.

Solution: Use the tangent ratio.

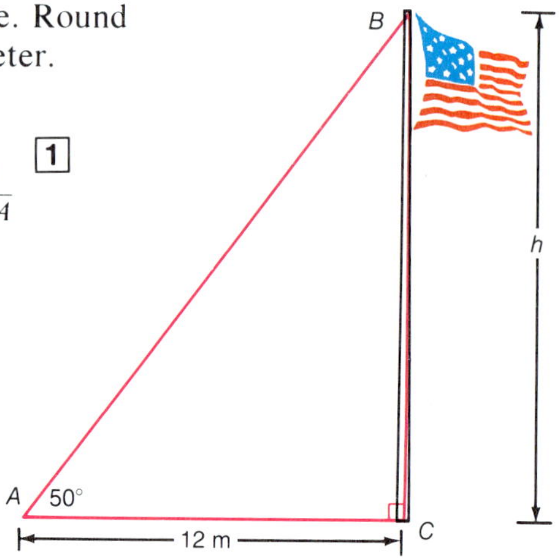

② $\tan A = \frac{\text{length of side opposite angle } A}{\text{length of side adjacent to angle } A}$

$\tan 50° = \frac{h}{12}$ ◀ Find tan 50° in the Tangent Table.

③ $1.19 = \frac{h}{12}$ ◀ Multiply each side by 12.

$1.19 \times 12 = \frac{h}{12} \times \frac{12}{1}$

$14.28 = h$ The flagpole is about **14 meters** high.

458   Chapter 15

Self-Check *Solve for x. Round your answer to the nearest whole number.*

3. $\tan 60° = \dfrac{x}{6}$ 4. $\tan 5° = \dfrac{x}{15}$ 5. $\tan 30° = \dfrac{x}{11}$

SELF-CHECK
3. 10   4. 1   5. 6

TANGENT TABLE

Angles	Tangents	Angles	Tangents
5°	0.087	50°	1.19
10°	0.176	55°	1.43
15°	0.268	60°	1.73
20°	0.364	65°	2.14
25°	0.466	70°	2.75
30°	0.577	75°	3.73
35°	0.700	80°	5.67
40°	0.839	85°	11.43
45°	1.00		

◀ $\tan 65° = 2.14$

◀ $\tan 80° = 5.67$

When you know the measure of angle *A* and the length of one leg of a right triangle, you can use the table and the definition of tan *A* to find the length of the other leg.

PROCEDURE

To use the tangent ratio to find a leg of a right triangle:

1. Draw and label a right triangle.
2. Use the known angle, the known leg, and the unknown leg to write the tangent ratio.
3. Solve for the length of the unknown leg.

CLASSROOM EXERCISES

In Exercises 1–5, use the Table of Tangents to find each of the following. (Table)

1. $\tan 10°$ 2. $\tan 85°$ 3. $\tan 25°$ 4. $\tan 75°$ 5. $\tan 45°$

In Exercises 6–9, use the Table of Tangents to solve each problem. Round each answer to the nearest whole number. (Example 2)

6. $\tan 10° = \dfrac{x}{20}$ 7. $\tan 35° = \dfrac{n}{9}$ 8. $\tan 20° = \dfrac{h}{15}$ 9. $\tan 85° = \dfrac{y}{32}$

CLASSROOM EXERCISES
1. 0.176   2. 11.43
3. 0.466   4. 3.73
5. 1.00

Special Triangles   **459**

CLASSROOM EXERCISES
6. 3.52   7. 6.30
8. 5.46   9. 365.76

ASSIGNMENT GUIDE
BASIC
p. 460: Omit
AVERAGE
p. 460: Omit
ABOVE AVERAGE
p. 460: 1-10

WRITTEN EXERCISES
1. 1.00   2. 3.33
3. 0.67   4. 0.80
5. 67.12 m   6. 201.36

WRITTEN EXERCISES

Goals: To use the tangent ratio to find the tangent of an acute angle of a right triangle

To use the tangent ratio and the tangent table to find the measure of a leg of a right triangle

Find tan A. Write a two-place decimal for your answer. (Example 1)

1. [Triangle with B, C (right angle), A; legs 4 and 4]
2. [Triangle with A, B, C (right angle); legs 3 and 10]
3. [Triangle with A, B, C (right angle); legs 3 and 2]
4. [Triangle with B, C (right angle), A; legs 4 and 5]

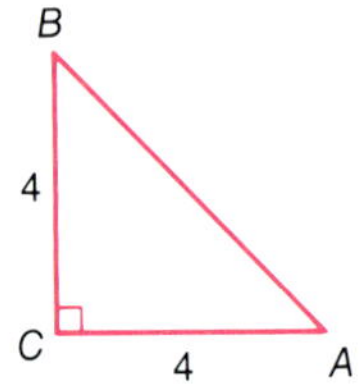

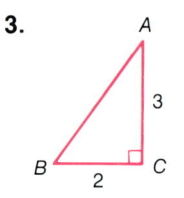

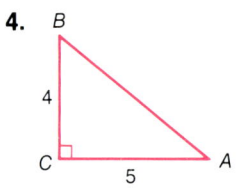

Solve. Use the tangent ratio and the Table of Tangents on page 459. Round each answer to the nearest whole number. (Example 2)

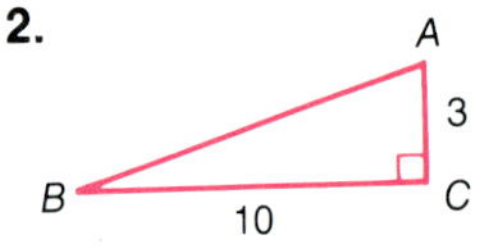

5. Find the distance, d.

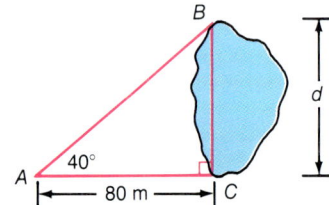

   [Figure: triangle with 40° angle at A, 80 m base to C, d vertical]

6. Find the width, w.

   [Figure: triangle with 40° angle at A, 240 m to C, w]

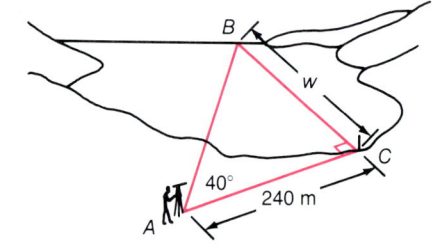

7. Find the height, h, of the plane above the ground.

   $\left(\text{HINT: } \tan 15° = \dfrac{h}{?}\right)$

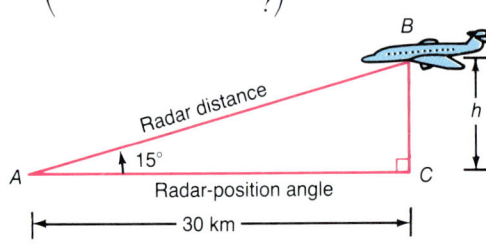

   [Figure: plane at B, 15° at A, radar distance AB, 30 km base AC, height h]

8. Find the height, h, of the television tower.

   $\left(\text{HINT: } \tan \underline{} = \dfrac{h}{40}\right)$

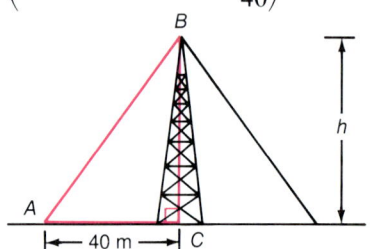

   [Figure: tower with B at top, A to C = 40 m, height h]

9. Find the height, h, of the mountain shown at the right.

   $\left(\text{HINT: } \tan 30° = \dfrac{h}{?}\right)$

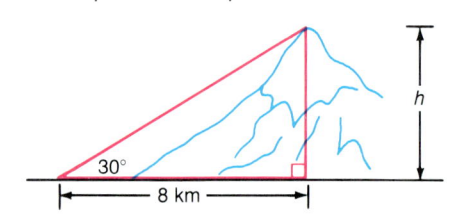

   [Figure: mountain with 30° angle, 8 km base, height h]

460   Chapter 15

TANGENT VALUES

You can use a scientific calculator instead of a table to find tangent values, given the angle measures.

EXAMPLE Find tan 42°.

Solution 4 2 [tan]   ⟶   0.900404

EXERCISES

Use a scientific calculator to find each of the following.

1. tan 3°   2. tan 16°   3. tan 27°   4. tan 46°   5. tan 88°
6. Use a scientific calculator to check your answers to Exercises 1-5 on page 459.

REVIEW: SECTIONS 15-4 — 15-6

Use the formula $A = s^2$ to find the length, s, of one side of a square with the given area, A. Round each answer to the nearest tenth. (Section 15-4)

1. $A = 12$ square meters; $s = \underline{\ ?\ }$
2. $A = 37$ square millimeters; $s = \underline{\ ?\ }$

Solve. (Section 15-5)

3. The base of a ramp is 12 meters long and 5 meters high. Find the length of the ramp. Use the Tables of Squares and Square Roots on page 447.

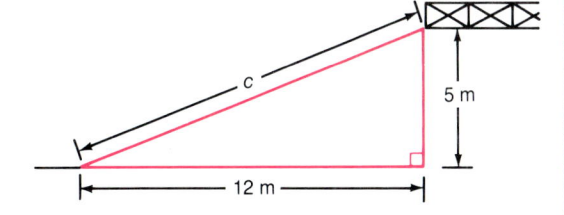

Solve. Use the Table of Tangents on page 459. Round each answer to the nearest whole number. (Section 15-6)

4. Find the height of the mountain shown below.
$$\left(\text{HINT: } \tan 25° = \frac{h}{?}\right)$$

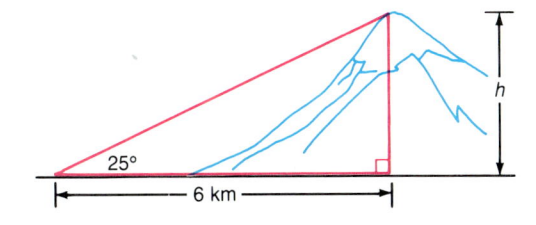

5. Find the height of the building shown below.
$$\left(\text{HINT: } \tan 30° = \frac{?}{42}\right)$$

WRITTEN EXERCISES
7. 4.82 mi   8. 57.2 m
9. 5 km

CALCULATOR EXERCISES
1. 0.0524078
2. 0.2867454
3. 0.5095254
4. 1.0355303
5. 28.636252

QUIZ: SECTIONS 15-4-15-6
After completing this Review, you may wish to administer a quiz covering the same sections. A Quiz is provided in the Teacher's Edition: Part II.

REVIEW: SECTIONS 15-4-15-6
1. 3.5 m   2. 6.1 mm
3. 13 m   4. 3 km
5. 18 m

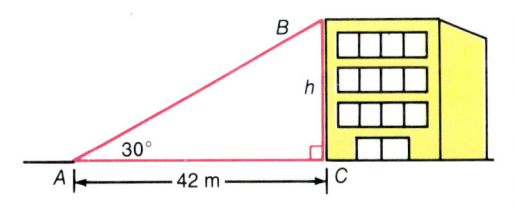

Special Triangles   **461**

CHAPTER REVIEW

1. Right triangle
2. Perfect square
3. Irrational
4. Square
5. Hypotenuse
6. Square root
7. 81   8. 225   9. 324
10. 576   11. 1764
12. 289   13. 4   14. 1
15. 9   16. 6   17. 10
18. 8   19. 11   20. 20
21. −2   22. −5
23. −12   24. −7
25. Rational
26. Rational
27. Irrational
28. Irrational
29. Rational
30. Irrational
31. Irrational
32. Rational
33. Rational
34. Irrational
35. Rational
36. Irrational
37. 324   38. 225
39. 441   40. 7056
41. 12,100   42. 4096
43. 8.660   44. 31
45. 10,677   46. 138
47. 9.747   48. 27
49. 3.5   50. 2.2
51. 5.2   52. 11.2
53. 2.8   54. 7.9

CHAPTER REVIEW

PART 1: VOCABULARY

For Exercises 1–6, choose from the box at the right the word(s) that best corresponds to each description.

1. A triangle with one right angle   ?
2. The square of a rational number   ?
3. The decimal equivalents of   ?   numbers are non-terminating and non-repeating.
4. The product of a number and itself is the   ?   of the number.
5. The longest side of a right triangle   ?
6. The inverse operation of squaring a number is taking its   ? .

> square
> irrational
> right triangle
> square root
> hypotenuse
> perfect square

PART 2: SKILLS

Square each of the following.   (Section 15-1)

7. 9   8. 15   9. 18   10. 24   11. 42   12. 17

Find each answer.   (Section 15-1)

13. $\sqrt{16}$ 14. $\sqrt{1}$ 15. $\sqrt{81}$ 16. $\sqrt{36}$ 17. $\sqrt{100}$ 18. $\sqrt{64}$
19. $\sqrt{121}$ 20. $\sqrt{400}$ 21. $-\sqrt{4}$ 22. $-\sqrt{25}$ 23. $-\sqrt{144}$ 24. $-\sqrt{49}$

Identify each number as rational or irrational.   (Section 15-2)

25. 4.5   26. $\sqrt{25}$ 27. $\sqrt{15}$ 28. 3.214···   29. $9.\overline{6}$ 30. $\sqrt{120}$
31. 8.562···   32. $3.\overline{13}$ 33. $\sqrt{9}$ 34. $\sqrt{90}$ 35. $4.5\overline{16}$ 36. $\sqrt{75}$

Refer to the table on page 447 to find each answer.   (Section 15-3)

37. 18^2 38. 15^2 39. 21^2 40. 84^2 41. 110^2 42. 64^2
43. $\sqrt{75}$ 44. $\sqrt{961}$ 45. $\sqrt{114}$ 46. $\sqrt{19{,}044}$ 47. $\sqrt{95}$ 48. $\sqrt{729}$

For Exercises 49–54, use $A = s^2$ to find s for the given values of A. Round each answer to the nearest tenth.   (Section 15-4)

49. $A = 12$ 50. $A = 5$ 51. $A = 27$ 52. $A = 125$ 53. $A = 8$ 54. $A = 63$

462   Chapter 15

For Exercises 55-59, use $D = 1.2\sqrt{A}$ to find D for the given values of A. Round each answer to the nearest whole number. (Section 15-4)

55. $A = 9604$ **56.** $A = 10{,}609$ **57.** $A = 6241$ **58.** $A = 18{,}496$ **59.** $A = 21{,}904$

For Exercises 60-63 find the unknown length for each right triangle. Round each answer to the nearest tenth. (Section 15-5)

60. $a = 12$, $b = 16$, $c = ?$

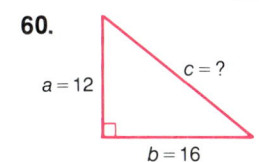

61. $a = 5$, $b = 10$, $c = ?$

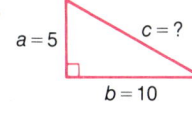

62. $a = ?$, $b = 25$, $c = 27$

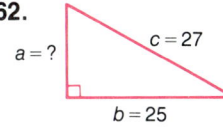

63. $a = 40$, $b = ?$, $c = 50$

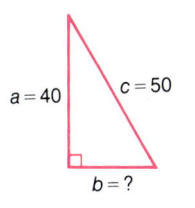

For Exercises 64-67, find tan A. Write a two-place decimal for each answer. (Section 15-6)

64. Triangle with B top, C right angle, A right; legs 4 and 4.

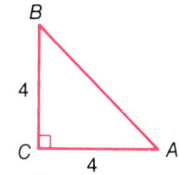

65. Triangle with A top, B right angle; legs 4 and 7.

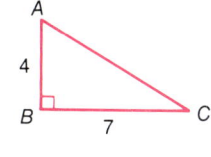

66. Triangle with B top, C right angle, A right; legs 4 and 8.

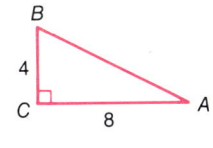

67. Triangle with B top, C right angle, A right; legs 7 and 9.

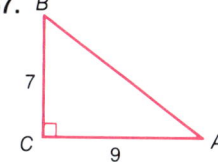

PART 3: APPLICATIONS

68. A ladder is placed against the side of a house as shown below. Find the length of the ladder. Use the table on page 447. Round the answer to the nearest tenth. (Section 15-5)

[Diagram: ladder with horizontal 2 m, vertical 7 m, hypotenuse c]

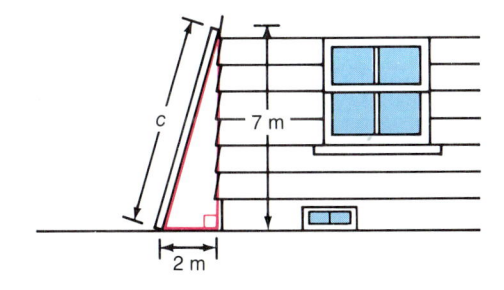

69. Find the height, h, of the telephone pole shown below. Use the Table of Tangents on page 459. Round the answer to the nearest whole number.

$\left(\text{HINT: } \tan 45° = \dfrac{h}{11}\right)$ (Section 15-6)

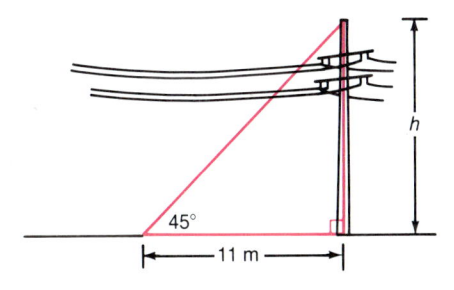

[Diagram: telephone pole, 45° angle, base 11 m, height h]

CHAPTER REVIEW

55. 117.6 **56.** 123.6
57. 94.8 **58.** 163.2
59. 177.6 **60.** 20
61. 11.2 **62.** 10.2
63. 30 **64.** 1.00
65. 1.75 **66.** 0.50
67. 0.78 **68.** 7.3 m
69. 11

Special Triangles 463

CHAPTER TEST

Two forms of a Chapter Test, Form A and From B, are provided on copying masters in the *Teacher's Edition: Part II*.

1. 25   2. 81   3. 7
4. 6   5. -12
6. Irrational
7. Rational
8. Rational
9. Irrational
10. Rational
11. 4096   12. 1369
13. 9.747 14. 10.100
15. 66   16. 15   17. 30
18. 12   19. 6.3 m
20. 43 m

A cumulative test covering the content of Chapters 1-15 is provided on copying masters in the *Teacher's Edition: Part II*. This cumulative test is primarily designed as a final examination for the Average course of study.

CHAPTER TEST

Find each answer.

1. 5^2
2. 9^2
3. $\sqrt{49}$
4. $\sqrt{36}$
5. $-\sqrt{144}$

Identify each number as rational or irrational.

6. $\sqrt{24}$
7. $\sqrt{81}$
8. $4.2\overline{6}$
9. $1.4132\cdots$
10. 0.18

Use the Table of Squares and Square Roots on page 447 to find each answer.

11. 64^2
12. 37^2
13. $\sqrt{95}$
14. $\sqrt{102}$
15. $\sqrt{4356}$

For Exercises 16-18, find the unknown length of each right triangle with hypotenuse c and legs a and b. Refer to the Table of Squares and Square Roots on page 447. Round each answer to the nearest tenth.

16. $a=9; b=12; c=\underline{}$
17. $b=16; c=34; a=\underline{}$
18. $a=5; c=13; b=\underline{}$

19. A ladder is placed against a shed as shown below. Find the length of the ladder. Use the Table of Squares and Square Roots on page 447. Round the answer to the nearest tenth.

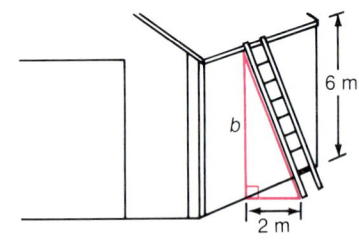

20. Find the height, h, of the smockstack shown below. Use the Table of Tangents on page 459. Round the answer to the nearest whole number.

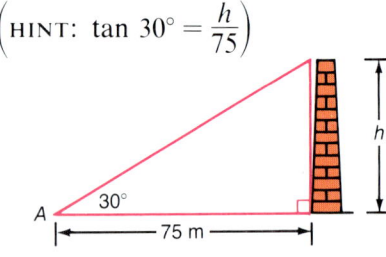

    $\left(\text{HINT: } \tan 30° = \dfrac{h}{75}\right)$

ANSWERS TO REVIEW CAPSULES

Page 442   1. 0.5 2. 0.67 3. 0.6 4. 0.25 5. 1.17 6. 2.25 7. 0.2 8. 0.75 9. $0.\overline{3}$ 10. $0.8\overline{3}$ 11. 0.15 12. 0.625 13. 0.875 14. 0.45 15. $1.\overline{1}$ 16. 3.8 17. 1.75 18. $3.\overline{6}$

Page 444   1. 36 2. 225 3. 2 4. 102 5. 1764 6. 5 7. 56 8. 3364 9. 13 10. 88 11. 8100 12. 13.924

Page 450   1. 43.2 2. 16.7 3. 27.8 4. 14.0 5. 82.9 6. 93.1 7. 12.5 8. 39.8 9. 55.5 10. 68.3 11. 10.9 12. 50.1 13. 4.249 14. 19.1562 15. 7.8125 16. 3.9936 17. 38 18. 28 19. 42 20. 17 21. 73 22. 25 23. 6 24. 18 25. 9 26. 2 27. 8 28. 124

Page 453   1. 9 2. 64 3. 100 4. 400 5. 1296 6. 361 7. 4 8. 13 9. 9 10. 13 11. 25 12. 45 13. 17 14. 18 15. 12.4 16. 21.7 17. 18.1 18. 32.2 19. 49.5 20. 62.4

Page 457   1. 0.50 2. 0.90 3. 2.50 4. 1.20 5. 2.25 6. 0.35 7. 12 8. 19 9. 37 10. 18 11. 100 12. 6 13. 70 14. 32 15. 1500 16. 81 17. 100 18. 125 19. 15 20. 47.6 21. 1.8 22. 3.6

464   Chapter 15

ENRICHMENT

Arithmetic Method for Finding $\sqrt{N}$

The flowchart at the left below outlines a procedure for approximating the square root of an irrational number N to a given number of decimal places. The Example at the right shows how to follow these steps to determine $\sqrt{38}$ to the nearest tenth.

Flowchart

1. DETERMINE TWO CONSECUTIVE INTEGERS A AND B SUCH THAT $A < \sqrt{N} < B$. THEN FIND THE AVERAGE OF A AND B.
2. DIVIDE N BY THE APPROXIMATION.
3. FIND THE AVERAGE OF THE QUOTIENT AND DIVISOR.
4. DO YOU WANT A MORE ACCURATE APPROXIMATION? — Yes → (return to step 2); No ↓
5. $\sqrt{N} \approx ?$

Example

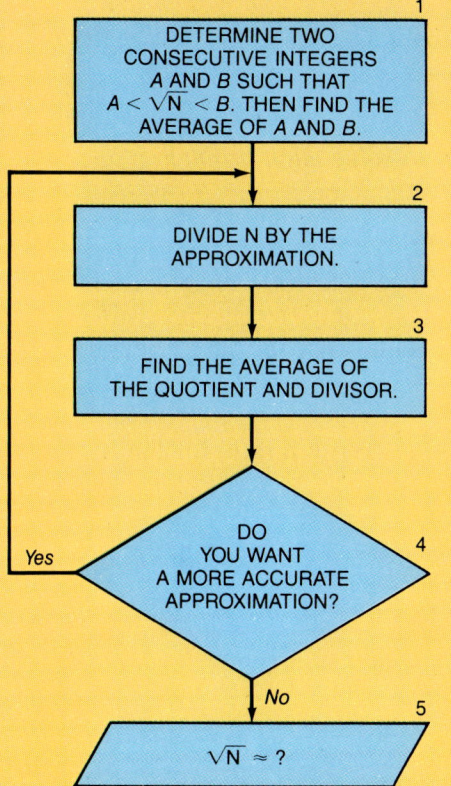

[1] $6 < \sqrt{38} < 7$ ◀ Find the average of 6 and 7.

First approximation: $\dfrac{6 + 7}{2} = 6.5$

[2] Divide N (38) by the approximation.

$6.5 \overline{)38.00}$ quotient 5.8 ◀ First division: Divide to one decimal place.

[3] Second approximation:

$\dfrac{5.8 + 6.5}{2} = 6.15$ ◀ Average of the quotient and divisor

[4] One loop is complete. To approximate $\sqrt{38}$ to the nearest tenth, you divide to two decimal places. Return to Step 2.

[2] $6.15 \overline{)38.0000}$ quotient 6.17 ◀ Second division: Divide to two decimal places.

[3] Third approximation:

$\dfrac{6.17 + 6.15}{2} = 6.16$ ◀ Average of the quotient and divisor

[4] Two loops are complete. Round 6.16 to the nearest tenth.

[5] $\sqrt{N} \approx 6.2$

EXERCISES

Follow the steps of the flowchart to approximate the square root of each number.

1. 6 (nearest tenth)
2. 8 (nearest tenth)
3. 15 (nearest tenth)
4. 5 (nearest hundredth)
5. 17 (nearest hundredth)
6. 60 (nearest hundredth)

ENRICHMENT
You may wish to use this lesson for students who performed well on the formal Chapter Test.

1. 2.4   2. 2.8   3. 3.8
4. 2.24   5. 4.12
6. 7.74

Special Triangles

ADDITIONAL PRACTICE

You may wish to use all or some of these exercises, depending on how well students performed on the formal Chapter Test.

1. 49   2. 121   3. 169
4. 5   5. 9   6. 11
7. Rational
8. Irrational
9. Rational
10. Rational
11. Rational
12. Irrational
13. 676   14. 1296
15. 2209 16. 71
17. 27   18. 91
19. 13; 0.42
20. 5.4; 0.40
21. 4; 0.75
22. 0.42; 0.4; 0.2
23. 40   24. 11   25. 93
26. 5   27. 23 m
28. 156 ft

ADDITIONAL PRACTICE

SKILLS

Square each number. (Pages 440–441)   *Find each answer.* (Pages 440–441)

1. 7   2. 11   3. 13   4. $\sqrt{25}$ 5. $\sqrt{81}$ 6. $\sqrt{121}$

Identify each number as rational or irrational. (Pages 442–444)

7. $\sqrt{64}$ 8. $\sqrt{3}$ 9. 4.12   10. $4.\overline{25}$ 11. $4.\overline{6}$ 12. 3.24135 . . .

Refer to the table on page 447 to find each answer. (Pages 445–446)

13. 26^2 14. 36^2 15. 47^2 16. $\sqrt{5041}$ 17. $\sqrt{729}$ 18. $\sqrt{8281}$

Use the Pythagorean Theorem to solve each problem. (Pages 454–457)

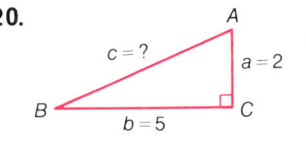

19. triangle with $c = ?$, $a = 5$, $b = 12$, right angle at B
20. triangle with $c = ?$, $a = 2$, $b = 5$, right angle at C
21. triangle with $c = 5$, $a = 3$, $b = ?$, right angle at B

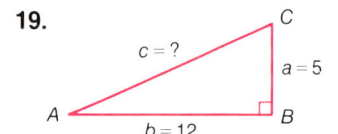

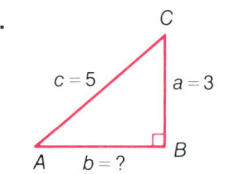

22. Find tan A for each triangle of Exercises 19–21. Write a ratio for tan A. (Pages 458–460)

Use the Table of Tangents on page 459 to find x.

23. $\tan 45° = \dfrac{x}{40}$ 24. $\tan 35° = \dfrac{x}{16}$ 25. $\tan 75° = \dfrac{x}{25}$ 26. $\tan 15° = \dfrac{x}{20}$

APPLICATIONS

27. The Ty family plans to build a square garage. The area of the floor will be 529 square meters. Find the length of one side of the garage. (Pages 451–453)

28. Find the height, h, of this radio tower. $\left(\text{HINT: } \tan \underline{\ ?\ } = \dfrac{h}{90}\right)$ (Pages 458–460)

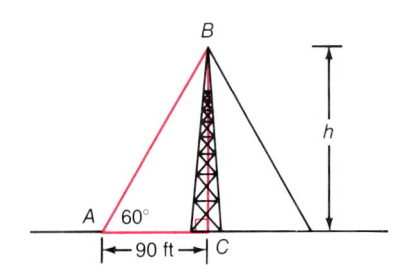

Radio tower diagram: angle $A = 60°$, horizontal distance from A to C is 90 ft, height h from C to B.

466   Chapter 15

COMMON ERRORS

Each of these problems contains a common error.

a. Find the correct answer.
b. Find the error.

Solve.

1. To graph the ordered pair $P(-3, -6)$, start at the origin.
   Move 3 units to the *right*.
   Then move 6 units *down*.

2. Complete the table of ordered pairs for the equation $y = 2x - 1$.

x	$2x - 1$	y
2	$2(2) - 1 = 3$	3
0	$2(0) - 1 = -1$	-1
-1	$2(-1) - 1 = 1$	1

3. Find the slope of the line containing the points $(3, 5)$ and $(2, 1)$.
   Vertical change: $3 - 2 = 1$
   Horizontal change: $5 - 1 = 4$
   Slope: $\frac{1}{4}$

4. Find the y intercept of $y = 3x + 6$.
   $y = 3x + 6$
   $0 = 3x + 6$
   $-6 = 3x$
   $-2 = x$ ◄ y intercept: -2

5. The area of a square is 10 meters. Find the length of one side.
   $A = s^2$
   $10 = s^2$
   $100 = s$
   The length of one side is **100 m**.

6. Use the Pythagorean Theorem to find the length of the hypotenuse of this right triangle.
   $a^2 + b^2 = c^2$
   $9^2 + 6^2 = c^2$
   $81 + 36 = c^2$
   $117 = c^2$
   $\sqrt{117} = c$
   $10.817 = c$ ◄ The unknown length

   (Triangle with legs 9 and 6, right angle at C, vertices A, B, C)

7. Find tan A. Refer to the figure at the right.
   $\tan A = \frac{\text{length of side opposite angle } A}{\text{length of side adjacent to angle } A}$
   $= \frac{12}{9} = \frac{4}{3} = \textbf{1.75}$

   (Right triangle with $a = 9$, $b = 12$, $c = ?$, right angle at C)

COMMON ERRORS

In preparation for the Cumulative Review, these exercises focus the student's attention on the most common errors to be avoided.

1. Start at the origin. Move 3 units to the left. Then move 6 units down.

2.
x	y
2	3
0	-1
-1	-3

3. 4   4. 6   5. 3.16
6. 10.817   7. 0.75

Special Triangles 467

CUMULATIVE REVIEW:
CHAPTERS 14-15

A cumulative test is provided on copying masters in the Teacher's Edition: Part II.

1. c   2. d   3. b
4. a   5. a   6. b
7. d   8. d   9. d
10. b

CUMULATIVE REVIEW: CHAPTERS 14-15

Choose the correct answer.
Choose a, b, c, or d.

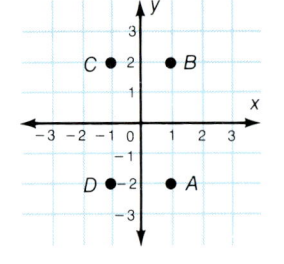

1. Name the point on the graph at the right that corresponds to the ordered pair $(-1, 2)$.
   a. A
   b. B
   c. C
   d. D

2. Find the slope of the line containing the points $A(3, 4)$ and $B(-1, 2)$.
   a. 1
   b. 2
   c. $\frac{1}{3}$
   d. $\frac{1}{2}$

3. Find the y-intercept of the linear equation $y = 4x + 16$.
   a. $y = 0$
   b. $y = 16$
   c. $y = 4$
   d. $y = -4$

4. Which ordered pair is a solution of this system? $\begin{cases} y = 2x + 6 \\ y = 3x + 1 \end{cases}$
   a. $(5, 16)$
   b. $(-3, 0)$
   c. $(3, 12)$
   d. $(2, 7)$

5. Which ordered pair is a solution to the inequality $y > -2x + 1$?
   a. $(4, -6)$
   b. $(-3, 6)$
   c. $(-100, 194)$
   d. $(0, 1)$

6. Which number is a perfect square?
   a. 24
   b. 81
   c. 18
   d. 28

7. Which of the following numbers is irrational?
   a. 3.1
   b. $3.\overline{3}$
   c. $\sqrt{25}$
   d. $\sqrt{17}$

8. Find the length in meters of a side of the square that has an area of 36 square meters.
   a. 9
   b. 1296
   c. 24
   d. 6

9. Use the Pythagorean Theorem to find the length in feet of the ladder shown below.
   a. 14
   b. 9
   c. 13
   d. 10

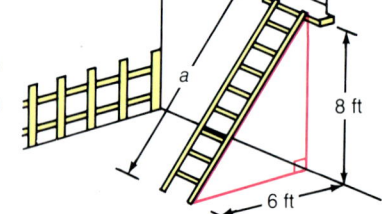

10. Use the tangent ratio to find the height in meters of the wall.
    (HINT: tan 15° = 0.27)
    a. 2962
    b. 216
    c. 21.6
    d. 292.2

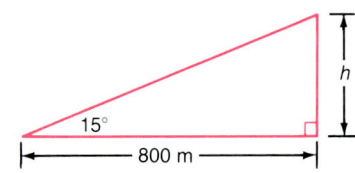

468 Cumulative Review: Chapters 14-15

CHAPTER 16 Statistics and Probability

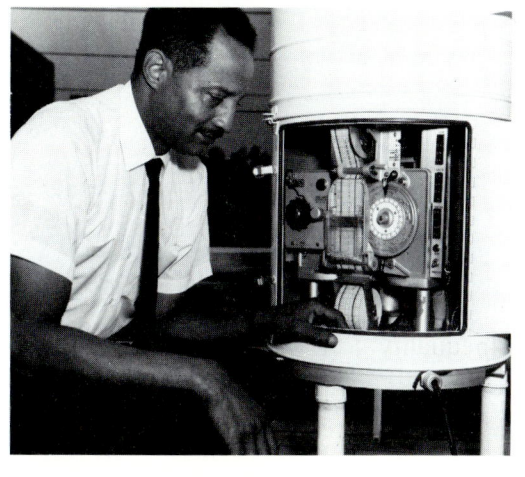

SECTIONS
- **16-1** Mean, Median, and Mode
- **16-2** Histograms
- **16-3** Probability
- **16-4** Probability and Tables
- **16-5** Tree Diagrams
- **16-6** Multiplying Probabilities

FEATURES
Calculator Application: *Multiplying Probabilities*
Career Application: *Statistical Clerk*
Enrichment: *Sampling*

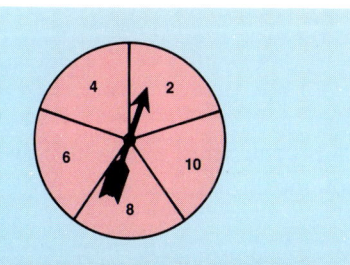

Teaching Suggestions
p. M-29

QUICK QUIZ

Find the average.

1. 3; 12; 8; 9; 23
   Ans: 11
2. 120; 175; 200; 145
   Ans: 160
3. $3\frac{1}{2}$; 7; $8\frac{1}{2}$; 6; 5
   Ans: 6
4. 1.7; 1.3; 1.5; 1.5; 1.6 Ans: 1.52
5. 3; 7; 7; 5; 4; 6; 7
   Ans: $5\frac{4}{7}$

ADDITIONAL EXAMPLE

Example 1
Find the mean.
Temperatures in °F

| 17 | 24 | 32 |
| 36 | 33 | 14 |

Ans: 26

SELF-CHECK
1. $21

16-1 Mean, Median, and Mode

George Blackwater records the amount of rainfall throughout the year for the Weather Service. He uses the data to keep records and spot trends. Every month, George calculates the *mean* rainfall.

PROCEDURE

To find the mean, divide the sum of the given measures by the number of measures.

$$\text{Mean} = \frac{\text{Sum of Measures}}{\text{Number of Measures}}$$

Mean is another word for average.

EXAMPLE 1 This table shows the rainfall for Louisville, Kentucky from May through September of a certain year. Find the mean rainfall for this 5-month period.

| Rainfall for Louisville, KY ||
Month	Rainfall (in)
May	4.2
June	4.1
July	3.8
August	3.0
September	2.9

Solution:

$$\text{Mean} = \frac{4.2 + 4.1 + 3.8 + 3.0 + 2.9}{5}$$

Mean = 3.6   The mean rainfall is **3.6 inches**.

Self-Check 1. Find the mean amount of savings.

Month	Jan.	Feb.	Mar.	Apr.	May	June	July	Aug.	Sept.
Amount	$20	$25	$17	$25	$21	$25	$17	$20	$19

In a listing of data, the **median** is the middle measure or score. When there are an even number of measures, the median is the mean of the two middle measures.

The **mode** is the measure that occurs most often. There may be more than one mode.

470 Chapter 16

> **PROCEDURE**
>
> **To find the median of a list of data:**
>
> 1. Arrange the data in order.
> 2. a. For an odd number of items, the median is the middle number listed.
>   b. For an even number of items, the median is the mean of the two middle numbers.

EXAMPLE 2 The table at the right shows the number of fiction books checked out of a library for 6 days.

a. Find the median number of books checked out.

b. Find the mode.

Number of Fiction Books Checked Out During Six Days

Day	Number
Monday	30
Tuesday	45
Wednesday	19
Thursday	30
Friday	36
Saturday	50

Solutions: a. 1 Arrange the numbers in order.

19   30   **30**
36   45   50

◀ There are two middle numbers.

2 Find the mean of 30 and 36.

$$\frac{30 + 36}{2} = \frac{66}{2} = 33$$

The median number of fiction books checked out is **33**.

b. Since the number of fiction books checked out most often is 30, the mode is **30**.

Self-Check 2. Find the median and mode of the prices for the television sets.

Brand	A	B	C	D	E	F
Price	$275	$672	$195	$450	$857	$672

CLASSROOM EXERCISES

For Exercises 1–2, find the mean. (Example 1)

1. **Passengers on Eight Buses**

   42   45   41   38
   40   42   32   40

2. **Prices of Six Trucks**

   $8439 $7680 $6755
   $7521 $6350 $5945

Statistics and Probability **471**

CLASSROOM EXERCISES
3. 85.5; 85   4. 16; 16

ASSIGNMENT GUIDE
BASIC
pp. 472-473: Omit
AVERAGE
pp. 472-473: Omit
ABOVE AVERAGE
pp. 472-473: 1-13

WRITTEN EXERCISES
1. 15,813   2. 284,527
3. $6.15   4. 7
5. 26, 26
6. $12,500; $12,500

For Exercises 3–4, find the median and mode. (Example 2)

3. **Math Scores for Ten Tests**

| 85 | 86 | 89 | 89 | 70 |
| 74 | 88 | 85 | 100 | 85 |

4. **Ages of Members of Golf Team**

| 16 | 15 | 15 | 17 | 16 |
| 17 | 14 | 16 | 18 | 14 |

WRITTEN EXERCISES

Goal: To find the mean, median, and mode of a list of data

For Exercises 1–4, find the mean. (Example 1)

1. **Heights in Feet of Six Mountains in the United States**

Mountain	Height
Blackburn	16,390
Churchill	15,638
McKinley	20,320
Pike's Peak	14,110
Rainier	14,410
San Luis	14,010

2. **Number of Telephones in Five Cities**

City	Number
Albany, NY	170,643
Cambridge, MA	111,498
Lexington, KY	163,114
Lubbock, TX	161,860
San Francisco, CA	815,520

3. **Earnings Per Hour for Eight Workers**

| $6.50 | $5.75 | $5.25 | $4.95 |
| $6.25 | $5.00 | $8.15 | $7.35 |

4. **Number of Hours of Sleep for Ten Adults**

| 8 | 6 | 6 | 8 | 7 |
| 7 | 7 | 8 | 5 | 8 |

For Exercises 5–8, find the median and mode. (Example 2)

5. **Ages of Twenty People at a Party**

22	31	27	26	19
23	26	28	26	21
24	27	31	27	29
25	22	28	30	26

6. **Yearly Salaries for Five Employees**

Manager	$40,000
Assistant	$25,000
Clerk	$12,500
Clerk	$12,500
Clerk	$12,500

472 Chapter 16

7. Rainfall for Omaha, Nebraska

Month	Rainfall (in)
September	3.3
October	1.9
November	1.1
December	0.8
January	0.8

8. Enrollment Per Grade

Grade	Enrollment
7	189
8	198
9	205
10	215
11	198
12	200

MORE CHALLENGING EXERCISES

9. Before ordering dresses for the summer, the owner of a dress shop examined the records of the number and sizes of dresses sold the previous summer. Would the mean, median, or mode be the most useful guide in determining the sizes to order this summer? Give a reason for your answer.

Mary and Bill own a small business. Of the twelve employees, one earns $16,500, one earns $14,000, two earn $9500 each, five earn $8100 each, and three earn $6000 each. Mary and Bill pay themselves $20,000 each.

10. Find the mean of the fourteen salaries.

11. Find the median and the mode of the fourteen salaries.

12. Which of these measures, the mean, median, or mode, gives the highest "average" salary?

13. Which of these measures, the mean, median, or mode, gives the lowest "average" salary?

REVIEW CAPSULE FOR SECTION 16-2

1. Construct a vertical bar graph for the following data. (Pages 270–273)

Personal Income for the First Six Months of a Certain Year

Month	January	February	March	April	May	June
Trillions of Dollars	2045	2055	2070	2070	2060	2080

The answers are on page 494.

Statistics and Probability 473

WRITTEN EXERCISES

7. 1.1; 0.8
8. 199; 198
9. mode; shows most frequent sales
10. $10,571.43
11. $8100; $8100
12. mean
13. median and mode

REVIEW CAPSULE
This Review Capsule reviews prior-taught skills used in Section 16-2. The reference is to the pages where the skills were taught.

Teaching Suggestions
p. M-29

QUICK QUIZ

Construct a horizontal bar graph for the following data.

Media Advertising Costs

Media	Millions
Newspapers:	$15,541
Magazines:	$ 3,149
Television:	$11,387
Radio:	$ 3,827
Mail:	$ 7,596

Ans:

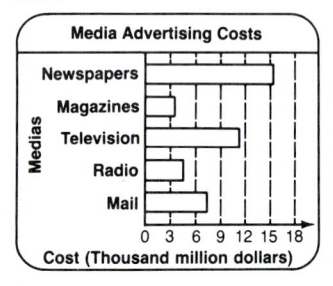

Media Advertising Costs
(horizontal bar graph with Newspapers, Magazines, Television, Radio, Mail on y-axis; Cost (Thousand million dollars) 0 3 6 9 12 15 18 on x-axis)

ADDITIONAL EXAMPLE

Example 1

Show the following data in a table. Use the given intervals.

Hourly Temperatures (°F)

11 11 13 14 17 18 18 18
20 23 24 25 27 27 30 26
24 23 21 20 16 14 14 13

Intervals: 10-15; 16-20; 21-25; 26-30

Ans:

Intervals	Tally	Count
10-15	|||| ||	7
16-20	|||| ||	7
21-25	||||		6
26-30	||||	4

See page 475 for answer to *SELF-CHECK*.

474

16-2 Histograms

Carol Davis runs the 100-yard dash for the school track team. At each meet, the coach records her time in seconds and lists the data in a table. Grouping the data in intervals makes the table shorter and easier to read.

PROCEDURE

To show grouped data in a table:
1. List the intervals.
2. Tally the number of measures in each interval. Count the measures.

EXAMPLE 1 Carol's times in seconds for the 100-yard dash are given at the right. List the data in a table.

Solution:

Intervals	Tally	Count
11.2–11.4	||||		6
11.5–11.7	||||	5
11.8–12.0	||||	4

Times in Seconds for the 100–yard Dash

11.6	11.2	11.5	12.0	12.0
11.2	11.5	11.6	11.8	11.4
12.0	11.4	11.2	11.6	11.4

Self-Check

1. Show the following data in a table. Use the given intervals.

Tractor Output Per Day for 24 Factories

31 40 35 67 85 90 32 39 68 87 55 50
33 43 64 78 82 70 54 60 41 85 32 68

Intervals: 31-40; 41-50; 51-60; 61-70; 71-80; 81-90

A **histogram** is a special kind of bar graph. The width of the bars of a histogram represent the size of the intervals.

PROCEDURE

To draw a histogram:
1. List the data in equal intervals in a table.
2. Draw a rectangle to represent the count for each interval.

474 Chapter 16

EXAMPLE 2 Pulse rates for 60 people are listed below. Draw a histogram to show the data.

Pulse Rates of 60 People

68	70	82	72	74	81
77	73	78	73	77	72
72	75	74	73	76	80
77	74	73	72	84	73
75	79	79	81	69	77
79	69	70	71	84	85
69	74	75	79	75	70
69	70	76	76	75	73
79	81	74	72	76	78
79	71	80	77	79	80

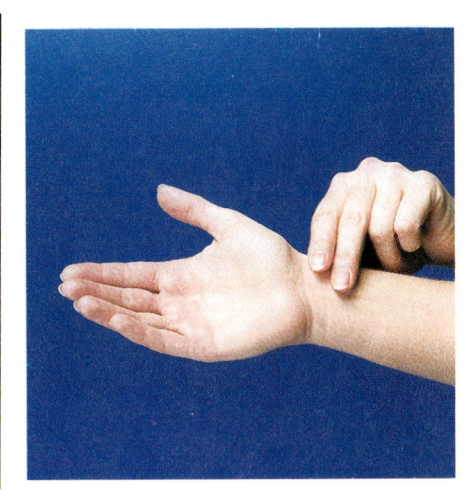

Solution:

1 List the data by intervals.

Pulse Rates	Count
65–69	5
70–74	22
75–79	23
80–84	9
85–89	1

2 Draw a rectangle for each interval.

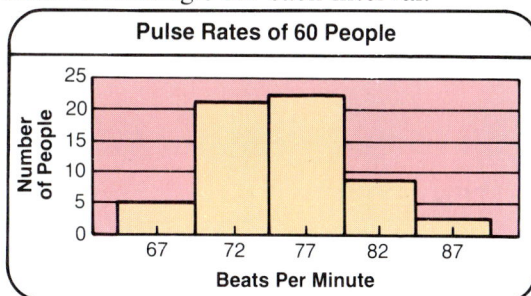

Pulse Rates of 60 People — histogram with Number of People (y-axis, 0–25) vs. Beats Per Minute (x-axis: 67, 72, 77, 82, 87).

Self-Check

2. Draw a histogram to show the data. Use the given intervals.

Cars	A	B	C	D	E	F	G	H	I	J	K	L	M	N
Gas Mileage (mi/gal)	28	32	25	33	30	24	35	26	31	34	27	38	36	37

Intervals: 21-25; 26-30; 31-35; 36-40

NOTE: In a histogram, the middle number of each interval is at the center of the base of the corresponding bar on the graph.

Statistics and Probability **475**

SELF-CHECK

1.

Intervals	Tally	Count
31–40	below	6
41–50	||||	4
51–60	|||	3
61–70	below	5
71–80			1
81–90	below	5

ADDITIONAL EXAMPLE

Example 2

Draw a histogram to show data given in Additional Example 1.

Ans:

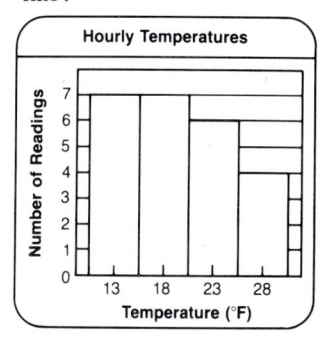

Hourly Temperatures — histogram with Number of Readings (y-axis, 0–7) vs. Temperature (°F) (x-axis: 13, 18, 23, 28).

SELF-CHECK

2.

Gas Mileage of 24 Cars — histogram with Cars (y-axis, 0–6) vs. Miles per gallon (x-axis: 23, 28, 33, 38).

475

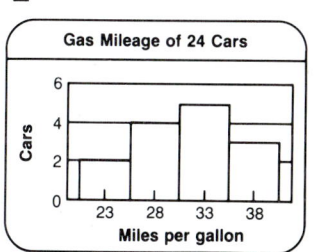

CLASSROOM EXERCISES
1. 3; 3; 6; 2; 2
2. 3; 4; 3; 6; 4

ASSIGNMENT GUIDE
BASIC
pp. 476-477: Omit
AVERAGE
pp. 476-477: Omit
ABOVE AVERAGE
pp. 476-477: 1-6

WRITTEN EXERCISES

1. | Heights | Count |
|---|---|
| 120-219 | 5 |
| 220-319 | 4 |
| 320-419 | 4 |
| 420-519 | 1 |
| 520-519 | 4 |
| 620-719 | 2 |

2. | Speed | Count |
|---|---|
| 170-174 | 2 |
| 175-179 | 3 |
| 180-184 | 5 |
| 185-189 | 7 |
| 190-194 | 3 |

3.

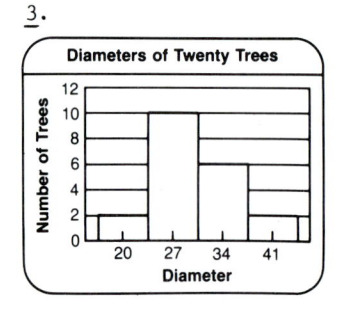

Diameters of Twenty Trees histogram — Number of Trees vs Diameter (20, 27, 34, 41)

CLASSROOM EXERCISES

Use the given information to complete each table. (Example 1)

1. **Capacity of 16 College Football Stadiums**

59,000	52,000	32,000	30,000
66,000	35,000	72,000	50,000
50,000	66,000	42,000	58,000
70,000	46,000	47,000	52,000

Capacity	Tally	Count
30,000–39,000	?	?
40,000–49,000	?	?
50,000–59,000	?	?
60,000–69,000	?	?
70,000–79,000	?	?

(Example 2)

2. **Average Class Size in Twenty Schools**

19	32	28	38	24
25	39	20	35	37
29	26	25	33	38
32	34	22	18	31

Class Size	Count
15–20	?
21–25	?
26–30	?
31–35	?
36–40	?

WRITTEN EXERCISES

Goals: To list data by using equal intervals in a table

To use a histogram to show data

For Exercises 1–2, use the given information to make a table. The intervals are also given. (Example 1)

1. **Heights in Feet of Twenty Major U.S. Dams**

121	340	250	325	519
585	121	250	160	210
550	564	402	637	625
245	382	610	154	239

Intervals: 120–219; 220–319; 320–419; 420–519; 520–619; 620–719

2. **Qualifying Speeds (mph) for a Stock Car Race**

175	186	183	171	182
190	184	187	188	178
191	173	182	186	177
185	193	186	182	189

Intervals: 170–174; 175–179; 180–184; 185–189; 190–194

476 Chapter 16

For Exercises 3–6, make a histogram to show the given data. Use the given intervals. (Example 2)

3. Diameters of Twenty Trees (cm)

25	32	36	28	20
26	35	40	27	30
36	23	29	31	24
29	32	43	25	30

Intervals: 17–23; 24–30; 31–37; 38–44

4. Bowling Scores for One Week

162	125	146	150
170	190	155	175
135	144	170	165
155	185	162	184

Intervals: 121–140; 141–160; 161–180; 181–200

5. Miles Bicycled Per Day for Triathlon Training

25	30	22	32	37
34	15	25	39	24
32	18	30	35	32
20	25	31	28	30

Intervals: 15–19; 20–24; 25–29; 30–34; 35–39

6. Hours Per Week of Evening Television Viewing by Teens

4	6	9	11	8
1	12	9	4	11
15	2	9	5	13
4	9	18	4	9

Intervals: 0–4; 5–9; 10–14; 15–19

REVIEW: SECTIONS 16-1–16-2

For Exercises 1–4, find the mean, median, and mode. (Section 16-1)

1. Science Test Scores (%)

| 92 | 85 | 94 | 81 |
| 83 | 92 | 89 | 92 |

2. Weight of a Wrestling Team (lbs)

| 160 | 177 | 122 | 151 |
| 143 | 156 | 160 | 147 |

3. Denver's High Temperatures for a Week

| 70 | 73 | 72 | 68 | 72 | 70 | 72 |

4. Ages of Golf Team Members

| 19 | 22 | 21 | 19 | 19 | 21 |
| 18 | 24 | 23 | 26 | 19 |

Make a histogram to show the given data. Use the given intervals. (Section 16-2)

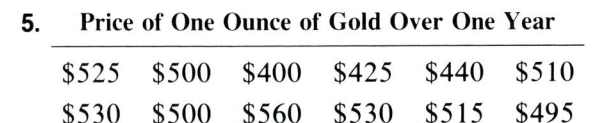

5. Price of One Ounce of Gold Over One Year

| $525 | $500 | $400 | $425 | $440 | $510 |
| $530 | $500 | $560 | $530 | $515 | $495 |

Intervals: $400–$439; $440–$479; $480–$519; $520–$559; $560–$599

Statistics and Probability **477**

WRITTEN EXERCISES

4. Bowling Scores for One Week (histogram)

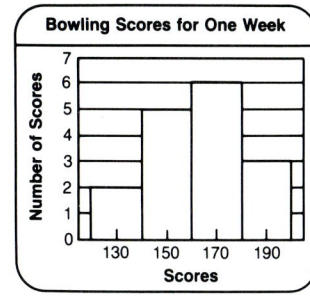

5. Miles Bicycled Per Day (histogram)

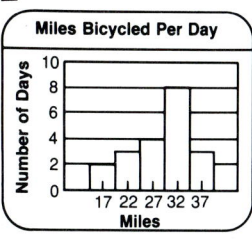

6. Television Viewing by Teens (histogram)

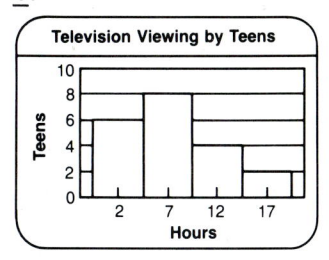

QUIZ: SECTIONS 16-1–16-2
After completing this Review, you may wish to administer a quiz covering the same sections. A Quiz is provided in the *Teacher's Edition: Part II.*

REVIEW: SECTIONS 16-1–16-2
See page 484 for the answers to Exercises 1–5.

CAREER APPLICATIONS

This feature introduces the statistical concept of standard deviation and illustrates a five-step procedure for calculating it. Because standard deviation relates to variations in collected data, it can be used after Section 16-2 (pages 474-477) is taught.

1. 6.1
2. 2.4
3. 5.3
4. 1.9

CAREER APPLICATIONS
Statistical Clerk

A **statistical clerk** often helps to collect and analyze data such as test scores. One measure that is computed is the **standard deviation.** The standard deviation shows how data spreads out or varies from the mean of the data.

EXAMPLE Find the standard deviation for these test scores.

80, 84, 87, 89, 90

SOLUTION

⃞1 Find the mean.

$$\frac{80 + 84 + 87 + 89 + 90}{5} = 86$$

⃞2 Find the difference between the mean and each score.

$86 - 80 = 6$
$86 - 84 = 2$
$86 - 87 = -1$
$86 - 89 = -3$
$86 - 90 = -4$

⃞3 Square the differences.

$6^2 = 36$
$2^2 = 4$
$(-1)^2 = 1$
$(-3)^2 = 9$
$(-4)^2 = 16$

⃞4 Find the average of the squares.

$$\frac{36 + 4 + 1 + 9 + 16}{5} = 13.2$$

⃞5 The standard deviation is the square root of the average. Use a calculator.

13.2 √ ⇒ 3.6331804

The standard deviation is about **3.6**.

Exercises

Find the standard deviation to the nearest tenth.

1. **Science Test Scores**

   75   78   84   91

2. **Heights of Players in Inches**

   76   74   77   70   73

3. **Miles Per Gallon For Six Cars**

   22   30   18   16   25   15

4. **Number of Hits in Six Baseball Games**

   1   1   3   5   6   2

478 Career Application

REVIEW CAPSULE FOR SECTION 16-3

Write each fraction in lowest terms. (Pages 124–125)

1. $\frac{4}{6}$ 2. $\frac{20}{24}$ 3. $\frac{6}{18}$ 4. $\frac{54}{60}$ 5. $\frac{15}{40}$ 6. $\frac{21}{36}$

Write a percent for each fraction. (Pages 227–230)

7. $\frac{1}{4}$ 8. $\frac{1}{10}$ 9. $\frac{3}{4}$ 10. $\frac{4}{5}$ 11. $\frac{37}{50}$ 12. $\frac{9}{20}$

The answers are on page 494.

16-3 Probability

Jason is one of 20 guests at a party. All the guests write their names on a card and place it in a box. The cards are shuffled and one card is drawn at random to choose a winner for a prize. Since each card is <u>as equally likely</u> to be drawn as any other, Jason can find the *probability* of winning the prize.

PROCEDURE

To find the **probability** of an event with equally likely outcomes, use this ratio.

$$P(\text{probability}) = \frac{\text{Number of Favorable Outcomes}}{\text{Number of Possible Outcomes}}$$

EXAMPLE 1 Find the probability that Jason will win.

Solution: Jason wins only if his card is drawn.

$\frac{\text{Number of Favorable Outcomes}}{\text{Number of Possible Outcomes}} = \frac{1}{20}$ ← Jason's card
← 20 cards in all

The probability that Jason will win the prize is $\frac{1}{20}$.

Self-Check Each of the 11 letters of the word "MATHEMATICS" is written on a separate card. The cards are placed face down and shuffled. A card is chosen at random. What is the probability that it will show each of the following?

1. The letter M   2. The letter E   3. A vowel

Statistics and Probability

ADDITIONAL EXAMPLE

Example 2

A regular 6-sided die is tossed. What is the probability of getting a number greater than four?

Ans: $\frac{1}{3}$

SELF-CHECK

4. $\frac{3}{10}$; or 30%
5. $\frac{2}{5}$; or 40%
6. 0; or 0%

CLASSROOM EXERCISES

1. $\frac{3}{5}$; or 60%
2. 12; $\frac{2}{5}$; or 40%
3. 18; 30; $\frac{3}{5}$; or 60%
4. 12; 30; $\frac{2}{5}$; or 40%
5. $\frac{1}{10}$; or 10%
6. $\frac{1}{5}$; or 20%
7. $\frac{2}{5}$; or 40%
8. $\frac{3}{10}$; or 30%
9. $\frac{3}{5}$; or 60%
10. $\frac{9}{10}$; or 90%
11. 0; or 0%
12. 1; or 100%

480

You can write a fraction or a percent for a probability.

EXAMPLE 2 Stan and Susan are playing a board game. Susan spins the spinner at the right once during her turn. Find the probability of the spinner stopping on each of the following.

a. A three
b. A two
c. A number greater than three
d. A number less than four

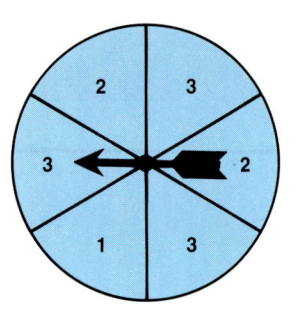

Solutions:

a. $\dfrac{\text{Number of Favorable Outcomes}}{\text{Number of Possible Outcomes}} = \dfrac{3}{6} = \dfrac{1}{2}$

$P = \dfrac{1}{2}$, or **50%**.

b. $\dfrac{\text{Number of Favorable Outcomes}}{\text{Number of Possible Outcomes}} = \dfrac{2}{6} = \dfrac{1}{3}$

$P = \dfrac{1}{3}$, or $33\dfrac{1}{3}\%$.

c. $\dfrac{\text{Number of Favorable Outcomes}}{\text{Number of Possible Outcomes}} = \dfrac{0}{6}$

$P = 0$, or **0%**.

d. $\dfrac{\text{Number of Favorable Outcomes}}{\text{Number of Possible Outcomes}} = \dfrac{6}{6}$

$P = 1$, or **100%**.

Self-Check There are 3 red pencils, 2 blue pencils, 1 green pencil, and 4 brown pencils in a desk drawer. A pencil is chosen at random. Find the probability of choosing each of the following.

4. A red pencil
5. A brown pencil
6. A purple pencil

The probability of an event that **cannot** happen is **0**.
The probability of an event that is **certain** to happen is **1**.

CLASSROOM EXERCISES

Suppose that there are 12 boys and 18 girls in a mathematics class. The teacher chooses a student at random to read the answers to the homework. Complete the table. (Example 1)

Event	Number of Favorable Outcomes	Number of Possible Outcomes	Probability
1. Choosing a girl	18	30	?
2. Choosing a boy	?	30	?
3. Not choosing a boy	?	?	?
4. Not choosing a girl	?	?	?

480 Chapter 16

There are 4 red marker pens, 3 blue marker pens, 2 brown marker pens, and 1 green marker pen in a box. Felipe chooses a pen at random. Find the probability of each of the following. (Example 2)

5. Choosing a green pen
6. Choosing a brown pen
7. Choosing a red pen
8. Choosing a blue pen
9. Not choosing a red pen
10. Not choosing a green pen
11. Choosing a yellow pen
12. Choosing a colored marker pen

WRITTEN EXERCISES

Goal: To find the probability of an event

These ten names are written on slips of paper and placed in a hat. One name is drawn at random. Find each probability. (Example 1)

Alice	Jennifer	Ronald	Anthony	Joyce
Philip	Anna	Sylvia	Carlos	Michael

1. Drawing a name that begins with the letter "A"
2. Drawing a name that begins with the letter "J"
3. Drawing a girl's name
4. Drawing a boy's name
5. Drawing a name that ends with the letter "e"
6. Drawing the name Rosa

A coin purse contains two pennies, a nickel, a dime, and a quarter. One coin is drawn at random. Find each probability. Write each answer as a fraction in lowest terms and as a percent. (Example 2)

7. Drawing a dime
8. Drawing a penny
9. Drawing a coin worth more than 1¢
10. Drawing a coin worth more than 5¢
11. Drawing a coin worth less than 25¢
12. Drawing a coin worth more than 25¢

Find the probability of the spinner at the right stopping on each of the following. Write each answer as a fraction in lowest terms and as a percent. (Example 2)

13. An even number
14. A three
15. An odd number
16. A number less than three
17. A number greater than five
18. A number less than nine

Statistics and Probability **481**

ASSIGNMENT GUIDE
BASIC
p. 481: Omit
AVERAGE
p. 481: Omit
ABOVE AVERAGE
p. 481: 1-18

WRITTEN EXERCISES
1. $\frac{3}{10}$ 2. $\frac{1}{5}$ 3. $\frac{1}{2}$
4. $\frac{1}{2}$ 5. $\frac{1}{5}$ 6. 0
7. $\frac{1}{5}$; or 20%
8. $\frac{2}{5}$; or 40%
9. $\frac{3}{5}$; or 60%
10. $\frac{2}{5}$; or 40%
11. $\frac{4}{5}$; or 80%
12. 0; or 0%
13. $\frac{1}{2}$; or 50%
14. $\frac{1}{8}$; or $12\frac{1}{2}$%
15. $\frac{1}{2}$; or 50%
16. $\frac{1}{4}$; or 25%
17. $\frac{3}{8}$; or $37\frac{1}{2}$%
18. 1; or 100%

Teaching Suggestions
P. M-29

QUICK QUIZ

Write in lowest terms.

1. $\frac{81}{108}$ Ans: $\frac{3}{4}$
2. $\frac{76}{100}$ Ans: $\frac{19}{25}$

Add. Write in lowest terms.

3. $\frac{2}{7} + \frac{3}{7}$ Ans: $\frac{5}{7}$
4. $\frac{1}{8} + \frac{5}{8}$ Ans: $\frac{3}{4}$
5. $\frac{7}{15} + \frac{3}{15}$ Ans: $\frac{2}{3}$

REVIEW CAPSULE FOR SECTION 16-4

Write each fraction in lowest terms. (Pages 124–125)

1. $\frac{10}{14}$ 2. $\frac{6}{10}$ 3. $\frac{13}{26}$ 4. $\frac{11}{33}$ 5. $\frac{15}{45}$ 6. $\frac{63}{72}$

Add. (Pages 158–160)

7. $\frac{1}{3} + \frac{1}{3}$ 8. $\frac{2}{5} + \frac{2}{5}$ 9. $\frac{3}{12} + \frac{4}{12}$ 10. $\frac{5}{36} + \frac{2}{36}$ 11. $\frac{5}{16} + \frac{7}{16}$ 12. $\frac{5}{10} + \frac{3}{10}$

The answers are on page 494.

16-4 Probability and Tables

Eric and Rosa are playing a baseball board game. The player at bat tosses a pair of dice and records the sum of the dice. The play for each sum is given in the table below.

Sum of Two Dice	Play
2	Homerun
3, 4, 5	Single
6, 7, 8	Out
9, 10, 11	Double
12	Triple

Sum of Two Dice

Second Die

First Die	1	2	3	4	5	6
1	2	3	4	5	6	7
2	3	4	5	6	7	8
3	4	5	6	7	8	9
4	5	6	7	8	9	10
5	6	7	8	9	10	11
6	7	8	9	10	11	12

TABLE 1

Table 1 shows the <u>sum</u> for each of the 36 possible ways that a pair of dice can land. You can get certain sums in more than one way. For example, there are 4 ways to get a sum of 9.

6 and 3   5 and 4   4 and 5   3 and 6

ADDITIONAL EXAMPLE

Example 1

A pair of dice is tossed. Use Table 1 to find the probability that the sum is 11. Ans: $\frac{1}{18}$

EXAMPLE 1   Rosa tosses the pair of dice. Use Table 1 to find the probability that the sum is 4 (a single). Write a fraction for the probability.

Solution: There are 3 ways to get a sum of 4.

$$P(\text{sum of 4}) = \frac{\text{Number of Favorable Outcomes}}{\text{Number of Possible Outcomes}} = \frac{3}{36} = \frac{1}{12}$$

Thus, the probability that the sum is 4 is $\frac{1}{12}$.

Chapter 16

Self-Check   *Use Table 1 to find each probability when a pair of dice is tossed. Write each answer as a fraction in lowest terms.*

1. Getting a sum of 5
2. Getting a sum of 12

Events that cannot happen at the same time are said to be **mutually exclusive**.

EXAMPLE 2   Mark tosses a pair of dice once. Use Table 1 to find the probability that the sum is 2 or 12 (homerun or triple). Write a fraction for the probability.

Solution: *Think:* Ways of getting a sum of 2:   1
             Ways of getting a sum of 12:   1

$P(\text{sum of 2 or 12}) = P(\text{sum of 2}) + P(\text{sum of 12})$ ◀ **Sum of the probabilities**

$= \frac{1}{36} + \frac{1}{36} = \frac{2}{36}$, or $\frac{1}{18}$

Self-Check   *A pair of dice is tossed. Find each probability. Write each answer as a fraction in lowest terms.*

3. Getting a sum of 2 or 11
4. Getting a sum of 3 or 5

CLASSROOM EXERCISES

A pair of dice is tossed. Refer to Table 1 on page 482 to write the number of favorable outcomes for each given sum of the two dice. (Example 1)

1. 1	**2.** 2	**3.** 3	**4.** 4	**5.** 5	**6.** 6
7. 7	**8.** 8	**9.** 9	**10.** 10	**11.** 11	**12.** 12

A pair of dice is tossed. Refer to Table 1 on page 482 to find each probability. Write each answer as a fraction in lowest terms. (Example 2)

13. $P(\text{sum of 2 or 4}) = \frac{1}{36} + \frac{?}{36} = \underline{}$
14. $P(\text{sum of 3 or 8}) = \frac{2}{36} + \frac{?}{36} = \underline{}$
15. $P(\text{sum of 10 or 11}) = \frac{?}{36} + \frac{?}{36} = \underline{}$
16. $P(\text{sum of 6 or 12}) = \frac{?}{36} + \frac{?}{36} = \underline{}$
17. $P(\text{sum of 2 or 3 or 4}) = \frac{1}{36} + \frac{?}{36} + \frac{?}{36} = \underline{}$
18. $P(\text{sum of 10 or 11 or 12}) = \frac{?}{36} + \frac{?}{36} + \frac{?}{36} = \underline{}$

SELF-CHECK

1. $\frac{1}{9}$ 2. $\frac{1}{36}$

ADDITIONAL EXAMPLE
Example 2
A pair of dice is tossed once. Use Table 1 to find the probability that the sum is 7 or 9. Write a fraction for the probability.
Ans: $\frac{5}{18}$

SELF-CHECK

3. $\frac{1}{12}$.   4. $\frac{1}{6}$

CLASSROOM EXERCISES

1. 0   2. $\frac{1}{36}$ 3. $\frac{1}{18}$
4. $\frac{1}{12}$ 5. $\frac{1}{9}$ 6. $\frac{5}{36}$
7. $\frac{1}{6}$ 8. $\frac{5}{36}$ 9. $\frac{1}{9}$
10. $\frac{1}{12}$ 11. $\frac{1}{18}$ 12. $\frac{1}{36}$
13. $\frac{1}{9}$ 14. $\frac{7}{36}$ 15. $\frac{5}{36}$
16. $\frac{1}{6}$ 17. $\frac{1}{6}$ 18. $\frac{1}{6}$

Statistics and Probability

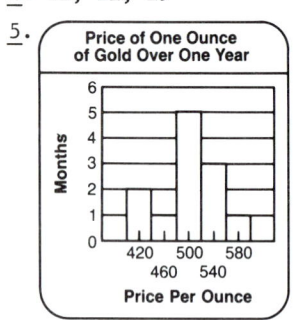

WRITTEN EXERCISES

Goals: To use a table to find the probability of an event
To find the probability of mutually exclusive events

A pair of dice is tossed once. Use Table 1 on page 482 to find each probability. Write each answer as a fraction in lowest terms.
(Example 1)

1. What is the probability of getting a sum of 3?
2. What is the probability of getting a sum of 6?
3. What is the probability of getting a sum of 7?
4. What is the probability of getting a sum of 9?
5. What is the probability of getting a sum of 10?
6. What is the probability of getting a sum of 8?

(Example 2)

7. What is the probability of getting a sum of 2 or 9?
8. What is the probability of getting a sum of 10 or 11?
9. What is the probability of getting a sum of 5 or 6?
10. What is the probability of getting a sum of 4 or 5?
11. What is the probability of getting a sum of 3 or 7?
12. What is the probability of getting a sum of 7 or 8?

MORE CHALLENGING EXERCISES

13. What is the probability of getting a single?
14. What is the probability of getting an out?
15. What is the probability of getting a double or a triple?
16. What is the probability of <u>not</u> getting an out?

REVIEW CAPSULE FOR SECTION 16-5

Write each fraction in lowest terms. (Pages 124–125)

1. $\frac{4}{36}$ 2. $\frac{12}{16}$ 3. $\frac{24}{60}$ 4. $\frac{18}{64}$ 5. $\frac{18}{32}$ 6. $\frac{15}{20}$

Write a percent for each fraction. (Pages 227–230)

7. $\frac{3}{5}$ 8. $\frac{1}{4}$ 9. $\frac{2}{3}$ 10. $\frac{1}{6}$ 11. $\frac{7}{8}$ 12. $\frac{1}{12}$

The answers are on page 494.

484 Chapter 16

ASSIGNMENT GUIDE
BASIC
p. 484: Omit
AVERAGE
p. 484: Omit
ABOVE AVERAGE
p. 484: 1–16

WRITTEN EXERCISES
1. $\frac{1}{18}$ 2. $\frac{5}{36}$ 3. $\frac{1}{6}$
4. $\frac{1}{9}$ 5. $\frac{1}{12}$ 6. $\frac{5}{36}$
7. $\frac{5}{36}$ 8. $\frac{5}{36}$ 9. $\frac{1}{4}$
10. $\frac{7}{36}$ 11. $\frac{2}{9}$ 12. $\frac{11}{36}$
13. $\frac{1}{4}$ 14. $\frac{4}{9}$ 15. $\frac{5}{18}$
16. $\frac{5}{9}$

REVIEW: SECTIONS 16-1–16-2
1. 88.5; 90.5; 92
2. 152; 153.5; 160
3. 71; 71; 72
4. 21; 21; 19
5. [Bar graph: Price of One Ounce of Gold Over One Year; x-axis Price Per Ounce 420, 460, 500, 540, 580; y-axis Months 0–6]

16-5 Tree Diagrams

There are two empty seats next to each other on a school bus. Two boys and two girls get on the bus. You can draw a **tree diagram** to show the number of ways the students can be seated.

CHOOSING TWO SEATS

First Seat	Second Seat	Possible Outcomes
G	G	GG
G	B	GB
G	B	GB
B	B	BB
B	G	BG
B	G	BG

PROCEDURE

To use a tree diagram to find the probability of an event:

1. Draw a tree diagram that shows all the possible ways the event can happen.
2. Use the tree diagram to find the probability.

EXAMPLE Two girls and two boys get on a school bus. There are only two empty seats next to each other. Refer to the tree diagram above to find the probability that a girl sits in the first seat and a boy sits in the second seat.

Solution: Think: There are six possible outcomes. In two of the six outcomes, a girl sits in the first seat and a boy sits in the second seat.

$$P(G, B) = \frac{\text{Number of Favorable Outcomes}}{\text{Number of Possible Outcomes}} = \frac{2}{6} = \frac{1}{3}, \text{ or } 33\frac{1}{3}\%$$

Self-Check 1. What is the probability that two boys will sit in the empty seats?

Statistics and Probability

CLASSROOM EXERCISES

1. $\frac{1}{3}$ 2. $\frac{2}{3}$ 3. $\frac{1}{6}$

4. $\frac{1}{6}$

ASSIGNMENT GUIDE
BASIC
p. 486: Omit
AVERAGE
p. 486: Omit
ABOVE AVERAGE
p. 486: 1-14

WRITTEN EXERCISES

1. Direct; Coastal;
   Central; Scenic;
   Direct; Overland;
   Scenic

2. $\frac{1}{6}$ 3. $\frac{1}{6}$

CLASSROOM EXERCISES

For Exercises 1–4, refer to the tree diagram on page 485 to find each probability. Write a fraction in lowest terms and a percent for each probability. (Example)

1. What is the probability that a boy sits in the first seat and a girl sits in the second?

2. What is the probability that a boy and girl sit next to each other?

3. What is the probability that both seats are occupied by girls?

4. What is the probability that both seats are occupied by boys?

WRITTEN EXERCISES

Goals: To draw a tree diagram

To use tree diagrams to find probabilities

For a trip from Miami to Chicago, Acme Tours offers a choice of three routes (Coastal, Central, or Overland) and two routes (Direct and Scenic) from Chicago to San Francisco.

1. Complete this tree diagram to show all possible choices of a route.

CHOOSING A ROUTE

Miami–Chicago	Chicago–San Francisco	Possible Outcomes
Coastal	Direct	Coastal, ___?___
	Scenic	___?___, Scenic
Central	Direct	___?___, Direct
	Scenic	Central, ___?___
Overland	Direct	Overland, ___?___
	Scenic	___?___, ___?___

For Exercises 2–7, refer to the tree diagram above to find each probability. Write a fraction in lowest terms and a percent for each probability. (Example)

2. What is the probability of choosing the coastal scenic route?

3. What is the probability of choosing the overland direct route?

486 Chapter 16

4. What is the probability of choosing a scenic route?

5. What is the probability of choosing a central route?

6. What is the probability of choosing a coastal <u>or</u> central route?

7. What is the probability of choosing the coastal direct <u>or</u> the overland scenic route?

MORE CHALLENGING EXERCISES

On a certain day, the menu at a school cafeteria offers a choice of two main courses — chicken or fish. Each of three students chooses one main course.

Lunch Menu — Main Courses —
CHICKEN
BROILED FISH

8. Draw a tree diagram that shows all the possible ways in which each of three students can order a main course.

For Exercises 9–14, refer to the tree diagram of Exercise 8 to find each probability. Write a fraction in lowest terms and a percent for each probability. (Example)

9. What is the probability that each student chooses chicken?

10. What is the probability that exactly two students choose fish?

11. What is the probability that only one student chooses chicken?

12. What is the probability that all three students chose fish?

13. What is the probability that <u>at least</u> two students choose chicken?

14. What is the probability that the first two students each choose fish?

WRITTEN EXERCISES

4. $\frac{1}{2}$ 5. $\frac{1}{3}$ 6. $\frac{2}{3}$

7. $\frac{1}{3}$

8.

9. $\frac{1}{8}$ 10. $\frac{3}{8}$

11. $\frac{3}{8}$ 12. $\frac{1}{8}$

13. $\frac{1}{2}$ 14. $\frac{1}{4}$

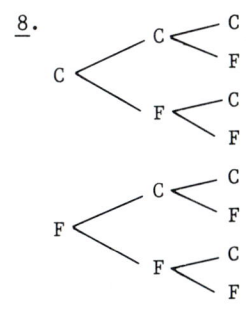

REVIEW CAPSULE FOR SECTION 16-6

Write a percent for each fraction. (Pages 227–230)

1. $\frac{3}{25}$ 2. $\frac{13}{50}$ 3. $\frac{5}{40}$ 4. $\frac{8}{40}$ 5. $\frac{8}{8}$ 6. $\frac{0}{9}$

Multiply. (Pages 129–131)

7. $\frac{1}{4} \times \frac{1}{4}$ 8. $\frac{1}{8} \times \frac{1}{8}$ 9. $\frac{1}{3} \times \frac{2}{5}$ 10. $\frac{3}{4} \times \frac{5}{8}$

11. $\frac{1}{2} \times \frac{6}{7}$ 12. $\frac{3}{5} \times \frac{5}{6}$ 13. $\frac{1}{3} \times \frac{2}{3}$ 14. $\frac{3}{8} \times \frac{3}{4}$

The answers are on page 494.

Statistics and Probability

Teaching Suggestions
p. M-29

QUICK QUIZ

Multiply. Write in lowest terms.

1. $\frac{2}{3} \times \frac{5}{8}$ Ans: $\frac{5}{12}$
2. $\frac{3}{6} \times \frac{4}{6}$ Ans: $\frac{1}{3}$

The letters D, I, V, I, D, E, N, D are written on separate cards. What is the probability of drawing

3. the letter D? Ans: $\frac{3}{8}$
4. the letter I? Ans: $\frac{1}{4}$

ADDITIONAL EXAMPLES

Example 1

Find each probability.

1. P(senior boy and junior girl) Ans: $\frac{40}{147}$
2. P(junior boy and junior girl) Ans: $\frac{12}{49}$

SELF-CHECK

1. $\frac{11}{42}$

Example 2

1. Find the probability of drawing a black sock; then a brown sock (see Example 2). Ans: $\frac{4}{15}$
2. Find the probability of choosing a senior girl; then a senior boy (see Table; page 488). Ans: $\frac{220}{861}$

16-6 Multiplying Probabilities

A student from each class at Smalltown High School is chosen by a random drawing for a community internship program. One name is drawn from each of four boxes labeled by class.

Smalltown High School

Class	Number of Boys	Number of Girls
Freshman	35	35
Sophomore	27	33
Junior	24	32
Senior	20	22

Since the outcome of the draw from one box **does not** affect the outcome of the draw from any other box, the events are **independent**.

You can find the probability of two or more independent events occurring by multiplying the probabilities of each event.

EXAMPLE 1 Find the probability of choosing a freshman boy and a junior girl.

Solution: P(freshman boy and junior girl) = P(freshman boy) × P(junior girl)

P(freshman boy) = $\frac{35}{70}$, or $\frac{1}{2}$ P(junior girl) = $\frac{32}{56} = \frac{4}{7}$

P(freshman boy and junior girl) = $\frac{1}{2} \times \frac{4}{7} = \frac{4}{14}$, or $\frac{2}{7}$

Self-Check 1. Find the probability of choosing a sophomore girl and a senior boy.

Sometimes what happens in the first event **does affect** what happens in following events. Events which affect each other are called **dependent events**.

EXAMPLE 2 A drawer contains 6 black socks and 4 brown socks. Two socks are drawn at random, one after the other, without replacement. Find the probability of drawing two black socks in a row.

Solution: P(black, then black) = P(black) × P(black)

First draw: P(black) = $\frac{6}{10}$, or $\frac{3}{5}$ ◀ The first sock is not replaced.

Second draw: P(black) = $\frac{5}{9}$ ◀ There are now 5 black socks out of 9 socks.

P(black, then black) = $\frac{3}{5} \times \frac{5}{9} = \frac{1}{3}$ ◀ The probability is $\frac{1}{3}$.

Chapter 16

Self-Check

2. Find the probability of drawing two brown socks in a row without replacement (see Example 2).

CLASSROOM EXERCISES

Refer to the table on page 488 to find each probability. Write each answer as a fraction in lowest terms. (Example 1)

1. P(freshman girl <u>and</u> junior boy)
 P(freshman girl) = __?__
 P(junior boy) = __?__
 P(freshman girl <u>and</u> junior boy) = __?__

2. P(sophomore boy <u>and</u> senior girl)
 P(sophomore boy) = __?__
 P(senior girl) = __?__
 P(sophomore boy <u>and</u> senior girl) = __?__

A drawer contains 8 black socks, 6 grey socks, and 2 white socks. Two socks are drawn at random, one after the other, without replacement. Complete to find each probability. (Example 2)

3. P(black, then black)
 First draw: P(black) = __?__
 Second draw: P(black) = __?__
 P(black, then black) = __?__

4. P(grey, then white)
 First draw: P(grey) = __?__
 Second draw: P(white) = __?__
 P(grey, then white) =

WRITTEN EXERCISES

Goals: To find the probability of two or more independent events
To find the probability of two or more dependent events

Connie has six scarves. Three are striped, one is blue, and two are white. She also has ten blouses. Three are blue, two are yellow, and five are plaid. She chooses a scarf and a blouse at random. For Exercises 1–6, find each probability. Write each probability as a fraction in lowest terms. (Example 1)

1. P(white scarf <u>and</u> plaid blouse)
2. P(blue scarf <u>and</u> yellow blouse)
3. P(striped scarf <u>and</u> blue blouse)
4. P(striped scarf <u>and</u> plaid blouse)
5. P(blue scarf <u>and</u> blue blouse)
6. P(white scarf <u>and</u> yellow blouse)

Statistics and Probability **489**

SELF-CHECK

2. $\frac{2}{15}$

CLASSROOM EXERCISES

1. $\frac{1}{2}$; $\frac{3}{7}$; $\frac{3}{14}$
2. $\frac{9}{20}$; $\frac{11}{21}$; $\frac{33}{140}$
3. $\frac{1}{2}$; $\frac{7}{15}$; $\frac{7}{30}$
4. $\frac{3}{8}$; $\frac{2}{15}$; $\frac{1}{20}$

ASSIGNMENT GUIDE
BASIC
p. 490: Omit
AVERAGE
p. 490: Omit
ABOVE AVERAGE
p. 490: 1–20

WRITTEN EXERCISES

1. $\frac{1}{6}$ 2. $\frac{1}{30}$ 3. $\frac{3}{20}$
4. $\frac{1}{4}$ 5. $\frac{1}{20}$ 6. $\frac{1}{15}$

WRITTEN EXERCISES

7. $\frac{2}{15}$ 8. $\frac{1}{15}$ 9. $\frac{1}{45}$
10. $\frac{1}{30}$ 11. $\frac{2}{15}$ 12. $\frac{4}{45}$
13. $\frac{6}{125}$ 14. $\frac{9}{500}$
15. $\frac{1}{125}$ 16. $\frac{3}{250}$
17. $\frac{1}{20}$ 18. $\frac{1}{60}$ 19. $\frac{1}{90}$
20. $\frac{1}{60}$

CALCULATOR EXERCISES

1. 0.9%   2. 12.5%
3. 28.6%   4. 55.9%
5. 25.4%   6. 17.6%

A box has 4 quarters, 3 dimes, 2 nickels, and 1 penny. Two coins are drawn at random, one after the other, without replacement. Write each probability as a fraction in lowest terms. (Example 2)

7. P(quarter, then quarter)   8. P(dime, then dime)   9. P(nickel, then penny)
10. P(penny, then dime)   11. P(dime, then quarter)   12. P(quarter, then nickel)

MORE CHALLENGING EXERCISES

In a bag there are 4 blue marbles, 3 yellow marbles, 2 white marbles, and 1 black marble. For Exercises 13–16, three marbles are drawn at random, one after the other, with replacement. Find each probability. Write each probability as a fraction in lowest terms.

13. P(blue and blue and yellow)   14. P(yellow and yellow and white)
15. P(blue and white and black)   16. P(yellow and blue and black)

For Exercises 17–20, three marbles are drawn at random, one after the other, without replacement. Find each probability. Write each probability as a fraction in lowest terms.

17. P(blue, then blue, then yellow)   18. P(yellow, then yellow, then white)
19. P(blue, then white, then black)   20. P(yellow, then blue, then black)

MULTIPLYING PROBABILITIES

A calculator with the $($ and $)$ keys enables you to multiply two or more probabilities written as fractions to obtain a probability written as a percent.

EXAMPLE   Multiply: $\frac{7}{14} \cdot \frac{3}{13} \cdot \frac{5}{12}$ ▶ The probabilities for 3 dependent events occurring in a row

Solution

$($ 7 $\times$ 3 $\times$ 5 $)$ $\div$ $($ 1 4 $\times$ 1 3 $\times$ 1 2 $)$ = $\times$ 1 0 0 =   `4.8076923`

The answer rounded to the nearest tenth is **4.8%**.

NOTE: Pressing the $)$ key completes the multiplication.

EXERCISES   *Multiply. Round each answer to the nearest tenth.*

1. $\frac{5}{14} \cdot \frac{4}{13} \cdot \frac{1}{12}$ 2. $\frac{9}{10} \cdot \frac{5}{9} \cdot \frac{2}{8}$ 3. $\frac{4}{7} \cdot \frac{5}{6} \cdot \frac{3}{5}$
4. $\frac{8}{13} \cdot \frac{2}{12} \cdot \frac{6}{11}$ 5. $\frac{11}{15} \cdot \frac{9}{14} \cdot \frac{7}{13}$ 6. $\frac{15}{17} \cdot \frac{12}{16} \cdot \frac{4}{15}$

490   Chapter 16

REVIEW: SECTIONS 16-3—16-6

Each of the 6 letters of the word "CHANCE" is written on a separate card. The cards are placed face down and shuffled. A card is chosen at random. What is the probability that it will contain each of the following? (Section 16-3)

1. The letter N?
2. The letter C?
3. A consonant?
4. A vowel?

A pair of dice is tossed once. For Exercises 5-8, refer to the table on page 482 to find each probability. Write each probability as a fraction in lowest terms. (Section 16-4)

5. What is the probability of getting a sum of 11?
6. What is the probability of getting a sum of 7?
7. What is the probability of getting a sum of 4 <u>or</u> 8?
8. What is the probability of getting a sum of 9 <u>or</u> 12?

9. Two coins are tossed at the same time. To find all the possible ways the coins can land, copy and complete the tree diagram below. (Section 16-5)

TOSSING TWO COINS

First Coin	Second Coin	Possible Outcomes
H	H	H H
	?	? ?
T	?	? ?
	?	? ?

For Exercises 10-13, refer to the tree diagram in Exercise 9 to find each probability. Write a fraction in lowest terms and a percent for each probability. (Section 16-5)

10. Getting two heads
11. Getting two tails
12. Getting a head and a tail
13. Getting a tail on the second coin

A box contains 12 eggs, two of which are cracked. (Section 16-6)

14. One egg is picked from the box and replaced. Then another egg is picked. Find P(good egg, then cracked egg.)
15. One egg is picked from the box and not replaced. Another egg is picked. Find P(good egg, then good egg).

Statistics and Probability   **491**

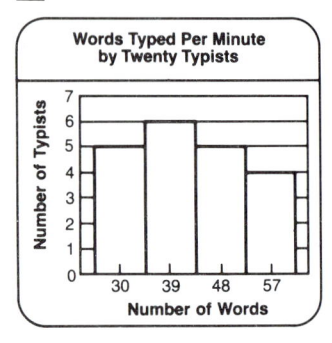

CHAPTER REVIEW

1. median 2. data
3. 0
4. dependent events
5. mean
6. mutually exclusive events
7. independent events
8. 1 9. mode
10. $.77 11. 237
12.

13. $\frac{1}{16}$; or $6\frac{1}{4}$%
14. $\frac{3}{16}$; or $18\frac{3}{4}$%
15. $\frac{1}{4}$; or 25%
16. $\frac{1}{8}$; or $12\frac{1}{2}$%

CHAPTER REVIEW

PART 1: VOCABULARY

For Exercises 1–9, choose from the box at the right below the word(s) or number(s) that best corresponds to each description.

1. The middle number in a list of data arranged in order of size ?
2. Another word for information ?
3. The probability of an event that cannot happen ?
4. Events which affect each other ?
5. The quotient of the sum of the given measures and the number of measures ?
6. Events that cannot happen at the same time ?
7. Events which do not affect each other ?
8. The probability of an event that is certain to happen ?
9. The number that occurs most often in a set of data ?

> mean
> mode
> median
> data
> 0
> 1
> mutually exclusive events
> independent events
> dependent events

PART 2: SKILLS AND APPLICATIONS

For Exercises 10–11, find the mean, median, and mode. (Section 16-1)

10. **Cost of Lettuce at Eight Stores**

    $.80 $.59 $.68 $.80
    $.96 $.97 $.59

11. **Miles Driven Per Day on a Trip**

    250   375   132   200
    132   400   182   225

12. Make a histogram to show the number of words typed per minute by 20 students. Use the data in the table. (Section 16–2)

Number of Words	26–34	35–43	44–52	53–61
Count	5	6	5	4

Each of the 16 letters of the words "MASSACHUSETTS BAY" is written on a separate card. A card is chosen at random. Find each probability. Write each probability as a fraction in lowest terms and as a percent. (Section 16-3)

13. What is the probability of drawing an M?
14. What is the probability of drawing an A?
15. What is the probability of drawing an S?
16. What is the probability of drawing a T?

492 Chapter 16

This table shows the possible outcomes for a game. In the game, each player spins the arrow twice and adds the scores.

Second Spin

+	2	4	6	8	10
2	4	6	8	10	12
4	6	8	10	12	14
6	8	10	12	14	16
8	10	12	14	16	18
10	12	14	16	18	20

First Spin

For Exercises 17–22, refer to the table above to find each probability. Write each answer as a fraction in lowest terms. (Section 16-4)

17. What is the probability of getting a sum of 6?

18. What is the probability of getting a sum of 10?

19. What is the probability of getting a sum of 20?

20. What is the probability of getting a sum of 12?

21. What is the probability of getting a sum of 6 or 8?

22. What is the probability of getting a sum of 10 or 12?

23. Three coins are tossed at the same time. Draw a tree diagram that shows all the possible ways for three coins to land. (Section 16-5)

For Exercises 24–27, refer to the tree diagram of Exercise 23 to find each probability. Write each probability as a fraction in lowest terms and as a percent. (Section 16-5).

24. What is the probability of getting two tails and a head?

25. What is the probability of getting three heads?

26. What is the probability of getting three tails?

27. What is the probability of getting a tail on the first coin?

A bag contains 3 green marbles, 5 white marbles, and 7 blue marbles. For Exercises 28–30, one marble is drawn and replaced. Then a second marble is drawn. Find each probability. Write each probability as a fraction in lowest terms and as a percent. (Section 16-6)

28. P(green, then green) **29.** P(green, then white) **30.** P(white, then blue)

For Exercises 31–33, one marble is drawn and not replaced. Then a second marble is drawn. Find each probability. Write each probability as a fraction in lowest terms and as a percent. (Section 16-6)

31. P(green, then green) **32.** P(green, then white) **33.** P(white, then blue)

Statistics and Probability

CHAPTER REVIEW

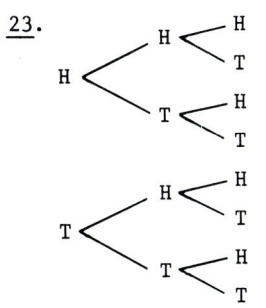

17. $\frac{2}{25}$ 18. $\frac{4}{25}$ 19. $\frac{1}{25}$

20. $\frac{1}{5}$ 21. $\frac{1}{5}$ 22. $\frac{9}{25}$

23.

```
       H — H
   H <
       T — H
           T
       H — H
   T <
       T — H
           T
```

24. $\frac{3}{8}$ 25. $\frac{1}{8}$ 26. $\frac{1}{8}$

27. $\frac{1}{2}$ 28. $\frac{1}{25}$

29. $\frac{1}{15}$ 30. $\frac{7}{45}$

31. $\frac{1}{35}$ 32. $\frac{1}{14}$

33. $\frac{1}{6}$

CHAPTER TEST

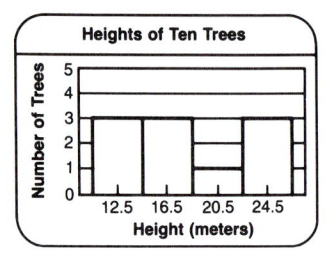

1. Find the mean, median, and mode.

Number of Weeks on the Chart for the Top Ten Records

14	6	8	14	15
9	7	3	7	7

2. Draw a histogram to show this data.

Heights in Meters of Ten Trees

23	14	17	25	16
23	19	18	12	11

Intervals: 11–14; 15–18; 19–22; 23–26

Each of the eleven letters of the word "MISSISSIPPI" is written on a separate card. The cards are placed face down and shuffled. A card is chosen at random. What is the probability that the card will show each of the following?

3. The letter S   **4.** A vowel   **5.** Not the letter S

6. There are two Yes–No questions on a test. Draw a tree diagram to show all the possible ways to answer the questions.

A box contains twenty light bulbs, four of which are broken. One light bulb is drawn from the box. Then a second bulb is drawn. For Exercises 7–8, find each probability if the first light bulb is replaced. Write each answer as a fraction in lowest terms.

7. What is the probability of drawing a broken bulb and a good bulb?

8. What is the probability of drawing a good bulb and a good bulb?

For Exercises 9–10, find each probability if the first light bulb is not replaced. Write each answer as a fraction in lowest terms.

9. What is the probability of drawing a broken bulb, then a good bulb?

10. What is the probability of drawing a good bulb, then a good bulb?

ANSWERS TO REVIEW CAPSULES

Page 473   **1.** Let 10 cm represent the scale from 2000 to 2100. Jan: 4.5 cm; Feb: 5.5 cm; Mar: 7.0 cm; Apr: 7.0 cm; May: 6.0 cm; Jun: 8.0 cm

Page 479   **1.** $\frac{2}{3}$ **2.** $\frac{5}{6}$ **3.** $\frac{1}{3}$ **4.** $\frac{9}{10}$ **5.** $\frac{3}{8}$ **6.** $\frac{7}{12}$ **7.** 25%   **8.** 10%   **9.** 75%   **10.** 80%   **11.** 74%   **12.** 45%

Page 482   **1.** $\frac{5}{7}$ **2.** $\frac{3}{5}$ **3.** $\frac{1}{2}$ **4.** $\frac{1}{3}$ **5.** $\frac{1}{3}$ **6.** $\frac{7}{8}$ **7.** $\frac{2}{3}$ **8.** $\frac{4}{5}$ **9.** $\frac{7}{12}$ **10.** $\frac{7}{36}$ **11.** $\frac{3}{4}$ **12.** $\frac{4}{5}$

Page 484   **1.** $\frac{1}{9}$ **2.** $\frac{3}{4}$ **3.** $\frac{2}{5}$ **4.** $\frac{9}{32}$ **5.** $\frac{9}{16}$ **6.** $\frac{3}{4}$ **7.** 60%   **8.** 25%   **9.** $66\frac{2}{3}$%   **10.** $16\frac{2}{3}$%   **11.** $87\frac{1}{2}$%   **12.** $8\frac{1}{3}$%

Page 487   **1.** 12%   **2.** 26%   **3.** $12\frac{1}{2}$%   **4.** 20%   **5.** 100%   **6.** 0%   **7.** $\frac{1}{16}$ **8.** $\frac{1}{64}$ **9.** $\frac{2}{15}$ **10.** $\frac{15}{32}$ **11.** $\frac{3}{7}$ **12.** $\frac{1}{2}$ **13.** $\frac{2}{9}$ **14.** $\frac{9}{32}$

494   Chapter 16

ENRICHMENT

Sampling

To find how many defective batteries there are in a large shipment, a quality control technician inspects and tests a certain number of batteries. The batteries that are actually tested are called a **sample**.

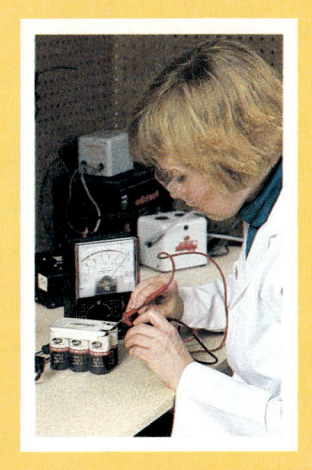

EXAMPLE A technician found that 6 out of every 120 batteries tested were defective. Find the probable number of defective batteries in a shipment of 80,000.

Solution:

1. Find the percent of defective batteries in the sample.

   6 out of $120 = \frac{6}{120} = \frac{1}{20} = 0.05 = 5\%$ ◄ **5% are defective.**

2. Since 5% of the batteries in the sample are defective, it is probable that 5% of the batteries in the shipment are also defective. Find 5% of 80,000.

   5% of $80,000 = 0.05 \times 80,000 =$ **4000** ◄ **Probable number of defective batteries**

EXERCISES

For Exercises 1–4, complete the table.

	Product	Defective Items in Sample	Percent of Defective Items in Sample	Total Number of Items in Shipment	Probable Number of Defective Items in Shipment
1.	Calculators	4 out of 100	?	800	?
2.	Light bulbs	6 out of 100	?	13,400	?
3.	Stereos	2 out of 50	?	4000	?
4.	Watches	3 out of 60	?	6800	?

5. A sample of 400 cameras contains 20 that are defective. Find the probable number of defective cameras in a shipment of 26,000.

6. In a shipment of 950 radios, the probability that any one is defective is 2%. Find the probable number of defective radios in the shipment.

Statistics and Probability **495**

ADDITIONAL PRACTICE

You may wish to use all or some of these exercises, depending on how well students performed on the formal Chapter Test.

1. $4\frac{3}{8}$; $4\frac{1}{2}$; 3

2.

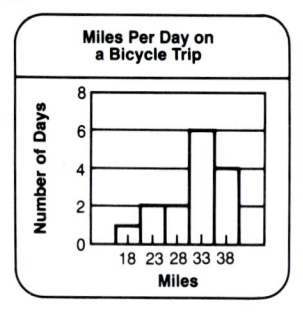

Miles Per Day on a Bicycle Trip

(histogram with Number of Days on y-axis, Miles on x-axis at 18, 23, 28, 33, 38)

3. $\frac{2}{5}$; 40%   4. $\frac{3}{10}$; 30%
5. $\frac{4}{5}$; 80%   6. $\frac{9}{10}$; 90%
7. $\frac{7}{36}$ 8. $\frac{2}{9}$ 9. $\frac{1}{6}$
10. $\frac{1}{18}$

11. (tree diagram for four coins)

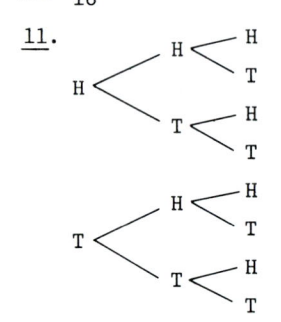

12. $\frac{4}{25}$ 13. $\frac{28}{45}$

496

ADDITIONAL PRACTICE

SKILLS

1. Find the mean, median, and mode. (Pages 470–473)

Number of Passengers in Eight Taxis

| 6 | 5 | 3 | 6 |
| 5 | 4 | 3 | 3 |

2. Construct a histogram to show the data. Use the given intervals. (Pages 474–477)

Miles Per Day on a Bicycle Trip

16	30	33	34	37
34	39	26	21	38
32	36	35	24	33

Intervals: 16–20; 21–25; 26–30; 31–35; 36–40

A box contains ten neckties. Three are green, two are white, one is black, and four are blue. One necktie is chosen at random.

Find each probability. Write each answer as a fraction in lowest terms and as a percent. (Pages 479–481)

3. Choosing a blue necktie
4. Choosing a green necktie
5. Not choosing a white necktie
6. Not choosing a black necktie

For Exercises 7–10, refer to Table 1 on page 482 to find each probability. Write each answer as a fraction in lowest terms. (Pages 482–484)

7. Getting a sum of 5 or 10
8. Getting a sum of 4 or 8
9. Getting a sum of 1 or 7
10. Getting a sum of 2 or 12

11. Four coins are tossed at the same time. Draw a tree diagram to find all possible ways the coins can land. (Pages 485–487)

APPLICATIONS

12. A basket contains 8 oranges and 2 grapefruit. One piece of fruit is picked from the basket and replaced. Then another piece of fruit is picked. What is the probability of picking one orange, then one grapefruit? (Pages 488–490)

13. In the basket of fruit for Exercise 12, one piece of fruit is picked and not replaced. Then another piece of fruit is picked. What is the probability of picking one orange, then one orange? (Pages 488–490)

496   Chapter 16

CHAPTER 17 Geometry

SECTIONS

17-1 Introduction to Geometry
17-2 Angles
17-3 Properties of Triangles
17-4 Problem Solving and Applications: Using Similar Triangles
17-5 Perpendicular and Parallel Lines
17-6 Properties of Quadrilaterals
17-7 Problem Solving and Applications: Surface Area and Rectangular Prisms
17-8 Volume: Cylinders
17-9 Volume: Pyramids and Cones

FEATURES

Calculator Application: *Similar Polygons*
Computer Application: *Angle Measures*
Enrichment: *Spheres*
Common Errors

Sidebar

Teaching Suggestions
p. M-30

QUICK QUIZ

Identify each as a line, ray, angle, or line segment.

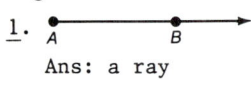

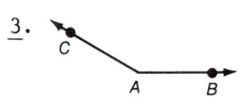

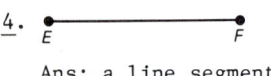

1. [figure: ray from A through B]
   Ans: a ray
2. [figure: line through C and D]
   Ans: a line
3. [figure: angle with vertex A, rays to C and B]
   Ans: an angle
4. [figure: segment EF]
   Ans: a line segment
5. [figure: angle with vertex B]
   Ans: an angle

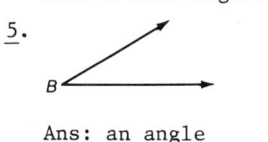

ADDITIONAL EXAMPLE

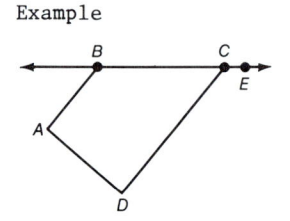

Example

[figure showing triangle with points A, B, C, D, E]

Use symbols and the figure above to name four different angles.
Ans: Any four of ∠A, ∠D, ∠ECD, ∠BCD, ∠ABC

17-1 Introduction to Geometry

Geometry is the study of points, lines, planes, and the figures they form. This table shows some basic geometric figures.

TABLE

Figure	Name	Symbol	Description
A	Point A	A	Shows location. All points are named with capital letters.
[line through A, B]	Line AB	$\overleftrightarrow{AB}$	Extends indefinitely in opposite directions.
[segment A to B]	Line segment AB	$\overline{AB}$	Part of a line that is named by its endpoints.
[ray from A through B]	Ray AB (In this figure, A is the endpoint)	$\overrightarrow{AB}$	Extends indefinitely in one direction from its endpoint.
[angle with vertex B, rays to A and C]	Angle ABC, angle CBA, or angle B (In this figure, B is the vertex)	$\angle ABC$, $\angle CBA$, or $\angle B$	A pair of rays that have a common endpoint called the vertex. The vertex (plural: vertices) is the middle point named.
[parallelogram ABCD labeled I]	Plane **I**, or Plane $ABCD$	Plane **I**, or Plane $ABCD$	A flat surface that extends indefinitely in all directions. A plane is represented by a four-sided figure. Points, lines, rays, and angles all lie in planes.

Use the symbols to name lines, line segments, rays, and angles.

EXAMPLE Use symbols and the given line to name each of the following.

a. Three different segments
b. Three different rays
c. Ray AB in two other ways
d. The line in three ways

[figure: line with points A, B, C, D]

Solutions:
a. $\overline{AB}$, $\overline{AC}$, $\overline{BC}$
b. $\overrightarrow{AD}$, $\overrightarrow{DA}$, $\overrightarrow{BD}$
c. $\overrightarrow{AC}$, $\overrightarrow{AD}$
d. $\overleftrightarrow{AD}$, $\overleftrightarrow{BC}$, $\overleftrightarrow{BD}$

Chapter 17

Self-Check

1. Use symbols and line AB in the Example to name two different segments and two different rays not already named.

   NOTE: $\overline{AD}$ and $\overline{DA}$ are the same segment. However, $\overrightarrow{AD}$ and $\overrightarrow{DA}$ are different rays.

SELF-CHECK
1. $\overline{AD}$, $\overline{CD}$, $\overrightarrow{CD}$, $\overrightarrow{BA}$

CLASSROOM EXERCISES

Replace the ? with the missing symbol or name. (Table)

1.  [figure: S T with ray pointing left]
   Name: Ray TS
   Symbol: ?

2.  [figure: M N with line]
   Name: ?
   Symbol: $\overleftrightarrow{MN}$

3.  [figure: angle with vertex E, sides to D and F]
   Name: ?
   Symbol: $\angle DEF$

4. X
   Name: Point X
   Symbol: ?

5.  [figure: Q R segment]
   Name: Line segment QR
   Symbol: ?

6. 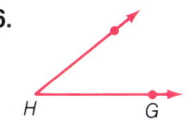 [figure: angle with vertex H, sides to G]
   Name: Angle H
   Symbol: ?

For Exercises 7–16, refer to lines MR and HK which intersect at E. Complete each statement. (Table)

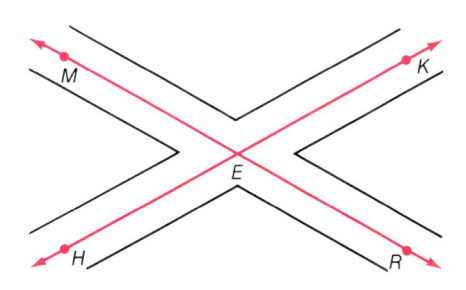

7. Point E is between points M and   ? .
8. Point E is between points H and   ? .
9. Point E is on lines   ?   and   ? .
10. Line segment EM has endpoints   ?   and   ? .
11. Line EM has   ?   endpoints.
12. The endpoint of $\overrightarrow{HE}$ is   ? .
13. Rays EH and ER form angle   ? .
14. Rays EM and EK form angle   ? .
15. Points M, E, and R are on the   ?   line.
16. Points K, E, and H are on the   ?   line.

CLASSROOM EXERCISES
1. $\overrightarrow{TS}$ 2. line MN
3. angle DEF   4. X
5. $\overline{QR}$ 6. $\angle H$ 7. R
8. K   9. MR; HK
10. E; M   11. No
12. H   13. HER; REH
14. MEK; KEM
15. same   16. same

Geometry

ASSIGNMENT GUIDE
BASIC
p. 500: Omit
AVERAGE
p. 500: Omit
ABOVE AVERAGE
p. 500: 1-14

WRITTEN EXERCISES
1. f; l   2. b; h   3. c
4. e; i   5. d; g; k
6. a; j   7. Answers may vary. Some answers are, $\overrightarrow{AB}$; $\overrightarrow{FA}$; $\overrightarrow{BF}$.
8. $\overrightarrow{BA}$; $\overrightarrow{BE}$; $\overrightarrow{BF}$; $\overrightarrow{BG}$
9. $\overrightarrow{BF}$ 10. $\overline{HG}$; $\overline{HE}$; $\overline{GE}$
11. ∠IGH   12. ∠HGI; ∠IGE; ∠EGB; ∠BGH
13. $\overleftrightarrow{HE}$; $\overleftrightarrow{BI}$ 14. B

REVIEW CAPSULE
This Review Capsule reviews prior-taught skills used in Section 17-2. The reference is to the pages where the skills were taught.

WRITTEN EXERCISES

Goals: To identify points, lines, segments, rays, angles, and planes

Match one or more names in Column A with each figure in Exercises 1-6. (Table)

1. [figure with points P, Q]
2. [figure with points R, S]
3. [figure with points T, U, V]
4. [figure with points W, X, Y]
5. [figure with points B, H, J]
6. [figure D, G, E, F with plane II]

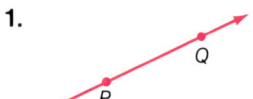

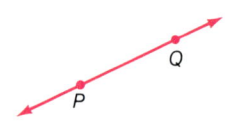

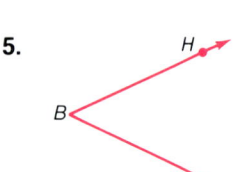

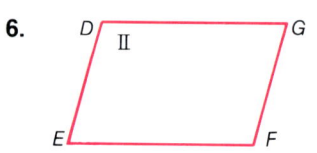

COLUMN A
a. Plane II
b. $\overline{RS}$
c. $\overrightarrow{TV}$
d. ∠HBJ
e. $\overrightarrow{WX}$
f. $\overleftrightarrow{PQ}$
g. ∠B
h. $\overline{SR}$
i. $\overrightarrow{WY}$
j. Plane DEFG
k. ∠JBH
l. $\overleftrightarrow{QP}$

For Exercises 7-14, refer to the figure at the right. (Example)

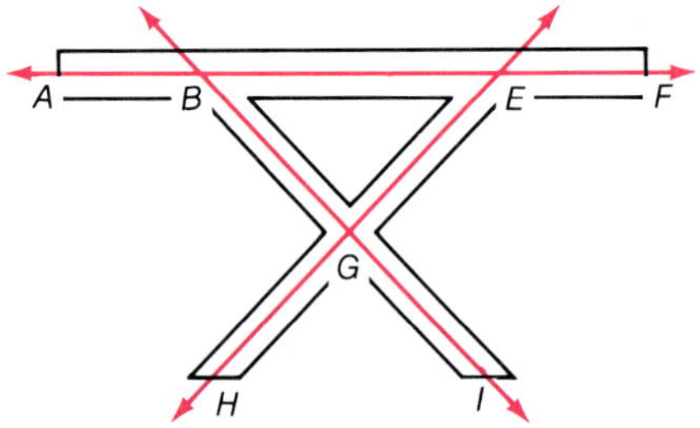

7. Name line *AF* in three different ways.
8. Name four rays that have *B* as an endpoint.
9. Name $\overrightarrow{BE}$ in a different way.
10. Name three different line segments on line *HE*.
11. Name angle *HGI* in one other way.
12. Name four angles with *G* as a vertex.
13. Name two lines that contain point *G*.
14. Name the point where $\overleftrightarrow{IG}$ and $\overleftrightarrow{FE}$ intersect.

REVIEW CAPSULE FOR SECTION 17-2

Replace the __?__ with > or <. (Pages 294-296)

1. 32 __?__ 90
2. 110 __?__ 90
3. 90 __?__ 60
4. 90 __?__ 15
5. 160 __?__ 90
6. 25 __?__ 90
7. 90 __?__ 89
8. 90 __?__ 179

The answers are on page 533.

500   Chapter 17

17-2 Angles

A ski jumper can increase the length of a jump by keeping the skis at the proper angle. If the angle is either too large or too small, the distance covered by the jump will be short.

The degree (symbol: "°") is the standard unit for measuring an angle. You can use a protractor to measure an angle.

PROCEDURE

To use a protractor to measure an angle:

1 Place the center of the base of the protractor at the vertex of the angle.

2 Place the protractor so that one ray of the angle passes through "0" on the protractor.

3 Read the measure of the angle from the second ray.

EXAMPLE 1 Use a protractor to find the measure of the given angles.

Solutions: a. ∠CBD: 30°   b. ∠CBE: 90°   c. ∠CBF: 130°   d. ∠ABD: 150°

Self-Check Use the protractor in Example 1 to find the measure of each angle.

1. Angle CBH   2. Angle CBG   3. Angle ABH

Geometry 501

Teaching Suggestions
p. M-30

QUICK QUIZ

Replace the ? with < or >.

1. 90 ? 27   Ans: >
2. 125 ? 90   Ans: >

Subtract.

3. 150 − 90   Ans: 60
4. 130 − 45   Ans: 85
5. 90 − 35   Ans: 55

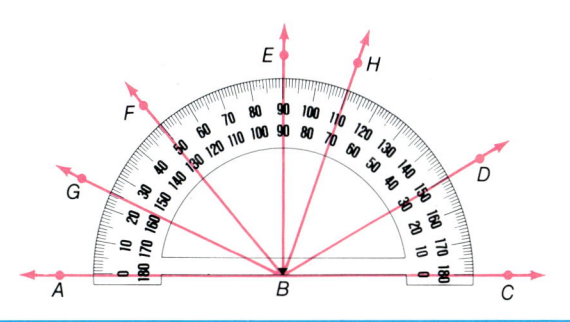

ADDITIONAL EXAMPLES
Example 1
Use the protractor in Example 1 to find the measure of each angle.
1. ∠ABG   Ans: 25°
2. ∠ABF   Ans: 50°

SELF-CHECK
1. 70°   2. 155°
3. 110°

501

Angles can be classified according to their measures.

TABLE

Right Angle	Acute Angle	Obtuse Angle	Straight Angle
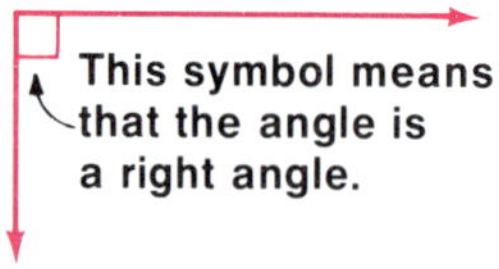 This symbol means that the angle is a right angle.	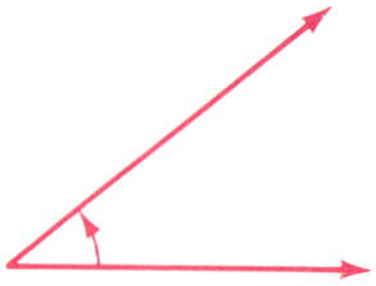	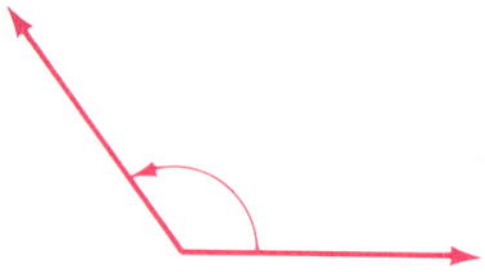	
Equals 90°	Less than 90°	Greater than 90°	Equals 180°

A protractor can also be used to draw angles.

EXAMPLE 2 Draw ∠RST so that its measure is 150.

Solution:

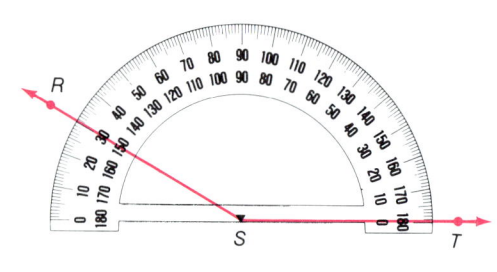

1. Label a point S and a point T. Then draw $\vec{ST}$. Place the center of the protractor at S.

2. Use the scale that begins with 0 on $\vec{ST}$. Find the 150 mark and label point R. Draw $\vec{SR}$.

The measure of ∠RST is **150°**.

Self-Check Use a protractor to draw angles having the given measures.

4. 45°   5. 90°   6. 115°   7. 135°

CLASSROOM EXERCISES

For Exercises 1–6, choose the angle that is closest to the given measure. Choose a or b.   (Example 1)

1. 30°
   a. 
   b.

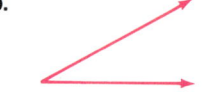

2. 120°
   a. 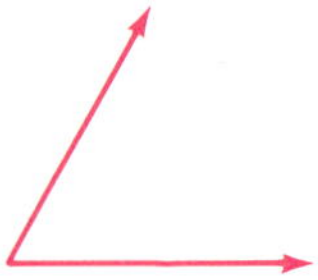
   b.

3. 60°
   a. 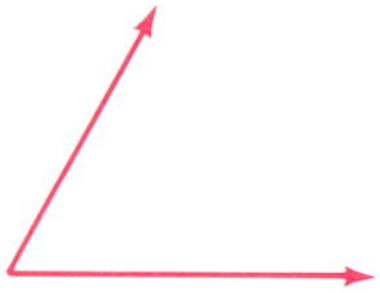
   b.

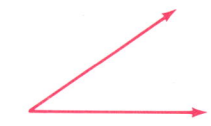

ADDITIONAL EXAMPLE

Example 2

Draw ∠ABC so that its measure is 85°.

Ans: Label a point B and a point C. Then draw $\overline{BC}$. Place the center of the protractor at B. Use the scale that begins with 0 on $\overline{AB}$. Find the 85 mark and label point A. Draw $\vec{BA}$. The measure of ∠ABC is 85°.

SELF-CHECK

For Exercises 4–7, check student's drawings.

CLASSROOM EXERCISES

1. b   2. b   3. a
4. a   5. a   6. b

502 Chapter 17

4. 170° **5.** 90° **6.** 45°

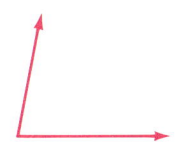

a.

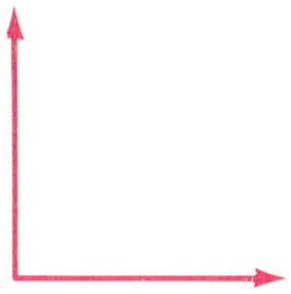

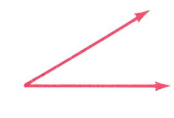

b.

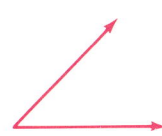

WRITTEN EXERCISES

Goals: To use a protractor to measure a given angle
To classify an angle according to its measure
To use a protractor to draw a given angle

For Exercises 1–9:

a. *Use the protractor to find the measure of each angle.*

b. *Identify each angle as right, acute, obtuse, or straight.*
(Example 1 and Table)

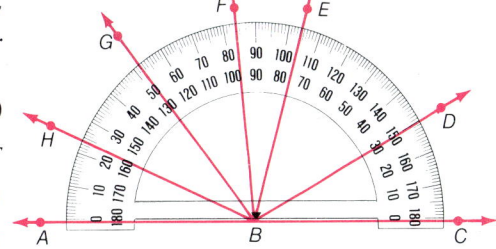

1. ∠CBD	2. ∠CBE	3. ∠CBF
4. ∠CBG	5. ∠CBH	6. ∠CBA
7. ∠ABH	8. ∠ABD	9. ∠EBA

For each of Exercises 10–19, use a protractor to draw an angle with the given measure. (Example 2)

| 10. 32° | 11. 47° | 12. 85° | 13. 90° | 14. 55° |
| 15. 105° | 16. 122° | 17. 137° | 18. 155° | 19. 175° |

MORE CHALLENGING EXERCISES

Use the protractor above to find the measure of each angle.

| 20. ∠DBE | 21. ∠FBG | 22. ∠EBF | 23. ∠GBH | 24. ∠DBF |

Geometry **503**

ASSIGNMENT GUIDE
BASIC
p. 503: Omit
AVERAGE
p. 503: Omit
ABOVE AVERAGE
p. 503: 1–24

WRITTEN EXERCISES
1. 30°; acute
2. 75°; acute
3. 95°; obtuse
4. 128°; obtuse
5. 155°; obtuse
6. 180°; straight
7. 25°; acute
8. 150°; obtuse
9. 105°; obtuse
For Exercises 10–19, teachers should check each student's drawings.
20. 45°   21. 33°
22. 20°   23. 27°
24. 65°

503

Teaching Suggestions
p. M-30

QUICK QUIZ
Classify each triangle.

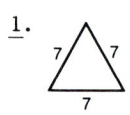

1. (triangle with sides 7, 7, 7)

Ans: equilateral; acute

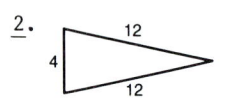

2. (triangle with sides 12, 4, 12)

Ans: isosceles; acute

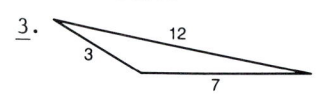

3. (triangle with sides 12, 3, 7)

Ans: scalene; obtuse

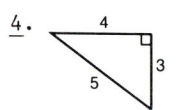

4. (right triangle with sides 4, 3, 5)

Ans: right; scalene

17-3 Properties of Triangles

The figures formed by the wood beams on the gable at the right are **closed plane figures**. Closed plane figures formed by line segments are called **polygons**.

A **triangle** is a polygon with three sides and three angles. The gable at the right is in the shape of a triangle. The triangle formed by the gable can be named as follows.

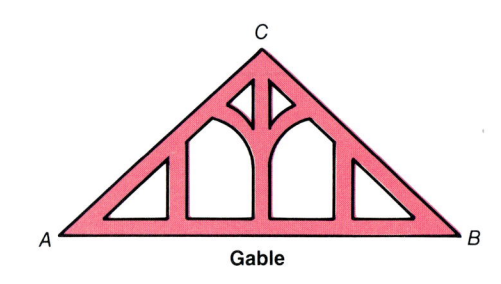

Gable

$\triangle ABC$, $\triangle BCA$, or $\triangle CAB$

Read: "Triangle ABC, triangle BCA, or triangle CAB."

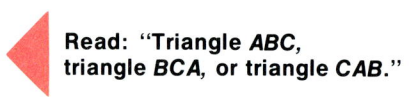

Triangles can be classified by the lengths of their sides.

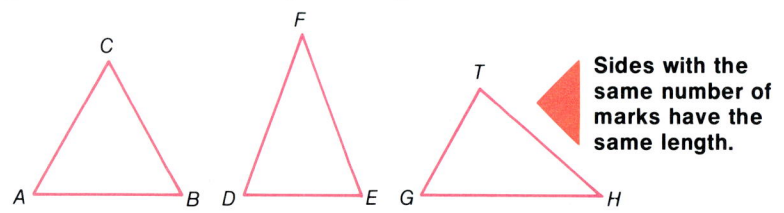

Sides with the same number of marks have the same length.

TABLE 1

Triangle	Name	Description
$\triangle ABC$	Equilateral	All sides have the same length.
$\triangle DEF$	Isosceles	Two sides have the same length.
$\triangle RST$	Scalene	No sides have the same length.

Triangles can also be classified by the measures of their angles.

TABLE 2

Right Triangle	Acute Triangle	Obtuse Triangle
One angle is a right angle.	All three angles are acute angles.	One angle is an obtuse angle.

504   Chapter 17

Congruent triangles are triangles that have the same shape and size.

Triangles *ABC* and *DEF* shown on the gate at the right are congruent triangles.

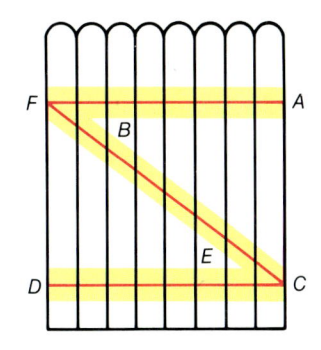

In symbols,
$\triangle ABC \cong \triangle DEF$

▶ Read: "△ *ABC* is congruent to △ *DEF*."

In congruent triangles, **corresponding angles** have the same measure and **corresponding sides** have the same length.
Note that corresponding congruent sides are **opposite** corresponding congruent angles.

TABLE 3

Corresponding Congruent Angles	Corresponding Congruent Sides
$\angle A \cong \angle D$	$\overline{BC} \cong \overline{FE}$
$\angle B \cong \angle E$	$\overline{AC} \cong \overline{FD}$
$\angle C \cong \angle F$	$\overline{AB} \cong \overline{DE}$

To show that two triangles are congruent, it is *not* necessary to know that all six corresponding parts are congruent.

Side–Angle–Side (SAS)

If two corresponding sides and their included angles are congruent, then the triangles are congruent.

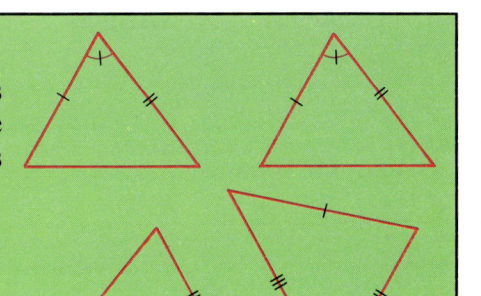

Side–Side–Side (SSS)

If three corresponding sides are congruent, then the triangles are congruent.

Angle–Side–Angle (ASA)

If two corresponding angles and their included side are congruent, then the triangles are congruent.

NOTE: Tick marks show corresponding congruent parts.

Geometry

ADDITIONAL EXAMPLES

Example

Determine whether each pair of triangles is congruent by SAS, SSS, or ASA. Give a reason.

1. 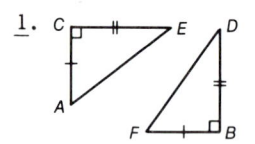 [figure]

Ans: SAS; $\overline{CE} \cong \overline{BD}$, $\angle C \cong \angle B$, $\overline{CA} \cong \overline{BF}$

2. 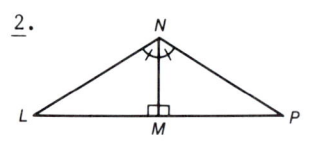 [figure]

Ans: ASA; $\angle LNM \cong \angle PNM$, $\overline{NM} \cong \overline{NM}$, $\angle LMN \cong \angle PMN$

SELF-CHECK

For Exercises 1–3, check student's drawings

CLASSROOM EXERCISES

1. isosceles
2. equilateral
3. scalene
4. isosceles
5. acute
6. obtuse
7. right
8. acute
9. $\angle M$ 10. $\overline{PM}$ 11. $\overline{PQ}$
12. $\angle P$ 13. $\angle T$
14. $\overline{RT}$

506

EXAMPLE 1 *Determine whether each pair of triangles is congruent by SAS, SSS, or ASA. Give a reason for each answer.*

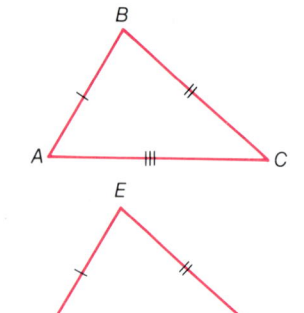

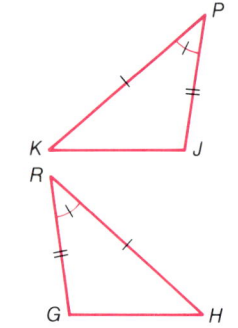

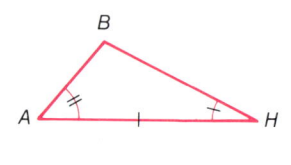

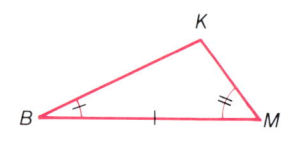

[figures of triangle pairs]

Solutions:
a. **SSS**, because $\overline{AB} \cong \overline{ED}$, $\overline{BC} \cong \overline{EF}$, and $\overline{CA} \cong \overline{FD}$.
b. **SAS**, because $\overline{KP} \cong \overline{HR}$, $\angle P \cong \angle R$, and $\overline{PJ} \cong \overline{RG}$.
c. **ASA**, because $\angle A \cong \angle M$, $\overline{AH} \cong \overline{MB}$, and $\angle H \cong \angle B$.

Self-Check *Draw a pair of triangles to illustrate each kind of congruence. Mark the corresponding congruent parts.*

1. SSS   2. SAS   3. ASA

CLASSROOM EXERCISES

The sides of a triangle have the lengths shown. Classify each triangle as equilateral, isosceles, or scalene. (Table 1)

1. 4 in; 5 in; 4 in
2. 6 cm; 6 cm; 6 cm
3. 7 in; 9 in; 12 in
4. 8 cm; 10 cm; 8 cm

The angles of a triangle have the measures shown. Classify each triangle as acute, right, or obtuse. (Table 2)

5. 89°; 43°; 48°
6. 25°; 20°; 135°
7. 30°; 60°; 90°
8. 60°; 60°; 60°

For Exercises 9–14, it is given that triangles RST and MPQ are congruent. Complete each statement. (Table 3)

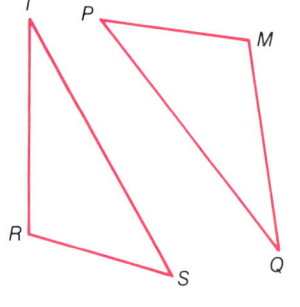

9. $\angle R \cong$ _?_
10. $\overline{RS} \cong$ _?_
11. $\overline{ST} \cong$ _?_
12. $\angle S \cong$ _?_
13. _?_ $\cong \angle Q$
14. _?_ $\cong \overline{QM}$

506 Chapter 17

WRITTEN EXERCISES

Goals: To classify triangles according to the lengths of their sides

To classify triangles according to the measures of their angles

To determine whether triangles are congruent by SSS, SAS, or ASA

Classify each triangle as equilateral, isosceles, or scalene. (Table 1)

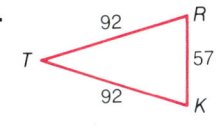

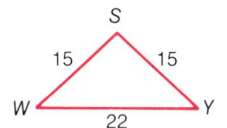

1. Triangle CED with sides 8, 8, 8.
2. Triangle TRK with sides 92, 57, 92.
3. Triangle WSY with sides 15, 15, 22.
4. Triangle NMP with sides 15, 17, 8.

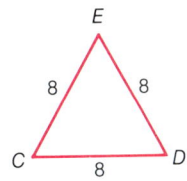

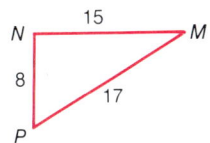

Determine whether each pair of triangles is congruent by SSS, SAS, or ASA. Name the corresponding congruent angles and sides. (Example 1)

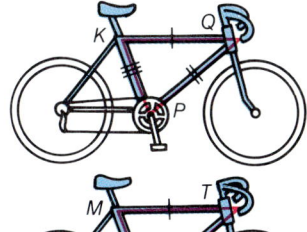

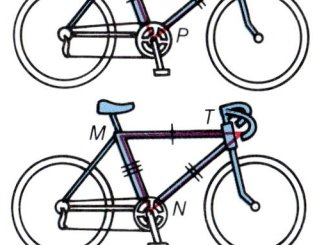

5. Coat Hangers 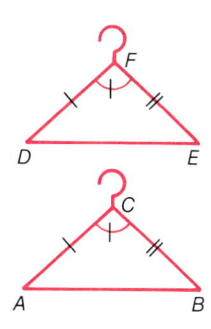
6. Bicycle Frames
7. Tent Flaps 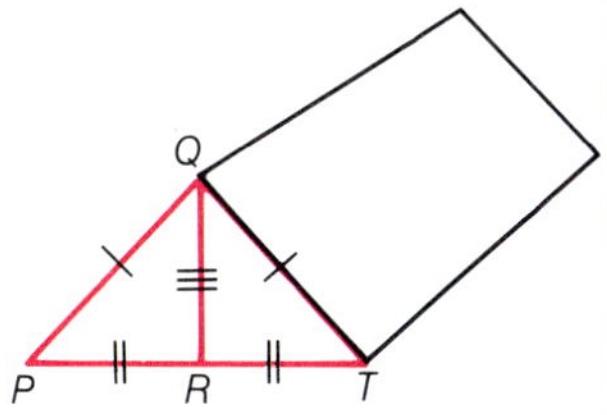
8. Antenna 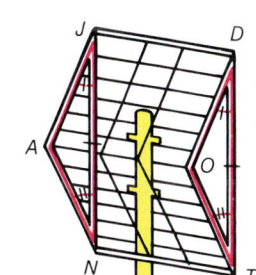
9. Lean-to 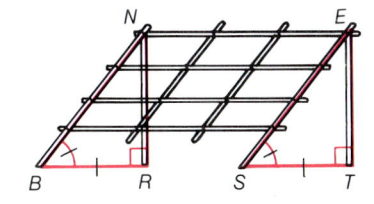
10. Table Legs

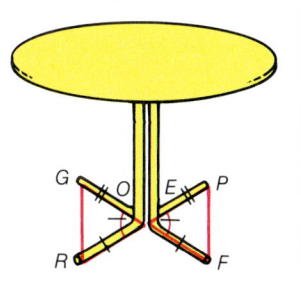

ASSIGNMENT GUIDE

BASIC
p. 507: Omit

AVERAGE
p. 507: Omit

ABOVE AVERAGE
P. 507: 1–10

WRITTEN EXERCISES

1. equilateral
2. isosceles
3. isosceles
4. scalene
5. $\overline{DF} \cong \overline{AC}$; $\angle F \cong \angle C$; $\overline{FE} \cong \overline{CB}$
6. $\overline{KQ} \cong \overline{MT}$; $\overline{QP} \cong \overline{TN}$; $\overline{PK} \cong \overline{NM}$
7. $\overline{PQ} \cong \overline{TQ}$; $\overline{PR} \cong \overline{RT}$; $\overline{QR} \cong \overline{QR}$
8. $\angle J \cong \angle D$; $\overline{JN} \cong \overline{DT}$; $\overline{LN} \cong \overline{LT}$
9. $\angle B \cong \angle S$; $\overline{BR} \cong \overline{ST}$; $\overline{LR} \cong \overline{LT}$
10. $\overline{GO} \cong \overline{EP}$; $\angle O \cong \angle E$; $\overline{OR} \cong \overline{EF}$

Geometry **507**

REVIEW CAPSULE FOR SECTION 17-4

Solve each proportion. (Pages 219–221)

1. $\dfrac{n}{18} = \dfrac{15}{20}$
2. $\dfrac{10}{9} \times \dfrac{40}{x}$
3. $\dfrac{16}{p} = \dfrac{4}{2}$
4. $\dfrac{a}{12} = \dfrac{2}{1}$
5. $\dfrac{11}{8} = \dfrac{b}{24}$
6. $\dfrac{25}{t} = \dfrac{5}{3}$

Round each number to the nearest tenth. (Pages 32–34)

7. 9.83
8. 11.88
9. 11.58
10. 12.85
11. 9.05
12. 11.51

The answers are on page 533.

PROBLEM SOLVING AND APPLICATIONS

17-4 Using Similar Triangles

Similar triangles are triangles that have the same shape.

> In similar triangles, angles that correspond have **equal measures**.
>
> In similar triangles, sides that cprrespond are **opposite equal angles**.
>
> Sides that correspond have **equivalent ratios**.

When you know that triangles are similar, you can write a proportion to find the length of an unknown side.

TABLE	SIMILAR TRIANGLES	EQUIVALENT RATIOS

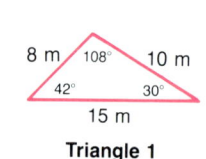

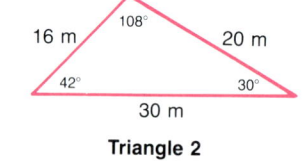

Triangle 1: 8 m, 10 m, 15 m; angles 108°, 42°, 30°
Triangle 2: 16 m, 20 m, 30 m; angles 108°, 42°, 30°

Triangle 1 | Triangle 2
$\dfrac{8}{15} = \dfrac{16}{30}$
$\dfrac{8}{10} = \dfrac{16}{20}$
$\dfrac{10}{15} = \dfrac{20}{30}$

> **PROCEDURE**
>
> **To find the length of an unknown side:**
>
> 1 Use the unknown side and a known side in one triangle to write a ratio.
>
> 2 Use sides that correspond in the second triangle to write another ratio.
>
> 3 Use the two ratios to write a proportion.
>
> 4 Solve the proportion.

508 Chapter 17

Teaching Suggestions p. M-30

QUICK QUIZ

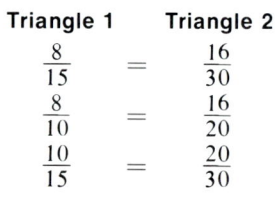

Solve.

1. $\dfrac{x}{10} = \dfrac{21}{35}$ Ans: x = 6
2. $\dfrac{7}{14} = \dfrac{1}{y}$ Ans: y = 2
3. $\dfrac{6}{a} = \dfrac{10}{45}$ Ans: a = 27

Round to the nearest tenth.

4. 21.632   Ans: 21.6
5. 1.78   Ans: 1.8
6. 7.182   Ans: 7.2

EXAMPLE Jill is 1.8 meters tall. On a sunny day, her shadow was 2 meters long at the same time that the shadow of a water tower was 40 meters long. Find the height of the tower.

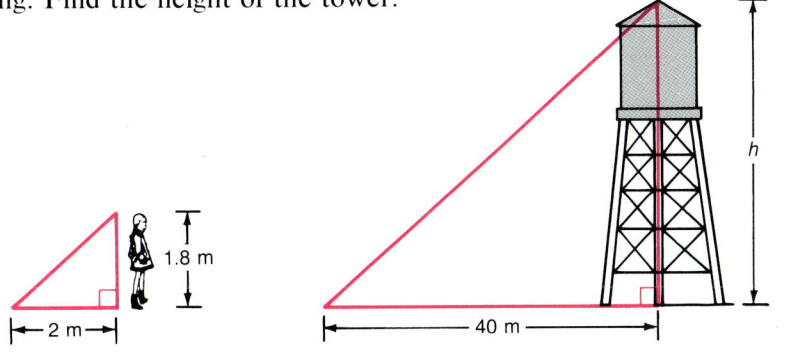

Solution:

1. Use the triangle with the unknown side to write a ratio. $\dfrac{h}{40}$ ← Tower's height / Shadow length

2. Use sides that correspond in the second triangle to write another ratio. $\dfrac{1.8}{2}$ ← Jill's height / Shadow length

3. Use the ratios to write a proportion. $\dfrac{h}{40} = \dfrac{1.8}{2}$

4. Solve the proportion.
$$h \times 2 = 40 \times 1.8$$
$$2h = 72$$
$$h = 36$$

The tower is 36 meters high.

Self-Check

1. Bill is 6 feet tall. On a sunny day, Bill's shadow is 8 feet long and the shadow of a high diving platform is 52 feet long. Find the height of the diving platform.

CLASSROOM EXERCISES

For Exercises 1–4, each pair of triangles is similar. Complete the proportion. (Table)

1. $\dfrac{n}{10} = \dfrac{?}{20}$

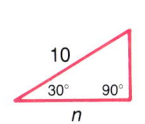

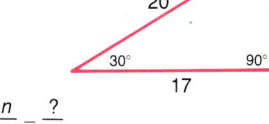

2. $\dfrac{n}{56} = \dfrac{20}{?}$

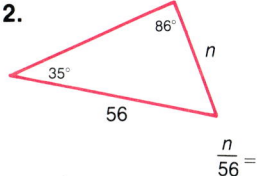

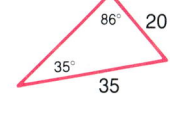

ADDITIONAL EXAMPLE

Example
Solve using a proportion.
A flag pole casts a shadow of 9 ft at the same time a shadow of 1.5 ft is cast by a 6-foot man. How high is the top of the flag pole?
Ans: $\dfrac{9}{x} = \dfrac{1.5}{6}$; 36 ft

SELF-CHECK
39 ft

CLASSROOM EXERCISES
1. 17   2. 35

Geometry

CLASSROOM EXERCISES

3. $\dfrac{1.8}{1.2}$ 4. $\dfrac{85}{60}$

ASSIGNMENT GUIDE
BASIC
p. 510-511: Omit
AVERAGE
p. 510-511: Omit
ABOVE AVERAGE
p. 510-511: 1-7

WRITTEN EXERCISES
1. 13   2. 60   3. 20.3
4. 133.3   5. 6.4 m

(Example)

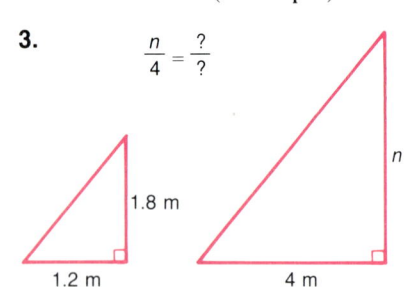

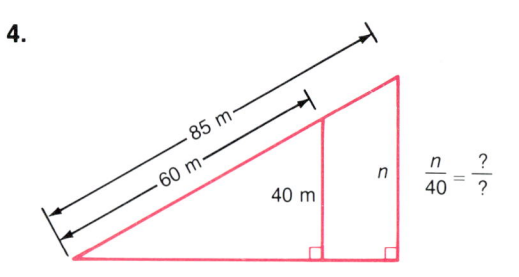

3. $\dfrac{n}{4} = \dfrac{?}{?}$

1.8 m, 1.2 m, n, 4 m

4. 85 m, 60 m, n, 40 m $\dfrac{n}{40} = \dfrac{?}{?}$

WRITTEN EXERCISES

Goals: To use a proportion to solve problems involving similar triangles
To apply this skill to solving word problems

Each pair of triangles is similar. Write a proportion to find the unknown length. Then solve the proportion. When necessary, round your answer to the nearest tenth. (Example)

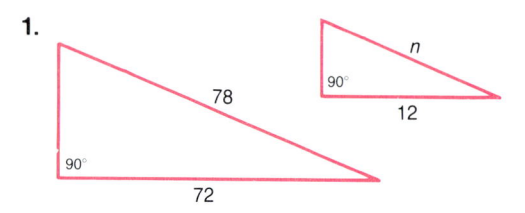

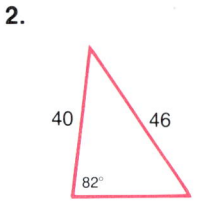

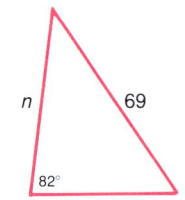

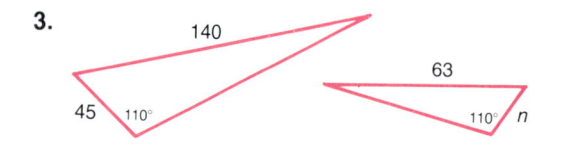

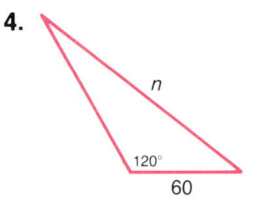

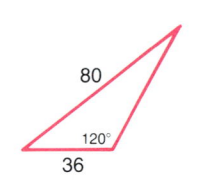

1. 78, 72, 90°; n, 12, 90°

2. 40, 46, 82°; n, 69, 82°

3. 140, 45, 110°; 63, n, 110°

4. n, 120°, 60; 80, 120°, 36

For Exercises 5–7, use a proportion to solve each problem. (Example)

5. Norman is 1.8 meters tall. On a sunny day, Patty measured Norman's shadow and the shadow of the school. Use the similar triangles shown below to find the height of the school. Round the answer to the nearest tenth.

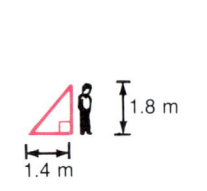

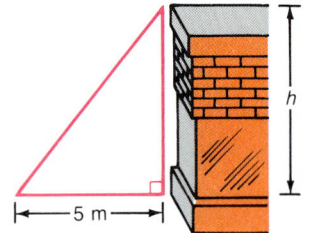

1.8 m, 1.4 m; h, 5 m

510   Chapter 17

6. Jaimie placed a mirror on the sidewalk. Then he walked backwards until he saw the top of the streetlight in the mirror. The figure below shows how he formed similar triangles. Find the height of the street lamp. Round your answer to the nearest tenth of a meter.

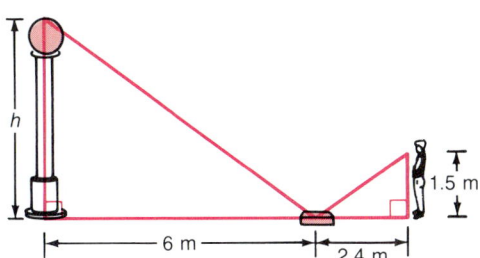

7. Find the distance across this lake. Round your answer to the nearest tenth of a meter.

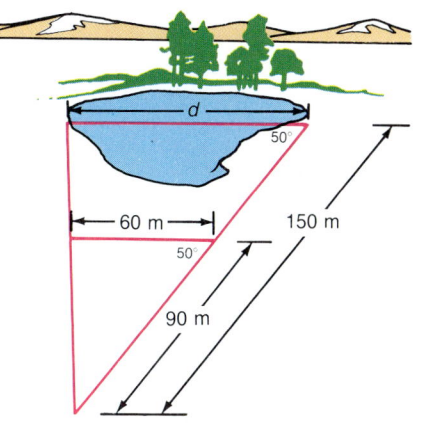

6. 3.75 m   7. 100 m

SIMILAR POLYGONS

You can use a scientific calculator and the ratio of corresponding sides of two similar triangles to find the lengths of unknown sides.

EXAMPLE The ratio of corresponding sides of similar triangles ABC and DEF is $\frac{11}{8}$. Find EF and DE when CB = 14 and AB = 17.

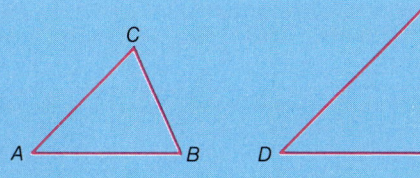

Solution Use the constant and multiplication keys (and) to store the ratio $\frac{11}{8}$. Then enter each known value and press

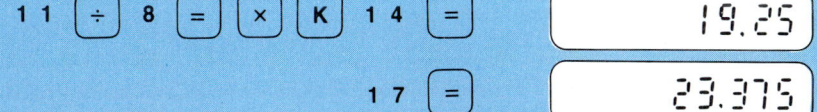

1 1 ÷ 8 = × K 1 4 =   19.25

1 7 =   23.375

EXERCISES

Use a scientific calculator to check the solutions to Exercises 1–4 on page 510.

CALCULATOR EXERCISES

1. 13   2. 60   3. 20.3
4. 133.3

Geometry 511

511

COMPUTER APPLICATIONS

This feature illustrates the use of a computer program for determining whether two angles are supplementary and for finding the measure of each base angle of an isosceles triangle when the measure of the vertex angle is known. It can be used after Section 17-3 (pages 504-507) is taught.

1. Angles A and B are supplementary.
2. Angles A and B are supplementary.
3. Angles A and B are not supplementary.
4. Angles A and B are supplementary.
5. Angles A and B are not supplementary.
6. Angles A and B are not supplementary.
7. 70°
8. 55°
9. 30°
10. 73.75°
11. 82.1°
12. 43.85°

COMPUTER APPLICATIONS Angle Measures

If you are given the measures of two angles greater than 0 and less than 180, you can use the following program to determine if they are supplementary.

Program 1:

```
100 PRINT "ENTER THE ANGLE MEASURES"
110 INPUT A, B          ← These are the angle measures.
120 IF A + B = 180 THEN 150  ← If A + B ≠ 180, then 130.
130 PRINT "ANGLES A AND B ARE NOT SUPPLEMENTARY."
140 GOTO 160
150 PRINT "ANGLES A AND B ARE SUPPLEMENTARY."
160 PRINT "ANY MORE ANGLES (1 = YES, 0 = NO)";
170 INPUT X             ← X = 1 or X = 0
180 IF X = 1 THEN 100
190 END
```

If you know the measure of the vertex angle of an isosceles triangle, you can use the following program to compute the measure of each base angle.

Program 2:

```
10 PRINT "ENTER THE MEASURE OF THE VERTEX ANGLE."
20 INPUT V              ← V is the measure of the vertex angle.
30 LET B = (180 - V) / 2  ← B is the measure of a base angle.
40 PRINT "THE MEASURE OF EACH ANGLE IS"; B
50 PRINT "ANY MORE ANGLES (1 = YES, 0 = NO)";
60 INPUT X
70 IF X = 1 THEN 10
80 END
```

NOTE: Statement 30 gives the formula for finding the measure of the base angle. It stores this value in memory location B.

Exercises

Run Program 1 for the following values.

1. $A = 40$; $B = 140$
2. $A = 60$; $B = 120$
3. $A = 82$; $B = 96$
4. $A = 95$; $B = 85$
5. $A = 62.3$; $B = 97.6$
6. $A = 42.5$; $B = 138.5$

Run Program 2 for the following values of V.

7. 40
8. 70
9. 120
10. 32.5
11. 15.8
12. 92.3

Computer Application

REVIEW: SECTIONS 17-1—17-4

For Exercises 1–3, refer to the figure at the right.
(Section 17-1)

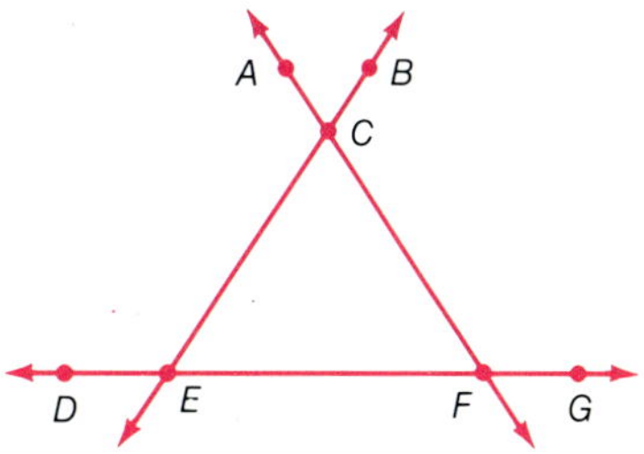

1. Name three different rays that have E as an endpoint.
2. Name two lines that contain point C.
3. Name four angles with C as a vertex.

Draw an angle with the given measure. (Section 17-2)

4. 30°
5. 87°
6. 112°
7. 90°
8. 45°

Determine whether each pair of triangles is congruent by SSS, SAS, or ASA. Name the corresponding congruent parts. (Section 17-3)

9. 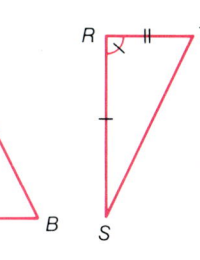 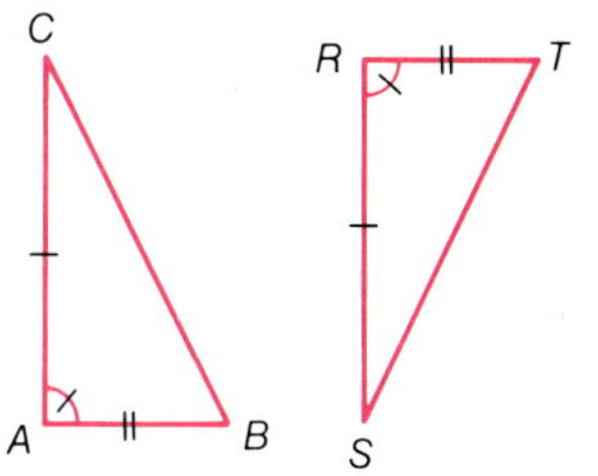
10. 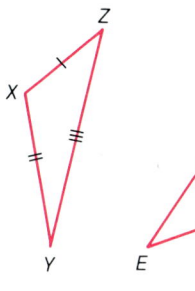 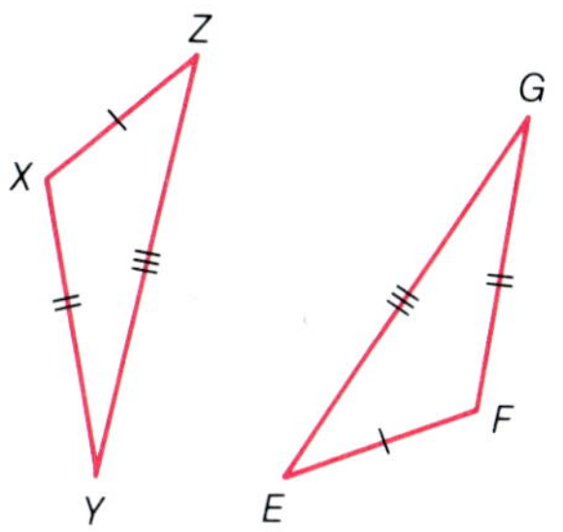
11.

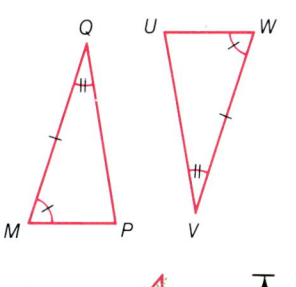

12. On a sunny day, the shadow of a tree was 12 meters long. Use the similar triangles at the right to find the height of the tree. Round your answer to the nearest tenth of a meter.

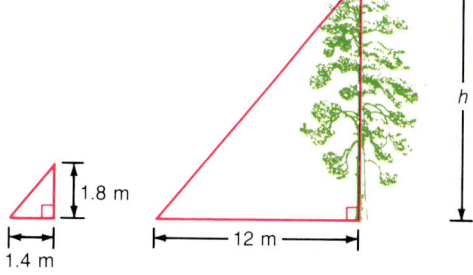

REVIEW CAPSULE FOR SECTION 17-5

Match one or more symbols in Column A with each figure in Exercises 1–4. (Pages 498–500)

The answers are on page 533.

1. 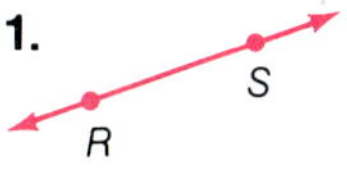
2. 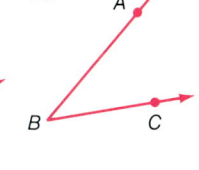
3. 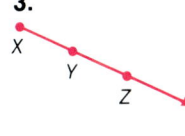
4.

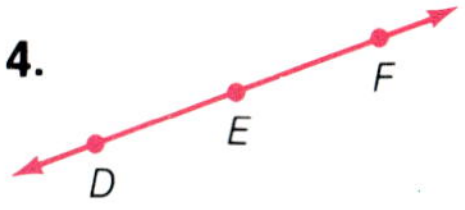

COLUMN A

a. $\angle ABC$ e. $\overrightarrow{XY}$
b. $\overleftrightarrow{RS}$ f. $\overleftrightarrow{DF}$
c. $\overleftrightarrow{FD}$ g. $\overrightarrow{SR}$
d. $\angle CBA$ h. $\overleftrightarrow{XZ}$

Geometry **513**

QUIZ: SECTIONS 17-1-17-4
After completing this Review, you may wish to administer a quiz covering the same sections. A Quiz is provided in the Teacher's Edition: Part II.

REVIEW: SECTIONS 17-1-17-4
1. $\overrightarrow{ED}$; $\overrightarrow{EG}$; $\overrightarrow{EB}$
2. $\overleftrightarrow{EB}$; $\overleftrightarrow{AF}$ 3. $\angle ACB$; $\angle BCF$; $\angle FCE$; $\angle ECA$
For Exercises 4-8, check student's drawings.
9. SAS; $\overline{AC} \cong \overline{RS}$; $\angle A \cong \angle R$; $\overline{AB} \cong \overline{RT}$
10. SSS; $\overline{XY} \cong \overline{EG}$; $\overline{XZ} \cong \overline{GF}$; $\overline{YZ} \cong \overline{EF}$
11. ASA; $\angle M \cong \angle V$; $\overline{MQ} \cong \overline{VW}$; $\angle Q \cong \angle W$
12. 15.43

Teaching Suggestions
p. M-30

QUICK QUIZ

Give the proper symbol for each figure.

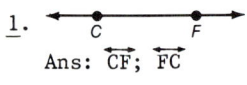

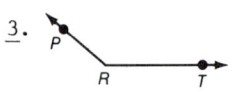

1. •——————•
     C   F
   Ans: $\overleftrightarrow{CF}$; $\overleftrightarrow{FC}$

2. •——————→
    A   B
   Ans: $\overrightarrow{AB}$

3.
   Ans: ∠PRT; ∠R; ∠TRP

Name the corresponding sides and angles

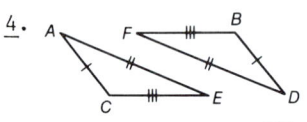

4.
   Ans: $\overline{AC} \cong \overline{DB}$; $\overline{AE} \cong \overline{DF}$;
   $\overline{CE} \cong \overline{BF}$; ∠A ≅ ∠D;
   ∠E ≅ ∠F; ∠C ≅ ∠B

ADDITIONAL EXAMPLES

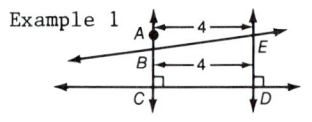

Example 1

Identify each of the following from the figure.

1. 2 parallel lines
   Ans: $\overleftrightarrow{AB}$ and $\overleftrightarrow{ED}$

2. 2 sets of perpendicular lines
   Ans: $\overleftrightarrow{BC}$ and $\overleftrightarrow{CD}$;
   $\overleftrightarrow{ED}$ and $\overleftrightarrow{CD}$

SELF-CHECK

1. $\overleftrightarrow{BC}$ and $\overleftrightarrow{CD}$
2. $\overleftrightarrow{CD} \perp \overleftrightarrow{AC}$

514

17-5 Perpendicular and Parallel Lines

In the figure at the right, boats 1 and 2 are sailing on parallel courses (lines). **Parallel lines** are lines in a plane that are always the same distance apart.

$\overleftrightarrow{AC} \parallel \overleftrightarrow{DF}$ Read: "Line *AC* is parallel to line *DF*."

The course of boat 3 meets (intersects) the courses of boats 1 and 2 at right angles. In a plane, **intersecting lines** have exactly one point in common. When two intersecting lines form a right angle, the lines are **perpendicular lines**.

$\overleftrightarrow{DF} \perp \overleftrightarrow{BE}$ Read: "Line *DF* is perpendicular to line *BE*."

EXAMPLE 1   Identify each of the following. Refer to the figure at the right.

  a. A pair of parallel lines

  b. A pair of intersecting lines that are not perpendicular

  c. A pair of perpendicular lines

Solutions: **a.** $\overleftrightarrow{AB} \parallel \overleftrightarrow{CD}$.   **b.** $\overleftrightarrow{BC}$ intersects $\overleftrightarrow{AB}$ at *B*.   **c.** $\overleftrightarrow{AB} \perp \overleftrightarrow{AC}$.

Self-Check   *Identify each of the following. Refer to the figure of Example 1.*

  1. A second pair of intersecting lines

  2. A second pair of perpendicular lines

A **transversal** is a line that intersects two other lines. When parallel lines are cut by a transversal, pairs of **congruent** angles are formed.

Recall that congruent angles have equal measures.

∠1 ≅ ∠3   Read: "Angle 1 is congruent to angle 3."

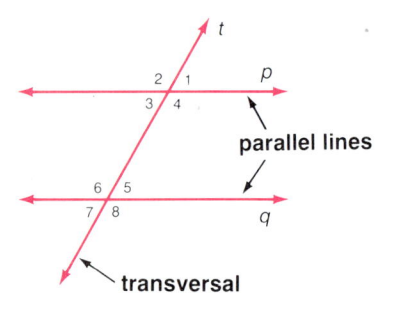

514   Chapter 17

The following table classifies the pairs of congruent angles formed when a transversal intersects two parallel lines. Refer to the figure at the bottom of page 514.

TABLE

Pairs of Congruent Angles	Name
∠1 ≅ ∠3, ∠2 ≅ ∠4, ∠5 ≅ ∠7, ∠6 ≅ ∠8	Vertical Angles
∠3 ≅ ∠5, ∠4 ≅ ∠6	Alternate interior angles
∠1 ≅ ∠7, ∠2 ≅ ∠8	Alternate exterior angles
∠1 ≅ ∠5, ∠2 ≅ ∠6, ∠3 ≅ ∠7, ∠4 ≅ ∠8	Corresponding angles

NOTE: A line can be also named with a single letter. Thus, in the figure at the bottom of page 514, the transversal is line t and lines p and q are parallel.

EXAMPLE 2 In the figure at the right, lines m, n, and p are parallel to each other. Line t is a transversal. Classify each pair of congruent angles as vertical, alternate interior, alternate exterior, or corresponding.

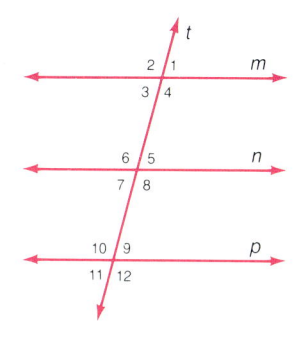

    a. ∠1 and ∠3   **b.** ∠3 and ∠5
    c. ∠5 and ∠9   **d.** ∠5 and ∠11

Solutions: **a.** Vertical   **b.** Alternate interior
              c. Corresponding   **d.** Alternate exterior

Self-Check Classify each pair of congruent angles in Example 2 as corresponding, alternate interior, alternate exterior, or vertical.

    3. ∠4 and ∠2   **4.** ∠7 and ∠9   **5.** ∠6 and ∠4   **6.** ∠9 and ∠11

CLASSROOM EXERCISES

For Exercises 1–6, identify each pair of lines as intersecting, parallel, or perpendicular. Some exercises may have more than one answer. (Example 1)

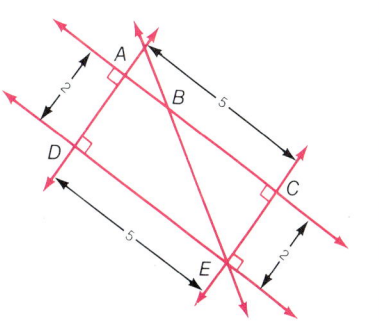

1. $\overleftrightarrow{AB}$ and $\overleftrightarrow{AD}$ **2.** $\overleftrightarrow{BE}$ and $\overleftrightarrow{DE}$ **3.** $\overleftrightarrow{AD}$ and $\overleftrightarrow{AC}$
4. $\overleftrightarrow{AD}$ and $\overleftrightarrow{DE}$ **5.** $\overleftrightarrow{AB}$ and $\overleftrightarrow{DE}$ **6.** $\overleftrightarrow{AD}$ and $\overleftrightarrow{CE}$

Geometry 515

ADDITIONAL EXAMPLES
Example 2

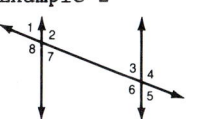

Classify as vertical, alternate interior, alternate exterior, or corresponding.
1. ∠1 and ∠3
   Ans: corresponding
2. ∠8 and ∠4
   Ans: alternate exterior
3. ∠3 and ∠5
   Ans: vertical
4. ∠6 and ∠2
   Ans: alternate interior

SELF-CHECK
3. vertical
4. alternate interior
5. alternate interior
6. vertical

CLASSROOM EXERCISES
1. intersecting; perpendicular
2. intersecting
3. intersecting; perpendicular
4. intersecting; perpendicular
5. parallel
6. parallel

CLASSROOM EXERCISES

7. b **8.** a **9.** a
10. c **11.** c **12.** d

ASSIGNMENT GUIDE
BASIC
p. 516: Omit
AVERAGE
p. 516: Omit
ABOVE AVERAGE
p. 516: 1–12

WRITTEN EXERCISES
1. $\overleftrightarrow{BA} \perp \overleftrightarrow{AC}$; $\overleftrightarrow{CD} \perp \overleftrightarrow{AC}$; $\overleftrightarrow{AF} \perp \overleftrightarrow{FD}$; $\overleftrightarrow{CD} \perp \overleftrightarrow{FD}$
2. $\overleftrightarrow{AC} \parallel \overleftrightarrow{FD}$; $\overleftrightarrow{AF} \parallel \overleftrightarrow{CD}$
3. $\overleftrightarrow{BE}$ and $\overleftrightarrow{BF}$; $\overleftrightarrow{BE}$ and $\overleftrightarrow{CD}$; $\overleftrightarrow{BE}$ and $\overleftrightarrow{AC}$; $\overleftrightarrow{BE}$ and $\overleftrightarrow{FD}$
4. corresponding
5. vertical
6. corresponding
7. alternate interior
8. vertical
9. corresponding
10. alternate exterior
11. alternate exterior
12. alternate interior

For Exercises 7–12, parallel lines AB and CD intersect transversals EF and GH. Match each pair of congruent angles with the correct name. Choose a, b, c, or d. Example 2)

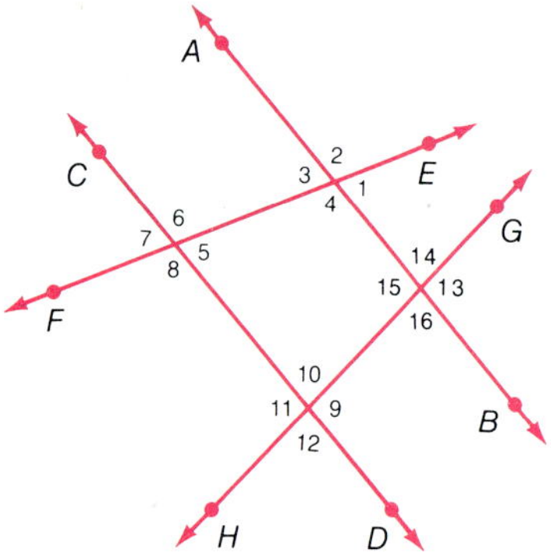

7. ∠3 and ∠5
8. ∠10 and ∠12
9. ∠5 and ∠7
10. ∠2 and ∠8
11. ∠13 and ∠11
12. ∠14 and ∠10

a. Vertical angles
b. Alternate interior angles
c. Alternate exterior angles
d. Corresponding angles

WRITTEN EXERCISES

Goals: To identify intersecting, perpendicular, and parallel lines

To identify the angle relationships that occur when parallel lines are cut by a transversal

Identify each of the following. Refer to the figure at the right.
(Example 1)

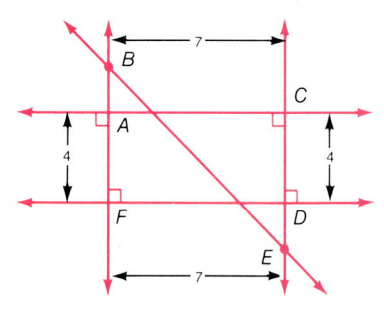

1. Four pairs of perpendicular lines
2. Two pairs of parallel lines
3. Four pairs of intersecting lines that are not perpendicular

In the figure at the right, $\overleftrightarrow{AB}$ is parallel to $\overleftrightarrow{CD}$ and $\overleftrightarrow{EF}$ is parallel to $\overleftrightarrow{GH}$. Classify each pair of congruent angles as corresponding, alternate interior, alternate exterior, or vertical (Example 2)

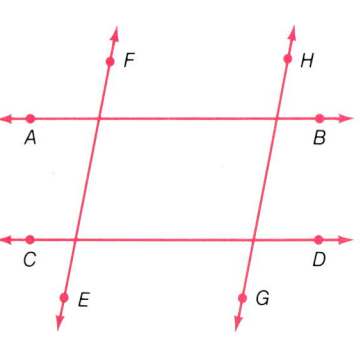

4. ∠1 and ∠13   **5.** ∠13 and ∠15   **6.** ∠15 and ∠3
7. ∠3 and ∠5   **8.** ∠5 and ∠7   **9.** ∠7 and ∠11
10. ∠11 and ∠13 **11.** ∠2 and ∠16   **12.** ∠13 and ∠3

516 Chapter 17

17-6 Properties of Quadrilaterals

The wings of the model plane at the right are in the shape of quadrilaterals. A **quadrilateral** is a polygon having four sides. The following table lists some quadrilaterals having special names and properties.

Quadrilateral	Figure	Definition
Trapezoid ABCD		Quadrilateral with exactly one pair of opposite sides parallel ($\overline{DC}$ and $\overline{AB}$ in this figure)
Parallelogram EFGH		Quadrilateral with both pairs of of opposite sides parallel
Rectangle KLIJ		Parallelogram with four right angles
Rhombus MPQN		Parallelogram with four congruent sides
Square WXYZ		Parallelogram with four congruent sides and four right angles

NOTE: Rectangles, rhombuses, and squares are all parallelograms. In all parallelograms, opposite sides are equal and opposite angles are equal.

Geometry 517

Teaching Suggestions
p. M-30

QUICK QUIZ

Name each polygon.

1.
Ans: square

2.
Ans: rectangle

3.
Ans: parallelogram

4.
Ans: octagon

5.
Ans: equilateral triangle

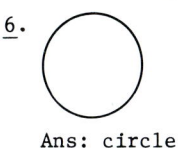

6.
Ans: circle

ADDITIONAL EXAMPLES

Example

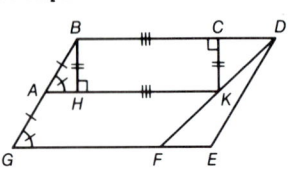

Refer to the diagram and classify the following.

1. AKFG

    Ans: trapezoid

2. CKHB

    Ans: parallelogram; rectangle

SELF-CHECK

1. trapezoid
2. parallelogram; rectangle
3. parallelogram; rectangle

CLASSROOM EXERCISES

1. Yes; Yes
2. No; No; No; Yes; Yes
3. No; Yes; Yes; Yes; Yes
4. No; Yes; Yes; Yes; Yes
5. No; No; Yes; No; Yes
6. Yes; No; No; No; No

EXAMPLE 1 In the figure at the right, line segments *AD*, *GE*, and *MH* are parallel, segments *MA*, *KG*, *JF*, and *HD* are parallel, and segments *MB*, *LC*, and *KD* are parallel. Refer to the figure to classify each figure as a trapezoid, parallelogram, rectangle, rhombus, or square.

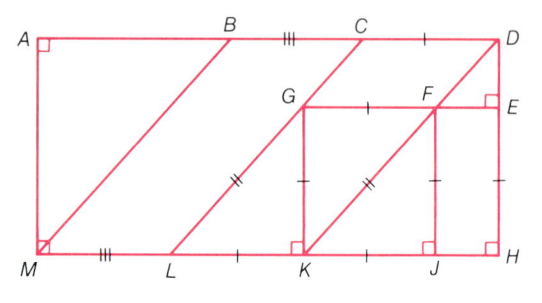

a. FEHJ   b. ACLM   c. GFJK   d. CDFG

Solutions: Refer to the definitions in the Table on page 517 as needed.

	Figure	Trapezoid	Parallelogram	Rectangle	Rhombus	Square
a.	FEHJ	No	Yes	Yes	No	No
b.	ACLM	Yes	No	No	No	No
c.	GFJK	No	Yes	Yes	Yes	Yes
d.	CDFG	No	Yes	No	No	No

Self-Check Refer to the figure in Example 1 to classify each figure as a trapezoid, parallelogram, rectangle, rhombus, or square.

1. GFJL   2. AMHD   3. GEHK

CLASSROOM EXERCISES

Complete. Answer <u>Yes</u> *or* <u>No</u>.   (Table)

Property	Figure				
	Trapezoid	Parallelogram	Rectangle	Rhombus	Square
1. Opposite sides parallel	No	Yes	Yes	?	?
2. All sides congruent	?	?	?	?	?
3. Opposite sides congruent	?	?	?	?	?
4. Opposite angles congruent	?	?	?	?	?
5. Four right angles	?	?	?	?	?
6. Only one pair of parallel sides	?	?	?	?	?

518   Chapter 17

WRITTEN EXERCISES

Goal: To identify the properties of quadrilaterals

Determine the unknown measure. (Table)

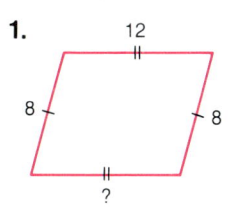

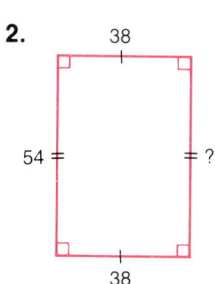

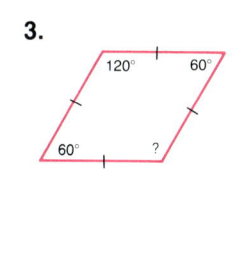

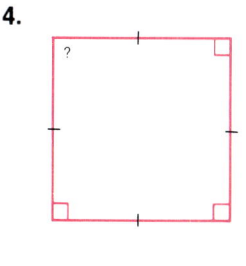

1. [rhombus with sides 12, 8, 8, ?]
2. [rectangle with sides 38, 54, ?, 38]
3. [parallelogram with angles 120°, 60°, 60°, ?]
4. [square with side ?]

In the figure below, line segments LA, HB, NK, and PC are parallel, segments EJ, RF, and AC are parallel, and segments PL and MG are parallel. Refer to the figure to classify each given quadrilateral as a trapezoid, parallelogram, rectangle, rhombus, or square. Some figures may have more than one name. (Example 2)

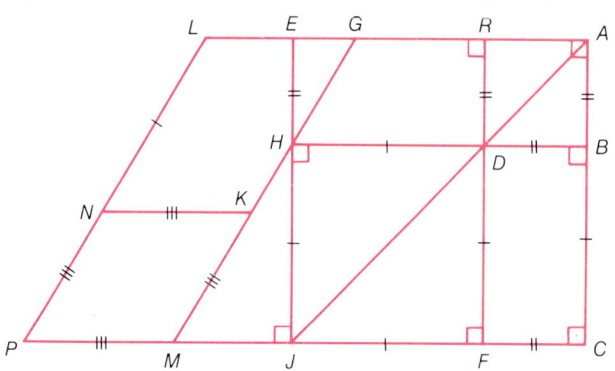

9. BCFD
10. BCJH
11. DFJH
12. GMPL
13. ACFD
14. ACPL
15. KMPN
16. ACJE
17. JELP
18. AJMG
19. ABHE
20. ABDR

MORE CHALLENGING EXERCISES

Complete. Use the words parallelogram, rhombus, rectangle, or square.

21. Every rectangle is also a __?__.
22. Every rhombus is also a __?__.
23. Every square is also a __?__, a __?__, and a __?__.
24. A parallelogram with four congruent angles is a __?__ or a __?__.
25. A parallelogram with four congruent sides is a __?__ or a __?__.

Geometry 519

ASSIGNMENT GUIDE
BASIC
p. 519: Omit
AVERAGE
p. 519: Omit
ABOVE AVERAGE
p. 519: 1-25

WRITTEN EXERCISES
1. 12   2. 54   3. 120°
4. 90° 5. 4.2 6. 90°
7. 115°; 65°
8. 2.3; 5.6
9. parallelogram; rectangle
10. parallelogram; rectangle
11. parallelogram; rectangle; rhombus; square
12. parallelogram
13. trapezoid
14. trapezoid
15. parallelogram; rhombus
16. parallelogram; rectangle; rhombus; square
17. trapezoid
18. trapezoid
19. parallelogram; rectangle
20. parallelogram; rectangle; rhombus; square

See page 520 for the answers to exercises 21-25.

519

Teaching Suggestions
p. M-30

QUICK QUIZ
Evaluate.
1. 2(21) + 2(79)
   Ans: 200
2. 3 · 4 + 4 · 5 + 5 · 6
   Ans: 62
3. 2(1.6) + 3(2.7)
   Ans: 11.3

Find the area of the rectangle.
4. l = 8 in; w = $5\frac{1}{2}$ in
   Ans: 44 in^2
5. l = 7.3 cm; w = 9.2 cm
   Ans: 67.16 cm^2

Find the area of the square.
6. s = 17 ft
   Ans: 289 ft^2
7. s = 3.6 mm
   Ans: 12.96 mm^2

WRITTEN EXERCISES
p. 519
21. parallelogram
22. parallelogram
23. rhombus; parallelogram; rectangle
24. rectangle; square
25. rhombus; square

REVIEW CAPSULE FOR SECTION 17-7

Evaluate each expression. (Pages 18–21)

1. 2 · 5 + 2 · 8
2. 3 · 5 + 2 · 9
3. 2(3.4) + 2(6.5)
4. 4(1.8) + 3(2.2)
5. 2 · $\frac{1}{2}$ + 2 · $\frac{1}{3}$
6. 3 · $1\frac{1}{2}$ + 2 · $1\frac{1}{4}$

Find the area of each rectangle. (Pages 70–73)

7. l = 5 m; w = 3 m
8. l = 1.4 cm; w = 0.5 cm
9. l = 13 ft; w = 8 ft

Find the area of each square. (Pages 70–73)

10. s = 7 in
11. s = $8\frac{1}{2}$ ft
12. s = 12.5 m

The answers are on page 533.

PROBLEM SOLVING AND APPLICATIONS

17-7 Surface Area and Rectangular Prisms

Each of the six sides or **faces** of a rectangular prism is a rectangle. The surface area is the combined areas of these six rectangles.

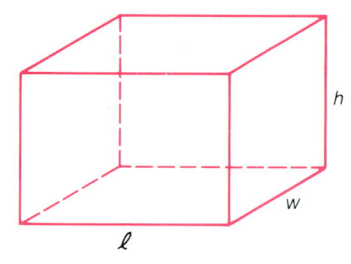

Word Rule: The **surface area** of a rectangular prism is the sum of the areas of the six faces.

The opposite faces of a rectangular prism, such as the top and bottom, have equal areas.

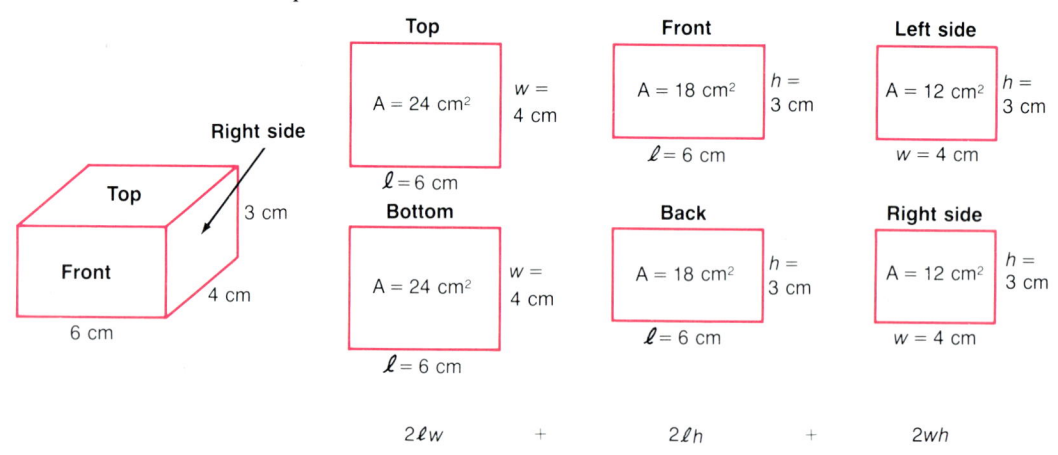

2ℓw + 2ℓh + 2wh

520 Chapter 17

520

Thus, the surface area of a rectangular prism can be found as follows.

Surface Area	=	Area of Top and Bottom	+	Area of Front and Back	+	Area of Left and Right Sides

Formula: $S = 2lw + 2lh + 2wh$

EXAMPLE Find the surface area of the box of paper clips shown at the right.

Solution:
1. Formula: $S = 2lw + 2lh + 2wh$
2. Known values: $l = 6$ cm; $w = 4$ cm; $h = 3$ cm
3. Replace l with 6, w with 4, and h with 3. $\quad S = 2(6)(4) + 2(6)(3) + 2(4)(3)$
4. Multiply. $\quad S = 48 + 36 + 24$
   $S = 108$

The surface area is **108 cm²**.

Self-Check 1. Find the surface area of a rectangular prism with a length of 6 centimeters, a width of 3 centimeters, and a height of 2 centimeters.

A **cube** is a special kind of rectangular prism. The length, width, and height are the same. The formula for the surface area of a cube can be written as follows.

$S = 6 \cdot s \cdot s$
$S = 6s^2$

s = length of any side

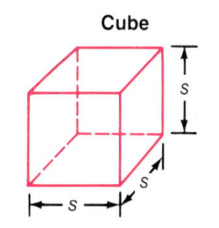

CLASSROOM EXERCISES

Complete. (Example)

	Length	Width	Height	2lw	2lh	2wh	Surface Area: 2lw + 2lh + 2wh
1.	4 m	3 m	5 m	?	?	?	?
2.	2.5 cm	4 cm	1 cm	?	?	?	?
3.	3 ft	4 ft	2 ft	?	?	?	?
4.	8 in	6 in	4 in	?	?	?	?

ADDITIONAL EXAMPLES
Example
Find the surface area of the rectangular prism.
1. $l = 5\frac{1}{2}$ ft; w = 3 ft; h = 2 ft
   Ans: 67 ft²
2. l = 3.7 cm; w = 1.6 cm; h = 5.8 cm
   Ans: 73.32 cm²

SELF-CHECK
1. 72 cm²

CLASSROOM EXERCISES
1. 24 m²; 40 m²; 30 m²; 94 m²
2. 20 cm²; 5.0 cm²; 8 cm²; 33 cm²
3. 24 ft²; 12 ft²; 16 ft²; 52 ft²
4. 96 in²; 64 in²; 48 in²; 208 in²

Geometry

ASSIGNMENT GUIDE
BASIC
pp. 522-523: Omit
AVERAGE
pp. 522-523: Omit
ABOVE AVERAGE
pp. 522-523: 1-20

WRITTEN EXERCISES
1. 2255 cm^2 2. 10.6 m^2
3. 54 cm^2 4. 108 m^2
5. 512 cm^2
6. 3727.6 cm^2
7. 47.28 m^2
8. 268 in^2
9. 72 ft^2
10. 225 in^2
11. 580 ft^2
12. 700 in^2
13. $414\frac{2}{3}$ in^2
14. $13\frac{1}{2}$ yd^2
15. 1400 in^2

WRITTEN EXERCISES

Goals: To find the surface area of a rectangular solid
To apply the skill of finding surface area to solving word problems

METRIC MEASURES

Find the surface area of each rectangular solid. (Example)

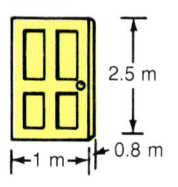

1. Book — 45 cm, 3.5 cm, 20 cm
2. Door — 1 m, 0.8 m, 2.5 m
3. Package — 3 cm, 3 cm, 3 cm

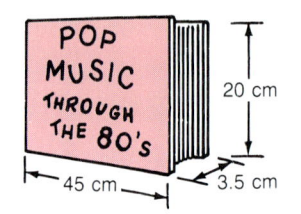

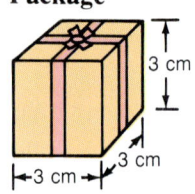

	Length	Width	Height
4.	3 m	4 m	6 m
6.	40 cm	29 cm	10.2 cm

	Length	Width	Height
5.	12 cm	8 cm	8 cm
7.	1.1 m	2.4 m	6 m

CUSTOMARY MEASURES

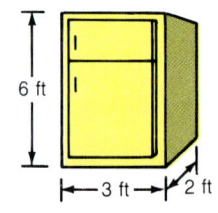

8. Cereal Box — 8 in, 3 in, 10 in
9. Refrigerator — 3 ft, 2 ft, 6 ft
10. Radio — 10 in, $2\frac{1}{2}$ in, 7 in

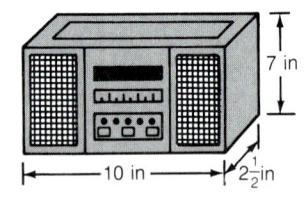

	Length	Width	Height
11.	5 ft	10 ft	16 ft
13.	$6\frac{2}{3}$ in	8 in	$10\frac{1}{2}$ in

	Length	Width	Height
12.	10 in	15 in	8 in
14.	$1\frac{1}{2}$ yd	$1\frac{1}{2}$ yd	$1\frac{1}{2}$ yd

15. Howard is wrapping a present. The box he is using is a rectangular prism with a length of 24 inches, a width of 15 inches, and a height of 4 inches. Find how many square inches of paper he needs to wrap the entire box. Add 368 square inches to the answer to allow for overlapping the paper.

522 Chapter 17

16. Lucy is painting a toy chest that is a rectangular prism. The chest is 4 feet long, 2 feet wide, and $1\frac{1}{2}$ feet high. What is the surface area of the chest?

17. A certain music box has the shape of a cube. Each side of the music box is 12 centimeters long. What is the surface area of the music box?

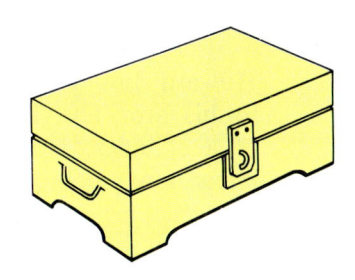

MORE CHALLENGING EXERCISES

Find the cost of roofing each building when the cost of shingles is $0.35 per square foot. Both sides of the roof have the same shape.

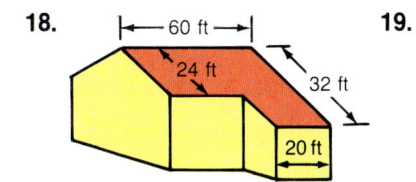

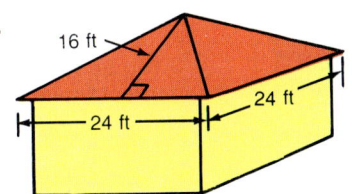

18.

19.

20. The roofs of 6 houses in a development are shaped as shown in the figure. Both sides of the roof are the same. The builder pays $0.27 per square foot for shingles and $8.50 per hour for labor. Five roofers shingle the 6 roofs in 40 hours. Find the total cost to the builder.

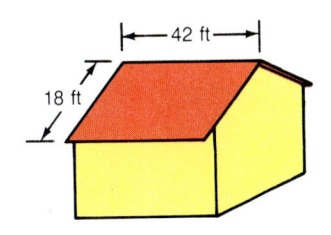

REVIEW CAPSULE FOR SECTION 17-8

Find the volume of each rectangular prism. (Pages 145-147)

1. $l = 4$ cm; $w = 6$ cm; $h = 3$ cm
2. $l = 5$ m; $w = 2.5$ m; $h = 7$ m
3. $l = 3$ ft; $w = 5$ ft; $h = 2$ ft
4. $l = 10$ in; $w = 8$ in; $h = 9$ in

Evaluate each expression. (Pages 176-179)

5. $3p$ when $p = 3.14$
6. $9t$ when $t = 3.14$
7. r^2 when $r = 0.6$
8. a^2 when $a = 5\frac{1}{2}$
9. $\frac{22}{7}p$ when $p = 5$
10. $\frac{22}{7}t$ when $t = 2\frac{1}{3}$

Round each decimal to the nearest whole number. (Pages 32-34)

11. 0.62
12. 1.98
13. 25.31
14. 36.81
15. 46.54
16. 99.5

The answers are on page 533.

Geometry **523**

WRITTEN EXERCISES
p. 522
16. 34 ft^2
17. 864 cm^2
18. $1120
19. $268.80
20. $2789.44

Teaching Suggestions
p. M-30

QUICK QUIZ
Find the volume of the rectangular prism.

1. l = 7 yd; w = 3 yd; h = 4 yd
   Ans: 84 yd³
2. l = 2.7 m; w = 1.6 m; h = 0.8 m
   Ans: 3.456 m³

Evaluate.

3. 14x when $x = \frac{22}{7}$
   Ans: 44
4. 8y when y = 3.14
   Ans: 25.12

Round to the nearest whole number.

5. 17.86   Ans: 18
6. 9.21   Ans: 9

ADDITIONAL EXAMPLES
Example
Find the volume of the cylinder. Use $\frac{22}{7}$ for π. Round the answer to the nearest whole number.

1. r = 8 ft; h = 23 ft
   Ans: 4626 ft³
2. r = 11 in; h = 35 in
   Ans: 13,310 in³

SELF-CHECK
849 cm³

524

17-8 Volume: Cylinders

Howard Lopez built a new silo for his farm. The silo has the shape of a cylinder. Howard wants to know how much corn the silo will hold: that is, he wants to know the **volume** of the silo.

Finding the volume of a cylinder is similar to finding the volume of a rectangular prism. Volume is measured in **cubic units**.

Rectangular Prism — The base is a rectangle.

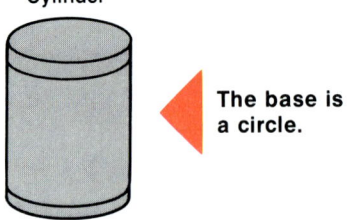

Cylinder — The base is a circle.

Volume = Area of Base × Height

$V = l \times w \times h$
$V = lwh$

Volume = Area of Base × Height

$V = \pi \times r^2 \times h$
$V = \pi r^2 h$

PROCEDURE

To find the volume of a cylinder, multiply the area of the base and the height.

EXAMPLE Howard's silo has a radius of 6.5 meters. The height is 18 meters. Find the volume of the silo. Use 3.14 for π. Round the answer to the nearest meter.

Solution: $V = \pi r^2 h$ ← r^2 means $r \cdot r$.

$V = 3.14(6.5)(6.5)(18)$

$V = 2387.97$ m³   The volume is about **2388 m³**.

Self-Check 1. Find the volume of a cylinder with a radius of 5.2 centimeters and a height of 10 centimeters. Use 3.14 for π. Round the answer to the nearest whole number.

524 Chapter 17

CLASSROOM EXERCISES

METRIC MEASURES

For Exercises 1–4, find the area of the base and the volume of each cylinder. Use 3.14 for π. Round each answer to the nearest whole number. (Example)

	Radius	Height	Area of Base	Volume of Cylinder
1.	3 m	6 m	28.26 m²	? m³
2.	6 cm	8 cm	? cm²	? cm³
3.	2.4 m	5 m	? m²	? m³
4.	25 mm	8.5 mm	? mm²	? mm³

CUSTOMARY MEASURES

For Exercises 5–8, find the area of the base and the volume of each cylinder. Use $\frac{22}{7}$ for π. Round each answer to the nearest whole number. (Example)

	Radius	Height	Area of Base	Volume of Cylinder
5.	7 ft	10 ft	154 ft²	? ft³
6.	14 yd	8 yd	? yd²	? yd³
7.	$2\frac{1}{3}$ ft	4 ft	? ft²	? ft³
8.	42 in	$3\frac{1}{3}$ in	? in²	? in³

WRITTEN EXERCISES

Goal: To find the volume of a cylinder

METRIC MEASURES

For Exercises 1–7, find the volume of each cylinder. Use 3.14 for π. Round each answer to the nearest whole number. (Example)

1. Mug 
$r = 4$ cm; $h = 8$ cm

2. Water Tower 
$r = 5$ m; $h = 16.6$ m

3. Trash Can 
$r = 23.5$ cm; $h = 56$ cm

Geometry 525

CLASSROOM EXERCISES
1. 170 m³
2. 113.04 cm²; 904 cm³
3. 18.09 m²; 90 m³
4. 1962.5 mm²; 16,681 mm³
5. 1540 ft³
6. 616 yd²; 4938 yd³
7. 17 ft²; 68 ft³
8. 5544 in²; 18,480 in³

ASSIGNMENT GUIDE
BASIC
pp. 525–526: Omit
AVERAGE
pp. 525–526: Omit
ABOVE AVERAGE
pp. 525–526: 1–18

WRITTEN EXERCISES
1. 402 cm²
2. 1303 m³
3. 97,108 cm³

WRITTEN EXERCISES

<u>4.</u> 1413 cm^3 <u>5.</u> 170 cm^3
<u>6.</u> 2 m^3 <u>7.</u> 661 cm^3
<u>8.</u> 308 in^3 <u>9.</u> 69 in^3
<u>10.</u> 1386 ft^3
<u>11.</u> 4928 in^3
<u>12.</u> 1697 in^3
<u>13.</u> 4505 in^3
<u>14.</u> 212 ft^2 <u>15.</u> 1 m^3
<u>16.</u> 35 in^3 <u>17.</u> 629 in^3
<u>18.</u> 322 cm^3

	Radius	Height
4.	5 cm	18 cm
6.	0.8 m	1 m

	Radius	Height
5.	3 cm	6 cm
7.	4.5 cm	10.4 cm

CUSTOMARY MEASURES

For Exercises 8–14, find the volume of each cylinder. Use $\frac{22}{7}$ for π. Round each answer to the nearest whole number. (Example)

8. Can of Varnish

$r = 3\frac{1}{2}$ in; $h = 8$ in

9. Glass

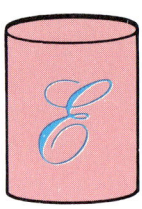

$r = 2$ in; $h = 5\frac{1}{2}$ in

10. Wading Pool

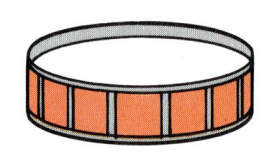

$r = 10\frac{1}{2}$ ft; $h = 4$ ft

	Radius	Height
11.	14 in	8 in
13.	$10\frac{1}{2}$ in	13 in

	Radius	Height
12.	6 in	15 in
14.	3 ft	$7\frac{1}{2}$ ft

APPLICATIONS: USING VOLUME OF A CYLINDER

Solve. Round each answer to the nearest whole number. (Example)

15. The radius of a hot water tank is 0.4 meters and its height is 1.5 meters. Find the volume of the tank. Use 3.14 for π.

16. A box of salt has the shape of a cylinder. The radius of the box is $1\frac{1}{2}$ inches and the height is 5 inches. Find the volume of the box. Use $\frac{22}{7}$ for π.

17. A cylinder-shaped water pitcher has a radius of 4 inches and a height of $12\frac{1}{2}$ inches. Find the volume of the water pitcher. Use $\frac{22}{7}$ for π.

18. The radius of a can of soup is 3.2 centimeters and the height is 10 centimeters. Find the volume of the can. Use 3.14 for π.

REVIEW CAPSULE FOR SECTION 17-9

Evaluate each expression. (Pages 22–24, 129–131)

1. $\frac{1}{3}y$ when $y = 18$
2. $\frac{1}{3}c$ when $c = 7.2$
3. $\frac{1}{3}mp$ when $m = 3.14$ and $p = 6$
4. $\frac{1}{3}nq$ when $n = \frac{22}{7}$ and $q = 36$

The answers are on page 533.

17-9 Volume: Pyramids and Cones

The pyramid at the right is a **rectangular pyramid**. The base of a pyramid determines its name.

When a pyramid and a rectangular prism have equal heights and bases that are equal in area, the volume of the pyramid is $\frac{1}{3}$ of the volume of the rectangular prism.

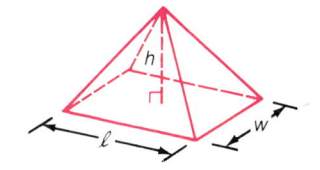

PROCEDURE

To find the volume of a pyramid, use this formula.

$$V = \frac{1}{3}Bh$$

B = area of base
h = height

EXAMPLE 1 The Pyramid of the Sun in Mexico has a rectangular base. The base is 229 meters long and 137 meters wide. The pyramid is 60 meters high. Find the volume.

Solution: Since the base is a rectangle,
$B = (229)(137) = 31{,}373$ m².

$V = \frac{1}{3}Bh$ Replace B with 31,373 and h with 60.

$V = \frac{1}{3}(31{,}373)(60)$

$V = 627{,}460$ The volume is **627,460 m³**.

Self-Check *Find the volume of each rectangular pyramid.*

1. $l = 3$ m; $w = 3$ m; $h = 6$ m
2. $l = 6$ ft; $w = 9$ ft; $h = 10$ ft

Geometry **527**

Teaching Suggestions
p. M-30

QUICK QUIZ
Evaluate.
1. $\frac{1}{3}y$ when y = 138
   Ans: 46
2. $\frac{b}{3}$ when b = 17.22
   Ans: 5.74
3. $\frac{ac}{3}$ when a = 6; b = 8
   Ans: 16
4. $\frac{hjk}{3}$ when h = 7; j = 8; k = 9
   Ans: 168
5. $\frac{1}{3}xy$ when x = $\frac{22}{7}$; y = 15   Ans: 15.7

ADDITIONAL EXAMPLES
Example 1
Find the volume of the pyramid.
1. l = 21 cm; w = 6 cm; h = 9 cm
   Ans: 378 cm³
2. l = 17 yd; w = 19 yd; h = 6 yd
   Ans: 646 yd³

SELF-CHECK
1. 18 m³   2. 180 ft³

527

The base of a cylinder is a circle. The base of a cone is also a circle. When a cone and a cylinder have equal radii (plural of radius) and equal heights, the volume of the cone is $\frac{1}{3}$ of the volume of the cylinder.

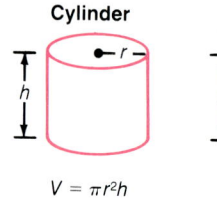

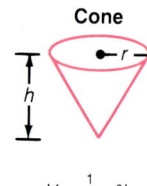

Cylinder: $V = \pi r^2 h$

Cone: $V = \frac{1}{3}\pi r^2 h$

PROCEDURE

To find the volume of a cone, use this formula.

$$V = \frac{1}{3} \cdot B(\text{area of base}) \cdot h$$ ◀ The base is a circle.

$$V = \frac{1}{3} \cdot \pi \cdot r^2 \cdot h, \text{ or } V = \frac{1}{3}\pi r^2 h$$

EXAMPLE 2 Find the volume of this paper cup. Use 3.14 for π. Round your answer to the nearest whole number.

Solution: $V = \frac{1}{3}\pi r^2 h$ ◀ $r = 3$, $h = 8$

$V = \frac{1}{3}(3.14)(3)(3)(8)$

$V = 75.36$ The volume is about **75 cm³**.

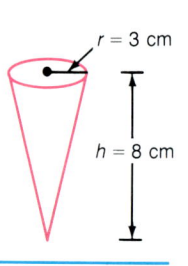

Self-Check Find the volume of each cone. Round each answer to the nearest whole number.

3. $r = 4$ m; $h = 10$ m; $\pi = 3.14$
4. $r = 6$ ft; $h = 15$ ft; $\pi = \frac{22}{7}$

CLASSROOM EXERCISES

Complete to find the volume of each rectangular pyramid. (Example 1)

1. $l = 5$ ft; $w = 5$ ft; $h = 12$ ft
$V = \frac{1}{3}Bh = \frac{1}{3}(5)(5)(12) = \underline{}$

2. $l = 2.4$ m; $w = 1.8$ m; $h = 6$ m
$V = \frac{1}{3}Bh = \frac{1}{3}(2.4)(1.8)(6) = \underline{}$

Complete to find the volume of each cone. Round each answer to the nearest whole number. (Example 2)

3. $r = 6$ m; $h = 20$ m; $\pi = 3.14$
$V = \frac{1}{3}\pi r^2 h = \frac{1}{3}(3.14)(6)(6)(20) = \underline{}$

4. $r = 5$ ft; $h = 12$ ft; $\pi = \frac{22}{7}$
$V = \frac{1}{3}\pi r^2 h = \frac{1}{3}\left(\frac{22}{7}\right)(5)(5)(12) = \underline{}$

ADDITIONAL EXAMPLES

Example 2

Find the volume of the cone. Round each answer to the nearest whole number.

1. $r = 18$ cm; $h = 5$ cm; use 3.14 for π.
Ans: 5087 cm³

2. $r = 5$ in; $h = 7$ in; use $\frac{22}{7}$ for π.
Ans: 550 in³

SELF-CHECK
3. 167 m³   4. 566 ft³

CLASSROOM EXERCISES
1. 100 ft³   2. 8.64 m³
3. 754 m³   4. 314 ft³

WRITTEN EXERCISES

Goals: To find the volume of a pyramid
To find the volume of a cone

METRIC MEASURES

Find each volume. Use $\pi = 3.14$. When necessary, round the answer to the nearest whole number. (Examples 1 and 2)

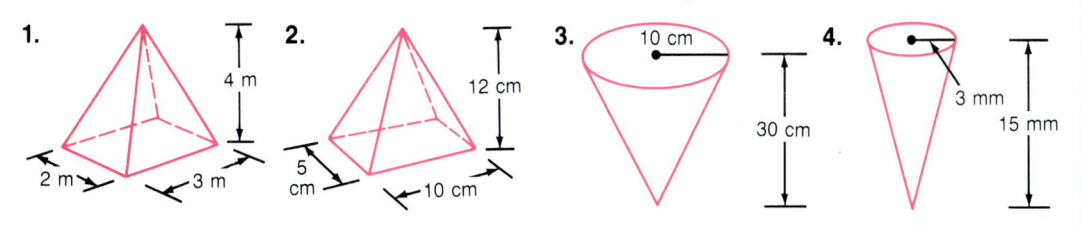

1. [pyramid: 4 m high, 2 m × 3 m base]
2. [pyramid: 12 cm high, 5 cm × 10 cm base]
3. [cone: radius 10 cm, height 30 cm]
4. [cone: radius 3 mm, height 15 mm]

CUSTOMARY MEASURES

Find each volume. Use $\pi = \frac{22}{7}$. When necessary, round the answer to the nearest whole number. (Examples 1 and 2)

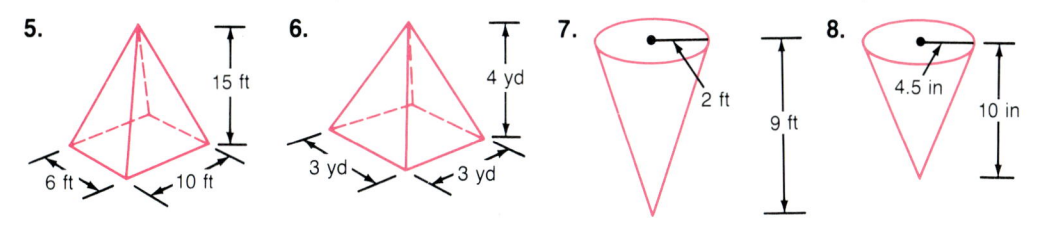

5. [pyramid: 15 ft high, 6 ft × 10 ft base]
6. [pyramid: 4 yd high, 3 yd × 3 yd base]
7. [cone: radius 2 ft, height 9 ft]
8. [cone: radius 4.5 in, height 10 in]

APPLICATIONS: USING VOLUME

9. Martha and Cherie pitched a tent that has the shape of a pyramid. The base of the tent is a rectangle that is 2.1 meters wide and 2.4 meters long. The tent is 2 meters high. Find the volume.

10. A pyramid-shaped container has a rectangular base. The base is 13 centimeters long and 9 centimeters wide. The container is 22.5 centimeters high. Find the volume.

11. A cone-shaped hanging planter has a radius of 7 centimeters and a height of 15 centimeters. Find the volume. Use 3.14 for π.

12. A cone-shaped papercup has a radius of 4 inches and a height of 8 inches. Find the volume. Use $\frac{22}{7}$ for π. Round your answer to the nearest whole number.

Geometry

REVIEW: SECTIONS 17-5–17-9

Identify each of the following. Refer to the figure at the right. (Section 17-5)

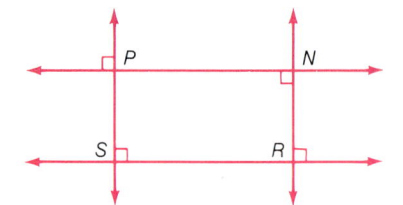

1. Four pairs of perpendicular lines.
2. Two pairs of parallel lines.

For Exercises 3-6, find each missing measure. (Section 17-6)

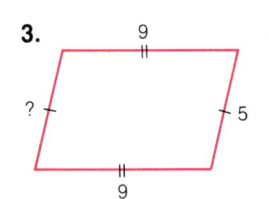

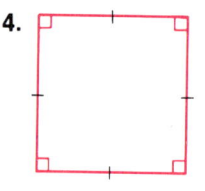

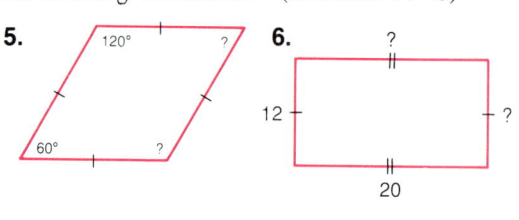

3.
4.
5.
6.

Find the surface area of each rectangular solid. (Section 17-7)

7. $l = 2$ m; $w = 5$ m; $h = 8$ m
8. $l = 2\frac{1}{2}$ ft; $w = 8$ ft; $h = 3\frac{1}{4}$ ft

Find the volume of each cylinder. Round each answer to the nearest whole number. (Section 17-8)

9. $r = 4$ cm; $h = 12$ cm; $\pi = 3.14$
10. $r = 1\frac{1}{2}$ in; $h = 7$ in; $\pi = \frac{22}{7}$

Find the volume of each pyramid. (Section 17-9)

11. $l = 2$ m; $w = 4$ m; $h = 9$ m
12. $l = 6$ cm; $w = 3.5$ cm; $h = 4.5$ cm

Find the volume of each cone. Round each answer to the nearest whole number. (Section 17-9)

13. $r = 7$ m; $h = 9$ m; $\pi = 3.14$
14. $r = 9$ cm; $h = 21$ cm; $\pi = 3.14$
15. $r = 12$ ft; $h = 14$ ft; $\pi = \frac{22}{7}$
16. $r = 3\frac{1}{2}$ yd; $h = 16$ yd; $\pi = \frac{22}{7}$

17. Each side of a cube is 22 inches long. What is the surface area of the cube? (Section 17-7)

18. A cone-shaped candle has a radius of 2 inches and a height of 7 inches. Find the volume. Use $\frac{22}{7}$ for π. (Section 17-9)

19. A can of frozen juice has the shape of a cylinder. The radius of the cylinder is 2 centimeters and the height is 9 centimeters. Find the volume. Use 3.14 for π. (Section 17-8)

20. A parking lot has pyramid-shaped markers with rectangular bases. The markers are 0.3 meters wide, 0.4 meters long, and 0.8 meters high. Find the volume of one marker. (Section 17-9)

530 Chapter 17

CHAPTER REVIEW

PART 1: VOCABULARY

For Exercises 1–18, choose from the box at the right the word(s) that best corresponds to each description.

1. The standard unit for measuring an angle __?__
2. A polygon with three sides __?__
3. A __?__ extends indefinitely in opposite directions.
4. A rectangular prism in which the length, width, and height are the same __?__
5. A closed plane figure formed by line segments __?__
6. Triangles that have the same size and shape __?__
7. A polygon with four sides __?__
8. A __?__ extends indefinitely in one direction from its endpoint.
9. An angle that has a measure of 90° __?__
10. Volume is measured in __?__.
11. Triangles that have the same shape but not the same size __?__
12. A pair of rays having a common endpoint __?__
13. An angle having a measure less than 90° __?__
14. A triangle having no sides the same length __?__
15. An angle having a measure greater than 90° __?__
16. A triangle having two sides the same length __?__
17. A part of a line that is named by its endpoints __?__
18. A triangle having all sides the same length __?__

> line
> line segment
> ray
> angle
> degree
> right angle
> acute angle
> obtuse angle
> polygon
> triangle
> equilateral triangle
> isosceles triangle
> scalene triangle
> congruent triangles
> similar triangles
> quadrilateral
> cube
> cubic units

PART 2: SKILLS

For Exercises 19–22, refer to the figure at the right. (Section 17-1)

19. Name line BE in three different ways.
20. Name $\overrightarrow{EB}$ in a different way.
21. Name four angles with B as a vertex.
22. Name two lines that contain point E.

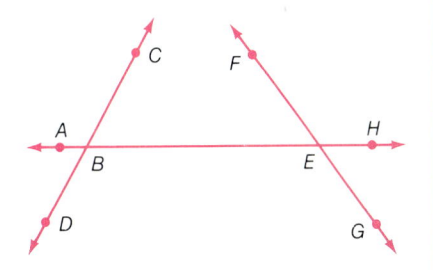

CHAPTER REVIEW
1. degree 2. triangle
3. line 4. cube
5. polygon
6. congruent triangles
7. quadrilateral
8. ray 9. right angle
10. cubic units
11. similar triangles
12. angle
13. acute angle
14. scalene triangle
15. obtuse angle
16. isosceles triangle
17. line segment
18. equilateral triangle
19. Answers may vary.
    Some answers are:
    $\overrightarrow{BH}$; $\overrightarrow{AH}$; and $\overleftrightarrow{AE}$
20. $\overrightarrow{EA}$
21. $\angle ABC$; $\angle CBE$; $\angle EBD$; $\angle DBA$
22. $\overleftrightarrow{FG}$; $\overleftrightarrow{BH}$

Geometry

CHAPTER REVIEW

For Exercises 23-26, teacher should check each student's drawings.

27. isosceles
28. equilateral
29. scalene
30. isosceles
31. 81   32. 15
33. alternate interior
34. alternate exterior
35. vertical
36. corresponding
37. parallelogram
38. parallelogram
39. parallelogram; rhombus

For each of Exercises 23-26, use a protractor to draw an angle with the given measure. (Section 17-2)

23. 20°   **24.** 50°   **25.** 90°   **26.** 165°

Classify each triangle as equilateral, isosceles, or scalene. (Section 17-3)

27. (97, 62, 97)   **28.** (6, 6, 6)   **29.** (26, 19, 28)   **30.** (18, 18, 25)

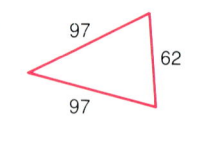

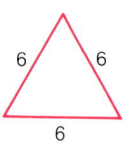

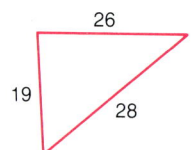

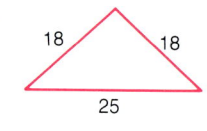

For Exercises 31-32, each pair of triangles is similar. Use a proportion to find the unknown length. (Section 17-4)

31. (45, 36; 20, n)   **32.** (13, 10, 84°; 19.5, n, 84°)

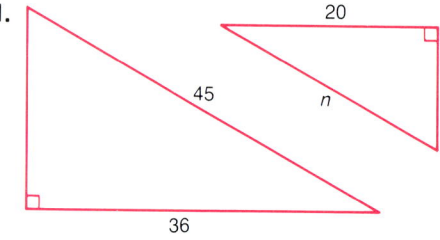

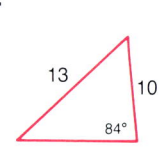

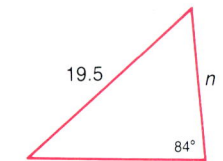

In the figure at the right, $\overleftrightarrow{EF}$ is parallel to $\overleftrightarrow{GH}$, and $\overleftrightarrow{IJ}$ is a transversal. Classify each pair of angles as alternate exterior, alternate interior, corresponding, or vertical. (Section 17-5)

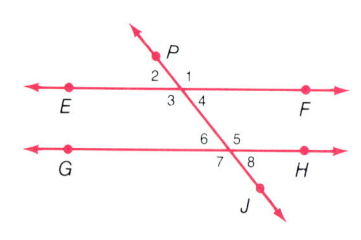

33. ∠6 ≅ ∠4   **34.** ∠1 ≅ ∠7   **35.** ∠5 ≅ ∠7   **36.** ∠2 ≅ ∠6

For Exercises 37-39, classify each quadrilateral as a trapezoid, parallelogram, rectangle, rhombus, or square. Some figures may have more than one name. (Section 17-6)

37.   **38.** 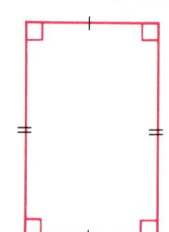   **39.** (105°, 75°, 75°, 105°)

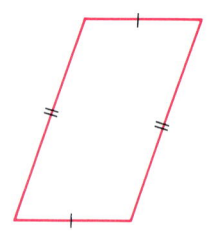

532   Chapter 17

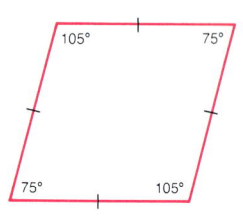

Find the surface area of each rectangular solid. (Section 17-7)

	Length	Width	Height
40.	$4\frac{1}{2}$ ft	6 ft	10 ft
41.	12 m	10.5 m	20.2 m
42.	$1\frac{1}{2}$ yd	8 yd	$5\frac{1}{5}$ yd

Find the volume of each cylinder. Round each answer to the nearest whole number. (Section 17-8)

	Radius	Height	π
43.	6 m	12 m	3.14
44.	$5\frac{1}{4}$ ft	9 ft	$\frac{22}{7}$
45.	14.4 cm	10.5 cm	3.14

PART 3: APPLICATIONS

46. Sue is 1.6 meters tall. On a sunny day, her shadow is 2 meters long at the same time as the shadow of a TV tower is 70 meters long. Find the height of the tower. (Section 17-4)

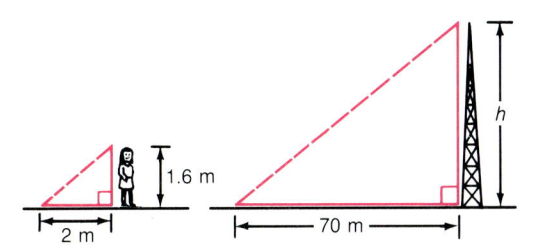

47. The radius of a cylindrical silo is 7 meters and its height is 28.5 meters. Find the volume of the silo. Use 3.14 for π. Round the answer to the nearest whole number. (Section 17-8)

48. Ramona is buying screen for a hamster cage that has the shape of a rectangular prism. The cage is 2 feet wide, $2\frac{1}{2}$ feet long, and $1\frac{1}{2}$ feet high. Find how many square feet of screen she needs to cover the entire cage. (Section 17-7)

49. The Great Pyramid of Egypt has a square base. Each side of the base is 230 meters long. The height of the pyramid is 147 meters. Find the volume. (Section 17-9)

CHAPTER REVIEW
40. 264 ft^2
41. 1161 m^2
42. $122\frac{4}{5}$ yd^2
43. 1356 m^3
44. 780 m^3
45. 6837 cm^3
46. 56 m
47. 4385 m^3
48. $23\frac{1}{2}$ ft^2
49. 2,592,100 m^3

ANSWERS TO REVIEW CAPSULES

Page 500 **1.** < **2.** > **3.** > **4.** > **5.** > **6.** < **7.** > **8.** <

Page 508 **1.** $13\frac{1}{2}$ **2.** 36 **3.** 8 **4.** 24 **5.** 33 **6.** 15 **7.** 9.8 **8.** 11.9 **9.** 11.6 **10.** 12.9 **11.** 9.1 **12.** 11.5

Page 513 **1.** b; g **2.** a; d **3.** e **4.** c; f

Page 520 **1.** 26 **2.** 33 **3.** 19.8 **4.** 13.8 **5.** $1\frac{2}{3}$ **6.** 7 **7.** 15 m^2 **8.** 0.7 cm^2 **9.** 104 ft^2 **10.** 49 in^2 **11.** $75\frac{1}{4}$ ft^2 **12.** 156.25 m^2

Page 523 **1.** 72 cm^3 **2.** 87.5 m^3 **3.** 30 ft^3 **4.** 720 in^3 **5.** 9.42 **6.** 28.26 **7.** 0.36 **8.** $30\frac{1}{4}$ **9.** $15\frac{5}{7}$ **10.** $7\frac{1}{3}$ **11.** 1 **12.** 2 **13.** 25 **14.** 37 **15.** 47 **16.** 100

Page 527 **1.** 6 **2.** 2.4 **3.** 6.28 **4.** $37\frac{5}{7}$

Geometry 533

CHAPTER TEST

Two forms of a chapter test, Form A and Form B, are provided on copying masters in the *Teacher's Edition: Part II*.

1. $\overrightarrow{GH}$; $\overrightarrow{GB}$; $\overrightarrow{GJ}$ 2. ∠GEF; ∠FED; ∠DEC; ∠CEG
3. acute
4. $\overleftrightarrow{BJ}$; $\overleftrightarrow{HD}$
5. $\overleftrightarrow{JG} \perp \overleftrightarrow{HE}$
6. trapezoid
7. $\overleftrightarrow{JF} \parallel \overleftrightarrow{BC}$; $\overleftrightarrow{JB} \parallel \overleftrightarrow{FC}$
8. SAS; $\overline{AC} \cong \overline{EF}$; ∠A ≅ ∠E; $\overline{AB} \cong \overline{DE}$
9. SSS; $\overline{YZ} \cong \overline{RS}$; $\overline{XY} \cong \overline{ST}$; $\overline{XZ} \cong \overline{RT}$
10. 3   11. 16.8
12. 52 ft²
13. 4537.3 m³
14. 259,200 m³
15. 198 in³

A cumulative test covering the content of Chapters 1–17 is provided on copying masters in the *Teacher's Edition: Part II*. This cumulative test is primarily designed as a final examination for the Above Average course of study.

CHAPTER TEST

For Exercises 1–7, refer to the figure at the right below.

1. Name three different rays that have *G* as an endpoint.
2. Name four angles having *E* as a vertex.
3. Is angle *ABJ* a right angle, an obtuse angle, or an acute angle?
4. Name a pair of intersecting lines.
5. Name a pair of perpendicular lines.
6. Lines *HD* and *AC* are not parallel. Which type of quadrilateral is quadrilateral *BCEG*?
7. Quadrilateral *BCFJ* is a parallelogram. List two properties of this quadrilateral.

Write whether each pair of triangles is congruent by SAS, SSS, or ASA. Then name the corresponding congruent sides and angles.

8.

9.

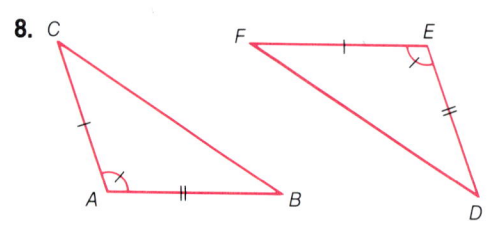

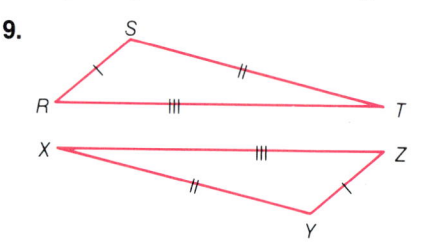

For Exercises 10–11, each pair of triangles is similar. Use a proportion to find the unknown length.

10.

11.

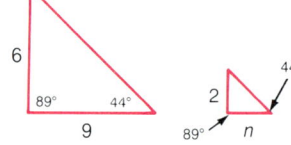

12. Eduardo is painting a box that is a rectangular prism. The box is 3 feet wide, 4 feet long, and 2 feet high. What is the surface area?

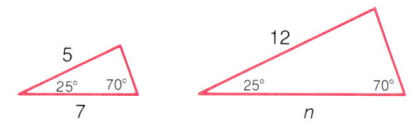

13. A cylinder-shaped water tower has a radius of 8.5 meters and a height of 20 meters. Find the volume. Use 3.14 for π.

14. The Step Pyramid in Egypt has a rectangular base, 120 meters long and 108 meters wide. The height is 60 meters. Find the volume.

15. A cone-shaped cup has a radius of 3 inches and a height of 7 inches. Find the volume. Use $\frac{22}{7}$ for π.

534   Chapter 17

ENRICHMENT

Spheres

You can use this formula to find the volume of a sphere.

$$V = \frac{4}{3}\pi r^3 \qquad r = \text{radius}$$

EXAMPLE The radius of a basketball is 12 centimeters. Find the volume of the basketball. Use 3.14 for π. Round your answer to the nearest whole number.

Solution: Use paper and pencil or use a calculator.

$$V = \frac{4}{3}\pi r^3 \qquad r^3 = r \times r \times r$$

$$V = \frac{4}{3} \times 3.14 \times 12 \times 12 \times 12$$

$V = 4$ ÷ 3 × 3.14 × 12 = $\boxed{7234.56}$

The volume is about **7235 cm³**. ◀ Nearest whole number

EXERCISES

For Exercises 1–6, find the volume of each sphere. Round each answer to the nearest whole number.

Sphere	Radius	π	Sphere	Radius	π
1. Golf ball	2.1 cm	3.14	2. Storage tank	7.1 m	3.14
3. Volleyball	4.2 in	$\frac{22}{7}$	4. Tennis ball	1.5 in	$\frac{22}{7}$
5. Beachball	10.2 cm	3.14	6. Marble	0.6 cm	3.14

7. Large aluminum spheres are used to transport liquid natural gas on tankers. One of these spheres has a diameter of 36 meters. Find its volume. (Use $\pi = 3.14$.)

8. How much greater is the volume of a grapefruit with a radius of 4 inches than the volume of an orange with a radius of 2 inches? (Use $\pi = \frac{22}{7}$.)

Geometry 535

ENRICHMENT

You may wish to use this lesson for students who performed well on the formal Chapter Test.

1. 39 cm³   2. 1498 m³
3. 310 in³   4. 14 in³
5. 4443 cm³   6. 1 cm³
7. 24,417 m³
8. 234 in³

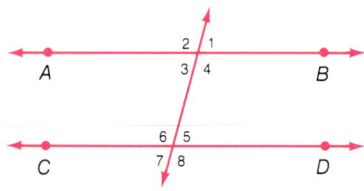

ADDITIONAL PRACTICE

You may wish to use all or some of these exercises, depending on how well students performed on the formal chapter test.

For Exercises 1-3, answers may vary.

1. Some answers are:
$\overrightarrow{DE}$; $\overrightarrow{EF}$; $\overrightarrow{GF}$

2. Some answers are:
$\overline{DE}$; $\overline{EF}$; $\overline{EG}$

3. Some answers are:
$\overleftrightarrow{DG}$; $\overleftrightarrow{EF}$; $\overleftrightarrow{FD}$

4. straight   5. acute
6. right   7. obtuse
8. acute   9. obtuse
10. ASA   11. C   12. C
13. NC   14. C
15. trapezoid
16. square
17. parallelogram; rectangle; square
18. square   19. 20 m
20. 295.2 in^3
21. 1538.6 cm^3
22. 94 ft^2

536

ADDITIONAL PRACTICE

SKILLS *Use symbols and the given line to name each of the following.* (Pages 498–500)

1. Three different rays.
2. Three different segments.
3. The line in three ways.

Identify each angle as right, acute, obtuse, or straight. (Pages 501–503)

4. 180°   5. 45°   6. 90°   7. 105°   8. 85°   9. 165°

10. Refer to the triangles at the right to determine whether they are congruent by SSS, SAS, or ASA. (Pages 504–507)

In the figure at the right, $\overleftrightarrow{AB}$ is parallel to $\overleftrightarrow{CD}$. Identify each pair of angles as congruent (C) or not congruent (NC). (Pages 514–516)

11. ∠6 and ∠8   12. ∠2 and ∠6   13. ∠3 and ∠6   14. ∠2 and ∠8

Identify each quadrilateral as a parallelogram, square, rectangle, or trapezoid. (Pages 517–519)

15. Exactly one pair of opposite sides parallel
16. Parallelogram with four congruent sides and four right angles
17. Both pairs of opposite sides parallel
18. Parallelogram with four right angles

APPLICATIONS

19. A fence post is 2 meters tall. On a sunny day, the shadow of the post was 2.2 meters long at the same time that the shadow of a light pole was 22 meters long. Find the height of the light pole. (Pages 508–511)

20. A planter shaped like a pyramid has a rectangular base that is 9 inches wide and 12.3 inches long. The planter is 8 inches high. Find the volume of the planter. (Pages 527–529)

21. The radius of a paint can is 7 centimeters and its height is 10 centimeters. Find the volume of the can. Use 3.14 for π. (Pages 524–526)

22. Find the surface area of a rectangular prism that has a length of 5 feet, a width of 4 feet, and a height of 3 feet. (Pages 520–523)

536   Chapter 17

COMMON ERRORS

Each of these problems contains a common error.

a. Find the correct answer.
b. Find the error.

For Exercises 1–2, refer to the table at the right.

Scores on Six Tests
85 82 90 76 87 90

1. Find the median.
Middle Scores: 90 and 76
$$\frac{90+76}{2} = \frac{166}{2} = 83$$

2. Find the mean.
$$\frac{85+82+90+76+87}{5} = 84$$

3. Janet is one of 25 guests at a party. A drawing for one prize is held. What is the probability that she will win the prize?
$$P = \frac{\text{Number of Favorable Outcomes}}{\text{Number of Possible Outcomes}}$$
$$P(\text{winning}) = \frac{1}{24}$$

4. A bag contains 5 red marbles, 3 green marbles, and one black marble. Two marbles are drawn at random, one after the other, without replacement. Find the probability of drawing a red marble, then a green marble.
$$P(\text{red, then green}) = \frac{5}{9} + \frac{3}{9} = \frac{8}{9}$$

5. These triangles are similar. Use a proportion to find the unknown length.
$$\frac{6}{8} = \frac{n}{4}$$
$$8n = 24$$
$$n = 3$$

6. Find the surface area of this rectangular prism.
$l = 4$ m; $w = 3$ m; $h = 2$ m
$$S = 2lw + 2lh + 2wh$$
$$= 4 \cdot 3 + 4 \cdot 2 + 3 \cdot 2$$
$$= 12 + 8 + 6$$
$$= 26 \text{ m}^3$$

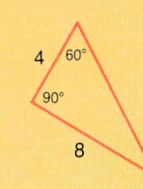

7. Find the volume of this cylinder.
$r = 5$ cm; $h = 6$ cm; $\pi = 3.14$
$$V = \pi r^2 h$$
$$= 3.14(2)(5)(10)$$
$$= 3.14(100)$$
$$= 314 \text{ cm}^3$$

8. Find the volume of this cone.
$r = 3$ in; $h = 7$ in; $\pi = \frac{22}{7}$
$$V = \frac{1}{3}\pi r^2 h$$
$$= \frac{22}{7} \cdot 3 \cdot 3 \cdot 7$$
$$= 198 \text{ in}^3$$

Geometry **537**

COMMON ERRORS
In preparation for the Cumulative Review, these exercises focus the student's attention on the most common errors to be avoided.

1. 86   2. 85   3. $\frac{1}{25}$
4. $\frac{5}{24}$ 5. 12   6. 52 m^2
7. 486 cm^3 8. 66 in^3

CUMULATIVE REVIEW

A cumulative test is provided on copying masters in the Teacher's Edition: Part II.

1. a 2. c 3. b
4. b 5. a 6. d
7. c 8. d 9. b

CUMULATIVE REVIEW: CHAPTERS 16–17

Choose the correct answer. Choose a, b, c, or d.
Refer to the table at the right below for Exercises 1–3.

Points Scored by a Basketball Team

65	42	81	53	64
66	72	51	64	79

1. Find the mean.
   a. 63.7   b. 64   c. 64.5   d. 63

2. Find the median.
   a. 63.7   b. 64   c. 64.5   d. 63

3. Find the mode.
   a. 63.7   b. 64   c. 64.5   d. 63

4. A box contains six quarters, eight dimes, and six nickels. One coin is chosen at random. What is the probability of choosing a quarter?
   a. $\frac{1}{3}$ b. $\frac{3}{10}$ c. $\frac{1}{2}$ d. $\frac{2}{5}$

5. The numbers 1 through 5 are written on separate slips of paper and placed into a hat. A number is chosen at random. Find the probability of choosing an odd number.
   a. 60%   b. 50%   c. 40%   d. 30%

6. A pair of dice is tossed. What is the probability that the sum is 2 or 12?
   a. 0   b. $\frac{1}{12}$ c. $\frac{1}{9}$ d. $\frac{1}{18}$

7. The tree diagram at the right shows all the possible ways two coins can land when tossed at the same time. What is the probability that both coins will be heads?
   a. $\frac{1}{2}$ b. $\frac{1}{6}$
   c. $\frac{1}{4}$ d. $\frac{2}{6}$

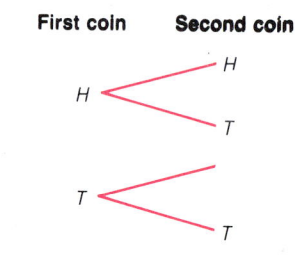

8. Three nickels and two dimes are placed into a coin purse. One coin is drawn at random and replaced. Then a second coin is drawn. What is the probability that both coins are nickels?
   a. $\frac{6}{5}$ b. $\frac{3}{5}$ c. $\frac{6}{25}$ d. $\frac{9}{25}$

9. Classify an angle of 95° as one of the following.
   a. acute angle   b. obtuse angle   c. right angle   d. straight angle

10. Determine whether the pair of triangles shown at the right are congruent by SSS, SAS, ASA, or AAA.
    a. SAS   b. SSS
    c. ASA   d. AAA

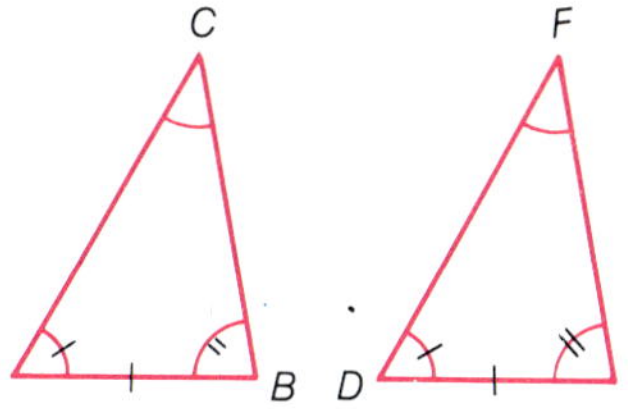

11. Two similar triangles are shown in the figure at the right. Use a proportion to find the height of the larger triangle.
    a. 16.2 m   b. 18.2 m
    c. 162 m   d. 182 m

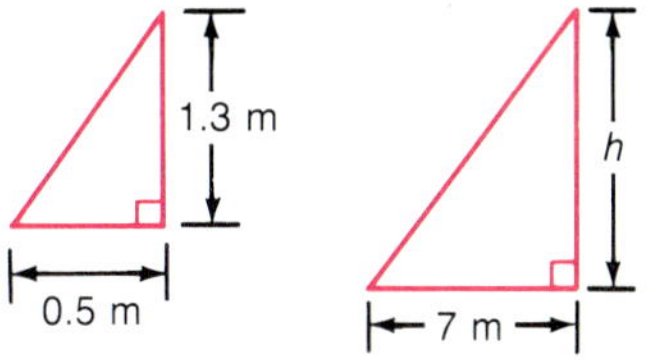

12. In the figure at the right lines AB and CD are parallel. Which of the following name a pair of corresponding angles?
    a. ∠2 and ∠5   b. ∠4 and ∠5
    b. ∠1 and ∠3   d. ∠8 and ∠2

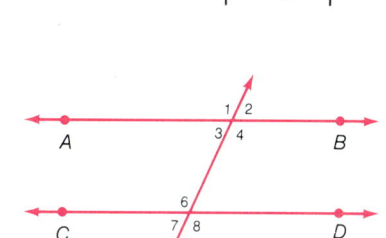

13. Refer to the parallelogram at the right to find the measure of the unknown angle.
    a. 70°   b. 180°
    c. 110°   d. 40°

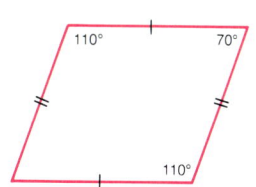

14. Find the surface area of a rectangular prism with a length of 20 centimeters, a width of 20 centimeters, and a height of 25 centimeters.
    a. 1400 cm²   b. 10,000 cm²   c. 2800 cm²   d. 65 cm²

15. A cylinder-shaped salt container has a diameter of 8 inches and a height of 14 inches. Find the volume of the container. Use $\frac{22}{7}$ for π.
    a. 176 in³   b. 1408 in³   c. 2816 in³   d. 704 in³

16. The pyramid in the figure at the right has a base 300 feet long and 300 feet wide. The height is 150 feet. Find the volume in cubic feet.
    a. 13,500,000   b. 4,500,000
    c. 45,000   d. 6,750,000

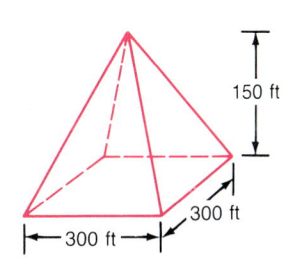

CUMULATIVE REVIEW
10. c   11. b   12. a
13. a   14. c   15. d
16. b

Cumulative Review: Chapters 16–17   **539**

GLOSSARY

The following definitions and statements reflect the usage of terms in this textbook.

Absolute Value To indicate distance, but not direction, from zero, you use *absolute value*. (Page 304)

Acute Angle An *acute angle* is an angle whose measure is less than 90°. (Page 502)

Algebraic Expression An *algebraic expression* contains at least one variable. (Page 22)

Angle An *angle* is a pair of rays that have a common endpoint. (Page 498)

Area The *area* is the number of square units needed to cover a surface. (Page 70)

Average The *average* or *mean* of two or more numbers or measures is the sum of the measures divided by the number of measures. (Page 10)

Binomial A *binomial* is the sum or difference of two monomials. (Page 315)

Circumference The *circumference* of a circle is the distance around the circle. (Page 176)

Combining Like Terms When you add or subtract like terms, you are *combining like terms*. (Page 76)

Composite Number A *composite number* is a counting number greater than 1 that is not a prime number. (Page 103)

Congruent Triangles Triangles that have the same size and shape are *congruent triangles*. (Page 505)

Coordinate Plane The *coordinate plane* has four regions, or quadrants, which are separated by the *x* and *y* axes. (Page 408)

Discount The amount that an item is reduced in price is the *discount*. (Page 244)

Divisible A number is *divisible* by another number if the remainder is 0. (Page 96)

Equation An *equation* is a sentence that contains the equality symbol "=." (Page 190)

Equivalent Equations Equations that have the same solution(s) are *equivalent equations*. (Page 193)

Equivalent Fractions *Equivalent fractions* name the same number. (Page 161)

Evaluate *Evaluate* means to find the value. (Page 18)

Even Number A number that is divisible by 2 is an *even number*. (Page 100)

Exponent The second power of 3 is written as 3^2. The raised two is called an *exponent*. The exponent indicates how many times a number is multiplied by itself. (Page 8)

Factor Since 24 is divisible by 8, 8 is a *factor* of 24. (Page 96)

Factoring Writing a sum or difference as a product is called *factoring*. (Page 364)

Formula A *formula* is a shorthand way of writing a word rule. In a formula, variables and symbols are used to represent words. (Page 2)

Greatest Common Factor The *greatest common factor* (GCF) of two or more counting numbers is the greatest number that is a factor of each number. (Page 115)

540   Glossary

Hypotenuse The *hypotenuse* of a right triangle is the side opposite the right angle. (Page 454)

Integers *Integers* are made up of the positive integers, zero, and the negative integers. (Page 294)

Interest *Interest* is the money charged for the use of borrowed money. (Page 240)

Interest Rate Interest is usually expressed as a percent of the principal. This percent is called the *interest rate*. (Page 240)

Kilowatt-hour You use one *kilowatt-hour* when you use 1 kilowatt (1000 watts) for one hour. (Page 5)

Least Common Multiple The *least common multiple* (LCM) of two or more counting numbers is the smallest counting number that is divisible by the given numbers. (Page 111)

Like Fractions *Like fractions* have a common denominator. (Page 158)

Like Terms Two or more *like terms* have the same variables and the same powers of these variables. (Page 76)

Linear Equation A *linear equation* is an equation whose graph is a straight line. (Page 410)

List Price The regular price of an article is called the *list price*. (Page 244)

Mean See Average.

Median In a listing of data, the *median* is the middle measure or score. (Page 470)

Mode In a listing of data, the *mode* is the measure that occurs most often. (Page 470)

Multiple A *multiple* of a given number is a product that has the given number as one of its factors. (Page 96)

Net Price The price of an article after a discount is deducted is the *net price*. (Page 245)

Numerical Expression A *numerical expression* includes at least one of the operations of addition, subtraction, multiplication, or division. (Page 18)

Obtuse Angle An *obtuse angle* is an angle whose measure is greater than 90°. (Page 502)

Odd Number A number that is not divisible by 2 is an *odd number*. (Page 100)

Opposites Two numbers that are the same distance from 0 and are in opposite directions from 0 are *opposites*. (Page 297)

Ordered Pair An *ordered pair* is a pair of numbers one of which is designated as the first and the other is the second. (Page 408)

Origin In the coordinate plane, the point with coordinates (0, 0) is called the *origin*. (Page 408)

Parallel Lines *Parallel lines* are lines in a plane that are always the same distance apart. (Page 514)

Parallelogram A *parallelogram* is a quadrilateral with both pairs of opposite sides parallel. (Page 517)

Percent A *percent* is a ratio in which the second term is 100. (Page 227)

Perfect Square A rational number is a *perfect square* if it is the square of a rational number. (Page 442)

Perimeter The *perimeter* of a figure is the distance around it. To find perimeter, add the lengths of the sides. (Page 81)

Perpendicular Lines When two intersecting lines form a right angle, the lines are *perpendicular lines*. (Page 514)

Polygon A *polygon* is a closed plane figure formed by line segments. (Page 504)

Polynomial A *polynomial* is a monomial or the sum or difference of two or more monomials. (Page 315)

Prime Factorization Every composite number can be expressed as a product of prime factors. Such a product is called the *prime factorization* of the number. (Page 108)

Prime Number A *prime number* is a counting number greater than 1 that has exactly two counting number factors, 1 and the number itself. (Page 103)

Principal The amount of money borrowed on which interest is paid is called the *principal*. (Page 240)

Probability The ratio of the number of favorable outcomes to the number of possible outcomes is the *probability* that an event will occur. (Page 479)

Proportion A *proportion* is an equation that states that two ratios are equal. (Page 219)

Pythagorean Theorem In any right triangle, the square of the length of the hypotenuse is equal to the sum of the squares of the lengths of the legs, or $c^2 = a^2 + b^2$. (Page 454)

Rate of Discount Discounts are often expressed as a certain percent of the list price. This percent is called the *rate of discount*. (Page 244)

Ratio A *ratio* is a quotient that compares the two numbers. (Page 216)

Rational Numbers A number that can be written as a fraction is a *rational number*. (Page 297)

Reciprocals Two numbers having 1 as their product are called *reciprocals*. (Page 148)

Retail Price The public pays the *retail price* for items sold in stores. (Page 261)

Right Angle A *right angle* is an angle whose measure is 90°. (Page 502)

Similar Triangles Triangles that have the same shape are *similar triangles*. (Page 508)

Slope The *slope* of a line is the ratio of the vertical change to the horizontal change between any two points on a line. (Page 417)

Solve an Equation To *solve an equation* means to find the numbers that can replace the variable to make the equation true. (Page 190)

System of Equations Two equations in the same two variables form a *system of equations*. (Page 427)

Tangent Ratio In a right triangle, the *tangent* of an acute angle, A, is the ratio

$$\frac{\text{length of side opposite angle } A}{\text{length of side adjacent to angle } A}$$

(Page 458)

Triangle A *triangle* is a polygon with three sides. (Page 504)

Trinomial A *trinomial* is the sum or difference of three monomials. (Page 315)

Variable A *variable* is a letter representing one or more numbers. (Page 22)

Volume The number of cubic units contained in a solid is the *volume* of the solid. (Page 145)

Wholesale Price *Wholesale price* is the price that a store pays for items it will sell. (Page 261)

X Intercept The *x intercept* of a line is the x value of its point on the x axis. (Page 422)

Y Intercept The *y intercept* of a line is the y value of its points on the y axis. (Page 422)

INDEX

Boldface numerals indicate the pages that contain formal or informal definitions.

Absolute value, 304-305
Acute angle, **502**-504
Addition
  Associative Property of, **79**, 312
  Commutative Property of, **79**, 312
  of decimals, 35
  of fractions having unlike denominators, 161
  of like fractions, 158
  of mixed numerals, 158
  of mixed numerals with unlike denominators, 165
  of polynomials, 315
  of positive and negative numbers, 300, 304-305
  of rational numbers, 304-305
  on a number line, 300
  Property for Equations, **324**, 326, 329
  Property for Inequalities, **340**
  Property of Opposites, **312**
  Property of Zero, **64**, 312
  to solve equations, 193-194
Algebraic expression, **22**, 334, 377
Angle(s), **498**, 501-502
  congruent, **514**-515
  of a triangle, 388
Annual percentage rate, 332
Applications
  addition and subtraction in equations, 196
  algebraic expressions, 334-335, 377-378
  angles in a triangle, 388
  area, 70-71, 140-141
  broken line graphs, 414-415
  career, 138, 311, 448, 478
  circumference and area, 176-178
  computer, 43, 75, 168, 226, 333, 387, 421, 449, 512
  consecutive numbers, 390-391
  consumer, 13, 43, 107, 139, 199, 269, 303, 332, 359
  discount, 244-245
  estimating with decimals, 47-48
  estimating with whole numbers, 14, 15
  inequalities, 396-397
  multiplication and division in equations, 206-207

  of scientific notation, 55-56
  percent, 257-258, 264-266, 274-275, 279-280
  perimeter, 81-82
  proportion, 222-223
  ratio, 222, 417, 424, 479
  simple interest, 240-241
  square root, **451**
  surface area and rectangular prisms, 520-521
  temperature, 132-133
  time cards, 169
  using similar triangles, 508-509
  volume, 145
Area, **70**
  of a circle, 177
  of a parallelogram, 140, 177
  of a rectangle, 70
  of a square, 70
  of a triangle, 141
Associative Property
  of Addition, **79**, 312
  of Multiplication, **67**, 356
Average, **10**

Bar graph, **270**
Batting average, **207**
Binomial, **315**
Break-even, **427**
Broken line graph, **414**-415

Calculator Applications
  adding rational numbers, 314
  annual percentage rate, 332
  area of triangles, 449
  checking equations, 331
  checking inequalities, 342
  computing y values for equations, 413
  cost of wallpaper, 199
  distributive property, 367
  estimation, 50
  evaluating powers, 164
  finding discount, 282
  greatest safe load, 138
  multiplying probabilities, 491
  multiplying rational numbers, 355
  Order of Operations, 21
  prime number/exponents, 106
  safe stopping distance, 303
  similar polygons, 511

  simple interest, 243
  standard deviation, 478
  tangent values, 461
  wind chill temperature, 448
Career Applications
  Engineering Technician, 138
  Meteorologist, 448
  Statistical Clerk, 478
  Travel Agent, 311
Celsius, **132**
Centigram, 52
Centiliter, 52
Centimeter, **51**
Circumference, **176**, 177
Clock arithmetic, 321
Commission, **236**
Common factors, **364**
Commutative Property
  of Addition, **79**, 312
  of Multiplication, **67**, 356
Composite number, **103**, 108
Computer Applications
  Adding Fractions, 333
  Angle Measures, 512
  Area of Triangles, 449
  Evaluating Expressions, 75
  Evaluating Powers, 168
  Problem Solving, 226
  Slope of a Line, 421
  Solving Equations, 387
  Unit Price, 43
Congruent angles, **514**-515
Congruent triangles, **505**-506
Consecutive number problems, 390-391
Consumer Applications
  Banking, 332
  Conserving Energy, 359
  Driving Safety, 303
  Nutrition, 269
  Painting and Estimation, 107
  Saving $ on Cooling Costs, 139
  Saving $ on Housing Costs, 13
  Unit Price, 43
  Wallpaper and Estimation, 199
Coordinate plane, **408**
Counting numbers, **96**, 97
Cube, **521**
Cubic units, **524**
Customary measures, 126-127

Data, **270**, 474

Index   543

Decigram, 52
Deciliter, 52
Decimals, 32
  as percent, 231-232
  comparing, 33
  non-repeating, 443
  place value, 32
  repeating, **174**, 213, 443
  terminating, **173**, 443
Decimeter, 51
Degree, **501**
Dekagram, 52
Dekaliter, 52
Dekameter, 51
Dependent events, **488**
Diameter of a circle, **176**
Digits, **99**
Direct variation, **424**
Discount, **244**, 245, 282
Distributive Property, **356**
Divisible by, **96**, 99-100
Division
  of decimals, **44**
  of fractions, 148
  of mixed numerals, 149
  of rational numbers, 360-361
  Property for Equations, **374**-375
  Property for Inequalities, **394**
  to solve equations, 200-201, 374

Equation(s), **190**-191
  Addition Property for, **324**, 326, 329
  Division Property for, **374**-375
  equivalent, **193**, 201, 203, 324
  for finding percent of a number, 235
  linear, **410**
  Multiplication Property for, **374**
  Solving, 190, 193, 196, 200, 203
  Subtraction Property for, **324**, 326, 329
  to solve word problems, 335
  with like terms, 326
  with more than one operation, 326, 381
  with parentheses, 383
  with variable on both sides, 329
Equilateral triangle, **504**
Equivalent equations, **193**, 201, 203, 324
Equivalent fractions, **161**, 235
Equivalent inequalities, **340**
Equivalent ratios, **216**-217, 508
Estimation, 14
Evaluate, **18**, 22

Even number, **100**
Exponent(s), **8**

Factor, **96**, 97, 108, 124
Factoring, **364**-365
Fahrenheit, **132**
Formula, **2**
  for area of a circle, 177
  for area of a parallelogram, 140, 177
  for area of a rectangle, 70
  for area of a square, 70
  for area of a triangle, 141
  for compound interest, 251
  for converting temperatures, 132, 133
  for discount, 244-245
  for distance in feet an object will fall, 8
  for distance in miles to horizon from aircraft above the ground, 451
  for distance/rate/time, 7, 206
  for finding airspeed in miles per hour, 196
  for finding an average, 10
  for finding angles of a triangle, 388
  for finding batting average, 207
  for finding certain prime numbers, 106
  for finding kilowatt-hours, 5
  for finding net pay, 196
  for finding sticker price, 2-3
  for greatest safe load, 138
  for heat transfer, 359
  for length of one side of a square, 451
  for perimeter of a rectangle, 81
  for perimeter of a square, 82
  for perimeter of a triangle, 82
  for prime numbers, 106, 117
  for sale price, 245
  for simple interest, 240
  for stopping distance, 303
  for sum of n numbers, 86
  for surface area of a cube, 521
  for surface area of a rectangular prism, 521
  for tangent ratio, 458
  for volume of a cone, 528
  for volume of a cylinder, 524
  for volume of a pyramid, 527
  for volume of a rectangular prism, 145
  for volume of a sphere, 535

Pythagorean Theorem, 454-455
Fraction(s), **124**
  as a decimal, 173-174
  as percent, 231
  equivalent, **161**

Geometry, **498**
Gram, **52**
Graph(s)
  of an ordered pair, 408-409
  of integers, 294, 300
  of linear equations, 410-411
  of linear inequalities, 430
  on a number line, 294, 300, 338, 340
  to show data as percents, 274
Graphical method of solving a system of equations, 427-428
Greatest common factor, **115**, 124
Gross earnings, **196**
Groundspeed, **196**

Head wind, **196**
Hecto-, **51**, 52
Hectogram, 52
Hectoliter, 52
Hectometer, 51
Hidden question, **3**, 170
Histogram, **474**-475
Hypotenuse, **454**

Independent events, **488**
Indirect variation, **425**
Inequalities, **295**, 338, 340
  Addition Property for, **340**
  applications of, 396-397
  Compound, **347**
  Division Property for, **394**
  equivalent, **340**
  Multiplication Property for, **393**
  on a number line, 338-339
  Subtraction Property for, **340**
Integers, **294**
  graphs of, 294, 300
  negative, 294
  positive, 294
Intercepts of a graph, 422-423
Interest, **240**
  compound, 251
  simple, **240**, 243
Interest rate, **240**
Inverse operation, **190**
  addition as inverse of subtraction, 193
  division as inverse of multiplication, 200

multiplication as inverse of division, 203
subtraction as inverse of addition, 190, 193
Irrational square root, **442**
Isosceles triangle, **504**

Kilo-, 51, 52
Kilogram, **52**
Kiloliter, **52**
Kilometer, **51**
Kilowatt-hour, **5**, 38

Least common multiple, 111
Like denominators, 158
Like fractions, **158**
Like terms, **76**, 326
  combining, 76, 326
  in solving equations, 326
Line(s), **498**
  parallel, **514**
  perpendicular, **514**
Linear equation, **410**
Linear inequalities, **430**
  graph of, 430
Line segment, **498**
List price, **244**
Liter, **52**

Mathematical sentences, 64
Mean, **10**, 470
Median, **470**-471
Meter, **51**
Metric ton, **52**
Milli-, **51**, 52
Milligram, **52**
Milliliter, **52**
Millimeter, **51**
Mixed numeral(s), **126**, 135, 158, 165
  multiplication of, 135-136
  division of, 149
Mode, **470**
Monomial, **315**
Multiple, **96**
Multiplication
  Associative Property of, **67**, 356
  Commutative Property of, **67**, 356
  of decimals, 38-39
  of fractions, 129
  of mixed numerals, 135-136
  of numbers with like signs, 353
  of probabilities, 488, 491
  of rational numbers, 353, 356

of two numbers with unlike signs, 350
Property for Equations, **374**
Property for Inequalities, **393**
Property of One, **64**, 351
Property of Zero, **64**, 351
to solve equations, 203-204, 374
Multiplicative inverses, **148**
Mutually exclusive events, **483**

Negative integers, 294
Negative rational numbers, **297**
Net pay, **196**
Net price, **245**
Non-repeating decimal, 443
Number(s)
  composite, 103, 108
  counting, 96, 97
  even, **100**
  geometric, 29
  odd, **100**
  prime, **103**, 108
  rational, **297**-298, 304
  whole, **99**
Number line, 294, 300
  inequalities, 338, 340
Number sequences, **288**
Numerical expressions, **18**

Obtuse angle, 502, 504
Odd number, **100**
Operation(s)
  inverse, **190**-191, 193-194, 200-201, 203-204
  order of, **18**, 19, 21
Opposites, **297**-298, 308
Order of Operations, **18**, 19, 21
Ordered pair, 408
Origin, **408**

Parallel lines, 514
Percent, **227**
  applications, 257-279
  and equivalent fractions, 235
  as a decimal, 231-232
  as a fraction, 231
  in circle graphs, 274
  of decrease, **258**
  of increase, **257**
  of what one number is of another, 254
  to find a number, 261
Perfect square, **442**
Perimeter, **22**, 81
  of a rectangle, 81

of a square, **22**, 82
of a triangle, 82
Perpendicular lines, **514**
Pi, **176**, 177
Place name, 32
Plane, **498**
Point, **498**
Point of intersection, 427-428
Polygon, **504**
Positive integers, 294
Positive rational numbers, 297, 304-305, 308
Price
  list, **244**
  net, **245**
  retail, 261
  sale, **245**
  wholesale, 261
Prime factorization, **108**-109, 111 115, 124
Prime number, **103**, 108
Principal, **240**
Probability
  multiplication of, 488, 491
  ratio, 479-480
  tables, 482-483
  using tree diagram, 485
Problem Solving and Applications (see Applications)
Property
  Addition, for Equations, **324**, 326, 329
  Addition, for Inequalities, **340**
  Addition, of Opposites, **312**
  Addition, of Zero, **64**, 312
  Associative, of Addition, 79, 312
  Associative, of Multiplication, 67, 356
  Commutative, of Addition, 79, 312
  Commutative, of Multiplication, 67, 356
  Distributive, **356**
  Division, for Equations, 374-375
  Division, for Inequalities, **394**
  Multiplication, for Equations, **374**
  Multiplication, for Inequalities, **393**
  Multiplication, of One, **64**, 351
  Multiplication, of Zero, **64**, 351
  Subtraction, for Equations, 324, 326, 329

Subtraction, for Inequalities, 340
Proportion(s), 219-220
    applications of, 222-223
Pythagorean Theorem, 454-455

Quadrilateral, 517-518

Radius of a circle, 176-177
Raised to a power, 8
Ratio, 216-217
    applications of, 222, 417, 424, 479
    equivalent, 216-217, 516
    probability, 479-480
    tangent, 458-459
Rational numbers, 297-298
    addition of, 304-305
    as fractions, 297
    as perfect square, 442
    negative, 297-298, 304-305
    positive, 297, 304-305
    subtraction of, 308-309
Ray, 498
Reciprocals, 148
Rectangle, 70, 81, 140
Rectangular prism, 145, 520-521
Repeating decimal, 174, 213, 443
Retail price, 261
Right angle, 502

Sale price, 245
Sample, 495
Scale, 270
Scalene triangle, 504
Scientific notation, 55-56
Set(s), 402
    element of, 402, 437
    empty, 402
    intersection of, 437
    subset of, 437
Similar triangles, 508-509

union of, 437
Slope, 417-418
Square, 70
Square of a number, 440-441
Square root, 440-441
    irrational, 442
Square units, 70
Standard deviation, 478
Sticker price, 2, 3
Straight angle, 502
Subtraction
    of decimals, 35
    of fractions having unlike denominators, 161
    of like fractions, 158
    of mixed numerals, 158
    of mixed numerals with unlike denominators, 165
    of rational numbers, 308-309
    Property for Equations, 324, 326, 329
    Property for Inequalities, 340
    to solve equations, 190-191
System(s) of equations, 427
    solved by graphing, 427-428

Table of Squares and Square Roots, 445-446, 447
Table of Tangents, 459
Tangent ratio, 458-459
Tangent value, 461
Temperature
    Celsius, 132
    conversions, 132-133
    Fahrenheit, 132
Terminating decimal, 173, 443
Total cost of the options, 3
Transversal, 514
Tree diagram, 485
Triangle(s)
    congruent, 505-506

equilateral, 504
isosceles, 504
scalene, 504
similar, 508-509
right, 454, 458, 504

Unlike denominators, 161-162, 165

Value of numerical expression, 18
Variable, 22, 235
    on both sides of equation, 329
Variation
    direct, 424
    indirect, 425
Venn diagram(s), 437
Vertex, 498
Volume, 145
    of a cone, 528
    of a cylinder, 524
    of a pyramid, 527
    of a rectangular prism, 145
    of a sphere, 535

Whole numbers, 99
Wholesale price, 261
Word expression, 334, 377

x axis, 408
x coordinate, 408
x intercept, 422-423

y axis, 408
y coordinate, 408
y intercept, 422-423

Zero, Addition Property of, 64, 312
Zero, Multiplication Property of, 64, 351

ANSWERS TO SELECTED EXERCISES

ANSWERS TO SELECTED EXERCISES

CHAPTER 1

Pages 2-3 Self-Check 1. $9875 **2.** $8308

Page 3 Classroom Exercises 1. 20 **3.** 743
5. 3875 **7.** 8409

Page 4 Written Exercises 1. $10,235 **3.** $9135
5. $10,640 **7.** $9880 **9.** $8828 **11.** $9667

Page 5 Self-Check 1. 176

Page 5 Classroom Exercises 1. 1212 **3.** 8317
5. 2526 **7.** 1249

Page 6 Written Exercises 1. 817 **3.** 4204 **5.** 878
7. 56 **9.** 8196 **11.** 14,543

Pages 7-8 Self-Check 1. 568 miles **2.** 2000 meters
3. 144 feet

Page 8 Classroom Exercises 1. 15 **3.** 522 **5.** 81
7. 64 **9.** 1

Page 9 Written Exercises 1. 2400 feet **3.** 375 kilometers **5.** 465 feet **7.** 2058 meters **9.** 4 **11.** 16
13. 1 **15.** 9 **17.** 36 **19.** 1000 **21.** 576 feet
23. 256 feet

Page 10 Self-Check 1. 143 **2.** 122 **3.** 5°

Page 11 Classroom Exercises 1. 12 **3.** 42 **5.** 123
7. 45 **9.** 7

Page 11 Written Exercises 1. 143 **3.** 159 **5.** 147
7. 45 kilograms **9.** 83 **11.** $150,600 **13.** 90
15. 22 **17.** 76,329 **19.** 25 **21.** Yes

Page 13 Special Topic 1. 140 **3.** 1 **5.** 4
7. 50 **9.** $86

Pages 14-15 Self-Check 1. 2300 kilometers **2.** 300 kilometers **3.** 900 rentals **4.** 8 hours **5.** 26 hours

Page 16 Classroom Exercises 1. a **3.** c

Page 16 Written Exercises 1. a **3.** a **5.** a **7.** b
9. 5000 meters **11.** 2700 seats

Page 17 Review: Sections 1-1—1-5 1. $7653
2. $13,792 **3.** 1084 kilowatt-hours **4.** 1628 kilowatt-hours **5.** 186 meters **6.** 784 feet **7.** 59
8. 85 **9.** 130 seconds **10.** 20 points

Page 19 Classroom Exercises 1. Rule 1 **3.** Rule 3
5. Rule 4 **7.** Rule 1 **9.** Rule 2 **11.** Rule 3
13. Rule 3 **15.** Rule 3

Page 20 Written Exercises 1. addition **3.** multiplication **5.** division **7.** multiplication **9.** 8
11. 27 **13.** 28 **15.** 8 **17.** 24 **19.** 54 **21.** 100
23. 4 **25.** 4 **27.** 47 **29.** 60 **31.** 27 **33.** 36
35. 96 **37.** 57 **39.** 50 **41.** 35 **43.** 3 **45.** 19
47. 23 **49.** 18 **51.** 56 **53.** 4 **55.** 90 **57.** 21
59. 1 **61.** 2 **63.** 24 **65.** 38 **67.** 49 **69.** 10
71. 8 **73.** 8 **75.** 19 **77.** 0 **79.** 25 **81.** 6 + 7 · 3 = 27 **83.** (9 + 9) · 4 = 72 **85.** 4 · 4 · 4 = 64
87. 8 · 9 + 4 − 9 + 6 = 73 **89.** 8 · 2 + 9 · 3 = 43
91. (12 + 3) · (21 ÷ 7) = 45 **93.** (6 · 9) − 14 = 40
95. 18 − (35 ÷ 7) = 13 **97.** (9 − 9) · (33 ÷ 11) = 0

Page 21 Calculator Exercises 1. 51 **3.** 5 **5.** 35
7. 35 **9.** 13

Pages 22-23 Self-Check 1. 108 **2.** 5 **3.** 1
4. 19 **5.** 51 **6.** 10 **7.** 3

Page 23 Classroom Exercises 1. 15 · 5 + 32
3. 6 · 4 + 5 · 7 **5.** 24 · 19 + 18 · 15 − 30
7. 2 · 3 · 4 − 4 **9.** 2 **11.** 3 **13.** 4 **15.** 9 **17.** 4
19. 6 **21.** 8 **23.** 22 **25.** 12 **27.** 120 **29.** 46
31. 216

Page 24 Written Exercises 1. 5 **3.** 8 **5.** 4 **7.** 4
9. 7 **11.** 9 **13.** 36 **15.** 51 **17.** 46 **19.** 12
21. 29 **23.** 100 **25.** 42 **27.** 4 **29.** 42 **31.** 15
33. 48

Page 24 Review: Sections 1-6—1-7 1. 6 **2.** 14
3. 10 **4.** 16 **5.** 43 **6.** 3 **7.** 5 **8.** 4 **9.** 4
10. 42 **11.** 13 **12.** 54 **13.** 0 **14.** 28 **15.** 22
16. 25

Page 25 Chapter Review 1. variable **3.** formula
5. kilowatt-hour **7.** evaluate **9.** $5382 **11.** $9104
13. 219 **15.** 667 **17.** 160 hours **19.** 90 meters
21. 1 **23.** 1 **25.** 49 **27.** 64 **29.** 400 **31.** 3600
33. 14° **35.** 83° **37.** c **39.** b **41.** 12 **43.** 20
45. 90 **47.** 30 **49.** 51 **51.** 21 **53.** 8 **55.** 1
57. 4 **59.** $10,166 **61.** 3331 miles **63.** 140 cases

Page 28 Chapter Test 1. $9100 **3.** 1440 meters
5. 1936 feet **7.** b **9.** b **11.** 20 **13.** 7 **15.** 18
17. 35 **19.** $5528

Page 29 Enrichment 1. a. A triangle with 1 dot in the first row, 2 dots in the second row, 3 dots in the third row, and so on. The last (fifth) row has 5 dots.
b. A triangle with 1 dot in the first row, 2 dots in the second row, and so on. The last (sixth) row has 6 dots.
c. A triangle with 1 dot in the first row, 2 dots in the second row, and so on. The last (seventh) row has 7 dots. **3. a.** A square having 5 rows and 5 columns; **b.** A square having 6 rows and 6 columns; **c.** A square having 7 rows and 7 columns.

Page 30 Additional Practice 1. $9810 **3.** 1527
kilowatt-hours **5.** 1425 feet **7.** 576 feet **9.** 2304
feet **11.** 78 **13.** 374 **15.** a **17.** a **19.** 11
21. 40 **23.** 48 **25.** 203 meters

CHAPTER 2

Pages 32-33 Self-Check 1. 50 **2.** 0.03 **3.** 0.4
4. 0 **5.** > **6.** < **7.** >

548 Answers to Selected Exercises

Page 33 Classroom Exercises 1. ones 3. thousands 5. thousandths 7. > 9. >

Page 34 Written Exercises 1. 0.5 3. 5 5. 20 7. 200 9. 2000 11. 0.007 13. 400 15. 0.3 17. 0.07 19. > 21. < 23. < 25. > 27. > 29. < 31. Jane 33. 22.00124

Page 35 Self-Check 1. 31.88 2. 225.653

Page 36 Classroom Exercises 1. 0.58 3. 19.94 5. 214.3 7. 115.263 9. 133.706 11. 2.32 13. 2.043 15. 0.40 17. 5.884 19. 6.8 21. 23.7

Page 36 Written Exercises 1. 24.92 3. 38.542 5. 242.921 7. 27.74 9. 946.66 11. 2.593 13. 0.83 15. 10.5 17. 8.952 19. 7.17 21. 135.44 23. 25.3 25. 71.21 27. 116 seconds 29. 0.09 meters

Pages 38-39 Self-Check 1. 73¢ 2. 2¢

Page 39 Classroom Exercises 1. 20.0 3. 20.0 5. 0.200 7. 0.00090 9. 0.000090 11. 27.0 13. 15.73 15. 18.200 17. 0.014 19. 0.002848

Page 40 Written Exercises 1. 9.6 3. 97.20 5. 13.6890 7. 28.084 9. 0.58357 11. 1.0224 13. 0.087528 15. 0.018 17. 0.00800 19. 0.0244 21. 0.0456 23. 0.00175 25. 0.00728 27. $3.08, $19.25, $1.54, $3.96 29. $5.47, $34.16, $2.73, $7.03 31. $.42, $2.63, $.21, $.54 33. $8.53, $53.34, $4.27, $10.97 35. $.02 37. $.05 39. $6.45 41. $11.96 43. $10.95

Page 42 Review: Sections 2-1—2-3 1. hundredths 2. tens 3. < 4. < 5. < 6. > 7. 19.94 8. 192.7 9. 2.19 inches 10. $3.47 11. $.06 12. 2.05 inches

Page 43 Special Topic 1. $0.045 3. $0.11 5. $0.017 7. $0.075 9. $0.15

Pages 44-45 Self-Check 1. 26.2 miles 2. 35 3. 730 4. 2900

Page 45 Classroom Exercises 1. 3.48 3. 0.716 5. 28 7. 200 9. 9.4 11. 5470

Page 45 Written Exercises 1. 2.84 3. 9.81 5. 0.065 7. 8.07 9. 1.47 11. 0.056 13. 6.2 15. 89 17. 0.35 19. 7060 21. 3670 23. 8620 25. 16.7 miles 27. 84.5 miles 29. first car

Pages 47-48 Self-Check 1. $13.00 2. 50 pounds

Page 48 Classroom Exercises 1. c 3. a 5. b 7. b 9. b 11. c

Page 49 Written Exercises 1. b 3. b 5. c 7. b 9. $11 11. $66 13. 50 pennants

Page 50 Calculator Exercises 1. 280, 281 3. 12,000, 12,139 5. 70, 68.6

Pages 52-53 Self-Check 1. 25,000 m 2. 1.5 cm 3. 790 cm 4. 3600 cg 5. 0.350 g 6. 53.1 dg 7. 0.022 L 8. 37,000 mL 9. 792,000 L

Page 53 Classroom Exercises 1. divide 3. multiply 5. divide 7. 10 9. 1000

Page 53 Written Exercises 1. 30 mm 3. 54,000 m 5. 3500 m, 350,000 cm 7. 0.256 Km, 25,600 cm 9. 5.100 Kg 11. 900 Kg 13. 3500 g, 3,500,000 mg 15. 0.72 Kg, 720,000 mg 17. 1000 L 19. 450 ml 21. 23,000 ml 23. 620 L 25. 4200 m 27. 0.946 27. 0.946 ml 29. 1000 mg

Pages 55-56 Self-Check 1. 2.6×10^3 2. 864,000

Page 56 Classroom Exercises 1. 1 3. 3 5. 4 7. 4×10 9. 4×10^2 11. 4×10^3 13. 160 15. 38,000

Page 56 Written Exercises 1. 3.4×10^3 3. 9×10^4 5. 1.05×10^6 7. 8.75×10^2 9. 8×10^7 11. 7.6×10^4 13. 3200 15. 251,000 17. 9,123,000 19. 52,000,000 21. 14.2 23. 2×10^6

Page 57 Review: Sections 2-4—2-7 1. 0.7 m 2. 18.4 miles 3. b 4. a 5. c 6. a 7. 0.0043 8. 400 9. 0.3204 10. 3 11. 5 12. 8.25 13. 4.5

Page 58 Chapter Review 1. liter 3. deka- 5. meter 7. metric ton 9. scientific notation 11. gram 13. 0.4 15. 0.04 17. 0.005 19. > 21. < 23. > 25. 10.84 27. 1.669 29. 23.595 31. 105.792 33. 40.81 35. $14.94, $11.43, $7.16 37. 5.89 39. 122 41. c 43. b 45. 20 47. 4510 49. 550 51. 2.5×10^2 53. 5×10^6 55. c 57. $299.64

Page 60 Chapter Test 1. hundredths 3. > 5. < 7. 191.83 9. 37.35 11. 476.37 13. 3.26 15. 0.002832 17. b 19. 2600 21. 3 23. 6 25. 28.2 miles

Page 61 Enrichment 1. 500 sec 3. a. 16,292 h b. 679 days

Page 62 Additional Practice 1. 0.8 3. 60 5. 0.0003 7. < 9. > 11. 82.993 13. 1.26 15. 0.000912 17. 13.2 19. 660 21. b 23. b 25. 4,300,000 27. 14,000 29. 7.7×10^4 31. 1.3×10^2 33. 310 35. 35,000,000 37. $.01

CHAPTER 3

Pages 64-65 Self-Check 1. $t = 0$, Add. Prop. of Zero 2. $s = 0$, Mult. Prop. of Zero 3. $r = 58$, Mult. Prop. of One 4. $r = 21$ 5. $r = 15$ 6. $r = 21$

Page 65 Classroom Exercises 1. Add. Prop. of Zero 3. Mult. Prop. of Zero 5. $x = 0$ 7. $c = 7$ 9. $p = 1$ 11. $t = 0$ 13. $m = 24$ 15. $k = 0.3$ 17. $x = 7$ 19. $a = 2.5$

Answers to Selected Exercises 549

Page 65 Written Exercises 1. $n = 0$ 3. $x = 0.45$ 5. $y = 0$ 7. $m = 0$ 9. $n = 1$ 11. $k = 4.5$ 13. 13. $x = 13$ 15. $z = 19$ 17. $n = 21.3$ 19. $p = 23$ 21. $c = 3$ 23. $m = 5$ 25. $n = 1$ 27. $p = 0$ 29. $t = 0$ 31. $c = 1$ 33. $n = 16$ 35. $x = 0$ 37. $n = 3$ 39. $n = 7$ 41. $n = 3$ 43. $x = 13$ 45. $b = 5$ 47. $z = 0$ 49. $n = 6$ 51. $r = 36$ 53. $n = 5$

Pages 67-68 Self-Check 1. $k = 2$ 2. $d = 7$ 3. $30x$ 4. $62xy$ 5. $33cz$ 6. $11xy$ 7. $50c^2n$ 8. $30a^3b$ 9. $48xy^3$

Page 68 Classroom Exercises 1. $x = 5$ 3. $n = 30$ 5. $p = 2.5$ 7. $y = 7$ 9. $t = 3$ 11. $48t$ 13. $5.5k$ 15. $8ab$ 17. $8rs$ 19. $6x^2y$ 21. $21p^2q^2r$ 23. $90f^2g^2h$ 25. $49k^2n^3$

Page 69 Written Exercises 1. $a = 7$ 3. $n = 2$ 5. $n = 6.2$ 7. $k = 5$ 9. $x = 3$ 11. $24y$ 13. $90b$ 15. $8.4n$ 17. $96r$ 19. $96st$ 21. $8.4pq$ 23. $19.2ab$ 25. $6.08rst$ 27. $84y^2$ 29. $36k^2t$ 31. $42m^3n$ 33. $6st^3$ 35. $80m^2n^3$ 37. $81a^2d^2p^2$ 39. $72mn$ 41. $297p$ 43. $6m^2n^2$ 45. $5.6a$ 47. $24abc$ 49. $6m^3np^2$ 51. $24x^2y^2$; 96 53. x^2y^4; 4 55. $90x^3y^5$; 720

Pages 70-71 Self-Check 1. 1.25 m^2 2. 9 yd^2

Page 71 Classroom Exercises 1. 50 cm^2 3. 5 km^2 5. 21 in^2 7. 64 cm^2 9. 0.25 m^2 11. 225 in^2 13. 625 ft^2

Page 71 Written Exercises 1. 1134 m^2 3. 12 m^2 5. 3 m^2 7. 36.4 mm^2 9. 21 m^2 11. 308 yd^2 13. 104 in^2 15. 48 yd^2 17. 3 mi^2 19. 961 m^2 21. 1.21 m^2 23. 100 m^2 25. 190.44 mm^2 27. 4 mi^2 29. 9 ft^2 31. 256 yd^2 33. 16 mi^2 35. $12,500 \text{ cm}^2$ 37. $13,120 \text{ ft}^2$ 39. 225 in^2 41. $20,908.4 \text{ m}^2$ 43. 8.99 m^2

Page 74 Review: Sections 3-1—3-3 1. Mult. Prop. of Zero 2. Add. Prop. of Zero 3. Mult. Prop. of One 4. $t = 3.8$ 5. $r = 32$ 6. $x = 0$ 7. Assoc. Prop. of Mult. 8. Comm. Prop. of Mult. 9. Assoc. Prop. of Mult. 10. $x = 8$ 11. $g = 6$ 12. $y = 9$ 13. $24y$ 14. $24t$ 15. $12a^2b$ 16. 5670.09 m^2 17. 1008 in^2 18. 2914 cm^2 19. 324 in^2

Page 75 Special Topic 1. 9 3. 15 5. 27.5 7. 7 9. 20 11. 36 13. 84 15. 108

Pages 76-77 Self-Check 1. yes 2. yes 3. no 4. no 5. $10y$ 6. $13rs$ 7. $9.5z^2$ 8. $6p$ 9. $12r$ 10. $4.1t^2$

Page 77 Classroom Exercises 1. like; same variables 3. unlike; different exponents 5. like; same variables 7. unlike; different variables 9. $7 + 15 = 22$ 11. $0.9 - 0.4 = 0.5$ 13. $5 - 4 = 1$ 15. $32 - 19 = 13$ 17. $11x$ 19. $0.9d$ 21. $21r^2$ 23. $34ad$ 25. $5y$ 27. $25b$ 29. $3.3m^2$ 31. $20st$

Page 78 Written Exercises 1. H 3. D, R 5. O 7. C 9. Q 11. $16a$ 13. $13b$ 15. $39y$ 17. $3.9x$ 19. $12m^2$ 21. $14xy$ 23. $7s$ 25. $5a$ 27. $5.9t$ 29. $3.8a$ 31. $2x^2$ 33. $3ab$ 35. $2n$ 37. $45st$ 39. $8.2f^2$ 41. 16.02 m

Pages 79-80 Self-Check 1. $q = 11.5$ 2. $t = 3$ 3. $17a + 7$ 4. $11t$ 5. $3.9x + 13$

Page 80 Classroom Exercises 1. $r = 11$, Comm. and Assoc. Prop. of Add. 3. $x = 5$, Comm. Prop. of Add. 5. $52r + 10$ 7. $42st + 25$ 9. $1.5t + 1.8$ 11. $87n^2 + 9$

Page 80 Written Exercises 1. $r = 2.7$ 3. $x = 4$ 5. $x = 6$ 7. $5a + 4$ 9. $29a + 15$ 11. $3.9n + 4.2$ 13. $55k + 7$ 15. $18x$ 17. $9.5t$ 19. $13a^2 + 4$ 21. $10xy + 4$ 23. $3.9p + 1.9$ 25. $15p + 16$ 27. $91r$ 29. $190 + 254p$

Pages 81-82 Self-Check 1. 82 m 2. 74 ft 3. 20 m 4. 13 yd

Page 82 Classroom Exercises 1. 22 cm 3. 54 m 5. 24 in 7. 19 cm 9. 33.8 cm 11. 22 in

Page 83 Written Exercises 1. 130 cm 3. 7.2 m 5. 69.8 m 7. 10.2 cm 9. 64 mm 11. 16 ft 13. 42 in 15. 24 yd 17. 20 ft 19. 2.9 m 21. 25.1 m 23. 9.8 m 25. 32 cm 27. 9 m 29. 36 ft 31. 12 ft 33. 15 yd 35. 49 in 37. 2256 ft 39. 104 yd 41. $8b$ 43. $3a$ 45. $2a + 2b + c$ 47. $12x$

Page 86 Review: Sections 3-4—3-6 1. yes 2. no 3. yes 4. $5x$ 5. $7.2b$ 6. $1.9ef$ 7. $8n^2$ 8. $a = 3.4$ 9. $c = 4$ 10. $m = 3$ 11. $12b$ 12. $14g + 17$ 13. $11.3t + 10$ 14. $8.2x^2 + 2.5$ 15. 180 mi 16. 18.4 km

Page 86 Calculator Exercises 1. 153 3. 2346 5. 183,315

Page 87 Chapter Review 1. area 3. $4 + 3 = 3 + 4$ 5. square 7. $25 \cdot 0 = 0$ 9. $5 \cdot 12 = 12 \cdot 5$ 11. perimeter 13. Mult. Prop. of Zero 15. Comm. Prop. of Mult. 17. Assoc. Prop. of Mult. 19. $y = 1.9$ 21. $x = 3$ 23. $k = 4$ 25. $25.2m$ 27. $17rt$ 29. $45a^2$ 31. $68p^2q$ 33. $100a^2b^3$ 35. $54x^2y^3$ 37. 70.56 cm^2 39. 441 m^2 41. like 43. unlike 45. unlike 47. $13n$ 49. $11a$ 51. $12.1t$ 53. $2.8n^2$ 55. $y = 3$ 57. $6.7a + 3$ 59. $13ab + 5$ 61. $9t + 8$ 63. $9x^2$ 65. 84 in 67. 29.5 cm 69. 108 ft^2 71. 136.6 m

Page 89 Chapter Test 1. $x = 5.2$ 3. $a = 1$ 5. $14.76x$ 7. $36x^2$ 9. 11.56 m^2 11. 328 in^2 13. $15x$ 15. $14a + 3$ 17. 40.4 m 19. 27 cm 21. 111 ft 23. 21.9 cm 25. 8.3 m

Page 90 Enrichment 1. 190 3. 21,000 5. 37,100 7. 162 9. 500 11. 16,100

Page 91 Additional Practice 1. $x = 0$ 3. $x = 1$

5. $x = 6$ **7.** $y = 13$ **9.** $36x$ **11.** $20p^3$
13. $20x^3y^3$ **15.** 84.64 cm² **17.** 137.64 m²
19. $30d$ **21.** $11c + 12$ **23.** $5xy + 3$ **25.** $12.8x + 4$
27. $18.3m^2n + 13$ **29.** 208 yd **31.** 72 cm
33. square; 1 yd greater **35.** 23.32 yd²

Page 92 Common Errors **1.** 9 **3.** 11 **5.** 21.39
7. 4000 mm **9.** $9b$ **11.** 32 cm

Page 93 Cumulative Review: Chapters 1-3 **1.** c
3. c **5.** a **7.** b **9.** b **11.** a **13.** b **15.** c
17. 4 **19.** d **21.** c **23.** d **25.** d

CHAPTER 4

Pages 96-97 Self-Check **1.** False; since $5 \cdot 7 = 35$
2. True; since $28 \div 4 = 7$ **3.** 1, 2, 3, 4, 6, 12 **4.** 1, 2, 3, 4, 6, 8, 12, 16, 24, 48 **5.** 1, 3 **6.** 1, 11 **7.** 1, 2, 4, 19, 38, 76 **8.** 1, 3, 5, 7, 15, 21, 35, 105

Page 97 Classroom Exercises **1.** 40 is divisible by 20. **3.** 6 is divisible by 2. **5.** 40 is divisible by 5.
7. 7 is divisible by 7. **9.** 11 is a factor of 55. **11.** 8 is a factor of 40. **13.** 1 is a factor of 7. **15.** 33 is a factor of 33. **17.** 24 is a multiple of 8. **19.** 6 is a multiple of 6. **21.** 9 is a multiple of 1. **23.** 15 is a multiple of 3. **25.** $1 \cdot 13$ **27.** $1 \cdot 18; 2 \cdot 9; 3 \cdot 6$
29. $1 \cdot 30; 2 \cdot 15; 3 \cdot 10; 5 \cdot 6$

Page 98 Written Exercises **1.** True, since $18 \div 6 = 3$
3. False, since $21 \div 8 = 2R5$ **5.** True, since $29.1 = 29$
7. True, since $5 \cdot 15 = 75$ **9.** True, since $143 \div 13 = 11$ **11.** 1, 13 **13.** 1, 23 **15.** 1, 31 **17.** 1, 2, 3, 4, 5, 6, 10, 12, 15, 20, 30, 60 **19.** 1, 2, 23, 46
21. 1, 7, 49 **23.** No; No; Yes; 1, 3, 5, 15 **25.** Yes; Yes; Yes; 1, 2, 4, 7, 8, 14, 28, 56

Page 100 Self-Check **1.** 2, 4 **2.** 2, 4 **3.** 2, 4, 5, 10 **4.** 5 **5.** 2, 4, 5, 10 **6.** 3, 9 **7.** 3 **8.** 3
9. neither **10.** No **11.** 2

Page 101 Classroom Exercises **1.** False; 71 does not end in 2 **3.** True; 16 is divisible by 4 **5.** False; 385 does not end in 0 **7.** True; 4130 ends in 0 **9.** True; 9 is divisible by 3 **11.** False; 19 is not divisible by 9
13. True; 18 is divisible by 9 **15.** True; 18 is divisible by 9 **17.** True; 15 is divisible by 3 **19.** True; 8 is divisible by 8

Page 101 Written Exercises **1.** True; 428 ends in 8
3. True; 95 ends in 5 **5.** False; 17 is not divisible by 4
7. True; 28 is divisible by 4 **9.** False; 926 does not end in 0 **11.** True; 15 is divisible by 3 **13.** False; 10 is not divisible by 3 **15.** True; 21 is divisible by 3
17. False; 21 is not divisible by 9 **19.** True; 27 is divisible by 9 **21.** False; 52,374 does not end in 0
23. True; 18 is divisible by 3 **25.** True; 44 is divisible by 4 **27.** False; 67,237 does not end in 0, 2, 4, 6, or 8
29. True; 675 ends in 5 **31.** False; 16 is not divisible by 9 **33.** No; 8 **35.** No; 2

Page 104 Self-Check **1.** 3 **2.** 2 **3.** 2 **4.** 5
5. 31 **6.** $3^2 \cdot 5^2 \cdot 7^2 \cdot 11$ **7.** $2 \cdot 3^2 \cdot 5^3 \cdot 19$

Page 104 Classroom Exercises **1.** prime **3.** composite **5.** prime **7.** 2 **9.** 3 **11.** 7 **13.** $2 \cdot 3 \cdot 5^2$
15. $2^3 \cdot 3^2 \cdot 7 \cdot 11^2$ **17.** $2^4 \cdot 11^2$ **19.** $3^4 \cdot 4^2 \cdot 7^3$

Page 105 Written Exercises **1.** 2 **3.** 3 **5.** 5
7. 5 **9.** 11 **11.** 13 **13.** 29 **15.** 3 **17.** 11
19. $2^3 \cdot 3^2$ **21.** $2^3 \cdot 3^2 \cdot 5 \cdot 11^2$ **23.** $2^2 \cdot 11^3 \cdot 13^2$
25. $2^3 \cdot 3^3 \cdot 5^2 \cdot 7$ **27.** $2^6 \cdot 3^3$

Page 105 Review: Sections 4-1–4-3 **1.** 1, 3, 11, 33
2. 1, 2, 5, 10 **3.** 1, 2, 4, 5, 8, 10, 20, 40 **4.** 1, 2, 13, 26 **5.** 1, 3, 5, 15 **6.** 1, 2, 3, 4, 6, 8, 12, 24 **7.** 1, 2, 4, 5, 10, 20 **8.** 1, 2, 3, 4, 6, 12 **9.** 1, 7 **10.** 1, 13 **11.** 1, 2, 3, 5, 6, 10, 15, 30 **12.** 1, 2, 4, 8, 16, 32
13. True; 12 is divisible by 3 **14.** True; 24 is divisible by 3 **15.** False; 25 is not divisible by 9 **16.** True; 24 is divisible by 4 **17.** True; 44 is divisible by 4
18. False; 1379 does not end in 0, 2, 4, 6, or 8
19. True; 2160 ends in 0 **20.** True; 13,370 ends in 0
21. 2 **22.** 5 **23.** 3 **24.** 7 **25.** $3^3 \cdot 5^2$
26. $2^3 \cdot 3 \cdot 7^2$ **27.** $2^2 \cdot 3^2 \cdot 11^3$ **28.** $2 \cdot 3^3 \cdot 5^3 \cdot 7^2$
29. $3^3 \cdot 5^2$ **30.** $2^2 \cdot 3^2 \cdot 4^2 \cdot 5^2$ **31.** No; 3
32. No; 1

Page 106 Calculator Exercises **1.** 7 **3.** 1023
5. 32,767

Page 107 Special Topic **1.** 6 L **3.** 5 L

Pages 108-109 Self-Check **1.** $7 \cdot 11$ **2.** $3 \cdot 11$
3. $2 \cdot 3 \cdot 5$ **4.** $2 \cdot 3 \cdot 5 \cdot 7$ **5.** $2^2 \cdot 5$ **6.** $2 \cdot 3^2$
7. $2^2 \cdot 5^2$ **8.** $2 \cdot 11^2$

Page 109 Classroom Exercises **1.** $5 \cdot 7 \cdot 13$
3. $2 \cdot 3 \cdot 5^2$ **5.** $2^2 \cdot 3 \cdot 41$ **7.** $2 \cdot 3^2 \cdot 11$ **9.** $3^2 \cdot 29$ **11.** $2 \cdot 3^2 \cdot 5^2$ **13.** $5 \cdot 7$ **15.** $3 \cdot 5$ **17.** $2 \cdot 5 \cdot 13$ **19.** $2^2 \cdot 3^2$ **21.** $2^2 \cdot 5 \cdot 7$ **23.** $2 \cdot 3^3 \cdot 11$

Page 110 Written Exercises **1.** $2 \cdot 3$ **3.** $2 \cdot 17$
5. $5 \cdot 11$ **7.** $2 \cdot 5 \cdot 11$ **9.** $5 \cdot 17$ **11.** $7 \cdot 13$
13. $2^3 \cdot 7$ **15.** $2 \cdot 5^2$ **17.** $2 \cdot 3 \cdot 13$ **19.** $2^2 \cdot 3 \cdot 7$
21. 2^7 **23.** 2^6 **25.** $2 \cdot 3^3 \cdot 7$ **27.** $2^2 \cdot 3^4$
29. $2^2 \cdot 3^3 \cdot 5$ **31.** $2^2 \cdot 3^2 \cdot 5$ **33.** $7 \cdot 17$ **35.** $2^2 \cdot 3^2 \cdot 13$ **37.** $11 \cdot 17$ **39.** 2^9 **41.** $2^2 \cdot 11^2$
43. $2^3 \cdot 13 \cdot 17$ **45.** $11 \cdot 13 \cdot 17$ **47.** $17 \cdot 19$
49. $5 \cdot 7^2 \cdot 13$ **51.** $19 \cdot 23$ **53.** $2 \cdot 5^2 \cdot 23$
55. $5 \cdot 7^2 \cdot 19$ **57.** $2^2 \cdot 3^2 \cdot 7 \cdot 13^2$ **59.** $3 \cdot 5^2 \cdot 13^2$

Page 112 Self-Check **1.** 8 **2.** 84 **3.** 63 **4.** 12
5. 54 **6.** 300

Page 113 Classroom Exercises **1.** $2^2 \cdot 3^2$ **3.** $2^3 \cdot 3^3 \cdot 5^2 \cdot 7$ **5.** $2^3 \cdot 3^3 \cdot 5^3$ **7.** $2^3 \cdot 3^2 \cdot 5$ **9.** 20
11. 15 **13.** 189 **15.** 54 **17.** 21 **19.** 75
21. 240 **23.** 240 **25.** 15 **27.** 135 **29.** 27
31. 75 **33.** 42 **35.** 36

Page 113 Written Exercises **1.** 75 **3.** 60 **5.** 33
7. 280 **9.** 168 **11.** 135 **13.** 294 **15.** 672
17. 168 **19.** 288 **21.** 120 **23.** 240 **25.** 2025
27. 840 **29.** 200 **31.** 180 **33.** 2100 **35.** 216

Answers to Selected Exercises 551

37. 187 **39.** 319 **41.** 120 **43.** 630 **45.** 144
47. 210 **49.** twelfth day **51.** 360 in

Page 116 Self-Check 1. 2 **2.** 4 **3.** 3

Page 116 Classroom Exercises 1. 2 **3.** 2 **5.** 2 · 3 · 3; 2

Page 116 Written Exercises 1. 3; 5; 2 **3.** 3; 3; 6
5. 2 **7.** 3 **9.** 8 **11.** 4 **13.** 15 **15.** 15 **17.** 6
19. 5 **21.** 4 **23.** 9

Page 117 Review: Sections 4-4—4-6 1. $2^4 \cdot 3$
2. $2^3 \cdot 5$ **3.** $2^2 \cdot 3^2 \cdot 5$ **4.** $2^3 \cdot 5^2$ **5.** 5^3
6. $2 \cdot 3 \cdot 11$ **7.** $3 \cdot 5^2$ **8.** $2^2 \cdot 5 \cdot 11$ **9.** $5 \cdot 19$
10. $3 \cdot 11^2$ **11.** $2^5 \cdot 3^2$ **12.** $2 \cdot 5^2$ **13.** 40
14. 360 **15.** 336 **16.** 300 **17.** 1260 **18.** 210
19. 247 **20.** 140 **21.** 12 days **22.** 24 days
23. 18 **24.** 8 **25.** 12 **26.** 11 **27.** 3 **28.** 5
29. 4 **30.** 6

Page 117 Calculator Exercises 1. 131 **3.** 691
5. 1447

Page 118 Chapter Review 1. digits **3.** least common multiple **5.** composite numbers **7.** even number **9.** counting numbers **11.** multiple **13.** 1, 2, 7, 14 **15.** 1, 2, 3, 4, 6, 9, 12, 18, 36 **17.** 1, 2, 3, 5, 6, 9, 18 **19.** 1, 2, 5, 10 **21.** 1, 17 **23.** 1, 2, 3, 4, 6, 8, 9, 12, 18, 24, 36, 72 **25.** True; 435 ends in 5
27. False; 11 is not divisible by 9 **29.** False; 531 does not end in 0, 2, 4, 6, or 8 **31.** $2 \cdot 3^4 \cdot 5^2 \cdot 7 \cdot 11$
33. $3 \cdot 5^4 \cdot 27^2$ **35.** $3 \cdot 5 \cdot 7$ **37.** $2^2 \cdot 3 \cdot 5^2$
39. $2^3 \cdot 11$ **41.** $2 \cdot 3^2 \cdot 5^3 \cdot 13^2$ **43.** $2^4 \cdot 3^4 \cdot 5^3 \cdot 7 \cdot 11$ **45.** 120 **47.** 252 **49.** 7 **51.** 3 **53.** 5
55. 3 **57.** No; 2 **59.** 360

Page 120 Chapter Test 1. 1, 2, 3, 5, 6, 10, 15, 30
3. 1, 2, 3, 4, 6, 8, 12, 24 **5.** True; 27 is divisible by 3
7. False; 13,324 is not divisible by 15 **9.** 7 **11.** $2^4 \cdot 3^3 \cdot 5^2$ **13.** $5^5 \cdot 7^2 \cdot 11 \cdot 13$ **15.** $2 \cdot 3^2 \cdot 5$
17. $2^3 \cdot 3^2 \cdot 5$ **19.** 1 **21.** No; 1

Page 121 Enrichments 1. $2 \times 3 \times 7^2$ **3.** $2^2 \times 3^2$
5. $2^2 \times 5^2$ **7.** $2 \times 3 \times 5^2$ **9.** $2^3 \times 5^3$ **11.** $3^4 \times 5^2$

Page 122 Additional Practice 1. 1, 3, 9 **3.** 1, 2, 3, 4, 6, 8, 12, 16, 24, 48 **5.** 1, 2, 4, 8, 16, 32 **7.** True; 9 is divisible by 9 **9.** True; since 6 · 11 = 66
11. True; since 135 ends in 5 **13.** False; 1342 does not end in 0 **15.** 37 **17.** 3 **19.** 3 **21.** $2^3 \cdot 3^2 \cdot 5$
23. $2^3 \cdot 3^2 \cdot 11^2 \cdot 17$ **25.** $5 \cdot 13$ **27.** $2 \cdot 7 \cdot 13$
29. $2^3 \cdot 3^3$ **31.** 450; 2 **33.** 630; 1 **35.** No; 2

CHAPTER 5

Page 125 Classroom Exercises 1. No **3.** Yes
5. No

Page 125 Written Exercises 1. $\frac{2}{3}$ **3.** $\frac{4}{7}$ **5.** $\frac{1}{8}$
7. $\frac{7}{8}$ **9.** $\frac{2}{15}$ **11.** $\frac{1}{3}$ **13.** $\frac{9}{16}$ **15.** $\frac{4}{9}$ **17.** $\frac{19}{30}$
19. $\frac{2}{5}$ **21.** $\frac{t}{2}$ **23.** $\frac{2}{3y}$ **25.** $\frac{2}{3n}$ **27.** $\frac{3}{7y}$ **29.** $\frac{3n}{5p}$
31. $\frac{9c}{20f}$

Page 126-127 Self-Check 1. $3\frac{3}{4}$ lb **2.** $4\frac{2}{3}$ ft
3. $2\frac{2}{3}$ yd **4.** $8\frac{1}{3}$ yd **5.** $4\frac{7}{8}$ lb **6.** $4\frac{1}{2}$ gal

Page 127 Classroom Exercises 1. $4\frac{1}{2}$ gal **3.** $5\frac{1}{2}$ qt
5. $6\frac{1}{2}$ lb **7.** $3\frac{1}{2}$ qt **9.** $6\frac{2}{3}$ ft **11.** $21\frac{1}{4}$ lb
13. $4\frac{5}{6}$ ft **15.** $3\frac{2}{3}$ yd

Page 127 Written Exercises 1. $6\frac{3}{4}$ gal **3.** $6\frac{2}{3}$ yd
5. $4\frac{3}{4}$ gal **7.** $2\frac{1}{2}$ pt **9.** $1\frac{1}{10}$ mi **11.** $2\frac{2}{3}$ yd
13. $3\frac{11}{12}$ ft **15.** $10\frac{1}{3}$ yd **17.** $11\frac{3}{4}$ gal **19.** $13\frac{1}{4}$ yd
21. $2\frac{1}{4}$ ft **23.** $6\frac{2}{3}$ yd **25.** $7\frac{1}{4}$ gal **27.** $4\frac{1}{2}$ lb
29. $3\frac{1}{4}$ gal **31.** $1\frac{7}{12}$

Pages 129-130 Self-Check 1. $\frac{9}{40}$ **2.** $\frac{8}{15}$ **3.** $\frac{11}{18}$
4. $\frac{1}{5}$ **5.** $\frac{4}{5}$ **6.** $\frac{2}{3}$ **7.** $\frac{3}{5}$ **8.** $\frac{1}{4}$

Page 130 Classroom Exercises 1. $\frac{1}{14}$ **3.** $\frac{3}{16}$ **5.** $\frac{16}{45}$
7. $\frac{5}{21}$ **9.** $\frac{15}{7}$ **11.** $\frac{6}{7}$ **13.** $\frac{4}{7}$ **15.** $\frac{2}{5}$ **17.** $\frac{1}{5}$ **19.** $\frac{2}{5}$
21. $\frac{7}{18}$ **23.** $\frac{3}{10}$

Page 130 Written Exercises 1. $\frac{1}{8}$ **3.** $\frac{3}{10}$ **5.** $\frac{8}{45}$
7. $\frac{4}{5}$ **9.** $\frac{1}{2}$ **11.** $\frac{8}{15}$ **13.** $\frac{7}{15}$ **15.** $\frac{5}{21}$ **17.** $\frac{3}{5}$
19. $\frac{1}{4}$ **21.** $\frac{3}{4}$ **23.** $\frac{1}{6}$ **25.** $\frac{11}{12}$ **27.** $\frac{4}{9}$ **29.** 1
31. $\frac{3}{4}$ **33.** 1 **35.** $\frac{8}{15}$ **37.** $\frac{5}{6}$ hr **39.** $\frac{3}{10}$ gal **41.** $\frac{1}{3}$
43. $\frac{3}{5}$ **45.** st **47.** $\frac{6}{b}$ **49.** $\frac{3}{m}$

Pages 132-133 Self-Check 1. 41°F **2.** 221°F
3. 1472°F **4.** 2192°F **5.** 5°C **6.** 105°C
7. 530°C **8.** 85°C

Page 133 Classroom Exercises 1. 122°F **3.** 59°F
5. 392°F **7.** 20°C **9.** 60°C **11.** 55°C

Page 133 Written Exercises 1. 50°F **3.** 212°F
5. 437°F **7.** 10°C **9.** 45°C **11.** 180°C
13. 120°C **15.** 9977°F

Page 134 Review: Sections 5-1—5-4 1. $\frac{3}{4}$ **2.** $\frac{5}{8}$
3. $\frac{6}{7}$ **4.** $\frac{2}{5}$ **5.** $\frac{4}{5}$ **6.** $\frac{2}{3}$ **7.** $8\frac{11}{12}$ ft **8.** $19\frac{5}{16}$ lb
9. $3\frac{2}{3}$ yd **10.** $2\frac{3}{4}$ gal **11.** $\frac{5}{12}$ **12.** $\frac{1}{3}$ **13.** $\frac{6}{7}$
14. $\frac{2}{15}$ **15.** 20°C **16.** 122°F **17.** 180°C
18. 932°F **19.** $1\frac{7}{12}$ ft **20.** $\frac{3}{4}$ hr

Pages 135-136 Self-Check 1. $3\frac{2}{7}$ **2.** $8\frac{1}{5}$ **3.** $5\frac{1}{3}$
4. $13\frac{1}{2}$ **5.** 10 **6.** 60 **7.** $5\frac{1}{2}$ **8.** $5\frac{11}{16}$

Page 136 Classroom Exercises 1. $4\frac{1}{2}$ **3.** $5\frac{2}{5}$
5. $6\frac{4}{7}$ **7.** $1\frac{3}{5}$ **9.** $\frac{9}{4}$ **11.** $\frac{75}{8}$ **13.** $\frac{55}{16}$ **15.** $17\frac{1}{2}$
17. 4 **19.** 8

Page 136 Written Exercises 1. $1\frac{3}{5}$ **3.** $2\frac{2}{5}$ **5.** $1\frac{7}{8}$
7. $2\frac{7}{10}$ **9.** $\frac{11}{3}$ **11.** $\frac{37}{5}$ **13.** $\frac{31}{9}$ **15.** 7 **17.** 6 **19.** 9
21. $7\frac{7}{8}$ **23.** $11\frac{7}{10}$ **25.** $34\frac{5}{6}$ **27.** $18\frac{39}{40}$ **29.** $2\frac{1}{3}$

31. $13\frac{4}{9}$ **33.** $18\frac{1}{3}$ **35.** $116\frac{2}{3}$ mi **37.** $3\frac{3}{4}$ cups
39. $\frac{p}{2}$ **41.** $\frac{9}{2a}$ **43.** $7m$ **45.** $\frac{7t}{2}$ **47.** $\frac{3}{2p}$

Page 138 Special Topic **1.** 8975 lb **3.** 34,464 lb
5. 13,350 lb

Page 139 Special Topic **1.** 6500 **3.** 9000 **5.** 3500

Page 141 Self-Check **1.** 22.5 in² **2.** 71.68 cm²
3. $13\frac{1}{2}$ ft² **4.** 43 m²

Page 142 Classroom Exercises **1.** 28 in² **3.** 91 yd²
5. 98.8 m² **7.** 6 ft² **9.** 21 yd² **11.** 36.6 cm²

Page 142 Written Exercises **1.** 225 in² **3.** $1\frac{3}{4}$ ft²
5. 96 ft² **7.** 200 in² **9.** 227.5 cm² **11.** 12.8 cm²
13. 84 m² **15.** $4\frac{1}{2}$ ft² **17.** $26\frac{5}{6}$ in² **19.** $91\frac{1}{4}$ ft²
21. $7\frac{7}{9}$ yd² **23.** 5.75 m² **25.** 12.6 cm²
27. 4.55 cm² **29.** 135 ft² **31.** $15\frac{3}{4}$ in²

Page 145 Self-Check **1.** $43\frac{1}{3}$ ft³ **2.** 299.52 m³

Page 146 Classroom Exercises **1.** 135 ft³
3. 0.85 cm³ **5.** $59\frac{1}{2}$ yd³ **7.** 754.4 mm³

Page 146 Written Exercises **1.** 102 ft³ **3.** $67\frac{1}{2}$ ft³
5. 297 in³ **7.** 125 ft³ **9.** 47,424 mm³ **11.** 40 cm³
13. 275.4 m³ **15.** $13\frac{1}{8}$ ft³ **17.** 34,290 cm³
19. 161.4 m³ **21.** 3630 ft³

Pages 148-149 Self-Check **1.** $2\frac{1}{4}$ **2.** $\frac{1}{24}$ **3.** $\frac{2}{25}$
4. $\frac{5}{12}$ **5.** $\frac{1}{2}$ **6.** $4\frac{2}{3}$ **7.** $\frac{2}{3}$ **8.** $1\frac{7}{11}$

Page 149 Classroom Exercises **1.** $\frac{3}{2}$ **3.** $\frac{7}{9}$ **5.** $\frac{1}{4}$
7. $\frac{2}{3}$ **9.** $\frac{1}{8}$ **11.** $\frac{6}{7}$ **13.** $\frac{1}{20}$ **15.** $\frac{3}{50}$ **17.** $\frac{1}{20}$
19. $2\frac{8}{25}$ **21.** 1 **23.** $\frac{1}{2}$ **25.** 3 **27.** $6\frac{1}{8}$

Page 150 Written Exercises **1.** 3 **3.** $\frac{7}{4}$ **5.** $\frac{1}{8}$
7. $\frac{1}{12}$ **9.** $\frac{11}{18}$ **11.** $\frac{4}{13}$ **13.** $\frac{9}{7}$ **15.** $\frac{2}{3}$ **17.** $\frac{3}{4}$
19. $\frac{5}{6}$ **21.** $\frac{3}{5}$ **23.** $\frac{20}{33}$ **25.** $\frac{1}{30}$ **27.** 6 **29.** $\frac{2}{15}$
31. $3\frac{7}{25}$ **33.** $\frac{1}{2}$ **35.** $3\frac{1}{5}$ **37.** 4 **39.** $\frac{4}{5}$ **41.** $3\frac{17}{21}$
43. $\frac{31}{32}$ **45.** 12 **47.** $\frac{1}{2}$ **49.** $7\frac{1}{2}$ **51.** $1\frac{1}{2}$ **53.** $\frac{1}{2}$
55. $\frac{6}{11}$ **57.** $1\frac{1}{5}$ **59.** $1\frac{1}{10}$ **61.** $1\frac{7}{20}$ **63.** $\frac{2}{3}$ **65.** 16
67. 9 **69.** $\frac{1}{6}$ **71.** $\frac{1}{t}$ **73.** 2 **75.** $\frac{3}{2p}$ **77.** $\frac{5t}{4}$

Page 151 Review: Sections 5-5—5-8 **1.** 3 **2.** 5
3. $23\frac{1}{3}$ **4.** $6\frac{7}{9}$ **5.** 15 ft² **6.** 15.05 m² **7.** $12\frac{3}{8}$ in²
8. 102 yd³ **9.** 48.1 cm³ **10.** $1\frac{1}{2}$ **11.** $\frac{1}{5}$ **12.** $1\frac{2}{5}$
13. $\frac{15}{16}$ **14.** $7\frac{7}{8}$ in² **15.** $18\frac{3}{4}$ miles

Page 152 Chapter Review **1.** Fahrenheit **3.** mixed numeral **5.** lowest terms **7.** $\frac{1}{2}$ **9.** $\frac{1}{4}$ **11.** $\frac{7}{16}$
13. $3\frac{11}{12}$ ft **15.** $10\frac{1}{2}$ gal **17.** $4\frac{2}{3}$ yd **19.** $6\frac{1}{4}$ gal
21. $\frac{3}{16}$ **23.** $\frac{2}{5}$ **25.** $\frac{2}{3}$ **27.** $\frac{1}{4}$ **29.** $\frac{1}{2}$ **31.** 104°F
33. 50°F **35.** $3\frac{1}{2}$ **37.** $3\frac{3}{4}$ **39.** $2\frac{5}{8}$ **41.** $\frac{5}{4}$

43. $\frac{35}{8}$ **45.** $\frac{18}{7}$ **47.** 39 **49.** $2\frac{3}{4}$ **51.** $2\frac{1}{10}$
53. $16\frac{1}{3}$ **55.** 12 **57.** 10 in² **59.** 4.55 cm²
61. 127.1 m³ **63.** 455 in³ **65.** $2\frac{7}{9}$ **67.** $\frac{3}{8}$ **69.** $\frac{1}{2}$
71. $2\frac{1}{2}$ **73.** 17 **75.** 15°C **77.** 490 ft³

Page 154 Chapter Test **1.** $\frac{1}{2}$ **3.** $\frac{3}{8}$ **5.** $8\frac{3}{4}$ ft
7. $\frac{5}{12}$ **9.** 14 **11.** 130.5 m² **13.** $26\frac{2}{3}$ in² **15.** 2
17. $6\frac{3}{4}$ **19.** 55°C

Page 155 Enrichment **1.** $32\frac{1}{2}$ m² **3.** 66 m²
5. 28 ft²

Page 156 Additional Practice **1.** $\frac{1}{2}$ **3.** $\frac{2}{7}$ **5.** $\frac{3}{7}$
7. $28\frac{1}{4}$ lb **9.** $1\frac{3}{4}$ ft **11.** $\frac{1}{4}$ **13.** $\frac{6}{35}$ **15.** $\frac{6}{11}$
17. 95°F **19.** 0°C **21.** $22\frac{1}{2}$ **23.** 6 **25.** 28 cm²
27. 12 m² **29.** 60 m³ **31.** $1\frac{1}{4}$ **33.** $\frac{9}{20}$ **35.** $5\frac{1}{3}$
37. 7500 m³

CHAPTER 6

Pages 158-159 Self-Check **1.** $1\frac{1}{3}$ **2.** $\frac{2}{5}$ **3.** $\frac{2}{5}$
4. $1\frac{3}{5}$ **5.** 8 **6.** $3\frac{1}{2}$ **7.** $5\frac{1}{5}$ **8.** $7\frac{4}{5}$

Page 159 Classroom Exercises **1.** $\frac{1}{2}$ **3.** 1 **5.** $1\frac{1}{2}$
7. $2\frac{1}{2}$ **9.** $1\frac{1}{2}$

Page 159 Written Exercises **1.** $\frac{2}{3}$ **3.** $\frac{1}{3}$ **5.** $\frac{1}{2}$
7. 1 **9.** $1\frac{2}{5}$ **11.** $5\frac{2}{7}$ **13.** $9\frac{5}{9}$ **15.** $6\frac{2}{3}$ **17.** $5\frac{2}{3}$
19. $9\frac{5}{7}$ **21.** $10\frac{1}{11}$ **23.** $\frac{3}{5}$ **25.** $\frac{3}{25}$ **27.** $8\frac{2}{5}$ **29.** $2\frac{1}{4}$
31. $14\frac{1}{4}$ **33.** $13\frac{1}{4}$ **35.** $6\frac{1}{2}$ hrs **37.** $\frac{8}{a}$ **39.** $\frac{9}{x}$
41. $\frac{2x}{3}$ **43.** $\frac{2s}{9}$ **45.** $\frac{m}{5}$ **47.** $\frac{3x}{y}$ **49.** $\frac{5a}{b}$ **51.** $\frac{5a}{3f}$

Pages 161-162 Self-Check **1.** $\frac{5}{10}$ **2.** $\frac{8}{20}$ **3.** $\frac{35}{60}$
4. $\frac{20}{64}$ **5.** $1\frac{1}{2}$ **6.** $1\frac{1}{5}$ **7.** $\frac{19}{20}$ **8.** $\frac{19}{24}$ **9.** $\frac{1}{4}$ **10.** $\frac{1}{2}$
11. $\frac{1}{40}$ **12.** $\frac{1}{16}$

Page 162 Classroom Exercises **1.** $\frac{3}{9}$ **3.** $\frac{12}{16}$ **5.** $\frac{15}{35}$
7. $\frac{7}{8}$ **9.** $\frac{7}{18}$ **11.** $\frac{17}{24}$ **13.** $1\frac{3}{14}$ **15.** $1\frac{7}{40}$ **77.** $\frac{1}{2}$
19. $\frac{11}{24}$ **21.** $\frac{11}{20}$ **23.** $\frac{1}{6}$ **25.** $\frac{1}{8}$

Page 163 Written Exercises **1.** $\frac{2}{6}$ **3.** $\frac{25}{40}$ **5.** $\frac{30}{36}$
7. $\frac{18}{42}$ **9.** $\frac{10}{15}$ **11.** $\frac{21}{36}$ **13.** $\frac{7}{8}$ **15.** $\frac{4}{5}$ **17.** $\frac{7}{18}$
19. $\frac{22}{45}$ **21.** $\frac{14}{15}$ **23.** $\frac{23}{24}$ **25.** $1\frac{1}{2}$ **27.** $1\frac{1}{42}$ **29.** $\frac{1}{2}$
31. $\frac{2}{15}$ **33.** $\frac{7}{36}$ **35.** $\frac{11}{18}$ **37.** $\frac{7}{45}$ **39.** $\frac{7}{39}$ **41.** $\frac{1}{8}$
43. $1\frac{2}{3}$ **45.** $\frac{21}{40}$ **47.** $\frac{25}{44}$ **49.** $1\frac{1}{45}$ **51.** $\frac{13}{14}$
53. $1\frac{3}{8}$ in **55.** $\frac{7}{30}$ mi **57.** $\frac{6x}{25}$ **59.** $\frac{a}{8}$ **61.** $\frac{5x}{8}$
63. $\frac{13}{2y}$ **65.** $\frac{7}{3x}$ **67.** $\frac{7a}{2n}$ **69.** $\frac{9y}{16z}$ **71.** $\frac{61s}{15t}$

Page 165 Self-Check **1.** $14\frac{2}{15}$ **2.** $3\frac{1}{4}$ **3.** $2\frac{5}{12}$
4. $8\frac{1}{8}$ **5.** $1\frac{1}{2}$ **6.** $2\frac{1}{2}$ **7.** $5\frac{2}{2}$ **8.** $2\frac{2}{3}$

Page 166 Classroom Exercises **1.** $2\frac{2}{4}$ **3.** $6\frac{16}{20}$

Answers to Selected Exercises 553

5. $6\frac{7}{6}$ 7. $5\frac{4}{4}$

Page 166 Written Exercises 1. $7\frac{3}{4}$ 3. $1\frac{11}{12}$ 5. 8
7. $7\frac{3}{20}$ 9. $3\frac{1}{4}$ 11. $2\frac{1}{12}$ 13. $2\frac{1}{8}$ 15. $6\frac{2}{15}$
17. $4\frac{2}{3}$ 19. $1\frac{1}{2}$ 21. $10\frac{2}{3}$ 23. $4\frac{47}{60}$ 25. $5\frac{1}{10}$
27. $3\frac{7}{8}$ 29. $6\frac{11}{16}$ 31. $9\frac{14}{12}$ 33. $16\frac{1}{12}$ 35. $5\frac{7}{8}$
37. $1\frac{5}{12}$ 39. $12\frac{5}{12}$ 41. $\frac{11}{20}$ min 43. $16\frac{1}{6}$ ft

Page 167 Review: Sections 6-1—6-3 1. $\frac{3}{4}$ 2. $7\frac{2}{3}$
3. $\frac{3}{13}$ 4. $11\frac{3}{10}$ 5. $10\frac{1}{2}$ 6. $\frac{10}{20}$ 7. $\frac{12}{28}$ 8. $\frac{9}{12}$
9. $\frac{25}{30}$ 10. $\frac{42}{48}$ 11. $1\frac{1}{6}$ 12. $\frac{1}{2}$ 13. $\frac{17}{24}$ 14. $\frac{1}{2}$
15. $\frac{1}{3}$ 16. $4\frac{1}{12}$ 17. $9\frac{13}{15}$ 18. $4\frac{1}{2}$ 19. $2\frac{17}{20}$
20. $10\frac{7}{8}$ 21. $3\frac{4}{5}$ mi 22. $1\frac{5}{12}$ hrs

Page 168 Special Topic 1. 49 3. 4096 5. 1024
7. 19,487,171 9. 62,748,517 11. 4,100,625

Pages 169-170 Self-Check 1. $15\frac{5}{12}$ hrs 2. $63.98

Page 170 Classroom Exercises 1. $4\frac{1}{2}$; 7; 7; $6\frac{1}{2}$; $2\frac{3}{4}$; $27\frac{3}{4}$ hrs

Page 171 Written Exercises 1. $7\frac{1}{4}$; $7\frac{1}{2}$; $8\frac{3}{4}$; $7\frac{1}{2}$; $7\frac{1}{2}$; $38\frac{1}{2}$ hrs 3. $7\frac{1}{2}$; $7\frac{3}{4}$; $3\frac{3}{4}$; $7\frac{3}{4}$; $7\frac{1}{2}$; $34\frac{1}{4}$ hrs 5. $200.20
7. $212.35 9. $93.80 11. $134.20 13. $22.95
15. $12.90 per hour

Pages 173-174 Self-Check 1. 0.5 2. 0.4
3. 0.625 4. 2.5 5. 1.8 6. $0.\overline{333}\cdots$
7. $0.\overline{833}\cdots$ 8. $0.\overline{133}\cdots$ 9. $0.\overline{555}\cdots$
10. $0.\overline{277}\cdots$ 11. < 12. < 13. > 14. =

Page 174 Classroom Exercises 1. = 3. ≠ 5. ≠
7. = 9. T 11. R 13. T 15. < 17. <

Page 175 Written Exercises 1. 0.55 3. 3.8
5. 0.15 7. 0.625 9. 0.325 11. $0.\overline{3}$ 13. $0.\overline{7}$
15. $0.7\overline{3}$ 17. $1.\overline{1}$ 19. $0.58\overline{3}$ 21. $0.8\overline{3}$ 23. $2.\overline{3}$
25. $0.1\overline{6}$ 27. > 29. < 31. > 33. < 35. <
37. Earl 39. Phyllis

Pages 177-178 Self-Check 1. 19 in 2. 63 ft
3. 16 cm 4. 79 m 5. 314 cm² 6. 39 ft²

Page 178 Classroom Exercises 1. 88 in 3. 94 m
5. 28 yd²

Page 178 Written Exercises 1. 92 cm 3. 20 cm
5. 4 m 7. 76 cm 9. 47 ft 11. 42 in 13. 13 yd
15. 38 in 17. 1256 cm² 19. 2 m² 21. 408 cm²
23. 23 m² 25. 227 in² 27. 16 in² 29. 50 in²
31. 13 in²

Page 180 Review: Sections 6-4—6-6 1. a. $7\frac{1}{4}$; 7; $7\frac{1}{3}$; $7\frac{1}{4}$; $4\frac{2}{3}$ b. $33\frac{1}{2}$ 2. $180.90 3. < 4. > 5. =
6. < 7. $37\frac{5}{7}$ in; $113\frac{1}{7}$ in² 8. 26.38 m; 55.39 m²
9. Tara's 10. $234\frac{1}{7}$ in²

Page 180 Calculator Exercises 31. > 33. <
35. <

Page 181 Chapter Review 1. terminating 3. like
5. unlike 7. $\frac{6}{7}$ 9. $\frac{1}{2}$ 11. $3\frac{2}{3}$ 13. $4\frac{1}{2}$ 15. $\frac{1}{5}$
17. $\frac{8}{18}$ 19. $\frac{21}{36}$ 21. $\frac{14}{15}$ 23. $\frac{43}{48}$ 25. $\frac{5}{9}$ 27. $\frac{13}{30}$
29. $\frac{16}{21}$ 31. $3\frac{5}{6}$ 33. $15\frac{20}{21}$ 35. $7\frac{23}{24}$ 37. $5\frac{4}{9}$
39. $6\frac{7}{24}$ 41. $38\frac{5}{6}$; $186.40 43. 0.125 45. $0.\overline{54}$
47. 0.15 49. < 51. > 53. 24 m; 45 m² 55. 76 ft; 453 ft² 57. $\frac{1}{5}$ hr 59. $1\frac{17}{20}$ mi 61. 100 m

Page 183 Chapter Test 1. $\frac{8}{10}$ 3. $\frac{20}{48}$ 5. $\frac{1}{2}$
7. $1\frac{1}{18}$ 9. $3\frac{4}{7}$ 11. $13\frac{5}{24}$ 13. 0.6 15. $0.8\overline{3}$
17. 14 m; 17 m² 19. $93.10

Page 184 Enrichment 1. .1 cm; ± 0.05 cm; 4.35 cm; 4.25 cm 3. $\frac{1}{2}$ in; ± $\frac{1}{4}$ in; $2\frac{3}{4}$ in; $2\frac{1}{4}$ in

Page 185 Additional Practice 1. $\frac{3}{5}$ 3. $\frac{1}{2}$ 5. $\frac{3}{4}$
7. $\frac{17}{35}$ 9. $\frac{1}{2}$ 11. $\frac{9}{14}$ 13. $8\frac{1}{2}$ 15. $9\frac{5}{8}$ 17. $3\frac{5}{18}$
19. $5\frac{13}{15}$ 21. 0.7 23. $0.2\overline{6}$ 25. $0.41\overline{6}$ 27. >; 0.8 > 0.75 29. >; 0.4 > 0.375 31. >; 0.375 > $0.\overline{3}$
33. 79 m; 491 m² 35. 95 cm; 725 cm² 37. 13 mi; 13 mi² 39. 57 yd; 255 yd² 41. more computers

Page 186 Common Errors 1. 1, 2, 3, 6, 9, 18
3. 24 5. $\frac{12}{25}$ 7. $2\frac{4}{25}$ 9. $1\frac{5}{12}$ 11. $9\frac{1}{3}$ 13. $2\frac{1}{4}$ yd²

Page 187 Cumulative Review: Chapters 4-6 1. b
3. d 5. b 7. a 9. c 11. c 13. d 15. a
17. a 19. b 21. c 23. d

CHAPTER 7

Pages 190-191 Self-Check 1. 36 2. 3 3. 0
4. 1.8 5. $\frac{3}{8}$ 6. $1\frac{1}{4}$

Page 191 Classroom Exercises 1. 8 3. n 5. x
7. n 9. $x + 5 - 5 = 19 - 5$ 11. $n + 5.7 - 5.7 = 10.3 - 5.7$ 13. $b + \frac{1}{4} - \frac{1}{4} = \frac{1}{2} - \frac{1}{4}$ 15. $6\frac{7}{8} - 2\frac{3}{4} = y + 2\frac{3}{4} - 2\frac{3}{4}$ 17. $3.2 - 3.2 + r = 9 - 3.2$ 19. $6 - 6 + c = 9 - 6$

Page 192 Written Exercises 1. 7 3. 17 5. 24
7. 17 9. 27 11. 49 13. 1.6 15. 3.5 17. 2.9
19. 11.5 21. 0.38 23. 21.74 25. $1\frac{1}{4}$ 27. $2\frac{1}{4}$
29. $3\frac{1}{2}$ 31. $\frac{1}{3}$ 33. $3\frac{5}{9}$ 35. $2\frac{2}{15}$ 37. 2.54
39. $3\frac{5}{6}$ 41. 213 43. 1.07 45. 7 47. $1\frac{7}{8}$
49. 1.58 51. 0

Pages 193-194 Self-Check 1. 23 2. 18 3. 5
4. 5.9 5. 1 6. $4\frac{1}{2}$

Page 194 Classroom Exercises 1. a 3. r 5. x
7. a 9. $3\frac{1}{4}$ 11. $a - 12 + 12 = 5 + 12$ 13. $47 +$

554   *Answers to Selected Exercises*

$83 = t - 83 + 83$ **15.** $12.9 + 5.6 = b - 5.6 + 5.6$ **17.** $4\frac{1}{4} + 1\frac{7}{8} = n - 1\frac{7}{8} + 1\frac{7}{8}$

Page 194 Written Exercises 1. 58 **3.** 101 **5.** 203 **7.** 392 **9.** 338 **11.** 4107 **13.** 10.1 **15.** 1.13 **17.** 48 **19.** 24.37 **21.** 20.5 **23.** 110.18 **25.** 8 **27.** $14\frac{1}{2}$ **29.** $1\frac{5}{8}$ **31.** $5\frac{1}{2}$ **33.** 14 **35.** $3\frac{7}{8}$ **37.** 127 **39.** 368 **41.** $6\frac{1}{4}$ **43.** $3\frac{11}{16}$ **45.** 3.05 **47.** 334.4 **49.** 8 **51.** 2, 4, 6, 8, 10 **53.** 8

Pages 196-197 Self-Check 1. $336.20 **2.** 655 mi/h

Page 197 Classroom Exercises 1. $233 **3.** $462.11 **5.** 375 mi/h

Page 197 Written Exercises 1. $439.00 **3.** $491.68 **5.** $337.23 **7.** $291.77 **9.** 337 mi/h **11.** 875.4 mi/h **13.** 376.8 mi/h **15.** 475 mi/h **17.** $a + t = g$

Page 198 Review: Sections 7-1—7-3 1. 15 **2.** 22 **3.** 8.6 **4.** 0.33 **5.** $\frac{1}{3}$ **6.** $2\frac{5}{8}$ **7.** 4.3 **8.** $\frac{1}{6}$ **9.** $4\frac{3}{10}$ **10.** 38 **11.** 130 **12.** 13.6 **13.** 108 **14.** 8 **15.** $2\frac{1}{8}$ **16.** 10.0 **17.** $12\frac{7}{12}$ **18.** $5\frac{17}{20}$ **19.** $257.69 **20.** $148.80 **21.** 569.9 mi/h **22.** 597.5 mi/h

Page 199 Special Topic 1. 11; $63.69 **3.** 11; $87.89

Pages 200-201 Self-Check 1. 14 **2.** 86 **3.** 16 **4.** 32 **5.** 70 **6.** $15\frac{4}{5}$ **7.** 2 **8.** $12\frac{1}{2}$

Page 201 Classroom Exercises 1. x **3.** n **5.** $\frac{4n}{4} = \frac{32}{4}$ **7.** $\frac{44}{11} = \frac{11y}{11}$ **9.** $\frac{2.6a}{2.6} = \frac{33.8}{2.6}$ **11.** $\frac{96}{19.2} = \frac{19.2x}{19.2}$ **13.** No **15.** Yes **17.** Yes **19.** No

Page 202 Written Exercises 1. 9 **3.** 5 **5.** 32 **7.** 5 **9.** 14 **11.** 41 **13.** 20 **15.** 120 **17.** 40 **19.** $9\frac{1}{5}$ **21.** $16\frac{4}{5}$ **23.** $10\frac{5}{8}$ **25.** $\frac{3}{4}$ **27.** 32 **29.** 12 **31.** 53.5 **33.** 125 **35.** $12\frac{1}{2}$ **37.** 24.5 **39.** $\frac{2}{5}$ **41.** 22 **43.** $15\frac{1}{3}$ **45.** $\frac{5}{8}$ **47.** 12 **49.** $\frac{9}{10}$ **51.** $2\frac{1}{4}$ **53.** 28 **55.** $1\frac{1}{3}$

Pages 203-204 Self-Check 1. 27 **2.** 8 **3.** 336 **4.** 108 **5.** 126 **6.** 18 **7.** 10 **8.** 12.3

Page 204 Classroom Exercises 1. q **3.** a **5.** r **7.** p **9.** $\frac{n}{5} \cdot 5 = 6 \cdot 5$ **11.** $\frac{y}{10} \cdot 10 = 3 \cdot 10$ **13.** $\frac{s}{5} \cdot 5 = 1.1(5)$ **15.** $2\frac{1}{2} \cdot 4 = \frac{v}{4} \cdot 4$ **17.** $\frac{b}{17} \cdot 17 = 5 \cdot 17$ **19.** $3\frac{3}{8} \cdot 8 = \frac{d}{8} \cdot 8$ **21.** $\frac{f}{3} \cdot 3 = 4\frac{5}{6} \cdot 3$ **23.** $\frac{r}{7} \cdot 7 = 12 \cdot 7$

Page 205 Written Exercises 1. 136 **3.** 276 **5.** 861 **7.** 4498 **9.** 10 **11.** 16 **13.** 126 **15.** 7.47 **17.** 2.96 **19.** 56.4 **21.** 10.0 **23.** 9.9 **25.** 28 **27.** 70 **29.** 4 **31.** $162\frac{1}{2}$ **33.** 33.6 **35.** 910 **37.** 62 **39.** 4794 **41.** 4.8 **43.** 24 **45.** 9 **47.** $3\frac{1}{5}$ **49.** $\frac{2}{3}$ **51.** $3\frac{2}{3}$

Pages 206-207 Self-Check 1. 255 seconds **2.** 26 hits

Page 207 Classroom Exercises 1. 20 hours **3.** 24 km/h **5.** 28 hits

Page 208 Written Exercises 1. 6 h **3.** 5.24 m/sec **5.** $8\frac{1}{5}$ sec **7.** 1083.75 mi/h **9.** 15 h **11.** 50 hits **13.** 30 hits **15.** 39 hits **17.** 176 hits **19.** 26 hits **21.** $2\frac{3}{5}$ hours **23.** 200 times at bat

Page 209 Review: Sections 7-4—7-6 1. 7 **2.** 6 **3.** 6 **4.** 15 **5.** $4\frac{2}{7}$ **6.** $\frac{3}{8}$ **7.** 26 **8.** 11 **9.** 117 **10.** 100 **11.** 24 **12.** 16.87 **13.** $10\frac{2}{3}$ **14.** $39\frac{2}{3}$ **15.** 1664 **16.** 0.104 **17.** 1460 m/sec **18.** 17 hits

Page 210 Chapter Review 1. net pay **3.** addition **5.** divide **7.** equation **9.** equivalent equations **11.** 34 **13.** 12.8 **15.** $10\frac{3}{4}$ **17.** 22 **19.** 36.4 **21.** $11\frac{3}{8}$ **23.** $224.00 **25.** $239.45 **27.** 440 mi/h **29.** 277.4 mi/h **31.** 9 **33.** $8\frac{3}{4}$ **35.** 182 **37.** $126\frac{2}{3}$ **39.** 60 sec **41.** 52 km/h **43.** 102 hits **45.** $299.45 **47.** 5.1 mi/h

Page 212 Chapter Test 1. 25 **3.** 2.89 **5.** 77 **7.** $1\frac{4}{5}$ **9.** 11 **11.** $10\frac{1}{2}$ **13.** 4112 **15.** 219.24 **17.** $352.11 **19.** 156.25 mi/h

Page 213 Enrichment 1. $\frac{1}{3}$ **3.** $\frac{2}{11}$ **5.** $\frac{8}{11}$ **7.** $\frac{5}{3}$ **9.** $\frac{38}{11}$ **11.** $\frac{34}{37}$ **13.** $\frac{619}{99}$ **15.** $\frac{725}{99}$

Page 214 Additional Practice 1. 305 **3.** 1.12 **5.** $1\frac{3}{8}$ **7.** 89 **9.** 30.2 **11.** $2\frac{9}{16}$ **13.** $515.00 **15.** 75 km/h **17.** 16 **19.** 65 **21.** 12.3 **23.** 1.52 **25.** 448 **27.** 27.6 **29.** 26.88 **31.** 55 **33.** 195 hits **35.** $262.92

CHAPTER 8

Page 216-217 Self-Check 1. $\frac{1}{3}$ **2.** $\frac{4}{3}$ **3.** $\frac{1}{7}$ **4.** $\frac{5}{1}$ **5.** No **6.** Yes **7.** No **8.** No **9.** $\frac{2}{5}$ **10.** $\frac{5}{7}$ **11.** $\frac{1}{4n}$ **12.** $\frac{3}{7p}$

Page 218 Classroom Exercises 1. $\frac{3}{4}$ **3.** $\frac{5}{1}$ **5.** $\frac{4}{7}$ **7.** Yes **9.** No **11.** $\frac{3}{5}$ **13.** $\frac{7p}{2}$ **15.** $\frac{3}{2}$ **17.** $\frac{1}{2q}$ **19.** $\frac{1}{3}$ **21.** $\frac{1}{1}$

Page 218 Written Exercises 1. $\frac{1}{3}$ **3.** $\frac{3}{8}$ **5.** $\frac{1}{10}$ **7.** $\frac{1}{3}$ **9.** $\frac{1}{4}$ **11.** $\frac{1}{3}$ **13.** Yes **15.** Yee **17.** No **19.** No **21.** $\frac{5}{6}$ **23.** $\frac{5r}{2}$ **25.** $\frac{10}{7}$ **27.** $\frac{3}{7}$ **29.** $\frac{5}{12}$ **31.** $\frac{15a}{1}$ **33.** $\frac{1}{1}$ **35.** $\frac{4}{1}$ **37.** $\frac{r}{2}$ **39.** $\frac{r}{1}$ **41.** $\frac{1}{9}$ **43.** $\frac{8}{3}$

Page 220 Self-Check 1. 1 **2.** 1 **3.** 13 **4.** $2\frac{3}{5}$ **5.** $2\frac{2}{5}$ **6.** 4 **7.** 8 **8.** 16

Page 220 Classroom Exercises 1. $16n = 80$ **3.** $45 = 7x$ **5.** $24 = 4m$ **7.** $x = 3$ **9.** $p = 8$ **11.** $k = 2$

Page 220 Written Exercises 1. 18 **3.** 1 **5.** 2

Answers to Selected Exercises 555

7. 7 **9.** 27 **11.** 16 **13.** 54 **15.** 7 **17.** 18
19. 60 **21.** 11 **23.** 36 **25.** $3\frac{1}{5}$ **27.** $4\frac{1}{2}$ **29.** $11\frac{1}{4}$
31. 2 **33.** 400 **35.** $1\frac{1}{4}$

Page 222-223 Self-Check 1. $21\frac{1}{3}$ quarts **2.** 95 beats

Page 223 Classroom Exercises 1. $\frac{5}{x}$ **3.** $\frac{90}{x}$

Page 224 Written Exercises 1. 585 km **3.** 18
5. $77\frac{7}{9}$ **7.** 10.5 in **9.** 30 in **11.** $762\frac{1}{2}$ bu **13.** 300

Page 225 Review: Sections 8-1–8-3 1. $\frac{4}{5}$ **2.** $\frac{3}{2}$
3. $\frac{8}{5}$ **4.** $\frac{1}{5}$ **5.** $\frac{3}{5}$ **6.** $\frac{13}{25}$ **7.** $\frac{6}{25}$ **8.** $\frac{2k}{3}$ **9.** Yes
10. Yes **11.** No **12.** No **13.** 15 **14.** 6 **15.** 6
16. $16\frac{4}{5}$ **17.** 480 **18.** 50

Page 226 Special Topic
```
10 INPUT H, R        (H = Number of Hours)
20 LET P = H * R     (R = Pay per Hour)
30 PRINT P           (P = Total Pay)
40 END
```

Pages 227-228 Self-Check 1. 25% **2.** 60%
3. 70% **4.** 0% **5.** 45% **6.** 82% **7.** $33\frac{1}{3}$%
8. $83\frac{1}{3}$% **9.** $42\frac{6}{7}$% **10.** $22\frac{2}{9}$% **11.** $41\frac{2}{3}$%
12. $56\frac{1}{4}$% **13.** 100% **14.** 100% **15.** 350%
16. 625% **17.** $216\frac{2}{3}$% **18.** 325%

Page 229 Classroom Exercises 1. 9% **3.** 7.6%
5. $16\frac{2}{3}$% **7.** $\frac{n}{100} = \frac{3}{4}$ **9.** $\frac{n}{100} = \frac{7}{20}$ **11.** $\frac{n}{100} = \frac{1}{2}$
13. $\frac{n}{100} = \frac{2}{3}$ **15.** $\frac{n}{100} = \frac{4}{15}$ **17.** $\frac{n}{100} = \frac{1}{6}$ **19.** $\frac{n}{100} = \frac{16}{5}$ **21.** $\frac{n}{100} = \frac{20}{7}$ **23.** $\frac{n}{100} = \frac{20}{19}$

Page 229 Written Exercises 1. 25% **3.** 20%
5. 8% **7.** 80% **9.** 16% **11.** 85% **13.** $62\frac{1}{2}$%
15. $2\frac{1}{2}$% **17.** $71\frac{3}{7}$% **19.** 2% **21.** $63\frac{7}{11}$%
23. $32\frac{1}{2}$% **25.** 120% **27.** 150% **29.** $131\frac{1}{4}$%
31. 102% **33.** 220% **35.** 340% **37.** 40%
39. $77\frac{7}{9}$% **41.** $146\frac{2}{3}$% **43.** 18% **45.** $63\frac{7}{11}$%
47. $8\frac{1}{3}$% **49.** 610% **51.** 1190% **53.** 70%
55. 40% **57.** 88% **59.** 68% **61.** 62%

Page 232 Self-Check 1. 8% **2.** 63% **3.** 50.1%
4. 940% **5.** 358% **6.** 0.15 **7.** 0.06 **8.** 0.215
9. 0.004 **10.** 1.13

Page 233 Classroom Exercises 1. $\frac{41}{100}$ **3.** $\frac{97}{100}$
5. $\frac{790}{100}$ **7.** 0.21 **9.** 0.007 **11.** 0.063 **13.** 2.00
15. 0.015 **17.** 0.0005

Page 233 Written Exercises 1. 40% **3.** 940%
5. 20.3% **7.** 150% **9.** 26% **11.** 10% **13.** 860%
15. 4% **17.** 362% **19.** 0.18 **21.** 0.661 **23.** 0.03
25. 1.01 **27.** 0.005 **29.** 0.20 **31.** 0.0025
33. 4.00 **35.** 0.045 **37.** 14%; $\frac{7}{50}$ **39.** 0.33; $\frac{33}{100}$
41. 85%; 0.85 **43.** 0.60; $\frac{3}{5}$ **45.** 90%; 0.90

47. 0.47 **49.** 0.05 **51.** 0.025 **53.** 0.025
55. 0.52 **57.** 0.131

Pages 235-236 Self-Check 1. 0.3 **2.** 2.4 **3.** 2
4. 60 **5.** $3900

Page 237 Classroom Exercises 1. $n = 0.07 \cdot 121$
3. $r = 0.15 \cdot 144$ **5.** $w = \frac{3}{5} \cdot 75$ **7.** $r = \frac{1}{3} \cdot 96$
9. $c = \frac{3}{4} \cdot 4$ **11.** $n = \frac{3}{4} \cdot 30$

Page 237 Written Exercises 1. $n = 0.35 \cdot 400$
3. $n = 0.83 \cdot 100$ **5.** $n = \frac{3}{8} \cdot 16$ **7.** 51.6 **9.** 54
11. 0.312 **13.** 108 **15.** 20 **17.** 48 **19.** 93
21. 8 **23.** 34 **25.** 120 **27.** $7500 **29.** $3.00
31. 15 **33.** $255.36 **35.** 12 **37.** $5.45

Pages 240-241 Self-Check 1. 24 **2.** 27.50
3. 781.25

Page 242 Classroom Exercises 1. $i = (950)(0.08)(1)$
3. $i = (1956)(0.05)(\frac{1}{2})$ **5.** $i = (2575)(0.125)(2)$
7. $5.00 **9.** $16.00 **11.** $3.00 **13.** $8.00
15. $100.00 **17.** $19.00

Page 242 Written Exercises 1. $40.00 **3.** $12.00
5. $6.00 **7.** $22.50 **9.** $30.00 **11.** $75.00
13. $144.00 **15.** $2812.50 **17.** $725.00
19. $575.00 **21.** $3120.00

Page 243 Calculator Exercises 1. $76.00
3. $48.90 **5.** $643.75 **7.** $5.00 **9.** $16.00
11. $3.00 **13.** $8.00 **15.** $100.00 **17.** $19.00

Page 245 Self-Check 1. $12.50 **2.** $42.32
3. $12.99 **4.** $99.83

Page 246 Classroom Exercises 1. $6.00 **3.** $27.30
5. $15.80 **7.** $25.13

Page 246 Written Exercises 1. $57.00 **3.** $19.50
5. $52.80 **7.** $323.00 **9.** $13.00 **11.** $2144.00
13. $504.00

Page 247 Review: Sections 8-4–8-8 1. 50%
2. 15% **3.** 54% **4.** 84% **5.** $122\frac{1}{2}$% **6.** 620%
7. 90% **8.** 380% **9.** 17% **10.** 104% **11.** 10%
12. 1% **13.** 0.12 **14.** 0.02 **15.** 0.85 **16.** 0.70
17. 0.095 **18.** 0.153 **19.** 6.6 **20.** 1.28
21. $4130 **22.** $66.00 **23.** $480.00 **24.** $120.00
25. $14.41 **26.** $20.52

Page 248 Chapter Review 1. commission
3. discount **5.** proportion **7.** principal **9.** net price **11.** $\frac{3}{5}$ **13.** $\frac{3}{2}$ **15.** $\frac{12}{7}$ **17.** $\frac{1}{2}$ **19.** $\frac{3}{7}$
21. $\frac{3}{5}$ **23.** No **25.** No **27.** Yes **29.** Yes
31. 4 **33.** 3 **35.** 16 **37.** 12 **39.** $4\frac{4}{5}$ **41.** $6\frac{2}{3}$
43. $\frac{1}{20}$; 5% **45.** $\frac{1}{40}$; 0.025 **47.** 45 **49.** 73.08
51. $1840 **53.** $100 **55.** $8.95 **57.** $216
59. 60¢

Page 250 Chapter Test 1. Yes **3.** No **5.** 24
7. 42 **9.** 60% **11.** 70% **13.** 0.18 **15.** 0.085
17. 21 **19.** 0.81 **21.** 8 in **23.** $127.50
25. $3.75

556   *Answers to Selected Exercises*

Page 251 Enrichment 1. $1030.41 3. $944.78 5. $1464.10 7. $4207.66

Page 252 Additional Practice 1. $\frac{1}{5}$ 3. $\frac{4}{9}$ 5. $\frac{4}{1}$ 7. No 9. Yes 11. $\frac{2}{5}$ 13. $\frac{5t}{1}$ 15. $\frac{6}{25y}$ 17. $11\frac{2}{3}$ 19. 36 21. $8\frac{3}{4}$ 23. $36\frac{4}{11}$% 25. 275% 27. 14% 29. 25%; $\frac{1}{4}$ 31. 0.42; $\frac{21}{50}$ 33. 85%; 0.85 35. 52.91 37. $24 39. 132 beats

CHAPTER 9

Pages 254-255 Self-Check 1. 20% 2. 40% 3. 8%

Page 255 Classroom Exercises 1. $p \cdot 36 = 20$ 3. $p \cdot 108 = 81$ 5. $120 = p \cdot 12$ 7. $p \cdot 4000 = 200$

Page 256 Written Exercises 1. $p \cdot 20 = 15$ 3. $p \cdot 15 = 9$ 5. $56 = p \cdot 64$ 7. $37\frac{1}{2}$% 9. 250% 11. $33\frac{1}{3}$% 13. 20% 15. 30% 17. $83\frac{1}{3}$% 19. $66\frac{2}{3}$% 21. 60% 23. 80% 25. 16%

Pages 257-258 Self-Check 1. 40% 2. 15%

Page 258 Classroom Exercises 1. $19; \frac{19}{100}$; 19% 3. $16; \frac{16}{80}$; 20% 5. $12; \frac{12}{80}$; 15% 7. $9; \frac{9}{150}$; 6%

Page 259 Written Exercises 1. 10% 3. 11% 5. 14% 7. 19% 9. 9% 11. 6% 13. 18% 15. 21% 17. 12% 19. 12% 21. 26% 23. 20%

Pages 261-262 Self-Check 1. 37.5 2. 375 3. $53

Page 262 Classroom Exercises 1. $50 = 0.20n$ 3. $9 = 0.15t$ 5. $26 = 0.15p$ 7. $18 = 0.45n$ 9. $6 = 0.30n$ 11. $18,000; 20; 75

Page 262 Written Exercises 1. 75 3. 100 5. 168 7. 450 9. 20 11. 97.5 13. 120 15. 300 17. $109 19. $1944 21. $42; \frac{1}{6}$ 23. $640; \frac{1}{8}$ 25. 21 min; $\frac{1}{5}$

Page 265 Self-Check 1. 9.6 2. 350% 3. 20

Page 266 Classroom Exercises 1. 8% of 21 is n 3. n% of 85 is 17 5. 15% of n is 4 7. 100% of k is k; $\frac{100}{300} = \frac{k}{36}$ 9. 100% of t is t; $\frac{100}{128} = \frac{t}{45}$ 11. 100% of 126 is 126; $\frac{3}{100} = \frac{h}{126}$ 13. 100% of a is a; $\frac{100}{15} = \frac{a}{10}$

Page 266 Written Exercises 1. 7 3. 10.8 5. 105 7. 5.12 9. 75% 11. 25 13. 25 15. 43.75 17. 150% 19. 50 21. 25% 23. 560 25. 340 27. 85% 29. 3.12 oz 31. $2080 33. $2\frac{1}{2}$%

Page 267 Review: Sections 9-1—9-4 1. 90% 2. 120% 3. $41\frac{2}{3}$% 4. 140% 5. 64% 6. 75% 7. 6% 8. 52% 9. 30% 10. 9% 11. 90 12. 30 13. 80 14. 200 15. 60 16. 62.5 17. 6% 18. $300

Page 269 Special Topic 1. 150 I.U. 3. 9mg 5. 2% 7. 20mg

Pages 270-271 Self-Check 1. 40% 2. Let 4 cm represent 40°C. Phoenix: 3.4 cm; Chicago: 2.4 cm; Boston: 2.2 cm; Los Angeles: 2.2 cm; Cheyenne: 2.0 cm

Page 271 Classroom Exercises 1. 120 3. $33\frac{1}{3}$%

Page 272 Written Exercises 1. 185,000 3. 14% 5. 86 7. 52% 9. Let 6 cm represent 600 books. January: 1.5 cm; February: 2 cm; March: 3 cm; April: 4 cm; May: 1.75 cm; June: 5 cm 11. Let 5 cm represent 500 m. Yosemite Lower Falls: 1.0 cm; Yosemite Upper Falls: 4.4 cm; Nevada Falls: 1.8 cm; Great Falls: 2.6 cm; Ribbon Falls: 4.9 cm; Silver Strand Falls: 3.6 cm

Pages 274-275 Self-Check 1. 25 2. New York: 133°; Los Angeles: 68°; San Francisco: 61°; Washington, D.C.: 50°; Orlando: 47°

Page 276 Classroom Exercises 1. $375 3. $n = 0.20(900); $180 5. 15%; $n = 0.15(900); $135 7. 90% 9. 0.18; 65° 11. 0.21; 76°

Page 277 Written Exercises 1. $900 3. $1200 5. 1000 7. 2200 9. Schools: 126°; Public Works: 54°; Health and Safety: 144°; Other: 36° 11. Lightning: 122°; Children: 61°; Campers: 40°; Auto Passengers: 36°; Other: 101° 13. 1976: 79°; 1977: 83°; 1980: 90°; 1985: 108°

Pages 279-280 Self-Check 1. $12 2. $40 3. 20%

Page 280 Classroom Exercises 1. c 3. b 5. c 7. b 9. a 11. b 13. b 15. b 17. b

Page 281 Written Exercises 1. a 3. a 5. c 7. a 9. b 11. 25% 13. 20% 15. 10% 17. a 19. a 21. 25% 23. $16 25. 50%

Page 282 Calculator Exercises 1. $200; $208.00 3. $22; $21.98 5. $2; $2.13

Page 283 Review: Sections 9-5—9-7 1. Let 4 cm represent scale 900 — 1100 students. 1976: 2.6 cm; 1977: 2.4 cm; 1978: 3.9 cm; 1980: 3.5 cm 2. $100 3. $300 4. $200 5. 65 and over: 40°; 45-65: 72°; 20-44: 126°; 19 and under: 122° 6. a 7. c 8. b 9. c 10. $33\frac{1}{3}$% 11. $16\frac{2}{3}$% 12. $125 13. $24

Page 284 Chapter Review 1. percent of decrease 3. percent 5. percent of increase 7. 60% 9. $85\frac{5}{7}$% 11. $12\frac{1}{2}$% 13. 19% increase 15. 3% decrease 17. 24% decrease 19. 41% increase 21. 120 23. 60 25. 50 27. 189 29. 10% 31. 256 33. 25% 35. Pacific: 166°; Atlantic: 83°; Indian: 72°; Arctic: 14°; Other: 25° 37. a 39. c 41. $33\frac{1}{3}$% 43. 10% 45. 10% 47. $250

Page 287 Chapter Test 1. 8 3. 20 5. 14 7. 256 9. 19% 11. 30% 13. $150 15. Newspaper: 54°; TV: 198°; Radio: 72°; Other: 36°

Page 288 Enrichment 1. 9, 11, 13 3. 512, 2048,

Answers to Selected Exercises 557

8192 **5.** 16, 8, 4 **7.** 21, 34, 55 **9.** 35, 57, 92

Page 289 Additional Practice **1.** 50% **3.** 50%
5. 50% **7.** 48 **9.** 680 **11.** Suits: 144°;
Dresses: 126°; Skirts: 54°; Shoes: 36° **13.** b
15. 1200

Page 290 Common Errors **1.** 11 **3.** 2 **5.** $16.\overline{6}$
7. 6 **9.** 300 **11.** 25%

Page 291 Cumulative Review: Chapters 7-9 **1.** d
3. c **5.** c **7.** d **9.** b **11.** b **13.** b **15.** a
17. d **19.** a **21.** a

CHAPTER 10

Pages 294-295 Self-Check **1.** +200; or 200
2. −120 **3.** −7; −6; −5; 3; 4 **4.** −9; −3; 0; 8; 10
5. > **6.** > **7.** < **8.** <

Page 295 Classroom Exercises **1.** −8 **3.** −20
5. −3; 1; 2; 3; 7 **7.** −8; −1; 0; 2; 11 **9.** −12; 0; 6;
12; 13 **11.** < **13.** >

Page 296 Written Exercises **1.** 10 **3.** −2000
5. −2 **7.** −3; 0; 3; 6 **9.** −2; 0; 1; 3 **11.** −4; −2;
1; 3; 8 **13.** < **15.** > **17.** < **19.** < **21.** >
23. < **25.** > **27.** <

Pages 297-298 Self-Check **1.** $\frac{4}{1}$ **2.** $\frac{-9}{2}$ **3.** $\frac{38}{10}$
4. $\frac{-5}{1}$ **5.** $\frac{9}{4}$ **6.** 15 **7.** $\frac{8}{3}$ **8.** 0 **9.** 2.7 **10.** $\frac{-4}{5}$

Page 298 Classroom Exercises **1.** $\frac{4}{1}$ **3.** $\frac{10}{1}$
5. $\frac{32}{10}$ **7.** $-\frac{11}{5}$ **9.** $\frac{9}{4}$ **11.** $-\frac{18}{10}$ **13.** −12 **15.** 6
17. −3.2 **19.** 1.5 **21.** $-\frac{1}{4}$ **23.** $1\frac{4}{5}$

Page 298 Written Exercises **1.** $\frac{17}{1}$ **3.** $-\frac{7}{1}$ **5.** $\frac{38}{10}$
7. $-\frac{10}{3}$ **9.** $\frac{13}{4}$ **11.** $\frac{27}{5}$ **13.** Answers may vary.
15. $-\frac{93}{10}$ **17.** $\frac{75}{10}$ **19.** −12 **21.** 20 **23.** $-2\frac{1}{5}$
25. $\frac{3}{4}$ **27.** $-\frac{2}{5}$ **29.** $8\frac{1}{2}$ **31.** −4.8 **33.** 8.9
35. 0 **37.** $\frac{5}{2}$ **39.** $-\frac{326}{10}$ **41.** $\frac{435}{100}$ **43.** $\frac{1325}{10}$
45. $-\frac{55}{10}$ **47.** $\frac{125}{10}$ **49.** $-5\frac{1}{5}$; $-2\frac{1}{5}$; $-\frac{1}{4}$; $\frac{1}{4}$; 1; $3\frac{1}{3}$
51. $-1\frac{1}{2}$; $-\frac{3}{4}$; $-\frac{2}{3}$; $\frac{4}{3}$; 1; 2 **53.** −1.875; −1.83;
−0.687; 0; 1.5; 2.25 **55.** −3; −2.875; −2.5; −1.125;
−1; −0.25

Pages 300-301 Self-Check **1.** −4 **2.** −5 **3.** −9
4. −14 **5.** 1 **6.** −3 **7.** 0 **8.** −7 **9.** −3 **10.** 0
11. 3 **12.** 5

Page 301 Classroom Exercises **1.** −6 **3.** −10
5. −10 **7.** −7 **9.** 3 **11.** 3 **13.** −4 **15.** 3
17. −5 **19.** −3 **21.** 1 **23.** 5

Page 301 Written Exercises **1.** −9 **3.** −7 **5.** −10
7. −4 **9.** −15 **11.** −5 **13.** 3 **15.** 5 **17.** 6
19. 4 **21.** −4 **23.** 0 **25.** −2 **27.** 2 **29.** −5
31. 5 **33.** −6 **35.** 7 **37.** −3 **39.** −3 **41.** −5
43. 6 **45.** −6 **47.** 3 **49.** −5 **51.** loss of 4 yd

Page 302 Review: Sections 10-1–10-3 **1.** −3; −1;
0; 2; 5 **2.** −8; −4; 1; 7; 9 **3.** −4; −2; 0; 1; 4 **4.** $\frac{15}{1}$

5. $-\frac{8}{1}$ **6.** $-\frac{58}{10}$ **7.** $\frac{17}{8}$ **8.** $-\frac{22}{10}$ **9.** Answers may
vary. **10.** −3 **11.** −13 **12.** 8 **13.** −8 **14.** −11
15. 2 **16.** 0 **17.** 0 **18.** 2°C

Page 303 Special Topic **1.** 121 ft **3.** 162 ft **5.** 101 ft

Page 305 Self-Check **1.** −18 **2.** −22 **3.** −13
4. 6 **5.** 8 **6.** 7 **7.** −1 **8.** −6 **9.** −6

Page 306 Classroom Exercises **1.** 8 **3.** 6 **5.** 15
7. −18 **9.** −22 **11.** 29 **13.** −22 **15.** 2 **17.** 11
19. 4 **21.** 12 **23.** −3 **25.** −8 **27.** −4 **29.** −15

Page 306 Written Exercises **1.** 7 **3.** 5 **5.** 15
7. 8 **9.** 32 **11.** 0 **13.** 1.3 **15.** 6.7 **17.** $\frac{1}{3}$
19. −16 **21.** 28 **23.** −36 **25.** −18 **27.** −5.5
29. −5.2 **31.** $-\frac{2}{3}$ **33.** $-\frac{3}{4}$ **35.** 4 **37.** 8 **39.** 9
41. 5 **43.** 0.7 **45.** 3.9 **47.** $\frac{1}{5}$ **49.** $\frac{1}{6}$ **51.** −5
53. −12 **55.** 21 **58.** −8 **59.** −1.9 **61.** −1.1
63. $-\frac{1}{4}$ **65.** $-\frac{1}{8}$ **67.** 5 **69.** −34 **71.** −5
73. 15 **75.** −25 **77.** 6 **79.** −5.7 **81.** 3.9
83. $-\frac{3}{5}$ **85.** $\frac{1}{4}$ **87.** 185 m **89.** −8°C

Pages 308-309 Self-Check **1.** 4 + (−9) **2.** −10 +
(−18) **3.** −1 + 12 **4.** 17 + 21 **5.** −6 **6.** −85
7. −2.6 **8.** −5.9 **9.** 29 **10.** 63 **11.** 15 **12.** 7

Page 309 Classroom Exercises **1.** (−5) **3.** 14
5. (−17) **7.** −3 **9.** −22 **11.** −3 **13.** −25
15. 9 **17.** −3 **19.** 21 **21.** 6

Page 309 Written Exercises **1.** 11 + (−3) **3.** 14 + 6
5. −6 + (−1) **7.** −8 + 5 **9.** −7 + (−7) **11.** 4.7 +
(−8.6) **13.** −4 **15.** −41 **17.** −24 **19.** −121
21. −0.9 **23.** $-1\frac{1}{2}$ **25.** 4 **27.** 8 **29.** 25 **31.** 57
33. 23.4 **35.** $5\frac{1}{2}$ **37.** 0 **39.** 7 **41.** −188
43. −351 **45.** 12 **47.** −81 **49.** 11.5 **51.** $2\frac{7}{8}$
53. 56°C **55.** 14,776 ft

Page 311 Special Topic **1.** 12:05 P.M. **3.** 12:05 P.M.
5. 11:05 P.M.

Pages 312-313 Self-Check **1.** −10 **2.** 12 **3.** 6
4. 4

Page 313 Classroom Exercises **1.** (−1); 11 **3.** 4;
−10 **5.** (−2); 1 **7.** 0 **9.** −3 **11.** 0

Page 313 Written Exercises **1.** −5 **3.** 2 **5.** −29
7. −24 **9.** 16 **11.** −64 **13.** −10 **15.** 13 **17.** 5
19. 0 **21.** −6 **23.** 8 **25.** −9 **27.** −9 **29.** 26
31. −4 **33.** −12 **35.** 2 **37.** −10.7 **39.** $-\frac{1}{4}$
41. −0.7 **43.** $1\frac{1}{4}$

Page 314 Calculator Exercises **1.** −97 **3.** −0.67
5. −6.99

Pages 315-316 Self-Check **1.** $4x + 3$ **2.** $-5p^2 - 3p$ **3.** $-3y - 2$ **4.** $5n^2 + 3n + 7$

Page 316 Classroom Exercises **1.** monomial
3. binomial **5.** 2 **7.** 3

Answers to Selected Exercises

Page 316 Written Exercises 1. $7p + 3$ **3.** $6c^2 - 6c$
5. $7d - 2$ **7.** $5n^2 - 6n + 5$ **9.** $3x^2 - 2x - 2$
11. $4r^2 + 4r - 2$ **13.** $11x$ **15.** $-9p^2$ **17.** $-6y$
19. $-9t^2$ **21.** $8x^2$ **23.** $-2y$

Page 317 Review: Sections 10-4–10-7 1. 6
2. -21 **3.** -5 **4.** 11 **5.** 11 **6.** 50 **7.** -14
8. -39 **9.** 7 **10.** -24 **11.** 24 **12.** -17
13. 39 **14.** -16 **15.** 6 **16.** -12 **17.** $-9°$F
18. $50°$C **19.** -18 **20.** -26 **21.** -14 **22.** 3
23. 3 **24.** 3 **25.** $-x + 8$ **26.** $-6y^2 - y - 3$
27. $20t + 1$ **28.** $3p^2 + 4p$ **29.** $5y^2 - 4y + 2$
30. $3n^2 + 4n - 7$

Page 318 Chapter Review 1. $4 + (-7) = -7 + 4$
3. opposites **5.** polynomial **7.** negative integers
9. $3 + (-3) = 0$ **11.** positive integers **13.** The points on the numberline from left to right in this order: -5; $1; 2; 4$ **15.** The points on the numberline from left to right in this order: $-3; -1; 0; 5$ **17.** $>$ **19.** $<$
21. $<$ **23.** $>$ **25.** $-\frac{5}{1}$ **27.** $-\frac{10}{3}$ **29.** $-\frac{23}{10}$
31. 6 **33.** -8 **35.** 7 **37.** 8 **39.** -5 **41.** -22
43. 16 **45.** -33 **47.** 0 **49.** -17 **51.** 5
53. -12 **55.** 16 **57.** 13 **59.** -17 **61.** -7
63. -23 **65.** 36 **67.** $-4x^2 - 3x$ **69.** $6b^2 + b - 1$
71. $-\frac{65}{2}$ **73.** $\frac{245}{1}$ **75.** gain of 2 yd **77.** $130°$F

Page 320 Chapter Test 1. $>$ **3.** $>$ **5.** 7 **7.** -72
9. -9 **11.** -28 **13.** -25 **15.** -13 **17.** -15
19. -20 **21.** -13 **23.** 21 **25.** $34°$F

Page 321 Enrichment 1. 1 **3.** 6 **5.** 2 **7.** 3
9. 6 **11.** 6 **13.** 1 **15.** 1 **17.** 1

Page 322 Additional Practice 1. $-6; -4; 0; 2$
3. $-1; 2; 3; 4; 5$ **5.** $>$ **7.** $<$ **9.** $\frac{6}{10}$ **11.** $-\frac{35}{10}$
13. $\frac{10}{3}$ **15.** -4 **17.** 7 **19.** 2.5 **21.** -4 **23.** -6
25. 3 **27.** 32 **29.** 0 **31.** 9 **33.** -78 **35.** -7.9
37. 2 **39.** -14 **41.** 9 **43.** $\frac{1}{8}$ **45.** $\frac{1}{2}$ **47.** 0
49. 1 **51.** gain of 2 yd

CHAPTER 11

Page 324 Self-Check 1. -4 **2.** 21 **3.** -3
4. -15 **5.** -4 **6.** -17

Page 325 Classroom Exercises 1. $y + 18 - 18 = 7 - 18$ **3.** $a - 5 + 5 = 22 + 5$ **5.** $-15 - 23 = 23 + b - 23$ **7.** -1 **9.** 29 **11.** -2 **13.** -12 **15.** 3
17. -7

Page 325 Written Exercises 1. 26 **3.** 2 **5.** -13
7. 39 **9.** 31 **11.** -1.8 **13.** 6 **15.** -9 **17.** -21
19. -13 **21.** -35 **23.** 5.9 **25.** -25 **27.** 18
29. 22 **31.** -102 **33.** 0 **35.** -7.5 **37.** 10.5
39. $-1\frac{2}{5}$

Pages 326-327 Self-Check 1. 4 **2.** 5 **3.** 6 **4.** 3

Page 327 Classroom Exercises 1. $8x = 32$ **3.** $c = 19$ **5.** $7x - 5 = 16$ **7.** $2a + 4 = 6$ **9.** 6 **11.** 8
13. 7 **15.** 4

Page 327 Written Exercises 1. 9 **3.** 4 **5.** 9
7. 3 **9.** 23 **11.** 2 **13.** 4 **15.** 6 **17.** 2 **19.** 5
21. 6 **23.** 19 **25.** 33 **27.** $5\frac{3}{5}$ **29.** $3\frac{1}{2}$ **31.** 13
33. $6\frac{1}{2}$ **35.** 6 **37.** 8 **39.** 10 **41.** 5 **43.** 10
45. 4 **47.** 5.5

Page 329 Self-Check 1. 5 **2.** 3 **3.** 12 **4.** 4
5. 4

Page 330 Classroom Exercises 1. $v; v; 2v; 3$ **3.** $d; d; 6d; 12$ **5.** $5h; 5h; h; 12; 7$

Page 330 Written Exercises 1. 10 **3.** 7 **5.** 16
7. $7\frac{1}{2}$ **9.** $3\frac{1}{2}$ **11.** 50 **13.** 7 **15.** 10 **17.** 8
19. 7 **21.** 3 **23.** 5 **25.** 3 **27.** 2 **29.** 30 **31.** 4

Page 331 Calculator Exercises 1. Does not check
3. Does not check

Page 331 Review: Sections 11-1–11-3 1. 7
3. -24 **5.** 2 **7.** -3 **9.** 1 **11.** 6 **13.** 3 **15.** 8

Page 332 Special Topic 1. 14.7% **3.** 17.2% **5.** 15.3%

Page 333 Special Topic 1. $\frac{7}{10}$ **3.** $\frac{1}{9}$ **5.** $-\frac{3}{6}$
7. $-\frac{12}{5}$

Page 335 Self-Check 1. $53 **2.** 104 lb

Page 336 Classroom Exercises 1. $y - 6$ **3.** $y + 6$
5. $6 + y$ **7.** $6 + y; y + 6$ **9.** $y - 6$ **11.** $c + 12$
13. $60 - d = 24$ **15.** $k - 100 = 425$

Page 336 Written Exercises 1. $25 - n$ **3.** $x + 5$, or $5 + x$ **5.** $e - 6$ **7.** $32 + b$ **9.** $85 + c$, or $c + 85$
11. $d - 14.5$ **13.** $y + 25$ **15.** $42 - (-p)$ **17.** $t + 19 = 42$ **19.** $t + 32 = 78$ **21.** $24 - t = 15$ **23.** $t - 12 = 52$ **25.** 682 km **27.** $36.75 **29.** $165
31. 8 **33.** 9 cm **35.** 34

Pages 338-339 Self-Check 1. All points to the right of and not including 2 **2.** All points to the left of and not including -1 **3.** All points to the left of and not including 0 **4.** All points to the left of and including 3
5. All points to the right and including -5 **6.** All points to the right of and including 0

Page 339 Classroom Exercises 1. No **3.** Yes
5. Yes **7.** Yes **9.** No **11.** Yes **13.** Yes **15.** No
17. Yes **19.** Yes **21.** No **23.** Yes

Page 339 Written Exercises 1. All points to the right of and not including 2 **3.** All points to the left of and not including -1 **5.** All points to the left of and not including 4 **7.** All points to the right of and not including 0 **9.** All points to the right of and not including 1 **11.** All points to the right of and including 1 **13.** All points to the right of and including -5
15. All points to the left of and including 0 **17.** All points to the left of and including -2 **19.** All points to the right of and including -2 **21.** All points to the right of and not including 3 **23.** All points to the right of and including -3 **25.** All points to the left of and including -1 **27.** All points to the right of and includ-

Answers to Selected Exercises 559

ing 5 **29.** All points to the right of and not including −5 **31.** All points to the left of and not including −3 **33.** All points to the left of and including 4 **35.** All points to the right of and including −4

Pages 340-341 Self-Check 1. $p > 7$; All points to the right of and not including 7 **2.** $b < 5$; All points to the left of and not including 5 **3.** $t < 1$; All points to the left of and not including 1 **4.** $x < 3$; All points to the left of and not including 3 **5.** $c > -6$; All points to the right of and not including −6 **6.** $a > -5$; All points to the right of and not including −5

Page 341 Classroom Exercises 1. $a - 5 + 5 > 3 + 5$ **3.** $w - 4 + 4 < 8 + 4$ **5.** $y - 2 + 2 > 6 + 2$ **7.** $m + 12 - 12 \leq 3 - 12$ **9.** $t + 15 - 15 \geq -5 - 15$ **11.** $c + 5 - 5 < 12 - 5$ **13.** $g > 12$; All points to the right of and not including 12 **15.** $w \leq -3$; All points to the left of and including −3 **17.** $b < 5$; All points to the left of and not including 5 **19.** $p < 6$; All points to the left of and not including 6 **21.** $y \geq 4$; All points to the right of and including 4 **23.** $b > 10$; All points to the left of and not including 10

Page 341 Written Exercises 1. $t > 5$; All points to the right of and not including 5 **3.** $p < 2$; All points to the left of and not including 2 **5.** $x > 16$; All points to the right of and not including 16 **7.** $c > 17$; All points to the right of and not including 17 **9.** $b \leq -6$; All points to the left of and including −6 **11.** $z < 1$; All points to the left of and not including 1 **13.** $n < 4$; All points to the left of and not including 4 **15.** $k > -2$; All points to the right of and not including −2 **17.** $w < -6$; All points to the left of and not including −6 **19.** $d \leq 11$; All points to the left of and including 11 **21.** $b \geq -15$; All points to the right of and including −15 **23.** $n < 1$; All points to the left of and not including 1 **25.** $x \geq -7$; All points to the right of and including −7 **27.** $a > 16$; All points to the right of and not including 16 **29.** $t \leq 50$; All points to the left of and including 50 **31.** $x \geq -9$; All points to the right of and including −9 **33.** $y > -1$; All points to the right of and not including −1 **35.** $x < 4$; All points to the left of and not including 4 **37.** $n < -4$; All points to the left of and not including −4 **39.** $r < 2$; All points to the left of and not including 2 **41.** $w < 43$; All points to the left of and not including 43 **43.** $p \geq -9$; All points to the right of and including −9 **45.** $x > -7$; All points to the right of and not including −7 **47.** $c \geq -9$; All points to the right of and including −9

Page 342 Calculator Exercises 1. Checks **3.** Does not check **5.** Does not check

Page 343 Review: Sections 11-4—11-6 1. $n + 5$ **2.** $d - 7$ **3.** $w - 5$ **4.** $b + 31$ **5.** $c - 5$ **6.** $y - 27$ **7.** $x + 14$ **8.** $n + 3$ **9.** $c + 8 = 33$ **10.** $12 + p = 56$ **11.** $a - 3 = 13$ **12.** $d - 10 = 36$ **13.** $x + 11 = 34$ **14.** $y - 5 = 19$ **15.** $20 - w = 13$ **16.** $s - 28 = 103$ **17.** $65 **18.** $141 **19.** $48\frac{1}{2}$ in **20.** 608 km

21. All points to the right of and not including −8 **22.** All points to the left of and not including 5 **23.** All points to the left of and including −3 **24.** All points to the right of and including 6 **25.** $y < 9$; All points to the left of and not including 9 **26.** $c > 2$; All points to the right of and not including 2 **27.** $t > -1$; All points to the right of and not including −1 **28.** $k \geq 4$; All points to the right of and including 4

Page 344 Chapter Review 1. inequalities **3.** word **5.** algebraic **7.** 4 **9.** −44 **11.** −4 **13.** −3 **15.** 9 **17.** 6 **19.** 3 **21.** 4 **23.** 9 **25.** 5 **27.** 1 **29.** $35 + t$ **31.** $c - 6$ **33.** $d + 5$ **35.** $r - 15 = 42$ **37.** $w + 16 = 84$ **39.** All points to the left of and not including 5 **41.** All points to the right of and not including −4 **43.** All points to the left and including −4 **45.** All points to the right of and including 0 **47.** $t < 4$; All points to the left of and not including 4 **49.** $y \geq -3$; All points to the right of and including −3 **51.** $t < -5$; All points to the left of and not including −5 **53.** $k > -3$; All points to the right of and not including −3 **55.** $16 **57.** $88

Page 346 Chapter Test 1. 27 **3.** −15 **5.** 2 **7.** 3 **9.** 4 **11.** $40 - d$ **13.** $11 + n$ **15.** $24.50 **17.** $153.20 **19.** $z > 8$; All points to the right of and not including 8

Page 347 Enrichment 1. All points between 3 and 4 and not including 3 and 4 **3.** All points between −2 and 3 including −2, but not including 3 **5.** All points to the right of, and not including 0, and all points to the left of, and including −3 **7.** $x > 150$ and $x \leq 220$

Page 348 Additional Practice 1. 23 **3.** −93 **5.** −41 **7.** −44.6 **9.** −40.7 **11.** 3 **13.** −13 **15.** 3 **17.** 7 **19.** 7 **21.** 50 **23.** 3 **25.** 19 **27.** 33 **29.** All points to the right of and not including 5 **31.** All points to the right of and including −4 **33.** $n < 3$; All points to the left of and not including 3 **35.** $a > 14$; All points to the right of and not including 14 **37.** $3 + n = 84$ **39.** $c - 24 = 36$ **41.** $c - 55 = 326$; $381

CHAPTER 12

Page 350 Self-Check 1. −70 **2.** −225 **3.** −96 **4.** −36 **5.** −72 **6.** −54

Page 351 Classroom Exercises 1. −16 **3.** −10; −15 **5.** 0 **7.** −65 **9.** −231 **11.** −3456 **13.** −6 **15.** −20 **17.** 0 **19.** −384 **21.** −425 **23.** −1107

Page 351 Written Exercises 1. −30 **3.** −180 **5.** −665 **7.** −2100 **9.** −3.15 **11.** −3 **13.** 0 **15.** −81 **17.** −156 **19.** −840 **21.** −32.8 **23.** $-\frac{1}{12}$ **25.** 0 **27.** −2016 **29.** −6272 **31.** −4402 **33.** −86.8 **35.** $-\frac{4}{5}$ **37.** −3(5); −15 **39.** 6(−15); −90 **41.** 4(−6); −24 **43.** −2(78); −156 **45.** −145 **47.** −102 **49.** −27 **51.** −50y **53.** −8ab **55.** −56ab **57.** −6m^3 **59.** −2v^5 **61.** −30n^3

560   Answers to Selected Exercises

Page 353 Self-Check 1. 14 2. 130 3. 300
4. 77

Page 354 Classroom Exercises 1. 12 3. 9; 18

Page 354 Written Excerises 1. 35 3. 9 5. 735
7. 810 9. 625 11. 2992 13. 1498 15. 7812
17. 6.6 19. 8.28 21. 5.7 23. 21.294 25. $\frac{1}{2}$
27. $10\frac{1}{5}$ 29. $\frac{1}{8}$ 31. $4\frac{1}{6}$ 33. 13 35. 36 37. 52
39. 1 41. $18a$ 43. $15x^2$ 45. $10a^2b^2$ 47. $12x^3y$

Page 355 Calculator Exercises 1. $-26,752$
3. 5044.48 5. -0.1623792 7. 51,504
9. 46,818.486

Pages 356-357 Self-Check 1. $10n + 8$ 2. $-24y + 18$ 3. $4p + 28$ 4. $-12p - 21$ 5. $-4x - 48$
6. $-8x - 2$ 7. $24x^2 - 12x$ 8. $15t^2 - 6t$
9. $-6r^2 - 2r$

Page 357 Classroom Exercises 1. $2x$ 3. -8
5. $6(-3); 6x$ 7. $2n^2$

Page 357 Written Exercises 1. $15s + 50$ 3. $8x + 14$ 5. $14n + 21$ 7. $24q + 6$ 9. $-27c + 18$
11. $-2r - 14$ 13. $6p - 3$ 15. $4c - 3$ 17. $-5z + 10$ 19. $a^2 - a$ 21. $q^2 - 2q$ 23. $3y^2 - 12y$
25. $6p^2 - 12p$ 27. $12x^2 - 28x$ 29. $8x + 20$
31. $2t + 18$ 33. $-3p - 15$ 35. $-12n - 4$
37. $40b + 70$ 39. $16d^2 - 8d$ 41. $4n + 12$
43. $-b - 6$ 45. $-14p + 12$ 47. $m^2 + 3md$

Page 358 Review: Sections 12-1–12-3 1. -40
2. -60 3. -135 4. -357 5. -672 6. -840
7. -3150 8. $-11,742$ 9. 32 10. 15 11. 63
12. 512 13. 1000 14. 4484 15. 3696
16. 28,743 17. $10a + 6$ 18. $6b + 48$ 19. $-4x - 8$ 20. $-15t - 5$ 21. $y^2 - 5y$ 22. $8r^2 - 4r$
23. $-20x^2 - 15c$ 24. $-18v^2 + 42v$

Page 359 Special Topic 1. -678 BTU's 3. -2236 BTU's

Page 361 Self-Check 1. 9 2. -7 3. -6 4. -8
5. -12 6. $-\frac{1}{10}$ 7. $-3\frac{1}{2}$

Page 362 Classroom Exercises 1. 8; 6 3. 7; -8
5. 6 7. -9 9. -10 11. 11 13. -6 15. -4
17. -12 19. 10 21. $-\frac{1}{5}$ 23. $1\frac{1}{2}$

Page 362 Written Exercises 1. 9 3. 7 5. -3
7. -7 9. -3 11. -18 13. 5 15. -15 17. -8
19. -23 21. -5 23. -7 25. -48 27. $\frac{1}{27}$
29. -6 31. $-1\frac{1}{2}$ 33. $1\frac{2}{9}$ 35. -10 37. $-\frac{a}{8}$; or $-\frac{1}{8}a$ 39. $\frac{2e}{3}$; or $\frac{2}{3}e$ 41. $\frac{b}{18}$; or $\frac{1}{18}b$ 43. $-\frac{5h}{2}$; or $-\frac{5}{2}h$ 45. $-\frac{1}{x}$ 47. 1 49. $\frac{c}{4}$; or $\frac{1}{4}c$ 51. $-f$
53. No; difference may be a negative integer 55. Yes

Pages 364-365 Self-Check 1. $3(a + 1)$ 2. $7(x + 1)$
3. $2(2 + p)$ 4. $5(3 + n)$ 5. $3(x - 2)$ 6. $5(m - 2)$
7. $3(4n - 1)$ 8. $t(t + 3)$ 9. $p(p + 5)$ 10. $y(4 - y)$

Page 365 Classroom Exercises 1. 2 3. 2 5. 3

7. m 9. 3 11. 2 13. $r + 7$ 15. $9 + t$ 17. 3
19. 3 21. $r - 1$ 23. $2x - 1$ 25. n 27. $t + 10$
29. x 31. $12 - c$

Page 366 Written Exercises 1. $2(n + 1)$ 3. $3(r + 1)$
5. $3(y + 5)$ 7. $7(3 + c)$ 9. $7(n - 2)$ 11. $5(3a - 1)$
13. $7(n - 4)$ 15. $3(p - 4)$ 17. $y(y + 5)$
19. $c(c - 4)$ 21. $x(13 + x)$ 23. $r(r + 1)$
25. $r(r - 1)$ 27. $t(t + 2)$ 29. $13(y + 1)$
31. $7(n + 6)$ 33. $3(t + 4)$ 35. $5(a - 4b)$
37. $x(x - 12)$ 39. $y(9 - y)$ 41. $3(x^2 + 3x + 5)$
43. $5(a^2 + 2a + 3)$ 45. $7(d^2 + 2d + 4)$ 47. $3(6m^2 + 5m + 9)$ 49. $2(x^2 - x - 1)$

Page 367 Calculator Exercises 1. 31.2 3. 1184.08
5. 5205.6 7. 15,713.88 9. 71.34 11. 1,684,956

Page 367 Review: Sections 12-4–12-5 1. 8 2. 7
3. -18 4. -27 5. -3 6. -8 7. 16 8. 5
9. -31 10. -5 11. -5 12. -12 13. -10
14. -16 15. 5 16. -7 17. -16 18. -15
19. -10 20. -25 21. $1\frac{1}{2}$ 22. $\frac{2}{3}$ 23. -1
24. $-\frac{1}{5}$ 25. $2(x + 1)$ 26. $5(x - 2)$ 27. $3(x - 1)$
28. $5(x + 3)$ 29. $x(x - 2)$ 30. $x(x + 12)$
31. $2(x + 7)$ 32. $x(x - 3)$ 33. $7(x - 6)$
34. $3(5 + x)$ 35. $6(3 + x)$ 36. $4(3 + x)$
37. $x(16 - x)$ 38. $x(3 + x)$ 39. $x(7 + x)$
40. $x(13 - x)$

Page 368 Chapter Review 1. $(6)(-3) = (-3)(6)$
3. $5 \cdot 0 = 0$ 5. $(-2 \cdot 5)4 = -2(5 \cdot 4)$ 7. factoring
9. -108 11. -84 13. -391 15. -960
17. -2394 19. -4536 21. -10.44 23. $-\frac{1}{24}$
25. 24 27. 144 29. 1598 31. 3654 33. 1526
35. 2856 37. 8.64 39. $3\frac{3}{4}$ 41. $2x + 6$
43. $-4b - 20$ 45. $r^2 - 6r$ 47. $-8n + 16$
49. $2p - 3$ 51. $6y^2 - 6y$ 53. 10 55. 3 57. -4
59. -4 61. -17 63. -13 65. -16 67. $1\frac{1}{2}$
69. $-\frac{2}{3}$ 71. $\frac{1}{20}$ 73. $2(x + 1)$ 75. $k(k + 1)$
77. $13(3r - 1)$ 79. $2(1 + h)$ 81. $7(2 - f)$
83. $y(13 - y)$ 85. $2(-10); -20$ 87. $2(-6); -12$
89. $5(-12); -60$

Page 370 Chapter Test 1. -384 3. -306
5. -7.75 7. 900 9. $2a + 2$ 11. $4y^2 - 8y$
13. $-2r - 12$ 15. 9 17. -8 19. $-\frac{1}{3}$ 21. -5
23. $5(x + 1)$ 25. $2(y - 8)$ 27. $m(m + 2)$
29. $5(y - 4)$

Page 371 Enrichment 1. $x^2 + 5x + 6$ 3. $w^2 + 9w + 20$ 5. $m^2 + 15m + 56$ 7. $a^2 - 5a + 4$ 9. $p^2 - 4p + 3$ 11. $w^2 - 11w + 28$ 13. $c^2 + 4c - 21$
15. $b^2 + b - 20$ 17. $y^2 - 2y - 8$ 19. $t^2 + 19t + 84$

Page 372 Additional Practice 1. -156 3. -189
5. 0 7. -2520 9. -8.40 11. $-\frac{5}{8}$ 13. 24
15. 66 17. 1404 19. 6405 21. 1.12 23. 2
25. $15x + 20$ 27. $-30b - 50$ 29. $-12n - 12$
31. $4y^2 - 2y$ 33. $28p^2 - 35p$ 35. $12f^2 - 24f$
37. 10 39. 4 41. -4 43. -3 45. -4
47. -6 49. $\frac{2}{3}$ 51. -10 53. $14(x + 1)$

55. $7(3m + 1)$ **57.** $7(p - 5)$ **59.** $4(c - 1)$
61. $4(9 - v)$ **63.** $x(x - 6)$ **65.** $4(-15); -60$

CHAPTER 13

Page 374 Self-Check 1. 42 **2.** −10 **3.** −112
4. 20 **5.** −9 **6.** −7

Page 375 Classroom Exercises 1. −18 **3.** −7
5. 15 **7.** 96 **9.** −4 **11.** 7 **13.** −3 **15.** −2
17. −36 **19.** 63 **21.** −28 **23.** −44 **25.** −3
27. −9 **29.** 11 **31.** −6 **33.** −3 **35.** 3

Page 376 Written Exercises 1. −96 **3.** −60
5. 24 **7.** −72 **9.** −114 **11.** −516 **13.** −4
15. −11 **17.** 24 **19.** −3 **21.** $-72\frac{1}{2}$ **23.** $16\frac{2}{3}$
25. −28 **27.** 29 **29.** 175 **31.** −140 **33.** −8
35. 0 **37.** −1.4 **39.** 12.5

Page 378 Self-Check 1. 6 **2.** 120

Page 378 Classroom Exercises 1. $5t$ **3.** $\frac{q}{4}$ **5.** $4q$
7. $2t$ **9.** $7m$ **11.** $\frac{c}{12} = 25$ **13.** $\frac{h}{7} = 6$

Page 379 Written Exercises 1. $\frac{x}{9}$ **3.** $2q$ **5.** $3r$
7. $-4a$ **9.** $3p$ **11.** $3w$ **13.** $\frac{t}{14}$ **15.** $4s = 28$
17. $6h = 132$ **19.** $\frac{k}{78} = 4$ **21.** $\frac{m}{7} = 10$ **23.** 3
25. 38 **27.** 400 **29.** $675 **31.** 71 **33.** 270

Page 381 Self-Check 1. −1 **2.** 6 **3.** −15
4. −24 **5.** 18 **6.** −30

Page 382 Classroom Exercises 1. $3k + 5 - 5 = 2 - 5$
3. $7b + 1 - 1 = 15 - 1$ **5.** $12 - 8 = -2y + 8 - 8$
7. $4x + 7 - 7 = -13 - 7$ **9.** $\frac{x}{3} - 1 + 1 = -2 + 1$
11. $\frac{v}{-2} + 3 - 3 = 7 - 3$ **13.** $2 - 4 = \frac{y}{2} + 4 - 4$
15. $\frac{r}{-2} - 3 + 3 = 4 + 3$

Page 382 Written Exercises 1. −2 **3.** 2 **5.** −4
7. −4 **9.** −5 **11.** −8 **13.** −6 **15.** −8 **17.** 20
19. −66 **21.** 20 **23.** −160 **25.** −2 **27.** 5
29. −35 **31.** −12 **33.** 60 **35.** 21 **37.** −3 **39.** 3

Pages 383-384 Self-Check 1. 2 **2.** −1 **3.** −6
4. 2 **5.** −6

Page 384 Classroom Exercises 1. 21 **3.** $9e$; 72
5. — **7.** 35; $-3z$ **9.** 12; — **11.** Yes **13.** No
15. Yes **17.** Yes

Page 385 Written Exercises 1. 4 **3.** $3\frac{2}{3}$ **5.** −3
7. −16 **9.** 11 **11.** 10 **13.** −3 **15.** $\frac{2}{3}$ **17.** 5
19. $\frac{1}{2}$ **21.** −3 **23.** $-2\frac{1}{4}$ **25.** 15 **27.** −8
29. −5 **31.** −8 **33.** 10 **35.** 3 **37.** −2 **39.** 4

Page 386 Review: Sections 13-1–13-4 1. −30
2. −56 **3.** −108 **4.** −45 **5.** −7 **6.** −5 **7.** 6
8. 3 **9.** 128 **10.** 15 **11.** 4 **12.** 336 **13.** 1
14. −2 **15.** −3 **16.** −2 **17.** −8 **18.** −6
19. −30 **20.** 18 **21.** 4 **22.** 1 **23.** −20 **24.** −5
25. −6 **26.** −10

Page 387 Special Topic 1. 11.5 **3.** −0.5 **5.** −4
7. −6.5 **9.** 5

Page 388 Self-Check 1. 15°; 75°; 90°

Page 388 Classroom Exercises 1. 30; 60; 90
3. 60; 100; 0

Page 389 Written Exercises 1. $x = 40°; 2x = 80°$;
$x + 20 = 60°$ **3.** $y = 45°; 2y = 90°$ **5.** $y = 50°; y + 15 = 65°$ **7.** $d = 75°; d - 15 = 60°; d - 30 = 45°$
9. $x = 50°; x + 10 = 60°; x + 20 = 70°$

Pages 390-391 Self-Check 1. 7; 8 **2.** 8; 9; 10

Page 391 Classroom Exercises 1. Yes **3.** No
5. $n + n + 1 = 5$ **7.** $n + n + 1 + n + 2 = 21$

Page 392 Written Exercises 1. Louise; 14;
Reynaldo: 15 **3.** William: 85; Juan: 84 **5.** 104;
105; 106 **7.** Morris: 21; Eric: 20; Maria: 22 **9.** Hans:
17; Rosaline: 16 **11.** 6; 7; 8 **13.** 64; 65

Page 394 Self-Check 1. $c < 8$; All points to the left
of and not including 8 **2.** $x > -3$; All points to the
right of and not including -3 **3.** $d > -18$; All points
to the right of and not including -18 **4.** $r < 20$; All
points to the left of and not including 20 **5.** $m > -5$;
All points to the right of and not including -5 **6.** $b < -6$; All points to the left of and not including -6
7. $q \leq -5$; All points to the left of and including -5

Page 394 Classroom Exercises 1. 5 **3.** −2 **5.** 9
7. 7 **9.** −8 **11.** −3 **13.** −6 **15.** 5 **17.** −7

Page 395 Written Exercises 1. $y < 9$; All points to
the left of and not including 9 **3.** $a > -70$; All points
to the right of and not including -70 **5.** $w \leq -6$; All
points to the left of and including -6 **7.** $k < 50$; All
points to the left of and not including 50 **9.** $n > -7$;
All points to the right of and not including -7
11. $b > 4$; All points to the right of and not including 4
13. $m < -10$; All points to the left of and not including
-10 **15.** $t \leq -5$; All points to the left of and including
-5 **17.** $x > 96$; All points to the right of and not
including 96 **19.** $a < -40$; All points to the left of and
not including -40 **21.** $j > 10$; All points to the right
of and not including 10 **23.** $n > -2$; All points to the
right of and not including -2 **25.** $b < -15$; All points
to the left of and not including -15 **27.** $d < 32$; All
points to the left of and not including 32 **29.** $p \leq -11$; All points to the left of and including -11
31. $n \geq -11$; All points to the left of and including -11
31. $n \geq -16$; All points to the right of and including -16

Pages 396-397 Self-Check 1. $27.99 **2.** 95

Page 397 Classroom Exercises 1. $b + 8 < 45$
3. $b + 15 < 32$ **5.** $\frac{87 + 92 + y}{3}$ **7.** $\frac{65 + 70 + 68 + x}{4} > 66$

Page 397 Written Exercises 1. $27.99 **3.** $44.99
5. 89 **7.** 101

Page 398 Review: Sections 13-5–13-8 1. 8; 9
2. 12; 13; 14 **3.** $x < -8$; All points to the left of and
not including -8 **4.** $c < -6$; All points to the left of
and not including -6 **5.** $n < -2$; All points to the left

562   *Answers to Selected Exercises*

sales **11.** $8100; $8100 **13.** median and mode

Page 474-475 Self-Check 1. 31-40, 7; 41-50, 3; 51-60, 3; 61-70, 5; 81-90, 5 **2.** The height of the bar for each interval is as given: 21-25: 2 units; 26-30: 4 units; 31-35: 5 units; 36-40: 3 units

Page 476 Classroom Exercises 1. 3; 3; 6; 2; 2

Page 476 Written Exercises 1. 120-219, 5; 220-319, 4; 320-419, 4; 420-519, 1; 520-619, 4; 620-719, 2 For Exercises 3-5, the height of the bar for each interval is as given. **3.** 17-23: 2 units; 24-30: 10 units; 31-37: 6 units; 38-44: 2 units **5.** 15-19: 3 units; 20-24: 3 units; 25-29: 4 units; 30-34: 8 units; 35-39: 3 units

Page 477 Review: Sections 16-1 — 16-2 1. 88.5; 90.5; 92 **2.** 152; 153.5; 160 **3.** 71; 71; 72 **4.** 21; 21; 19 **5.** The height of the bar for each interval is as given: 400-439: 2 units; 440-479: 2 units; 480-519: 5 units; 520-559: 3 units; 560-599: 1 unit

Page 478 Special Topic 1. 6.1 **3.** 5.3

Pages 479-480 Self-Check 1. $\frac{2}{11}$ **2.** $\frac{2}{11}$ **3.** $\frac{4}{11}$ **4.** $\frac{3}{10}$, or 30% **5.** $\frac{2}{5}$, or 40% **6.** 0, or 0%

Page 480 Classroom Exercises 1. $\frac{1}{30}$, or $3\frac{1}{3}$% **3.** 18; 30; $\frac{3}{5}$, or 60% **5.** $\frac{1}{10}$, or 10% **7.** $\frac{2}{5}$, or 40% **9.** $\frac{3}{5}$, or 60% **11.** 0, or 0%

Page 481 Written Exercises 1. $\frac{3}{10}$ **3.** $\frac{1}{2}$ **5.** $\frac{1}{5}$ **7.** $\frac{1}{5}$, or 20% **9.** $\frac{3}{5}$, or 60% **11.** $\frac{4}{5}$, or 80% **13.** $\frac{1}{2}$, or 50% **15.** $\frac{1}{2}$, or 50% **17.** $\frac{3}{8}$, or $37\frac{1}{2}$%

Page 483 Self-Check 1. $\frac{1}{9}$ **2.** $\frac{1}{36}$ **3.** $\frac{1}{12}$ **4.** $\frac{1}{6}$

Page 483 Classroom Exercises 1. 0 **3.** $\frac{1}{18}$ **5.** $\frac{1}{9}$ **7.** $\frac{1}{6}$ **9.** $\frac{1}{9}$ **11.** $\frac{1}{18}$ **13.** $\frac{1}{9}$ **15.** $\frac{5}{36}$ **17.** $\frac{1}{6}$

Page 484 Written Exercises 1. $\frac{1}{18}$ **3.** $\frac{1}{6}$ **5.** $\frac{1}{12}$ **7.** $\frac{5}{36}$ **9.** $\frac{1}{4}$ **11.** $\frac{2}{9}$ **13.** $\frac{1}{4}$ **15.** $\frac{5}{18}$

Page 485 Self-Check 1. $\frac{1}{6}$

Page 486 Classroom Exercises 1. $\frac{1}{3}$ **3.** $\frac{1}{6}$

Page 486 Written Exercises 1. Direct; Coastal; Central; Scenic; Overland; Scenic **3.** $\frac{1}{6}$ **5.** $\frac{1}{3}$ **7.** $\frac{1}{3}$ **9.** $\frac{1}{8}$ **11.** $\frac{3}{8}$ **13.** $\frac{1}{2}$

Pages 488-489 Self-Check 1. $\frac{11}{42}$ **2.** $\frac{2}{15}$

Page 489 Classroom Exercises 1. $\frac{1}{2}$; $\frac{3}{7}$; $\frac{3}{14}$ **3.** $\frac{1}{2}$; $\frac{7}{15}$, $\frac{7}{30}$

Page 490 Written Exercises 1. $\frac{1}{6}$ **3.** $\frac{3}{20}$ **5.** $\frac{1}{20}$ **7.** $\frac{2}{15}$ **9.** $\frac{1}{45}$ **11.** $\frac{1}{15}$ **13.** $\frac{6}{125}$ **15.** $\frac{1}{125}$ **17.** $\frac{1}{20}$ **19.** $\frac{1}{90}$

Page 491 Calculator Exercises 1. 0.9% **3.** 28.6% **5.** 25.4%

Page 491 Review: Sections 16-3 — 16-6 1. $\frac{1}{6}$ **2.** $\frac{1}{3}$ **3.** $\frac{2}{3}$ **4.** $\frac{1}{3}$ **5.** $\frac{1}{18}$ **6.** $\frac{1}{6}$ **7.** $\frac{2}{9}$ **8.** $\frac{5}{36}$ **9.** THT; HTH; TTT **10.** $\frac{1}{4}$ **11.** $\frac{1}{4}$ **12.** $\frac{1}{2}$ **13.** $\frac{1}{2}$ **14.** $\frac{5}{36}$ **15.** $\frac{15}{22}$

Page 492 Chapter Review 1. median **3.** 0 **5.** mean **7.** independent events **9.** mode **11.** 237 **13.** $\frac{1}{16}$, or $6\frac{1}{4}$% **15.** $\frac{1}{4}$, or 25% **17.** $\frac{2}{25}$ **19.** $\frac{1}{25}$ **21.** $\frac{1}{5}$ **23.** HHH; HHT; HTH; HTT; THH; THT; TTH; TTT **25.** $\frac{1}{8}$ **27.** $\frac{1}{2}$ **29.** $\frac{1}{15}$ **31.** $\frac{1}{35}$ **33.** $\frac{1}{6}$

Page 494 Chapter Test 1. 9; 7.5; 7 **3.** $\frac{4}{11}$ **5.** $\frac{7}{11}$ **7.** $\frac{4}{25}$ **9.** $\frac{16}{95}$

Page 495 Enrichment 1. 4%; 32 **3.** 4%; 160 **5.** 1300

Page 496 Additional Practice 1. $4\frac{3}{8}$; $4\frac{1}{2}$; 3 **3.** $\frac{2}{5}$, 40% **5.** $\frac{4}{5}$, 80% **7.** $\frac{7}{36}$ **9.** $\frac{1}{6}$ **11.** HHHH; HHHT; HHTH; HHTT; HTHH; HTHT; HTTH; HTTT; THHH; THHT; THTH; THTT; TTHH; TTHT; TTTH; TTTT **13.** $\frac{28}{45}$

CHAPTER 17

Page 499 Self-Check 1. $\overline{AD}$, $\overline{CD}$; $\overrightarrow{CD}$, $\overrightarrow{BA}$

Page 499 Classroom Exercises 1. $\overrightarrow{TS}$ **3.** angle DEF **5.** $\overline{QR}$ **7.** R **9.** MR; HK **11.** No **13.** HER or REH **15.** same

Page 500 Written Exercises 1. f or l **3.** c **5.** d, g, or k **7.** Answers may vary. Some answers are: $\overleftrightarrow{AB}$; $\overleftrightarrow{FA}$; $\overrightarrow{BF}$ **9.** $\overrightarrow{BF}$ **11.** $\angle IGH$ **13.** $\overleftrightarrow{HE}$; $\overleftrightarrow{BI}$

Pages 501-502 Self-Check 1. 70° **2.** 155° **3.** 110° For Exercises 4-7, teacher should check each student's drawings.

Page 502 Classroom Exercises 1. b **3.** a **5.** a

Page 503 Written Exercises 1. 30°; acute **3.** 95°; obtuse **5.** 155°; obtuse **7.** 25°; acute **9.** 105°; obtuse For Exercises 11-19, teacher should check each student's drawings. **21.** 33° **23.** 27°

Page 506 Self-Check For Exercises 1-3, teacher should check each student's drawings.

Page 506 Classroom Exercises 1. isosceles **3.** scalene **5.** acute **7.** right **9.** $\angle LM$ **11.** $\overline{PQ}$ **13.** $\angle T$

Page 507 Written Exercises 1. equilateral **3.** isosceles **5.** $\overline{DF} \cong \overline{AC}$; $\angle F \cong \angle C$; $\overline{FE} \cong \overline{CB}$ **7.** $\overline{PQ} \cong \overline{TO}$; $\overline{PR} \cong \overline{RT}$; $\overline{QR} \cong \overline{QR}$ **9.** $\angle B \cong \angle S$; $\overline{BR} \cong \overline{ST}$; $\angle R \cong \angle T$

17. 324 **19.** 3600 **21.** 576 **23.** 10,000 **25.** 2
27. 7 **29.** 3 **31.** 6 **33.** 10 **35.** 13 **37.** −3
39. −7 **41.** −6 **43.** 14 **45.** 40 **47.** 60

Pages 442-443 Self-Check 1. rational **2.** irrational
3. irrational **4.** rational **5.** irrational **6.** rational
7. rational **8.** irrational

Page 443 Classroom Exercises 1. No **3.** Yes
5. No **7.** Yes **9.** No **11.** No **13.** irrational
15. rational **17.** irrational **19.** rational **21.** rational
23. irrational **25.** rational **27.** irrational
29. rational **31.** irrational

Page 444 Written Exercises 1. rational; 25 is a perfect square **3.** rational; 100 is a perfect square
5. irrational; 18 is not a perfect square **7.** irrational; 22 is not a perfect square **9.** irrational; 7 is not a perfect square **11.** rational; 36 is a perfect square
13. irrational; 112 is not a perfect square **15.** rational; 1600 is a perfect square **17.** rational; 1000 is a perfect square **19.** rational; terminating **21.** irrational; non-terminating, non-repeating **23.** rational; repeating
25. rational; repeating **27.** irrational; non-terminating, non-repeating **29.** rational; terminating **31.** irrational; 19 is not a perfect square **33.** rational; terminating **35.** irrational; non-terminating, non-repeating **37.** rational; repeating **39.** irrational; 61 is not a perfect square **41.** rational; terminating
43. b; c **45.** c **47.** d

Pages 445-446 Self-Check 1. 1225 **2.** 8281
3. 4.583 **4.** 9.165 **5.** 29 **6.** 85 **7.** 20 **8.** 99

Page 446 Classroom Exercises 1. 3.742 **3.** 400; 4.472 **5.** 324; 4.243 **7.** 3025; 7.416

Page 446 Written Exercises 1. 169 **3.** 441
5. 1024 **7.** 6561 **9.** 14,400 **11.** 16,900
13. 2.828 **15.** 4.583 **17.** 7.483 **19.** 11
21. 6.708 **23.** 12.207 **25.** 15 **27.** 26 **29.** 39
31. 81 **33.** 60 **35.** 55

Page 448 Special Topic 1. −11.9°C **3.** −13.7°C

Page 449 Special Topic 1. 60 **3.** 862 **5.** 139

Page 450 Review: Sections 15-1 — 15-3 1. 16
2. 121 **3.** 361 **4.** 64 **5.** 576 **6.** 900 **7.** 7
8. 2 **9.** 10 **10.** 1 **11.** −6 **12.** −9 **13.** −4
14. 5 **15.** −8 **16.** −12 **17.** 3 **18.** 11 **19.** irrational **20.** rational **21.** rational **22.** irrational
23. rational **24.** rational **25.** irrational **26.** rational **27.** rational **28.** rational **29.** rational
30. irrational **31.** 324 **32.** 484 **33.** 12,544
34. 4225 **35.** 5.657 **36.** 8.660 **37.** 11.180
38. 11.790 **39.** 31 **40.** 97 **41.** 130 **42.** 62
43. 29 **44.** 96 **45.** 146 **46.** 71 **47.** 135
48. 149

Page 451 Self-Check 1. 12 ft **2.** 67 mi

Page 452 Classroom Exercises 1. 3.7 m **3.** 1.7 km
5. 6 in **7.** 12 yd **9.** 60 mi **11.** 71, 85 mi
13. 98, 118 mi **15.** 114, 137 mi

Page 452 Written Exercises 1. 4.8 m **3.** 5.9 cm
5. 11.7 cm **7.** 6.5 mi **9.** 6 ft **11.** 11 in **13.** 18 ft
15. 66 mi. **17.** 146 mi **19.** 74 mi **21.** 122 mi

Page 455 Self-Check 1. 8.6 in **2.** 10.6 m **3.** 3 m
4. 7.5 ft

Page 455 Classroom Exercises 1. 10 **3.** 16 **5.** 28

Page 456 Written Exercises 1. 9 m **3.** 5.4 m
5. 14.3 ft

Page 458-459 Self-Check 1. 1.5 **2.** 0.42 **3.** 10
4. 1 **5.** 6

Page 459 Classroom Exercises 1. 0.176 **3.** 0.466
5. 1.00 **7.** 6.30 **9.** 365.76

Page 460 Written Exercises 1. 1.00 **3.** 0.67
5. 67.12 m **7.** 4.82 mi **9.** 5 km

Page 461 Calculator Exercises 1. 0.0524078
3. 0.5095254 **5.** 28.636252

Page 461 Review: Sections 15-4 — 15-6 1. 3.5 m
3. 13 m **5.** 18 m

Page 462 Chapter Review 1. right triangle **3.** irrational **5.** hypotenuse **7.** 81 **9.** 324 **11.** 1764
13. 4 **15.** 9 **17.** 10 **19.** 11 **21.** −2 **23.** −12
25. rational **27.** irrational **29.** rational **31.** irrational **33.** rational **35.** rational **37.** 324 **39.** 441
41. 12,100 **43.** 8.660 **45.** 10.677 **47.** 9.747
49. 3.5 **51.** 5.2 **53.** 2.8 **55.** 117.6 **57.** 94.8
59. 177.6 **61.** 11.2 **63.** 30 **65.** 1.75 **67.** 0.78
69. 11

Page 464 Chapter Test 1. 25 **3.** 7 **5.** −12 **7.** 9
9. irrational **11.** 4096 **13.** 9.747 **15.** 66. **17.** 30
19. 6.3 m

Page 465 Enrichment 1. 2.4 **3.** 3.8 **5.** 4.12

Page 466 Additional Practice 1. 49 **3.** 169 **5.** 9
7. rational **9.** rational **11.** rational **13.** 676
15. 2209 **17.** 27 **19.** 13; 0.42 **21.** 4; 0.75
23. 11 **25.** 5 **27.** 156

Page 467 Common Errors 1. Start at the origin. Move 3 units to the left. Then move 6 units down. **3.** 4
5. 3.16 **7.** 0.75

Page 468 Cumulative Review: Chapters 14-15 1. c
3. b **5.** a **7.** d **9.** d

CHAPTER 16

Pages 470-471 Self-Check 1. $21 **2.** $561; $672

Page 471 Classroom Exercises 1. 40 **3.** 85.5; 85

Page 472 Written Exercises 1. 15,813 **3.** $6.15
5. 26; 26 **7.** 1.1; 0.8 **9.** mode; shows most frequent

21. (0, 4); (−2, 0) **23.** (0, −4); (1, 0) **25.** (0, 8); (4, 0) **27.** (0, −6); (−2, 0)

Page 425 Self-Check 1. $470 **2.** 68.6 km/h

Page 425 Classroom Exercises 1. $c = 5n$ **3.** $rt = 300$

Page 426 Written Exercises 1. $c = 2n$; $3 **3.** $lw = 400$; $3\frac{1}{5}$

Pages 427-428 Self-Check 1. 50 For Exercises 2-4, each graph is a pair of straight lines. The solution is given for each exercise. **2.** (−3, −4) **3.** (−1, 2) **4.** (0, 0)

Page 428 Classroom Exercises 1. Yes **3.** Yes **5.** Yes

Page 429 Written Exercises 1. 2000, 2500, 4000; 0, 1500, 4500; the graph is a pair of straight lines. The solution is (40, 3000). For Exercises 3-11, each graph is a pair of straight lines. The solution is given for each exercise. **3.** (−2, −1) **5.** (−1, −2) **7.** (0, 3) **9.** (1, −3)

Page 430 Self-Check For Exercises 1-2, each graph is a region that does not include its linear boundary. The location of the region with respect to the boundary is indicated for each exercise. **1.** Below the line containing (0, 3) and (−3, 0) **2.** Above the line containing (−3, −3) and (2, 2)

Page 431 Classroom Exercises 1. Yes **3.** No **5.** Yes **7.** a **9.** b

Page 431 Written Exercises 1. Yes **3.** Yes **5.** Yes **7.** No **9.** No For Exercises 11-21, each graph is a region that does not include its linear boundary. The location of the region with respect to the boundary is indicated for each exercise. **11.** Below the line containing (−3, −4) and (3, 2) **13.** To the left of the line containing (−1, −3) and (1, 3) **15.** To the right of the line containing (−1, −4) and (1, 2) **17.** To the left of the line containing (−1, 3) and (1, −3) **19.** Above the line containing (−3, 1) and (2, −4) **21.** To the right of the line containing (0, 3) and (2, −5)

Page 432 Review: Sections 14-5 − 14-8 1. 2; −2 **2.** −5; 5 **3.** −4; 2 **4.** 6; −2 For Exercises 5-8, each graph is a straight line containing the given points. **5.** (0, 3); (−3, 0) **6.** (0, −4); (4, 0) **7.** (0, 3); (−1, 0) **8.** (0, −8); (2, 0) **9.** $D = 60m$; 660 **10.** $dw = 24$; 6 For Exercises 11-14, each graph is a pair of straight lines. The solution is given for each exercise. **11.** (1, 3) **12.** (1, 0) **13.** (1, 1) **14.** (0, 0) For Exercises 15-18, each graph is a region that does not include its linear boundary. The location of the region with respect to the boundary is indicated for each exercise. **15.** Below the line containing (0, 3) and (−3, 0) **16.** Above the line containing (0, −4) and (4, 0) **17.** To the right of the line containing (−1, 3) and (1, −3) **18.** To the right of the line containing (0, −4) and (2, 0)

Page 433 Chapter Review 1. origin **3.** y axis **5.** line graph **7.** break-even **9.** slope **11.** indirect variation **13.** x coordinate **15.** point of intersection **17.** (1, −1) **19.** (−4, 2) **21.** (4, 3) **23.** (2, 1) **25.** (−4, −2) **27.** (1, 2) For Exercises 29-35, start at the origin. Then make the indicated moves for each point. **29.** Right 4; down 4 **31.** Right 3; down 2 **33.** Right 5 **35.** Right 5; up 4 For Exercises 37-39, each graph is a straight line containing the given points. **37.** (−1, −2); (2, 4) **39.** (3, 1); (−1, 5) **41.** 200 **43.** $\frac{1}{2}$ **45.** $-\frac{5}{3}$ **47.** $-\frac{6}{7}$ **49.** 7 **51.** 3 **53.** −6 **55.** −4 **57.** $T = 0.75a$; 4.50 For Exercises 59-61, each graph is a pair of straight lines. The solution is given for each exercise. **59.** (1, 0) **61.** (−1, −1) For Exercises 63-65, each graph is a region that does not include its linear boundary. The location of the region with respect to the boundary is indicated for each exercise. **63.** Below the line containing (0, 3) and (3, 0) **65.** To the left of the line containing (−3, −5) and (1, 3) **67.** The graph is a broken line connecting these points in order from left to right: (M, 2500); (T, 2400); (W, 3000); (Th, 2200); (F, 1900); (S, 2500); (S, 2200)

Page 436 Chapter Test 1. (4, 0) **3.** (2, 4) **5.** The graph is a straight line containing (0, −3) and (4, 1) **7.** $\frac{3}{2}$ **9.** 2; −2; the graph is a straight line containing (0, 2) and (−2, 0) **11.** The graph is the region below the line containing (0, 5) and (−3, 2) **13.** The graph is a pair of straight lines. The solution is (3, 9) **15.** $v = 9w$; 45

Page 437 Enrichment 1. $\{2, 3, 4, 5\}$ **3.** $\{3, 4, 5, 6, 9\}$ **5.** $\{3, 4\}$ **7.** ¢ **9.** ⊂ **11.** ⊂

Page 438 Additional Practice For Exercises 1-3, start at the origin. Then make the indicated moves for each point. **1.** Right 1; up 1 **3.** Left 2; down 2 For Exercises 5-7, the graph is a straight line containing the given points. **5.** (2, 2); (−3, −3) **7.** (−3, −1); (2, 4) **9.** $\frac{4}{3}$ **11.** $\frac{2}{3}$ **13.** 6; −2; the graph is a straight line containing (0, 6) and (3, 0) **15.** 9; 3; the graph is a straight line containing (0, 9) and (3, 0) **17.** $xy = 40$ **19.** The graph is a pair of straight lines. The solution is (0, 3) For Exercises 21-23, each graph is a region that does not include its linear boundary. The location of the region with respect to the boundary is indicated for each exercise. **21.** Above the line containing (−4, −1) and (4, 1) **23.** To the right of the line containing (−2, −3) and (1, 3) **25.** $Q = 4, 12, 20$; the graph is a straight line starting at (0, 0) and continuing through (5, 20)

CHAPTER 15

Pages 440-441 Self-Check 1. 9 **2.** 169 **3.** 324 **4.** 961 **5.** 7 **6.** 3 **7.** −4 **8.** −9

Page 441 Classroom Exercises 1. 16 **3.** 9; 81 **5.** 4 **7.** 7; 7

Page 441 Written Exercises 1. 4 **3.** 36 **5.** 100 **7.** 4 **9.** 225 **11.** 900 **13.** 441 **15.** 169

of and not including −2 **6.** $y < 4$; All points to the left of and not including 4 **7.** $x = 25°$; $2x = 50°$; $4x + 5 = 105°$ **8.** $349.99

Page 399 Chapter Review 1. multiplication **3.** consecutive **5.** reversed **7.** −40 **9.** 36 **11.** 8 **13.** −8 **15.** $3m = 42$ **17.** $\frac{p}{16} = 8$ **19.** −3 **21.** −6 **23.** 7 **25.** −36 **27.** 9 **29.** 12 **31.** $x = 60°$ **33.** $t = 30°$; $4t = 120°$ **35.** 17; 18 **37.** 12; 13; 14 **39.** $p < -135$; All points to the left of and not including −135 **51.** $b < 48$; All points to the left of and not including 48 **43.** $t > 11$; All points to the right of and not including 11 **45.** $q > -2$; All points to the right of and not including −2 **47.** 90 **49.** Kermit: 18; Vera: 19

Page 401 Chapter Test 1. −60 **3.** −8 **5.** 7 **7.** −75 **9.** 3 **11.** $b < 8$; All points to the left of and not including 8 **13.** $x = 18°$; $3x = 54°$; $6x = 108°$ **15.** $31.99

Page 402 Enrichment 1. Yes **3.** No **5.** Yes

Page 403 Additional Practice 1. 150 **3.** −36 **5.** −8 **7.** −3.2 **9.** 10 **11.** −5 **13.** −35 **15.** 48 **17.** 3 **19.** 1 **21.** −12 **23.** $x < 8$; All points to the left of and not including 8 **25.** $x > -4$; All points to the right of and not including −4 **27.** $3x = 33$ **29.** $\frac{w}{4} = 9$ **31.** $839.99 **33.** 108; 109; 110

Page 404 Common Errors 1. −13 **3.** −16 **5.** 3 **7.** 7 **9.** −18 **11.** $-16\frac{1}{2}$ **13.** $4m + 36$

Page 405 Cumulative Review: Chapters 10-13 1. c **3.** c **5.** a **7.** d **9.** a **11.** c **13.** c **15.** d **17.** b **19.** c **21.** b **23.** a **25.** d **27.** b

CHAPTER 14

Pages 408-409 Self-Check 1. (−2, 4) **2.** (−5, −5) **3.** (2, −2) **4.** (2, 3) For Exercises 5-7, start at the origin. Then make the indicated moves for each point. **5.** Left 1; down 4 **6.** Right 4 **7.** Down 2

Page 409 Classroom Exercises 1. N **3.** P **5.** Q **7.** J **9.** L **11.** S

Page 409 Written Exercises 1. (2, 3) **3.** (−3, 3) **5.** (−4, 1) **7.** (−5, −2) **9.** (4, −3) **11.** (0, −4) For Exercises 13-21, start at the origin. Then make the indicated moves for each point. **13.** Right 4; up 2 **15.** Left 2; down 4 **17.** Right 6; down 3 **19.** Left 2; up 6 **21.** Left 4

Pages 410-411 Self-Check 1. $\begin{array}{c|ccc} t & 1 & 3 & 5 \\ \hline d & 1100 & 3300 & 5500 \end{array}$ **2.** The graph is a straight line starting at (0, 0) and continuing through (5, 5500). **3.** The graph is a straight line containing (−2, −3) and (1, 0).

Page 411 Classroom Exercises 1. $\begin{array}{c|ccc} t & 0 & 2 & 4 \\ \hline d & 0 & 160 & 320 \end{array}$

3. $\begin{array}{c|ccc} t & 0 & 1 & 2 \\ \hline d & 0 & 500 & 1000 \end{array}$

Page 412 Written Exercises 1. $d = 0, 100, 200$; the graph is a straight line starting at (0, 0) and continuing through (4, 200). **3.** 1; 0; −2 **5.** 6; 0; −6 For Exercises 7-25, each graph is a straight line containing the given points. **7.** (1, 1); (0, 0); (−2, −2) **9.** (2, 6); (0, 0); (−2, −6) **11.** (0, 1); (−1, −2) **13.** (1, 1); (0, −3) **15.** (0, 4); (2, 0) **17.** (0, −3); (2, 1) **19.** (−2, 3); (4, −3) **21.** (2, 4); (−2, −4) **23.** (−2, 3); (2, −5) **25.** (−4, −2); (2, 1)

Page 413 Calculator Exercises 1. 1; −5; −8 **3.** 33; −15; −27

Pages 414-415 Self-Check 1. 2 in **2.** The graph is a broken line connecting these points in order from left to right: (J, 10); (F, 8); (M, 9); (A, 16); (M, 20); (J, 18)

Page 415 Classroom Exercises 1. 900 **3.** 200

Page 416 Written Exercises 1. 49 **3.** $12\frac{1}{2}$% **5.** The graph is a broken line connecting these points: (1940, 6.6); (1950, 8); (1960, 7.6); (1970, 7.6); (1980, 6.4)

Pages 417-418 Self-Check 1. 1 **2.** $\frac{5}{3}$ **3.** $-\frac{1}{2}$ **4.** $\frac{2}{3}$

Page 418 Classroom Exercises 1. negative **3.** zero **5.** $\frac{1}{4}$ **7.** $-1 - (-3) = 2$; $\frac{3}{2}$ **9.** −3 **11.** $4 - 4 = 0$; $-2 - (-6) = 4$; 0

Page 419 Written Exercises 1. 2 **3.** 1 **5.** $\frac{1}{6}$ **7.** $\frac{1}{2}$ **9.** $\frac{1}{5}$ **11.** 3 **13.** $-\frac{2}{3}$ **15.** $-\frac{3}{4}$ **17.** $-\frac{5}{4}$ **19.** 0 **21.** $-\frac{1}{2}$ **23.** 0 **25.** 7 **27.** $-\frac{3}{4}$ **29.** 3 **31.** $\frac{9}{4}$ **33.** $-\frac{4}{9}$ **35.** 0

Page 420 Review: Sections 14-1–14-4 1. (−4, 4) **2.** (1, 3) **3.** (0, −1) **4.** (3, −2) For Exercises 5-8, start at the origin and make the moves given for each point. **5.** Right 4; up 2 **6.** Left 3; up 1 **7.** Left 2; down 4 **8.** Right 1; down 3 **9.** 0; 80; 160 **10.** −1; 1; 3 **11.** 2; 4; 6 **12.** 0; 4; 8 For Exercises 13-16, each graph is a straight line containing the given points. **13.** (0, 4); (−4, 0) **14.** (−1, −6); (3, −2) **15.** (0, 3); (−2, −1) **16.** (−2, −5); (1, 4) **17.** The graph is a broken line connecting these points: (M, 2000); (T, 2500); (W, 1500); (Th, 2400); (F, 2100); (S, 3000); (S, 1200) **18.** $\frac{1}{3}$ **19.** $-\frac{2}{5}$ **20.** $-\frac{1}{5}$

Page 421 Special Topic 1. $-\frac{6}{5}$ **3.** Slope is Undefined **5.** $\frac{-3}{3}$ **7.** $\frac{4}{-1}$ **9.** $\frac{8}{-7}$

Pages 422-423 Self-Check 1. 4 **2.** −1 **3.** 2 **4.** 3 For Exercises 5-7, each graph is a straight line containing the given points. **5.** (0, 4); (−2, 0) **6.** (0, −10); (2, 0) **7.** (0, 9); (3, 0)

Page 423 Classroom Exercises 1. 2; −8 **3.** 1; 1

Page 423 Written Exercises 1. 2 **3.** 7 **5.** −3 **7.** 4 **9.** 3 **11.** −6 **13.** 3 **15.** −3 For Exercises 17-27, each graph is a straight line containing the given points. **17.** (0, 3); (−3, 0) **19.** (0, −7); (7, 0)

Answers to Selected Exercises

PICTURE CREDITS

Photos not listed below, Blaise Zito Associates, Inc.

Key: (t) top, (b) bottom, (l) left, (r) right, (c) center.

ILLUSTRATORS: Page 134, Tom Powers; all others, Blaise Zito Associates, Inc.

RESEARCH PHOTOS: CHAPTER ONE: Pages 5, Chris Reeberg; 7, © Marvin E. Newman/Woodfin Camp; 8, Focus on Sports; 10, Focus on Sports; 12, Will Blanche/DPI; 14, © Craig Aurness/Woodfin Camp; 15, Stephen Kraseman/© Peter Arnold, Inc.; 16, Charlie Ott/DPI; 22, © George Hall/Woodfin Camp. CHAPTER TWO: Pages 34, Katrina Thomas/Photo Researchers, Inc.; 35, Focus on Sports; 37, Focus on Sports; 42, Will Blanche/DPI; 46, DPI; 54, Focus on Sports; 55, Focus on Sports; 64, Chuck Solo/Focus on Sports; 73, © Randa Bishop/DPI; 85, Gianni Tortoli/Photo Researchers, Inc.; 88, DPI. CHAPTER FOUR: Pages 99, © Shelly Grossman/Woodfin Camp; 107, HBJ Photo. CHAPTER FIVE: Pages 128, Chris Reeberg/DPI; 134, Dorothy Affa/Focus on Sports; 138, H. Armstrong Roberts; 144, © Alastair Black/Focus on Sports; 151, Bob Witt/DPI; 153, NASA. CHAPTER SIX: Pages 164, Weiner Muller/© Peter Arnold, Inc.; 165, Focus on Sports; 167, Focus on Sports; 183, © Larry Smith/DPI. CHAPTER SEVEN: Pages 198, E. Simonsen/H. Armstrong Roberts; 206, © C.B. Jones/Taurus Photos; 209, © David Madison/Focus on Sports; 211, Adolph Rohrer/DPI; 212, Focus on Sports. CHAPTER EIGHT: Pages 223, DPI; 230 top, © Lawrence Fried/The Image Bank; 230 (b), DPI; 231, H. Armstrong Roberts; 237, Focus on Sports; 238 (b), S. Shackman/Monkmeyer; 243, © Robert McElroy/Woodfin Camp; 254, U.S. Dept. of Energy; 255, Chris Reeberg/DPI; 267, Joel Gordon/DPI; 269, Runk/Schoenberger from Grant Heilman; 287, © Jerry Wachter/Focus on Sports. CHAPTER TEN: Pages 296, DPI; 303, William Kelly/Shostal Associates; 307, © David Burnett/Contact Press Images/Woodfin Camp; 310 (t); © George Gerster/Rapho Division/Photo Researchers, Inc.; 310 (b), © L.L.T. Rhodes/Taurus Photos; 320, © Skyviews Survey, Inc./DPI. CHAPTER ELEVEN: Pages 332, H. Armstrong Roberts; 334, Focus on Sports; 337, Will Blanche/DPI; 338, © Bill Strode/Woodfin Camp. CHAPTER TWELVE: Pages 359, R. Krubner/H. Armstrong Roberts; 369, © Jonathan Blair/Woodfin Camp. CHAPTER THIRTEEN: Pages 377, DPI; 401, © Adam Woolfitt/Woodfin Camp. CHAPTER FOURTEEN: Pages 410, © John Blaustein/Woodfin Camp; 427, © John Lei/Omni Photo Communications. CHAPTER FIFTEEN: Pages 448, H. Armstrong Roberts. CHAPTER SIXTEEN: Pages 470, National Oceanic and Atmospheric Administration; 474, Focus on Sports. CHAPTER SEVENTEEN: Pages 501, Guiliano Bevilacqua/Focus on Sports; 524, Lizabeth Corlett/DPI; 526, Isaac Geib from Grant Heilman; 527, J. Alex Langley/DPI; 533, DPI. CHAPTER OPENERS: CHAPTER ONE: (l & tr) Blaise Zito Associates, Inc.; (br), © Charlie Ott/DPI. CHAPTER TWO: (tl), © DPI. CHAPTER THREE: (r), © Gianni Tortoli/Photo Researchers, Inc.; (tl), © HBJ Photo; (bl), Chuck Solo/Focus on Sports. CHAPTER FOUR: Blaise Zito Associates, Inc. CHAPTER FIVE: (l), NASA; (tr), Blaise Zito Associates, Inc.; (br), H. Armstrong Roberts. CHAPTER SIX: (l), Blaise Zito Associates, Inc.; (tr), © Larry Smith/DPI; (br), Blaise Zito Associates, Inc. CHAPTER SEVEN: (r), Blaise Zito Associates, Inc.; (tl), © C.B. Jones/Taurus Photos; (bl), Adolph Rohrer/DPI. CHAPTER EIGHT: Blaise Zito Associates, Inc. CHAPTER NINE: (r), Runk/Schoenberger from Grant Heilman; (tl), © David Burnett/Contact Press Images/Woodfin Camp. CHAPTER TEN: (r), © Skyviews Survey, Inc./DPI. CHAPTER ELEVEN: (l), H. Armstrong Roberts; (tr), Will Blanche/DPI; bottom right, © Bill Strode/Woodfin Camp. CHAPTER TWELVE: (r), © Adam Woolfitt/Woodfin Camp. CHAPTER THIRTEEN: (r), © Jonathan Blair/Woodfin Camp.(bl),Blaise Zito Associates, Inc. CHAPTER FOURTEEN: (l), © John Lei/Omni Photo Communications. CHAPTER FIFTEEN: Blaise Zito Associates, Inc. CHAPTER SIXTEEN: (r), Blaise Zito Associates, Inc. CHAPTER SEVENTEEN: (l), J. Alex Langley/DPI; (r), Blaise Zito Associates, Inc. and National Oceanic and Atmospheric Administration.

Answers to Selected Exercises 567

Page 509 Self-Check 1. 39 ft

Page 509 Classroom Exercises 1. 17 **3.** $\frac{1}{8}$; $\frac{1}{2}$

Page 510 Written Exercises 1. 13 **3.** 20.3 **5.** 6.4 m **7.** 100 m

Page 511 Calculator Exercises 1. 13 **3.** 20.3

Page 512 Special Topic 1-3. Angles A and B are supp **5.** Angles A and B are not supp **7.** 70° **9.** 30° **11.** 82.1°

Page 513 Review: Sections 17-1–17-4 1. $\overleftrightarrow{ED}$; $\overleftrightarrow{EG}$; $\overleftrightarrow{EB}$ **2.** $\overrightarrow{AF}$ **3.** ∠ACB; ∠BCF; ∠FCE; ∠ECA
For Exercises 4-8, teacher should check each student's drawings. **9.** SAS; $\overline{AC} \cong \overline{RS}$; ∠A ≅ ∠R; $\overline{AB} \cong \overline{RT}$ **10.** SSS; $\overline{XY} \cong \overline{EG}$; $\overline{XZ} \cong \overline{GF}$; $\overline{YZ} \cong \overline{EF}$ **11.** ASA; ∠M ≅ ∠V; $\overline{MO} \cong \overline{VW}$; ∠O ≅ ∠W **12.** 1.47

Pages 514-515 Self-Check 1. $\overleftrightarrow{BC}$ intersects $\overleftrightarrow{CD}$ at C **2.** $\overleftrightarrow{CD} \perp \overleftrightarrow{AC}$ **3.** vertical **4.** alternate interior **5.** alternate interior **6.** vertical

Page 515 Classroom Exercises 1. intersecting; perpendicular **3.** intersecting; perpendicular **5.** parallel **7.** b **9.** a **11.** c

Page 516 Written Exercises 1. $\overleftrightarrow{BA} \perp \overleftrightarrow{AC}$; $\overleftrightarrow{CD} \perp \overleftrightarrow{AC}$; $\overleftrightarrow{AF} \perp \overleftrightarrow{FD}$; $\overleftrightarrow{CD} \parallel \overleftrightarrow{FD}$ **3.** $\overleftrightarrow{BE}$ and $\overleftrightarrow{BF}$; $\overleftrightarrow{BE}$ and $\overleftrightarrow{CD}$; $\overleftrightarrow{BE}$ and $\overleftrightarrow{AC}$; $\overleftrightarrow{BE}$ and $\overleftrightarrow{FD}$ **5.** vertical **7.** alternate interior **9.** corresponding **11.** alternate exterior

Page 518 Self-Check 1. trapezoid **2.** parallelogram; rectangle **3.** parallelogram; rectangle

Page 518 Classroom Exercises 1. Yes; Yes **3.** No; Yes; Yes; Yes **5.** No; No; Yes; No; Yes

Page 519 Written Exercises 1. 12 **3.** 120° **5.** 4.2 **7.** 115°; 65° **9.** parallelogram; rectangle **11.** parallelogram; rectangle **13.** trapezoid **15.** parallelogram; rhombus **17.** parallelogram **19.** parallelogram; rectangle **21.** parallelogram; rhombus; rectangle **23.** rhombus; parallelogram; rectangle **25.** rhombus; square

Page 521 Self-Check 1. 72 cm²

Page 521 Classroom Exercises 1. 24 m²; 40 m²; 30 m²; 94 m² **3.** 24 ft²; 12 ft²; 16 ft²; 52 ft²

Page 522 Written Exercises 1. 2255 cm² **3.** 54 cm² **5.** 512 cm² **7.** 47.28 m² **9.** 72 ft² **11.** 580 ft² **13.** $414\frac{2}{3}$ in² **15.** 1400 in² **17.** 864 cm² **19.** $268.80

Page 524 Self-Check 1. 849 cm³

Page 525 Classroom Exercises 1. 170 m³

Page 525 Written Exercises 1. 402 m² **3.** 97,108 cm³ **5.** 170 cm³ **7.** 661 cm³ **9.** 69 in³ **11.** 4928 in³ **13.** 4505 in³ **15.** 1 m³ **17.** 629 in³

Pages 527-528 Self-Check 1. 18 m³ **2.** 180 ft³ **3.** 167 m³ **4.** 566 ft³

Page 528 Classroom Exercises 1. 100 ft³ **3.** 754 m³

Page 529 Written Exercises 1. 8 m³ **3.** 3140 cm³ **5.** 300 ft³ **7.** 38 ft³ **9.** 3.36 m³ **11.** 769.3 m³ **13.** 18.09 m²; 90 m³ **15.** 1540 ft³ **17.** 17 ft²; 68 ft³

Page 530 Review: Sections 17-5–17-9 1. $\overline{NP} \perp \overline{NR}$; $\overline{NR} \perp \overline{SR}$; $\overline{LPS} \perp \overline{SR}$; $\overline{LNR}$ **2.** $\overline{PN} \parallel \overline{SR}$; $\overline{PS} \parallel \overline{NR}$ **3.** 5 **4.** 90° **5.** 120°; 60° **6.** 20 **7.** 132 in² **8.** $108\frac{1}{4}$ ft² **9.** 603 m² **10.** 50 in³ **11.** 24 in³ **12.** 31.5 cm³ **13.** 462 m³ **14.** 1780 cm³ **15.** 2112 ft³ **16.** 205 yd³ **17.** 2904 in³ **18.** $29\frac{1}{3}$ in³ **19.** 113.04 cm³ **20.** 0.032 m³

Page 531 Chapter Review 1. degree **3.** line **5.** polygon **7.** quadrilateral **9.** right angle **11.** similar triangles **13.** acute angle **15.** obtuse angle **17.** line segment **19.** Answers may vary. Some answers are: $\overrightarrow{BH}$; $\overrightarrow{AH}$; $\overrightarrow{AE}$ **21.** ∠ABC; ∠CBE; ∠EBD; ∠DBA For Exercises 23-25, teacher should check each student's drawings. **27.** isosceles **29.** scalene **31.** 25 **33.** alternate interior **35.** vertical **37.** parallelogram; rhombus **39.** parallelogram **41.** 1161 m² **43.** 1356 m³ **45.** 6837 cm³ **47.** 4385 m³ **49.** 2,592,100 m³

Page 534 Chapter Test 1. acute **3.** $\overline{GH}$; $\overline{GB}$; $\overline{GJ}$ **5.** $\overline{JG} \perp \overline{HE}$ **7.** $\overline{JF} \parallel \overline{BC}$; $\overline{JB} \parallel \overline{FC}$ **9.** SSS; $\overline{YZ} \cong \overline{RS}$; $\overline{XY} \cong \overline{ST}$; $\overline{XZ} \cong \overline{RT}$ **11.** 16.8 **13.** 4537.3 m³ **15.** 198 in³

Page 535 Enrichment 1. 39 cm³ **3.** 310 in³ **5.** 4443 cm³ **7.** 24,417 m³

Page 536 Additional Practice For Exercises 1-3, answers may vary. **1.** Some answers are: $\overline{DE}$; $\overline{EF}$; $\overline{GF}$ **3.** Some answers are: $\overline{DG}$; $\overline{EF}$; $\overline{FD}$ **5.** acute **7.** obtuse **9.** obtuse **11.** C **13.** NC **15.** trapezoid **17.** parallelogram; rectangle; square **19.** 20 m **21.** 1538.6 cm³

Page 537 Common Errors 1. 86 **3.** $\frac{1}{25}$ **5.** 12 **7.** 486 cm³

Page 538 Cumulative Review: Chapters 16-17 1. a **3.** c **5.** a **7.** c **9.** b **11.** b **13.** a **15.** d